PHYSICS
AND
GEOLOGY

McGRAW-HILL INTERNATIONAL SERIES
IN THE EARTH AND PLANETARY SCIENCES

AGER: Principles of Paleoecology
BERNER: Principles of Chemical Sedimentology
BROECKER and OVERSBY: Chemical Equilibria in the Earth
DE SITTER: Structural Geology
DOMENICO: Concepts and Models in Groundwater Hydrology
EWING, JARDETZKY, and PRESS: Elastic Waves in Layered Media
GRANT and WEST: Interpretation Theory in Applied Geophysics
GRIFFITHS: Scientific Method in Analysis of Sediments
GRIM: Applied Clay Mineralogy
GRIM: Clay Mineralogy
HOWELL: Introduction to Geophysics
JACOBS, RUSSELL, and WILSON: Physics and Geology
KRAUSKOPF: Introduction to Geochemistry
KRUMBEIN and GRAYBILL: An Introduction to Statistical Models in Geology
LEGGET: Geology and Engineering
MENARD: Marine Geology of the Pacific
MILLER: Photogeology
OFFICER: Introduction to the Theory of Sound Transmission
RAMSAY: Folding and Fracturing of Rocks
ROBERTSON: The Nature of the Solid Earth
SHROCK and TWENHOFEL: Principles of Invertebrate Paleontology
STANTON: Ore Petrology
TOLSTOY: Wave Propagation
TURNER: Metamorphic Petrology
TURNER and VERHOOGEN: Igneous and Metamorphic Petrology
TURNER and WEISS: Structural Analysis of Metamorphic Tectonites
WHITE: Seismic Waves: Radiation, Transmission, and Attenuation

This book was set in Press Roman.
The editors were Jack L. Farnsworth and James W. Bradley;
cover design by Nicholas Krenitsky;
and the production supervisor was Joe Campanella.
The drawings were done by John Cordes, J & R Technical Services, Inc.
The printer and binder was Kingsport Press, Inc.

ary of Congress Cataloging in Publication Data

bs, John Arthur, 1916-
hysics and geology.

(International series in the earth and planetary sciences)
ibliography: p.
Geophysics. I. Russell, Richard Doncaster, 1929- joint author.
Wilson, J. Tuzo, 1908- joint author. III. Title. IV. Series.
1.J25 1973 551 73-6621
0-07-032148-5

ICS

OGY

5 6 7 8 9 0 K P K P 7 9 8 7 6 5 4

McGRAW-HILL BOOK COMPANY

New York
St. Louis
San Francisco
Düsseldorf
Johannesburg
Kuala Lumpur
London
Mexico
Montreal
New Delhi
Panama
Rio de Janeiro
Singapore
Sydney
Toronto

J. A. JACOBS

Killam Memorial Professor of S
Director, Institute of Earth and
University of Alberta

R. D. RUSSELL

Professor and Head
Department of Geophysics ar
University of British Columb

J. TUZO WILSON

Professor of Geophysics
Principal, Erindale College
University of Toronto
President, Royal Society of

Physics and Geolo

SECOND EDITIO

Lil
Jac
II.
QE5
ISBN

PHY
AND
GEO

1 2 3

In Memory of
the late Professor LACHLAN GILCHRIST
1875-1962
First Professor of Geophysics
Department of Physics
University of Toronto

CONTENTS

PREFACE

The first edition of this book was an outgrowth of courses on the physics of the earth given by the authors between 1946 and 1959 to senior undergraduate and graduate students in the Department of Physics and in the Department of Geology at the University of Toronto. As a consequence the book has two aims:

1 To give students of geology an introduction to the physics of the earth
2 To give scientists in other fields some knowledge of geology and its relation to geophysics

In the authors' view the earth should be regarded as an active body, the physiology as well as the anatomy of which can be studied. In the past, geology has been chiefly concerned with describing that part of the earth's surface exposed above the sea and with tracing the earth's later history as indexed by fossils. Methods developed and applied during the past few years, however, have made it possible to describe the whole earth, from its deep interior to its outer atmosphere. The earth's development has been put into better perspective on an absolute time scale, and it has been possible to suggest the physical nature and causes of some of its processes. It is this broader picture of terrestrial behavior which the authors have tried to sketch, blending the older outlines of geology with the newer colors of physics.

The objectives of the book have remained the same, but the development of the subject has been so rapid that almost every part of the book has had to be rewritten. The opportunity has been taken to add an extensive list of references.

When the first edition was being written, the three authors were working together in the physics laboratories of the University of Toronto, but their backgrounds were such that they could easily divide the preparation of the draft manuscript, for in addition to training in physics, Dr. Jacobs is a mathematician, Dr. Russell a chemist, and Dr. Wilson a geologist. Nevertheless they collaborated closely in writing and rewriting that edition.

Since then two of the authors have moved to other universities so that, although the preparation of a second edition has necessarily become a more difficult undertaking, the authors hope that this edition still reflects the blend of varied backgrounds that contributed to its development. All three authors continue to accept joint responsibility, although it seems fair to mention that Dr. Jacobs wrote the greater part of Chaps. 1, 2, 4, 7, 8, 9, and 10 and Secs. 11–4 to 11–6. Dr. Russell wrote Chaps. 5, 6, and part of 3, and Dr. Wilson the rest. Dr. Jacobs also undertook the task of assembling the whole.

The authors wish to thank their colleagues in their three universities and elsewhere for much sound advice and help. Mr. A. Aiken and Mr. K. Khan helped prepare many of the illustrations. Mrs. G. Dinwoodie played an important part in assembling the material and especially in collating the references.

In addition to the Universities of Alberta, British Columbia, and Toronto and the respective provincial agencies, the authors have received assistance in carrying out research, which formed the basis for parts of this book, from the National Research Council of Canada, the Defence Research Board of Canada, the Geological Survey of Canada, the Dominion Observatory (now the Earth Physics Branch), and other bodies and companies some of which were mentioned in the preface to the first edition. To all these and to their many patient secretaries who over the years have typed and retyped drafts we express our thanks.

J. A. JACOBS

R. D. RUSSELL

J. TUZO WILSON

PHYSICS
AND
GEOLOGY

1

THE UNIVERSE AND THE SOLAR SYSTEM

1-1 INTRODUCTION

Before discussing in detail any particular aspect of the surface features or interior constitution of the earth, it is well to consider it first in its proper setting as a member of the solar system. Sooner or later in any discussion of the physical processes which occur within the earth, questions about its origin and early history are bound to arise. These in turn lead to the problem of the origin of the solar system and to other far-reaching astrophysical questions. Such questions are, in their very nature, bound to be extremely controversial, but it is wrong to ignore them, for the answers are important in many matters connected with the later development of the earth. Detailed studies of the thermal history of the earth, for example, depend quite critically on what initial temperatures are chosen. The origin of the earth will therefore be discussed at some length in this chapter; this will help in assessing the validity of particular earth models which will be constructed in later chapters. Before doing so, it will be convenient to introduce some of the terms which will be used by giving a brief account of some of the main features of the earth's surface and interior.

The earth is almost spherical in form, with a diameter of slightly less than 8000 miles. To be more precise and to give figures in the cgs units which will be used throughout this book, the earth has the shape of a spheroid with a mean equatorial radius of 6378.388 km and a polar radius of 6356.912 km. These are the figures adopted internationally. The radius of a sphere having the same volume is 6371.2 km, and a value of 6371 km will be used in any calculation. The earth's mass is 5.975×10^{27} g, and its average density is just over 5.5 g/cm^3. The average density of surface rocks, on the other hand, is approximately 2.8 g/cm^3 so that there must be a density increase toward the earth's center, where the pressure exceeds 3.5×10^{12} dynes/cm^2 ($\approx$ 3½ million atm). The temperature also increases toward the center. The average value of the surface gradient is about 30°C/km, but this gradient is not maintained at depth. The temperature at the center of the earth is almost certainly less than 10,000°C, and probably is no more than 5000°C.

There are a number of peculiarities about the surface features of the earth. Less than 30 percent of the earth's surface is land, which is markedly concentrated in the Northern Hemisphere, the oceans being concentrated in the Southern Hemisphere. This contrast is reversed in polar regions. Moreover, 81 percent of all the land is concentrated in one hemisphere, with its pole in Brittany, the corresponding "water hemisphere" having its pole near New Zealand. There is also a curious antipodal relation between land and sea. Although about 45 percent of the surface has sea opposite sea, only 1.4 percent has land opposite land. T. Hatherton has shown that the probability of most of the land being opposite to sea (as at present) could be due to chance, although whether this arrangement of surface features has existed throughout geological time is an intriguing question, and one on which complete agreement has not yet been reached. (See Chaps. 13 and 14, where polar wandering and continental drift are discussed in some detail.) The oceans hold as many geophysical problems as, or more than, the continents do. The greatest ocean depth exceeds the greatest mountain height, and there are mighty mountain ranges on the ocean floor which rival any of the ranges visible on land today.

The topmost layers of the earth are called the *crust*. Its thickness and composition are not constant, but vary between the continental crust, which consists of 30 to 60 km of light rocks (such as gneiss, granodiorite, and granite), and the oceanic crust, made up of dark rocks (such as basalt) usually not more than 5 to 6 km thick. The upper part of the crust is a thin and discontinuous layer of sedimentary rocks and oceanic deposits. That part of the earth between the crust and a depth of approximately 2900 km is called the *mantle*, and for the remaining part–inside the mantle–the word *core* is used. The boundaries between these three main divisions of the earth, the crust, mantle, and core, are sharp and distinct and mark large changes in seismic velocities, the evidence for which will be discussed more fully in the next chapter. The boundary between the crust and mantle is called the Mohorovičić discontinuity, or Moho.

Another useful division is based on the ability of different layers to flow. The crust and uppermost mantle to a depth of about 70 km are cold enough to be rigid and brittle so that they fracture rather than flow and are called the *lithosphere*, or *tectosphere*. Below that, a layer called the *asthenosphere*, or *rheosphere*, is hot enough and is under sufficiently low pressure so that it is capable of slow deformation and flow. At a depth of a few hundred kilometers, pressure leads to increased rigidity again. Changes in phase and in composition, and particularly changes in water content, may lead to changes in viscosity. The terms *sial* and *sima* were introduced before the advent of seismic information to distinguish on chemical grounds between rocks rich in *si*lica and *al*umina and those rich in *si*lica and *ma*gnesia, but with increasingly precise knowledge, these and many other older concepts and terms have become obsolete.

1-2 THE SOLAR SYSTEM

Until the time of Copernicus (1473-1543) it was generally held that the earth was the center of the universe and that around it revolved the sun, the moon, the planets then known, and the stars. This geocentric theory, or Ptolemaic theory, of the universe became ever more complicated and artificial as it tried to take account of the increasing accuracy of the observations and the newly discovered astronomical phenomena. Copernicus reintroduced the idea which had occurred to Aristarchus in 200 B.C. that the sun was the central body around which the earth and the other planets revolved in circular paths. The earth was thus displaced from its position as the center of the universe, being relegated to the status of a mere planet of an undistinguished star in a galaxy which itself is but an ordinary member of an uncountable number of galaxies. The astronomical observations of Galileo (1564-1642), following the invention of the telescope, confirmed the Copernican theory. The researches of Kepler (1571-1630), based on more accurate observations than were available to Copernicus, showed that the planetary orbits were not exactly circular, but elliptical with the sun at a focus. The discovery of the law of universal gravitation by Newton (1643-1727) gave impetus to the theoretical study of planetary motions, and Kepler's laws, which before had seemed distinct and unconnected, were shown to follow as simple deductions from Newton's law. A most sensational achievement was the discovery of a new planet, Neptune, in 1846. Using the discrepancies between the calculated and observed positions of Uranus, Adams (1819-1892) and Le Verrier (1811-1877), independently and unknown to each other, showed that they could be accounted for on the assumption that they were caused by the attraction on Uranus of an unknown planet; moreover, they were able to locate the position of the unknown planet, which was in due course observed through the telescope.

The planets fall into two groups, called the inner and outer planets. In order of distance from the sun, the inner planets are Mercury, Venus, Earth, and Mars, while the outer planets are Jupiter, Saturn, Uranus, Neptune, and Pluto. Some information of a general character on the planets is given in Table 1-1, which also includes data on the sun and moon. It can be seen that the four inner planets are

Table 1-1 CHARACTERISTICS OF THE SOLAR SYSTEM

Body	Mean density, g/cm^3	Total mass (earth masses)	Mean radius, km	Period of rotation,[a] days	Number of identified satellites[b]	Sidereal period,[c] years	Mean distance from sun, AU[d]	Eccentricity of orbit	Inclination of orbit to ecliptic,[e] rad
Sun	1.4	333,441.	696,000	25.36	....	....	....	....	....
Mercury	6.03	0.0556	2434	59.7	0	0.241	0.3871	0.206	0.1221
Venus	5.11	0.8161	6056	−243.09	0	0.615	0.7233	0.007	0.0591
Earth	5.52	1.0123	6370	1.00	1	1.000	1.0000	0.017	
Mars	4.16	0.1076	3370	1.03	2	1.881	1.5237	0.093	0.0322
Jupiter	1.34	318.3637	69,900	0.40	12	11.865	5.2037	0.049	0.0228
Saturn	0.68	95.2254	58,500	0.43	10	29.650	9.5803	0.051	0.0434
Uranus	1.55	14.5805	23,300	0.89	5	83.744	19.1410	0.046	0.0135
Neptune	2.23	17.2642	22,100	0.53	2	165.451	30.1982	0.005	0.0309
Pluto	4.(?)	0.926(?)	3000	6.39	0	247.687	39.4387	0.250	0.2995
Moon	3.34	0.0123	1738	27.32	....	0.0748	....	0.055	0.0899

a The rotation periods for Mercury and Venus have been obtained by radar measurements. Until recently it was thought that the rotation period of Mercury was the same as its rate of revolution around the sun. Venus has a very dense atmosphere (almost entirely carbon dioxide) which conceals its surface and makes an optical estimate of its rotation period exceedingly difficult. The sun's period of rotation varies from 24.7 days at its equator to 26.6 days at latitude 35°.

b A tenth satellite of Saturn was discovered by A. Dollfus in 1966.

c The sidereal period is the time of one revolution with respect to the stars.

d One astronomical unit (AU) is the mean distance from the Earth to the sun. It is 92,960,000 miles, or 149,598,000 km. (The distance from the Earth to the sun varies between 91,500,000 and 94,500,000 miles.) The distance from the Earth to the moon varies between 252,710 and 221,463 miles, the mean distance being 238,857 miles, or 384,403 km.

e The ecliptic is the plane of the Earth's orbit about the sun.

small planets, Earth being the largest of the group. With the exception of Pluto, the outer planets are very much larger than the inner ones. Pluto was discovered in 1930, and at present little is known about it. It has been suggested that it is possibly an escaped satellite of Neptune.

In 1801 a small planet called Ceres was discovered moving in a path between the orbits of Mars and Jupiter. That was the first of the minor planets (or asteroids, as they are often called) to be discovered. There are probably at least 30,000 of them; Ceres, the largest, has a diameter of about 730 km. The orbits of the minor planets lie essentially between those of Mars and Jupiter, and it has been suggested that the minor planets are the remnants of a major planet which was shattered into fragments at some stage in its history. However, a number of arguments have been advanced against this theory (see, for example, E. Anders, 1964, 1965); the origin of the asteroids will be discussed briefly in Sec. 1-5. The planetary distances r from the sun in astronomical units (AU) have been expressed by an empirical relationship known as Titius Bode's law.

$$r = 0.4 + 0.3 \times 2^n \qquad (1\text{-}1)$$

where $n = -\infty$ for Mercury, 0 for Venus, 1 for Earth, etc. There is no planet at a distance corresponding to $n = 3$, although the orbits of the minor planets fall in the area between those of Mars ($n = 2$) and Jupiter ($n = 4$).

Two things about the planetary orbits are of particular interest: (1) all the planets revolve around the sun in the same direction, and (2) the orbital planes of the planets, with the exception of Pluto, differ but little from the plane of the ecliptic. Moreover, the sun, moon, and planets, with the exception of Venus and Uranus, all rotate about their axes in the same sense as the planets revolve around the sun. These very special features must be taken into account when formulating any theory of the origin of the solar system.

By the first quarter of the present century, the development of giant optical telescopes and spectrographic methods of analyzing light had revealed that some of the diffuse and luminous stellar clouds called nebulas are distant concentrations of many thousands of millions of stars called galaxies. The Milky Way system is an example of a spiral galaxy, consisting of some 100,000 million stars in the shape of a lens. The sun is about 26,000 light years from the center of the spiral, which rotates with a period of about 200 m yr. Galaxies tend to be separated in space by distances averaging several times their diameters. They are scattered through all space visible to us with the largest telescopes. The most remote are so distant that light from them takes about 2000 m yr to reach the earth and registers only as faint marks on photographic plates exposed for many hours in the great 200-in. reflecting telescope at Palomar Mountain, California. In the last few years, the invention of another tool, radio astronomy, has made possible the detection of galaxies at even greater distances. Surveys show that galaxies tend to cluster in groups, containing up to a thousand or more. The distribution of clusters is a problem whose solution is vital to the present conflicting views on the origin of the universe.

1-3 AGES OF THE EARTH AND THE UNIVERSE

The ages of the earth and the universe have always been most intriguing problems. Did the universe originate at some finite time in the past, or has it existed forever? Many of the older cosmologies assume that the universe was created in very much the state in which we find it now; in contrast, some recent theories assume that the universe had no beginning. If the first assumption is correct, it is difficult to account for the observed existence of irreversible processes in nature; the second assumption, on the other hand, presents difficulties in accounting for the continued existence of radioactivity.

That the earth has a finite age was suspected by philosophers long before scientists could support such a view, but now there is a wealth of scientific information bearing on the subject. For example, the rivers of the world are continually contributing sodium salts to the oceans at a rate which can be measured; despite this fact, the amount of salt in the oceans is finite. Separate stages in the evolutionary scale have been recognized and suggest a progressive development of life from some obscure but real beginning. The well-known red shift in the light received from distant nebulas shows that they are rushing away from our galaxy at speeds proportional to their distance. If these motions are real and if they have been going on continuously in the past, all the matter spread throughout the universe must some time ago have been compressed into a very small compass. This would imply a possible explosive origin for the universe, which, assuming that the velocities have not changed throughout time, has been estimated to have occurred some 5500 m yr or more ago.

Modern methods of radioactivity and geochronology have been brought to bear on the problem. The fact that radioactive elements having half-lives of 10^9 to 10^{10} years exist in nature shows that the age of the earth cannot be much greater than these figures; conversely, the absence of most radioactive isotopes of shorter half-life suggests an age not much less. Ages determined for specific rocks and minerals by measuring the total amount of end product produced by a radioactive parent exceed 3000 m yr (Chap. 5). Postulated explanations for the variations observed in lead-isotope abundance ratios demand ages for the earth of from 3000 to 5000 m yr. It has also been shown that lead from certain meteorites which contain negligible uranium and thorium could be converted into the lead found in modern rocks by the uranium and thorium in those rocks in a period of some 4500 m yr. (See Sec. 5-11 for more details.)

At present there seems to be little doubt that the earth is between 4000 and 5000 m yr old. The figure 4500 m yr is generally accepted and is satisfactory to all branches of science, being probably as accurate as the ages which can be determined for the time of formation of some minerals and rocks.

1-4 ORIGIN OF THE SOLAR SYSTEM

It must be appreciated that it is by no means certain a priori that the problem of the origin of the solar system can be given a scientific solution. Consider, for example, a vessel in which the air has been stirred. After some time there remains no clue to the nature or time of the stirring. All memory of the event within the system has been lost.

However, the solar system is not infinitely old and shows some properties that must reflect directly or indirectly its mode of formation. Most evidence indicates that the solar system came into being as the result of some single process, and one is led to inquire whether this was not the basic process of stellar formation. One must also consider the question of whether the solar system just after its formation was essentially the same physically and dynamically as it is today, or whether important evolution took place, after the formation period, some 4000 to 5000 m yr ago. G. P. Kuiper (1956) has given good reasons for believing that the planetary masses and compositions have not changed since that time. However, internal rearrangements must have taken place; that is apparent, for example, from the changing features of the earth's crust. Whether such changes are progressive or cyclic is one of the questions on which universal agreement has not been reached. Some of the problems which must be answered are these: Have the continents grown from nuclei during geological time or have they always been roughly the same size and have they merely been reworked? What is the origin of the hydrosphere? While most evidence favors development from the earth's interior during geological time, no direct evidence for the evolution of water from the interior has been found. An attempt to answer these and other questions will be made in the following chapters.

Returning to the problem of the origin of the solar system, all that one can do is to attempt to derive its present state from an assumed event or set of circumstances which occurred in the distant past. In a sense the method is one of trial and error. The various processes which have been suggested can be roughly divided into two classes: those which regard the origin as the result of a gradual evolutionary process and those which attribute it to some cataclysmic action—usually associated with the hypothetical encounter of the sun with a star in the distant past. An example of the first type is the nebular theory of Laplace which he published as long ago as 1796, the main features being originally due to Kant (1755). Laplace supposed that far back in time the sun was a rotating, gaseous nebula. Under the general gravitational attraction, the nebula would gradually contract and its rotation would become more rapid. Laplace then supposed that, when the centrifugal force in the outer layers of the nebula exceeded the gravitational attraction of the nebula as a whole, gaseous matter was thrown off, just as mud is thrown off the rim of a rotating wheel when the rotation is sufficiently rapid. The expelled material later formed a ring, like Saturn's ring, revolving in the equatorial plane of the nebula. The nebula continued to contract while the material of the ring slowly collected into a single aggregation of gaseous matter which, on further condensation and cooling, developed into a planet revolving around the central body. As a result of further contraction of the nebula and its increased rotation, additional material was thrown off, to become another planet by the method already described.

Laplace's hypothesis in its original form was generally abandoned at the beginning of the present century. There are two main difficulties in his scheme:

1 In 1859 J. C. Maxwell, and again in 1900 F. R. Moulton, pointed out that if the mass of the present planets were spread out along Laplacian rings, such rings could never coalesce into planets. In this respect it is interesting to recall Laplace's appeal to the analogy of Saturn's rings. Research on the stability of the rings by Maxwell showed that, however constituted, they could never be collected into a planet as Laplace supposed.

2 Although the sun has 0.999 of the total mass of the solar system, it possesses less than 0.02 of the angular momentum. Yet as Fouché in 1884, and later Moulton in 1900, pointed out, one would expect, on the basis of Laplace's hypothesis, that the sun would rotate with its maximum possible angular velocity compatible with stability. If the whole of the mass and of the angular momentum of the solar system were concentrated in the sun, its period of rotation would be about 12h. The centrifugal force at its equator would then be only about 5 percent of the force of gravity, and rings of matter could not possibly be thrown off in the manner suggested by Laplace.

To answer these difficulties F. R. Moulton and T. C. Chamberlin in 1905 proposed a return to a cataclysmic theory of the origin of the solar system such as the collision of the sun with another body, as had been proposed some 200 years earlier by Buffon. Their ideas were later modified by J. H. Jeans (1919) and H. Jeffreys (1918), who attributed the formation of the solar system to the interaction between the sun and a star during a close encounter. Just as the moon by its gravitational attraction raises tides on the earth, so the star raised immense gaseous tides on the sun, and matter was pulled away from the sun, roughly in the direction of the star, to form a long gaseous filament. The filament, being unstable lengthwise, would soon break into several parts, each forming a distinct aggregation of matter, later to develop by cooling and contraction into a planet.

To explain how planetary matter was removed from the immediate neighborhood of the sun to the present great distances of the planets and set in motion in nearly circular orbits, H. N. Russell suggested in 1935 that perhaps the sun was at one time not a single star but a binary, and that another star was involved in a close encounter with the sun's companion, causing the results already described. This suggestion raises the problem of the removal from the sun's control of its companion while allowing for the retention of the tidal filament which is supposed later to condense into planets. This difficulty was explained by R. A. Lyttleton in 1936, but even so the binary-star hypothesis is still open to many objections and cannot be accepted. For example, if the planets were all formed at much the same distance from the sun, what process has supervened since then to bring about the immense changes in their distances from the sun as required by their present orbits? Moreover, F. Nölcke (1930) had already shown that a filament of matter drawn out from the sun would rapidly disperse, its density being well below the Roche limit.[1] And further, if the matter had been drawn out of the sun by a near encounter with another star, its temperature must have been more than 1 million °K and its pressure more than 1 million atm. At 1 million °K the mean velocity of hydrogen atoms exceeds 150 km/sec, and L. Spitzer (1939) showed that the velocity of escape would in fact be reached within a few hours and that the filament would then dissipate. Some of the material would escape into interstellar space; the rest would form an extended gaseous nebula around one or more of the stars involved. These arguments of Nölcke and Spitzer virtually rule out the possibility of a catastrophic origin of the solar system. P. L. Bhatnagar (1940) also showed that in the two-body problem, the formation of a planetary ribbon was impossible. Lyttleton (1938, 1941a,b) attempted to rescue the theory by supposing that the planets were formed in a two-stage process or that the sun was originally the principal component of a triple system, the other two components being initially close together. However, no plausible explanation can be given of many features of the solar system, and theories of a catastrophic nature have generally been abandoned today.

One is thus led to reconsider the Kant-Laplace hypothesis and in particular the reason for its failure. Any model based on the assumption that the angular momentum and the total mass of the initial solar system have remained unchanged cannot explain the present planetary system. The composition of the planets is so highly selective that the original mass must have greatly exceeded its present value. Advances in astrophysics, particularly in our knowledge of the abundances of the elements, indicate that the composition of the planets is exceptional but that the earth could have formed from matter of approximately solar composition by the loss of most of the gaseous elements. C. V. von Weizsäcker in 1944 returned to the nebular hypothesis. He considered that the sun at some stage in its history passed through a comparatively dense interstellar cloud of gas and dust particles. Galactic space is well filled with clouds, known as diffuse nebulas, which are very extensive. If the sun passed into one of them, it would remain within it for hundreds of

[1] At the Roche density the self-gravitation of the gas cloud will just balance the solar tidal force. Gravitational stability will exist, and hence planetary condensation will proceed, when the critical density is well exceeded.

thousands of years, and owing to its predominant gravitational attraction, it would gather great quantities of nebular material into a vast solar envelope. The mass of the nebula is estimated to have been about one-tenth that of the sun. The envelope would develop slowly as a result of frictional forces into a disklike shape. In 1944 von Weizsäcker attempted to explain the breakup of the solar nebula and the formation of the protoplanets from condensations in this nebular disk. The first attempt was based upon hydrodynamic considerations and was largely qualitative. G. P. Kuiper (1949, 1951a,b) has developed a modified form of the theory in which an important role is played by the process of gravitational instability. Kuiper showed that somewhat less than 1 percent of the mass of the protoplanets condensed into the planets themselves, the larger planets collecting about 100 times as large a fraction of protoplanetary material as did the smaller ones. At the time of von Weizsäcker's work, a school of cosmogony was growing up around the Russian astronomer O. J. Schmidt. Schmidt (1944, 1959) also assumed that the sun passed through a cloud of dust and gas, capturing some of it to form the solar nebula. Since this cloud possesses angular momentum, collisions will in time reduce it to a disk. Since this disk contains dust, the parts nearer the sun will absorb the solar radiation and warm up, while parts farther away will remain cool. This heating results in the dispersal of all volatile material from near the sun. The remaining nonvolatile material then agglomerates to produce the terrestrial planets. In the cooler parts farther away from the sun and its heating effects, all the material will agglomerate to produce the major planets. The theory, however, makes no attempt to explain the slow rotation of the sun.

A completely different type of theory is that proposed by H. Alfvén (1942, 1945, 1954). He assumes that the sun had a general magnetic field and that the gaseous cloud which surrounded it was ionized by radiation from the sun. The cloud was thus electrically conducting, and Alfvén considers that electromagnetic forces are of prime importance. The degree of ionization of any particular element will depend upon the temperature, and thus upon distance from the sun. The gaseous cloud falls in toward the sun under its gravitational attraction, but the magnetic field around the sun impedes the fall of the ionized constituents. Thus there is a gradual diffusion of the nonionized constituents through those that are ionized. Any given constituent, as it falls toward the sun, sooner or later reaches a distance at which, because of its increasing temperature, thermal ionization sets in and the magnetic field begins to act as a brake. Alfvén believes that four main clouds (A, B, C, and D), differing in composition, are produced by this process. Angular momentum is transferred from the sun to the gas clouds by electromagnetic forces, causing a concentration of gas in the equatorial planes. Then through condensation the gas is transformed into small solid or liquid bodies, and the planets are formed by the agglomeration of these bodies. The moon and Mars are supposed to have condensed out of impurities in the A cloud, which consisted mainly of helium. The Earth, Venus, and Mercury condensed from impurities in the B cloud, which consisted mainly of hydrogen. The C cloud consisted mainly of carbon, and it is from this cloud that the four major planets were formed. Pluto, and perhaps the Neptune satellite Triton, formed from the *D* cloud, which consisted mainly of iron and silicon. One of the main difficulties in the theory is that the postulated magnetic fields are required to exceed certain critical values which are larger than seem plausible. In particular, for the sun, the surface field that corresponds to the required dipole moment is 300,000 gauss, whereas the field at the present time is less than 2 gauss. Moreover, some of the mechanisms have not been worked out in sufficient detail.

Theories of the origin of the solar system may also be divided into two other categories:

1 Those that regard the origin and formation of the planets as being essentially related to the formation of the sun.

2 Those that regard the formation of the planets as being independent of the formation process of the sun, the planets forming after the sun had become a normal star.

Category 2 may be subdivided into two subcategories:

2a Where the material for the formation of the planets is extracted either from the sun or from another star.

2b Where the material is acquired from interstellar space.

We have seen that theories in category *2a* cannot offer a satisfactory solution to the origin of the solar system, while in category *2b* only that of Alfvén seems possible. In category 1, two recent theories seem promising, that of F. Hoyle (1955, 1960) and that of W. H. McCrea (1960). Hoyle begins with a cloud of gas possessing angular momentum, approximately that which it would have if it rotated with the general galactic rotation. For a star of the sun's mass, rotational instability sets in when its radius is roughly equal to Mercury's orbital radius. A ring of material is ejected, allowing the star to contract further. This is reminiscent of Laplace's nebular theory. However, unlike Laplace, Hoyle postulated that a magnetic torque comes into existence between the disk and the star. The effect of this torque is to transfer angular momentum from the sun to the disk, resulting in the disk moving outward from the central condensation while the condensation itself contracts further, now with roughly constant angular velocity. The temperature of the solar condensation at the time when the disk was ejected is of the order of 1000°K, and so a number of nonvolatile materials must be in solid form, probably as fine smoke particles. These particles will grow in size as a result of further condensation and accretion. Hoyle showed that these particles will be swept outward with the disk only if their diameter at the Earth's orbit is less than 100 cm. Thus, as the main disk moves out, a subsidiary disk consisting only of nonvolatile material remains behind. Hoyle suggests that the formation of the terrestrial planets takes place in this subsidiary disk, and thus they are different in composition from the sun. When the disk reaches the Jupiter-Saturn region, the gravitational field of the sun is weaker, and particles of diameter less than about 10 m are now swept along. The temperature is also much lower, and ice particles can form. By the time the condensations here are large enough to capture the gas by their own gravitational field, most of the gas will have escaped from the system. Hence Jupiter and Saturn, forming first, each have a large amount of gas, whereas Uranus and Neptune do not. Hoyle thus obtained very good agreement with observations for the mass and composition of the planets. The slow rotation of the sun is also explained.

W. H. McCrea (1960) considered the formation of the sun to have taken place at the same time as the formation of a number of other stars, these stars eventually forming a stellar cluster. Thus he envisaged an initial cloud of several hundred solar masses. However, he did not assume homogeneity and isotropy within the cloud; instead he assumed that the cloud was broken up into many cloudlets, or *floccules*, in random motion among themselves. When two of these floccules collide, parts will coalesce, while the rest proceed as smaller floccules. By a series of collisions, minor condensations are formed throughout the cloud. A reasonably large condensation will have a gravitational field which attracts further material, resulting in a growing condensation. Material entering the condensation must initially have been traveling toward the region, and so material joins it from all directions, so that very little angular momentum is carried into the condensation.

McCrea showed that each major condensation is about one solar mass and that about 1000 floccules will be captured in orbit about it. These floccules will tend to flatten into a plane with a tendency for collisions. In this way floccules orbiting in opposite directions will lose all their angular momentum and fall into the sun. There remain about 200 floccules, all orbiting in the same direction; these will tend to condense in the same manner as that which operated in the formation of the sun. Since a condensation must be about 20 floccule masses in order to hold together, about the correct number of planets will be formed by this process. All the protoplanets will be roughly similar, having about the same mass and composition as the present-day Saturn and Jupiter. The major planets form out of these protoplanets by direct condensation. Another process, that of separating the heavy elements from the hydrogen, plays a part in the formation of the terrestrial planets. McCrea showed that the major planets are outside the Roche limit and can condense, while the terrestrial planets are inside the limit and so will be torn apart. He suggested that a heavy element core is formed in all the protoplanets. Inside the Roche limit the core is all that remains after the tidal action of the sun has torn away the outer layers. This forms planets of the same mass and composition as the terrestrial planets.

Recent developments in astronomy have given much more information about the conditions under which the solar system must have evolved. Thus a number of authors have shown that some of the elements and compounds found on the earth and in meteorites are not compatible with the earth having condensed out of a hot gas cloud of solar composition. Hence the separation of the elements of the earth from the surrounding hydrogen must have taken place at low temperature. It is beyond the scope of this book, however, to pursue the question of the origin of the solar system any further. The origin of the elements is discussed briefly in Sec. 6-8.

Attention must be called, however, to H. C. Urey's (1952b, 1963) investigations of how the planets might have developed from the protoplanets. His work is based to a large extent upon chemical considerations and serves as a timely reminder that in all this work chemistry cannot be ignored, but must be considered equally with the physical arguments. It is more than possible that evidence no longer exists which would enable us to determine beyond a shadow of doubt how the solar system originated. Perhaps the most we can hope for is a theory which is in full accord with established physical principles, which is within the bounds of probability, and which is capable of accounting for the general features of the solar system as known at present. Some recent reviews on the origin of the solar system have been given by D. ter Haar (1967), T. Herczeg (1968), and I. P. Williams and A. W. Cremin (1968).

1-5 METEORS AND METEORITES

Before closing this brief discussion of the solar system, some mention must be made of two intruders, meteors and meteorites. The word *meteor* has now acquired a very broad meaning, describing the whole phenomenon associated with the entry into the earth's atmosphere of a solid particle from space; in particular, it describes the flash of light produced by the interaction and also the ionization generated in the upper atmosphere. Extremely small pieces of cosmic matter enter the earth's atmosphere at great speed (about 10 to 70 km/sec), and friction causes the solid material to be vaporized. High-speed collisions produce luminosity, and the meteor leavcs a bright but highly evanescent trail of light across the night sky. Visual observations of the tracks of meteors, supplemented by photographic and radio techniques, have in recent years added much to our knowledge of the physics of the upper atmosphere.

Occasionally a piece of cosmic matter of very much greater size succeeds in reaching the earth's surface before it is completely vaporized as it passes through the atmosphere. Such a body is called a *meteorite*, and a few thousand meteorities, weighing anything up to a few tons, have been picked up from all over the earth, although large ones are exceedingly rare. Meteorites are extremely valuable in any cosmogonic studies because, apart from the lunar samples brought back by the Apollo and Luna spacecraft, they are the only samples of extraterrestrial material we have, and they may give us information about the composition of the earth's interior. It is believed that meteors are small solid particles imbedded in the icy nucleus of a comet that have become freed as solar radiation vaporizes the ices when the comet approaches the sun. Meteorites, on the other hand, probably come from the asteroidal zone.

It has been estimated that about 1500 meteorites with a mass greater than 100 kg reach the neighborhood of the earth each year, although a mass of 100 kg would be reduced to about 10 kg by the time it hit the ground (G. S. Hawkins, 1964). The recovery rate, however, is only about 5 to 10 per year, most of them landing in the oceans and in uninhabited regions of the earth. Meteorites are classified according to their composition and structure (see, for example, B. Mason, 1962). Those that consist almost entirely of nickel iron are called *siderites*, or simply iron meteorites; those that are chiefly composed of iron and magnesium silicates (mainly olivine and pyroxene) with comparatively little nickel iron are called *aerolites*, or stony meteorites. There is also a third class called *siderolites*, or stony irons, which contain about 50 percent nickel iron and 50 percent silicates. None resembling sedimentary or metamorphic rocks have ever been found, but many meteorites contain small quantities of a peculiar iron-sulfide mineral (FeS) called troilite, and a few contain carbonaceous material.

Although many analyses of meteorites have been carried out, it is difficult to combine them in such a manner as to give an average composition for meteoritic matter as a whole. It is not sufficient to take the proportions of nickel iron, silicate, and troilite from the average content of all meteorites collected. Since about two-thirds of all meteorites found are irons, it would appear at first sight that irons are more abundant than other types. However, of all the meteorites observed to fall, over 90 percent are stones. The reason for this apparent paradox is that irons are easily recognized as meteorites, whereas stony meteorites, unless observed to fall, would easily be overlooked as such, and iron meteorites are much more resistant to weathering than stones. Thus a truer indication of the relative abundance of the different meteorite types is given from those observed to fall. Table 1-2 gives the elemental abundances in meteoritic matter according to V. M.

Table 1-2 THE AVERAGE COMPOSITION OF METEORITIC MATTER

		Goldschmidt (1937), ppm	Urey (1952a), ppm	Levin et al. (1956), ppm	Mason (1962)	
					ppm	atoms per 10^6 atoms Si
3	Li	4	5	3.2	2	47
4	Be	1	1	0.09	1?	18?
5	B	1.5	1.5	2.6	2?	30?
8	O	323,000	....	346,000	330,000	3,420,000
9	F	28	40	40	30?	260?
11	Na	5950	7500	7000	6800	49,000
12	Mg	123,000	135,500	139,000	138,600	945,000
13	Al	13,800	14,300	14,000	11,000	68,000
14	Si	163,000	179,600	178,000	169,500	1,000,000
15	P	1050	1500	1600	1300	6950
16	S	21,200	20,100	20,000	20,700	107,000
17	Cl	1000–1500?	470	800	100?	500?
19	K	1540	900	900	1000	4300
20	Ca	13,300	14,300	16,000	13,900	57,600
21	Sc	4	5	5	9	33
22	Ti	1320	580	700	800	2800
23	V	30	50	80	65	210
24	Cr	3430	2700	2500	3000	9600
25	Mn	2080	2400	2000	2000	6000
26	Fe	288,000	241,000	256,000	286,000	849,000
27	Co	1200	1100	900	1000	2800
28	Ni	15,680	14,500	14,000	16,800	47,400

Table 1-2 THE AVERAGE COMPOSITION OF METEORITIC MATTER (*continued*)

		Goldschmidt (1937), ppm	Urey (1952a) ppm	Levin et al. (1956), ppm	Mason (1962)	
					ppm	atoms per 10^6 atoms Si
29	Cu	170	170	(40)	100	260
30	Zn	138	76	20	50	125
31	Ga	4.2	4.6	8	5	12
32	Ge	79	53	(40)	10	23
33	As	....	18	70	2	4
34	Se	7	6.7	9	9	19
35	Br	20	25	(22)	10?	21?
37	Rb	3.5	8	8	3	6
38	Sr	20	23	22	11	21
39	Y	4.72	5.5	5	4	7
40	Zr	73	80	90	33	75
41	Nb	....	0.41	0.5	0.5	0.9
42	Mo	5.3	3.6	5	1.6	2.8
44	Ru	2.23	1.4	2	1	1.6
45	Rh	0.80	0.47	0.6	0.2	0.3
46	Pd	1.54	0.92	0.5	1?	1.6?
47	Ag	2.0	1.35	0.5	0.1	0.15
48	Cd	....	1.6	2	0.5?	0.7?
49	In	0.15	0.2	0.2	0.001	0.001
50	Sn	20	14	20	1	1.4
51	Sb	....	0.64	0.4	0.1	0.14
52	Te	0.1?	0.13	0.14	1	1.3
53	I	1	1.25	1	0.04	0.05
55	Cs	0.08	1.1	0.08	0.1	0.12
56	Ba	6.9	2.9	7	3.4	4.1
57	La	1.58	1.9	200	0.33	0.40
58	Ce	1.77?	2.1	2	0.51	0.62
59	Pr	0.75	0.88	0.8	0.12	0.15
60	Nd	2.59	3.0	3	0.63	0.74
62	Sm	0.95	1.1	1	0.22	0.25
63	Eu	0.25	0.27	0.3	0.083	0.097
64	Gd	1.42	1.7	1.6	0.34	0.36
65	Tb	0.45	0.52	0.5	0.051	0.056
66	Dy	1.80	2.1	2	0.37	0.39
67	Ho	0.51	0.60	0.6	0.075	0.078
68	Er	1.48	1.7	1.7	0.21	0.21
69	Tm	0.26	0.31	0.3	0.038	0.039
70	Yb	1.42	1.7	1.6	0.19	0.19
71	Lu	0.46	0.54	0.5	0.036	0.036
72	Hf	1.6	1.6	0.8	1.4	1.4
73	Ta	....	0.28	0.3	0.02	0.019
74	W	15	16	17	0.14	0.13
75	Re	0.0020	0.08	0.0018	0.08	0.08
76	Os	1.92	1.2	1.1	1.0	1.0
77	Ir	0.65	0.38	0.6	0.5	0.43
78	Pt	3.25	1.9	3	5?	4.3?
79	Au	0.65	0.25	0.26	0.3	0.25
80	Hg	....	<0.01	(0.009)	0.1?	0.8?
81	Tl	....	0.15	(0.14)	0.0004?	0.0003?
82	Pb	11	....	(2)	0.15	0.12
83	Bi	....	0.02	(0.16)	0.003?	0.002?
90	Th	0.8	....	0.2	0.04	0.028
92	U	0.36	....	0.05	0.014	0.097

FIGURE 1-1
Barringer meteorite crater, Arizona. Looking northwest from a height of 1400 m above the plateau. (*After J. S. Shelton, 1966.*)

Goldschmidt (1937), H. C. Urey (1952a), B. Y. Levin et al. (1956), and B. Mason (1962). The proportions of silicate, troilite, and nickel iron in average meteoritic matter are about 13:1:3.

A meteorite with a mass less than about 1000 kg approaches the earth with an average velocity in the range of 15 to 20 km/sec. By the time it has reached a height of 20 km, the object falls freely at its terminal velocity—usually about 0.1 km/sec—and this makes a fairly soft landing. If the original mass is greater than about 10^6 kg, the meteorite is not decelerated to any appreciable extent, and strikes the surface of the earth with its original cosmic velocity, with catastrophic effects.

Many features of the earth's surface are attributed to the impact of large meteorites. The most famous is the Barringer crater in Arizona, which is about 1300 m in diameter and 175 m deep (Fig. 1-1). It has been estimated that the mass of the object that formed this crater was about 10^9 kg and that its age is about 25,000 years. Detailed study of the earth's surface features has revealed many more ancient craters, such as the great Ashanti crater in Ghana, now filled by Lake Bosumtwi, which has a diameter of 10 km. At an estimated rate of one great fall

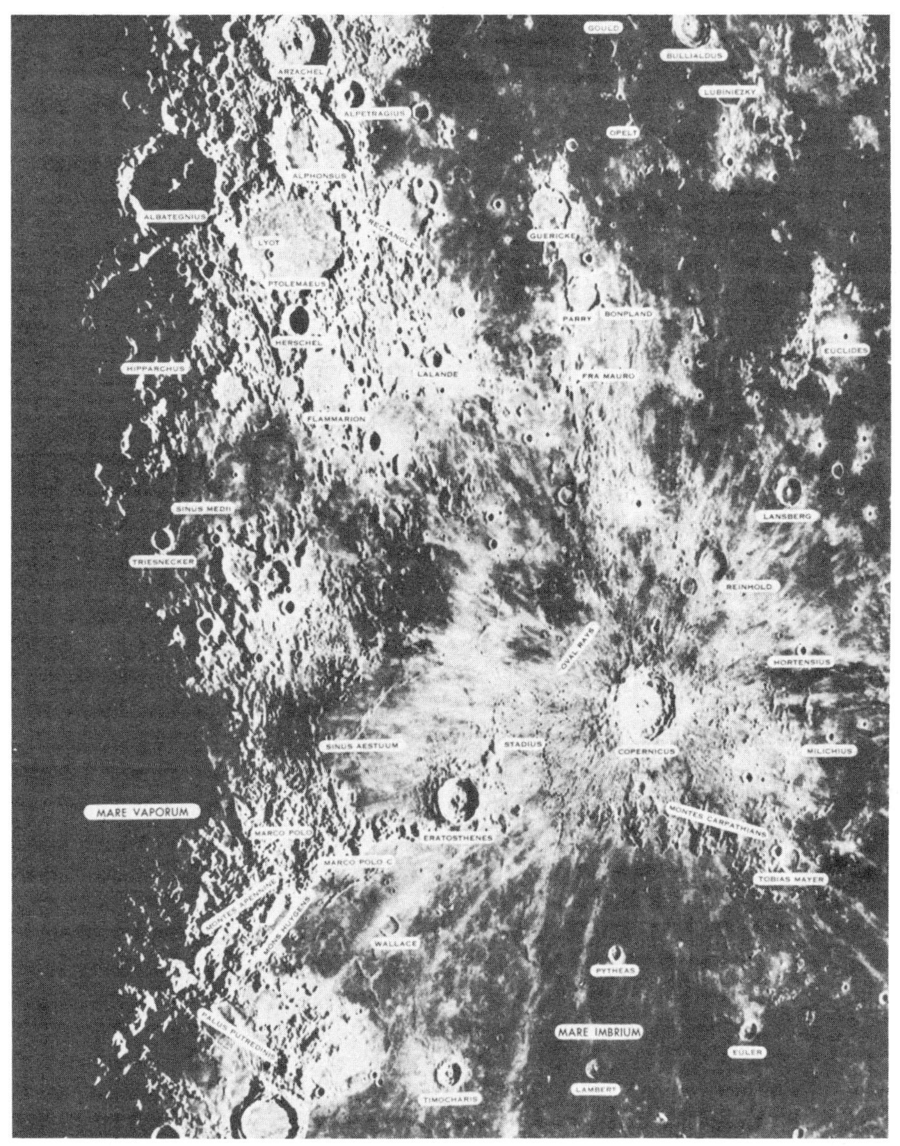

FIGURE 1-2
Craters on the moon. Satellite photographs have revealed a great amount of extra detail on the moon's surface features. In particular, there are many more craters too small to be seen from the earth. (*Plate 32, Lunar Atlas, edited by Dinsmore Alter, Dover Publishers, Inc., 1968.*)

FIGURE 1-3
Brent crater in central Ontario, a shallow depression about 3 km across. Its circular form, never observed at ground level, shows up clearly from the air. Geophysical investigations and borings show that it is a meteorite crater filled with sedimentary rocks at least 400 m yr old. (*Dominion Observatory, Ottawa, Canada, photograph 5015.*)

every 10,000 years, some 50,000 giant meteorites must have struck the earth since the Precambrian. However, on the earth the craters they made would not persist as visible features. Tectonic processes would alter their round shapes, erosion wear away their rims, and sedimentation fill them up. On the moon, however, which lacks water and an atmosphere, meteorite craters have remained unchanged, except through the impact of later meteorites (Fig. 1-2). Craters clearly visible on the earth today must all have been formed by impact during the last few million years.

A search of air photographs has turned up a number of faint circular features which further investigation has identified as fossil meteorite craters, or *astroblemes*, as they have been called by R. S. Dietz (1963a). C. S. Beals and his colleagues (1963) have been particularly active in this field. After careful ground examination of many circular structures in Canada, they have identified 16 probable and 11 possible craters (see Fig. 1-3 as an example). This number is to some extent a matter of personal opinion and varies considerably from one worker to another.

The world list contains about 55 probable meteorite craters. The speed of a meteorite large enough to produce a huge crater would hardly be reduced at all as it passed through the earth's atmosphere. The shock upon impact would generate pressures of the order of millions of atmospheres (volcanic explosions, in contrast, involve pressures of hundreds of atmospheres). The explosive force of the meteorite that caused the Barringer meteorite crater in Arizona must have been of the order of a 30-megaton hydrogen bomb; the meteorite itself would not have survived the impact.

R. S. Dietz distinguishes between meteorite craters and astroblemes on morphological grounds. In his usage astroblemes are old structures attributable to meteorite impact but lacking crater form. Although the details of rim deformation in young impact craters are distinctive, the topography and gross structure of older impact craters can be approximated by volcanic features (hence the arguments about the origin of lunar craters), so that other criteria are necessary. These include the following.

1-5.1 Discovery of Associated Meteorite Fragments

These are only preserved at young sites ($\approx 10^5$ years old or less) in surficial deposits. Deeper layers exposed by erosion or drilling probably contain some meteoritic material, but it defies recognition due to intense shock metamorphism, fragmentation, and mixing with pulverized or melted rock.

1-5.2 Shatter Cones

Shatter cones are conical fragments of rock characterized by striations radiating from an apex (Fig. 1-4*a* and *b*). They vary in size up to a meter or more in length, but complete cones are rare. The fracture surfaces are tensional in character and do not involve shear. They form when a rock is compressed by a shock pulse of the necessary energy density and then rapidly unloaded. The cones are probably generated at minor inhomogeneities in the rock. The best cones are formed in fine-grained carbonate rocks at shock pressures estimated at several tens of thousands of atmospheres. Shatter cones have identified the Vredefort Ring in South Africa as the remains of an old crater some 40 km in diameter.

1-5.3 Coesite and Stishovite

Coesite and stishovite are high-pressure forms of silica. Coesite was first formed artificially by L. Coes, Jr. (1953), and stishovite by S. M. Stishov and S. V. Popova (1961). The stability field for coesite lies above about 20,000 atm, and for stishovite above about 90,000 atm. [See, for example, P. D. Gigl and F. Dachille (1968), who examined the stability of stishovite and concluded that natural stishovite can only come from high-pressure impacts.] The coesite and stishovite

(a)

(b)

FIGURE 1-4
(a) Shatter cone from Gosses Bluff, Northern Territory, Australia. (*Personal photograph from D. J. Milton, 1969.*) (b) Two samples of shatter cones from the Charlevoix site northeast of Quebec City. The left-hand sample is fine-grained Ordovician limestone; the right-hand sample is granodioritic gneiss. P. B. Robertson's (1968) studies of these cones provided the first link between shatter coning and microscopic evidence of shock deformation. (*Dominion Observatory, Ottawa, Canada, photograph 6633.*)

seen in naturally shocked rocks are the products of strong shock metamorphism and probably formed at shock pressures of about 400,000 atm (E. C. T. Chao, 1968). Coesite has now been found in the highly sheared and fused sandstone at the Barringer meteorite crater, and in the Ries Kessel, an ancient basin some 25 km in diameter in southern Germany, which had long been supposed to be of volcanic origin.

1-5.4 Shock Metamorphism

Breccias and fractured rocks from the central regions of impact craters show evidence of shock metamorphism, including planar structures in crystals of quartz and feldspar, kinking in mica, vitrification without melting (by destruction of long-range order in crystals) to form maskelynite and similar phases, and ultimately complete fusion (B. M. French and N. M. Short, 1968).

1-5.5 Lack of Volcanic Roots

Drilling and geophysical surveys at a number of craters (six have been drilled in Canada) have demonstrated that crater deformation dies out at depth. This is difficult to demonstrate unambiguously in large craters, except by very extensive drilling programs. However, the igneous rocks found in large craters can be accounted for by shock melting (extreme shock metamorphism) and are not necessarily an indication of volcanic intrusion. This has led to controversy at Sudbury, Ontario, where R. S. Dietz (1963a,b) has suggested that the numerous shatter cones found there are evidence that the eruptive is an astrobleme, or ancient meteorite crater. In this case the intrusive sheet is 55 by 30 km and about 2 km thick. If this view is supported, it raises the question of whether even the larger isolated intrusive lopoliths like the Bushveldt complex in South Africa could not also be due to meteoritic impact.

1-6 ORIGIN OF METEORITES

Evidence presented in later chapters indicates that the earth has a nickel-iron core surrounded by a rocky mantle, and it is thus natural to suppose that iron meteorites come from the core of a fair-sized planet and that stony meteorites come from the surrounding mantle. Astronomical evidence indicates that meteorites may be asteroids that have been deflected from the asteroidal zone between Mars and Jupiter, and that the asteroids themselves may be fragments of a former planet. It is not possible in this book to consider in detail the problem of the formation of the asteroids. It is unlikely that they were formed almost as they are now—a swarm of irregular fragments—although some authors (for example, E. Anders, 1964, 1965) believe that the asteroidal belt is remarkably like its primitive condition, despite many collisions and some breakup of members up to 250 km in diameter. It is also unlikely that they were formed by the breakup of a single original planet–the amount of energy required to explode a planet would be enormous. It is more likely that two or more small planets, like Ceres, were formed in the asteroidal zone and suffered collision. The planets must have been of sufficient size for the interior to have become heated by radioactivity and for a chemical separation of iron and silicon to have taken place. However, the planets must have been somewhat smaller than the moon since, if an object the size of the moon had become molten in its interior, there would not have been sufficient time in the history of the solar system for the object to have cooled down and solidified. Iron meteorites show crystal patterns and were undoubtedly formed from a solid, not a liquid, core. The original planet was thus about 1000 km in diameter. Most recent estimates favor several (four or more) parent bodies for the meteorites to account for the chemical differences. The sizes of the parent bodies were probably between 50 and 250 km in diameter, at least for the iron meteorites.

Stony meteorites may be subdivided into two classes: *chondrites* and *achondrites*. Eighty-five percent of all witnessed meteorite falls and 95 percent of all stones are chondrites. Meteorites generally contain a few major constituents, although nearly fifty different minerals have been identified in them. The chondrites are composed of olivine, pyroxene, oligoclase, troilite, and flecks of nickel iron. When viewed through a microscope, a thin section of a chondrite shows these minerals in a disorderly array, quite unlike any terrestrial rock, with the fragments no more than a fraction of a millimeter across. The distinguishing feature is the *chondrule*, a round inclusion, typically 1 mm in diameter. J. A. Wood (1963a,b) gives an excellent description of chondrules and presents some evidence for glass in many of them. They appear to be frozen liquid droplets, mainly of silicate, with some metallic iron, and may be primordial planetary matter, although other origins have been suggested.

The *carbonaceous chondrites* form a particularly interesting subgroup. They are extremely rare, less than 2 percent of the total number of stones. They are very dark in color throughout because of the inclusion of about 5 percent of a black, tarlike material presumed to be a complex polymer of high molecular weight. They are fragile and porous and have a low density ($\approx 2.2\ g/cm^3$). The low crushing strength may be responsible for their scarcity. The porous structure may account for the discovery of microscopic structures within the meteorite, raising the intriguing question of the possibility of extraterrestrial life.

The *achondrites* do not contain chondrules. The most common minerals are pyroxene and plagioclase. They are very similar to terrestrial rocks such as gabbro. The achondrites are considered to show the effect of metamorphism and are thought to have crystallized from a magma in the same way that terrestrial rocks were produced. The achondrites could be derived from chondritic material that has been melted and subsequently recrystallized. As well as showing igneous textures, most achondrites have been shocked and brecciated.

It has already been suggested that nickel-iron meteorites probably crystallized very slowly in the molten core of a small planet. Two processes can take place in a nickel-iron mixture: first a crystal can grow within the melt, then slow changes can take place in the solid crystal as it cools. Iron with a low nickel content crystallizes at a high temperature, and so a nickel-iron mineral (*taenite*) is the first to form within the melt. As the temperature falls, this mineral transforms to *kamacite*, a body-centered cubic crystal with a nickel content between the narrow limits of 5 to 6 percent. The shape of the crystal is a hexahedron, a six-sided figure; this mineral is found in iron meteorites called *hexahedrites. Neumann bands*, caused by a slippage of the crystal structure, like the sliding of a pack of cards, are found in all hexahedrites and are most readily produced when the crystal is subjected to a violent impact between temperatures of 300 to 600°C. The bands were probably produced when the solidified core of the small planet was smashed by encounter with another body.

As the kamacite crystals form, the mother liquid becomes richer in nickel. Crystals of taenite again form when the nickel content is greater than 6 percent, but nickel-rich taenite does not degenerate during the annealing process. Taenite forms in octahedral crystals and is the basic constituent of an octahedrite, the most common type of iron meteorite. A typical polished and etched surface of an octahedrite shows the *Widmanstatten pattern*, which is caused by a minor degenerative process in the octahedral crystal during the annealing process. The bands which are poor in nickel are kamacite.

As crystallization proceeded in the core, the mother liquid would be further enriched in nickel. When the nickel content reached about 25 percent, the octahedral crystals of taenite would form as before. However, with the high percentage of nickel, no degeneration would take place in the solid crystal as it annealed, which accounts for the nickel-rich *ataxites.* Approximately 1 iron meteorite in 100 contains more than 25 percent nickel, and then the inclusion of kamacite is very rare. In the laboratory, the melting point of a nickel-iron alloy becomes lower as more nickel is added. The lowest melting point occurs at 1430°C at normal atmospheric pressure when the nickel content is 68 percent. This means that the mother liquid of the nickel-iron core would be expected to solidify as a whole when the nickel content reached about 68 percent. The maximum nickel content observed so far in a meteorite is 62 percent—a remarkable confirmation of the hypothesis that all nickel-iron meteorites were formed by solidification from a nickel-iron melt in the core of a small planet. However, while the nickel-iron compositions of iron meteorites can be explained by phase chemistry in terms of differentiation within a single parent body, other data favor several parent bodies (see E. Anders, 1964, for a full discussion).

SUGGESTIONS FOR FURTHER READING

The problem of the origin of the solar system has attracted much attention over the years, and the amount of literature on the subject is enormous. No attempt has been made to compile a comprehensive review or bibliography. The same applies to a lesser extent to meteorites. Short up-to-date reviews on these topics have been presented by:

BERLAGE, H. P.: "The Origin of the Solar System," Pergamon Press, New York, 1968.

HAWKINS, G. S.: "Meteors, Comets and Meteorites," McGraw-Hill Book Company, New York, 1964.

WOOD, J. A.: "Meteorites and the Origin of Planets," McGraw-Hill Book Company, New York, 1968.

Older references include:

ALFVÉN, H.: "On the Origin of the Solar System," Clarendon Press, Oxford, 1954.

HOYLE, F.: "Frontiers of Astronomy," William Heinemann, Ltd., London, 1955.

MASON, B.: "Meteorites," John Wiley & Sons, Inc., New York, 1962.

SCHMIDT, O.: "A Theory of the Origin of the Earth: Four Lectures," Lawrence and Wishart, 1959.

UREY, H. C.: "The Planets: Their Origin and Development," Yale University Press, New Haven, Conn., 1952.

2
SEISMOLOGY AND THE INTERIOR OF THE EARTH

2-1 STRESS AND STRAIN

Many of the problems of the physics of the earth are mechanical, and hence an understanding of the mechanics of solids and fluids is necessary. The fundamental problem is the relationship between stress and strain. Broadly speaking, *stress* specifies the nature of the internal forces acting within a material, while *strain* defines the changes of size and shape arising from those forces.

Stress is measured by the force **F** per unit area across a surface within the material. Since **F** may be at any angle to the surface, its components in any three perpendicular directions are needed to specify it completely. Again, since the surface may be drawn in any direction, it would seem that the specification of stress is extremely complex. However, it can be shown that it is only necessary to consider three mutually perpendicular directions for the surface, the stress components across any other surface being expressible in terms of those across those three directions. Although it would thus appear that nine components are necessary to completely define the stress at a point P, it can be proved that in fact only six are necessary. Taking mutually perpendicular axes $P(x,y,z)$ at the point P, the stresses on the yz plane may be denoted by p_{xx}, p_{xy}, and p_{xz} (Fig. 2-1). Here the first subscript x indicates that the plane is perpendicular to the x direction, i.e.,

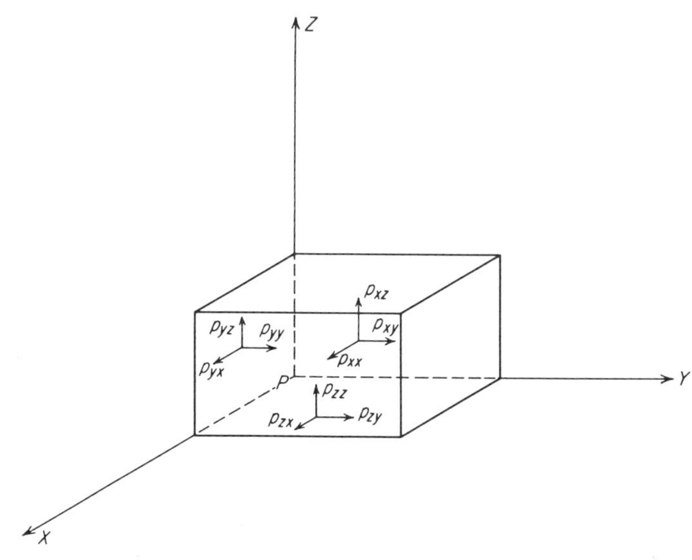

FIGURE 2-1
Representation of the stresses acting at a given point.

that it is the yz plane, while the second subscript denotes the direction of the stress component. We can similarly define the stress components across the zx and xy planes. Each of the components p_{xx}, p_{yy}, and p_{zz} is normal to the plane across which it acts. If the three components are directed toward the material on which they act, they are thrusts; if away from the material, they are tensions. p_{xy}, p_{yz}, ... are called shearing stresses, p_{xy} denoting the shear across the plane perpendicular to x resolved in the direction y. It can be shown that $p_{xy} = p_{yx}$, $p_{yz} = p_{zy}$, and $p_{zx} = p_{xz}$. It is the equality of these cross-shears that reduces the nine components of stress at a point to six. A further simplification is possible if the directions of the axes are suitably chosen. No matter what the state of stress is, three perpendicular directions such that a stress across a plane containing any two of them is wholly along the third can always be found; i.e., the state of stress can be represented by normal stresses only, the shearing stresses across the planes being zero. These stresses are called the principal stresses, and the directions of the axes, the principal axes. In a perfect, i.e., nonviscous, fluid or a real fluid at rest, the principal stresses are all equal; in solids and real fluids in motion, that is not in general so. If the principal stresses happen to be equal, the stress is called hydrostatic.

There are two kinds of displacement, namely, rigid-body displacement and deformation. A rigid-body displacement consists of a translation and a rotation without change of size or shape. It is change in size or shape that characterizes deformation and is called strain. More precisely, a strain which causes only a change in shape with no change in volume will be called a *distortion*; a change in volume without change in shape will be referred to as a simple expansion or contraction. Just as the state of stress at a point can be represented by nine components p_{ij} $(i,j = x,y,z)$, so the strain can be defined by nine components ϵ_{ij} $(i,j = x,y,z)$. ϵ_{xx}, ϵ_{yy}, ϵ_{zz} are simple expansions or contractions; ϵ_{xy}, ϵ_{yz}, . . . are shear strains. These two arrays of nine numbers which specify the state of stress and strain at any point are examples of tensor quantities; they are discussed in more detail in Sec. 11-4. Both the stress and strain tensors are symmetrical tensors, that is, $p_{xy} = p_{yx}$, etc., and $\epsilon_{xy} = \epsilon_{yx}$, etc., so that of the nine components of stress and strain at any point, only six are independent.

All the above definitions apply equally well to solids and fluids, the difference between the two states of matter being the stress-strain relations. We will consider first a perfectly elastic isotropic material. In *perfect elasticity*, the state of strain at any point (under specified thermodynamic conditions) is determined by the state of stress. By *isotropic* is meant that the elastic behavior is independent of any particular direction.

If a weight W is hung on the end of a wire of length l and cross-sectional area S, the wire is stretched and becomes thinner. The stress is W/S, and the strain is defined as $\delta l/l$, where δl is the elongation. Hooke's law shows that there is a linear relationship between stress and strain, the ratio stress/strain being defined as Young's modulus E. Thus $E = Wl/(S\,\delta l)$. If the reduction in diameter d is δd, then the ratio $(\delta d/d)/(\delta l/l)$ is called Poisson's ratio σ.

Other simple experiments may be carried out; for example, the above wire may be twisted by a torque Q, producing this time a change of shape but no change of volume. The torque Q will set up shearing stresses q proportional to the angle $\delta\theta$ through which the wire has been twisted. The relationship is linear and may be written $q = C\mu\,\delta\theta$, where C depends only on the dimensions of the wire and μ is called the modulus of rigidity and, like E, is different for different materials. Again, a sphere may be compressed by a uniform pressure p, thus causing a change in volume but no change in shape. If the volume V is reduced by an amount δV, the relationship between the stress p and the strain $\delta V/V$ is again linear and is written $p = k\,\delta V/V$, where k is the bulk modulus, or incompressibility. The numbers E, σ, μ, and k are known as elastic constants and are different for different materials. There are many other notations for the stresses, strains, and some of the elastic constants. No useful purpose would be served by giving all of them, and the above notation will be used consistently throughout this book. Only two of the four elastic constants are independent, however, and all four may be expressed in terms of any two.

The above experiments are but simple illustrations of a generalization of Hooke's law, namely, that the components of stress are homogeneous linear functions of the components of strain, and vice versa. This is true only if the displacements are small. Mathematically speaking, this means that the displacements must be sufficiently small for their second powers to be neglected in equations containing significant first-order terms. The theory developed with this restriction is known as the infinitesimal-strain theory, and in many cases it agrees well with what happens in practice. However, sometimes when the stresses are high, the theory is inadequate and finite-strain theory must be used, although the mathematical difficulties in such a treatment are very great. It has been shown that in seismology the infinitesimal-strain theory will, in general, be sufficiently accurate.

For a liquid it is found that under a given stress the shear is not constant but is proportional to time, the ratio of stress to the shear per unit time being called the *viscosity*. Whether the materials which compose the earth exhibit viscous or plastic flow will be considered in Chap. 11.

2-2 WAVE MOTION

The earth is continually undergoing deformation owing to the stresses which are set up within it. If the stresses are not too great, elastic or plastic deformation will occur. However, if the stresses continue to build up over a long time, fracture may eventually take place. This involves a sudden release of stress, and the disturbance will set up elastic waves which will travel through the earth. These waves emanate from a confined region below the surface of the earth, called the *focus* of an earthquake.

The energy released by a large earthquake is of the order of 10^{25} ergs. However, the time during which this main energy is released does not exceed the order of a second or so, while the linear dimensions of the focal region may be of the order of several kilometers. It can be shown that three types of waves can be propagated: P and S waves, both of which are body waves, and surface waves. P waves are longitudinal waves; i.e., as the waves advance, each particle of the solid is displaced in the direction of travel of the waves, as is the case in sound waves. S waves are transverse waves, as are light waves. The velocity of P waves is given by

$$V_p = \sqrt{\frac{k + \frac{4}{3}\mu}{\rho}} \tag{2-1}$$

and the velocity of S waves by

$$V_s = \sqrt{\frac{\mu}{\rho}} \tag{2-2}$$

where ρ is the density. Both wave velocities thus depend only on the elastic constants and the density of the medium. In particular, if the rigidity μ is zero, then V_s is zero; i.e., shear waves cannot be transmitted through a material of zero rigidity. It follows from the above equations that

$$V_p^2 - \tfrac{4}{3}V_s^2 = \frac{k}{\rho} \tag{2-3}$$

These relations will be used later (Secs. 2-5 and 2-6) to find the variation of physical properties, such as density, throughout the earth.

If a stone is thrown into a pond or a strong wind blows across an open stretch of water, the surface of the water is agitated and waves that travel away from the source of disturbance are set up. On deep water the amplitude of the wave is greatest at the surface; at a depth of a wavelength, the displacement at any point is but a fraction of what it is at the surface. Such waves, called surface waves, can also occur in a solid; in that case they are controlled by elasticity, whereas the surface waves on water in the example given above are controlled by gravity. There are two types of surface waves in solids, Rayleigh waves and Love waves. In Rayleigh waves the displacement of the surface is partly in the direction of propagation and partly vertical. Rayleigh waves can be generated on a uniform solid, whereas Love waves are possible only if the material is nonuniform. For Love waves there must be a superficial layer resting on another layer, the velocity of *S* waves being less in the upper layer than in the lower. The displacement in these waves is entirely horizontal and at right angles to the direction of propagation. The velocity of Love waves depends on the ratio of the wavelength to the thickness of the layer; thus, unlike *P* and *S* waves, Love waves show the phenomenon of dispersion. If the velocity is found by observation for a number of wavelengths, it is possible to determine the thickness of the layer. For more details about surface waves, the reader is referred to the works of H. Jeffreys, B. Gutenberg, K. E. Bullen, W. M. Ewing, and F. Press.

When a wave meets a surface of discontinuity such as the surface of the earth or the boundary between the core and the mantle, part of it will be reflected and part refracted, and laws of reflection and refraction analogous to those of geometrical optics apply. The case of elastic waves is more complicated, however, because waves of both *P* and *S* type may be reflected and refracted. Thus an incident *P* wave can be reflected to give both a *P* wave and an *S* wave (called *PP* waves and *PS* waves). Likewise, an incident *S* wave can be reflected to give both a *P* and an *S* wave (called *SP* waves and *SS* waves). The symbol *c* is used to indicate an upward reflection at the core boundary, and the symbol *K* to denote the part of the path of a wave which is refracted through the outer core. Thus *ScS* is an *S* wave that has traveled down to the core boundary and been reflected as an *S* wave. Similarly, *SKP* is an *S* wave that has been refracted into the core and refracted back into the mantle as a *P* wave. *S* waves have never been observed in the core, which is the main reason for believing that the core is liquid, since a liquid cannot transmit shear waves. However, there is now evidence that the innermost region of the earth, called the inner core, may have substantial rigidity and be able to transmit *S* waves. Since *S* waves have never been found in the outer core, the symbol *K* is associated with *P* waves. The symbol *I* is used for a segment of *P* type in the inner core, and the symbol *J* has been proposed for paths of *S* type in the inner core. Thus *PKIKP* is a *P* wave which has passed through the inner core, remaining a *P* wave throughout its path. Figure 2-2 illustrates some of the

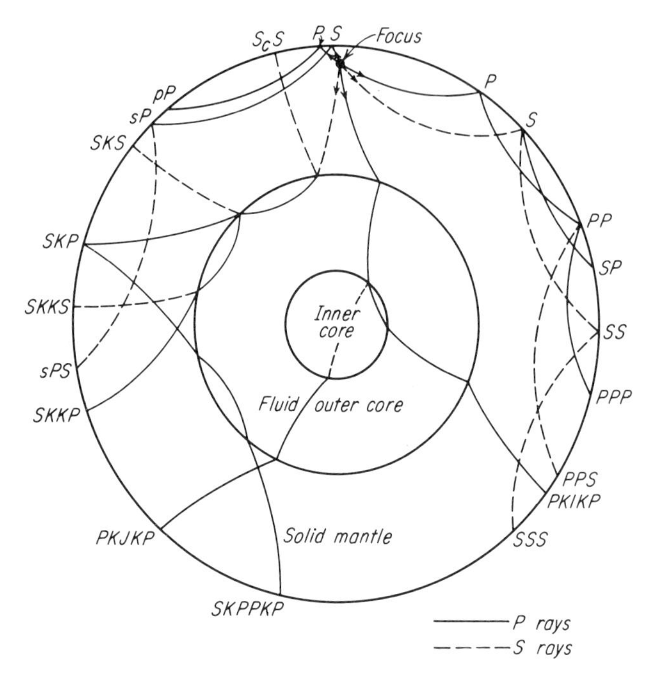

FIGURE 2-2
Representative seismic rays. (*After K. E. Bullen, 1954.*)

great number of possible reflections and refractions that may result from any disturbance.

2-3 TRAVEL–TIME TABLES AND VELOCITY–DEPTH CURVES

P and *S* waves and surface waves were first distinguished in actual records of earthquakes by R. D. Oldham, in 1900. His studies showed that at any station the first indication of an earthquake was the sudden arrival of a *P* wave. This was followed by a train of waves, and at some later time another sharp movement announced the arrival of *S* waves. *S* waves were followed by a series of smooth waves of increasing amplitude that reached a maximum before gradually dying down. These were at first identified as Rayleigh waves; the interpretation, however, was inadequate. Rayleigh waves should give no horizontal displacement at right angles to the direction of travel; yet actual records showed strong movements in that direction. The surface waves have been explained by A. E. H. Love when there is a superficial layer resting on another layer. Both Love and Rayleigh waves have now been identified; they arrive at different times, and in any time interval have different periods. Figure 2-3 shows two typical seismograms and illustrates the arrival of the different types of waves.

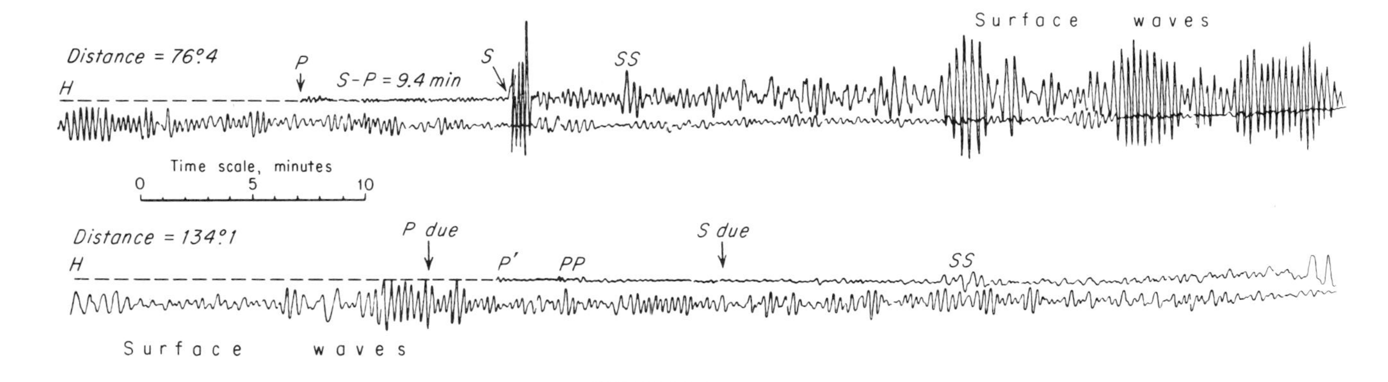

FIGURE 2-3
Two earthquake records. The first is a copy of the Milne-Shaw EW component, Harvard Seismograph Station, Dec. 1, 1928. Earthquake in Chile at 34° S, 73° W. The second is a copy of the Milne-Shaw NS component, Harvard Seismograph Station, June 16, 1929. Earthquake in New Zealand at 41.8° S, 172.2° E. H on each record is the time at which the earthquake occurred. In each case the lower curve is a continuation of the upper one. (*After L. Don Leet.*)

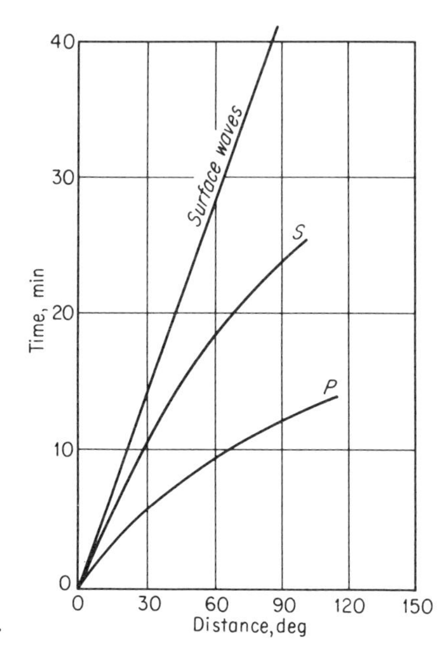

FIGURE 2-4
Travel-time curves.

The times at which signals from the same seismic shock arrive at different stations can be recorded, so that it is possible to determine the travel time of the disturbance as a function of the distance. Figure 2-4 illustrates travel-time curves for *P*, *S*, and surface waves, where the distance Δ is measured in degrees along the surface of the earth. As would be expected, the graph for the first arrival of surface waves is a straight line, while the graphs for the *P* and *S* waves show curves with a downward concavity. At a distance of about 105° both *P* and *S* waves seem to fade out, although waves, apparently of *P* type, are found for the range 142 to 180°. The range $105° < \Delta < 142°$ is referred to as a "shadow zone" (Fig. 2-5). However, some waves are recorded in this range, so that it is not a true shadow zone. The amplitudes of such waves are much reduced, and for many years their presence was attributed to diffraction around the boundary of the core. Miss I. Lehmann suggested in 1936 that the waves recorded in the shadow zone had passed through an inner core in which the *P* velocity is significantly greater than that in the outer core. Later work has corroborated her hypothesis, and the existence of an inner core seems fairly well established now.

Recent studies by the Carnegie Institution of Washington (S. Sacks, 1966, 1967) of the spectral behavior of direct and diffracted *P* waves at distances between 70 and 167° have indicated that the shadow boundary of the core is near 96° rather than at 105°, the uncertainty in this newer determination being about 1°. A consequence of this result is that the earth's core is larger than previously thought; also, the amplitude decrease in the shadow zone puts an upper limit of about 10^8

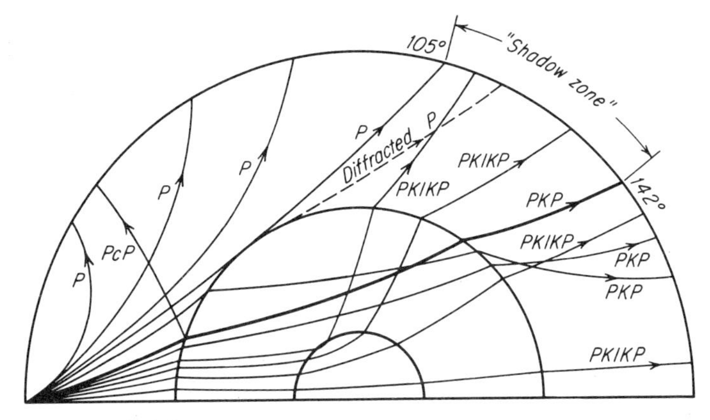

FIGURE 2-5
P, PcP, PKP, PKIKP, and diffracted *P* rays. (*After K. E. Bullen, 1954.*)

dynes/cm^2 on the rigidity in the core. An increase in the core radius has also been proposed by other workers; see L. R. Johnson (1969) for a summary of this work. A radius of 3481 km has been suggested—an increase of about 8 km over earlier estimates. In a series of new earth models (Sec. 2-5) R. A. W. Haddon and K. E. Bullen (1969) have suggested a larger increase of some 15 to 20 km. A larger increase (13 to 17 km) has also been proposed by A. L. Hales and J. L. Roberts (1970).

From the travel-time tables it is possible to calculate the velocity at any depth within the earth, although the calculations are rather complex (Appendix A). Many estimates of the relation between velocity and depth have been given from 1910 onward. An excellent account of this early work, with a full bibliography, has been given by J. B. Macelwane (1951). The extent to which individual judgment and choice of material have affected the construction of travel-time tables, and hence velocity-depth curves, is illustrated by Macelwane in a figure in which half a dozen different solutions are plotted on the same graph. There is general agreement upon the major features of the curves, the differences for the most part being concerned with minor inflections. In the later work of Jeffreys and Gutenberg (see, for example, Jeffreys, 1962, and Bullen, 1963a) most of these inflections have disappeared. Table 2-1 gives the velocities of both *P* and *S* waves at different depths within the earth according to Jeffreys, and Fig. 2-6 shows the distribution of seismic velocities in both the core and mantle according to Jeffreys and Gutenberg. Gutenberg has given an alternative estimate of the velocities in the upper part of the mantle, but his results do not differ appreciably from those of Jeffreys. However, there is some difference of interpretation at the boundary of the inner core (at a depth of approximately 5000 km). Jeffreys postulates a major discontinuity, whereas Gutenberg concludes that there is neither a decrease of velocity nor a strictly vertical segment, although there is a sharp change of slope in the

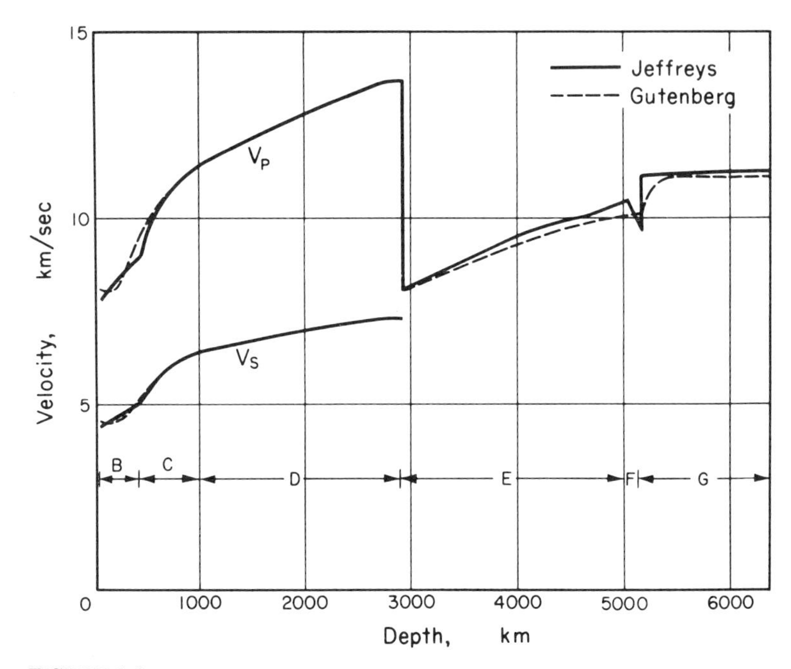

FIGURE 2-6
Seismic velocities V_p (longitudinal) and V_s (transverse) as a function of depth. (*After F. Birch, 1952.*)

Table 2-1 VELOCITIES OF *P* AND *S* WAVES AT DIFFERENT DEPTHS IN THE EARTH

Depth, km	Velocity of longitudinal (*P*) waves, km/sec	Velocity of transverse (*S*) waves, km/sec	Depth, km	Velocity of longitudinal (*P*) waves, km/sec
33	7.76	4.36	2898	8.10
100	7.95	4.45	3000	8.22
200	8.26	4.60	3200	8.47
300	8.58	4.76	3400	8.76
413	8.97	4.96	3600	9.04
600	10.25	5.66	3800	9.28
800	11.00	6.13	4000	9.51
1000	11.42	6.36	4200	9.70
1200	11.71	6.50	4400	9.88
1400	11.99	6.62	4600	10.06
1600	12.26	6.73	4800	10.25
1800	12.53	6.83	4892	10.44
2000	12.79	6.93	5121	(9.7)
2200	13.03	7.02	5121	11.16
2400	13.27	7.12	5700	11.26
2600	13.50	7.21	6371	11.31
2800	13.64	7.30		
2898	13.64	7.30		

(*After K. E. Bullen, 1963a.*)

velocity-depth curve. The velocity-depth curves in the upper mantle and inner core are discussed in greater detail in Chap. 10.

2-4 MAJOR SUBDIVISIONS OF THE EARTH

From an analysis of the velocity-depth curves, it is possible to designate a number of subdivisions of the interior of the earth. These are listed in Table 2-2 and are taken from Bullen (1963a), whose analysis is based on the velocity-depth curves of Jeffreys. The most notable feature of the velocity-depth curves is the discontinuity at a depth of about 2898 km. At that depth, there is a discontinuous drop in the velocity of the compressional P waves; below that depth, the passage of transverse S waves has never been detected. The depth marks a major structural division within the earth and defines the boundary between the core and the mantle.

Solids are distinguished from liquids in that they are able to support shear stresses as a consequence of their finite rigidity. Equation (2-2) shows that the velocity of transverse S waves is proportional to the square root of the modulus of rigidity of the medium. Since no transverse S waves have ever been observed to travel through the core of the earth, it follows that the core has negligible rigidity and is therefore fluid. H. Takeuchi (1950), using tidal and seismic data in the mantle, showed that the rigidity of the outer core E is less than one-fortieth of that of the rocks of the outer mantle, again indicating that the E region is essentially fluid, i.e, a material of zero or very small rigidity. The physical state of the inner core is still in doubt, however.

Table 2-2 DIMENSIONS AND DESCRIPTIONS OF INTERNAL REGIONS OF THE EARTH (*After K. E. Bullen, 1963a.*)

Region		Depth to boundaries, km	Fraction of volume	Features of regions
Crust†	*A*	0	0.0155	Conditions fairly heterogeneous
		33		
	B		0.1667	Probably homogeneous (?)
		413		
Mantle	*C*		0.2131	Transition region
		984		
	D		0.4428	Probably homogeneous
		2898		
	E		0.1516	Homogeneous fluid
		4982		
Core	*F*		0.0028	Transition layer
		5121		
	G		0.0076	Inner core (solid?)
		6371		

† The thickness of the crust under the continents is not constant; it averages about 33 km. Under the oceans, the crust is much thinner, being only about 5 or 6 km thick.

B. A. Bolt (1962) has revised the Jeffreys distribution of the velocity of *P* waves in the earth's core, obtaining two discontinuous increases in V_p at the boundaries of an intermediate region *F* separating an inner core *G* from an outer core *E*. The velocities throughout *F* and *G* are taken as constant. Bolt made this revision in an attempt to explain the first arrivals *PKP* at epicentral distances between about 125 and 140°. At the shorter end of this range such waves arrive some 15 to 20 sec earlier than the normal larger-amplitude *PKIKP* waves. Gutenberg had already called attention to these early onsets of *PKP* waves and put forward as a tentative explanation wave dispersion in the transition region *F*. Bolt's interpretation, with its revised velocity-depth curve, has an important bearing on the physical properties of the earth's core (Sec. 10-10). Further observational studies of the *PKP* phase by R. D. Adams and M. J. Randall (1964) appear to indicate three sharp increases in V_p near the boundary of the inner core, suggesting that the outer parts of the inner core may have a layered structure.

There is a second major discontinuity in the velocity-depth curves just below the earth's surface which marks the boundary between the crust and the mantle. This boundary is called the Mohorovičić discontinuity, after its discoverer, who identified it on a seismogram of an earthquake in Croatia in 1909. The nature of the Mohorovičić discontinuity is discussed in some detail in Sec. 10-2. The depth to the "Moho" is about 33 km under the continents and about 5 km under the ocean floors. Although the detailed structure of the crust is extremely complex, it is less than 1 percent of the mass of the whole earth, and the details of its structure do not affect the interpretation of the physical properties of the deeper parts of the mantle and core. In calculations involving the deep interior of the earth the crust has been conventionally taken to consist of a layer 15 km thick and of density 2.65 g/cm^3, resting on a second layer 18 km thick and of density 2.87 g/cm^3. This model is now out of date, but since it is not possible to give a complete description of the crust, it is advantageous to keep it for comparison purposes in calculations relating to the whole earth. Errors in a model of the crustal layers have little effect on results in the deep interior.

Until about 1940 our knowledge of the crustal layers rested on body-wave data from near earthquakes and on surface-wave data. Seismic data of 1953 indicate that, apart from the sedimentary layers, there are two fairly well-defined crustal layers under the continental regions with velocities of *P* and *S* waves of 6.2 and 3.6 km/sec in the upper part, increasing with depth to about 7.0 and 3.8 km/sec, respectively. Estimates of the velocities of *Pn* waves and *Sn* waves (i.e., waves in the mantle just below the crust) are 8.15 and 4.7 km/sec, respectively.

Important evidence supplementing the data from near earthquakes has been obtained since 1940 from explosions. The results of the Burton-on-Trent explosion of 1944 showed that in northern Europe the velocity of *Pn* waves is approximately 8.1 km/sec. This was confirmed by the Heligoland explosion of 1947, and the most recent analyses of near-earthquake data indicate that the *Pn* velocity is within 1 percent of the value 8.1 km/sec in most parts of the world.

Explosions of atom bombs may also be used to investigate the structure of the mantle. The Bikini atom bomb explosion of 1946 was the first to yield valuable information; in that case the energy produced was comparable with that of moderately large earthquakes, and P waves were recorded at distances up to 80° from the source. Controlled explosions have a great advantage over natural earthquakes in that the exact location of the focus and the time of origin are both known. The Bikini explosion has since been supplemented by a large number of additional explosions which have led to new travel-time curves (see, for example, 1968 Seismological Tables for P Phases, *Bulletin of the Seismological Society of America*, Special Number, vol. 58, no. 4, August 1968). Recent seismic-refraction studies of crustal structure have indicated significant lateral variations in crustal thickness and mantle velocity. These will be discussed in Chaps. 15 and 16, where the structures of the mid-ocean ridges, the deep-sea trenches, tectonic belts, shield areas, and continental margins are considered in some detail.

Other discontinuities within the earth are indicated from seismic data; none of them, however, is so striking as either the Mohorovičić discontinuity or the core-mantle boundary. Two have been placed by Bullen within the mantle at depths of 413 and 984 km, where there are second-order discontinuities in the velocity-depth curves, i.e., discontinuities in the slopes of the curves. Although it is unlikely that there will be any major revisions in the velocity-depth curves, their detailed structure is not well established. L. R. Johnson (1967) has estimated the velocity structure for P waves in the upper mantle using $dT/d\Delta$ measurements with the extended array at the Tonto Forest Seismological Observatory in Arizona. Short period P waves from earthquakes in the distance range 0 to 30° were used in his analysis. He found two regions with high velocity gradients located near depths of 400 and 650 km. Both these zones are between 50 and 100 km thick and involve a velocity increase of 9 to 10 percent. They are interpreted as phase transitions and are discussed in more detail in Sec. 10-5.

There is also a growing body of evidence that there is a low-velocity (LV) layer in the upper mantle, thicker under ocean bottoms than under continents. Recent seismic results also indicate that there are significant regional differences. Beneath the oceans there is a definite LV zone for S, and possibly one for P as well. Beneath Precambrian shields the LV zone for S is less pronounced and the LV zone for P seems to be absent. Other continental regions are intermediate between these two cases. The LV zone is considered to be due to high thermal gradients with mineralogical and chemical heterogeneity superimposed. The effect of the LV layer on the constitution of the mantle will be discussed later (Sec. 10-3).

Using the extended array at the Tonto Forest Seismological Observatory to measure $dT/d\Delta$ of direct P waves in the distance range 30 to 100°, L. R. Johnson (1969) has estimated P velocities in the lower mantle. Although he finds no large discrepancies with the traditional models of Gutenberg and Jeffreys, increased velocity gradients near depths of 830, 1000, 1230, 1540, 1910, and 2370 km are indicated. These anomalies in the lower mantle are spread over depth intervals of at least 50 km, and are an order of magnitude smaller than those which have been found for the upper mantle.

2-5 VARIATION OF DENSITY WITHIN THE EARTH

Since the mean density of the earth is about 5.5 g/cm^3, the material deep within the earth must have a density considerably greater than that of typical surface rocks. The density ρ will depend on the pressure p, the temperature, and an indefinite number of parameters specifying the chemical composition. If m is the mass of the material within a sphere of radius r, then, since the stress in the earth's interior is essentially equivalent to a hydrostatic pressure,

$$\frac{dp}{dr} = -g\rho \tag{2-4}$$

where

$$g = \frac{Gm}{r^2} \tag{2-5}$$

and G is the constant of gravitation. The assumption of hydrostatic pressure would be a poor approximation for the stress in the crust where the strength of the rocks is of the same order as the mean pressure. But the mean pressure steadily increases with depth, while the strength, or maximum stress difference, decreases, so that a depth at which the approximation is satisfactory is soon reached. Considering for the moment a chemically homogeneous layer in which the temperature variation is adiabatic, it follows from Eq. (2-4) that

$$\frac{d\rho}{dr} = \frac{d\rho}{dp}\frac{dp}{dr} = -\frac{g\rho^2}{k} \tag{2-6}$$

where k is the adiabatic incompressibility defined by the equation

$$\frac{1}{k} = \frac{1}{\rho}\left(\frac{\partial\rho}{\partial p}\right)_S$$

and S is the entropy. By a homogeneous region is meant one in which there are no significant changes of phase or chemical composition. If the temperature variation is not adiabatic, the right-hand side of Eq. (2-6) may be modified by including a factor $(1 - \delta)$; that is,

$$\frac{d\rho}{dr} = \frac{-g\rho^2}{k}(1 - \delta) \tag{2-7}$$

F. Birch (1952) has estimated that a superadiabatic temperature gradient of 2°C/km reduces the density gradient by perhaps 20 percent, and this reduces the density by 0.2 g/cm^3 at the base of the mantle (approximately 4 percent). There is no direct evidence of the variation of the temperature gradient in the earth's deep interior, although more recent calculations tend to reduce earlier estimates of differences from the adiabatic gradient. Since the effect of the term δ in Eq. (2-7) is to increase $d\rho/dr$, whereas the effect of any variance from chemical homogeneity is in the opposite direction, δ is taken as zero; i.e., Eq. (2-6) is applied in those regions where there is no evidence of inhomogeneity. The detailed analysis of Birch (1952) is compatible with neglecting δ at depths greater than about 1000 km, although the term may be important in the upper mantle. Bullen has estimated that neglecting δ should not lead to errors of more than 0.1 g/cm^3 in the computed values of the density.

Combining Eqs. (2-5) and (2-6), we have

$$\frac{d\rho}{dr} = \frac{-Gm\rho^2}{kr^2} \tag{2-8}$$

Since $dm/dr = 4\pi\rho r^2$ and k/ρ is known from the velocity-depth curves [Eq. (2-3)], Eq. (2-8) can be integrated numerically to obtain the density distribution in those regions of the earth where chemical and nonadiabatic temperature variations may be neglected. This equation was first obtained by L. H. Adams and E. D. Williamson in 1923 and has been extensively used by Bullen and other workers to obtain density distributions.

Any density distribution must satisfy two conditions: it must yield the correct total mass of the earth and the correct moment of inertia about its rotational axis. Using these two conditions and a value of ρ_1 of 3.32 g/cm^3 for ρ at the top of layer B of the mantle, Bullen applied Eq. (2-8) throughout layers B, C, and D. He then found that this solution led to a value of the moment of inertia I_c of the core greater than that of a uniform sphere of the same size and mass. This would entail the density decreasing inward, which would be an unstable state in a fluid. Allowance for the term δ [i.e., the use of Eq. (2-7) instead of (2-8)] increased the value of I_c. A reasonable value could be obtained by increasing the value of ρ_1, but only if an impossibly high value (at least 3.7 g/cm^3) is taken. Hence the assumption of chemical homogeneity must be in error; i.e., the Adams-Williamson equation (2-8) cannot be applied throughout the mantle. The most likely layer where this assumption breaks down is C, where the velocity-depth curves indicate changes in slope. In his earth model A, Bullen thus used Eq. (2-8) in layers B and D while in C he fitted a quadratic expression in r for $\rho = \rho(r)$.

In the outer core (region E) Eq. (2-8) is likely to apply, and values of ρ down to a depth of about 5000 km can be obtained with some confidence, although it is not easy to determine the density in regions F and G with any certainty. However, since these regions constitute only about 1 percent of the earth's total volume, the density distribution within E can be estimated fairly precisely. In the core one boundary condition is $m = 0$ at $r = 0$, but lack of evidence on the value of the density ρ_2 at the center of the earth leads to some indeterminacy regarding the density distribution in the core. Bullen showed, however, that strong controls on permissible density values were exercised by various moment of inertia criteria. In particular, he showed that the minimum possible value of ρ_2 was 12.3 g/cm^3 and that increasing the value of ρ_2 by 5 g/cm^3 affected the formally computed densities elsewhere by maximum amounts of only 0.03 g/cm^3 in the mantle and 0.4 g/cm^3 in the outer core. Bullen derived density distributions on two fairly extreme hypotheses: (1) ρ_2 = 12.3 g/cm^3 and (2) ρ_2 = 22.3 g/cm^3 (this value being taken quite arbitrarily). A model with density values midway between those of these two hypotheses has been called model A. More recent evidence (to be discussed later, in Chap. 10) indicates that ρ_2 is probably much nearer its minimum value and that a model based on ρ_2 = 12.3 g/cm^3 (model A-i) is more likely to be correct.

A new distribution of the density in the earth has been determined by M. Landisman, Y. Sato, and J. Nafe (1965) using a method which is independent of the assumptions of homogeneity and an adiabatic temperature gradient, except in the region of the outer core. Torsional and spheroidal observations of the longer period modes of the free vibrations of the earth excited by the Chilean earthquake of May 22, 1960, provide an additional source of data for determining the physical properties of its deep interior. When combined with data from the travel-time curves, they indicate that a revision of the density distribution in the mantle is required. The agreement between the *P* and *S* wave observations of Jeffreys and Gutenberg has led to nearly identical velocity distributions in the middle and lower mantle. In general, the more precise observations of nuclear explosions by S. D. Kogan (1960) strongly support the validity of these earlier earthquake body wave studies. However, neither the Jeffreys nor the Gutenberg velocity distributions, combined with Bullen's A-i densities, are able to satisfy the longer period torsional and spheroidal free oscillation data from the Chilean earthquake.

Landisman et al. (1965) considered the inverse problem, that of determining the radial distribution of ρ, V_p, and V_s from the travel-time curves for *P* and *S* waves and the free periods of the earth. They investigated a number of earth models subject to the constraints fixed by a density of 3.32 g/cm^3 at the top of the mantle, the total mass of the earth, and its mean moment of inertia as revealed by perturbations of artificial satellite orbits (fractional moment of inertia 0.33089 instead of 0.3335). The outer core is taken to be chemically homogeneous with an adiabatic temperature gradient, and in it the Adams-Williamson equation is used to find the density. In their model M1, the lower core is chemically inhomogeneous, following the model proposed by Bullen (1962, 1963b) and B. Bolt (1962). In their model M3, the entire core is taken to be a homogeneous adiabatic fluid. As with all models, Hooke's law is assumed to be valid throughout the earth and no density inversions are permitted at any depth.

The density for both models M1 and M3 is about 10 percent lower near the base of the mantle than that predicted by the Bullen A-i distribution. Constant densities are found for both models at depths between about 1600 and 2800 km. A superadiabatic temperature gradient of 4 to 5°/km would be required to explain this result—this would lead to excessive temperatures at the core-mantle boundary. If the vanishing density gradient is the result of a compositional change such as the depletion of iron, a concomitant increase of almost 10 percent would be expected in the shear velocity gradient. From depths of 500 to 1100 km, densities for models M1 and M3 are slightly lower and from 1100 to 1800 km slightly higher than those of the Bullen A-i distribution. For model M1, the density of the outer core is found to be about 2 percent higher than that for Bullen A-i, while for model M3 it is about 4 percent higher. The central density for model M1 is about 15.42 g/cm^3, and that for model M3 about 12.63 g/cm^3.

C. L. Pekeris (1966) has also obtained density distributions in the Earth using the observed periods of the free oscillations of the Earth, without using the Adams-Williamson assumption of homogeneity and adiabaticity for any region. The density distribution $\rho(r)$ was represented by 50 pivotal values $\rho(r_k)$ with linear variations in between and with discontinuities at the Moho and at the core-mantle boundary. The ρ_k were varied by the method of steepest descent so as to minimize the sum of the squares of the residuals of all the observed periods. Pekeris found that a somewhat better fit to the observed spectral data is obtained for a Gutenberg type of velocity distribution (with a low-velocity layer) than for the Jeffreys distribution. The final density distributions for the various models converge in the depth range 400 to 1200 km, to a narrow band close to that of Bullen's model A-i. In the first 400 km, the density is nearly constant for Gutenberg-type models, with a tendency toward negative density gradients in the first 200 km. All models give a density at the outer boundary of the core close to 10.0 g/cm^3; as was to be expected, the density distribution in the inner core has little effect on the spectrum as a whole. The same degree of improvement in the spectral fit obtained by varying the density alone can also be achieved by varying the shear velocity alone. The new shear-velocity distribution shows a tendency to be constant or minimum in the depth range 100 to 200 km.

F. Birch (1964) has determined the density distribution in the earth based on the empirical observation that for silicates and oxides of about the same iron content, there is an approximate linear relationship between the density ρ and the velocity V_p of compressional waves; that is,

$$\rho = a + bV_p \qquad (2\text{-}9)$$

In the lower mantle and core the density distribution is found from the Adams-Williamson equation, and the empirical relation (2-9) is used only for the upper mantle and transition zone; i.e., the Adams-Williamson equation is used where the change of density is most probably determined by compression alone, and the empirical rule for the transition layer where the principal effect is produced by phase changes. Two models were considered. In the first (solution I) the constant b was given the value 0.328 (g/cm^3)/(km/sec) as found for rocks and crystals of low mean atomic weight. The constant a and the mass of the core are then the only adjustable parameters and are determined by the total mass and moment of inertia of the earth. For the second model (solution II) Birch chose the

Table 2-3 DENSITY, GRAVITY, AND PRESSURE DISTRIBUTION IN THE MANTLE

Depth, km	Density, g/cm^3	Gravity, cm/sec^2	Pressure, $\times 10^{12}$ dynes/cm^2
33	3.32	985	0.009
100	3.38	989	0.031
200	3.47	992	0.065
300	3.55	995	0.100
400	3.63	997	0.136
413	3.64	998	0.141
500	3.89	1000	0.173
600	4.13	1001	0.213
700	4.33	1000	0.256
800	4.49	999	0.300
900	4.60	997	0.346
1000	4.68	995	0.392
1200	4.80	991	0.49
1400	4.91	988	0.58
1600	5.03	986	0.68
1800	5.13	985	0.78
2000	5.24	986	0.88
2200	5.34	990	0.99
2400	5.44	998	1.09
2600	5.54	1009	1.20
2800	5.63	1026	1.32
2898	5.68	1037	1.37

Table 2-4 DENSITY, GRAVITY, AND PRESSURE DISTRIBUTION IN THE CORE

Depth, km	Density, g/cm^3	Gravity, cm/sec^2	Pressure, $\times 10^{12}$ dynes/cm^2
2898	9.43	1037	1.37
3000	9.57	1019	1.47
3200	9.85	979	1.67
3400	10.11	936	1.85
3600	10.35	892	2.04
3800	10.56	848	2.22
4000	10.76	803	2.40
4200	10.94	758	2.57
4400	11.11	716	2.73
4600	11.27	677	2.88
4800	11.41	646	3.03
4982	11.54	626	3.17
5121	(14.2)	585	3.27
5121	(16.8)	....	....
5400	....	460	3.41
5700	....	320	3.53
6000	....	177	3.60
6371	(17.2)	0	3.64

constant a so that the density at 33 km is 3.32 g/cm^3, and the adjustable parameters are then b and the mass of the core. The density in the lower mantle is about 1 percent higher and in the core about 1 percent lower in the second than in the first model, and the models are very similar to Bullen's model A-i. The densities in the lower mantle are in good agreement with shock-wave measurements on rocks having FeO contents in the range 10 ± 2 percent by weight and may be accounted for in terms of mixtures of close-packed oxides, silica transforming to stishovite in the transition layer. A constant density in the lower mantle, combined with rising velocities, as suggested by Landisman et al. (1965), is not compatible with Birch's results. However, the value of the central density for Landisman's model M3 is almost identical with Birch's. Tables 2-3 and 2-4 give the density distribution in the mantle and the core for Bullen's model A.

C.-Y. Wang (1970) has suggested that better estimates of the density in the upper mantle (to a depth of ≈1000 km) may be obtained by using an empirical relationship between the bulk sound velocity $C = (V_p{}^2 - \frac{4}{3}V_s{}^2)^{1/2}$ and density ρ, rather than the linear relation between V_p and ρ proposed by Birch. Such a relationship was first suggested by R. G. McQueen et al. (1964) and later discussed by Wang (1968, 1969). Wang (1970) finds that the density in the upper 200 km of the mantle is 3.3 to 3.4 g/cm^3, this lower value lending support to the suggestion that the upper mantle may be composed of some variations of peridotite rather than eclogite, which requires densities in the range 3.5 to 3.6 g/cm^3. Wang also finds that the mean atomic weight of the lower mantle is about 21.3 to 21.5 and suggests the possibility that the entire mantle may have a uniform iron content.

S. P. Clark and A. E. Ringwood (1964) have estimated densities in the earth using petrological models of the upper mantle constructed on the assumption of an overall pyrolite (ultrabasic) composition and an eclogite composition (Sec. 10-3). There is an almost exact agreement between their pyrolite model and Bullen's model A-i in the lower mantle; densities in the upper mantle and transition zone are systematically lower than Bullen's because of their use of petrological arguments in those regions. Figure 2-7*a*, *b*, and *c* shows the density distribution for some of the more recently proposed earth models, together with Bullen's model A-i for comparison. The HB_1 model (R. A. W. Haddon and K. E. Bullen, 1969) is discussed later in this section.

F. Press (1968a,b) has used a Monte Carlo inversion method to obtain a number of earth models, using as data 97 eigenperiods, travel times of P and S waves, and the mass and moment of inertia of the earth. The Monte Carlo method uses random selection to generate large numbers of models in a computer, subjecting each model to a test against geophysical data. Only those models whose properties fit the data within prescribed limits are retained. This procedure has the advantage of finding models without bias from preconceived or oversimplified notions of earth structure. Monte Carlo methods also offer the advantage of exploring the range of possible solutions, and indicate the degree of uniqueness obtainable with currently available geophysical data.

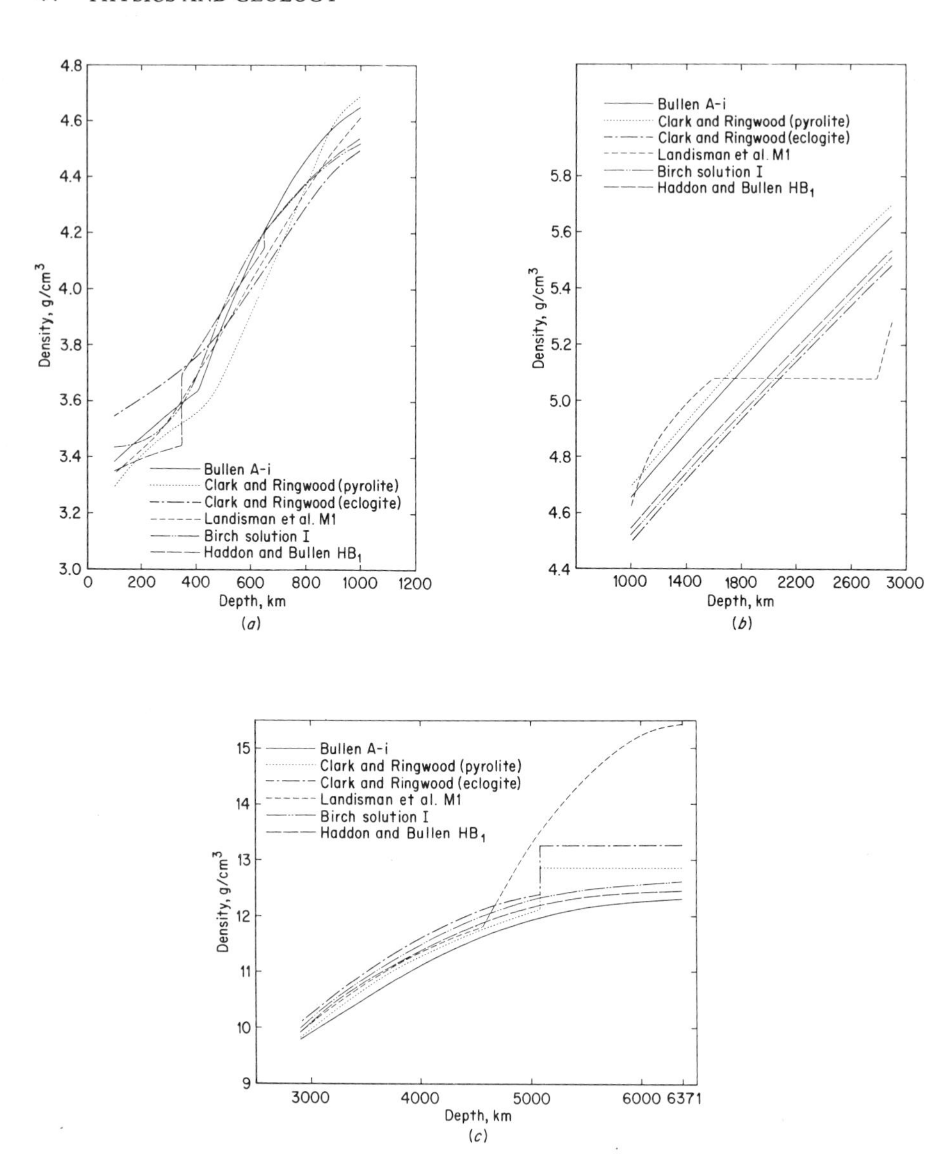

FIGURE 2-7
Variation of density in the earth's interior for different earth models.

Press was able later (1970*a*) to speed up considerably his Monte Carlo procedures. Using new, more extensive, and more accurate data, he was able to find a larger number of successful models. Figure 2-8 shows 27 successful shear velocity distributions for the mantle. Figures 2-9 and 2-10 show the corresponding density distributions in the mantle and core, respectively. Of the millions of earth models

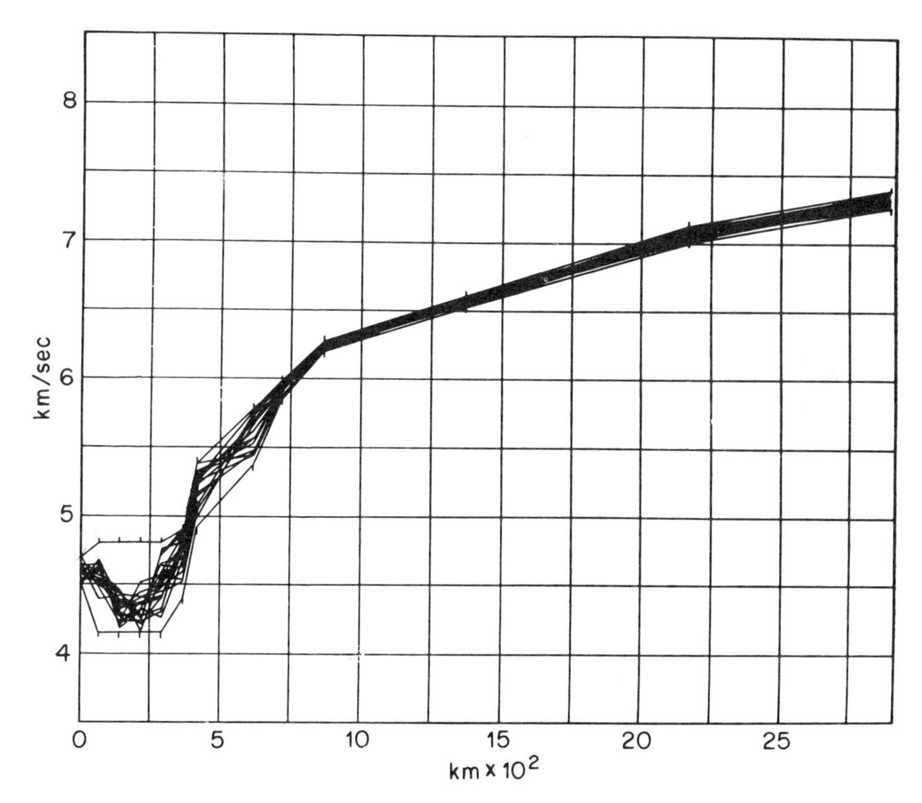

FIGURE 2-8
Twenty-seven successful shear-velocity models for the mantle. Ticks on the upper and lower bounds show where parameter was randomly varied. (*After F. Press, 1970.*)

generated and examined, every successful model showed a low-velocity zone for shear waves in the suboceanic mantle centered at depths between 150 and 250 km. Again, although density values in the mantle just below the Moho cover the entire permissible range, in the vicinity of 100 km all values fell in the narrow band 3.5 to 3.6 g/cm^3. This value of the density in the lithosphere near 100 km is so high as to narrow the range of its possible composition to an eclogitic facies (see also Sec. 10-3 and Fig. 10-5).

Large density and velocity gradients were found in the transition zone in the mantle without prior assumption of an equation of state. In particular, high density gradients are localized near 350 to 450, and 550 to 800 km, the details differing between models. These are interpreted as phase transitions and will be discussed in more detail in Sec. 10-5. Upper-mantle models fit better than "standard" models when they are more complex, with large fluctuations in the shear velocity and density. Such complexity can be expected if the mantle is chemically and

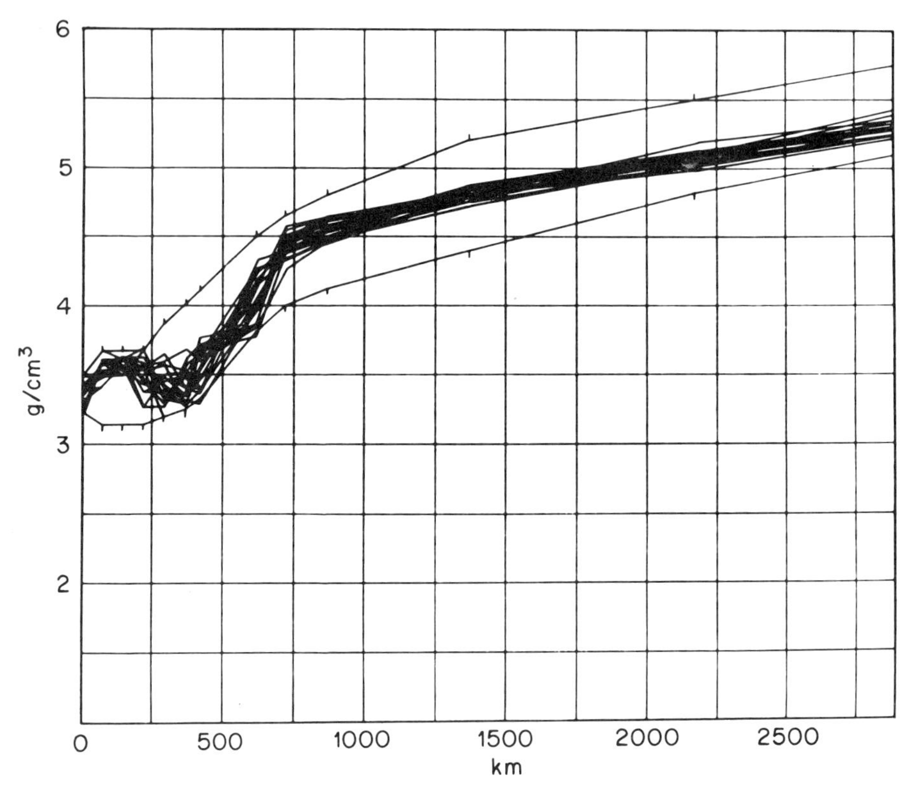

FIGURE 2-9
Successful density distributions in the mantle. (*After F. Press, 1970.*)

mineralogically zoned and if high thermal gradients and partial melting take place. The magnitude of the fluctuations suggests that the zoning is lateral and that the mantle is variable in composition laterally, ranging from pyrolite to eclogite (see also Sec. 10-3). In his earlier results Press (1968a,b) found an increase in the FeO:FeO + MgO ratio in the lower mantle by a factor of 2 compared with its value in the upper mantle. The enrichment of iron in the lower mantle would imply that the mantle as a whole has achieved a measure of gravitational stability that would inhibit mantle-wide convection. However, J. W. Fairborn's new $dT/d\Delta$ data (1969) for the lower mantle show a higher shear velocity gradient than previously supposed. This requires a compensatory reduction in the density gradient in order to fit the spheroidal eigenperiod data. Press (1970) found a change of mean atomic weight from 22 to 23 at the top of the lower mantle to 20 to 22 at the bottom of the mantle, implying a depletion of iron with depth. Press also confirmed that the earth's core is inhomogeneous. The density at the top of the core is constrained to

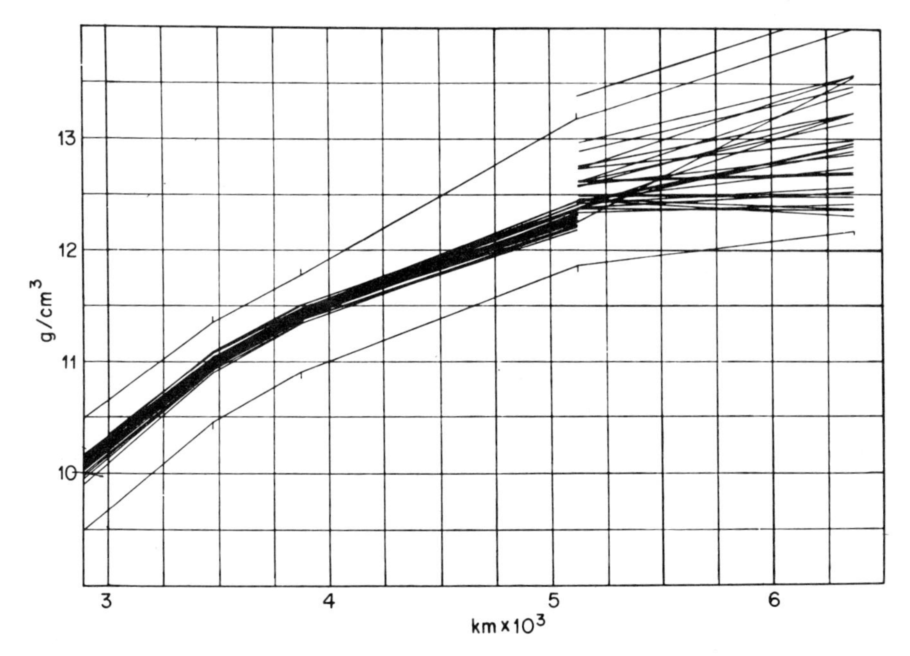

FIGURE 2-10
Successful density distributions in the core. (*After F. Press, 1970.*)

the narrow range 9.9 to 10.2 g/cm^3, a value appropriate for a mixture of iron with about 15 wt % silicon. Changes in the core radius ranged from −3 to +10 km.

R. A. W. Haddon and K. E. Bullen (1969) have constructed a series of new earth models (HB) incorporating free oscillation data, consisting of the observed periods of fundamental spheroidal and torsional oscillations for $0 \leqslant n \leqslant 48$ and $2 \leqslant n \leqslant 44$ respectively, where n is the order number, and of first and second spheroidal overtones for $n \leqslant 20$. The data were taken from the records of both the Chilean (May 1960) and Alaskan (March 1964) earthquakes. Their procedure was to start from models derived independently of the oscillation data and to produce a sequence of models showing improved agreement with all the available data. In passing from one model to its successor, a guiding principle was to introduce and vary one or more parameters in the model description at any stage in order to satisfy the oscillation data. They thus tried to establish models described in terms of the minimum number of parameters demanded by the data. Thus a major difference in principle between their method and F. Press's (1968a,b, 1970a) Monte Carlo procedure is the comparatively large number of parameters that Press has permitted to be randomly varied. Haddon and Bullen point out that the predominance of complex models found by Press is inherent in his method and that a simple random walk would automatically have a low probability. Press's results, however, broadly confirm Haddon and Bullen's findings. Haddon and Bullen also point out that the "average" earth to which average periods of free oscillation modes relate is not necessarily the same as an earth model to which the currently available average seismic body wave travel times apply, earthquake epicenters and recording stations not being randomly distributed over the earth's surface.

One of the intentions of Haddon and Bullen was to examine the conclusion of M. Landisman et al. (1965) that the oscillation data require $d\rho/dr$ to be abnormally low (≈ 0) throughout much of the lower mantle. They can accept this conclusion only if the core radius is kept unchanged; however, an increase in the core radius of some 15 to 20 km not only satisfies the oscillation data, but also leads to reduced density gradients in layer B and normal gradients throughout most of the lower mantle (layer D). Additional evidence indicates that the effect of damping on the oscillation periods may be significant; if so, the increase needed in the core radius may be somewhat less than 15 km. This evidence would allow some reduction in the density gradient in the lower mantle, but not nearly so much as that demanded by Landisman et al. Another result of Haddon and Bullen's work is a reduction in the thickness of the crustal layer A from 33 to 15 km. This is a result of the changes made in V_s and ρ in layer B in the course of fitting the oscillation data. Such a reduction is reasonable since it involves a change to a crust that may be more representative of average conditions in the earth.

It should be emphasized that the overall density distribution is not drastically changed by taking into account the combined effects of the revised moment of inertia of the earth (as determined by satellite data) and the observational data on the free vibrations of the earth. Inside the mantle the largest difference in ρ between models A-i and HB_1 is only 0.15 g/cm^3, and inside the core the values of ρ in model HB_1 exceed those in model A-i at all levels by amounts between 0.2 and 0.3 g/cm^3.

2-6 PRESSURE DISTRIBUTION, VARIATION OF ACCELERATION DUE TO GRAVITY, AND ELASTIC CONSTANTS WITHIN THE EARTH

From Eqs. (2-4) and (2-5), it follows that

$$\frac{dp}{dr} = \frac{-Gm\rho}{r^2} \qquad (2\text{-}10)$$

Hence, by numerical integration, the pressure distribution may be obtained once the density distribution has been determined. Since the density is used only to determine the pressure gradient, the pressure distribution is insensitive to small changes in the density distribution and may be determined quite accurately. The results (based on Bullen's model A density distribution) are given in Tables 2-3 and 2-4 and are shown graphically in Fig. 2-11. The uncertainty in the values for the

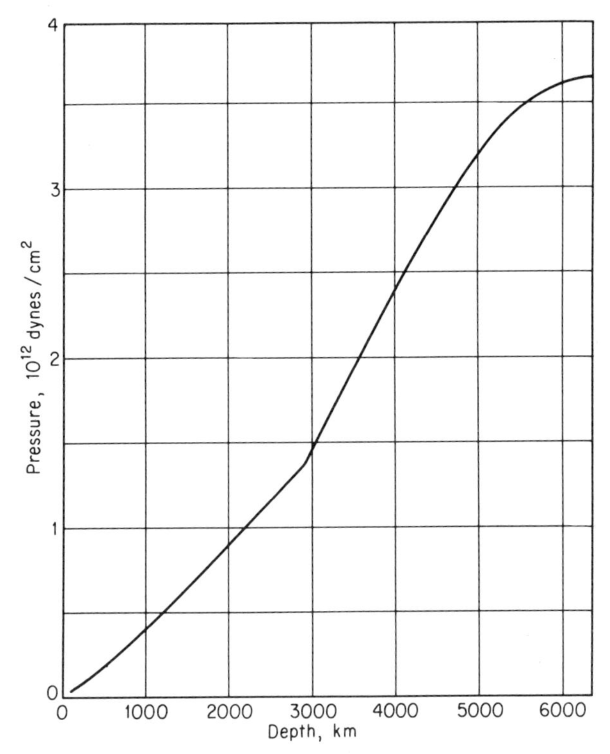

FIGURE 2-11
Pressure distribution in the earth's interior – Bullen model A. (*After K. E. Bullen, 1963a.*)

regions F and G is probably of the order of 3 percent, and for all other regions less than 1 percent.

The variation of g can be calculated from Eq. (2-5); the values of g are also given in Tables 2-3 and 2-4. The value of g does not differ by more than 1 percent from the value 990 cm/sec^2 until a depth of over 2400 km is reached. On the other hand, the values of g deep within the earth are sensitive to changes in density, and values below 4000 km may be in error by as much as 5 percent.

From a knowledge of the density distribution, it is easy to compute values of the elastic constants. Thus Eqs. (2-2) and (2-3) give μ and k directly. From the known relationships between the elastic constants, it is possible to compute the distribution of Young's modulus E and Poisson's ratio σ. In particular,

$$\sigma = \frac{3k - 2\mu}{6k + 2\mu} = \frac{V_p^{\,2} - 2V_s^{\,2}}{2(V_p^{\,2} - V_s^{\,2})} \tag{2-11}$$

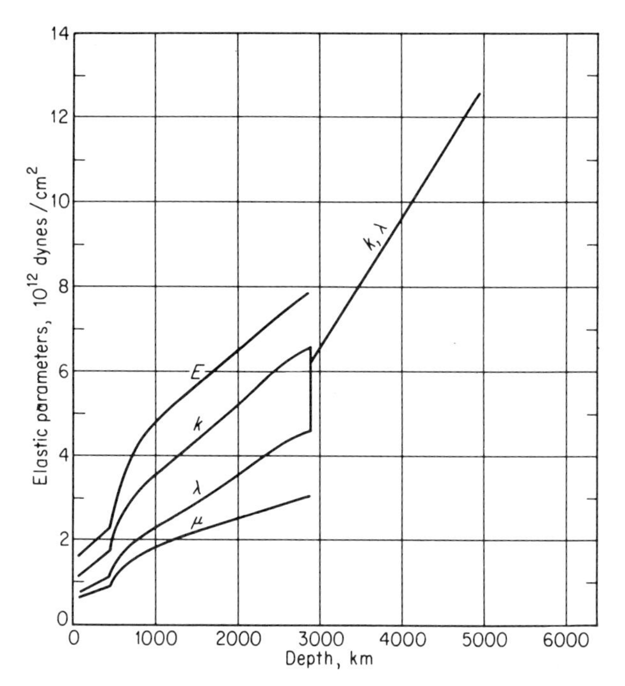

FIGURE 2-12
Variation of the elastic constants in the earth's interior – Bullen model A. (*After K. E. Bullen, 1963a.*)

and is thus independent of the density ρ. Figure 2-12 illustrates the variation of μ and k throughout the earth. Values in the inner core are not included because they are subject to much uncertainty.

2-7 MAGNITUDE, INTENSITY, AND ENERGY OF EARTHQUAKES

There is often a great discrepancy between the significance of earthquakes as disasters (in terms of loss of life and property) and the amplitudes shown on seismograms at distant stations. Thus on Feb. 29, 1960, an earthquake occurred with focus at shallow depth immediately beneath the town of Agadir, Morocco. (See F. Duffaud et al., 1962, for details.) It caused great destruction, about 10,000 deaths and 25,000 injuries, but European seismologists assigned only a very moderate magnitude (5.7 or 5.6) to this earthquake. Conversely, earthquakes recorded as major events on seismographs all over the world often receive little notice because most of them occur in remote places or under ocean basins. In an average year there are twenty or more major earthquakes (magnitude >7), only a few of which occur in populated areas and many of which are submarine.

Much of the early literature in seismology consisted of descriptive reports of individual earthquakes; later attempts were made to classify earthquakes in some manner according to size. The early efforts to assign a "bigness" to earthquakes depended of necessity on personal observations of the severity of their effects. The greater variety of observable effects led to the development of numerical scales, called *intensity* scales. The intensity of an earthquake at a particular location is a number denoting the severity of the earthquake in terms of its effect on buildings and natural surface features of the earth. Intensity scales are arbitrary and empirical, and are not based on actual measurements of any kind, but on judgment applied to effects observed in the field. Inexperienced investigators may make serious errors in attempting to estimate intensities.

The intensity scale generally used in the United States is the Modified Mercalli scale of 1931. Prior to 1931, the Rossi-Forel scale was widely used. This scale was developed in 1883 by M. S. De Rossi in Italy and F. A. Forel in Switzerland and is still in general use in much of Europe. The description of the effects given in the Rossi-Forel scale were too European to be universally applicable, and this led G. Mercalli to publish in 1902 an improved scale bearing his name, and this was the basis for the Modified Mercalli scale of 1931.

To classify an earthquake in a manner more nearly representative of its tectonic importance and to clarify the statistics of earthquake occurrence, C. F. Richter introduced in 1935 the concept of earthquake magnitude. At that time all stations in Southern California were provided with the same instrumentation (a pair of Wood-Anderson torsion seismometers). Because of the close similarity in instrumental response, it was practical to work with the trace amplitudes measured directly on the seismograms without measuring in addition the corresponding periods and calculating ground amplitudes. Richter found that when amplitudes were plotted on a logarithmic scale against the distance from the station to the earthquake, the plots for a number of Southern California earthquakes were roughly parallel. The vertical distance between the parallel lines was used as a measure of the *local* magnitude M_L. The zero of the M_L scale was placed at a level below that of the smallest earthquakes then being recorded. An earthquake recorded with a maximum *trace* amplitude of 1 mm by a standard torsion seismometer at a station 100 km from the epicenter was assigned magnitude 3. This would give for magnitude zero a corresponding trace amplitude of 0.001 mm (1 μ) which is unobservable under normal recording and measuring conditions. In this form, the scale is of very restricted application, since it applies to local earthquakes in California observed by a particular type of seismograph.

It is clearly desirable to have a magnitude scale of broader applicability, as well as one for which the values are simply related to energy release. The energy release should be obtained by multiplying the kinetic energy per unit volume by factors which include the geometry and geology of the path taken by the seismic wave; i.e., the energy is expected to be proportional to $\rho v^2/2$, where ρ is the density of the material on which the seismograph station rests, and v is the observed ground velocity. The logarithm of the energy will therefore contain a term proportional to $\log v$, or in terms of amplitude and period, to $\log A/T$. When T can be regarded as fixed, the amplitude dependent term is simply proportional to $\log A$. For earthquakes more distant than 600 km, to which the local magnitude scale cannot be extended, surface waves are often of largest amplitude in a period range near 20 sec. This leads to the surface wave scale

$$M_S = \text{const} + \log A \qquad (2\text{-}12)$$

where A is the *ground* amplitude in microns, and the constant, which is a function of the distance to the earthquake, is usually taken from tables such as those given by Richter (1958).

Deep-focus earthquakes give weak surface waves, so that a magnitude scale must depend on body waves. The period of the largest P wave recorded on seismograms is quite variable, so that the form $\log A/T$ must be used for the amplitude dependent term. This leads to a body wave magnitude, often called the *unified magnitude*, given by

$$m = \text{const} + \log A/T \qquad (2\text{-}13)$$

where this constant, too, is empirically determined.

While values for M_L and M_S are reasonably consistent, those for m differ from them by substantial amounts, the relationship being

$$m = 2.5 + 0.63M \qquad (2\text{-}14)$$

There is poor correlation in general between the maximum reported intensities of earthquakes and the amplitudes of their traces on seismograms. Magnitude is characteristic of the earthquake as a whole, while intensity describes the effects at a particular place. The magnitude of an earthquake is related to the energy radiated by the shock in the form of elastic waves, although many other factors besides the total energy influence the size of the record. Because of its logarithmic scale the term magnitude as applied to earthquakes is similar to the term order of magnitude as used by the physicist or astronomer. The actual range of magnitudes is enormous, the largest events representing motions a hundred million times as large as those of small events (8 magnitudes).

Most attempts to estimate the elastic wave energy radiated during particular earthquakes use one of two methods. The first involves estimating the elastic strain energy (initially stored in the rocks around the focus) expended in producing the displacements of blocks of the earth's crust observed after the earthquake. The second involves estimating the elastic wave energy more directly from seismograms recorded at a distant seismic station. Neither method can be routinely applied to many earthquakes, and the number of energy estimates is small. Richter has suggested the empirical formula

$$\log E = A + BM \qquad (2\text{-}15)$$

where M is the Richter magnitude, E is the total radiated energy, and A and B are constants. The most recent values of the constants are $A = 11.4$ and $B = 1.5$ when the energy E is measured in ergs.

SUGGESTIONS FOR FURTHER READING

For information on earthquakes in general, the reader is referred to Hodgson and Richter. The theory of elastic waves with particular reference to the earth is covered in Jeffreys, Ewing et al., and Bullen. The two books by Bullen also contain detailed accounts of the application of seismology to our knowledge of the physical properties of the earth's interior. The Handbook of Physical Constants is an extremely valuable source of information on the properties of the earth.

HODGSON, J. H.: "Earthquakes and Earth Structure," Prentice-Hall, Inc., Englewood Cliffs, N.J., 1964.

RICHTER, C. F.: "Elementary Seismology," W. H. Freeman and Co., San Francisco, 1958.

JEFFREYS, H.: "The Earth," 4th ed., Cambridge University Press, New York, 1962.

EWING, W. M., W. S. JARDETZKY, and F. PRESS: "Elastic Waves in Layered Media," McGraw-Hill Book Company, New York, 1957.

BULLEN, K. E.: "Seismology," Methuen & Co., Ltd., London, 1954.

——: "An Introduction to the Theory of Seismology," 3d ed., Cambridge University Press, New York, 1963.

Handbook of Physical Constants (S. P. Clark, Jr., ed.), Geological Society of America Memoir 97, 1966.

3
COMPOSITION OF THE EARTH

3-1 INTRODUCTION

This chapter presents a very generalized outline of the variety and distribution of the rocks which constitute the crust of the earth. Thus it deals with aspects of *geochemistry*, which is the study of the composition of the earth and of the chemical processes these materials undergo. This branch of science is an offshoot of geology, begun in large part by V. I. Vernadsky (1863-1945) and V. M. Goldschmidt (1888-1947), who developed the subject in the field, and by A. L. Day (1869-1960) of the Carnegie Institution of Washington, who started experimental geochemistry upon silicate and sulfide systems under high temperatures and pressures. Laboratory work has contributed much to our understanding of the processes which are possible in the inaccessible mantle.

This chapter summarizes our knowledge of the chemical composition of the atmosphere, hydrosphere, and crust, while Chap. 10 presents conclusions about the core and mantle.

3-2 COMPOSITION OF STELLAR AND COSMIC MATTER

Before considering the earth in detail, it is well to broaden the field and consider the chemical composition of the universe as a whole. Our knowledge is obtained from spectroscopic examination of the radiation from the sun, the stars, and interstellar gas (Table 3-1), and from the direct chemical analysis of lunar samples and meteorites. A general picture emerges from all these investigations. Although the abundances of the different elements vary considerably, it may be seen from Fig. 3-1 that only about a dozen elements are quantitatively important, and of these the lightest elements, hydrogen and helium, far outweigh the rest. With the exception of the lightest gases, the chemical composition of the universe is everywhere much the same. Moreover, the abundances of the elements, with certain exceptions, show a rapid exponential decrease with increasing atomic number up to about $Z = 40$, followed by an almost constant value for the heavier elements. Terrestrial matter, like meteoritic matter (Table 1-2), differs from solar material chiefly in the rarity of the gaseous elements. This difference is not surprising and

Table 3-1 RELATIVE ABUNDANCES OF THE ELEMENTS IN STELLAR ATMOSPHERES OF THE SUN AND TWO STARS
Tabulated figures are $12 + \log_{10} N/N_H$

Star Spectral type	αLyra A 0 V	Sun G2 V	ϵVirgo G8 III
1 H	12.0	12.00	12.00
2 He	11.4		
6 C	...	8.72	8.60
7 N	8.8	7.98	
8 O	9.3	8.96	
10 Ne			
11 Na	7.3	6.30	6.60
12 Mg	7.7	7.36	7.40
13 Al	5.7	6.20	6.34
14 Si	8.2	7.45	7.58
16 S	...	7.30	7.39
20 Ca	6.3	6.15	6.25
21 Sc	3.4	2.82	2.76
22 Ti	4.8	4.68	4.67
23 V	4.0	3.70	3.67
24 Cr	5.6	5.36	5.36
25 Mn	5.3	4.90	4.97
26 Fe	6.5	6.47	6.47
27 Co	...	4.64	4.61
28 Ni	7.0	5.91	5.94
38 Sr	2.8	2.60	2.62
39 Y	2.1	2.25	2.08
40 Zr	2.9	2.23	2.08
56 Ba	...	2.10	2.01

SOURCE: Unsöld, 1969a,b.

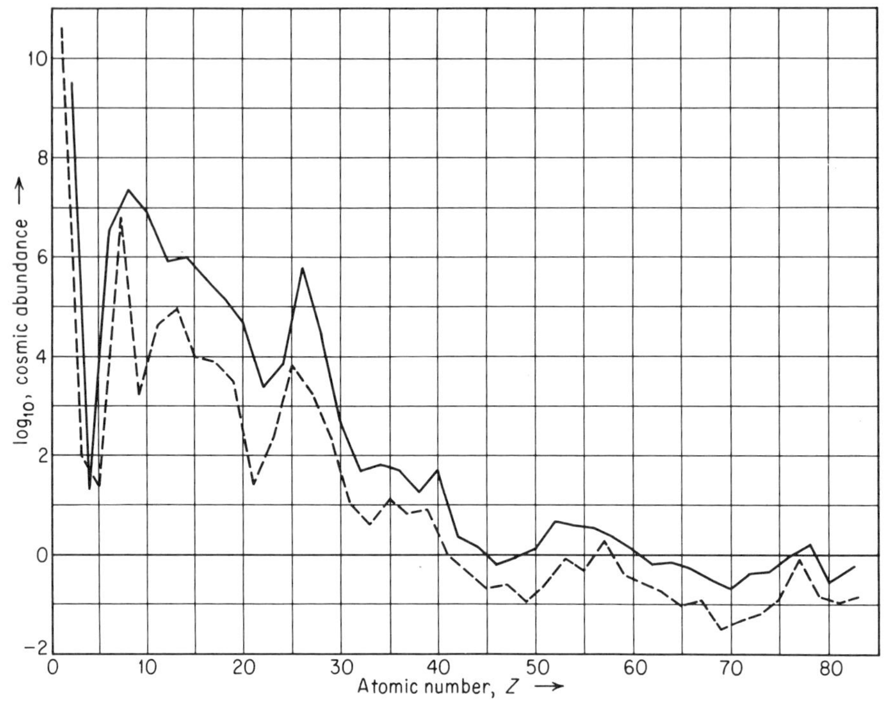

FIGURE 3-1
Cosmic abundances of the elements. The solid line shows elements of even atomic number, and the broken line shows elements of odd atomic number.

can be explained by the conditions which are supposed to have accompanied the birth of the solar system.

The proportion of hydrogen is greatest in the oldest stars (ages $\approx 10^{10}$ years) and least in the youngest stars and in the interstellar material out of which they are being formed. The evidence is consistent with the view that hydrogen and helium are processed to heavier elements in the interior of stars, some of which subsequently explode, to enrich the interstellar matter with heavy elements. Young stars are second- or third-generation stars which have condensed out of the enriched material.

The return of samples of the lunar surface by the Apollo missions has provided a new approach to the knowledge of our surroundings (P. H. Abelson et al., 1970; A. A. Levinson, 1970; B. Mason and W. G. Melson, 1970). The early studies have shown mechanical and chemical features different from those at the earth's surface or in meteorites. The occurrence of metallic iron and the absence of water and hydrated minerals are preliminary observations of considerable interest.

Isotopic studies and age determinations have also been valuable. Samples at the lunar surface have been found to have ages in excess of any known terrestrial rocks, but the preliminary estimates of the moon's age are comparable with the age of the earth. Anomalies in the isotopic abundance of gadolinium, of the order of 1/10 percent, are interpreted to indicate the effect of particle bombardment of the materials.

3-3 COMPOSITION OF THE ATMOSPHERE AND HYDROSPHERE

The size of the earth and its distance from the sun are such that the earth can retain a thin protective atmosphere which moderates the extremes of temperature. Moreover, the mean temperature of the earth is such that liquid water (a substance extremely rare in the universe) can exist on its surface. The presence of water in the liquid state plays an important part in determining the nature of the earth's surface. The oceans act as an enormous heat reservoir further moderating the surface temperature, while the weathering action of rainfall and the powerful transporting and sorting effects of water are largely responsible for the structure and distribution of surface rocks.

The total mass of the oceans is 1.41×10^{24} g, or about 0.03 percent of the mass of the earth, and the mass of the atmosphere is about 5.1×10^{21} g, or about 0.0001 percent of the mass of the earth. Thus the materials of the oceans and atmosphere represent an insignificant fraction of the whole earth. However, they provide important reservoirs for such elements as hydrogen, oxygen, and nitrogen, and their importance in determining the surface conditions of the earth cannot be overestimated.

Until recent years it had been supposed that the atmosphere was stratified, in the sense that gravitational forces should concentrate the heavier gases toward the bottom and that the lighter gases should rise. Sampling of the upper atmosphere by high-altitude rockets has shown this idea to be wrong, and the composition of the atmosphere is now known to vary only slightly with altitude, until altitudes of about 100 km are reached. Exceptions to this are water vapor, which varies greatly from place to place; ozone, which is formed in layers at high altitudes; and rare injections of dust from great volcanic eruptions (H. H. Lamb, 1970). The presence of carbon dioxide in the air, and hence in rain water, provides the principal means of chemically weathering silicate minerals (notably feldspars) and forming the clay minerals. The composition of the atmosphere is summarized in Table 3-2.

The hydrosphere consists of the oceans, with their connected seas and gulfs; continental ice, glaciers, and snow; surface water in rivers and lakes; and ground water. H. W. Menard and S. M. Smith (1966) have recently published new figures for the volume and depths of the ocean basins and found about 2 percent less water than that given by E. Kossinna in an older estimate. A. Bauer (1967) has estimated the volumes and masses of water in various forms, including ice. Table 3-3 combines these results.

Table 3-2 AVERAGE COMPOSITION OF THE ATMOSPHERE

Gas	Composition by volume, ppm	Composition by weight, ppm	Total mass, 10^{20} g
N_2	780,900	755,100	38.648
O_2	209,500	231,500	11.841
Ar	9,300	12,800	0.655
CO_2	300	460	0.0233
Ne	18	12.5	0.000636
He	5.2	0.72	0.000037
CH_4	1.5	0.94	0.000043
Kr	1	2.9	0.000146
N_2O	0.5	0.8	0.000040
H_2	0.5	0.035	0.000002
O_3†	0.4	0.7	0.000035
Xe	0.08	0.36	0.000018

†Variable, increases with height.
SOURCE: B. Mason, 1966.

The oceans constitute 98 percent of the hydrosphere. They cover 361 x 10^6 km^2, or 70.8 percent of the earth's surface, to an estimated average depth of 3729 m. Recent measurements of the thickness of ice sheets in Greenland and Antarctica show that in places the ice extends below sea level and that earlier estimates of its volume were too low. If all the glaciers and ice sheets melted, sea level would rise by about 80 m.

Ocean water varies greatly in salinity, or content of dissolved salts, but the composition of the dissolved salts is remarkably constant. Their relative proportions are given in Table 3-4. Detailed consideration of the material balance of the oceans, which is closely related to studies of weathering of rocks, is too large a subject to be included in this book.

There are two major problems regarding the hydrosphere and atmosphere which have not been solved. The first is whether the atmosphere and hydrosphere are original, having existed since the formation of the earth, or have formed during

Table 3-3 AVERAGE COMPOSITION OF THE HYDROSPHERE

Component	Volume, 10^6 km^3	Mass, 10^{21} g
Sea water†	1349.9	1380.4
Continental ice‡	29.1	29.1
Fresh water‡	0.5	0.5
Total	1379.5	1410.0

†From H. W. Menard and S. M. Smith, 1966.
‡From A. Bauer, 1967.

geological time by the accumulation of gases from within the earth. In this connection, materials newly arriving at the surface of the earth from the mantle for the first time are referred to as *juvenile.* W. W. Rubey (1951) has suggested that the oceans could have been formed during the lifetime of the earth from juvenile waters. It has also been demonstrated by H. A. Shillibeer and R. D. Russell et al. (1955) that the amount of argon-40 in the atmosphere is in reasonable agreement with the amount which continental material, arriving at the surface from the mantle, might be expected to have contained from the decay of radioactive potassium-40 and might be expected to have released if the continents were formed in that fashion. K. I. Mayne (1956) has attempted to extend the arguments to terrestrial helium-3 and helium-4. Hydrogen and helium escape from the earth's atmosphere, but neon, argon-36, krypton, and xenon, which are relatively abundant in the universe, should be more abundant unless the earth had at one time lost its whole atmosphere and generated a new one from within. It should be noted that these arguments only demonstrate the feasibility of the hypothesis of the growth of the atmosphere and oceans and do not show that the growth actually took place.

The second problem is whether the atmosphere has always been an oxidizing one as it is now or was originally reducing and has been altered to its present state by chemical and biological processes. The idea of a reducing atmosphere in the early part of the earth's history has been put forward by H. C. Urey (1952c), who has also considered the consequences in some detail. S. L. Miller and H. C. Urey (1959) have shown that amino acids could form spontaneously from the compounds which Urey considers to have existed in the early atmosphere and oceans. Amino acids are related to the far more complicated proteins which are basic to living materials. Thus the idea of an early reducing atmosphere seems to offer some advantage in trying to explain the existence of life on the earth. Nitrogen in the atmosphere could have been formed by the dissociation of ammonia in the upper atmosphere. The release of free oxygen was probably closely associated with the development of early life forms.

Some direct evidence on the nature of the early atmosphere can be obtained

Table 3-4 PRINCIPAL CONSTITUENTS PRESENT IN SOLUTION IN SEA WATER

	Concentration, ppm†		Concentration, ppm†
Cl	18,980	Br	65
Na	10,556	Sr	8
SO_4	2,649	H_3BO_3	26
Mg	1,272	B	4.6
Ca	400	F	1.3
K	380	Si	3.0
HCO_3	140	Al	0.01

†The concentration of all other elements is less than 1.0 ppm.
SOURCE: B. Mason, 1966.

by studying the oxidation state of iron in certain hematite ores. A paper published by P. Geijer (1956) reports evidence from Swedish rocks showing that an oxidizing atmosphere existed as long ago as 2000 m yr. Since other isotopic arguments seem to suggest the presence of a reducing atmosphere in more recent times, there is clearly need for more research of this kind.

3-4 MINERALS AND ROCKS

The crust of the earth is made up of rocks; rocks, in turn, are formed of minerals. The distinction between rocks and minerals is that *minerals* are crystalline chemical compounds, whereas *rocks* are aggregates. Over 2000 mineral species are known, and a few additional rare minerals are identified and named each year. The crystal form, composition, and classification of minerals is a precise aspect of geology upon which there is general agreement. Many common minerals can be recognized upon inspection. All minerals, even in the form of small specimens, can be identified by x-ray analysis.

Many readily available textbooks describe the various species of minerals (W. A. Deer, R. A. Howie, and J. Zussman, 1966). This discussion is no substitute and will only attempt an elementary description of some of the commonest forms.

A few minerals are simple elements, e.g., gold and sulfur; some are oxides, carbonates, chlorides, sulfides, and other salts; but most are complex silicates. The commonest oxide is quartz, SiO_2, a colorless, hard, glassy mineral which is the chief constituent of sand and sandstone and which can be seen as gray grains in granite. Chert is an amorphous form. Iron oxides and hydroxides are also common and produce the brown, yellow, or red colors of many rocks and soils. Magnetite, Fe_3O_4, and hematite, Fe_2O_3, are important in paleomagnetism because of their high magnetic susceptibility. Calcite, $CaCO_3$, is the soft, white to gray mineral which in finely crystalline form or as fossil fragments constitutes limestone. Rock salt, NaCl, crystallizes with other salts when sea water is evaporated, and by this means it is frequently collected in thick beds, some of which become buried by younger strata and preserved from solution. Pyrite, FeS_2, is the commonest of the sulfide minerals, many of which form important ores of the base metals.

The most abundant minerals are complex silicates. W. H. and W. L. Bragg (1962) showed that x-rays can be used to analyze and identify minerals by exploring their crystal lattices. The fundamental unit of all silicates is a group of four oxygen ions surrounding one silicon ion. The effective radius of the oxygen ions is large, and that of the silicon small, so that the arrangement is like that of four tennis balls piled in a tetrahedral pyramid about a marble. Each of the four oxygen ions, O^{--}, has two negative charges, one of which balances one of the four positive charges on the silicon ion, Si^{4+}, leaving four surplus negative charges on each tetrahedron. These can be neutralized in two ways, either by the addition of metal ions or by extending the structure in a repetitive pattern. Many combinations

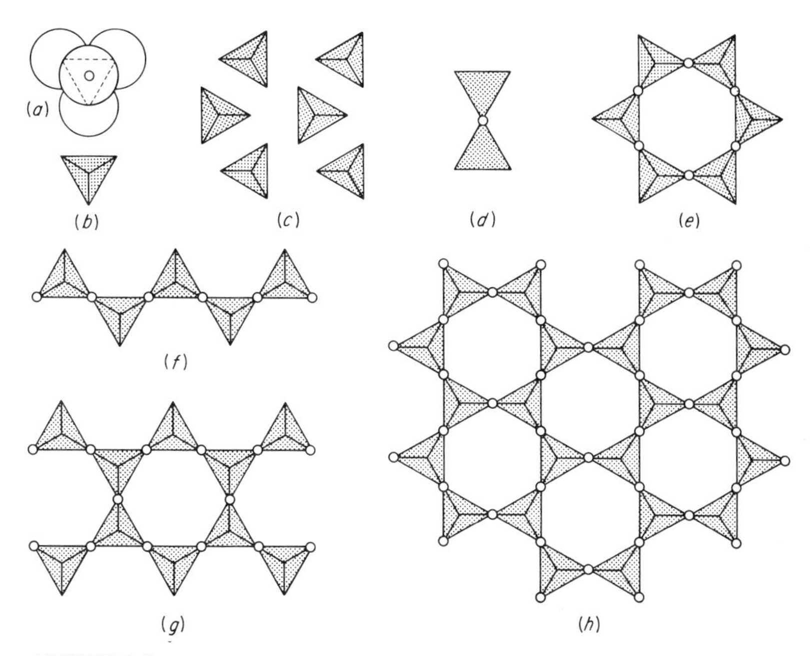

FIGURE 3-2
Some of the chief structural arrangements of the SiO_4 tetrahedron in crystals. (*a*) The SiO_4 tetrahedron, with the ions approximately to scale. (*b*) Conventional representations of the SiO_4 tetrahedron, shown by broken lines in (*a*). In (*d*) to (*h*) shared oxygens are indicated by open circles. (*c*) No oxygens shared (e.g., olivine). (*d*) A pair of tetrahedra sharing one oxygen (e.g., melilite). (*e*) A ring of six tetrahedra, each sharing two oxygens (e.g., beryl). (*f*) A single chain of tetrahedra, each sharing two oxygens (e.g., pyroxenes). (*g*) A double chain of tetrahedra; the outward-pointing tetrahedra share two oxygens, as in (*f*), while those pointing inward share three oxygens and so produce a succession of hexagonal "holes" large enough to accommodate ions of hydroxyl (OH) or fluorine (F) (e.g., amphiboles). (*h*) A sheet of tetrahedra, each sharing three oxygens and forming a continuous network with hexagonal "holes" as in (*g*) (e.g., mica). (*After A. Holmes, 1965.*)

of these two structural arrangements are possible, and they determine the properties of silicates.

If all the oxygen ions are neutralized by metal ions, no links form and the structure of the mineral is based on isolated tetrahedra. When two tetrahedra share one oxygen ion, a paired structure results. As tetrahedra share increasing numbers of oxygen ions, chains, rings, sheets, or solid frameworks are the result. This is illustrated in Fig. 3-2 and Table 3-5.

The complexity of minerals and rocks is increased by the fact that many silicates form solid-solution series. These are defined in terms of the end members which may combine in any ratio. For example, olivine, $(Mg, Fe)_2SiO_4$, consists of any combination of forsterite, Mg_2SiO_4, and fayalite, Fe_2SiO_4. Olivine is a yellow-green mineral forming an important constituent of the mantle, ultrabasic rocks, and some basalts. Other examples are the feldspars, which form the bulk of igneous rocks and are the most abundant minerals in the crust. They are hard, porcelanous minerals, pink or gray in color, with flat and glistening cleavage faces. The most familiar is orthoclase, a potassium aluminum silicate, which is the pink or white opaque mineral shining conspicuously in granite. Albite and anorthite are sodium and calcium aluminum silicates, which combine in all proportions to form the solid-solution series of plagioclase feldspars. Near the surface of the earth the action of water containing dissolved CO_2 slowly decomposes feldspars to clay minerals and fine-grained white micas called sericite. This is an important part of the process of breaking down igneous rocks into sedimentary ones.

Table 3-5 CLASSIFICATION OF SILICATE STRUCTURES

Number of shared oxygen ions	Structure	Si:O or (SiAl):O ratio	Common examples
None	Separate SiO_4 tetrahedra, each neutralized by metal ions	1:4	Olivine: Forsterite, $MgSiO_4$ Fayalite, Fe_2SiO_4
One	Paired tetrahedra	1:3½	Rare (e.g., melilite)
Two	Closed rings of 3, 4, or 6 tetrahedra	1:3	Rare (e.g., emerald)
Two	Single chains of many tetrahedra	1:3	Pyroxenes: Hypersthene $(Mg,Fe)SiO_3$ Diopside, $Ca(Mg,Fe)Si_2O_6$ Augite (complex)
Two and three alternatively	Double chains linked laterally	1:2¾	Amphiboles: Tremolite, $Ca_2Mg_5Si_8O_{22}(OH)_2$ Hornblende (complex)
Three	Network of tetrahedra forming plane sheets	1:2½	Serpentine, $Mg_3Si_2O_5(OH)_4$ Clay minerals, e.g., $Al_2Si_20_5(OH)_4$ Micas: Muscovite, $KAl_2(AlSiO_{10})(OH,F)_2$ Biotite, e.g., $K(Mg,Fe)_3(Si_2Al)O_{10}(OH)_2$
Four	Solid frameworks of tetrahedra	1:2	Quartz SiO_2 Feldspars: Orthoclase, $K(AlSi_3)O_8$ Albite, $Na(AlSi_3)O_8$ Anorthite, $Ca(Al_2Si_2)O_8$

3-5 CLASSIFICATION OF ROCKS

3-5.1 Classification of Sedimentary, Volcanic, Metamorphic, and Plutonic Rocks

Unlike minerals, rocks have neither definite composition nor crystal structures, but are aggregates varying widely in many respects—texture, composition, origin, and association—so that their classification is a complex problem to which there is no simple solution. Different authorities have used different variables as the basis for subdivision; for example the C.I.P.W.[1] classification depends solely upon chemical analyses. S. J. Shand (1943) based his classification upon constituent minerals. In the field, all geologists must use readily observable characteristics such as structure, color, grain size, and the presence or absence of a few distinctive minerals. Even after a basis for subdivision has been selected, many difficulties remain. Scarcely any two rocks have exactly the same characteristics, and the placing of boundaries between types is arbitrary. If mineral composition is chosen as the basis for classification, petrographic study of thin sections of rock under a microscope is required. Although some minerals such as quartz and calcite are readily identifiable, the composition of others which form members of series cannot be established without careful measurements. If chemical composition is selected, time-consuming and expensive silicate analyses are required, and H. W. Fairbairn et al. (1951) have shown that even the most careful analyses are liable to considerable error. Many of the older published analyses, therefore, cannot be relied upon, although new methods promise greater accuracy in future (J. A. Maxwell, 1968; J. G. Jeffrey, 1970).

Since the methods of classification vary and boundaries are indefinite, many rock types have been variously defined by different authors. Unfortunately, this has been the case with some of the commonest rocks.

In only two respects do all authors and all systems agree: that *volcanic rocks* have crystallized from hot molten lava or from *magma*, as molten rock is called before it reaches the surface and loses its volatile constituents, and that *sedimentary rocks* have been formed by accumulation of rock fragments and by precipitation, generally in the sea.

Difficulties in classification arise in the case of rocks whose formation cannot be observed. Unfortunately, they include most of the coarsely crystalline rocks of the *basement*, or *fundamental complex*, which constitutes the bulk of the crust in *shields*, the stable platforms forming the central part of all continents. For most of the past century geologists have debated the relative importance of the different processes by which the basement rocks are formed. On the one hand, there is no doubt that some such rocks are volcanic rocks which were trapped and which solidified slowly beneath the surface. In doing so, they formed crystals larger than those found in extruded lavas, which cooled quickly. On the other hand, it is true that other coarse rocks, especially some foliated gneisses, can be traced laterally

[1] C.I.P.W. after W. Cross, J. P. Iddings, L. V. Pirsson, and H. S. Washington, the petrologists who devised the method.

Table 3-6 DIAGRAM TO ILLUSTRATE THE RELATION BETWEEN TWO METHODS OF CLASSIFYING ROCKS. THE FIRST COLUMN IS NOW GENERALLY PREFERRED TO THE TRADITIONAL METHOD OF THE SECOND COLUMN

<table>
<tr><td>Volcanic</td><td rowspan="2">Igneous</td></tr>
<tr><td rowspan="2">Plutonic</td></tr>
<tr><td>Metamorphic</td></tr>
<tr><td>Sedimentary</td><td>Sedimentary</td></tr>
</table>

into sedimentary strata and are the result of widespread *regional metamorphism*, perhaps with the addition of chemical replacement, a process which is often called *metasomatism.* With *ultrametamorphism*, rocks may become plastic, mobile, and even partially molten (J. J. Sederholm, 1967; K. R. Mehnert, 1968; R. W. Kistler et al., 1971).

Some geologists put the extrusive volcanic rocks and most of the coarse basement rocks together in one group called *igneous* rocks and restrict "metamorphic rocks" to those whose origin from pre-existing strata can be clearly established. Other geologists regard most of the crustal rocks as the products of extreme metamorphism. H. H. Read (1947, 1957) introduced the term *plutonic rocks* to describe the great bulk of basement rocks. Table 3-6 shows the relationships between the two methods of classification.

Field work and chemical analyses provide no conclusive arguments, but consideration of abundance, rates of formation, and isotopic studies suggests that many acid and intermediate gneisses and plutonic rocks are metamorphosed sediments and did not cool from a molten state (Sec. 17-4).

3-5.2 Classification of Igneous Rocks

Table 3-7 presents in short form the usual classification of igneous and volcanic rocks according to color, texture, and essential minerals, characteristics observable in the field and equally applicable whether the coarse-grained rocks are volcanic in origin or the end products of extreme metamorphism.

Volcanic rocks are emitted from *central volcanoes* about whose vents or craters the erupted material builds cones (Fig. 3-3) or from *fissure volcanoes* in the form of linear cracks in the crust. Eruptions may be either *effusive*, in which *lava flows*, usually of basalt, predominate, or *explosive*, in which lavas, often siliceous and viscid, are thrown upward as spray to solidify and fall as ash or as froth which forms *cinders* and *pumice.* In some eruptions dense clouds, called *nuées ardentes* of

Table 3-7 CLASSIFICATION OF IGNEOUS ROCKS ACCORDING TO COMPOSITION AND TEXTURE

Characteristic	Kindreds, or classes, of igneous rocks by composition				
	Ultrabasic	Basic	Intermediate		Acid
Approximate silica content	<45%	45-55%	55-65%		>65%
Color	Dark (green or brown)	Dark (green or brown)	Light or dark (gray or green)		Light (buff or pink)
Essential minerals	Olivine or pyroxene	Calcium feldspar, pyroxene	Sodium feldspar	Potassium feldspar	Potassium feldspar and quartz
Common minerals	Calcium feldspar	Olivine	Quartz, ferromagnesian minerals	Quartz, ferromagnesian minerals	Ferromagnesian minerals
Texture:					
Fine-grained (volcanic rocks)	. . .	Basalt	Andesite	Trachyte	Rhyolite
Coarse-grained nonfoliated (plutonic rocks)	Peridotite, dunite, pyroxenite	Gabbro†	Granodiorite, diorite	Syenite	Granite
Coarse-grained foliated (plutonic rocks)	. . .	. . .	Gneiss	Gneiss	Gneiss

†Most geologists who regard granites and granodiorites as the end products of metamorphism would regard most gabbros as accidental intrusives and so place them in the class of volcanic rocks in spite of their coarse grain.

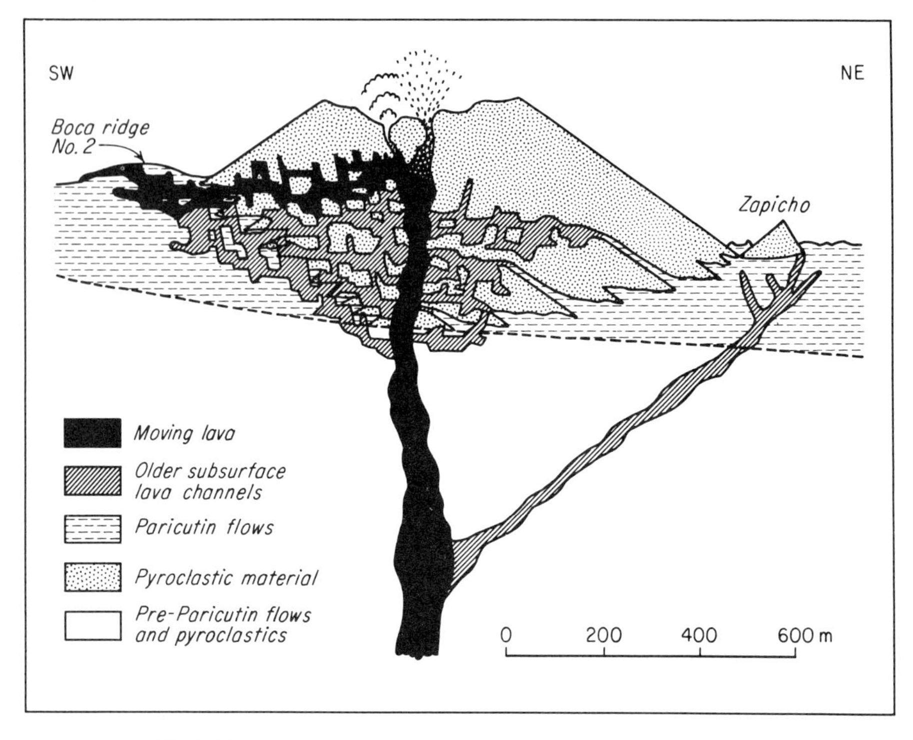

FIGURE 3-3
Inferred vertical section through Paricutin, a volcano, December 1945. (*After K. B. Krauskopf, 1948.*)

superheated steam, red-hot spray, and ash may pour down mountainsides at great speed, as in Martinique in 1902. If hot enough, the ash may congeal on settling to form *welded tuffs,* or *ignimbrites,* often mistaken for flows of rhyolite. The products of explosive eruption are collectively called *pyroclastics*, or *tuff*, whether deposited in water or subaerially.

The greatest historical eruptions have been at Laki, Iceland, where in 1783 a fissure 25 km long emitted 12 km^3 of dust and between 12 and 27 km^3 of basalt which flowed out to cover 525 km^2, and at Tambora, Indonesia, where in 1815 the explosion of Gunung volcano ejected between 100 and 300 km^3 of solid materials (H. H. Lamb, 1970). During the late Tertiary several flows of the Columbia River basalts each covered over 10,000 km^2 with more than 100 km^3 of lava (I. L. Gibson, 1969).

Dikes are nearly vertical intrusions tending to occur in parallel swarms of similar dikes (E. M. Anderson, 1951). *Sills* are horizontal intrusive bodies, while *laccoliths* are bun-shaped. Lists of the world's active volcanoes, great historical eruptions, and natural disasters in general have been published by M. N. van Padang (1951-67), H. H. Lamb (1970), and J. H. Latter (1968, 1969), respectively.

Plutonic rocks usually take the form of great *batholiths*, or mantled gneiss domes, often tens of kilometers in diameter with unseen roots; geophysical work suggests they are relatively shallow and within the crust (P. E. Eskola, 1948; W. B. Hamilton and W. B. Myers, 1967). Batholiths are usually considered to be due to the rise of rocks of low density. Such upwellings, often called *diapirs*, occur in many forms, including *salt domes*. The largest upwellings seem likely to be those in the mantle that give rise to mid-ocean ridges and related rifted domes.

3-5.3 Classification of Sedimentary Rocks

The great bulk of sedimentary rocks have been deposited by water, although three peculiar types, loess, till, and tuff, are the result of accumulation by wind, ice, and volcanic eruptions, respectively. *Loess* is consolidated dust formed around the borders of deserts. *Till* is an accumulation of rocks ground by glacial action, and it is characteristically undecomposed and a mixture of fragments of all sizes from rock flour to large boulders.

Water-laid sedimentary rocks have traditionally been identified by texture and composition, and a system of classifying them by those properties, devised by P. D. Krynine (1945, 1948) and followed by F. J. Pettijohn (1957), is given in Table 3-8. The system is genetic and has the advantage that rocks in any column tend to occur together under particular conditions, and are called a *sedimentary facies.*

The classification devised by Krynine included only the first three facies, which are those found on land, but recent exploration of the ocean floor by seismic and coring devices has shown that the abyssal beds are also quantitatively important.

Sedimentary rocks which have been formed slowly, like those of the platform facies, contain highly resistant minerals, notably quartz, which remain after igneous rocks have been weathered and eroded, whereas the presence of feldspar is a sign of rapid accumulation because it is readily altered to kaolin, $H_4Si_2Al_2O_9$, and calcite, $CaCO_3$. Pettijohn points out that the mineralogical differences between the fine-grained shale members of the different facies are as yet unknown.

3-5.4 Classification of Metamorphic and Plutonic Rocks

Metamorphic rocks are the most difficult to identify and classify (B. Bayly, 1968). Metamorphism may be thermal, as a result of change in temperature, or dynamic, as a result of change in pressure, or both. It may be local or regional. The intrusion of a hot igneous body into sedimentary rocks frequently causes *local metamorphism,* which is then easily recognized, because the cause is obvious. Burial deep in the roots of mountains followed by uplift, erosion, and exposure may cause *regional metamorphism* in ancient rocks, but this is harder to recognize. It is sometimes possible to make use of the presence of certain diagnostic minerals (including staurolite, sillimanite, and zeolite). Tracing metamorphosed rocks into unaltered

Table 3-8 CLASSIFICATION OF SEDIMENTARY ROCKS ACCORDING TO COMPOSITION AND TEXTURE

	Facies of sedimentary rocks by composition and location			
Characteristic	Platform orthoquartzite-carbonate	Greywacke	Arkose	Abyssal
Composition	Quartz ±chert† ±limestone	Quartz + chert +mica +chlorite ±feldspar	Quartz + feldspar +clay	(See Sec. 3-6.4)
Predominant color	Light	Dark	Red	Dark
Location	Covered shields and shelves	Island arcs	Intermontane basins	Ocean basins
Texture:				
Coarse (conglomerate)	Quartzose conglomerate	Greywacke conglomerate	Arkose conglomerate	Lacking
Medium (sandstone)	Quartzose sandstone	Greywacke	Arkose	Graded sands
Fine (shale)	Quartzose shale	Micaceous or chloritic shale	Feldspathic shale	Lutite
Precipitate	Limestone, dolomite, evaporites	Chert	. . .	Chert

SOURCE: P. D. Krynine, 1945, and F. J. Pettijohn, 1957.
†The sign ± should be read "containing more or less."

rocks, and the identification of the altered remnants or premetamorphic structures, can also be useful.

It is also hard to classify metamorphic rocks. The variety of rocks is wide, and any may be metamorphosed. It might be supposed that the mineral constitution would depend on the composition, but different degrees of pressure and temperature produce different minerals in the same rock, and even those depend on the rate of change of conditions, the time of exposure, and whether equilibrium was achieved. The effects of the presence, addition, or extraction of water are considerable. Once a rock has been changed by being heated and compressed, it is likely to retain the minerals characteristic of those conditions and not to undergo *retrograde metamorphism*.

The first attempt at classification depended upon the recognition of metamorphic zones, or *isograds*. At the end of the last century, while working in Scotland, G. Barrow (1893, 1912) recognized that across a broad belt of country a series of rocks which had once been uniform shales had been transformed without alteration in bulk chemical composition into rocks made of different minerals. He divided the altered rocks into zones, each zone characterized by the first appearance of a key mineral. The lines which separate zones are called *isograds* and join those places where a key mineral first appears (F. J. Turner and J. Verhoogen, 1960). The zones are listed in Table 3-9 and illustrated in Fig. 3-4. The minerals found in any one zone vary with the composition of the rock. The metamorphic changes are best seen in shales, which first pass into shiny micaceous schists and then into coarse feldspathic gneisses, with considerable changes in mineralogical but not in bulk composition. The metamorphic zones are harder to distinguish in other rocks. For example, pure quartz sandstone does not change in mineral composition, but only recrystallizes.

Barrow found bodies of granite within his highest zone, and he believed that the whole series had developed as a result of heat due to the intrusion of the granite into the deeply buried rocks. Other workers have noticed the wide differences in the degree of regional metamorphism existing in different areas intruded by granite. In some areas all the rocks have been regionally metamorphosed, whereas in other

Table 3-9 TYPICAL INDEX MINERALS OF ZONES OF INCREASING METAMORPHISM

Index mineral of zone	Rock type	Grade of metamorphism
Chlorite, albite	Shale	Low grade
Brown biotite	Schist	↓
Almandine garnet	Schist	
Staurolite, calcium-feldspar	Gneiss	Increasing temperature and pressure
Kyanite	Gneiss	↓
Sillimanite, augite	Gneiss	High grade

SOURCE: G. Barrow, 1893 and 1912.

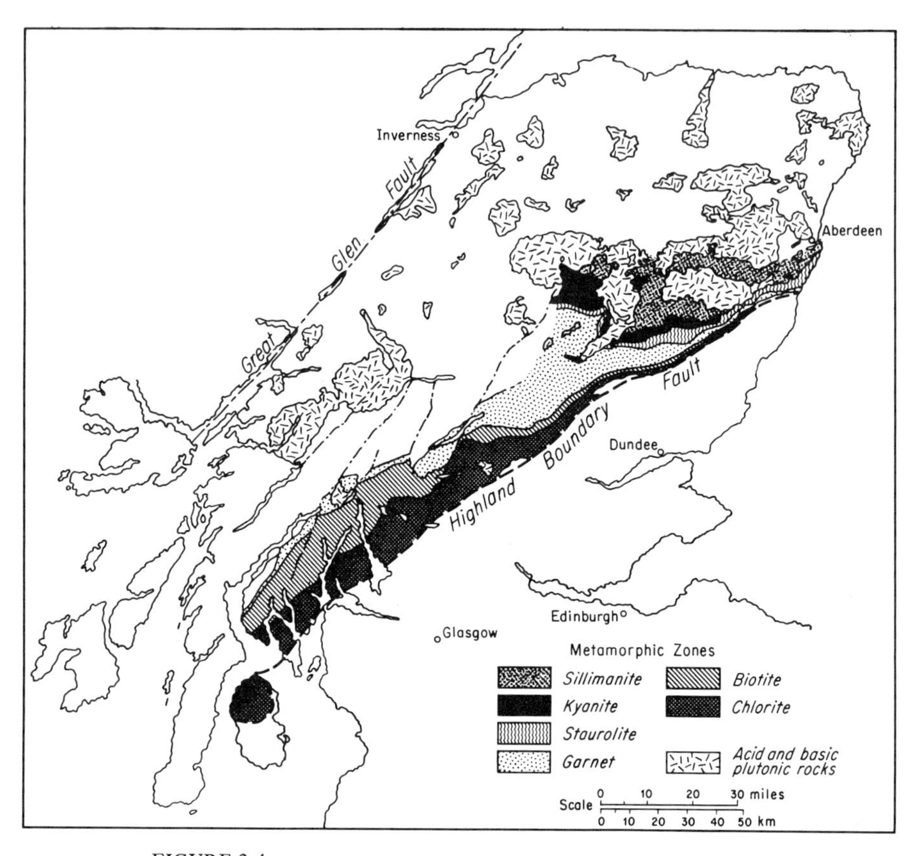

FIGURE 3-4
Metamorphic zones and intrusive bodies of the Southern Highlands of Scotland. (*After G. Barrow, 1893, 1912.*)

places there is only local metamorphism surrounding volcanic intrusives. Therefore they have held that intrusion of granite is not the cause of regional metamorphism, but may be its end product. If so, differences in composition show that in many places granite has either undergone changes in composition *in situ* or that it has been mobilized and moved upward from a deeper and different source.

Another method is to define those rocks which have been metamorphosed to a particular degree of temperature and pressure as belonging to a certain *metamorphic facies* named after minerals common to those conditions (Fig. 3-5). The concept is useful in spite of being somewhat ill-defined.

A new form of metamorphism has recently been recognized on the ocean floors. Along the crest of mid-ocean ridges, high temperature gradients, frequent volcanic intrusions, burial, and the action of circulating sea water produce an

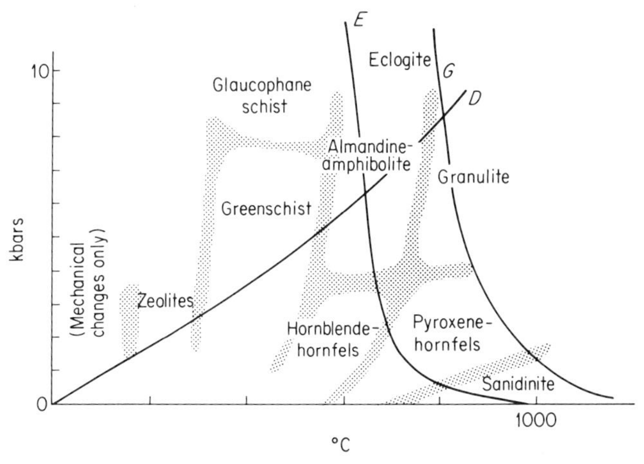

FIGURE 3-5
Nomenclature of metamorphic facies. Line *OD* indicates approximately the common trend of conditions, while line *E* is the melting range of granite in equilibrium with water vapor. Line *G* is the range at which basalts begin to melt. (*After F. J. Turner and J. Verhoogen, 1960; H. G. F. Winkler, 1965; B. Bayly, 1968.*)

unfoliated, retrograde metamorphism in which the originally igneous rocks are converted to green schists and amphibolites, with loss of calcium and potassium and gain in sodium and magnesium (K. S. Deffeyes, 1970).

3-6 ROCK ASSOCIATIONS

Section 3-5 described the principal varieties of rocks; this section will deal with their distribution. Most works on geology are written from the point of view of field geologists, whose task it is to identify, map, and describe the particular rock types occurring in limited areas. On the other hand, in broad discussions about the earth, it is more important to realize that certain groups of rocks are commonly found together in particular environments. Such assemblages will be called *rock associations.* F. J. Turner and J. Verhoogen (1960) have already used this method for classifying igneous rocks, and P. D. Krynine (1948) and F. J. Pettijohn (1957) have used it for sedimentary rocks. It is considered valid and useful to try to combine these classifications and relate them to specific environments.

All the rocks related to a particular environment will be called an *association.* Related igneous rocks tending to occur together within any association will be called a *kindred,* a name introduced by G. W. Tyrrell (1926), while the term *facies* will be used to refer to similar assemblages of sedimentary rocks. Thus a rock

Table 3-10 ROCK ASSOCIATIONS AND THEIR CHARACTERISTICS

Association	Igneous kindred	Sedimentary facies	Metamorphism	Location	Stages of ocean basin–mountain cycle (Sec. 15-1)
Rift valley	Basalt (tholeiite and alkali)	Arkose	Contact	Plateaus and rift valleys	1
Continental plain	Minor	Platform	Slight to low-grade regional	Interior plains	Chiefly 3
Continental margin	Minor basalts	Platform grading to abyssal	Negligible	Continental margins	2,3,4,5,
Ocean floor	Basalt (chiefly tholeiite)	Abyssal: particle by particle (pelagic), geostrophic current, turbidity current	Due to heat, burial, and action of sea water	Ocean floors	2,3,4,5
Oceanic island	Basalt (tholeiite, alkali trachytic)	Modified platform	Contact	Remote ocean islands	2,3,4,5
Island arc	Andesite and plutonic granodiorite	Greywacke	Regional and contact	Island arcs	4,5
Primary mountain	Andesite and plutonic granodiorite	Minor arkose, erosion predominant	Regional and contact	Active mountains	4,5,6
Exposed shield	Basalt and plutonic granodiorite	Erosion predominant	Regional	Central part of continents	6

association is normally made up of sedimentary rocks of a particular facies plus one or more kindreds of igneous rocks. Each association is also characterized by the degree to which most of the rocks have been metamorphosed. Table 3-10 lists the components of some common rock associations. The igneous kindreds and sedimentary associations are those described in Tables 3-7 and 3-8. Of the igneous kindreds the plutonic consists essentially of granite and gneiss, the basaltic of basalt, trachyte, and minor gabbro, and the andesitic of andesite, granodiorite, and minor basalt. Ultrabasic rocks of a wide variety occur as minor constituents in most associations (P. J. Wyllie, 1967).

In Chaps. 15 and 16 the associations are tentatively linked to stages in the major processes of the earth. Each association includes only the rocks formed during one cycle of activity. Thus, if an association of young sedimentary and volcanic rocks rests upon or beside the basement of a shield, the young association is regarded as belonging to a later cycle than the shield, which is made up of the eroded roots of mountains formed in earlier cycles.

3-6.1 The Rift-valley Association

In East Africa, the Colorado Plateau, along the Rhine Valley and elsewhere, domes or plateaus have been uplifted. As H. Cloos (1939) pointed out, rifting and volcanism have frequently accompanied or followed uplift. The East African rift valleys have been chosen as the type locality of this association. The vicinity of Lake Baikal and the Basin and Range province of the western United States may be other examples (Secs. 15-2, 15-3, and 16-6).

The igneous rocks belong to the *basalt kindred* and are commonly divided into two major classes, tholeiites (after Tholey, France) and *alkali basalts* (H. H. Hess and A. Poldervaart, 1967). Some authors distinguish high-alumina basalts also. Tholeiites are essentially an assemblage of pyroxene, calcic plagioclase, and glass, whereas alkali basalts have in addition alkali feldspars, or feldspathoids. There has been much debate whether all basalts are derived from a common source or different sources. Section 17-5 suggests that there are two dominant sources at different depths in the mantle, but it appears that, due to fractionation and mixing, basalts form a gradational sequence, although the types present in any one area may be distinctive. Table 3-11 shows average compositions and illustrates that continental-flood basalts have less titania and more silica and potash than oceanic basalts, perhaps because of contamination (V. Manson, 1967). In this association volcanic eruptions are either basalt flows, predominantly fed from fissures and often extensive, reaching a thickness of 3500 m in Ethiopia (P. A. Mohr, 1962), or much more varied eruptions from central volcanoes. The latter include basaltic piles like Kilimanjaro, smaller volcanoes of alkali basalt, and others of more highly differentiated rocks such as nephelinites and carbonatites.

Although the formation of sedimentary rocks depends upon the rainfall and the latitude, many uplifts are in arid regions in the heart of continents where rapid

Table 3-11 AVERAGE VALUES FOR BASALT ANALYSES

	SiO_2	TiO_2	Al_2O_3	Fe_2O_3	FeO	MnO	MgO	CaO	NA_2O	K_2O	H_2O	P_2O_5
Continental tholeiite	51.5	1.2	16.3	2.8	7.9	0.17	5.9	9.8	2.5	0.86	0.81	0.21
Oceanic tholeiite	49.3	2.4	14.6	3.2	8.5	0.17	7.4	10.6	2.2	0.53	0.79	0.26
Continental alkali basalt	47.1	2.2	15.7	3.4	7.8	0.16	7.1	10.1	3.3	1.5	1.1	0.47
Oceanic alkali basalt	46.9	3.0	15.5	3.1	8.6	0.16	6.9	10.4	3.0	1.3	0.8	0.39

SOURCE: V. Manson, 1967.

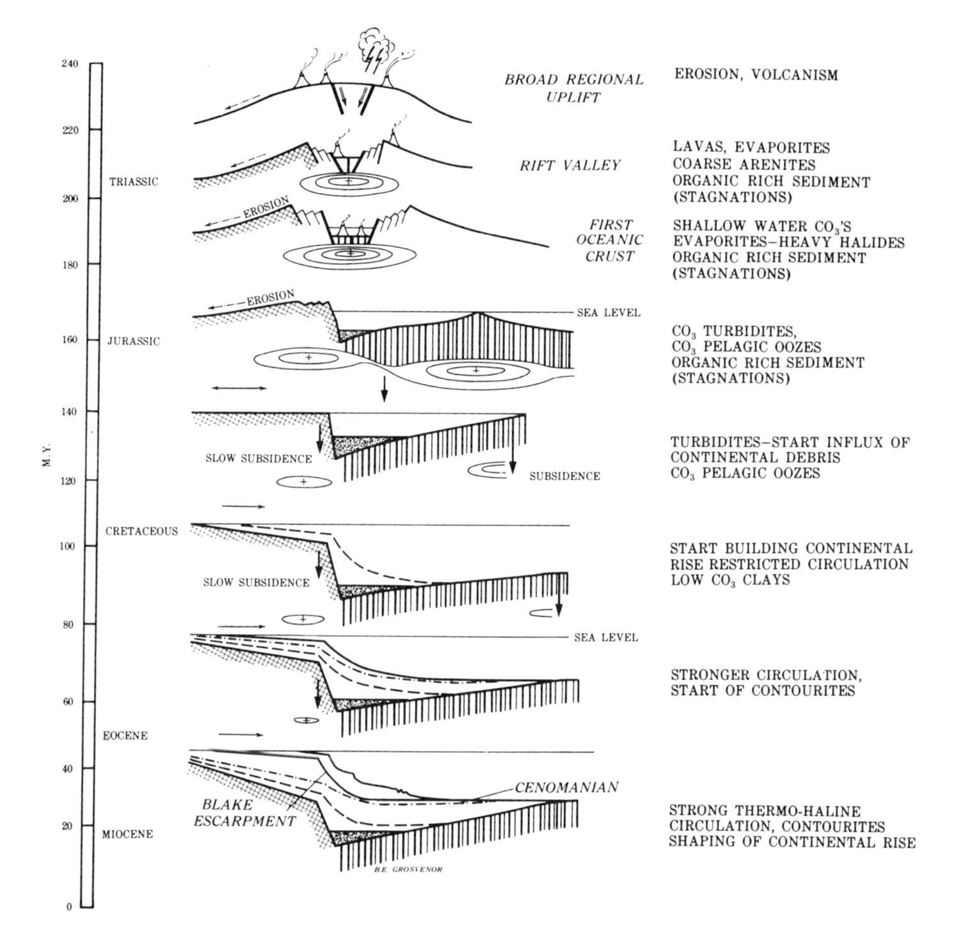

FIGURE 3-6
Sketch cross sections of the history of the development of the East Coast of the United States. (*After E. D. Schneider, 1969, and P. R. Vogt, 1970.*)

mechanical disintegration, without chemical alteration of the feldspars, produces arkose. In rift valleys, evaporites form, but few ores. In East Africa, deposits of lacustrine, alluvial, and pyroclastic sediments contain mammalian and archeological assemblages of great importance for the study of human evolution (G. Kurth, 1968; L. S. B. Leakey, 1969; F. C. Howell, 1969; B. Patterson et al., 1970; W. W. Bishop and G. R. Chapman, 1970).

It has been suggested that there may be a cycle of ocean building of which this association may mark the first stage (Fig. 3-6 and Table 15-1). If so, this would explain why dike swarms follow some coasts and why the largest areas of flood basalts are adjacent to others. For example, it seems likely that the Palisades sill in

Table 3-12 CHEMICAL COMPOSITION OF TYPICAL ROCKS OF PLATFORM SEDIMENTARY FACIES

Rock	SiO_2	TiO_2	Al_2O_3	Fe_3O_3	FeO	MnO	MgO	CaO	Na_2O	K_2O	H_2O	CO_2	Total
Average shale	58.1	0.7	15.4	4.0	2.5	. . .	2.4	3.1	1.3	3.2	5.0	2.6	98.3
Average sandstone	93.2	0.0	1.3	0.4		. . .	0.1	3.1	0.4		0.7	2.0	101.2
Average limestone	5.2	0.1	0.8	0.5		. . .	7.9	42.6	0.1	0.3	0.8	41.6	99.9
Average platform rock†	55.1	0.4	9.7	2.5	1.6	. . .	3.2	11.0	0.8	2.0	3.2	10.3	99.9
Average of Mississippi delta	70.0	0.6	10.5	3.5		0.1	1.4	2.2	1.5	2.3	5.7	1.4	99.2

SOURCE: F. J. Pettijohn, 1957, pp. 10, 298, 344, 384.

†The composition of average platform rock is calculated on the basis of a ratio of shale: sandstone:limestone = 58:22:20.

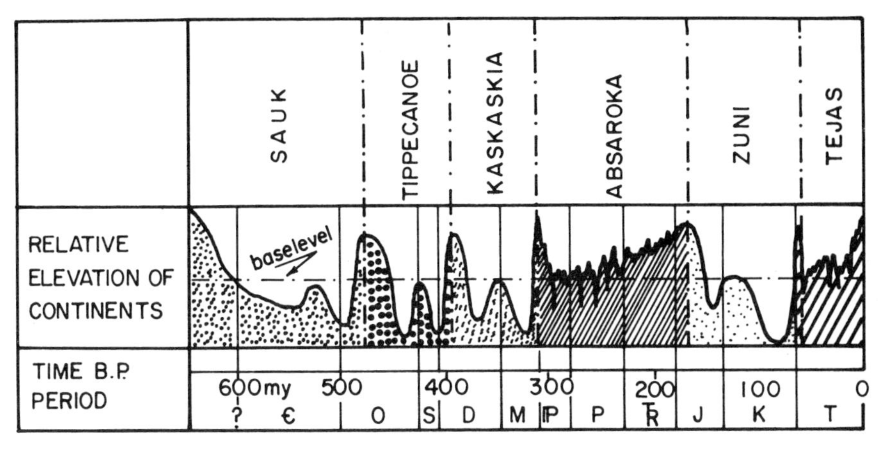

FIGURE 3-7
Postulated relative vertical movement between a stable North American shield and base level during Phanerozoic time. The names are of periods of major invasion by the sea separated by withdrawals. The Absaroka and Tejas invasions were less important than the others and covered only coastal regions. (*After L. L. Sloss, 1966.*)

New Jersey, the Deccan traps in India, the Serra Geral basalts in South Africa, and the Ferrar dolerites of Antarctica may have been intruded when the adjacent oceans first rifted open.

3-6.2 Continental-plain Association

The stability of shields allows them to be eroded almost to sea level and renders them susceptible to widespread flooding. Rocks of the platform facies are formed in these shallow seas and are well exposed on every continent (Table 3-12).

L. L. Sloss (1963) claims that over North America the sea has *transgressed* and *regressed* in several great cycles at intervals of about 100 m yr (Fig. 3-7). Two are dominant, and there is evidence that major changes in sea level are worldwide. This suggests that the early Paleozoic and early Mesozoic withdrawals were more important than the others. Since the volume of the present mid-ocean ridges is approximately 1.8×10^8 km^3 ($1/2 \times 60{,}000 \times 3{,}000 \times 2$ km) and the area of the ocean basins is 2.4×10^8 km^2, major changes in the mid-ocean-ridge system can change sea level by hundreds of meters (R. L. Grasty, 1967). In Permo-Triassic time when the continents were together in Pangea, the mid-ocean ridges were diminished and the supercontinent was high (E. M. Moores, 1970a).

In every transgression shallow seas encroached over old and deeply weathered land, slowly forming their beds (7000 m in the Michigan basin is 10 times the usual thickness). Good aeration supported abundant life, the remains of which are preserved in fossils, reefs, and limestones. Slight changes in level produced frequent interruptions called *hiatuses* and *disconformities.* Strong wave action sorted the rocks well, depositing conglomerates and sandstones of nearby pure quartz on the old basement and washing the finer grains of clay and mica from disintegrated feldspars into deeper water, to be laid on conglomerates and sandstones already formed. A common sequence is thus as follows:

Shales with local or reef limestones
Quartzose sandstones and transported limestones
Quartzose sandstones
Basal conglomerates

Slow transgressions deposit beds of the same formation or facies at progressively younger times as the sea creeps inland (M. Kay, 1951). It was at one time believed that shallow-water deposits so well exposed on continents are the commonest sedimentary rocks, but drilling for petroleum has demonstrated the greater abundance of other sedimentary rocks in deep basins, on coastal plains and offshore under continental shelves, and in island arcs. In these places the beds thicken and change their characteristics, becoming less well sorted and generally finer. With increasing thickness the chance of tectonic disturbance increases.

In arid regions evaporation of sea water in enclosed basins has produced saline deposits called *evaporites*. Under some other coastal and lacustrine conditions *coal* has formed from vegetation which has escaped oxidation and has been preserved.

Usually the rocks of the association are little disturbed, and any metamorphism is of low grade, forming hydrated minerals called *zeolites* (D. S. Coombs, 1961; A. Miyashiro and F. Shido, 1970). Igneous rocks are of negligible importance, for only a few volcanoes of strongly fractionated rocks break through, as for example the Monteregian Hills of alkali basalts which extend from northern Vermont to Montreal and beyond.

Cryptoexplosion craters are characteristic, if rare, disturbances. Which of these fascinating structures are due to purely volcanic causes and which to meteoric impact has been the subject of much debate (Sec. 1-5). They are of interest here because, in several supposed astroblemes, isolated patches of sedimentary rocks have been preserved far within the borders of the Canadian shield. One at Gypsumville, Manitoba, is the site of an important evaporite deposit (C. S. Beals and I. Halliday, 1967; H. R. McCabe and B. B. Bannatyre, 1970).

3-6.3 Continental-margin Association

The continental shelves and slopes where rivers bring detritus to the edge of ocean basins 5 km deep are favorable places for the accumulation of great thicknesses of sediments, and in recent years offshore drilling and geophysical methods have revealed their structure (Secs. 15-9 and 15-10). *Continental shelves* extend to the

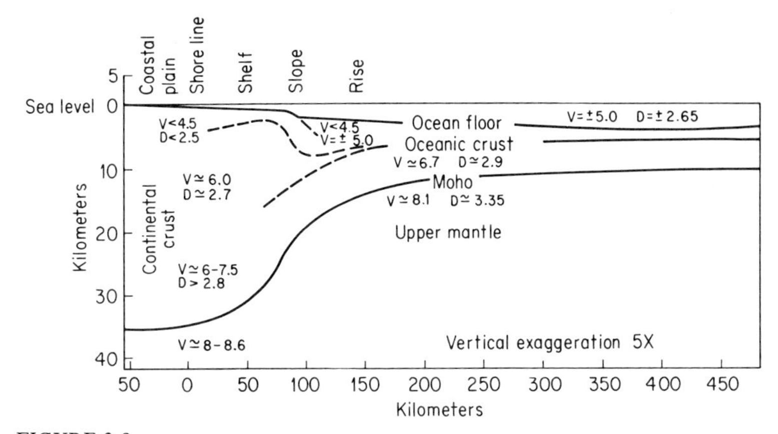

FIGURE 3-8
Generalized crustal structure at continental margin. (*After H. D. Hedberg, 1970.*)

shelf break, usually at a depth of about 100 fathoms, or 200 m, where they give place to the steep *continental slopes*. In general, the thickness of sediments increases under the shelf and beyond it for some distance to form a great basin or wedge of coalesced deltas several kilometers thick along the coast and over the oceanic crust (H. D. Hedberg, 1970). Along the Texas Gulf Coast G. E. Murray (1961) found that Jurassic to Recent sediments have a volume of 13.7×10^6 km^3 in the form of a succession of lenses. The maximum thickness of the lenses is 130,000 m, although they are not laid one upon another, but overlap like shingles and gradually extend out into deeper water (Figs. 3-8 and 3-9). The outer slopes of the continental shelves built of these accumulations are steep and have frequently slumped. This load has depressed the sea floor and coast, allowing more and more sediments to accumulate at sea level as the coast is tilted.

The predominant rocks are shale, alternating with some sandy beds. Toward the shore coarser beds increase and the rocks are better sorted. Table 3-12 includes one estimate of the composition of the sediments in the Mississippi delta. Near the mouths of large and muddy rivers limestones are scarce, but away from them in tropical waters coral reefs may be important, as in the Great Barrier Reef off Queensland, Australia.

Metamorphism of these rocks is slight, and igneous rocks are rare. Occasional dikes are found, and volcanic rocks can become incorporated where volcanoes are active close to the coast.

Salt domes have recently been discovered in water over 3500 m deep close to the shores of the Gulf of Mexico and Africa (Sec. 15-10). They indicate the presence of evaporite beds near the base of the sedimentary succession at the continental margin. Ocean basins are too large to be evaporite basins, so that either

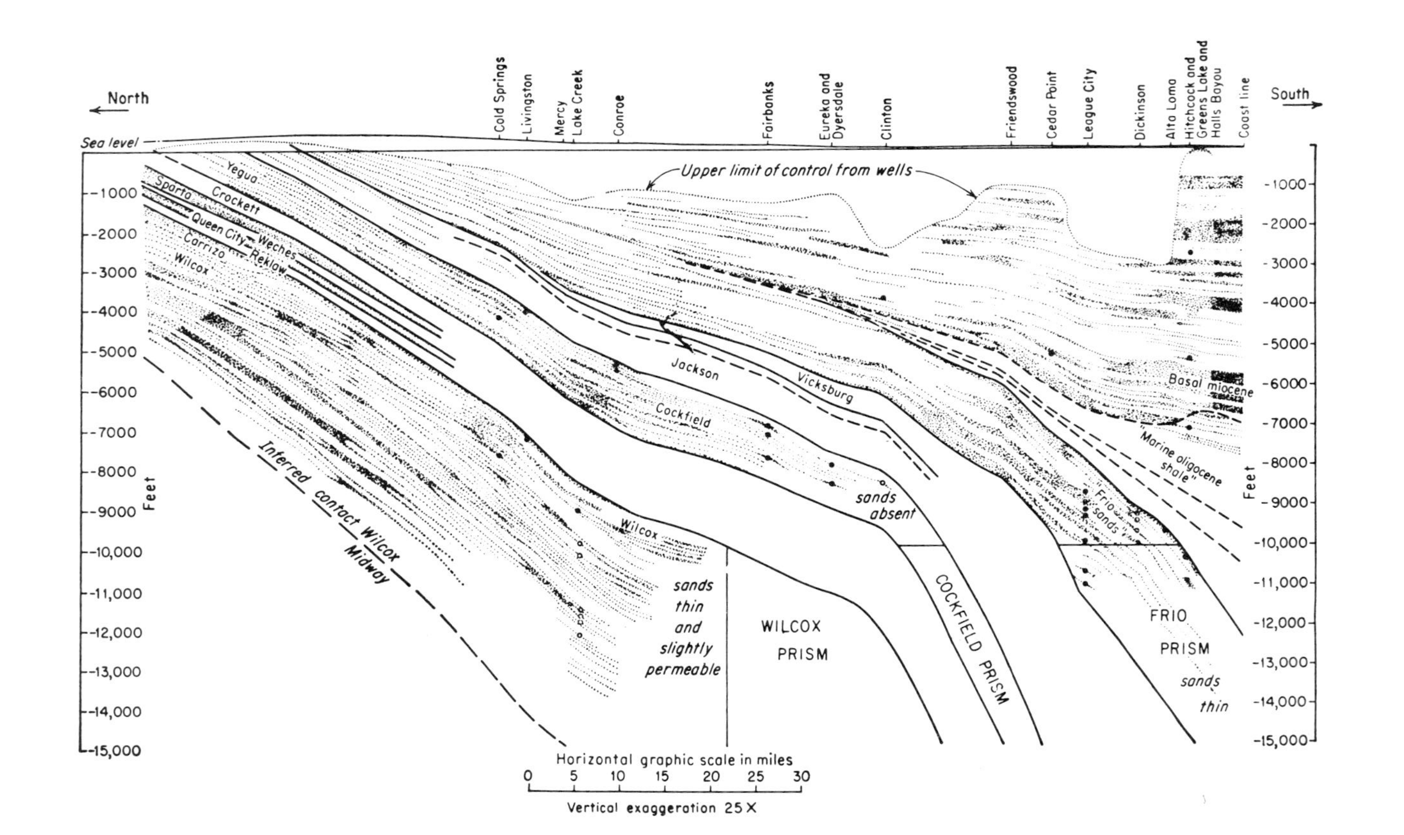

FIGURE 3-9
Cross section of Texas Gulf Coast shelf. (*After J. Colle et al., 1952.*)

the beds formed above ocean level and sunk by several kilometers or the Atlantic and other ocean basins have opened from an initially narrow rift like the Red Sea, in which evaporites are known to have formed.

In 1857 J. Hall (1859) drew attention to the presence in the western flank of the Appalachian Mountains of shallow-water deposits of Paleozoic age which thicken toward the mountains to more than 10,000 m and become increasingly disturbed. J. D. Dana (1873) gave the name "geosynclinal," later changed to *geosyncline*, to elongated basins filled with sedimentary rocks, and the name has been widely used in a variety of ways. This matter and the question of the relationship between geosynclines and shelf deposits are discussed in Sec. 17-2.

3-6.4 Ocean Floor

Deep oceans cover two-thirds of the earth's surface. Until recently little was known of their floors, but modern geophysical and drilling techniques have provided information about the thickness, age, and physical properties of the layers which underlie the ocean floor (Sec. 14-2).

It has not been easy to correlate results obtained by different methods. Drilling and coring have thus far penetrated only the uppermost 1000 m of sediments. Dredging can recover samples of deeper layers exposed on fault scarps, but identification of the precise layers involved has been uncertain, as has the question of whether disturbed regions are typical.

The interpretation of seismic surveys depends upon the assumption that the sea floor is underlaid by a series of uniform layers with velocities increasing downward. Sampling has shown that the two top layers are composed of sediments and basalt, respectively. From 1957 to 1970 there was general agreement that only three uniform layers constitute the oceanic crust between the floor of the ocean and the mantle, but the nature of the third layer was the subject of much debate until G. L. Maynard (1970) partly resolved the issue by showing that layer 3 is in reality two separate layers. Table 3-13 gives some estimates of the nature of the layers on the ocean floor.

Layer 1 is the only layer of the oceanic crust which varies markedly in thickness from place to place. Over the crests of mid-ocean ridges it is either absent or consists of discontinuous patches of very young sediments. Away from the crests, both the thickness and the maximum ages found increase. Some increases are sudden. This was at first attributed to irregularities in the motion of global plates (J. Ewing and M. Ewing, 1967), but the patterns of magnetic anomalies and the results of JOIDES drilling indicate steady motion (J. R. Heirtzler et al., 1968; J. D. Phillips and B. P. Luyendyk, 1970; A. E. Maxwell et al., 1970a, b). On the other hand, E. D. Schneider et al. (1967) have shown that periods of abnormally high sedimentation by bottom-contour currents can explain the sudden increases (Fig. 3-10). Basalt sills from layer 2 often intrude into the base of the sediments (J. D. Hays, 1970). In the Argentine basin and other large basins off continental margins

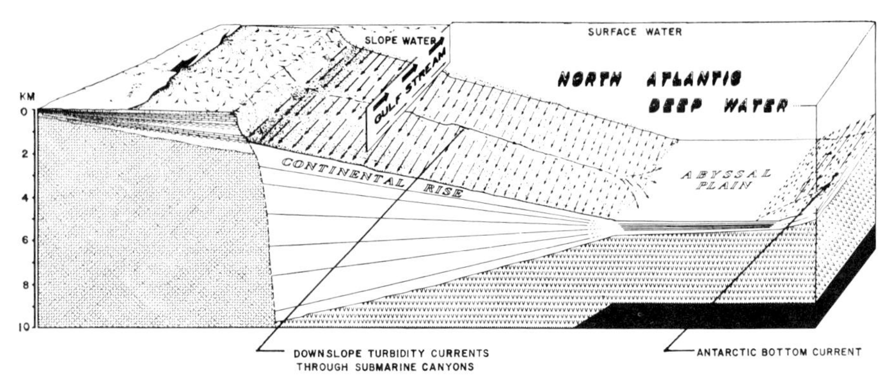

FIGURE 3-10
Shaping the continental rise by geostrophic contour currents shown by arrows. (*After B. C. Heezen et al., 1966a.*)

Table 3-13 ESTIMATES OF AVERAGE THICKNESSES, VELOCITIES AND COMPOSITIONS OF LAYERS OF OCEANIC CRUST (ESTIMATES GIVEN FOR THREE- AND FOUR-LAYER CRUST)

	Average thickness, km		Range of velocity, km/sec		Nature of material	
	Refs. 1,2,3	Ref. 4	Refs. 1,3	Ref. 4	Ref. 5	Ref. 6
Layer 1	0.3	0.3	1.5-3.4	1.5-3.4	Bottom sediments	Bottom sediments
Layer 2	1.3	1.3	5.0	5.0	Dikes, breccias, and sediments	Basalt flows
Layer 3	5.2	2.0	8.8	6.8	Serpentinite or gabbro or amphibolite	Dikes
Layer 4	0	3.2	. . .	7.3	-	Serpentinite, gabbro, and amphibolite
Total	6.8	6.8				

References: 1. G. G. Shor, Jr., and R. W. Raitt (1969).
2. A. B. Ronov and A. A. Yaroshevsky (1969).
3. J. I. Ewing (1969).
4. G. L. Maynard (1970).
5. P. R. Vogt et al. (1969b).
6. J. R. Cann (1970).

the sediments may be as much as 7 km thick and the base as old as Upper Jurassic (J. J. Zambrano and C. M. Urien, 1970). In the basins unconsolidated sediments up to 3 km thick with velocities of about 2 km/sec overlie an acoustically more opaque layer up to 3 or 4 km thick with velocities of about 3 km/sec (R. E. Houtz et al., 1968). In the North Atlantic the upper layer contains three prominent reflecting horizons A, β, and B. JOIDES drilling has shown that the widespread horizon A is a layer of early middle Eocene chert (M. Ewing et al., 1969a; J. Ewing et al., 1970a). Horizon β may be middle or lower Cretaceous in age. Like the changes in rates of sedimentation, these horizons may mark changes in ocean currents due to stages in the opening of the oceans and may be widespread. Horizon B appears to be a basalt flow of Late Jurassic age and is thus among the oldest rocks yet found in any ocean basin.

The sediments of layer 1 vary widely in composition, physical properties, and manner of deposition. Investigations begun by dredging on the *Challenger* expedition (1872-1875) showed that the chief materials are small grains of minerals, shells of such minute organisms as forminifera ($CaCO_3$), diatoms, and radiolaria ($SiO_2 \cdot H_2O$), bones and tests of other marine creatures, and various chemical precipitates. At that time all were considered to have accumulated by *particle-by-particle* (pelagic) *deposition.* Deep currents were not suspected. Two major factors in determining the nature of this type of sedimentation are the depth of the water and its productivity. Where the water is deep and productivity low, calcite tests are dissolved and only a fine deposit remains called *lutite,* or red clay (although brown in color). Where life is abundant over the equatorial upwelling and in shallow water the deposits contain much fossil material.

In 1952, B. C. Heezen and M. Ewing demonstrated the importance of a second method of deposition by slumping and *turbidity currents* (B. C. Heezen and C. L. Drake, 1964; W. B. F. Ryan and B. C. Heezen, 1965; D. C. Krause et al., 1970). These vast but infrequent flows of muddy water sweep down continental slopes, eroding *submarine canyons* by their passage and carrying coarse and fine sediments for hundreds of kilometers over the ocean floor (Fig. 3-11). After each flow the coarser fractions settle faster than the fine, forming a *graded bed.* These fill some of the deepest and flattest parts in the oceans. Those areas which have slopes of less than 1:1000 are by definition called *abyssal plains* (B. C. Heezen and A. S. Laughton, 1963). Turbidity currents were probably much more frequent during the lowered sea level of the Pleistocene.

More recently B. C. Heezen et al. (1966a) and E. D. Schneider et al. (1967) have shown that *contour currents* form a third agent of deposition. These are deep geostrophic currents created by global forces but constrained to follow the bottom contours around the margins of ocean basins (Fig. 3-10). They have contributed material to help build the *continental rises,* which are the gently sloping parts of the continental margins below the continental slopes.

N. D. Opdyke et al. (1966) found both normal and reverse magnetization in abyssal sediments and applied the geomagnetic time scale to determine rates of

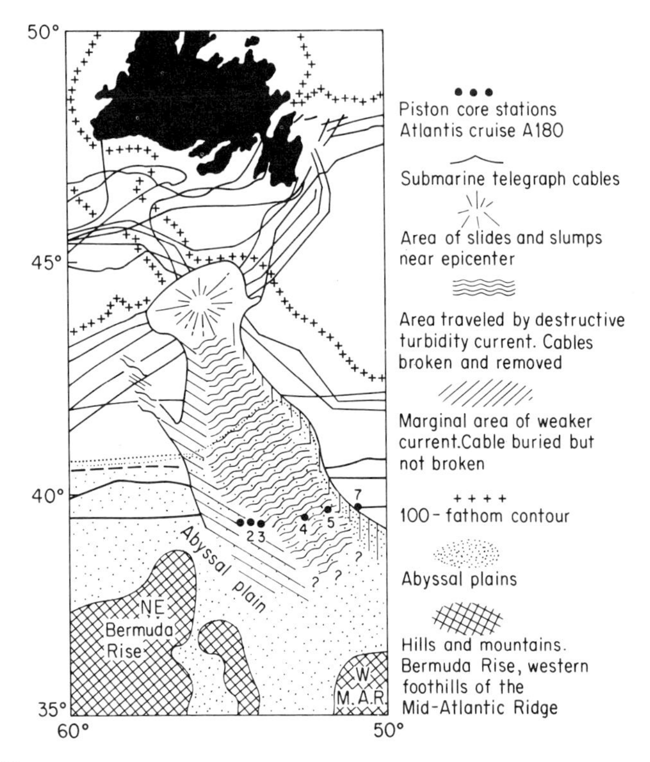

FIGURE 3-11
Sketch map of 1929 Grand Banks turbidity current. (*After B. C. Heezen, D. B. Ericson, and W. M. Ewing, 1954.*)

deposition for the past several million years (Sec. 8-9). In deep oceans they found the average to be 2 m/m yr of $CaCO_3$-free sediment, with no rates less than 0.5 m/m yr, agreeing with rates measured by radiometric dating of cores (T.-L. Ku and W. S. Broecker, 1967). These rates can be accommodated only if the ocean floor is young. Many people now believe that the ocean floors behave like paired conveyor belts placed back to back. Fresh floors are generated at mid-ocean ridges and move away, slowly accumulating sediments as they go, until in the ocean trenches they return into the mantle. This motion, plus the effect of the proximity to continental sources, can explain why sediments increase in thickness and maximum age from the crests of mid-ocean ridges to coasts, and why no rocks older than Jurassic have yet been found on the ocean floors.

The average composition of abyssal sediments has not been the subject of much study, but since most of the sediments are carried by turbidity and contour currents from the shelves of continents, the composition is probably similar to that in shelves.

Layer 2 crops out widely along mid-ocean ridges and consists of tholeiitic basalt. Where not altered, it is of relatively constant composition, close to that given in Table 3-11. Alumina content tends to be high, while potash content may be less than 0.1 percent. It is often in the form of pillow lava, whose structure is due to the rapid chilling of the surface of a succession of sack-shaped spurts or blobs of liquid lava as they flow out on the sea floor. Dikes and small intrusives are also found. Tholeiite is presumed to be a product of partial melting of the mantle beneath mid-ocean ridges but shows sufficient variation in chemical and particularly in isotopic composition to demonstrate that the mantle cannot be homogeneous (D. H. Green, 1968 and S. R. Hart, 1969). Layer 2 incorporates some sediments, but not nearly enough have been found to meet the requirements of the advocates of permanent ocean basins. This layer appears to be slightly thinner in deep water and where the spreading rate is fast, as beneath the Pacific Ocean (P. R. Vogt et al., 1969a).

Until 1970 *layer 3* was regarded as a single layer of uniform thickness, and its composition has been a subject of controversy. H. H. Hess (1962) proposed that it is peridotite from the mantle which has been 70 percent serpentinized in the upper part of the mantle, where temperatures fall below about 500°C along the ridge axis. C. A. Burk (1964) and others showed that serpentinite often crops out on fracture zones and on the walls of ocean trenches. J. R. Cann (1968) advanced several arguments against this view and proposed that the layer is basalt altered to amphibolite, which also crops out on the ocean floor (J. R. Cann and B. M. Funnell, 1967). E. R. Oxburgh and D. L. Turcotte (1968) supporting this alternative view, as did P. R. Vogt et al. (1969b), who further suggested that layer 3 is made of an amphibolitized sheet of contiguous dikes.

The matter was complicated by the discovery first of diorite and then of thick differentiated sills of basic rock in the lower part of the crust where it had been exposed on fault scarps in the sea floor near the crest of the ridge (F. Aumento, 1969; A. Miyashiro et al., 1970; W. G. Melson and G. Thompson, 1970) and by G. L. Maynard's (1970) demonstration that there are in reality two layers with different velocities over great stretches of the Pacific Ocean floor and probably in all oceans (Table 3-13).

At almost the same time J. R. Cann (1970) proposed a model of the sea floor with four layers above the Mohorovičić discontinuity, the two lower layers corresponding to the former layer 3. Of these he considers the upper to be a sheet of dikes and the lower to be differentiated gabbro sills intruded by some serpentinite diapirs from the upper mantle (Fig. 3-12). This model, derived by considering the possible effects of plate tectonic theory on the rocks of the ocean floor, is similar to the ophiolite complexes seen on continents and interpreted as slices of ocean floor thrust up in mountains during mountain building, for example in the Troodos complex in Cyprus (I. G. Gass, 1968; Fig. 3-13), the south Oman complex in Arabia (B. M. Reinhardt, 1969), in Newfoundland (R. K. Stevens, 1970), in California (E. H. Bailey et al., 1970), and in the Apennines (A. Bezzi and G. B. Picardo, 1971).

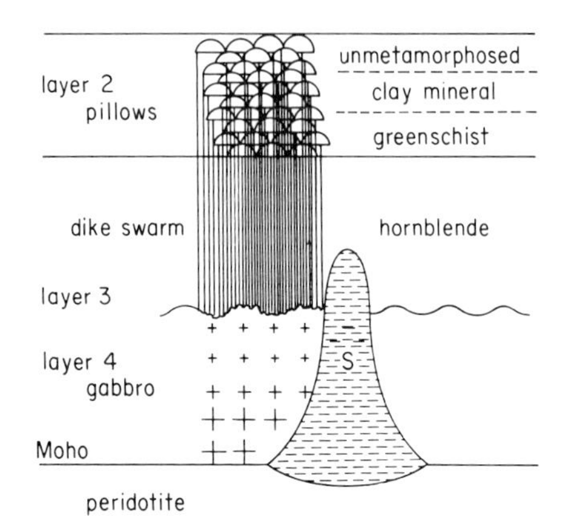

FIGURE 3-12
Section through the ocean crust with layer 1 of sediments omitted. *S* is a serpentinite diapir. (*After J. R. Cann, 1970.*)

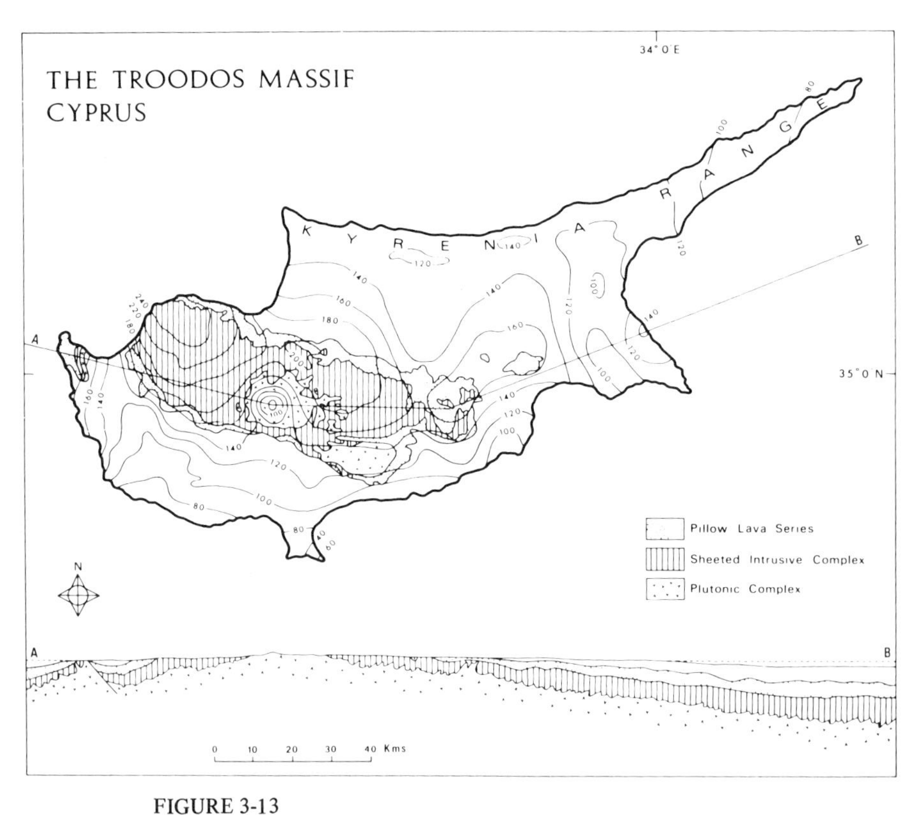

FIGURE 3-13
Map and cross section of the Troodos massif, Cyprus. It is considered that the pillow lavas, sheeted complex, and at least part of the plutonic complex may be upthrust ocean floor. Bouguer gravity anomalies at 20-mgal intervals. (*After I. G. Gass, 1968.*)

3-6.5 Oceanic-island Association

It is convenient to divide islands lying within the ocean basins (and not merely submerged parts of continents) into two chief groups. Those which form island arcs and chains toward the borders of the basins are dealt with in the next section. It is convenient to refer to isolated islands and seamounts which are not part of those chains as *oceanic*. Chapters 15 and 16 give descriptions and references to some of them. Nearly all are volcanoes or have a volcanic core (A. R. McBirney and I. G. Gass, 1967). Notable exceptions are St. Paul Rocks in the Atlantic, consisting of ultrabasic rocks presumed to be faulted up from the mantle; the Seychelles, a microcontinent of granite in the Indian Ocean; and Macquarie Island south of New Zealand, which appears to be uplifted ocean floor.

All the volcanic oceanic islands are basaltic, but the suites vary from rather uniform tholeiites, abundant in flows as on the Faroe Islands, to more variable alkali basalts associated with central volcanic pipes on many islands such as Hawaii, and more siliceous lavas, including trachybasalts and soda rhyolites found in small quantities on a few islands and abundantly on Iceland.

The different suites grade into one another and are found mixed on the same islands, so that it was at first thought that the variations were due to variations in the depths from which the lavas rose, in the extent of partial melting of the source rocks (T. H. Green et al., 1967) and to later fractionation of the melt, particularly in the conduits of central volcanoes and in chambers within volcanic cones (J. P. Eaton and K. Murata, 1960).

Although these factors certainly cause variety, there has been a growing realization that the distinctions are greater than was at first realized, and that two if not three suites arise in different ways without complete gradation between them. A. E. J. Engel et al. (1965) observed a difference between oceanic-island lavas and ocean-ridge basalts which P. W. Gast (1970), R. Kay et al. (1970), F. A. Frey (1970), S. R. Hart (1971), and others have defined more closely. The igneous rocks originating along mid-ocean ridges away from islands, and hence forming most of those found on the sea floor, are characterized relative to the basaltic rocks found in oceanic islands by (1) rare-earth-element patterns depleted in light REE, (2) occasional anomalously low values of europium, (3) lower contents of K, Rb, U, Th, and Ba, (4) lower contents of Sr ranging from 80 to 180 ppm, (5) lower Sr^{87}/Sr^{86} ratios, (6) K/Rb ratios of between 300 and 1500, (7) covariation of Al and Ti contents, and (8) covariation of Ni and Mg contents.

These authors conclude that ocean-ridge basalts originate by partial melting of mantle material at shallow depths of perhaps 15 to 25 km. W. J. Morgan (1971) suggests that, on the other hand, the sources of lavas for the principal oceanic islands are plumes rising from much greater depths in the mantle. A further distinction seems to be that the most siliceous lavas occur on those islands which lie both over a plume and on a mid-ocean ridge. Thus rhyolites are found on Iceland and Easter Island, but not in the Hawaiian Islands (Sec. 17-5).

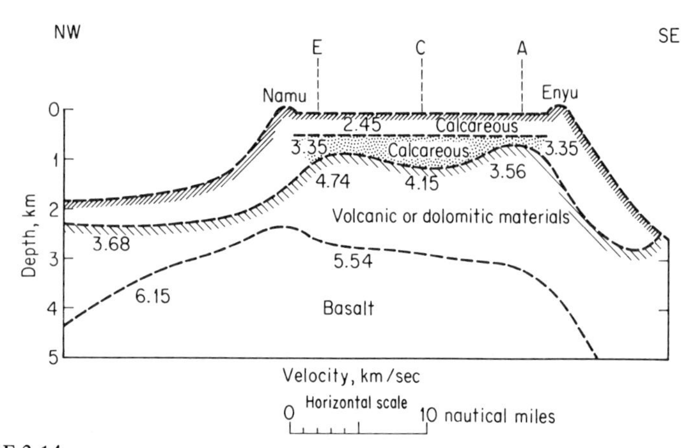

FIGURE 3-14
Cross section of Bikini atoll showing geological interpretation of zones of different seismic velocities. (*After M. B. Dobrin, 1954.*)

The movement of tectonic plates over the plumes forms chains of islands (J. T. Wilson, 1965a; W. J. Morgan, 1971). There is also a tendency for alkali lavas to become more abundant in younger flows and with elevation (A. E. J. Engel and C. G. Engel, 1964; A. R. McBirney and I. G. Gass, 1967). Thus all the Hawaiian Islands started as tholeiite shield volcanoes which turned to alkali-basalt volcanoes, with late parasitic cones, like Diamond Head, of nepheline-bearing lavas. In the Western Pacific many seamounts have flat tops and are then called tablemounts, or guyots. On some of them the discovery of shallow-water fossils and beach deposits has shown that they were islands in Cretaceous time and have sunk (H. W. Menard, 1964; H. H. Hess, 1965). This confirms C. Darwin's (1962) proposal that volcanoes subside to provide a base for coral islands and atolls (Fig. 3-14).

The sedimentary rocks on all islands in the deep oceans are highly specialized, being solely volcanic detritus, coralline and algal limestone, laterite, glacial ice, and calcium phosphate rock formed by the reaction of guano on limestone.

3-6.6 Island-arc Association

The Pacific Ocean is rimmed along the east coast of Asia with island arcs (Fig. 12-1). Apparently related are the straight to irregular chains of islands bordering the Southwestern Pacific, the Andes and Central American mountain ranges, and the West Indian and Scotia arcs. Associated with all these features are ocean trenches, intermediate to deep earthquakes, particular patterns of geophysical phenomena, and volcanoes which chiefly emit rocks of the *andesitic kindred* (A. R. McBirney, 1969; Table 3-14). Basalts are universal, but andesites are confined to arcs and are erupted by few volcanoes in the main ocean. This led P. Marshall and I. Born to

Table 3-14 CHEMICAL COMPOSITION OF ROCKS OF THE ANDESITE KINDRED

Rock	SiO_2	TiO_2	Al_2O_2	Fe_2O_3	FeO	MnO	MgO	CaO	Na_2O	K_2O	H_2O	P_2O_5	Total
Pyroxene andesite	63.2	0.5	18.2	1.4	3.3	. . .	2.3	5.2	4.1	1.2	0.5	0.1	100.0
Average andesite (Daly)	59.6	0.8	17.3	3.3	3.1	0.2	2.7	5.8	3.6	2.0	1.3	0.3	100.0
Average rhyolite (Daly)	72.8	0.3	13.5	1.4	0.9	0.1	0.4	1.2	3.4	4.4	1.5	0.1	100.0
Cenozoic andesite (Chayes)	58.2	0.8	17.3	3.1	4.1	. . .	3.2	6.9	3.2	1.6	1.2	0.2	99.8

SOURCE: R. A. Daly, 1933; F. J. Turner and J. Verhoogen, 1960; and F. Chayes, 1969.

postulate a boundary called the *Andesite Line* which followed the trenches in the Southwest Pacific (J. Gilluly, 1955), but the concept is somewhat misleading because andesites are not erupted in the seas behind arcs or on continents but are confined to the vicinity of arcs and similar structures (C. B. Raleigh and W. H. K. Lee, 1969).

The finest examples of island arcs lie off the whole coast of East Asia from the Aleutian Islands to Indonesia (Fig. 12-1). They differ in age, and the greater the age of the oldest rocks found in an arc, the larger are the islands. This suggests that the islands grow (J. T. Wilson, 1967).

Acid magmas are more viscous than basalt and do not produce such durable rocks. Viscosity causes them to pile up into steep peaks, and the escaping gases produce ash and pumice, which are readily eroded, so that andesites are not well preserved and their importance is disguised. Of the world's 1050 centers of volcanic activity, the great majority are andesitic. They are dangerous because the lavas contain enough water in solution to build up great steam pressures and, being viscous, tend to freeze in the conduits. Sudden release produces eruptions, of which some of the greatest have been in Santorini Island about 1400 B.C., which may have destroyed the Minoan civilization (Sec. 16-9), in the southern Yukon or Alaska between 500 B.C. and A.D. 500, and in Vesuvius in A.D. 79 (H. H. Lamb, 1970).

As andesite volcanics are eroded they form sediments of the greywacke facies (Table 3-15). On the ocean side of arcs these sediments mix with abyssal types in ocean trenches, while in the marginal seas on the inner side of arcs they mingle with sediments of deltaic facies from the continent. For example, deposition has almost filled the East China Sea, but the seas farther north are only partly filled, and large deep basins remain.

All the sediments formed under these conditions tend to be of the greywacke facies in which the breakdown of rocks and minerals is not complete, so that feldspars, micas, and ferromagnesian minerals have commonly not disintegrated. Sedimentation is rapid and on steep slopes so that slumps, mud flows, and turbidity currents are common. The latter settle at the bottom of trenches to form deep narrow abyssal plains floored by graded beds, but the slumps produce clasts of mud and other peculiar forms called turbidites, which have been mistaken for conglomerates. These rocks, unlike those common on continents, are poorly sorted and not well aerated, lack ripple marks and mud cracks, and have few fossils (J. R. L. Allen, 1970). Shale and greywacke are the commonest rocks, often present in the ratio of 2 or 3 parts to 1.

Along the axes of many island arcs and mountain belts, *ultrabasic bodies* are found in swarms and linear series. They are frequently, but not always, the oldest intrusives. Usually they are only a few hundreds or thousands of meters in length. It has been established that some of the larger bodies, which measure tens of kilometers across and of which the Lizard intrusive in Cornwall, England, is a good example, were emplaced by infaulting when they were in a warm but solid state. C. B. Raleigh and M. S. Paterson (1965) have supported this view by showing

Table 3-15 CHEMICAL COMPOSITION OF TYPICAL ROCKS OF GREYWACKE AND ARKOSE FACIES

Rock	SiO_2	TiO_2	Al_2O_3	Fe_2O_3	FeO	MnO	MgO	CaO	Na_2O	K_2O	H_2O	P_2O_5	CO_2	Total
Average greywacke	64.2	0.5	14.1	1.0	4.2	0.1	2.9	3.5	3.4	2.0	2.1	0.1	1.6	100.0
Average shale	58.1	0.7	15.4	4.0	2.5	. . .	2.4	3.1	1.3	3.2	5.0	0.2	2.6	98.5
1 greywacke +2 shale	60.1	0.7	15.0	3.1	3.0	. . .	2.5	3.2	2.0	2.8	4.0	0.2	2.2	98.8
Average arkose	75.5	. . .	11.4	2.4	. . .	0.2	0.1	1.6	2.0	5.6	0.6	. . .	0.4	99.8

SOURCE: F. J. Pettijohn, 1957.

experimentally that serpentinite is weakened by partial dehydration when heated to temperatures in the 300 to 600°C range, but is strong and brittle at higher and lower temperatures. Olivine is also ductile. Such bodies are known as Alpine type ultramafics, to distinguish them from the ultramafic parts of layered complexes like the Bushveld lopolith and from rare ultrabasic complexes which were injected as very hot liquid magmas or are conceivably due to alteration (P. J. Wyllie, 1967). Some of these ultrabasic bodies may have been upthrust from the mantle during mountain building, but along the shores of some islands like Cyprus, New Caledonia, and New Guinea such bodies are larger, as long as 100 km, and grade upward in gabbro intrusives, basalt flows, and sedimentary deposits typical of the ocean floors. There is a growing body of opinion that these masses are slices of old sea floor, including places in the whole oceanic crust and some upper mantle as well, which have been thrust on to islands or embodied in mountain systems (Secs. 16-9 and 16-11).

Some of the smaller island arcs have no exposed rocks older than Cretaceous or Tertiary, but the larger the islands in arcs, the older they are found to be, suggesting that the islands have grown by the addition of lavas escaping from the interior of the earth. This observation coupled with the close correlation of intermediate and deep earthquakes with andesitic volcanoes is entirely compatible with the concepts of global plate tectonics, according to which old ocean floor is being returned to the mantle by being thrust down under trenches and island arcs. Experimental work by T. H. Green and A. E. Ringwood (1969) shows that andesites can be produced from ultrabasic rocks like the mantle in two stages, of which the first might be partial melting to produce tholeiitic basalt on the sea floor and the second might be the remelting of this basalt when forced down under island arcs. H. Kuno (1966a) and W. R. Dickinson and T. Hatherton (1967) find that the composition of andesites varies across the width of Japan and with depth to underlying volcanoes in such a way as to demand a source that is constantly renewed, which again lends support to the concept of moving plates (Fig. 12-1).

Volcanism, intrusion, and tectonic activity combine to produce much metamorphism in island arcs.

3-6.7 Primary-mountain Association

All the associations so far discussed have accumulated by the extrusion of volcanic rocks and by the deposition of sedimentary rocks. Deformation and metamorphism have played a minor role.

The primary-mountain association, on the other hand, marks the stages of uplift and erosion in which the rocks accumulated in earlier associations and stages are deformed and metamorphosed. Deposition plays a minor part in this association because the products of erosion are chiefly carried beyond its borders. The dominant rocks are those of the *plutonic kindred,* which form great stocks and batholiths as well as minor associated intrusives in the form of pegmatites. They are

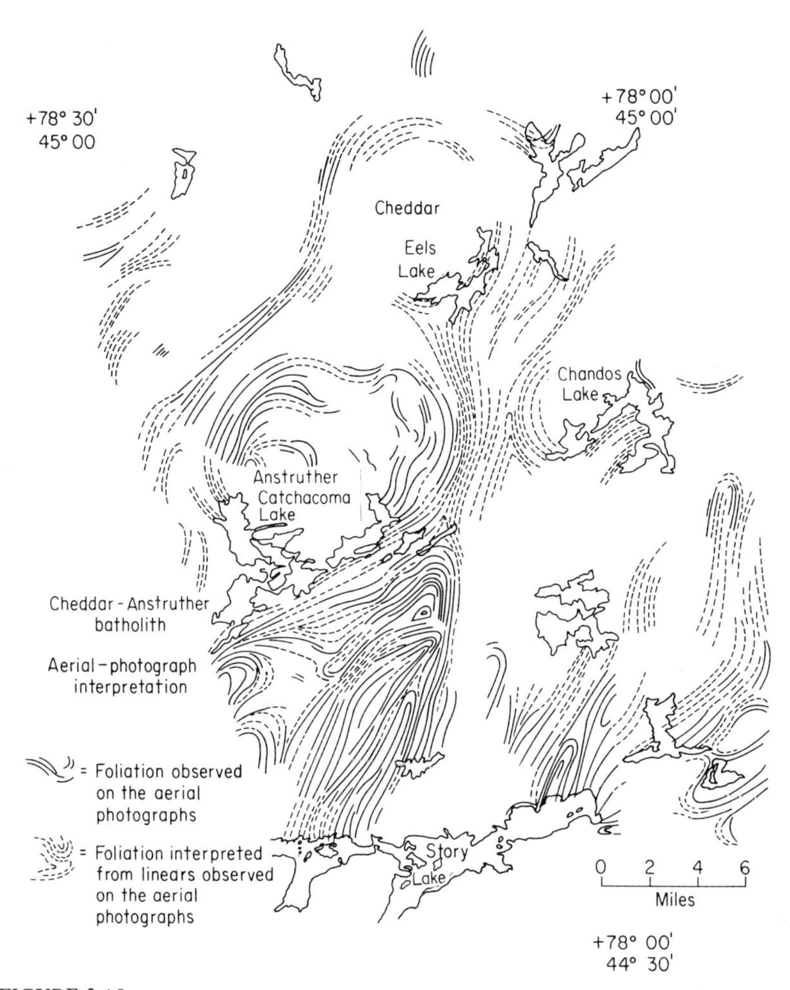

FIGURE 3-15
Aerial-photograph interpretation of area of Cheddar and Anstruther batholiths, Haliburton, Ontario.

either granodiorites or granites grading into gneisses and granulites (Fig. 3-15 and Table 3-16).

The term *primary* is used to distinguish these mountains from others, including volcanoes, uplifted blocks along rifts, mid-ocean ridges, and one other important class, here called secondary mountains, often associated with primary mountains and frequently lying parallel to their inland margin. Unlike primary mountains, secondary mountains are little metamorphosed and have no deep roots. They are built of sheets, folds, and nappes of sedimentary rocks which have been

Table 3-16 CHEMICAL COMPOSITION OF ROCKS OF THE PLUTONIC AND ULTRABASIC KINDREDS

Rock	SiO_2	TiO_2	Al_2O_3	Fe_2O_3	FeO	MnO	MgO	CaO	Na_2O	K_2O	H_2O	P_2O_5	CO_2	Total
Average granite	70.2	0.4	14.4	1.6	1.8	0.1	0.9	2.0	3.5	4.1	0.8	0.2	...	100.0
Average granodiorite	65.0	0.6	15.9	1.7	2.7	0.1	1.9	4.4	3.7	2.8	1.0	0.2	...	100.0
Average dunite	40.5	0.2	0.8	2.8	5.5	0.1	46.3	0.7	0.1	0.04	2.9	0.05	...	100.0
Average Canadian Shield	65.3	0.5	15.9	1.4	3.1	0.1	2.2	3.4	3.9	2.9	0.8	0.2	0.2	100.0

SOURCE: R. A. Daly, 1933, and W. F. Fahrig and K. E. Eade, 1968.

piled up by sliding and compression alongside the primary-mountain uplift. Good examples of such pairs are the Rocky Mountains of Canada, secondary mountains formed by compression and by sliding off the rising primary Monashee and Purcell Mountains to the west (A. W. Bally et al., 1966; J. O. Wheeler, 1970), and the secondary Valley and Ridge and primary Piedmont provinces of the Appalachian Mountains (M. Kay, 1951; P. B. King, 1959).

During the first half of the last century, as geologists worked out the stratigraphy of Phanerozoic time, they increasingly turned their attention to the older rocks, which they lumped together as pre-Cambrian, later to become Precambrian. They found relatively unaltered sedimentary sequences overlying metamorphic rocks, and some considered that the latter also consisted of recrystallized sedimentary sequences (W. Logan, 1863), but there was a general tendency to believe that the earth had cooled rapidly and that the great batholiths and shields represented an early hotter stage which had largely crystallized from liquid magma. Early workers in both microscope petrography and in experimental petrology supported these simplistic views (F. D. Adams and A. E. Barlow, 1912; F. F. Grout, 1948; O. F. Tuttle and N. L. Bowen, 1958).

It took a long time and more sophisticated considerations, including isotopic ratios and rates of formation of sediments, to return to the older view that most batholiths and shields are at least partly remobilized sedimentary rocks (J. J. Sederholm, 1967; H. C. Cooke, 1948; H. H. Read, 1957; Sec. 17-4).

P. C. Bateman and J. P. Eaton (1967) have gone so far as to advocate from combined geological and geophysical studies that the Sierra Nevada batholith was generated in a great downfold of sedimentary strata. They hold that high radiogenic-heat production in a thick prism of sedimentary rocks led to partial melting at depths of 25 to 45 km in the lower crust and that these melts rose at different times and places to metamorphose higher strata.

On the other hand, W. B. Hamilton and W. B. Myers (1967) and R. W. Kistler et al. (1971) hold that the process is not so simple, that some batholiths form away from thick sedimentary sequences and that processes in the mantle play an important role.

D. V. Higgs (1949) pointed out that with increasing age the proportion of sedimentary rocks preserved decreases. A corollary which is also true is that the relative proportion of metamorphic rocks increases in abundance with the age.

A comparison of Tables 3-14, 3-15, and 3-16 shows that the average compositions of andesite lavas, greywacke, and granodiorite are very similar, supporting the view that andesites erode to form greywackes and that greywackes are metamorphosed into granodiorites and gneisses. The preceding section showed that andesites could form by a two-stage process from the mantle, a view which isotopic studies strongly support (S. R. Taylor and A. J. R. White, 1965; S. R. Taylor, 1969). This provides a possible mechanism for growing continents during geological time.

It was at one time believed that shields grew by simple marginal additions, but this view is too simple. Not only have older parts been reworked, but if continental drift has operated for a long time, each shield may have been torn apart repeatedly and reassembled, so that additions are not necessarily now at the margins, and the margins are not necessarily young. Nevertheless, many Precambrian geologists remain skeptical, and the problem is complex and far from being completely solved (Chap. 16).

3-6.8 Exposed-shield Association

Next to ocean basins, continents are the largest of the earth's surface features and they are the most stable. Over great areas their basement structures and rocks, formed in earlier cycles, are exposed as shields. For the most part these are stable, low-lying areas reduced nearly to sea level and subject to but slight erosion, deposition, and igneous activity. Different latitudes and climates produce very different surface effects. Much of Australia and Central Africa offer good examples. For the past many tens of millions of years these surfaces have been exposed to weathering, which has produced a widespread crust of residual clays and related minerals called laterite. Where the climate is dry, as in the Sahara, deserts form with surfaces of blown sand, exposed rock, or pavements of pebbles from which the smaller particles have been blown away. In cold climates continental ice sheets may form, such as that which covers Antarctica. On retreating, these leave behind great areas of bare rock (Fig. 11-1) from which the weathered soils and fragments have been scraped and piled around the ice margins as moraines and sheets of till. Thus much soil and many boulders from the Canadian shield were carried to southern Canada and northern United States by the Pleistocene ice sheets.

The thin sedimentary cover is matched by an even greater lack of igneous activity. At rare intervals of time and distance small volcanic eruptions have produced pipes of very highly differentiated rocks, kimberlites, carbonatites, and alkali basalts. The concentration of rare elements and minerals in some of these makes them economically important in the production of diamonds, apatite (for phosphorus), columbium, rare earths, and base metals.

These rocks continue under later strata far beyond their exposures, and these covered shield areas are known from samples obtained at the bottom of many drill holes (W. R. Muehlberger et al., 1967; R. A. Burwash and J. Krupička, 1970).

Precambrian rocks have traditionally been subdivided into two, the older Archean era and the younger Proterozoic era (Sec. 17-4). All these rocks have an average composition very close to granodiorite and constitute a major part of the continental crust (D. M. Shaw et al., 1957; W. R. A. Baragar, 1968; Table 3-17). W. F. Fahrig and K. E. Eade (1968) found the Proterozoic rocks to have more K_2O, TiO_2, U, and Th than the Archean.

Table 3-17 VOLUMES, MASSES, AND AVERAGE CHEMICAL COMPOSITION OF THE CRUST

Types of crust	Shell	Volume 10^6 km^3	Average thick-ness, km	Mass 10^{24} g	Components, wt % and mass (10^{24} g)									
					SiO_2	TiO_2	Al_2O_3	Fe_2O_3	FeO	MnO	MgO	CaO	Na_2O	K_2O
Continental	Sedimentary	500	3.4	1.29	49.95	0.65	13.01	2.98	2.82	0.11	3.10	11.67	1.57	2.04
	Granitic	3000	20.1	8.20	63.94	0.57	15.18	2.00	2.86	0.10	2.21	3.98	3.06	3.29
	Basaltic	3000	20.1	8.70	58.23	0.90	15.49	2.86	4.78	0.19	3.85	6.05	3.10	2.58
Total		6500	43.6	18.19	60.22	0.73	15.18	2.48	3.77	0.14	3.06	5.51	2.97	–2.86
Subcon-tinental		190	2.9	0.48	49.95	0.65	13.01	2.98	2.82	0.11	3.10	11.67	1.57	2.04
	Granitic	590	9.1	1.61	63.94	0.57	15.18	2.00	2.86	0.10	2.21	3.98	3.06	3.29
	Basaltic	760	11.7	2.21	58.23	0.90	15.49	2.86	4.78	0.19	3.85	6.05	3.10	2.58
Total		1540	23.7	4.30	59.45	0.74	15.08	2.53	3.85	0.16	3.17	5.91	2.89	2.79
Oceanic	Sedimentary (layer 1)	120	0.4	0.19	40.63	0.62	11.31	4.62	0.97	0.34	2.95	16.70	1.13	2.03
	Volcanic sedimentary (layer 2)	350	1.2	0.96	45.50	1.09	14.46	3.20	4.15	0.25	5.27	14.03	2.00	1.02
	Basaltic (layer 3)	1700	5.7	4.92	49.58	1.51	17.13	2.02	6.84	0.17	7.21	11.75	2.75	0.18
Total		2170	7.3	6.07	48.65	1.40	16.52	2.29	6.23	0.18	6.79	12.28	2.57	0.37
Total crust		10,210	20.0	28.56	57.64	0.88	15.45	2.43	4.30	0.15	3.87	7.01	2.87	2.32

SOURCE: A. B. Ronov and A. A. Yaroshevsky, 1969.

3-7 AVERAGE COMPOSITION OF THE CRUST

Several authors have estimated the average composition of the surface of Precambrian shields, of the ocean floors, and of the crust as a whole. Their results are very consistent, which is encouraging because they collected data independently on different continents and separate ocean cruises. This section relies chiefly on a recent, thorough paper by A. B. Ronov and A. A. Yaroshevsky (1969) containing references to older work (Table 3-17).

To estimate the composition of the surface of the continental crust, F. W. Clarke and H. S. Washington averaged over 5000 analyses of igneous and plutonic rocks mostly from North America, V. M. Goldschmidt used the average of 77 samples of glacial clay scraped off the Precambrian shield of Scandinavia, and D. M. Shaw et al. (1967) and Ronov and Yaroshevsky (1969) diligently sampled the Canadian and Soviet basement rocks, respectively. The results agree, and B. Mason adopted the value of 60 percent silica for the continental crust.

On the other hand, A. Poldervaart (1955) adopted a value of 66 percent silica as the best value obtained from the work of a group of petrologists. This result is almost certainly too siliceous, for three reasons. The density of 2.7 to 2.75 g/cm^3 assumed for such rocks is at least 0.1 too low according to modern gravity data, so that the crustal rocks must be heavier, and hence more basic. Poldervaart's averages are based upon rocks collected from the glaciated Canadian and Scandinavian shields, where acid plutonic rocks tend to weather out and to be well exposed. It also seems likely that petrologists tend to collect the better-defined acid rocks in preference to basic schists, which are more nondescript and variable. In obtaining a value for the whole thickness of the continental crust, some workers make a correction for the sedimentary cover rocks, but their volume is small in relation to the crust and their average composition not far different, so that this correction is minor. On the other hand, seismic-refraction studies show that velocities increase in the lower half of the crust, indicating an increase in density and in basic elements. The low heat flow over continents suggests that the whole crust cannot contain as much radioactive elements as the outcrops, and this is compatible with a downward increase in basic elements. A. B. Ronov and A. A. Yaroshevsky (1969) divide the crust into two equal halves and support the view that the lower half is basaltic.

Seismic studies and dredging show that the oceanic crust is divided into three or four layers, of which the uppermost is minor, being a thin and discontinuous layer of sediments (Sec. 3-6.4). Most authorities regard layer 2 as tholeiitic basalt, but because Ronov and Yaroshevsky do not believe in continental drift, they are embarrassed that the volume of sediments in layer 1 is smaller than the rate of their accumulation, which would suggest that layer 2 is half sediments, and to obviate this difficulty they make this assumption. This lowers their estimate of its silica and magnesia content, but greatly increases the potash.

Opinion is divided about whether layer 3 is largely basalt, amphibolite rock, or serpentinite, but modern opinions, including those of Ronov and Yaroshevsky (1969) and J. R. Cann (1968), tend to favor basaltic composition. If so, it seems possible that this layer is worldwide and that the upper part of the continental crust has accumulated upon a deformed and altered oceanic crust (J. T. Wilson, 1951). Table 3-17 summarizes the conclusions of Ronov and Yaroshevsky.

4

THE FIGURE OF THE EARTH AND GRAVITY

4-1 THE FIGURE OF THE EARTH

The problem of determining the shape of the earth is one which has attracted the attention of scientists ever since early times, when the earth was considered flat. If the Pole Star is observed by means of a plumb line or a level, its altitude is found to be approximately 80° in northern Spitsbergen, 49° at Paris, and 0° at the equator, while in the Southern Hemisphere Polaris is invisible. This variation in altitude is not due to any difference in the actual direction of the Pole Star as seen from different places on the earth; it can only mean that the direction of the plumb line is different at different places. It thus follows that the earth is at any rate an isolated body with a surface which completely surrounds it. The approximate form was determined by Galileo, who observed the shadow cast by the earth on the moon during a lunar eclipse. Actually, Eratosthenes in the third century B.C. estimated the circumference of the earth by measuring the elevation of the sun and obtained a result correct to within about 1 percent. It is now known that the shape of the earth is approximately that of a sphere. The departures from the true spherical shape and the ruggedness of the surface due to the varying topography are very small when compared with the size of the earth. If an 18-in. globe is taken as a model of the earth, the greatest and least diameters differ by $\frac{1}{16}$ in., the highest mountains are less than $\frac{1}{75}$ in., while the average elevation of the continents above the ocean floors is no more than a layer of varnish.

If the earth is regarded as a sphere, it is easy to estimate its diameter by determining the length of a degree of latitude. This can be done by measuring the separation of two stations both in degrees and in distance along the surface. The first measurement is simplified if the two stations are on the same terrestrial meridian; their separation in degrees is then merely the difference in their latitudes, which may be found by a zenith telescope or by other astronomical means. The second measurement can be found by triangulation. The error in such geodetic observations is less than 1 ppm. It is found that the degree of latitude is shorter near the equator than near the poles by about 1 part in 100, the average distance being about 111 km. The only possible interpretation of this result is that the earth is approximately a spheroid more strongly curved near the equator than in high latitudes; i.e., it is flattened along the polar axis. If a is the equatorial and b the polar radius,

$$f = \frac{a - b}{a} \tag{4-1}$$

is defined as the oblateness of the spheroid. Its value is approximately $\frac{1}{298}$. The shape (but not the size) of the earth may also be determined by gravity surveys at different latitudes, by purely astronomical methods such as an analysis of the irregularity of the motion of the moon, and by artificial earth satellites (Sec. 4-4).

4-2 ROTATION OF THE EARTH

The earth rotates on its axis with an angular velocity

$$\Omega = \frac{2\pi}{86{,}164} = 7.292 \times 10^{-5} \text{ rad/sec}$$

This is extremely small, and the Coriolis force and the centrifugal force arising from the earth's rotation are not noticeable in our daily lives, although, as will be seen later, they are important from a geophysical standpoint. Broadly speaking, the centrifugal force is responsible for the earth's equatorial bulge, while the Coriolis force modifies the atmospheric and oceanic circulations, causing, for example, the trade winds.

Foucault in 1851 was the first to show that a pendulum could be used to demonstrate the earth's rotation. Suppose a pendulum at the North Pole is set vibrating as a simple pendulum in a vertical plane which is fixed in a Newtonian frame of reference. Then, as the earth turns under the pendulum with angular velocity Ω, the plane of vibration of the pendulum appears to an observer on the

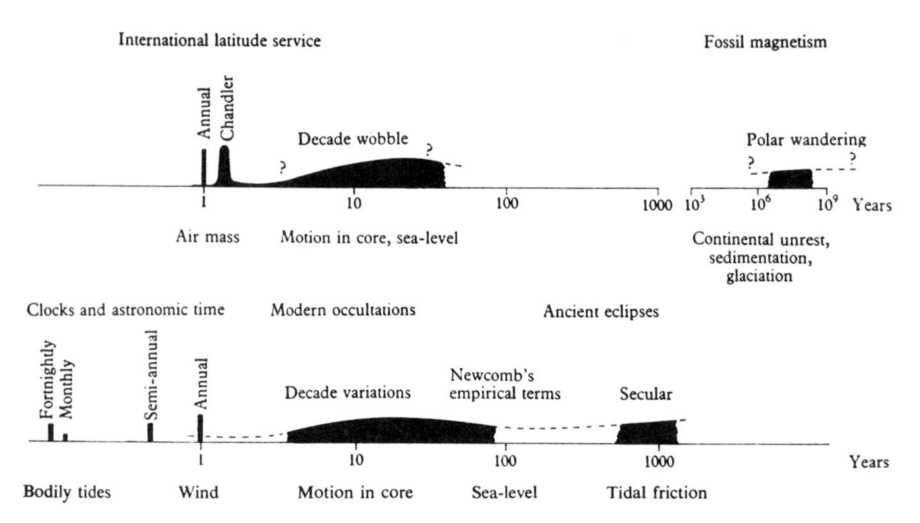

FIGURE 4-1
The spectrum of rotation. The wobble components (top) and length-of-day components (bottom) are schematically arranged according to their time scale in years. Vertical lines indicate discrete frequencies; shaded portions indicate a continuous, or noisy, spectrum. Principal source of the observations is shown above lines; presumable geophysical cause beneath lines. (*After W. H. Munk and G. J. F. MacDonald, 1960.*)

earth to turn with angular velocity Ω. Actually it is not necessary that the pendulum be situated at one of the earth's poles. It can be shown that the effect of the earth's rotation on the elliptical path of a spherical pendulum is to cause the ellipse to rotate with an angular velocity $-\Omega$ sin (latitude), that is, -15 sin (latitude) deg/hour, which is approximately 10 deg/hour in New York. The rotation is clockwise in the Northern Hemisphere and counterclockwise in the Southern Hemisphere.

The angular velocity of the earth is not constant. There are small fluctuations, not only in the rate of spin but also in the direction of the axis of rotation; i.e., there is a "wobble" of the earth. Figure 4-1 shows for different parts of the frequency spectrum the methods of observing the fluctuations and what are believed to be the principal causes.

Rapid changes of either sign in the rate of rotation are superposed on a substantially constant deceleration which is attributed to tidal friction in shallow seas. It is difficult to estimate the rate at which this deceleration has taken place, since our knowledge of paleogeography is scant and the result is so dependent on a few shallow seas. On the other hand, H. C. Urey (1952b) has suggested that, by differentiation of the materials of the earth and the growth of the core, the moment of inertia of the earth about its axis of rotation may have been reduced, and a decrease in the length of the day from 30 to 24 hours may have resulted from such a process. The changes caused by a growing core are considerably smaller (and of course of opposite sign) than those due to lunar tidal friction. A method of determining the rate of the earth's rotation in the geological past, and hence of testing the idea of a growing core, has been suggested by J. W. Wells (1963). He has shown that daily growth rings in the epitheca of coral can be recognized, and finds that the number of solar days in Middle Devonian time was about 400. The question of the growth of the earth's core is considered in more detail in Sec. 10-8. E. R. R. Holmberg (1952) believes that atmospheric solar tides may exert a controlling influence on the speed of rotation, the decelerating oceanic torque being balanced on the average by an accelerating atmospheric torque. Holmberg's theory, which is based on a sharp resonance in atmospheric oscillations at 12 hours, predicts that the length of the day, apart from minor fluctuations, will be held indefinitely to its present value. There is now, however, considerable doubt concerning the existence of this resonance peak.

The rapid irregular fluctuations in the length of the day over a decade cannot be explained by surface phenomena. No transport of mass at the surface which would alter the earth's moment of inertia is able to account for such large changes. It is believed that these irregular changes are due to electromagnetic torques adjusting the balance of angular momentum between the earth's core and mantle. Although the changes are so minute that they might seem to be of no consequence, modern views on the origin of the earth's main magnetic field and, in particular, on its westward drift are intimately connected with these irregular fluctuations in the speed of the earth's rotation (Secs. 8-5 and 8-6).

The annual wobble is principally due to the seasonal shifts in air masses, but this is unlikely to be the reason for the changes observed over decades. Astronomical evidence indicates that the position of the North Pole has not changed by more than 5 m between 1900 and 1950 (Fig. 4-2), and there is some evidence of a shift of 3 m toward 40°W. Such a shift could be caused by a rise in sea level of 3 cm due to the melting of the Greenland ice sheet. Paleomagnetic studies suggest that during geological time the magnetic poles may have wandered thousands of kilometers. A system can turn itself over in space in any way without violating the principle of the conservation of angular momentum. T. Gold has given as an example the ability of a cat, on being released upside down and without angular momentum, to land on all four paws. In the case of the earth, the equatorial bulge is the stabilizing influence. If the earth had complete stability of shape, it could maintain indefinitely the same axis of rotation within close limits, despite geological or other events occurring on its surface; but if an additional mass were placed on the earth (not at one of the poles or at the equator), so that the instantaneous axis of rotation were changed by a small angle, then if the earth is capable of yielding, the axis of figure would slowly follow. The equatorial bulge would be under a stress, tending to align it symmetrically with respect to the new instantaneous axis of rotation. Calculations indicate that large angles of polar

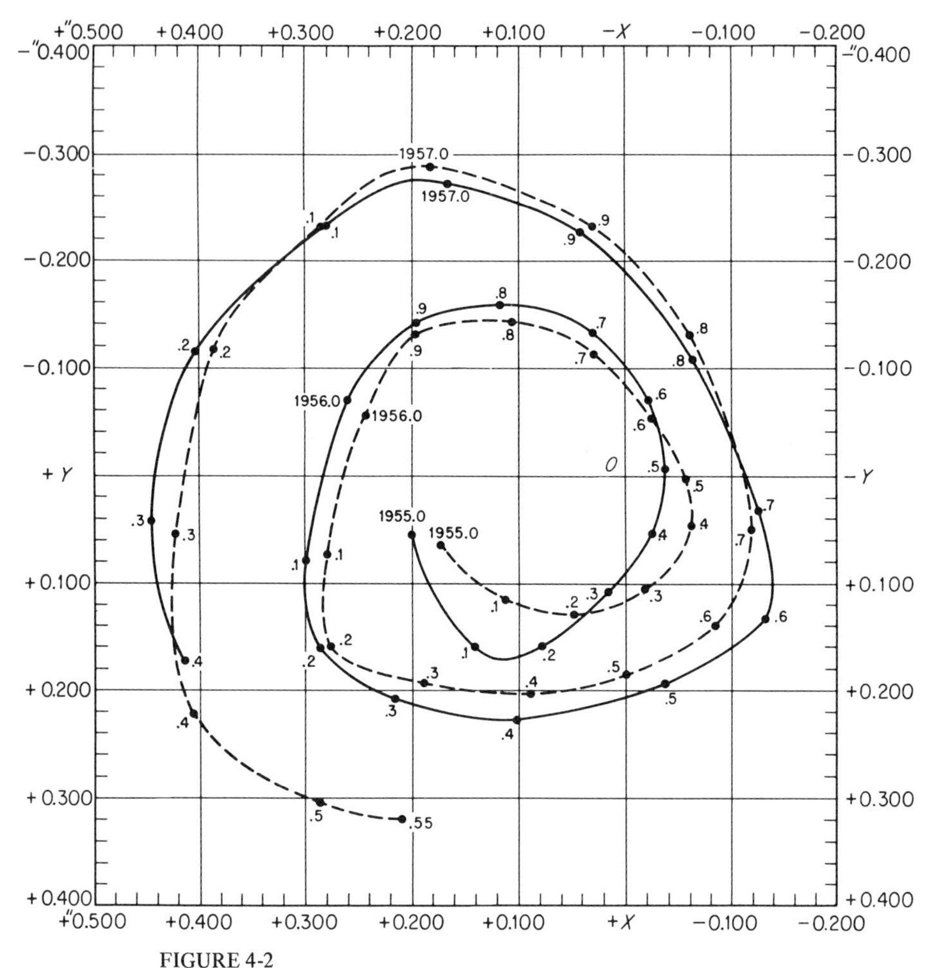

FIGURE 4-2
Wandering of the geographical pole, 1955–1957. The two curves show the results of two independent determinations. The letter *O* near the origin of coordinates shows the mean position of the pole 1910–1911. One large square represents 0.100 second of arc, or about 3.1 m. (*Communicated by the Director, Astronomical Observatory, Turin.*)

wandering are possible. This will be discussed in more detail in Sec. 8-10 after the results of paleomagnetism have been reviewed.

In 1891 S. C. Chandler isolated a component with a 14-month period in the wobble of the earth about its rotation axis. Rigid-body dynamics gives a 10-month period for the earth's natural wobble, although the longer period Chandler wobble can be explained if the earth does not behave as a rigid body (Sec. 8-10). The rotational deformation implies that the Chandler wobble must be subject to damping, and therefore a more or less continuous excitation is required to maintain it. The source of this excitation has for a long time been a matter of conjecture, and no agreement has as yet been reached on this question. A possible connection with seismic activity has often been suggested. Until recently the displacement fields of even the largest earthquakes were thought to extend no more than a few hundred kilometers from the focus, so that estimates of the contribution of earthquakes to the excitation of the Chandler wobble were several orders of magnitude too small. The work of F. Press (1965), however, has indicated that for great earthquakes a measurable displacement field may extend several thousand kilometers from the epicenter. Using dislocation theory, L. Mansinha and D. E. Smylie (1967, 1968) estimated the changes in the products of inertia of the earth arising from several large faults associated with major earthquakes and hence calculated their contribution to the excitation of the Chandler wobble. They found that the cumulative effect (based on earthquake statistics) could account for both the excitation of the Chandler wobble and a slow secular shift of the mean pole of rotation. One very interesting result of their investigations is that there is some indication of changes in the pole path 5 to 10 days before many of the largest earthquakes. If this is confirmed, a change in the pole path might be a signal that a very large strain release was about to occur and hence a warning of impending large seismic activity.

4-3 GRAVITATIONAL ATTRACTION

Newton's law of gravitation states that if two particles of masses m_1 and m_2 are at a distance r apart, each attracts the other with a force of magnitude Gm_1m_2/r^2, where G is a universal constant called the constant of gravitation. The most recent determination of G (in cgs units) is 6.674 ($\pm$ 0.004) x 10^{-8} (R. D. Rose et al., 1969). Thus, if M is the mass of the earth (of radius r), the weight W of a body of mass m at its surface is given by

$$W = mg = \frac{GMm}{r^2}$$

and hence

$$g = \frac{GM}{r^2} \tag{4-2}$$

where g is the weight of a unit mass or the acceleration due to gravity. At different points on the earth's surface g will be different, since r is not constant and (as will be shown later) centrifugal effects are not negligible. Since g can be measured very accurately and the size of the earth can be determined from geodetic and astronomical observations, the mass of the earth M can be inferred from a knowledge of G. Details of the laboratory techniques for measuring G or M directly

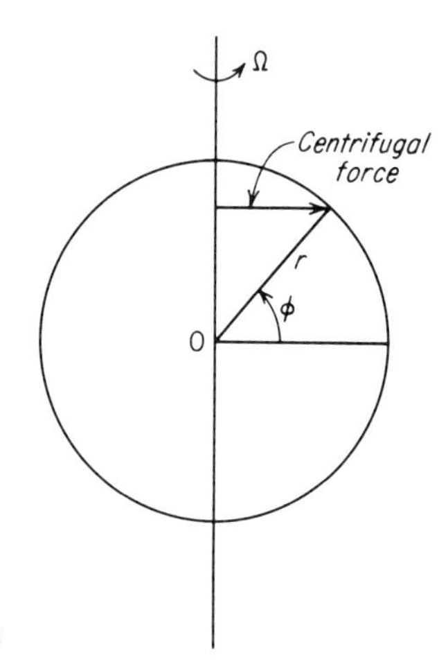

FIGURE 4-3

(by comparing the forces exerted by a known mass on another mass with the attraction of the earth on the second mass) will not be given here, nor will the instrumental details of the most modern methods of measuring g and its variations be given. Present-day absolute measurements of gravity have an accuracy better than 1 ppm. Relative changes in gravity, on the other hand, have been measured by means of spring gravimeters to 1 part in 1000 million (Sec. 4-5).

Equation (4-2) applies only to a nonrotating earth. The centripetal acceleration a for motion in a circle of radius r is $r\Omega^2$, where Ω is the angular velocity. Since $\Omega = 2\pi/T$, where T is the period,

$$a = \frac{4\pi^2 r}{T^2} \tag{4-3}$$

At the equator $$a \approx 3.4 \text{ cm/sec}^2 \approx \frac{g}{289}$$

In latitude ϕ, r is replaced by $r \cos \phi$, and hence the acceleration is $a \cos \phi$ (Fig. 4-3). This has a vertical component $a \cos^2 \phi$, which is a maximum at the equator and zero at the poles. The force of gravity is the resultant of the attraction of the earth and the centrifugal force due to the earth's rotation. It is this resultant force which determines the weight of a body at rest, or its velocity and direction of fall if falling freely. The time of vibration of an invariable pendulum depends on g, and if corrected for the effect of centrifugal force, pendulum observations can give relative distances from the center of the earth. It is found that g (pole) is greater than g (equator) by $^1/_{189}$. Centrifugal forces account for $^1/_{289}$, which leaves $^1/_{546}$ to be explained by the difference between the polar and equatorial radii, a matter of some 21 km. However, since the earth is not a perfect sphere, the inverse-square law is not strictly applicable, and also there are station corrections to be applied to the measurements of g (Sec. 4-6). An acceleration of 1 cm/sec^2 has been called a gal in honor of Galileo. A more convenient unit to work with in discussing gravity variations over the earth is the milligal (mgal), defined as 0.001 gal, or the gravity unit (gu), which is 0.0001 gal, or 0.1 mgal.

Pendulum or gravimetric measurements are far easier than geodetic at many places on the earth's surface, as in the polar regions for example. The land surface of the earth is now fairly well covered by a network of gravity stations, and in 1923 F. A. Vening Meinesz devised a technique for measuring gravity in a submarine, and in this way many oceanic regions have been sparsely covered by a network of about 5000 stations (F. A. Vening Meinesz, 1948a; J. L. Worzel, 1965). More recently Anton Graf developed a new submarine gravimeter that J. L. Worzel was able to adapt for use on a stabilized platform aboard a surface ship, and now gravity measurements can be made successfully at sea anywhere in the world. A detailed description of the instrument, operational procedures, and computation and correction methods have been given by Worzel (1959). Continuous gravity measurements can be made at sea with a precision better than 5 mgal. L. Lacoste et al. (1967) have recently redesigned their surface-ship gravimeter for stabilized platform operation. Repeated tests over the known gravity field of the San Luis Pass salt dome, about 30 km southeast of Freeport, Texas, by T. R. Lafehr and L. L. Nettleton (1967) gave an average difference of only 0.5 mgal. The greatest uncertainties arise from the difficulty of determining the ship's precise location and speed, which are needed for computing the Eötvös correction (Sec. 4-6). The continuous curves produced at sea greatly increase the value of gravity observations; they allow more representative values, which are required for many geodetic studies, to be chosen for oceanic areas. Also, the interpretation of gravity anomalies depends almost entirely on the shape of the observed curve; and a continuous curve is much better for this purpose than a curve drawn through a number of discrete points. Further experimental determinations of the reliability of surface-ship gravimeters have been given by P. Dehlinger and S. H. Yungul (1962) and by C. Gantar et al. (1962). More recently L. A. Gutarenko et al. (1967) have given details of gravity measurements at sea made in a towed gondola at a depth of 100 m and speeds of 4 to 5 knots. Gravity surveys have now been carried out over all the oceans [e.g., the Atlantic Ocean (M. Talwani and X. Le Pichon, 1969), the Indian Ocean (X. Le Pichon and M. Talwani, 1969), and the Pacific Ocean (P. Dehlinger, 1969a)].

More recently still, it has been possible to take gravity readings from an aircraft. The chief consideration in attempting to measure gravity in the air is its possible application for geodetic purposes. In geodetic determinations of the details of the shape of the earth, calculations are made using average gravity values for $1° \times 1°$ rectangles where available. The airborne gravity meter offers the hope of rapidly filling in wide gaps over many parts of the earth's surface. To achieve this would require a very precise navigation system since the Eötvös correction (Sec 4-6) is extremely large. A detailed evaluation of the precision of airborne gravity readings has been given by L. L. Nettleton et al. (1962), and a review of gravity measurements at sea and in the air by L. J. B. Lacoste (1967).

4-4 GRAVITATIONAL THEORY

The first approximation to the figure of the earth is a sphere, and the second an oblate spheroid. For theoretical purposes a surface which is everywhere horizontal is chosen, i.e., a surface which is everywhere perpendicular to a plumb line. Such a surface is called a level surface, or an *equipotential.* The problem of the determination of the figure of the earth is that of the determination of all equipotentials. Two equipotentials cannot intersect, and to give a definite point on one is to fix it completely. By taking a standard point near sea level, an equipotential called the geoid is obtained. It is a reference level and is in no way different (except in dimensions) from any other equipotential. It is to be noted that an equipotential does not constitute a surface of equal gravity. The ocean surface, for example, is approximately an equipotential surface (being everywhere perpendicular to the force of gravity), but the gravitational attraction over the ocean is quite variable.

In 1743, the French mathematician A. C. Clairaut obtained an expression relating gravity measurements at different points on the earth's surface with the form of the surface (Appendix B). Assuming that the earth is an oblate spheroid whose polar equation is given by

$$r = a(1 - f \sin^2 \phi) \qquad (4\text{-}4)$$

where f is the oblateness as defined by Eq. (4-1), he obtained the result

$$g = g_e \left[1 + \left(\frac{5c}{2} - f\right) \sin^2 \phi\right] \qquad (4\text{-}5)$$

where g_e is the value of g at the equator, $\phi = 0$, and c is the ratio of the centrifugal force to gravity at the equator,

$$c = \frac{\Omega^2 a}{g_e} \qquad (4\text{-}6)$$

It is easy to see that the coefficient of $\sin^2 \phi$ in Eq. (4-5), namely, $5c/2 - f$, is a measure of the "gravitational flattening" of the earth defined as $(g_p - g_e)/g_e$, where g_p is the value of gravity at the poles. Clairaut's theorem was extended by Stokes in 1849 and again by Helmert in 1884. Taking $f = 1/297$, the sea-level value of gravity can be calculated at any place from the formula

$$g = 978.049(1 + 0.0052884 \sin^2 \phi - 0.0000059 \sin^2 2\phi) \qquad (4\text{-}7)$$

This is the *international gravity formula*. The constant 978.049 is a statistically determined value for g_e based upon observational data. The constant for the term $\sin^2 \phi$ incorporates the effect of geometrical flattening and centrifugal force [Eq. (4-5)]. The term with $\sin^2 2\phi$ is a correction for nonconformity to the spheroidal shape assumed for the rotating body.

Recent satellite orbital studies and new astrogeodetic determinations suggest that the values of the parameters a and f used in obtaining Eq. (4-7) are incorrect by significant amounts. In 1967, the International Association of Geodesy (IAG) adopted two resolutions concerning the best values of a and f and of the coefficients in Eq. (4-7) (H. Moritz, 1968). The official values have not yet been published, but the preliminary revised version of the international gravity formula is

$$g = 978.0318(1 + 0.0053024 \sin^2 \phi - 0.0000058 \sin^2 2\phi) \qquad (4\text{-}7a)$$

The new value for g_e also incorporates a correction (approximately −13 mgal) to the absolute value of gravity at Potsdam (Sec. 4-5). The changes in theoretical gravity values show a difference of about −17 mgal at the equator and about −4 mgal at the poles. Since about −13 mgal can be attributed to the change of the absolute standard at Potsdam, the change due to the other geometric and dynamic parameters is of the order −4 mgal at the equator and +9 mgal at the poles.

It has also been suggested that the earth more exactly approaches a triaxial ellipsoid with a difference in equatorial radii of 100 m. M. Caputo (1963) has shown, however, that no actual gain in accuracy is obtained by adopting a triaxial instead of a biaxial ellipsoid, and that the equatorial bulge of the gravitational field for a triaxial ellipsoid is almost an order of magnitude smaller than some of the actual gravity anomalies found in many regions near the equator. The accuracy and coverage of present gravity and satellite data suggest that a more accurate expression of the actual field can be obtained only by means of higher order spherical harmonics.

Additional information about the earth's gravitational field and the figure of the earth has been made possible by the use of artificial satellites. If the earth were strictly spherical and without an atmosphere, the path of a satellite, apart from very small perturbations due to the sun and other bodies of the solar system, would be an ellipse with one focus at the earth's center. It is fortunate that the first-order perturbations due to the atmosphere and the earth's oblateness are quite distinct. D. G. King-Hele (1958) has shown that there are three main perturbations which arise because the earth's gravitational field is not spherically symmetric. First, the plane of the orbit is no longer fixed in direction, but rotates about the earth's axis in the direction opposite to the satellite's motion, while still remaining inclined at a fixed angle α to the equator. The rate of rotation is about

$$10.00 \left(\frac{a}{\bar{r}}\right)^{3.5} \cos \alpha \qquad \text{deg/day} \qquad (4\text{-}8)$$

where a is the earth's equatorial radius, and $\bar{r}$ is the harmonic mean distance of the satellite from the earth's center.

The second main effect of the earth's oblateness is that the major axis of the orbit rotates slowly in the orbital plane so that the orbit is a near ellipse with a rotating major axis. The rate of rotation is about

$$5.00 \left(\frac{a}{\bar{r}}\right)^{3.5} (5 \cos^2 \alpha - 1) \qquad \text{deg/day} \qquad (4\text{-}9)$$

This vanishes when $\cos \alpha = 1/\sqrt{5}$, or $\alpha \approx 63\tfrac{1}{2}°$.

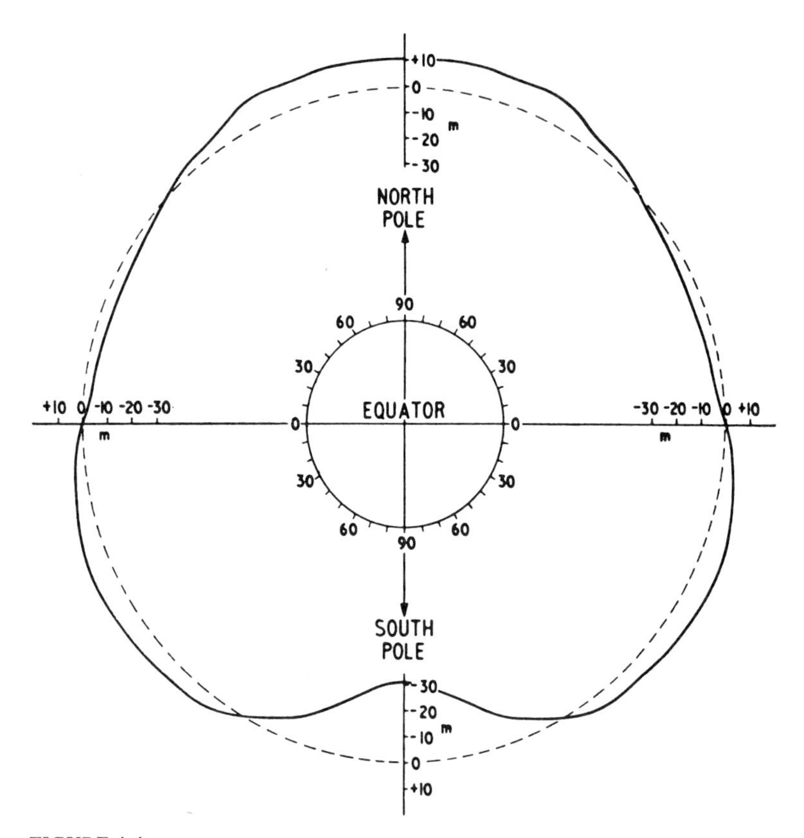

FIGURE 4-4
Height of the mean meridional section of the geoid (solid curve), relative to a spheroid of flattening $1/_{298.25}$ (dotted curve). (*After D. G. King-Hele, G. E. Cook, and Diana W. Scott, 1968.*)

In determining the orbit from observations, it is far easier to have a nonrotating major axis, and this was presumed one of the reasons why a value of α near 65° was chosen for the first Sputniks. At other inclinations the rate of rotation can be quite large—about 15°/day forward for near-equatorial orbits, and 4°/day backward for polar orbits. The third perturbation to the orbit is a change in the radial distance r of the satellite from the earth's center; this change is usually quite small, less than a kilometer.

The mean rate of rotation $\dot{\Omega}$ of the orbital plane can be found accurately from observations over a long period. Theoretical values of $\dot{\Omega}$ can be calculated using certain measured values of the orbital elements and the accepted values of geophysical constants relating to the earth's gravitational field. It was found, for example, that the ratio $\dot{\Omega}$ (theoretical)/$\dot{\Omega}$ (observed) remained almost constant during the lifetime of Sputnik 2 and had the value 1.0068 ± 0.0003. This difference between theory and observation is significant. Since the ratio remained nearly constant, it is extremely unlikely that it is due to atmospheric effects, which should increase as time goes on. The results suggest a lower value than that previously accepted for the ellipticity of the earth. Calculations from the orbit of satellite 1958 β_2 indicate a value of $^1/_{298.32}$.

Periodic variations in the eccentricity of the orbit of this satellite were also found which can be explained by the presence of a third zonal harmonic in the earth's gravitational field. This modifies the geoid, making it pear-shaped (Fig. 4-4). This figure greatly exaggerates the shape, the changes being only of the order of 30 m at the South Pole, 10 m at the North Pole, and 7.5 m in middle latitudes. However, the presence of a third harmonic of this magnitude indicates a very substantial load on the surface of the earth. The stresses involved must be supported either by a mechanical strength larger than that usually assumed for the interior of the earth or by large-scale convection currents in the mantle. The depressed areas of the geoid are those which have recently been covered with ice and are rapidly rising, and J. T. Wilson (1960) has pointed out that there is thus no evidence that the earth will long remain pear-shaped. A detailed discussion of the contribution of satellites to the determination of the earth's gravitational potential has been given by A. H. Cook (1963). To obtain a more complete picture of the gravitational field of the earth than can be obtained from satellite data alone, a combination of satellite and gravimetric data has been used by W. M. Kaula (1966) and R. H. Rapp (1968). Rapp used model anomalies for areas in which no gravity observations exist, whereas Kaula used statistical predictions for such areas.

4-5 MEASUREMENTS OF GRAVITY

The absolute determination of gravity at any point on the earth to a useful accuracy of 1 ppm is extremely difficult. The early absolute determinations were made by using some form of reversible pendulum of the type designed by Kater. The experimental difficulties are considerable if great accuracy is to be obtained, and all known sources of error, such as the yielding of the knife-edges under the weight of the pendulum, must be eliminated. Reversible pendulums were used to make a series of absolute determinations at Potsdam from 1898 to 1904 by L. Kühnen and P. Fürtwangler, and from these are derived the *Potsdam values* of g now used throughout the world. A number of later determinations of g have been made at various places, and these have shown that the Potsdam system gives values of g that are too high by an amount that is between 8 and 20 mgal. The discrepancies among these measurements are such, however, that they cannot provide an acceptable alternative to the Potsdam system, i.e., one that would be in error by an amount not greater than 1 or 2 mgal.

In 1946, C. Volet suggested that the timing of a rod in free fall could in principle give results that were comparable in accuracy with those obtained from compound pendulums and that such measurements would not be subject to the systematic experimental errors that affect pendulum experiments. Free-fall experiments have been made by P. N. Agaletzki and K. N. Egorov (1956), A. J. Martsinyak (1956) in Leningrad, B. A. Thulin (1958) in Sèvres, and H. Preston-Thomas et al. (1960) in Ottawa. The last paper contains references to the earlier work and gives a detailed account of the experimental techniques which measure the distances through which a nonmagnetic stainless-steel ruler falls *in vacuo* in discrete time intervals.

J. E. Faller (1965) has determined the absolute value of g at Princeton University, using one element of an optical interferometer as a freely falling body. A rotation-insensitive mirror was dropped approximately 10 cm, generating three sets of white light fringes as it fell. The times between the occurrence of these fringes were measured electronically, and their spacing interferometrically. The rms accuracy of the experiment was 7 parts in 10^7. The correction to the Potsdam value is -15.1 mgal, or -14.5 mgal, depending on the value adopted for g at the Washington Geophysical Laboratory, to which the value at Princeton was transferred. Work has begun on the construction of new apparatus, using a laser as the light source and a much longer dropping distance. It is hoped to obtain even greater accuracy.

Gravity differences, on the other hand, are far easier to determine than absolute values. If the properties of the pendulum are sufficiently stable, the ratio of the values of gravity at two stations is inversely proportional to the squares of the observed periods. Few pendulum instruments in use at the present time swing a single pendulum at a time, since that produces a motion of flexure of the support and may be disturbed by motions of the ground. Apparatus using two pendulums is sufficient for most observations on land. If two pendulums of equal period are swinging with equal amplitudes and phase difference π on the same support in the same vertical plane, both the adverse effects of a single pendulum are eliminated. The demands of exploration geophysics (chiefly for petroleum) have led to the development during the last 25 years of precise, portable gravimeters. Fundamentally they are designed to balance the force of a fixed mass against the elastic stresses in a system of springs or torsion fibers. The chief problem in their design is the achievement of the necessary sensitivity. For studies of the gravitational field over large areas, it is extremely important that the gravity readings all be consistent; otherwise discontinuities in the data between different regions may be interpreted as having a geological significance. Such discontinuities may arise from errors in connecting fundamental reference stations to the world base at Potsdam or from inaccurate calibration of the gravimeter. The observations of any survey should be made between stations of an interlocking network which includes pendulum stations, so that observed gravity differences around adjoining closed figures can be adjusted to give minimum closing errors, and the scale constant of the instrument can be determined from pendulum observations. The response of a gravimeter cannot be assumed to be linear over large variations in gravity, and it should not be used to extrapolate values from a limited calibration range. In recent years, calibration lines have been established with pendulums from north to south in North America and Europe, and numerous intercontinental ties have been made. Gravimeters can be calibrated by being flown between stations on these lines and may then be used to extend the network of gravity stations. G. P. Woollard (1969) has pointed out that much of the existing pendulum gravity data are in error and that there are several different gravity standards in use for calibrating the response of high-range gravimeters used for global studies.

4-6 GRAVITY ANOMALIES

Although the broad-scale variation of g over the earth's surface is intimately connected with the earth's figure, the more local departures from this smoothly varying field are related to the inhomogeneous density of the earth's crust. The range in density is, however, rather restricted, and the effect of structures in the earth's crust, even of continental proportions, amounts to only a small fraction of the total gravitational field. Any measured value of gravity is a composite of many effects, and these, in approximate order of importance, are:

1 Shape and size of the earth as a whole

2 Rotation of the earth

The sea-level value of g varies from approximately 978.049 cm/sec^2 at the equator to 985.221 cm/sec^2 at the poles. This change incorporates both effects 1 and 2 and represents by far the major changes present; the effect of all other factors is not likely to exceed 1 gal. An error in position of 1 mile in a north or south direction will have an effect of approximately 1 mgal in the anomaly.

3 Changes in elevation

Let g_0 be the acceleration due to gravity at some selected level and g_h the value at height h above this level. Then if r is the radius of the earth, it follows from Eq. (4-2) that

$$\frac{g_h}{g_0} = \left(\frac{r}{r+h}\right)^2 = \left(1 + \frac{h}{r}\right)^{-2} \approx 1 - \frac{2h}{r} \tag{4-10}$$

This correction is 0.3086 mgal/m, or 0.09406 mgal/ft. The sign of the correction is negative for increases in elevation above the reference level and positive for decreases below the reference level. The correction of an observed value to what it would be at sea level is known as the free-air reduction.

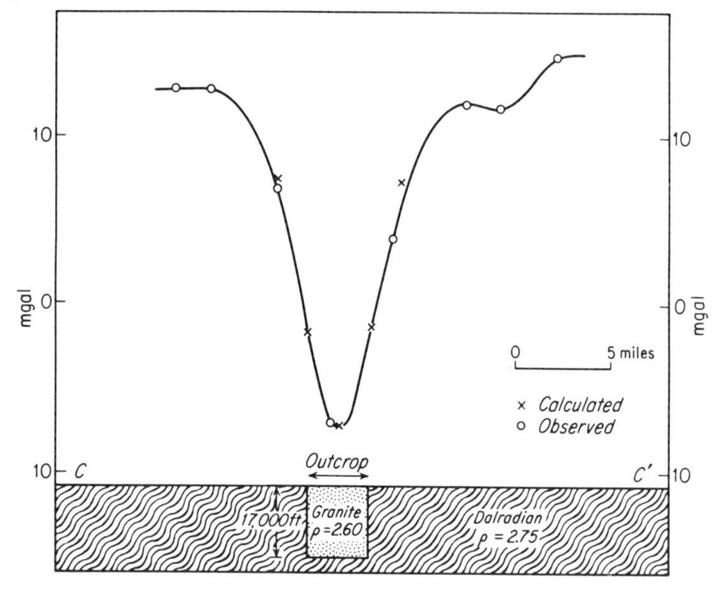

FIGURE 4-5
Gravity anomaly due to the granite at Barnesmore, Ireland. (*After A. H. Cook and T. Murphy, 1952.*)

4 Horizontal mass discontinuities caused by changes in the thickness of the outer crustal layers of the earth

This effect is given by $2\pi G\sigma h$, which is the gravitational attraction of a semi-infinite slab of thickness h and density σ. Its value is 0.1119 mgal/m, or 0.03410 mgal/ft, assuming normal crustal rocks with a density of 2.67 g/cm^3. The sign of the correction is always opposite to that for the elevation correction 3. The net effect for both corrections 3 and 4 for an increase in elevation of 1 m is –0.1967 mgal, or for an increase of 1 ft, –0.060 mgal. The effect of changes in crustal structure in the outer portions of the earth shows in the anomalies obtained when computed values of gravity are compared with observed values. Such anomalies (in combination with the free-air reduction) are called Bouguer anomalies. Figure 4-5 shows a typical anomaly due to a batholith. G. P. Woollard (1969) has pointed out that the value 2.67 g/cm^3 for the mean density σ of the crust used in calculating the Bouguer anomaly should be increased to 2.93 g/cm^3. He bases his higher value on data obtained from explosion seismological studies of the crust and laboratory studies of the physical properties of rocks under confining pressures. Since a difference of 0.1 g/cm^3 in mean crustal density results in about 4.2 mgal in the included mass correction for each 1000-m change in elevation, this change in σ would significantly affect Bouguer anomaly values.

5 Horizontal changes in mass caused by changes in density of the rocks in the crystalline complex beneath the surface sediments

This seldom amounts to as much as 100 mgal and is usually much less. Thus the effect of a volcanic pipe 3 km in diameter and of considerable depth may not exceed 30 mgal.

6 Horizontal changes in mass caused by changes in density of the surface and near-surface rocks

This usually amounts to less than 10 mgal, and in general these changes are removed in computing the Bouguer anomalies by changing the density σ in the mass-correction term.

7 Changes in topographical configuration of the buried crystalline rock surface

This seldom amounts to 20 mgal and is usually less than 5 mgal. Thus the Nemaha ridge, a buried granite ridge in Kansas, with over 1500 m relief, has a gravitational effect of only 2 mgal.

8 Changes in the relief of the surrounding surface topography

This depends more particularly on the actual relief at the point of observation. It is found that if the angle between the horizontal and the top or bottom of the adjacent topographical feature is less than 10°, the topographical effect does not exceed 1 mgal. Corrections for observations near a precipitous slope, however, may amount to 20 mgal or more. Regardless of whether the adjacent topographical feature is a hill or a valley, the sign of this correction is always negative.

9 The Eötvös correction

This is not an anomaly but a correction that must be applied to gravity measurements. For convenience it is discussed here.

As shown in Sec. 4-3, because of the rotation of the earth, there is in latitude ϕ an acceleration $r\Omega^2 \cos \phi$ directed toward the axis of rotation. This has a vertical component $r\Omega^2 \cos^2 \phi$, and hence for a stationary observer on a rotating spherical earth

$$g = \frac{GM}{r^2} - r\Omega^2 \cos^2 \phi \qquad (4\text{-}11)$$

This has already been allowed for in the international gravity formula (4-7a). Consider now a vehicle moving with constant speed V on a true course α on the surface of a rotating spherical earth; i.e., the observer has an additional velocity $V \cos \alpha$ northward and $V \sin \alpha$ eastward. The component $V \sin \alpha$ eastward is equivalent to an increase in the rotation of the earth by an amount $V \sin \alpha / r \cos \phi$. The component $V \cos \alpha$ northward will cause an additional acceleration $V^2 \cos^2 \alpha / r$ directed toward the center of the earth. Thus the value of gravity is

$$g_m = \frac{GM}{r^2} - r\left(\Omega + \frac{V \sin \alpha}{r \cos \phi}\right)^2 \cos^2 \phi - \frac{V^2 \cos^2 \alpha}{r}$$

and the Eötvös correction to observations made on a moving vehicle is

$$\Delta g = g - g_m = 2\Omega V \cos \phi \sin \alpha + V^2/r \qquad (4\text{-}12)$$

For gravity measurements at sea, the second term of the Eötvös correction in Eq. (4-12) is extremely small and can be neglected. For a speed V of 18 knots, it amounts to 1.3 mgal. The importance of the term rises rapidly, however, as the speed increases and is quite significant for measurements made on an aircraft, where the total correction may exceed 1000 mgal. At the equator errors in the Eötvös correction arising from inaccuracies in measuring the ground speed are, at 450 knots, 7.5 mgal/knot on a north-south course, 11 mgal/knot on an easterly course, and 3.7 mgal/knot on a westerly course. At current jet transport speeds account must also be taken of the ellipticity of the earth and the height of the aircraft. R. B. Harlan (1968) has obtained an expression for the Eötvös correction allowing for these effects; a distinction must also be made between ground speed and aircraft velocity.

4-7 ISOSTASY

If a mountain is considered merely as an excess mass superimposed on a uniform crust, then, ignoring the deformation of the interior, it is possible to estimate its effect on the gravitational attraction of bodies in its neighborhood. In particular, the attraction of a plumb bob toward the excess matter of the mountain would cause a deflection of the direction of the zenith. For example, two pendulum bobs at stations in Puerto Rico on the north and south coasts, 53 km apart and separated by a mountain range, were drawn together 56″. The observed results differ, however, from the predicted values. The first experiments were carried out by P. Bouguer, who found that the deflection due to the mountain Chimborazo, in the Andes, was much less than that calculated. The experiment was repeated by N. Maskelyne in 1774 in Scotland, and a repetition by F. Petit in 1849 in the Pyrenees showed that the attraction was not only small, but actually negative; i.e., the plumb line appeared to be deflected away from the mountains. Similar discrepancies were found by J. H. Pratt (1855) on analyzing the deflections observed by G. Everest in India, and by J. F. Hayford (1909, 1910) in America. These results can only mean that under the mountains the density increases more slowly than is normal. J. H. Pratt, G. B. Airy, J. F. Hayford, and F. R. Helmert found by gravity measurements

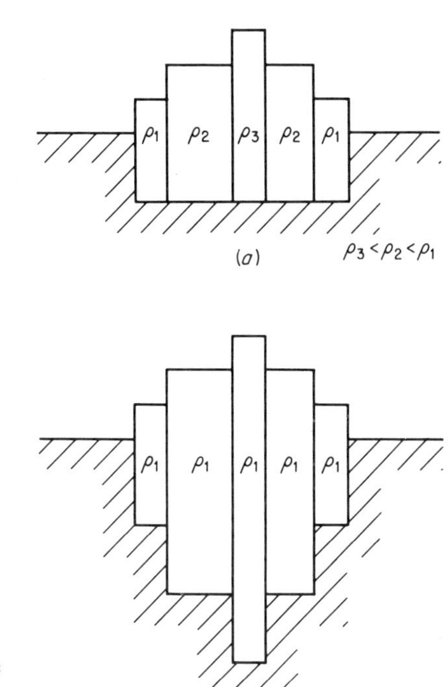

FIGURE 4-6
Schematic representation of isostatic equilibrium according to (*a*) Pratt and (*b*) Airy.

that the anomalous density changes extended down to about 120 km, although this is now regarded as too far. Two quite different physical explanations have been given to explain the result. The first is due to Pratt, who assumed that all crustal columns have the same mass above a uniform level. Hayford (1910) called this the *depth of compensation* and calculated that it is at a depth of 113.7 km below the surface. He took a crustal column extending from sea level to a depth of 113.7 km with a density of 2.67 g/cm^3 as standard. Any column with an elevation above sea level therefore has a proportionally lower density, depending upon its elevation. The alternative explanation, due to Airy, is to regard all crustal columns as having the same density and to suppose that differences in elevation reflect differences in the thickness of the outer crust, which is considered to be in hydrostatic equilibrium and floating in a denser subcrustal stratum analogous to blocks of ice floating in water. The depth of compensation is thus variable, being about 50 km under the continents. The continents, mountains, and other topographical features are said to be in "isostatic equilibrium," and according to Airy's theory, *isostasy* is nothing more than the Archimedian principle of hydrostatic balance between floating bodies.

Figure 4-6 illustrates schematically the difference between Pratt's and Airy's hypotheses. In practice it makes little difference which theory is assumed in

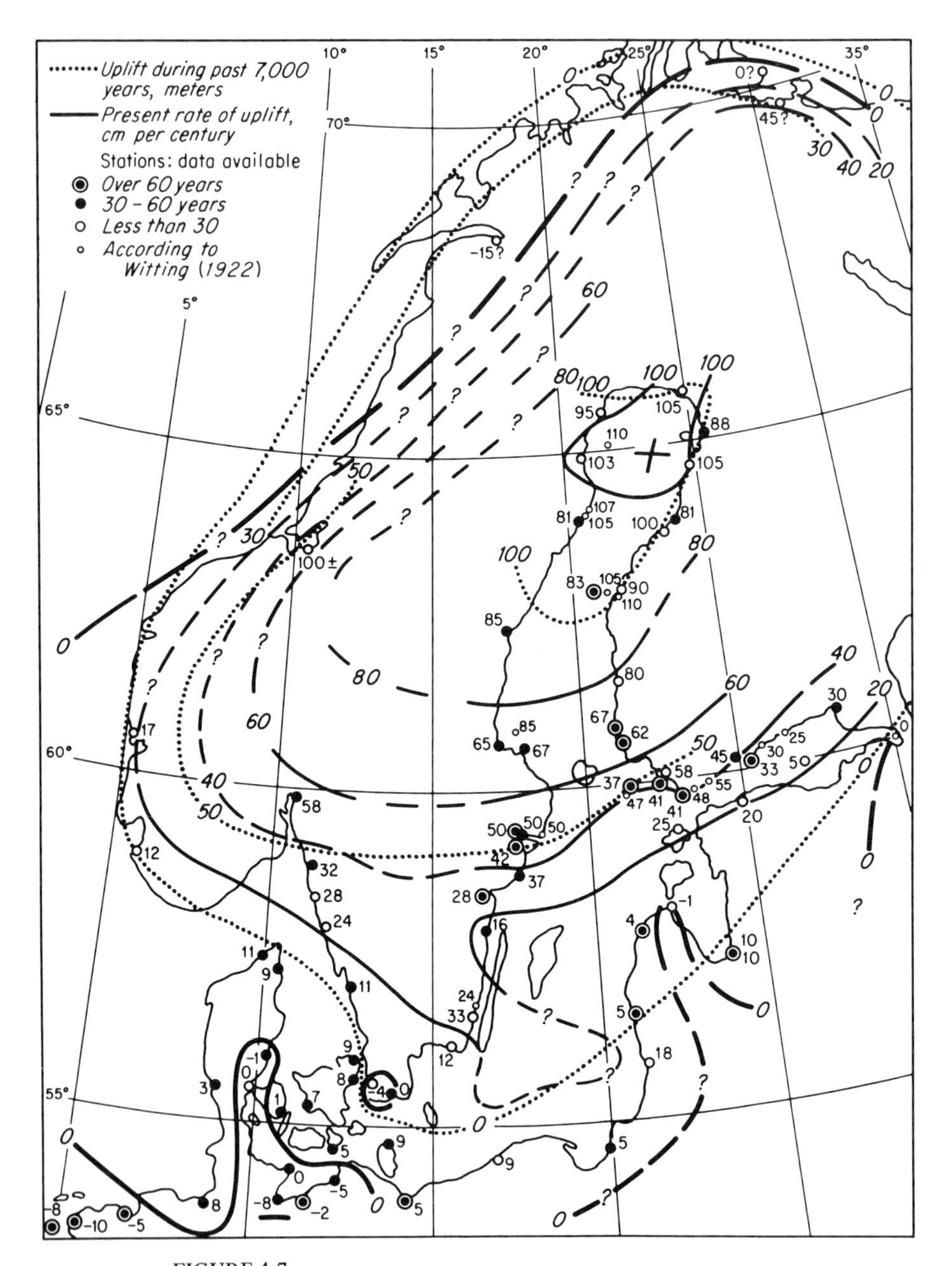

FIGURE 4-7
Postglacial uplift in Fennoscandia in centimeters per century, as shown by tide-gage stations. Data for past 7000 years, after Sauramo. (*From B. Gutenberg, 1941.*)

determining the isostatic correction to *g*. It is a laborious process because changes in elevation of all the surrounding terrain must be evaluated. Since the Airy theory is harder to apply in practice, the Pratt theory has often been adopted by geodesists. The correction is determined on the basis of the gravitational attraction of a vertical cylindrical column extending upward from −113.7 km with a density value determined by the proportional change in elevation of the actual column above the standard sea-level column.

The question of postglacial uplift may be considered as an example of isostasy. Fennoscandia has been rising considerably in recent times. The evidence for this is geological, geodetic, and tidal, as measured with tide-level gages (Fig. 4-7). In particular, M. Sauramo, by examining raised beaches, has concluded that the largest uplift took place near the Gulf of Bothnia and that since 6800 B.C. the land has risen 250 m. This is usually explained on the grounds of isostasy. Fennoscandia would have sunk after having been covered with the heavy, extensive load of ice during the Pleistocene ice age. After the melting of the ice, it should be rising again to achieve isostatic equilibrium corresponding to its present loading condition. However, there are large regions of the earth which are not isostatically compensated and which, in spite of this, do not show any signs of rising or subsiding, as for example India, with its large negative gravity anomalies. Moreover, E. N. Lyustikh (1956), in a recent analysis of geological data, maintains that the Fennoscandian shield was rising even before the last ice age.

Tilted beaches around the Great Lakes and raised marine features throughout arctic Canada which have been dated by radiocarbon analyses enable curves to be constructed of postglacial uplift versus time for a number of areas (W. R. Farrand, 1962). The mean pattern of uplift with respect to time at any given locality shows a strong exponential decrease from approximately the time of ice removal in that locality to the present day (Fig. 4-8). Curves for different areas are quite similar although displaced in time, and this displacement correlates with the time of deglaciation of each locality. These results give additional evidence that postglacial and recent uplift around the Canadian Shield (as in Scandinavia) are a result of glacial unloading. It also appears that far northern areas such as northwestern Victoria Island and northern Ellesmere Island probably had an ice cover comparable with that over the Great Lakes. Also, if the Wisconsin ice sheet achieved isostasy, the major part of isostatic recovery upon removal of the ice load took place in a given locality before complete deglaciation in the area. Isostatic recovery may be plotted as in Fig. 4-9 to show in a rough way the nature of crustal rebound. The dates and amounts of uplift are taken directly from Fig. 4-8. Further information on the postglacial uplift of the Canadian Arctic may be found in J. T. Andrews (1968) and J. T. Andrews and G. Falconer (1969).

Because an infinite number of mass distributions in space can create the same gravitational effect, no unique solution of the actual subsurface mass distribution can be determined from gravity data alone. Thus no more information would be gained by measuring gravity at different heights above the earth or from more than

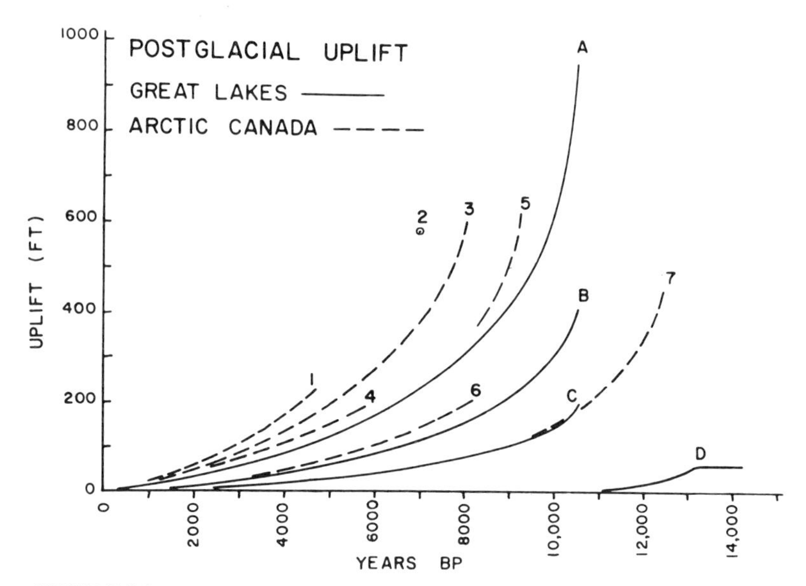

FIGURE 4-8
Postglacial uplift curves for North America. Solid curves are from Lake Huron; dashed curves are from arctic Canada. (*A*) North Bay, (*B*) Sault Ste. Marie, (*C*) Cape Rich, (*D*) Port Huron, (1) Igloolik, (2) Carr Lake, (3) James Bay, (4) Southampton Island, (5) Coronation Gulf, (6) northwest Ellesmere Island, (7) northwest Victoria Island. (*After W. R. Farrand, 1962.*)

one derivative of the potential. Although the distribution of the mass is indeterminate, the total disturbing mass is uniquely determined by the surface gravity field. Moreover, gravity data can be used to extend our knowledge when the key to the structure has been established from seismic data and can help, as in Puerto Rico, in making qualitative investigations where there are differences in crustal structure. Thus it is found that most young mountain ranges (such as the Rockies, Alps, Andes, and Himalayas) have a mass deficiency at depth which results in a Bouguer anomaly of about −300 mgal, and on the assumption of isostasy, this deficiency almost exactly compensates for the mass of the mountains above sea level. On the other hand, old eroded mountain ranges such as the Appalachians have a much smaller deficiency of mass associated with them, and although the Bouguer anomaly may be only 0 to 100 mgal in these areas, this mass deficiency may be much greater than is required for isostatic compensation. As a result, isostatic anomalies of approximately −50 mgal may be obtained, indicating that erosion and reduction in elevation in these areas has proceeded faster than the readjustment of the compensating mass at depth.

Figure 4-10 shows a simplified isostatic anomaly map of the United States. In the study of broad-scale effects, continental or subcontinental in scale, the question

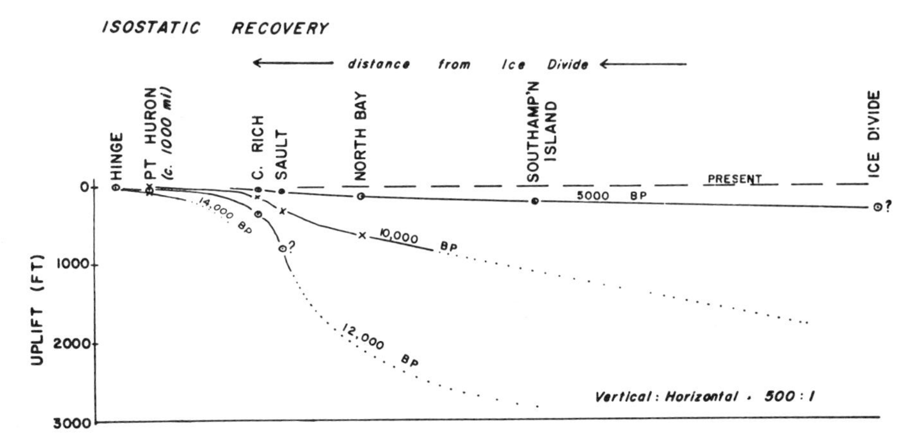

FIGURE 4-9
Schematic representation of isostatic recovery as interpreted from uplift curves in Fig. 4-8, assuming the Wisconsin ice sheet was in complete isostatic adjustment and depressed the crust approximately 3000 ft. The approximate horizontal scale is distance from a given station to the nearest ice divide, either in Labrador (Schefferville) or in Keewatin. The dotted portions of each curve have no control whatsoever. (*After W. R. Farrand, 1962.*)

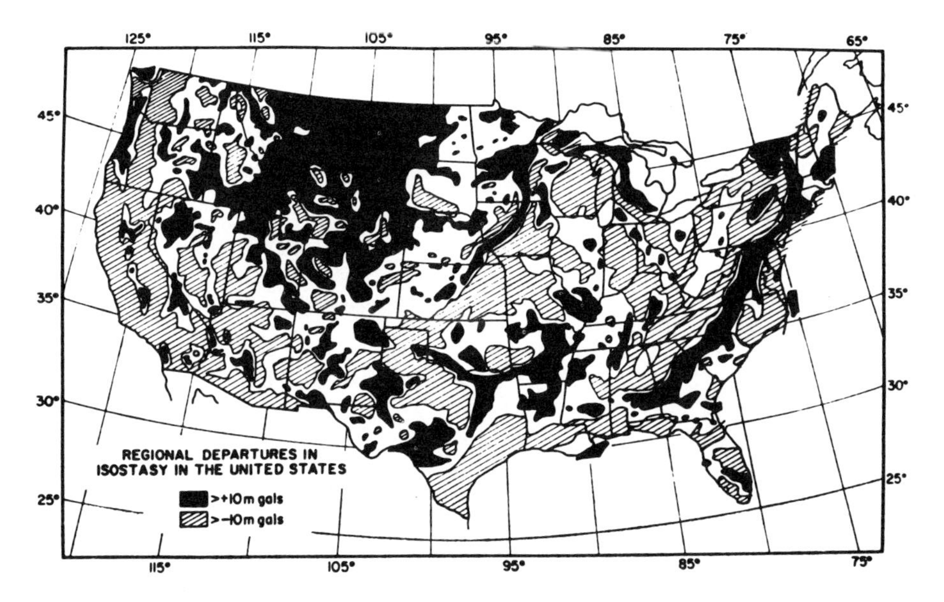

FIGURE 4-10
Simplified isostatic anomaly map of the United States. (*After G. P. Woollard, 1969.*)

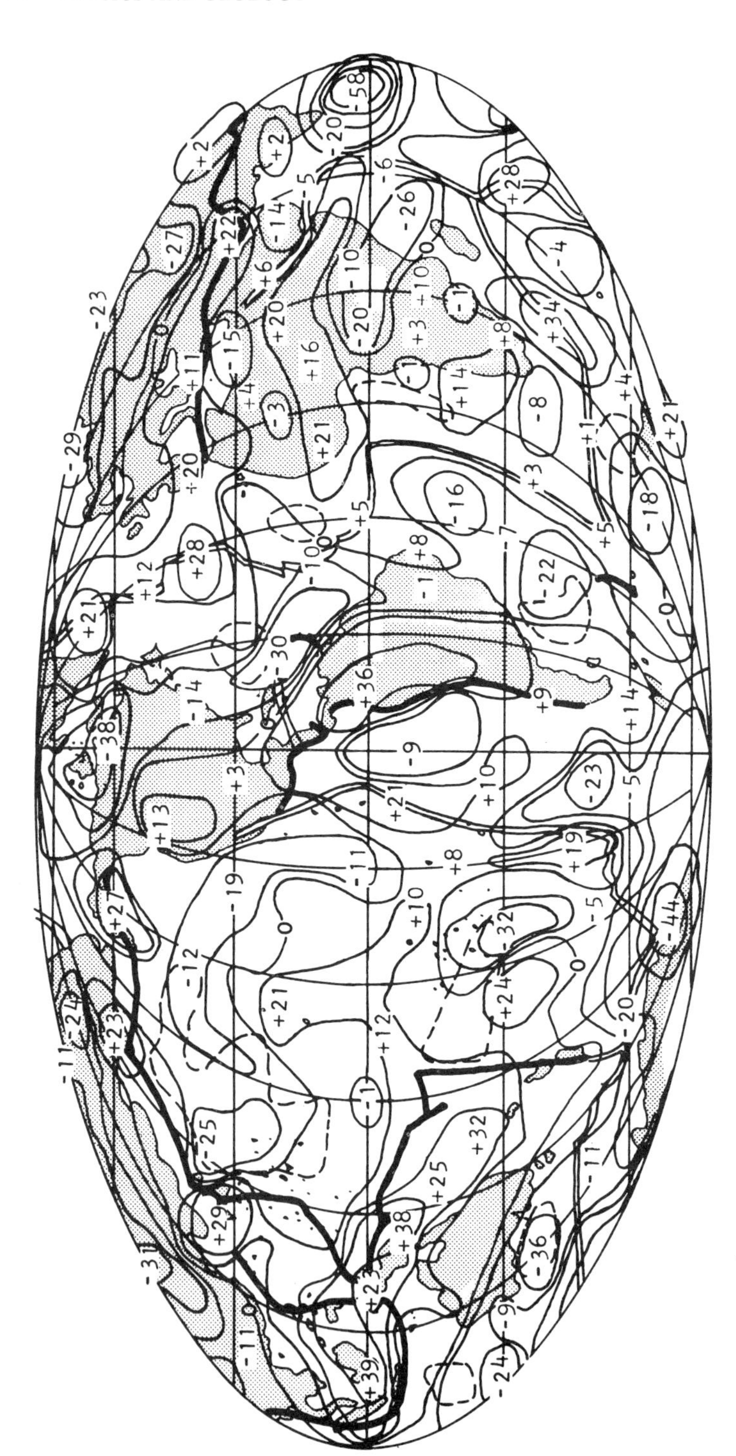
COMPRESSION
TENSION
APPROX. 3000 FATHOM

is one of smoothing out very local anomalies. Conversely, in the detailed study of a local area, a regional trend must often be removed before the smaller anomalies can be studied. However, the latest seismological and satellite-determined gravity evidence indicates that part of the compensation of young mountain belts must come from density differences in the mantle: All available evidence appears to indicate that the long wavelength components of the gravity field originate below the lithosphere.

W. M. Kaula (1969a) has attempted a tectonic classification of the earth's gravitational field based on a combination of satellite data and terrestrial gravimetry. He examined 43 areas: 19 markedly positive, 14 markedly negative, and 10 exceptionally "mild." The major features of the gravity field appear to be associated with relatively recent geological activity. This is particularly true for the dominant positive features. Thus Canada is negative because the recent glaciation dominates the character of the ancient shield, while Greenland is positive, possibly because it is part of the Iceland-North Atlantic oceanic flood basalt area which dominates the glaciation. Kaula suggests that the major tectonic disturbances are associated with positive gravity anomalies, whereas a greater part of the negative anomalies are the consequence of passive isostatic response. The major exceptions are the more rapidly spreading ocean rises, glaciated areas, and recent orogenies without extrusives (such as the Himalayas). The strongest correlation was found between positive-gravity anomalies and Quaternary volcanism; the lack of any systematic correlation between temperature and gravity anomalies would appear to indicate that horizontal variations in petrology are important. (See also Sec. 7-2.)

W. M. Kaula extended his work in a later paper (1970) using the most recent determination of the gravity field by E. M. Gaposhkin and K. Lambeck (1970), whose analysis is based primarily on 21 artificial satellite orbits and secondarily on mean gravity anomalies for $5^\circ \times 5^\circ$ squares covering 56 percent of the earth. Figure 4-11 shows a world map of the free-air anomalies. The improved resolution over Kaula's earlier (1969a) map results in significant changes in the relationship between gravity and tectonics in the southern oceans. Large positive anomalies are located along the ocean rises, the sole exception in the oceans being the anomaly lying over Hawaii. The trench and island-arc belts are also predominantly positive. The commonest negative anomaly features are the ocean basins, always located to the flanks of the rises. Kaula (1970) also offers an interpretation of these most recent gravity data which appears to confirm rather well the dependence of plate tectonics on mantle convection (see also Chap. 14). The greatest feature not

FIGURE 4-11
Free-air anomalies in milligals referred to an ellipsoid of flattening $1/_{299.8}$. Calculated from the spherical harmonic coefficients of the gravitational field of ◀ degrees 2 through 16 of E. M. Gaposhkin and K. Lambeck (1971). (Nonzero contours enclosing only one value have been omitted.) Global tectonic lines of compression and tension from B. Isacks et al. (1968) and major basins indicated by approximate 3000-fathom line. (*After W. M. Kaula, 1970.*)

FIGURE 4-12
Variations in the moon's gravitational field based upon Muller and Sjogren's (1968) analysis of Orbiter 5 results. Unit is 1 mgal. (*After W. M. Kaula, 1969b.*)

obviously related to the global tectonic system is the antarctic negative anomaly, which is much too large to be explained by glacial melting.

P. M. Muller and W. L. Sjogren (1968) have used tracking data of Orbiter 5 to obtain a gravimetric map of the near side of the moon (Fig. 4-12). They found several large positive gravity anomalies centered on the lunar ringed maria. These areas of high gravity indicate the presence of high density *mass con*centrations near the surface at these sites and have been called *mascons.* It has been estimated that they extend laterally from 50 to 200 km, are about 5 km thick, and lie at a depth of between about 50 and 100 km. In spite of their high density, they have not sunk into the moon to reach isostatic equilibrium. The obvious explanation of the existence of mascons is that large objects (iron meteorites?) of high density material collided with the moon and flattened out beneath the surface, where they have remained ever since. However, there has been much controversy over the interpretation of mascons, and their origin, like that of the maria themselves, is still uncertain.

SUGGESTIONS FOR FURTHER READING

CAPUTO, M.: "The Gravity Field of the Earth from Classical and Modern Methods," Academic Press, Inc., New York, 1967.

GARLAND, G. D.: "The Earth's Shape and Gravity," Pergamon Press, New York, 1965.

HEISKANEN, W. A., and F. A. VENING MEINESZ: "The Earth and Its Gravity Field," McGraw-Hill Book Company, New York, 1958.

MUNK, W. H., and G. J. F. MACDONALD: "The Rotation of the Earth," Cambridge University Press, New York, 1960.

RAMBERG, H.: "Gravity, Deformation and the Earth's Crust as Studied by Centrifuged Models," Academic Press, Inc., New York, 1967.

5
GEOCHRONOLOGY

5-1 INTRODUCTION

The study of the earth has provided a unique contribution to science in general. This is a splendid history of the past 3½ billion years preserved in the rocks with a detail unrivaled in other subjects. Two methods have contributed to the elucidation of this record. A century ago stratigraphers and paleontologists first established a precise scale of relative ages for the last half billion years. In recent years drilling for petroleum in deep basins, exploration of the ocean floors, and discoveries in remote regions have refined these ideas. Second, since 1930, radiometric methods have provided absolute dates for this stratigraphic scale; they have extended it to older rocks devoid of index fossils, and have established the probable age of the earth. (See Table 5-1 and Figs. 5-1 and 5-2.)

5-2 NOTE ON THE HISTORY OF STRATIGRAPHY

In 1785 when J. Hutton made the distinction clear between the nature and origin of volcanic and sedimentary rocks (Sec. 3-1), he was able to exclude from the successions younger intrusives of volcanic origin. He thus established the law of

Table 5-1 PRINCIPAL PALEONTOLOGICAL DIVISIONS OF THE LATTER PART OF GEOLOGIC TIME

Era	Period or system		Characteristic forms of life
Cenozoic	Quaternary Tertiary		Mammals
Mesozoic	Cretaceous Jurassic Triassic		Reptiles
Paleozoic	Permian		
	Pennsylvanian Mississippian	Carboniferous	Invertebrates and plants
	Devonian Silurian Ordovician Cambrian		Invertebrates
Precambrian time			No index fossils

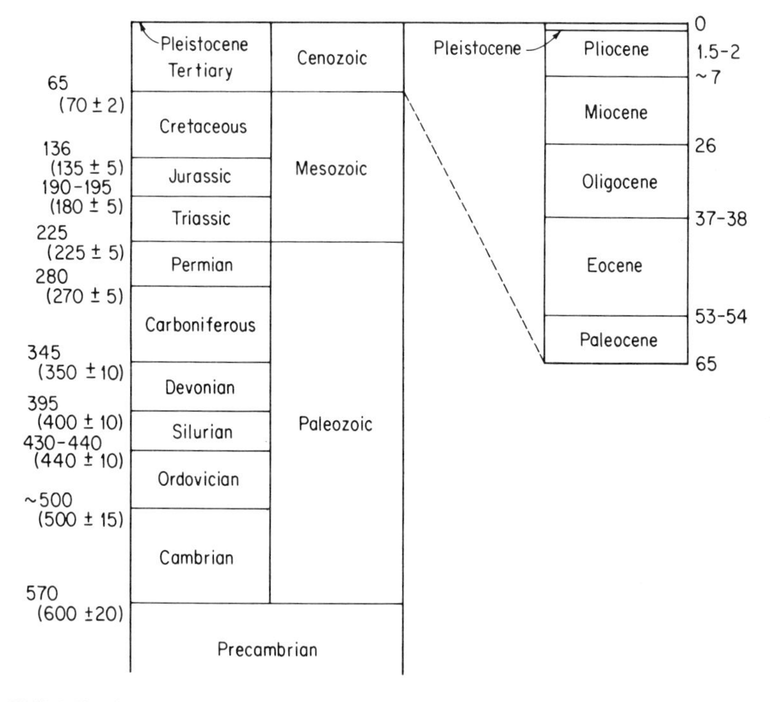

FIGURE 5-1
Time scale since the Precambrian. The dates are shown in millions of years before the present. (*Taken from W. B. Harland et al., 1964. Figures in parentheses are from A. Holmes, 1960.*)

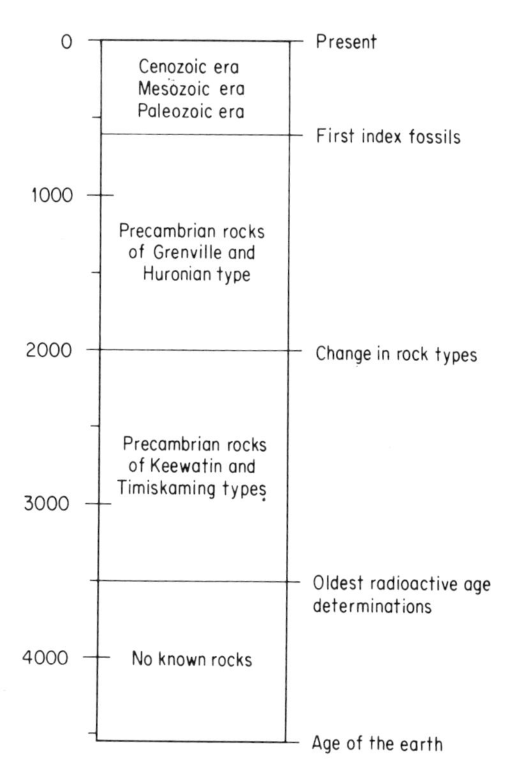

FIGURE 5-2
Gross divisions of the earth's history. The time scale is shown in millions of years. Dates shown for the various events are approximate and somewhat arbitrary.

superposition, which states that in undisturbed strata, the lowest beds are oldest. Paleontologists had already begun to relate particular fossils to individual strata and had found that certain forms were short-lived and always occurred in the same order in stratigraphic sections. These forms, called *index fossils*, have proved to be particularly useful in identifying horizons and in correlating them from place to place.

William Smith (1769-1839) developed these concepts and used them to map a succession of strata across Britain. He showed that the beds overlapped like rows of shingles across a roof, and that each formation preserved the same characteristics and contained many of the same fossils. This established the *law of faunal succession* and initiated the modern era of geological mapping.

In 1830, C. Lyell set forth these ideas in his great book on the principles of geology. With this as a guide, geologists could see what was to be done and enthusiastically began the great task of mapping the earth's land surface, in the firm belief that each geological stratum and system had its own life, lithological characteristics, and boundaries which represented universal dates. The general similarities between Europe and North America fixed the belief in a series of separate, successive creations until, in 1859, C. Darwin published "The Origin of

Species." The acceptance of evolution did not alter the opinion based upon the imperfect observations of that time that each formation had its own fauna and was separated from others by worldwide gaps in the record.

5-3 SUBDIVISIONS OF THE LATTER PART OF GEOLOGICAL TIME

The nineteenth-century geologists believed that they had established that the gaps in the geological record were worldwide and due to intermittent epochs of mountain building and to universal *transgressions* and *regressions* of the sea over the continents. They established a relative time scale based upon these ideas. Table 5-1 lists the principal divisions and shows how the rise to dominance of new forms of life distinguished the eras. Episodes of mountain building in Europe and North America mark, at least approximately, the boundaries between them. The geologists who first recognized the eras and periods regarded them as convenient structural and paleontological units, and this is generally the case in the type localities.

This was the first geological time scale. It was only a relative one, but geologists made reasonable estimates of its absolute length, better than did contemporary physicists who were biased by their ignorance of the heat produced by radioactive decay. It has proved useful and continues to provide the framework for subdivision today.

As geologists studied sections in greater detail, drilled deep basins for petroleum, and extended their work to distant countries, the concept of universal gaps and changes in the succession failed. In different places mountains have been built at different times. Although major transgressions of the sea have been universal, continents and smaller areas have also risen and fallen independently, and strata of the same age change in character from one place to another. A. W. Grabau (1936) emphasized the importance of these distinctions and their application to classify rocks according to their *facies*. These varying characteristics may be compared with those on a present-day coast with contemporaneous stretches of rock cliffs, sand beaches, coral reefs, and swamps. In these neither the materials nor the forms of life are identical, although common species may exist.

Even more puzzling are changes in *faunal realms*. This is well illustrated by the rocks of Cambrian age that border the North Atlantic Ocean. Over the greater part of Europe these strata, regardless of variation in facies, contain at least a few representatives of a fauna named *Paradoxides*, a characteristic genus of fossil trilobites. Over most of North America, strata occupying the same place in the succession contain another fauna, almost entirely different, named after the trilobite *Olenellus*.

That two different faunas have evolved is not surprising, but one might expect that each would be confined to a single continent. This is not the case. It is curious, but in some coastal areas along the margins of both continents the faunal realms are transposed. Thus in Norway and Scotland, the Cambrian faunas are of the North American *Olenellus* realm, while on the east coasts of Canada and New England and in drill holes in central Florida, European forms occur. This presents two problems. Why are some small regions separated by oceans from their main areas? Why on the landward side are these entirely different faunal realms closely juxtaposed without mingling? One possible explanation of this enigma is given in Sec. 13-6.

In seeking to establish a universal time scale it has become recognized that problems have arisen because three different methods have in fact been used to subdivide geological time and that it is a mistake to suppose that all give identical results. These three methods may be called the stratigraphical (or rock-stratigraphic), the paleontological (or time-stratigraphic), and the radiometric methods, according to the materials used (cf. Fig. 5-1 and Table 5-2).

From the viewpoint of a field geologist working in a quarry, on the outcrops of a hillside, or with core from a drill hole, the succession of strata may be most easily divided by changes in lithology, as between sandstone, gray shale, and black shale. The subdivisions so derived are known as stratigraphical, or more precisely as *rock-stratigraphic*, units. They can be traced laterally for distances of up to hundreds of kilometers, and the fundamental unit now recognized is the *formation*, which is defined as a mappable lithological unit.

A formation represents the occurrence of a particular environment; for example, the Potsdam sandstone records the transgression, or spread, of the Cambrian Sea over part of Eastern North America. The transgression did not occur simultaneously in all places, but as the sea advanced, the conditions were uniform for a long time, so that the formation is a rock-stratigraphic unit and not a unit of time. Formations may be divided into members and beds or combined into groups, as shown in Table 5-2. Figure 5-3 shows the relations of formations as recorded in two different areas.

Table 5-2 UNITS OF DIVISION OF TIME AND STRATA

1 Time units based on paleontology (geological time units)	*2* Rock units based on paleontology (time-stratigraphic units)	*3* Rock units based on lithology (rock-stratigraphic units)
Era	....	....
Period	*System*	....
Epoch	Series	Group
Age	Stage	*Formation*
....	Zone	Member, lentil, tongue, bed, stratum, layer

NOTE: In each column the units are in order of decreasing length. There is exact correspondence between units in columns 1 and 2, but only approximate correspondence with column 3. The fundamental unit in each column is in italic.
SOURCE: After H. G. Schenck and S. W. Muller (1941).

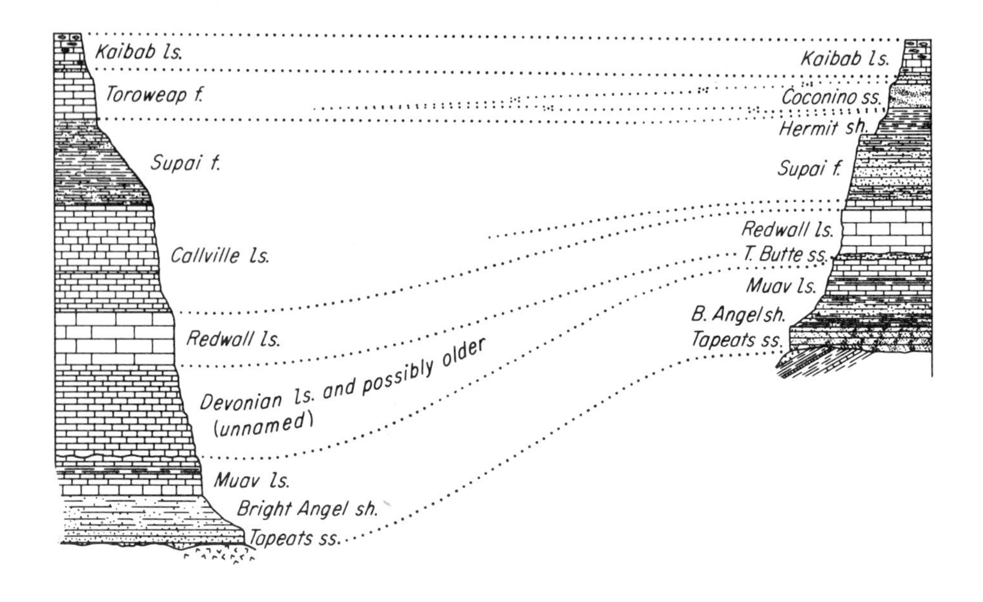

FIGURE 5-3
Two profile sections in the Grand Canyon, illustrating the principles used in correlation. The section at the left is near the mouth of the canyon and that at the right is over 100 miles farther east, near Bass Trail. (*After C. O. Dunbar, 1960.*)

These units provide the basis for local geological mapping because they are immediately recognizable in the field. Such units are not universal, for conditions of deposition change from place to place, altering the aspect, or facies, of contemporary deposits. Thus, while a sand beach is being formed in one place, coral reefs may be growing elsewhere.

The discovery that life has evolved through a succession of types which are never repeated and the realization that many species, especially free-floating, or pelagic, marine forms, have migrated rapidly about the earth relative to the rate of their evolution and the deposition of strata have led to a second method of subdividing rocks. This method has the value of linking together contemporary deposits differing in facies and occurring in places far apart and even on different continents. The subdivision of strata into those containing similar faunas and floras gives rise to paleontological, or *time-stratigraphic*, units. The fundamental world-wide unit is the *system*. A list of systems is given in Table 5-1. The systems are subdivided, but up to the present the smaller units have generally been correlated only within each continent or country.

Both of these methods of subdivision have been standardized in a Stratigraphic Code, and they have now been generally adopted, at least in North

America. They are administered by an American Committee on Stratigraphic Nomenclature, which includes representatives of the Geological Surveys and principal professional societies of the United States and Canada.

The relative geological time scale which was developed over the past century has also been formalized by this committee. Since radiometric ages based on radioactive decay are still few in number, since fossils provide worldwide correlations, and since it seems probable that pelagic marine species of fossils migrated far more rapidly and widely than transgressions of the sea, the paleontological, or time-stratigraphic, units have been chosen as the basis for the relative geological time scale. Corresponding to each time-stratigraphic unit there is a *geological time unit*, as shown in Table 5-2.

To give absolute dates by radioactive methods to this geological time scale has not been an easy task, and minor revisions to those accepted at present are to be expected. Basically, the difficulty is that those rocks which are most easily correlated with the geological succession are technically among the more difficult to date. Occasionally, added confusion has arisen through analytical errors and inadequate geological control of dated samples. Two widely used time scales are those of W. B. Harland (1964) and A. Holmes (1960); these are illustrated in Fig. 5-1.

5-4 ADVENT OF RADIOACTIVE METHODS

For establishing the chronology of younger rocks three methods are available: stratigraphical, paleontological, and radiometric, of which the second is dominant. When we turn to older and unfossiliferous rocks, only the first and third methods can be applied to sedimentary strata and only the last method to basement rocks. Before discussing the chronology of Precambrian rocks, it is therefore appropriate to describe the various radiometric methods for determining the age of rocks and of the earth.

The discovery of radioactivity by H. Becquerel in 1896 provided the means whereby absolute geological ages could be assigned. In 1906, E. Rutherford suggested that lead was produced from uranium and thorium by their decay. Applying this postulate, B. B. Boltwood in the following year used existing analyses of lead, uranium, and thorium in radioactive minerals to estimate the age of minerals. Many of the ages obtained were inaccurate, but they served to indicate that the history of the earth represented at least thousands of millions of years. This estimate was one order of magnitude greater than that which had been demanded by geologists for the formation of the earth's sedimentary rocks, for the accumulation of salt in the oceans, and for the evolution of modern life. It was two orders of magnitude greater than that deduced by Lord Kelvin for the time to exhaust the gravitational potential of the sun or to cool a molten earth to its present temperature. Of course the latter estimates were invalidated when it was discovered that the decay of radioactive elements produces heat. Thus the discovery of radioactivity completely revolutionized ideas about the early history of the earth (cf. J. T. Wilson, R. D. Russell, and R. M. Farquhar, 1956).

Since the early 1900's, the techniques of geochronology have been continually refined, until today ages of minerals or rocks are determined by specialized methods involving elaborate and expensive equipment. Any parent radioactive isotope having a half-life comparable with the age of Precambrian rocks is useful in this respect. Isotopes used are uranium-238, uranium-235, thorium-232, rubidium-87, and potassium-40. A few other long-lived isotopes exist, but they are so rare that their use has not proved practical. Isotopes having shorter half-lives are thorium-230 (ionium), lead-210, and other intermediate members of the uranium-238 decay series, and carbon-14, which is produced in the upper atmosphere by the interaction of cosmic rays with atmospheric nitrogen; all these isotopes are used in special applications. Other naturally occurring radioactive isotopes are either too short-lived or too long-lived to form the basis for geological-age determinations.

In considering age determinations it is important to emphasize that a good result depends first of all upon dependable chemical and isotopic analyses carried out on an unaltered mineral specimen, but this is far from the sole criterion. If the age obtained is to be useful, the exact location of the sample must be known, as well as its geological relationship to its surroundings. A precise age determination of a sample of dubious or unknown origin has no value, except insofar as it may provide an exercise in the application of techniques. In a technically adequate age determination, the full value of the results cannot be realized unless the age is interpreted with a proper understanding of the uses and limitations of radioactive dating methods. Equal attention to source of sample, the analysis, and the interpretation of the result is necessary.

5-5 CLOSED AND OPEN SYSTEMS

Fundamental to the interpretation of radiometric ages, and to the interpretation of common lead and common strontium, is the concept of a *closed system*. (The adjective "common" is used to indicate that the daughter element occurred in an environment free of its parent.) Such a system is understood to have existed during a well-defined interval of time, and during that interval no chemical transfer of parent or daughter across the boundaries of the system can have taken place. By using the equations of radioactive decay and physically determined constants, the change in the daughter/parent ratio may be calculated explicitly and precisely in terms of the bounding times.

Many real systems, minerals and rocks, approximate more or less to these conditions. Others may be equivalent if chemical exchanges do not change the daughter/parent ratio so that the closed-system formulation may apply. Substantial departures from the closed-system condition, such as through the loss of helium from uranium and thorium minerals, would preclude the application of closed-system equations to the interpretation. However, if the loss process is predictable, as in the postulated diffusion of lead from zircons, it may be handled analytically.

The most significant advance in this subject in the past ten years has been the realization that some geochemical systems which cannot be represented as a single closed system can be represented by a succession of two or more closed systems. This has greatly increased the power of geochronological investigations. In the discussion which follows, procedures applicable to closed systems will first be described. More complex models will then be introduced.

5-6 URANIUM METHODS

As well as being the first, age determinations based on the decay of uranium or thorium are still very important. These decay schemes have been well studied by physicists, and the constants are well known, and the presence of two radioactive uranium isotopes and one radioactive thorium isotope usually makes possible the determination of three independent ages for a uranium or thorium mineral. The details of the three decay schemes are shown in Tables 5-3 to 5-5.

Table 5-3 THE URANIUM (U^{238}) SERIES

Isotope†	Particle emitted	Particle energy, Mev	Half-life $T_{1/2}$
${}_{92}U^{238}$ (UI)	α	4.18	4.51 ± 0.01 × 10^9 years
${}_{90}Th^{234}$ (UX_1)	β	0.205, 0.111	24.101 ± 0.025 days
${}_{91}Pa^{234}$ (UX_2)	β	2.32, 1.50, 0.60	1.175 ± 0.003 min
${}_{92}U^{234}$ (UII)	α	4.763	2.475 ± 0.016 × 10^5 years
${}_{90}Th^{230}$ (Io)	α	4.68, 4.61	8.0 ± 0.3 × 10^4 years
${}_{88}Ra^{226}$ (Ra)	α	4.77	1622 ± 1 years
${}_{86}Rn^{222}$ (Rn)	α	5.486	3.825 ± 0.005 days
${}_{84}Po^{218}$ (RaA)	α	5.998	3.050 ± 0.009 min
${}_{82}Pb^{214}$ (RaB)	β	0.65	26.8 ± 0.1 min
${}_{83}Bi^{214}$ (RaC)	α 0.04% β 99.96%	5.46 1.65, 3.17	19.72 ± 0.04 min
${}_{84}Po^{214}$ (RaC′)	α	7.680	163.7 ± 0.2 μsec
${}_{81}Tl^{210}$ (RaC″)	β	1.8	1.32 ± 0.01 min
${}_{82}Pb^{210}$ (RaD)	β	0.018, 0.06	22.5 ± 0.4 years
${}_{83}Bi^{210}$ (RaE)	β	1.17	4.989 ± 0.013 days
${}_{84}Po^{210}$ (RaF)	α	5.298	138.374 ± 0.032 days
${}_{82}Pb^{206}$ (RaG)			Stable

† Original designations are given in parentheses.
SOURCE: J. L. Kulp, G. L. Bate, and W. S. Broecker, reported in J. T. Wilson et al. (1956), with revisions.

At the time when the first uranium-lead ages were made, mass-spectrometric analyses were not available so that it was not possible to distinguish between contamination from radiogenic lead formed within the mineral and common lead incorporated in the mineral when it was formed; nor was it possible to distinguish the leads formed as end products of the uranium and thorium series. The existence of the uranium-235 series as an independent decay series was not known. The theories of Rutherford and F. Soddy had predicted the relationships between the atomic weight of the parent and daughter isotopes and the number of α particles which should accompany the production of each daughter atom. Therefore, the rate of production of lead from both uranium and thorium could be estimated by counting the number of α particles given off by unit quantities of these elements. From this, an approximate formula can be written down for the age of a mineral containing uranium and thorium. With modern constants the equation is

$$t = \frac{7.37 \times 10^9 [Pb]}{[U] + 0.35[Th]} \quad \text{years} \qquad (5\text{-}1)$$

where [Pb], [U], and [Th] are masses of these elements found in a sample of the mineral.

This method is not very satisfactory, despite the fact that such pioneers as B. B. Boltwood, A. Holmes, A. C. Lane, and H. V. Ellsworth used it successfully to

Table 5-4 THE ACTINIUM (U^{235}) SERIES

Isotope†	Particle emitted	Particle energy, Mev	Half-life $T_{1/2}$
$_{92}U^{235}$ (AcU)	α	4.40, 4.58; 4.20, 4.47	$7.13 \pm 0.16 \times 10^8$ years
$_{90}Th^{231}$ (UY)	β	0.094, 0.302, 0.216	25.6 ± 0.1 hours
$_{91}Pa^{231}$ (Pa)	α	5.00	$3.43 \pm 0.03 \times 10^4$ years
$_{89}Ac^{227}$ (Ac)	β 99%; α 1%	0.04; 4.9	22.0 ± 0.3 years
$_{90}Th^{227}$ (RdAc)	α	6.00	18.6 ± 0.1 hours
$_{88}Ra^{223}$ (AcX)	α	5.70	11.2 ± 0.2 days
$_{87}Fr^{223}$	β 99+%	1.15	22 min
$_{86}Rn^{219}$ (An)	α	6.82	3.917 ± 0.015 sec
$_{84}Po^{215}$ (AcA)	α	7.37	$1.83 \pm 0.04 \times 10^{-3}$ sec
$_{82}Pb^{211}$ (AcB)	β	1.39, 0.50	36.1 ± 0.2 min
$_{83}Bi^{211}$ (AcC)	α 99.7%; β 0.3%	6.62, 6.27; 0.06	2.16 ± 0.03 min
$_{84}Po^{211}$ (AcC′)	α	7.43	0.52 ± 0.02 sec
$_{81}Tl^{207}$ (AcC″)	β	1.44	4.79 ± 0.02 min
$_{82}Pb^{207}$ (AcD)			Stable

† Original designations are given in parentheses.
SOURCE: J. L. Kulp, G. L. Bate, and W. S. Broecker, reported in J. T. Wilson et al. (1956), with revisions.

estimate the ages of many primary, unaltered uranium minerals; Eq. (5-1) not only neglects the exponential nature of radioactive decay and the possibility of the presence of common-lead contamination, but it also lumps together the effects of three independent decay series each of which can give an independent age.

For many purposes, the relatively short-lived, intermediate members of the series can be ignored, and the relationships

$$U^{238} \rightarrow Pb^{206} + 8He^4 + 6\beta$$
$$U^{235} \rightarrow Pb^{207} + 7He^4 + 4\beta$$
$$Th^{232} \rightarrow Pb^{208} + 6He^4 + 4\beta$$

are sufficient, because the half-lives ($T_{1/2}$) of the parent isotopes are so long compared with other members. The effect of this approximation is to underestimate ages by about 0.36 m yr and 0.05 m yr for the uranium-238 and uranium-235 ages, respectively. This is usually quite negligible.

As in any simple decay, the rate of disintegration of uranium-238, for example, is given by

$$\frac{d}{dt} N(U^{238}) = -\lambda N(U^{238}) \qquad (5\text{-}2)$$

where λ is used to denote the decay constant of uranium-238, and N denotes the number of atoms of the element shown in the parentheses. λ is obtained by dividing the natural logarithm of 2 by the half-life. Upon integration, Eq. (5-2) becomes

$$N(U^{238}) = N_0(U^{238})\ e^{-\lambda t} \qquad (5\text{-}3)$$

Table 5-5 THE THORIUM SERIES

Isotope†	Particle emitted	Particle energy, Mev	Half-life $T_{1/2}$	
$_{90}Th^{232}$ (Th)	α	3.98	1.39 ± 0.02 × 10^{10}	years
$_{88}Ra^{228}$ ($MsTh_1$)	β	0.012	6.7 ± 0.1	years
$_{89}Ac^{228}$ ($MsTh_2$)	β	1.15	6.13 ± 0.03	hours
$_{90}Th^{228}$ (RdTh)	α	5.42	1.90 ± 0.01	years
$_{88}Ra^{224}$ (ThX)	α	5.68	3.64 ± 0.01	days
$_{86}Rn^{220}$ (Tn)	α	6.28	54.53 ± 0.04	sec
$_{84}Po^{216}$ (ThA)	α	6.77	0.158 ± 0.008	sec
$_{82}Pb^{212}$ (ThB)	β	0.355	10.67 ± 0.05	hours
$_{83}Bi^{212}$ (ThC)	α 35%, β 65%	6.05, 2.25	60.48 ± 0.04	min
$_{84}Po^{212}$ (ThC′)	α	8.78	0.29 ± 0.01	μsec
$_{81}Ti^{208}$ (ThC″)	β	1.79	3.1 ± 0.1	min
$_{82}Pb^{208}$ (ThD)			Stable	

† Original designations are given in parentheses.
SOURCE: J. L. Kulp, G. L. Bate, and W. S. Broecker, reported in J. T. Wilson et al. (1956), with revisions.

Since each decaying uranium-238 atom gives rise to a single lead-206 atom, the number of lead-206 atoms found in a uranium mineral containing $N(U^{238})$ atoms of uranium-238 is readily seen to be

$$N(Pb^{206}) = N_0(U^{238}) - N(U^{238}) = N(U^{238})(e^{\lambda t} - 1) \tag{5-4}$$

Completely analogous equations can be written for the other two series, namely,

$$N(Pb^{207}) = N(U^{235})(e^{\lambda' t} - 1) \tag{5-5}$$

and

$$N(Pb^{208}) = N(Th^{232})(e^{\lambda'' t} - 1) \tag{5-6}$$

where λ' is the decay constant for uranium-235, and λ'' that for thorium-232. By solving these equations for t, substituting the appropriate numerical constants, and adapting them for the use of gram quantities rather than numbers of atoms, the following equations are obtained:

$$t_{206} = 6.50 \times 10^9 \log_e \left(1 + 1.158 \frac{[Pb^{206}]}{[U]}\right) \tag{5-7}$$

$$t_{207} = 1.03 \times 10^9 \log_e \left(1 + 159.6 \frac{[Pb^{207}]}{[U]}\right) \tag{5-8}$$

$$t_{208} = 20.0 \times 10^9 \log_e \left(1 + 1.121 \frac{[Pb^{208}]}{[Th]}\right) \tag{5-9}$$

$[Pb^{206}]$, $[Pb^{207}]$, and $[Pb^{208}]$ refer to the mass of lead multiplied by the atom fraction of the corresponding isotope. The total mass of uranium is $[U]$.

Equations (5-7) to (5-9) are approximate to the extent that an average value for the atomic weight of radiogenic lead is assumed. Dividing Eq. (5-5) by Eq. (5-4) yields an equation that relates only ratios of isotopes of the same elements. Thus

$$\frac{N(Pb^{207})}{N(Pb^{206})} = \frac{e^{\lambda' t} - 1}{e^{\lambda t} - 1} \frac{N(U^{235})}{N(U^{238})} \tag{5-10}$$

Ages obtained by Eqs. (5-7) to (5-9) are known as *uranium-lead* and *thorium-lead* ages, whereas those obtained from Eq. (5-10) are known as *lead-lead*, or *lead-ratio*, ages.

For a mineral formed initially with a reasonable concentration of uranium and thorium, and with an amount of common lead small enough so that a satisfactory correction can be made for its presence, the above methods yield a thorium age and three uranium ages. Of these the thorium age and any two of the uranium ages can be considered independent, while the third uranium age can be calculated if the other two are known.

The nature of Eq. (5-10), which is used for calculating lead-ratio ages, is such that the ratio varies very little with age for young minerals. This can be seen in

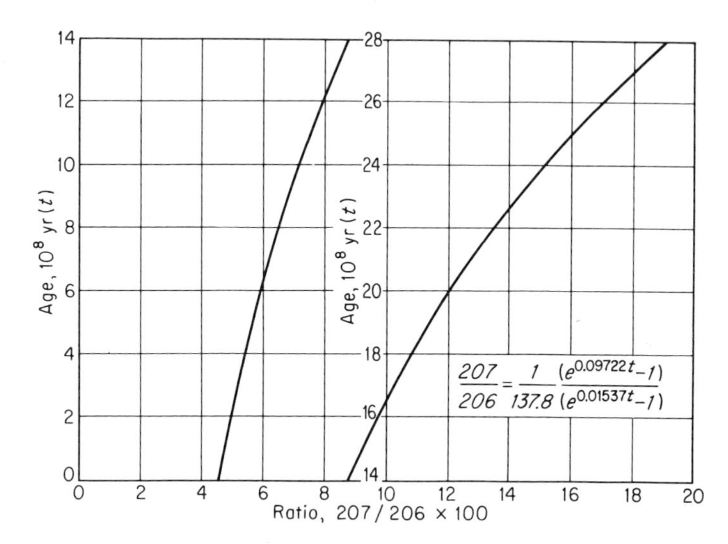

FIGURE 5-4
Curve for dating uranium minerals by the lead-ratio method.

Fig. 5-4, which shows the ratio of the radiogenic isotopes plotted against the ages of uranium minerals. In general, it is considered that this method is not very accurate for minerals younger than Cambrian (500 m yr), and that for minerals younger than Permian (200 m yr) the lead-ratio method is much less reliable than the other uranium-thorium-lead methods.

The following figures are an example of the calculation of the age of a uranium-thorium mineral, using data taken from an article by A. Holmes (1954) on age determinations from Rhodesia. It is a useful exercise to check the result of this calculation with that obtained by applying the approximate formulas (5-7) to (5-9).

Composition (wt % of sample):	*Lead-isotope abundances (atom % of total lead):*	*Assumed isotope ratios of nonradiogenic-lead contamination:*
U = 0.074	$Pb^{204} = 0.0066$	$Pb^{206}/Pb^{204} = 13.5$
Th = 2.39	$Pb^{206} = 9.49$	$Pb^{207}/Pb^{204} = 15.0$
Pb = 0.34	$Pb^{207} = 1.71$	$Pb^{208}/Pb^{204} = 33.5$
	$Pb^{208} = 88.80$	

Isotopic abundances corrected for nonradiogenic lead (%):

$Pb^{204} = 0$

$Pb^{206} = 9.49 - 13.5 \times 0.0066$
$= 9.40$

$Pb^{207} = 1.71 - 15.0 \times 0.0066$
$= 1.61$

$Pb^{208} = 88.8 - 33.5 \times 0.0066$
$= 88.6$

Atomic weight of radiogenic lead:

$$9.40 \times 206 = 1{,}936.4 \qquad \frac{88.6 \times 208 = 18{,}428.8}{99.6 \times \text{at. wt} = 20{,}698.5}$$

$$1.61 \times 207 = 333.3$$

$$\text{At wt} = \frac{20{,}698}{99.6} = 207.8$$

Number of moles of isotopes per 100 g:

$$U^{235} = \frac{0.074}{235}\,\frac{1}{138.8} = 2.27 \times 10^{-6}$$

$$U^{238} = \frac{0.074}{238}\,\frac{137.8}{138.8} = 3.08 \times 10^{-4}$$

$$Th^{232} = \frac{2.39}{232} = 1.03 \times 10^{-2}$$

$$Pb^{206} = \frac{0.34}{207.8}\,0.0940 = 1.54 \times 10^{-4}$$

$$Pb^{207} = \frac{0.34}{207.8}\,0.0161 = 2.63 \times 10^{-5}$$

$$Pb^{208} = \frac{0.34}{207.8}\,0.886 = 1.45 \times 10^{-3}$$

$$\frac{Pb^{206}}{U^{238}} = \frac{1.54 \times 10^{-4}}{3.08 \times 10^{-4}} = 0.500$$

$$\frac{Pb^{207}}{U^{235}} = \frac{2.63 \times 10^{-5}}{2.27 \times 10^{-6}} = 11.6$$

and

$$\frac{Pb^{207}}{Pb^{206}} = \frac{1.61}{9.40} = 0.171$$

$$\frac{Pb^{208}}{Th^{232}} = \frac{1.45 \times 10^{-3}}{1.03 \times 10^{-2}} = 0.141$$

Thus from Pb^{206}: $t = \dfrac{10^9}{0.1537} \log_e 1.500$ years = 2640 m yr

from Pb^{207}: $t = \dfrac{10^9}{0.9722} \log_e 12.6$ years = 2610 m yr

from Pb^{207}/Pb^{206}: t = 2610 m yr (graphically)

from Pb^{208}: $t = \dfrac{10^9}{0.0499} \log_e 1.141$ years = 2640 m yr

The development of methods of geochronology has in large part been made possible by the enormous advances in the design of the mass spectrometer, which is one of the most accurate and sensitive tools of modern geology. One of the most impressive advances in mass spectrometry has been the development of instruments capable of performing accurate isotopic analyses with extremely small samples. Some mass spectrometers are so sensitive that a microgram of certain elements represents an enormous sample in terms of their limits of sensitivity. Unfortunately, lead is a difficult element to analyze isotopically, and large samples are required for precise work. Even so, the technique makes possible precise measurements of the distribution and concentrations of uranium, thorium, and lead in common igneous rocks.

When igneous rocks crystallize, uranium, thorium, and lead will be present, along with other elements. As crystallization proceeds, the potassium-bearing minerals, including certain feldspars and most micas, are among the first to appear. Because of the similarity of the ionic radius of lead (1.3 A), it is possible for the lead present to be readily accommodated in the lattices of the potassium minerals, which therefore act as a sink for the element. However, uranium and thorium form smaller ions (1.0 and 1.1 A, respectively) and are rejected from the lattices of the common silicates; they are therefore concentrated into the accessory minerals zircon, sphene, and apatite. In rocks containing these minerals (i.e., the more acidic of the igneous rocks) there is a natural separation of minerals containing primary lead but no uranium or thorium, and minerals containing uranium and thorium but negligible primary lead. Some minerals separated from a granite can be used for the determination of excellent uranium-lead ages, while others can be used to study variations in the isotope ratios of common lead, a subject which will be discussed in Chap. 6. The first intensive study of separated minerals was completed in 1954 by G. R. Tilton et al. (1955).

Various modifications of the uranium and thorium methods have been proposed; some of them are quite useful. The measurement of helium in the minerals as an alternative to the measurement of lead has generally met with failure because of the almost universal loss of helium from radioactive minerals, but it may still prove useful for dating certain minerals having very-close-packed structures and low radioactivities. The measurement of an intermediate member of the decay series has been proposed as an alternative to measuring the parent isotope: A particularly interesting example of this approach is lead-210, a member of the uranium-238 series. Although generally subject to the same limitations as the usual uranium-238 method, the lead-210 method can provide additional information about the state of radioactive equilibrium in a mineral. Moreover, the method avoids the need for making chemical analyses, since the lead-210 can be measured by observing its radioactivity.

5-7 POTASSIUM METHODS

The potassium methods deserve special consideration, for the very important reason that potassium minerals are among the most common rock-forming minerals. The application of this method has been delayed until recent years because the decay constants were not known, and adequate techniques for extracting and measuring very small amounts of radiogenic argon from minerals and rocks had not been developed.

The potassium-40 nucleus is doubly unstable, there being two modes of decay which occur with relatively high frequency. These modes correspond to the emission of a β particle, with the consequent production of a calcium-40 nucleus, or the absorption into the nucleus of an orbital electron, with the consequent production of an argon-40 nucleus. In the latter case the energy balance is obtained by the emission of a penetrating γ ray and an X-ray which may be converted within the atom by interaction with a second orbital electron to produce a low-energy *Auger electron*. The equations for the decay can be written

$$K^{40} + e \rightarrow Ar^{40} + \gamma + \text{X-ray or Auger electron}$$

or

$$K^{40} \rightarrow Ca^{40} + \beta^- \tag{5-11}$$

Since there are two alternative ways by which potassium-40 atoms may decay, there are two decay constants λ_β and λ_e, which are proportional to the probabilities per year that a potassium-40 atom will decay to a calcium-40 nucleus or an argon-40 nucleus, respectively.

The ratio of the second decay constant to the first, which is also the ratio of the two daughters produced by the decaying parent, is designated the branching ratio. Any two of these three constants are sufficient to enable the calculation of the total amount of either daughter in a mineral of age t. The two which will be used here are the decay constants λ_β and λ_e. In terms of these constants, the production of the radiogenic daughters is given by

$$N(Ca^{40}) = \frac{\lambda_\beta}{\lambda} N(K^{40})(e^{\lambda t} - 1) \tag{5-12}$$

and

$$N(Ar^{40}) = \frac{\lambda_e}{\lambda} N(K^{40})(e^{\lambda t} - 1) \tag{5-13}$$

where

$$\lambda = \lambda_\beta + \lambda_e$$

L. T. Aldrich and G. W. Wetherill (1958) have reviewed the physical measurements and suggest values of $\lambda_\beta = 0.472 \times 10^{-9}$ year^{-1} and $\lambda_e = 0.0585 \times 10^{-9}$ year^{-1}. These values are also closely supported by A. G. Smith (1964) and have been widely adopted. R. D. Beckinsdale and N. H. Gale (1969) propose slightly different values of $\lambda_\beta = (0.4905 \pm 0.0009) \times 10^{-9}$ year^{-1} and $\lambda_e = (0.0567 \pm 0.00035) \times 10^{-9}$ year^{-1}. For young ages, less than a few hundred million years, the Ar^{40}/K^{40} ratio is substantially independent of the value of λ_β,

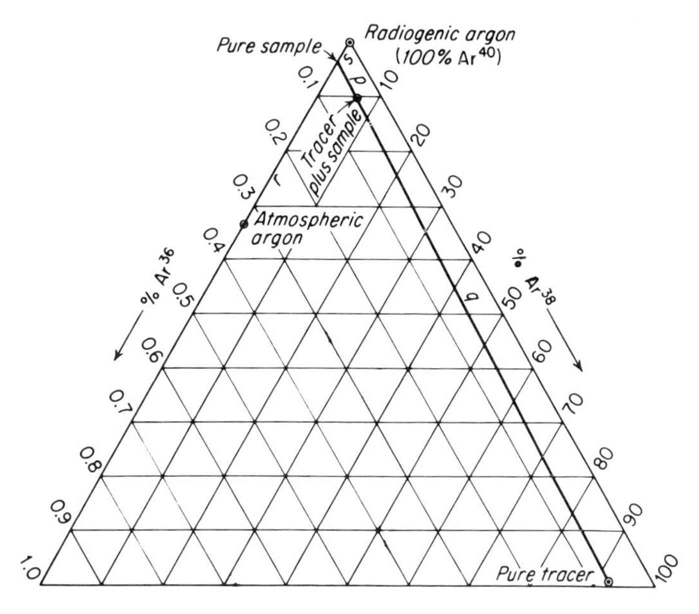

FIGURE 5-5
Plot showing the isotopic composition of argon samples analyzed in the course of a potassium-argon age determination.

and the age obtained is nearly inversely proportional to λ_e. Thus, the two sets of constants shown above give ages differing by 3 percent for young rocks and minerals but less than 3 percent for older materials.

The following analytical data obtained by M. Ozimà for a sample of mica separated from a boulder in the Bruce conglomerate in Ontario serve to illustrate the calculation of a potassium-argon age. Three grams of the mica were

Gas origin	Tracer	Sample plus tracer	Atmospheric argon
Ar^{36}%	0.0728	0.039	0.337
Ar^{38}%	91.822	6.22	0.063
Ar^{40}%	8.105	93.74	99.60

heated in an evacuated system until the mica was entirely molten and the entrapped gases released. The vacuum system also contained a tracer consisting of 1.197 x 10^{-4} cm^3 (STP) of argon. The tracer consisted primarily of argon-38. Isotopic analyses were carried out on the tracer and on the mixture of tracer with the gases from the mica, and the analyses were compared with the isotopic abundances of atmospheric argon. The compositions in the above table are shown plotted on the ternary diagram in Fig. 5-5. The distances *p, q, r*, and *s* measured on the diagram (in arbitrary units) can be used to calculate the volume of radiogenic argon-40. Thus

$$\text{Volume of ``pure sample''} = \frac{q}{p} \times 1.197 \times 10^{-4} \quad \text{cm}^3 \text{ (STP)}$$

$$\text{Volume of radiogenic argon} = \frac{r}{r+s} \times \text{vol. of ``pure sample''}$$

$$= \frac{rq}{p(r+s)} \times 1.197 \times 10^{-4} \quad \text{cm}^3 \text{ (STP)}$$

$$= \frac{30.1 \times 87.3}{33.7 \times 6.35} \times 1.197 \times 10^{-4} \quad \text{cm}^3 \text{ (STP)}$$

$$= 1.47 \times 10^{-3} \quad \text{cm}^3 \text{ (STP)}$$

$$\text{Mass of radiogenic argon} = \frac{1.47 \times 10^{-3}}{22{,}400} \times 40 \text{ g}$$

$$= 2.63 \times 10^{-6} \text{ g}$$

Therefore 1 g mica contains 8.76×10^{-7} g radiogenic Ar^{40}.

At the same time that the argon was released by melting the sample, the potassium was distilled out as the chloride. It was collected and analyzed with a flame photometer, giving a potassium content of 3.98 percent by weight. But the proportion of potassium-40 in potassium is 0.0119 atom % (A. O. Nier, 1950), which is 0.0122 wt %. Therefore, 1 g mica contains

$$0.0122 \times 3.98 \times 10^{-4} = 4.86 \times 10^{-6} \text{ g } K^{40}$$

Therefore

$$\frac{Ar^{40}}{K^{40}} = 0.180 \text{ atom per atom}$$

and

$$t = \frac{1}{\lambda} \log_e \left(1 + \frac{\lambda}{\lambda_e} \times 0.180\right)$$

Adopting the values

$$\lambda = 0.530 \times 10^{-9} \text{ year}^{-1}$$

$$\lambda_e = 0.0585 \times 10^{-9} \text{ year}^{-1}$$

results in

$$t = 1820 \text{ m yr}$$

Because heating of minerals and rocks to only a few hundred degrees centigrade will often result in the loss of substantial amounts of radiogenic argon, the method is often mistrusted in areas of complex history or where significant metamorphisms are thought to have followed the original crystallization. It is common practice to support potassium-argon age determinations by rubidium-strontium measurements, wherever this is practical.

This method is particularly valuable for young minerals, as young as 100,000 years or less. Its outstanding achievement has been the dating of reversals of the earth's magnetic field, a knowledge of which has led to dramatic proof of the spreading of the ocean floors outward from mid-ocean ridges (Chap. 14). The decay to calcium-40 is seldom used because of the difficulty of finding suitable minerals sufficiently poor in common calcium.

5-8 THE RUBIDIUM METHOD

After a rather uncertain beginning, age determinations based on the decay

$$Rb^{87} \rightarrow Sr^{87} + \beta^-$$

have rapidly become accurate and valuable. Early spectroscopic measurements were imprecise because of lack of suitable standards. Later measurements were made by adding to the sample known amounts of a less abundant isotope to serve as an internal standard (Sec. 5-7). Some of these early *isotope-dilution* measurements were accompanied by substantial analytical errors. However, rapid advances in modern solid-source mass spectrometry have brought simple and effective techniques for making this type of age measurement.

Of all the methods discussed, this decay is the simplest, and it may seem surprising that knowledge of a reliable decay constant has been a major difficulty. To obtain its value, one must make an absolute count of the rate at which β particles are emitted by a known mass of rubidium. This experiment is made very difficult by the shape of the energy spectrum of the particles. A large proportion of the particles have low energies, and extrapolation of experimental measurements to include these is difficult. As was also true for potassium, many of the early measurements were greatly in error, and values published as recently as 1961 show considerable scatter. The choice of the half-life for age-determination purposes usually lies between the value obtained by L. T. Aldrich et al. (1956), who demanded agreement between rubidium dates and concordant uranium dates, and the value obtained by counting experiments of K. F. Flynn and L. E. Glendenin (1959). These values are, in units of 10^{10} years, 5.0 ± 0.2 and 4.70 ± 0.05, respectively. The lower value claims the greater precision and is free from the geological uncertainty implicit in equating the ages of different minerals. In addition, it is supported by the value of $(4.77 \pm 0.10) \times 10^{10}$ years obtained by A. Kovach (1964), who used a similar counting technique, and by the value of $(4.72 \pm 0.04) \times 10^{10}$ years obtained by C. C. McMullen et al. (1966), who measured the rate of radiogenic accumulation of strontium-87 under laboratory conditions. The decay constant is obtained by dividing 0.693 by the half-life.

Often whole rocks are used for rubidium-strontium age determinations, because they are more likely than separated minerals to fulfill the closed-system requirement. The contamination by common strontium originally incorporated into the rock is often large, and the following graphical technique is frequently used to obtain the age. The observed ratio Sr^{87}/Sr^{86} is related to the original ratio

$(Sr^{87}/Sr^{86})_i$, under closed-system assumptions, by the equation

$$\frac{Sr^{87}}{Sr^{86}} = \left(\frac{Sr^{87}}{Sr^{86}}\right)_i + \frac{Rb^{87}}{Sr^{86}}(e^{\lambda t} - 1) \qquad (5\text{-}14)$$

since strontium-86 is nonradiogenic. Plotting the observed ratios Sr^{87}/Sr^{86} against Rb^{87}/Sr^{86} for a series of samples of common genesis should result in a straight line if all the assumptions are valid. The slope of the line is $(e^{\lambda t} - 1)$, from which t can be calculated; the intercept gives directly the original ratio of the strontium isotopes.

5-9 FISSION–TRACK DATING

Any natural process which results in a recognizable, unidirectional change at a predictable rate can, in principle, be used for age determination. Most of the physical methods of geochronology depend on radioactive decay, which is such a process. Another process which is important for the heavy elements is natural fission. The decay of a parent isotope by fission follows exponential laws of the same form as those used to describe radioactive decay. V. G. Khlopin and E. K. Gerling (1947), by determining the ratio xenon/uranium in uranium minerals, obtained the first geological ages based on the fission process.

The use of fossil tracks for geological-age determination was proposed by P. B. Price and R. M. Walker (1963). The method is based on the fact that the fission of a uranium-238 nucleus results in a production of energetic, massive particles which create damaged regions in mica and other minerals. The width of the tracks is about 10 mμ, and typical lengths are of the order of 10 μ. These can be observed with an electron microscope. However, a simpler technique is to etch the sample with hydrofluoric acid, whereupon the tracks grow in size until they become easily visible with an optical microscope. Only particles more massive than 30 amu (atomic mass units) leave visible tracks, and therefore the technique is insensitive to protons, α particles, and nuclei of small mass.

To determine the concentration of uranium, the samples are irradiated artificially with thermal neutrons. This produces additional tracks from the induced fission of uranium-235. The number of these tracks gives a measure of uranium concentration in the sample. Thus, to measure the age of a mineral by this technique, two counts must be made of the fission tracks, one before and the other after thermal-neutron irradiation. To avoid surface effects, the samples are cut and polished before etching and counting.

The rate of fission of uranium-238 atoms can be described by a rate constant λ_F, which is a quantity of the same type and dimensions as the normal radioactive-decay constant λ_D for that isotope. Since $\lambda_F << \lambda_D$, the number of U^{238} atoms which have decayed since time T is the usual expression U^{238} $(e^{\lambda_D T} - 1)$. The fraction that have decayed by fission is

$$\frac{\lambda_F}{\lambda_F + \lambda_D} \approx \frac{\lambda_F}{\lambda_D}$$

Thus the relationship between fission-track densities and the age of the sample is given by the equations

$$\rho = \frac{2\lambda_F}{\lambda_D} \phi(r_0) \cdot U^{238}(e^{\lambda_D T} - 1)$$

$$\Delta\rho = 2\phi(r_0) \cdot U^{235} \cdot n_t \cdot \sigma$$

and therefore

$$\frac{\rho}{\Delta\rho} = \frac{137.8\lambda_F(e^{\lambda_D T} - 1)}{\lambda_D n_t \sigma}$$

where ρ = natural-fission-track density

$\Delta\rho$ = increase in track density due to irradiation with slow neutrons

U^{235}, U^{238} = number densities of the appropriate atoms (U^{238}/U^{235} = 137.8)

$\phi(r_0)$ = a geometrical factor depending on the range r_0 of the particles and the characteristics of the etching process (r_0 = range)

σ = thermal-neutron cross section for uranium-235

n_t = thermal-neutron dose

λ_F = $(6.85 \pm 0.20) \times 10^{-17}$ year^{-1} (R. L. Fleischer and P. B. Price, 1964)

λ_D = 1.537×10^{-10} year^{-1} (Table 5-3)

The method is based on two important assumptions: that the uranium distribution in the sample is uniform and that spontaneous fission of uranium-238 is the dominant source of tracks. Price and Walker (1963) showed that fission induced by high-energy nucleons and π mesons in cosmic rays is unlikely to be significant for samples of geological importance. Also, they argued that fission of thorium is unlikely to be significant for the thorium/uranium ratios expected in the minerals to be dated. The tracks are annealed by quite short exposures to temperatures in the range 300 to 400°C, a fact which has to be kept in mind when considering ages obtained by this method.

The fission-track method is becoming a valuable tool for geochronology. The method can be applied to minerals in nearly every igneous rock in the age range from Precambrian to historic times.

5-10 RADIOCARBON AND TRITIUM METHODS

All the methods discussed so far have depended on the presence in the earth of radioactive isotopes with half-lives so long that the isotopes have not entirely decayed since the time when the earth was formed. However, there are also present in the earth isotopes which have much shorter half-lives but which are found in nature because they are continually being produced by natural processes. Obvious examples are the intermediate radioactive members of the uranium and thorium decay series which are found in minerals containing the radioactive parents. Others may be formed high in the earth's atmosphere as a result of the bombardment of gases by secondary cosmic rays.

Of great importance is the production of carbon-14 from nitrogen-14 by a neutron-proton reaction. Owing to the rapid circulation of the atmosphere, radioactive carbon is rapidly mixed into the active carbon reservoir, including the organic substances forming many living materials. As long as the animal or plant remains alive, the carbon is regularly exchanged with its surroundings and the supply of radioactive carbon-14 is kept replenished. Therefore the proportion of carbon-14 in the carbon in living materials is constant to within very close tolerances. However, when any living material dies, its exchange of carbon with the active reservoir stops, and the radiocarbon supply is cut off. The radioactivity of the carbon in the material, which is a measure of the carbon-14 content, no longer remains constant, but decays with time. The half-life of carbon-14 is 5730 years.

The age of the samples to which this method can be applied is determined by the sensitivity of the instruments used to measure the radioactivity. The techniques involved consist of very-low-level counting of β particles of very small energy. Discrimination against background is the major problem, and a number of procedures have been developed to solve it. Early measurements were made by counting β particles from a layer of elemental carbon deposited on the inside walls of a proportional counter. Later methods have used the solution of gases containing carbon, such as acetylene, in an organic solvent containing a liquid scintillator. The manufacture of organic liquids containing the carbon to be studied has also been attempted, and again liquid scintillators are used. The counters may be shielded with concrete, iron, lead, or mercury and may use anticoincidence circuits, with a ring of Geiger tubes surrounding the counting tube to greatly reduce the normal background. The oldest samples to which the method can be applied are about 50,000 years, so that the method is useful only for the study of extremely recent geological events. It has been of the greatest value in archeology.

Samples used for carbon-14 dating must satisfy the condition that the carbon now present is the same carbon which was present in the material at its death, i.e. that there has been no chemical or isotopic exchange of carbon with its surroundings. Experience has shown that substances of a chalky nature are generally poor choices. It is interesting to note that the carbon-14 content of the active reservoir has been significantly diluted by carbon dioxide produced by the combustion of large amounts of fuels since the Industrial Revolution. Examination of tree rings for the past 50 years shows a significant decrease in the relative abundance of radiocarbon. The abundance of carbon-14 is also modified through nuclear explosions and, possibly, through changes in the intensity of the earth's magnetic field.

Tritium, a radioactive isotope of hydrogen, is also produced in the upper atmosphere as the result of cosmic-ray activity. Even its very short half-life, 12.5 years, provides sufficient time for it to mix extensively with a considerable part of the active hydrogen reservoir. The activity of the hydrogen may be used to estimate the rate of mixing of lakes and oceans, to study the flow of underground rivers, and to solve other problems of a similar nature. However, tritium is a product of nuclear explosions, and this has severely limited its usefulness in geological-age studies, although by studying the distribution of bomb-produced tritium, much has been learned about the distribution of surface waters. The activity of this element is so weak that the proportion of tritium in a water sample must be artificially increased by distillation before it can be measured.

5-11 INTERPRETATION OF DISCORDANT AGES

If the ages obtained from the isotope ratios Pb^{206}/U^{238} and Pb^{207}/U^{235} agree (and preferably agree also with that calculated from Pb^{208}/Th^{232}), or if the age calculated from Ar^{40}/K^{40} agrees with that from Sr^{87}/Rb^{87}, the ages are termed *concordant.* Concordance provides convincing evidence for the validity of the closed-system assumption. When the ages are *discordant,* one or more of the systems have been open.

Episodic loss of parent or daughter is a likely reason for discordance. Particularly in the case of potassium-argon ages, heating of the minerals can result in loss of part or all of the daughter isotope. In this case the age will indicate a value intermediate between the time of primary crystallization and the time of loss. Zircons are very resistant to weathering, and their ages, as deduced from lead/uranium isotope ratios, are less likely to be modified by a metamorphic event, although this can happen. Figure 5-6 illustrates the effect on the calculated ages of a typical sample produced by losses of lead and uranium.

When the assumption of a single closed system fails to give concordant results, the assumption of consecutive closed systems may be useful. Suppose that a geologically coherent structure contains rocks deposited at a common initial time. Suppose, further, that there has been a subsequent metamorphism that *completely* homogenized the strontium in any individual rock sample but did not result in the exchange of strontium or rubidium between rock samples. A plot of Sr^{87}/Sr^{86} against Rb^{87}/Sr^{86} for those rocks which have remained closed systems throughout their history will give a straight line as before, and from the slope of the line, the

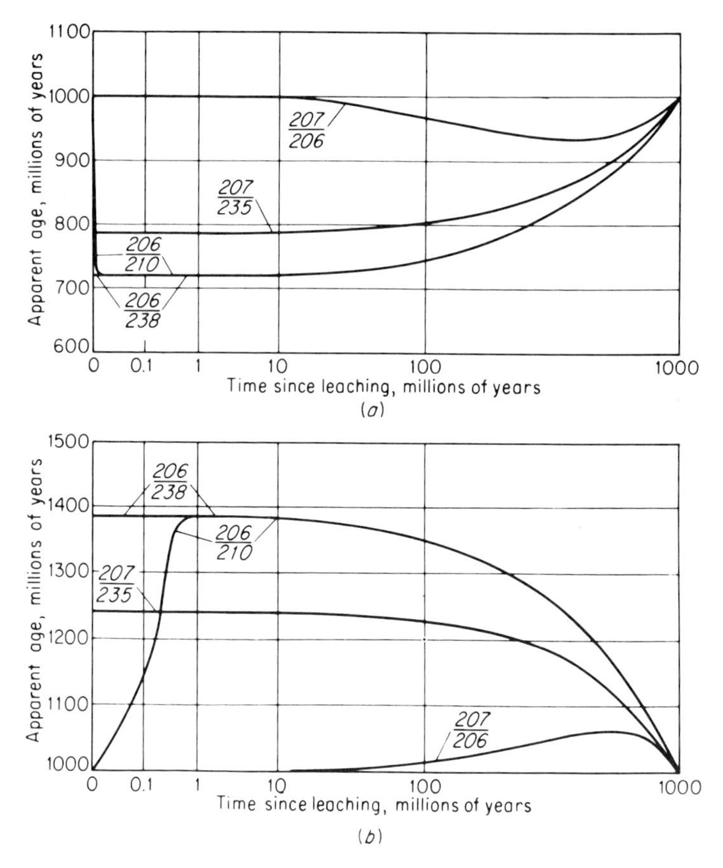

FIGURE 5-6
(*a*) The effect on apparent ages of leaching 30 percent of lead at various times from a uranium mineral 1000 m yr old. (*b*) The effect on apparent ages of leaching 30 percent of uranium at various times from a uranium mineral 1000 m yr old (assuming no thorium leaching).

age of the rock can be calculated (Fig. 5-7). The minerals separated from any one rock will give isotope ratios which will fall on a line of smaller slope; the time since the metamorphism can be calculated from this smaller slope. The correctness of the assumed two-stage model can be tested by comparing the results from several rocks, and if the ages of the metamorphic event agree, two events in the history of the rock have very probably been dated correctly. The method fails if the rock was involved in additional events, or if the metamorphic event did not thoroughly homogenize the strontium in each rock sample, or if the rock samples did not remain closed. That this can happen is clearly shown in Fig. 5-9.

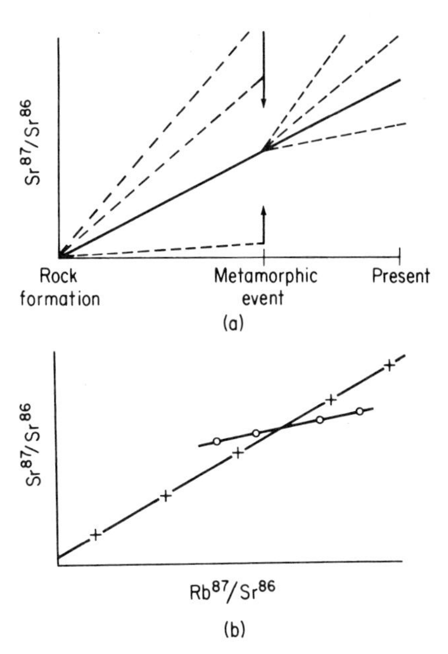

FIGURE 5-7
(*a*) Assumed variation with time of the ratio Sr^{87}/Sr^{86} for a whole rock (solid line) and separate minerals (broken lines). (*b*) Expected isotopic pattern for such rocks measured at the present time. The crosses indicate points for various whole rocks, and the circles show minerals separated from a single rock. (This plot is known as the BPI plot, after the Bernard Price Geophysical Institute, Johannesburg.)

T. J. Ulrych and P. H. Reynolds (1966) have reported an application of the same technique to uranium-lead age determinations from the Llano uplift, Texas. In this way, they were able to identify common-lead and radiogenic-lead components and to interpret each independently. The interpretation was internally consistent and in agreement with independent age evidence.

For the uranium methods, interpretations are more often based on the *concordia* plot devised by G. W. Wetherill (1956). This plot is the locus of Pb^{206}/U^{238} and Pb^{207}/U^{235} values giving concordant ages; age values can be indicated along the locus. Figure 5-8 shows an example of such a plot. A measured point for a discordant sample will lie off the curve. Lines projected from the point to the curve parallel to the axes will indicate two lead-uranium ages. A line drawn through the origin and through the point will intersect the concordia curve at the lead-ratio age.

This plot is a linear compositional diagram, and two-component mixtures will plot on it as straight lines. Suppose that a zircon has experienced a two-stage history; i.e., suppose that it existed as a closed system from its time of formation to a time of metamorphism. The resulting isotope ratio must lie along a line between the two points on the concordia corresponding to the two events in the mineral's history, or on its extension. This is necessarily so because the system is in effect a mixture of components which are the concordant ratios for the two times. A suite of zircons from the same rock unit should give isotope ratios which, if not all the same, will

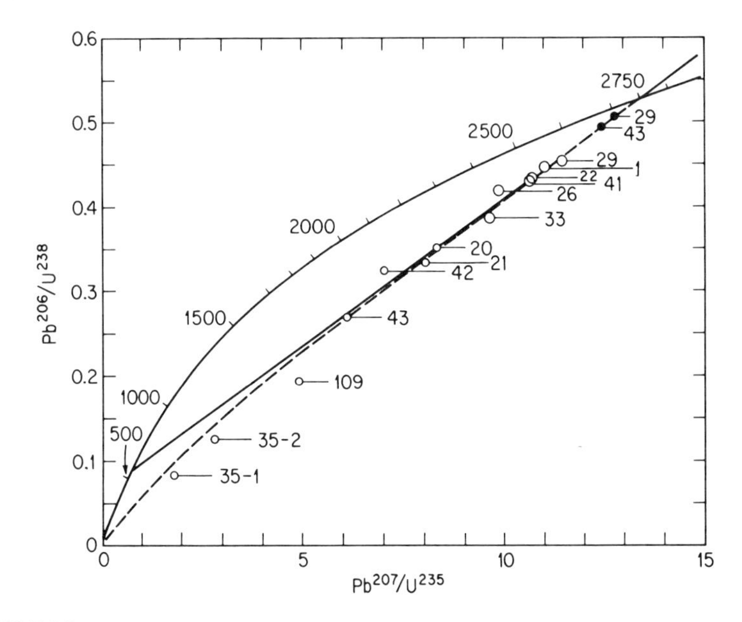

FIGURE 5-8
Example of a concordia plot for cogenetic zircons (open circles) and sphenes (solid circles) from Rainy Lake region of Ontario. (*After S. R. Hart and G. L. Davis, 1969.*)

indicate the position of the chord and, by the intersections with the concordia, the times of the two events.

It is clearly of major importance to establish whether the younger intercept with the concordia represents a real event. A word of caution on this point was given by R. D. Russell and L. H. Ahrens (1957), who pointed out that for uraninites and pitchblendes the two intercepts seemed closely correlated, which appears to be a strange coincidence. The possibility of a continuous loss mechanism was proposed, possibly associated with nuclear recoil following α-particle decay. Such a mechanism would explain the preferential loss of lead-207 which is usually observed. G. R. Tilton (1960) has shown that volume diffusion can give rise to loss patterns similar to those observed, the preferential loss of lead-207 arising from the greater age of the daughter atoms in the radioactive mineral. L. T. Silver (1963) has shown that in at least some cases a diffusion mechanism does not provide an adequate explanation for the experimental observations. More recently, R. H. Steiger and G. J. Wasserburg (1969) have made important contributions to the understanding of the concordia diagram. A full understanding of the processes involved has not yet been reached.

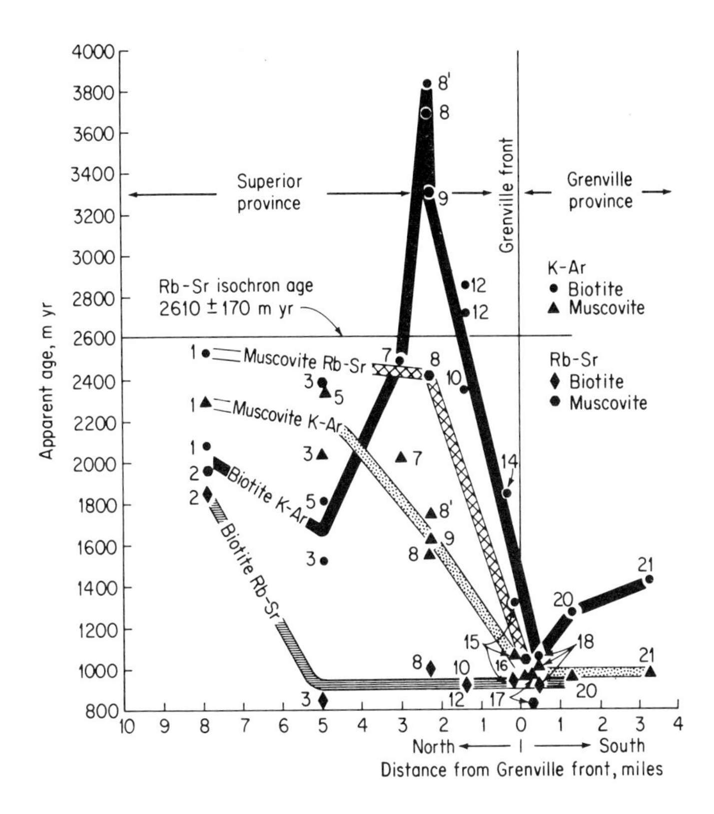

FIGURE 5-9
The distribution of apparent isotopic ages north and south of the Grenville Front, near Chibougameau, Quebec. (*After R. K. Wanless et al., 1969.*)

5-12 THE AGE OF THE EARTH

The "age of the earth," designated by the symbol t_0, is an important parameter in the discussion of common-lead-isotope abundance interpretations, which will be considered in the next chapter. Because of this, and because it is an interesting subject in its own right, it is worth discussing in some detail.

In order to be dated, any event must have had physical consequences, the effects of which can be studied today. Many such events have taken place on the earth, and these make possible the dating of individual rocks in minerals through radioactive-decay processes. To date the earth as a whole, one must seek to identify some event of much broader significance. Most successful have been attempts at dating the time when various meteorites were separated into individual isolated systems. Evidence suggests that meteorites are fragments of original members of our solar system (Sec. 1-5), and a reasonable assumption is made that the event which separated them is closely contemporaneous with the time when the earth took its present form. This assumption is supported by current studies of terrestrial common leads, which indicate that the isotopic composition of lead in the earth at that time resembles the lead now found in the troilite phase of iron meteorites. Lead from such meteorites is believed to represent isotopically unaltered samples of the lead present in the original solar system.

Most of the methods of geochronology are applicable to dating meteorites. The procedures are very much the same as those already described. Many meteorites contain significant uranium and thorium, and age determinations based on the decay of these elements are possible. However, as in any other mineral, it is first necessary to subtract from the abundances of the radiogenic isotopes the amount that was incorporated when the meteorite originally crystallized. Without a priori knowledge of the isotopic abundances of such initial lead contamination, an oblique attack on the problem must be made. Assuming that the meteorites evolved as closed systems, it is possible to relate the observed isotope ratios to the initial, or *primordial*, values by the equations

$$\frac{Pb^{206}}{Pb^{204}} = \left(\frac{Pb^{206}}{Pb^{204}}\right)_i + \frac{U^{238}}{Pb^{204}} (e^{\lambda t_0} - 1)$$

and

$$\frac{Pb^{207}}{Pb^{204}} = \left(\frac{Pb^{207}}{Pb^{204}}\right)_i + \frac{U^{238}}{Pb^{204}} (e^{\lambda' t_0} - 1) \qquad (5\text{-}15)$$

Since the ratio U^{238}/U^{235} is a constant, namely, 137.8, a plot of Pb^{207}/Pb^{204} against Pb^{206}/Pb^{204} should be a straight line. An exception occurs only if the chemical systems represented by the different meteorite samples do not all have the same uranium/lead ratio, in which case the sample abundances lie in a single point. The slope of the line is a function of only t_0 and the decay constants, and for presently available data gives 4550 m yr for t_0. Thus 4550 m yr ago an event common to the meteorites studied caused a gross dispersion of the uranium/lead ratios. (There was a similar dispersion in the thorium/lead ratios but not in the thorium/uranium ratios.)

The best estimate of the initial values of the lead-isotope ratios in meteorites is obtained by observing those meteorites in which the proportions of uranium and thorium are very small in comparison with the lead constant. The values usually accepted are the mean values obtained by V. R. Murthy and C. C. Patterson (1962) for the troilite phase of iron meteorites. Recently, Oversby (1969) has made similar measurements but using the technique of *double spiking* to eliminate systematic discrimination errors. The theory of double spiking has been reviewed by Russell (1971). The different values are summarized as follows:

	Murthy and Patterson	*Oversby*
$a_0 = \left(\frac{Pb^{206}}{Pb^{204}}\right)_i$	9.56	9.346
$b_0 = \left(\frac{Pb^{207}}{Pb^{204}}\right)_i$	10.42	10.218
$c_0 = \left(\frac{Pb^{208}}{Pb^{204}}\right)_i$	29.71	28.96

It is appropriate to ask whether the average lead-isotope ratios present in the outer parts of the earth today could have evolved from the same initial lead in the same time interval. A definitive answer is made more difficult because there is still significant uncertainty about the choice of values representative of average modern lead. However, most reasonable estimates lead to a time interval within a very few percent of 4550 m yr. Since this is unlikely to result from a coincidence, the age of the earth is assumed to be measured from the time when the earth's lead had the isotopic abundances of troilites. While it is very tempting to think that the earth was also involved in the event that resulted in the primary dispersion of the uranium/lead and thorium/lead ratios in meteorites, there is no direct evidence that these ratios were dispersed in the earth at that time.

The recent Apollo flights have returned substantial quantities of lunar material which have been subjected to extensive geochronometric and isotopic investigations. Papanastassiou and Wasserburg (1970, 1971), and others, have shown that there was a major event affecting lunar Rb/Sr ratios approximately 4600 m yr ago, in agreement with the supposed age of the earth. Soils from both Apollo 14 and Luna 16 missions yield apparently concordant U/Pb ages of about 4800–4900 m yr, but the significance of these older ages is still unclear. The lunar data clearly reflect several events in the age range 4600–3200 m yr, and the existence of volcanism on the moon at least as recent as the end of that interval.

Wetherill (1972) has attempted to reconcile this early chronological record for the moon, with the absence of many terrestrial materials having ages much older than 3300 m yr. (One age of 3900 m yr for southwest Greenland has been documented by L. P. Black et al., 1971.) He offers the hypothesis that accretion of the moon and earth were very rapid and approximately contemporaneous. The sizes of the two bodies are such that, under the conditions postulated, the earth would become entirely molten but the moon would become molten only at shallow depths of 100 to 200 km. The interior of the moon would never have melted, but the near surface material close to the melting point would result in early volcanism which would terminate as the temperature decreased. In the case of the earth, the molten planet would preserve little record of its history before the outer parts became cool and stable, about 3000 m yr ago.

5-13 NOTE ON THE HISTORY OF PRECAMBRIAN CHRONOLOGY

Wherever the early stratigraphers traced beds to the lowest horizons they found there either unfossiliferous strata to which they often gave the name Eocambrian or rocks so altered and confused that they were designated *fundamental complex*, or basement.

When large areas of the early unfossiliferous rocks were discovered exposed in shield areas in the hearts of continents and were first being mapped, some order was discovered in them. Unfortunately, this proved deceptive, because as each area was mapped, arguments arose about how the rocks should be correlated with those of other areas. Comparisons were really a matter of guesswork, and the same few names came to be used in many different places without any proof that they referred to rocks of similar age.

In 1902, in an endeavor to untangle the confusion which then prevailed regarding the Precambrian rocks around the Great Lakes, the Geological Surveys of the United States and Canada appointed a committee of their ablest geologists, which met in two subcommittees and examined the eastern and western areas.

In 1905, C. R. Van Hise published the report for the western areas, which divided the Precambrian rocks found in Minnesota into four parts. At that time Lord Kelvin had been emphasizing the short length of time which he then believed an initially hot earth would have taken to cool, and therefore the Precambrian was confined in an unrealistically short interval. Since the stratigraphers wanted as much time as possible for evolution, this apparently brief and simple classification for the Precambrian had a general appeal.

In 1906, in their "Textbook of Geology," T. C. Chamberlin and R. D. Salisbury adopted the classification as a general one for the Precambrian rather than as the local succession in Minnesota. They grouped the first two divisions of disturbed rocks into an Archean era, which they described as an early turbulent period in the history of a cooling earth and which was separated by a great unconformity from the later two divisions which formed a quieter Proterozoic era (Table 5-6). In a series of important papers Van Hise claimed that this twofold division of the Precambrian was universal. The report by the other subcommittee of the Geological Surveys which showed that the rocks east of the Great Lakes could not be fitted into the same classification was ignored. In 1907 the first age determinations were published and showed that Lord Kelvin was quite wrong and

Table 5-6 WIDELY ADOPTED BUT ERRONEOUS CLASSIFICATION OF THE PRECAMBRIAN

Keweenawan	Proterozoic era
Huronian	
Keewatin	Archean era
Laurentian	

that Precambrian time was a hundred times longer than had been supposed. This passed unnoticed because only a few reliable age determinations were published until 1950, and meanwhile the concept of a simple dual classification of the Precambrian had been officially adopted by geological surveys all over the world, so that it is deeply imbedded in the literature. This much greater tangle is only now beginning to be unraveled.

5-14 SUBDIVISION OF PRECAMBRIAN TIME

We have seen that the classification of the Precambrian which has come to be widely adopted as the standard was made at an unfortunate time and was based upon misconceptions then current. Furthermore, it was constructed upon principles which are well known not to apply to later rocks. Because it was promulgated before the discovery of radioactivity had made an impact on geology, it was based upon a concept of Precambrian time only one one-hundredth of the true length (tens instead of thousands of millions of years), and as a result much too simple a view of Precambrian time was taken.

Wherever stratified rocks have been investigated to sufficient depth, they are found to rest with profound unconformity upon a basement of altered rocks, but in the case of younger rocks no one supposes that the sequence is everywhere of the same age. For example, the overlap of Cretaceous strata upon metamorphosed Paleozoics in the Appalachian region cannot possibly be equated to the overlap of Paleozoic strata upon the Canadian shield. Around the Great Lakes, if all Precambrian time is divided into only Proterozoic and Archean eras, all such unconformities are lumped together without regard for the fact that they can differ widely in age, as is shown by radiometric age determinations.

The concept that all Archean rocks are highly altered and that Proterozoic rocks are not, and the use of degree of metamorphism as a criterion of age, run quite contrary to the discovery of metamorphic rocks of Cenozoic age in young mountains and radiometric determinations showing that little altered sediments can contain minerals of great age.

The concept of correlating rocks by type was carried even further. Thus the presence of gently folded red sandstones cut by diabase and basalt near the top of Precambrian sequences in several parts of the world has led to their correlation with similar rocks at Keweenawan Point on Lake Superior. They might as well be correlated with the Newark series of New Jersey, which is Triassic in age. There was no worldwide period for the formation of such rocks; rather, at the close of every orogeny there has been a tendency to form rocks of the Piedmont facies, which therefore in each area tend to lie at the top of the local column. This correlation of strata by lithology, or rock-stratigraphic units, might be possible even in the absence of fossils if the strata were continuous, but such is rarely the case over long distances for Precambrian rocks.

D. V. Higgs has pointed out that in the United States the proportion of sedimentary rocks preserved and exposed is less in each preceding period back to the Cambrian, and the same is probably also true for successively older sedimentary rocks in the Precambrian. Metamorphism and erosion are constantly reducing the volume of old unaltered sedimentary rocks, so that they tend to occur in small and isolated basins.

The converse, that the older the period the larger the proportion of metamorphic and igneous rocks, is also true. Thus in the Precambrian it is the basement rocks which are most widely and continuously exposed; since they also contain most of the minerals which can be used for radiometric dating, it is upon them that much of the system of Precambrian chronology will have to depend.

Fortunately, there is some correlation between the history of mountain and basement regions and that of adjacent basins, for uplift and erosion cannot proceed without accompanying deposition, and the time of uplift in one area may frequently be dated by a change in the abundance and facies of sediments deposited in an adjoining basin.

Regions which are uplifted are called *positive*; those which sink are *negative.* Changes in the relative level of land and sea are said to be due to *diastrophism*, which may be *orogenetic* if due to mountain building, *epeirogenic* if regional, and *eustatic* if worldwide.

The later part of geological time is divided into three eras, and many geologists who object to having only two eras in the Precambrian would divide that time into a larger but definite number of eras or cycles also. This concept, which has been advocated by A. Holmes and H. Stille, is based upon the idea that the earth undergoes intermittent pulsations of a worldwide nature which are generally thought to consist of brief orogenetic phases and longer intervening periods of quiescence.

On the other hand, the more carefully the uplift of mountains is examined and dated, the more drawn out and complex the process seems to become. It is found that the three young eras are not separated by universal tectonic division, but are in fact arbitrary units based upon fossils. In view of these findings, which have been strongly supported by A. Knopf, J. Gilluly, and L. M. R. Rutten, there seems to be no reason to attempt to divide Precambrian time into universal eras.

At the same time, it has been discovered that the basements of continents can be divided into great provinces, each apparently representing a recent or ancient belt of primary mountains. Each contains pegmatites of one fairly uniform date. These provinces are progressively younger toward the margins of continents. On the inner side of each one there are often basins of secondary mountains and of other cover rocks with which they can sometimes be correlated. Thus for eastern North America the Appalachian metamorphics with pegmatites about 300 million years old are related to the secondary Appalachian fold belts of the same age. The latter rest upon Grenville metamorphics with pegmatites about 1000 m yr old. Related to the Grenville and along their inner margin are secondary-fold belts of Huronian type, which in turn rest upon Keewatin basement with pegmatites about 2500 m yr old. For further information on the discussion of the Precambrian, see A. F. Trendall (1966).

SUGGESTIONS FOR FURTHER READING

FAUL, H.: "Ages of Rocks, Planets, and Stars," McGraw-Hill Book Company, New York, 1966.

HAMILTON, E. I.: "Applied Geochronology," Academic Press, Inc., New York, 1965.

—— and FARQUHAR, R. M.: "Radiometric Dating for Geologists," McGraw-Hill Book Company, New York, 1968.

HURLEY, P. M.: "How Old Is the Earth?" Anchor Books, Doubleday & Company, Inc., Garden City, N.Y.

SCHAEFFER, O. A., and J. ZAHRINGER: "Potassium Argon Dating," Springer-Verlag, New York, Inc., 1966.

YORK, D., and R. M. FARQUHAR: "The Earth's Age and Geochronology," Pergamon Press, New York, 1970.

6

ISOTOPE GEOLOGY

6-1 INTRODUCTION

Chapter 5 described the measurement of the geological age of rocks and minerals through the measurement of isotope ratios. There are other important studies which also depend on mass-spectrometer analyses but for which the principal objective is not geochronology. For example, the isotope ratios of lead and strontium vary in nature, even in minerals which contain negligible amounts of the radioactive parents. These minerals record the effects of radioactive decay processes before their formation. A correct interpretation of the observed isotopic composition of lead and strontium in such minerals will contribute to an understanding of the evolution of gross parts of the earth, as it involved the parent and daughter elements.

Since such isotope variations are the direct result of radioactive decay, they depend on the integrated effects of periods of association of lead or strontium with their parent elements prior to the formation of the final, nonradioactive mineral. Approximate dates for significant events in the history of these elements may be discovered through proper interpretation of the observed isotope ratios. However, it would be wrong to think of these common-lead and common-strontium methods as competing with the conventional dating techniques, since the radioactive decay processes occurred in systems which are unknown or which are not available for study. This is a substantial handicap in evaluating an age determination, but does illuminate the ultimate origin of the lead and strontium now found in common rocks and minerals.

Many elements, of course, are not daughter elements of naturally occurring radioactive decay processes, and many of these are of variable isotopic composition. These latter variations, which are very much smaller than those observed for radiogenic isotopes, are due to subtle changes in the physical and chemical properties of materials in which an isotopic substitution has been made.

Although detailed geochemical methods are beyond the scope of this book, methods dealing with the natural variations of the abundances of nonradiogenic isotopes are of great current interest and importance. Moreover, these studies show an interesting combination of purely physical measurements with interpretations which are based on chemical thermodynamics and statistical mechanics. Reactions involving only the exchange of isotopes between compounds are often determined by fewer variables than the ordinary chemical reactions, and therefore they may often be related to geological processes with more certainty. Studies of the abundances of the oxygen isotopes in marine fossils have provided information about the probable temperatures of ancient seas. Variations in the isotopic abundances of oxygen and hydrogen in waters help to interpret the relationships between rain, snow, and ground water. Carbon-isotope abundances vary widely in both organic and inorganic materials and provide important clues to the genesis of mineral species. Sulfur isotopes give information about the growth of life on our planet and are also used in correlating and studying organic-sulfur deposits associated with petroleum occurrences.

6-2 COMMON–LEAD INTERPRETATIONS

If the composition of a uranium mineral is carefully measured, it is found that the mineral contains isotopes of lead of masses 206, 207, and 208 that have been formed from the decay of the parent radioactive isotopes. The formation of this lead, called *radiogenic lead*, forms the basis of the uranium methods of age determination (Sec. 5-6). In addition, there are lead minerals, principally galenas, which are more or less pure lead compounds containing insignificant amounts of uranium and thorium. The lead in these minerals, which is distinguished by the name *common lead*, differs markedly from radiogenic lead in its isotopic composition. A most striking difference is the presence of a fourth isotope, lead-204, which is not generated by any radioactive process operating in the earth and which, therefore, is not present in purely radiogenic leads. Common leads are also characterized by the presence of much higher proportions of lead-207 than are found in radiogenic leads. There is no doubt that common leads are basically different from radiogenic leads.

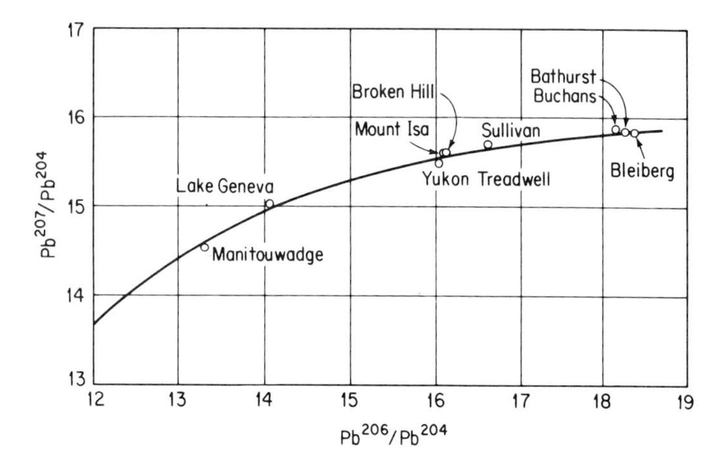

FIGURE 6-1
Primary-growth curve for terrestrial-lead samples. The points indicated along the curve are those from which the parameters of the curve were originally evaluated.

The isotopic composition of common leads has been thoroughly investigated, starting with the work of A. O. Nier (1938), and now analyses are available for thousands of samples from all parts of the earth and representing all parts of geological time. It is observed that the ratios of the isotopes in common leads vary enormously. The leads found in younger parts of the earth are generally richer in the isotopes which are produced by the decay of uranium and thorium. The nature of this variation is shown by the graphs plotted in Figs. 6-1 and 6-2.

When the variations were first observed, it was immediately suggested that they were closely related to the generation of lead isotopes by the uranium and thorium disseminated throughout the outer part of the earth. The process is visualized as follows: At the time when the earth reached its present state with a central metallic core and outer silicate mantle, it happened that the outer parts of the earth contained uranium, thorium, and lead in such proportions that the amount of lead could be increased substantially by the regular additions of lead-206, lead-207, and lead-208 from uranium and thorium as time went by. In this way the isotopic composition of lead slowly changed from a primeval lead, present in the early earth, to modern lead, which is now disseminated in the rocks of the outer part of the earth's mantle and crust. It is believed that modern lead is approximately two-thirds primeval lead and one-third radiogenic lead added within the earth's lifetime, so that the radiogenic component markedly influences the isotopic composition.

Just as the concept of a closed system was an essential consideration in the derivation of formulas for radiometric age determinations, it is basic to the interpretation of variations in common-lead-isotope abundance. By generalizing

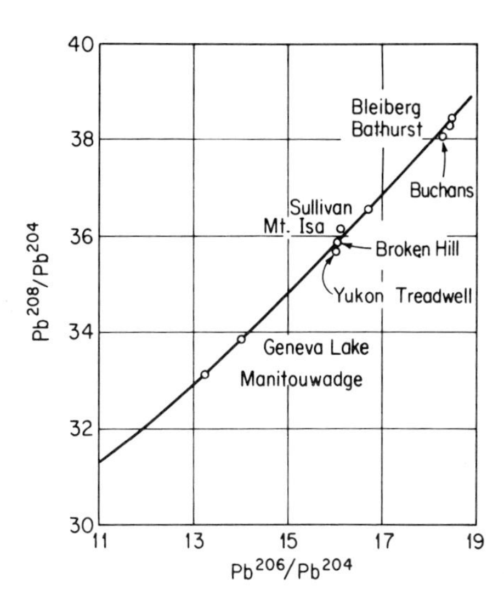

FIGURE 6-2
Primary-growth curve for terrestrial-lead samples. The points indicated along the curve are those from which the parameters of the curve were originally evaluated. (See also Fig. 6-1).

from formulas of the type given in Chap. 5, we can predict the change in an isotope ratio resulting from the existence of lead in a uranium-thorium environment between times t_1 and t_2. The results are as follows:

$$\Delta \frac{Pb^{206}}{Pb^{204}} = \frac{U^{238}}{Pb^{204}} (e^{\lambda t_1} - e^{\lambda t_2})$$

$$\Delta \frac{Pb^{207}}{Pb^{204}} = \frac{U^{235}}{Pb^{204}} (e^{\lambda' t_1} - e^{\lambda' t_2})$$

$$\Delta \frac{Pb^{208}}{Pb^{204}} = \frac{Th^{232}}{Pb^{204}} (e^{\lambda'' t_1} - e^{\lambda'' t_2}) \qquad (6\text{-}1)$$

$$\frac{U^{238}}{U^{235}} = 137.8$$

As before, the ratios U^{238}/Pb^{204}, etc., are understood to be extrapolated to the present time, since the effects of radioactive decay are taken care of in the exponential terms.

Interpretations of various complexity can be built up by adding together the contributions from a number of such closed systems. The total radiogenic increment is added to the ratios for primordial lead, which were shown in Sec. 5-12 to be 9.56, 10.42, and 29.71. The simplest model is that for which the lead is considered to have evolved in a single closed environment between the initial time t_0 and the time of mineralization t. Figures 6-1 and 6-2 show theoretical curves corresponding to evolution in such a system where U^{238}/Pb^{204} is taken as 9.0 and Th/U as 3.9. Lead in the primary system evolves with time from the bottom left of the curves at time t_0 to the top right at the present time. On the same figures are

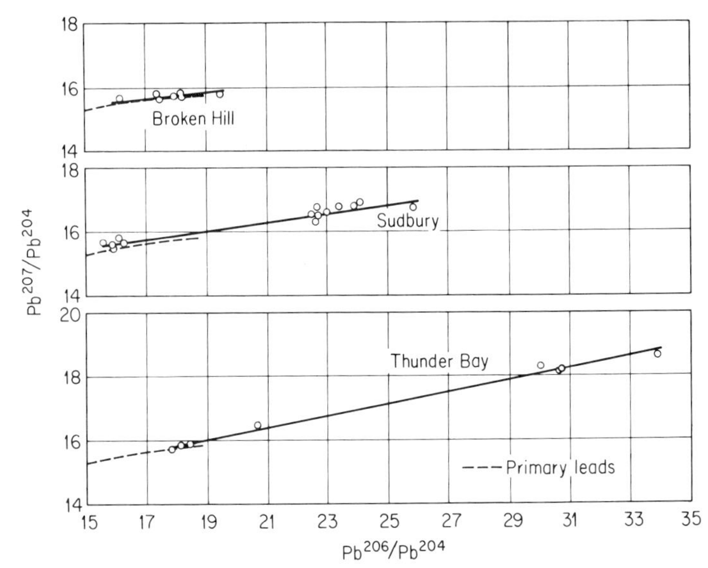

FIGURE 6-3
Isotopic patterns for samples which exhibit the effect of a multistage history of growth. The dashed line indicates the primary-lead curve, sometimes called the conformable-lead curve because the samples used to define it were found in lead deposits conformable in the geological sense. The curve shown by the dashed line is identical with that shown in Fig. 6-1. The solid lines are referred to as *anomalous lead lines*, or *secondary isochrons.*

plotted isotope ratios for lead from conformable ores selected according to geological criteria (R. L. Stanton and R. D. Russell, 1959; R. G. Ostic et al., 1967). According to a hypothesis of R. L. Stanton, the stratiform deposits in which the conformable ores occur formed in island-arc environments, as suggested by the occurrence of marine carbonates and volcanics, and were derived from subcrustal depths. The supposed mode of deposition makes it unlikely that these ores were much contaminated with crust-produced radiogenic lead, and therefore they should record only the effects of the subcrustal environment. It is seen that the simple model with U^{238}/Pb^{204} equal to 9.0 provides a good first-order theory for the interpretation of these samples.

In cases where, in addition to a subcrustal history, the leads have experienced a significant crustal history prior to ore formation, additional terms of the form of Eqs. (6-1) must be added for each additional distinct stage in the model. Figure 6-3 illustrates the pattern obtained for leads from Broken Hill, Australia, which are thought to correspond to a single primary (subcrustal) development stage and a single secondary (crustal) stage. A quantitative interpretation gives 1610 m yr for the time of incorporation of the lead into crustal rocks, and 510 m yr for the time

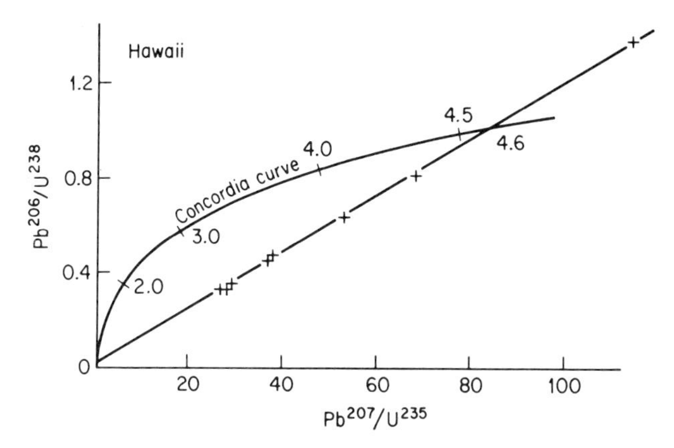

FIGURE 6-4
Example of Ulrych's adaptation of concordia plot to rock leads. (*After T. J. Ulrych, 1967.*)

of formation of the Thackaringa type of lead ores represented in the upper curve of Fig. 6-3. The first age is in excellent agreement with whole-rock rubidium-strontium ages, while the second agrees with potassium-argon age determinations.

Since ores are concentrated from rocks, it is natural to expect that knowledge of the isotope ratios in rocks would provide more fundamental information. Progress in interpreting rock-lead abundances has been disappointingly slow, partly because the analyses have been less precise and partly because the greater heterogeneity of rocks increases the difficulty of adequate sampling. Much of the available data suggest that points for rocks lie statistically below the curve in Fig. 6-1. These include samples from ocean sediments, beach sands, and oceanic basalts, as well as those from granitic rocks. P. H. Reynolds and R. D. Russell (1968) have carried out careful and precise analyses of lead from feldspars separated from granitic rocks near ore deposits at Balmat, New York, and at Nelson, British Columbia. Their findings indicate that these rock leads are unlikely to have evolved in the same primary system as that inferred from the conformable ores, but that the corresponding system for them had a uranium/lead ratio about 3 percent lower. However, mines in the immediate proximity of the Nelson batholith have isotope ratios simply related to those in the batholith.

T. J. Ulrych (1967) has adapted the concordia plot (Sec. 5-11) to the interpretation of rock leads. From his interpretation, the times of two events are found for the lead sample. One of these is the age of the earth, defined as the time when the earth's lead had the abundances of primordial lead, and the other is the time of a significant alteration of the uranium-lead environment of the sample (Fig. 6-4). He has attempted to apply the technique to interpret the observed isotope ratios of leads in rocks from oceanic islands, which Tatsumoto (1966) and others have shown to be products of a heterogeneous source. Recently, R. L. Armstrong (1968), R. D. Russell (1972), and others, have discussed bidirectional mixing models which may provide a more satisfactory understanding of global isotopic patterns. Table 6-1 shows some values of the ratio U^{238}/Pb^{204} found for conformable lead ores and for rocks (R. D. Russell et al., 1968).

The challenge of common-lead-isotope interpretations is to develop models of the form suggested above which are simple enough to permit numerical calculations but sophisticated enough to approximate adequately the complexities of geology.

Table 6-1 CALCULATED RATIOS OF U^{238}/Pb^{204} and Th^{232}/U FOR THE INITIAL EVOLUTIONARY STAGE OF LEAD OBSERVED IN ORES AND ROCKS†

Description	U^{238}/Pb^{204}	Th^{232}/U
Conformable ores:		
Manitouwadge, Ontario ‡	8.61	4.05
Broken Hill, Australia	8.98	3.86
Mount Isa, Australia	9.09	3.91
Captain's Flat, Australia	8.98	3.93
Cobar, Australia, lower horizon	8.97	3.90
Cobar, Australia, CSA	9.00	3.91
Bathurst, New Brunswick	9.00	3.85
Read Rosebery, Tasmania ‡	8.93	3.77
Hall's Peak, Australia	8.93	3.84
White Island, New Zealand ‡	8.89	3.75
Mean of all conformable ores	8.94 ± 0.13	3.88 ± 0.08
Mean omitting values indicated ‡	8.99 ± 0.05	3.89 ± 0.04
Continental rocks:		
Balmat, New York	8.75	
Llano uplift, Texas	8.60	
Oceanic basalts:		
Mid-Atlantic Ridge	8.66	
East Pacific rise	8.58	
Hawaii	8.63	
Easter Island	8.74	
Japan	8.74	
Mean of all rocks ‡	8.67 ± 0.06	

† All ratios were calculated for t_0 = 4550 m yr and for values of a_0, b_0, and c_0 = 9.56, 10.42, and 30.1. If a value of 29.71 were used for c_0, the thorium/uranium ratios would be increased by about 5 percent.

‡ It is possible that the difference between averages for rocks and ores is partly instrumental.

6-3 COMMON–STRONTIUM INTERPRETATIONS

Section 5-8 described the graphical interpretation of isotope ratios measured in rubidium-bearing minerals and showed that such interpretations lead to a knowledge of the ratio Sr^{87}/Sr^{86} in the minerals at the time of their formation. This

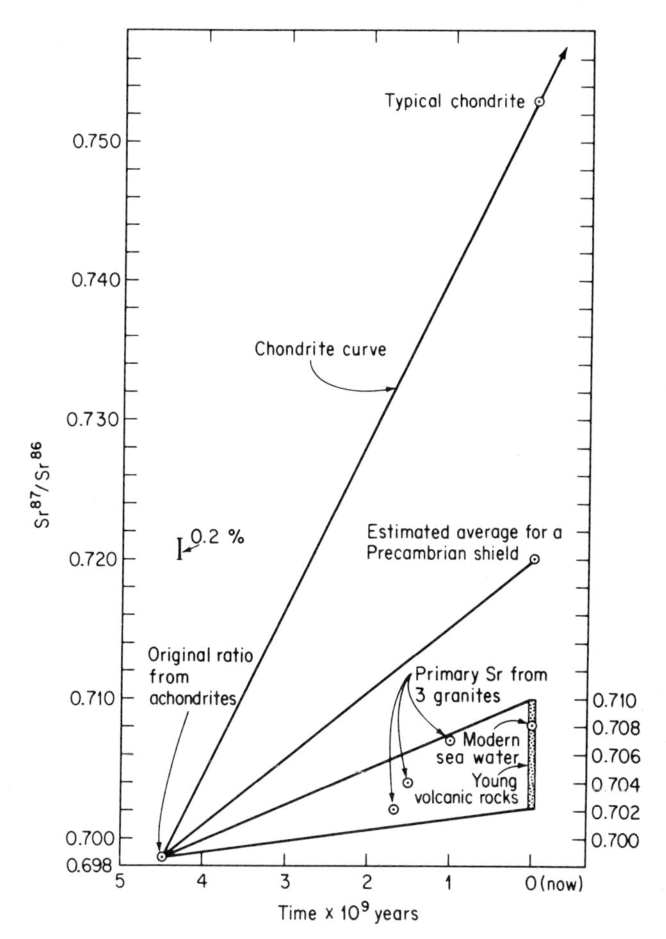

FIGURE 6-5
Schematic diagram for Sr^{87}/Sr^{86} growth with time for chondrites, showing the comparison with modern sea water, young volcanic rocks (continental and oceanic), and a present-day shield area. (*After G. J. Wasserburg, 1966.*)

ratio differs according to the history of the strontium in the earth before it was incorporated into the minerals now studied. In consequence, such strontium-isotope-abundance interpretations can provide valuable geochemical information on the abundances of rubidium and strontium in the outer parts of the earth.

The evolution of the ratio Sr^{87}/Sr^{86} differs significantly in different natural systems. The patterns of evolution are summarized by G. J. Wasserburg (1966), as shown in Fig. 6-5. Of particular interest is the pronounced difference in the patterns for basic volcanic rocks and some granites, on the one hand, and for an average Precambrian shield, on the other. Oceanic basalts are reported to occupy a still more restricted range than the band shown on the figure for young volcanic rocks. These differences make it possible, at least in principle, to detect the incorporation of old shield material in newly formed rock units.

For terrestrial materials the differences in the isotope ratio are small, because of the long half-life of the parent rubidium-87 and the rubidium/strontium ratios characteristic of earth materials. The precision available for solid-source analysis of strontium has been barely adequate for many of these studies. In some cases it has been found advantageous to normalize the observed isotope ratios to give a specified standard value of the ratio Sr^{86}/Sr^{88}, usually 0.1194. Since these isotopes are not formed in nature by radioactive decay, their ratio is expected to be the same in all samples. Some analyses of very high precision have recently been reported by D. A. Papanastassiou and G. J. Wasserburg (1969).

Progress in the interpretation of common-strontium-isotope abundances has been rapid and exciting. It has already contributed substantially to an understanding of the relationships between different parts of the surface of the earth.

6-4 THEORY OF ISOTOPIC EQUILIBRIUM

It might be anticipated that isotopes of the same element would not be found everywhere in identical proportions, because subtle differences in properties would lead to small discriminations in naturally occurring physical and chemical processes. Some discriminations arise purely from the differences in atomic mass of the isotopes of an element. An example of this occurs in the escape of helium from the earth's upper atmosphere, where the lighter helium-3 escapes more rapidly than the heavier helium-4. Another process is found in chemical equilibrium "exchange" reactions, where the chemical potential of many compounds is changed slightly by the substitution of one isotope of an element for another. Isotopic fractionations can also arise in nonequilibrium reactions in which the rate of reaction is not the same for compounds having different isotopic substitutions. Ordinary evaporations often come in this last category. The partial separation of isotopes through biological processes is so important in certain elements, notably sulfur and carbon, that it deserves special mention, although the actual mechanism of the exchange is probably a combination of those already listed.

Chance often plays an important part in scientific progress, and this was the case in the field of nonradiogenic isotopes. There exist in nature two stable isotopes of hydrogen (protium and deuterium) that have such grossly different properties that they could have been readily distinguished and isolated by techniques that were available many years ago. However, it happens that the abundance of the heavier hydrogen isotope in nature is only 0.015 percent of that of common hydrogen, and hence it was so diluted by the lighter isotope that its properties were not observed. If the hydrogen isotopes, like the isotopes of many other elements, had occurred in approximately equal amounts, the discovery of deuterium and the recognition of isotopic fractionation might have been advanced several decades. Actually, the presence of deuterium was first demonstrated in 1931 by H. C. Urey, F. G. Brickwedde, and G. M. Murphy (1932), who concentrated deuterium by allowing a large quantity of liquid hydrogen to evaporate until only a small residue remained, and then showed that the optical spectrum of the residue contained additional lines not observed in the ordinary hydrogen spectrum and having precisely the frequencies calculated for the Balmer series of hydrogen of atomic weight 2. The importance of the discovery of deuterium and the investigation of its properties was so great that Urey was awarded a Nobel prize in chemistry. Some of the properties of hydrogen and deuterium and their oxides are summarized in Table 6-2.

There are many other special cases of enormous differences in the properties of isotopes of a single element, such as differences in nuclear-scattering and nuclear-absorption cross sections, radioactive stability, and other primarily nuclear phenomena. To this could be added the example of helium, the heavier isotope of which exhibits the properties of a superfluid below 2.18°K, properties which have not been observed in the lighter isotope. These latter examples, however, take no part in fractionating the isotopes under terrestrial conditions.

One of the basic concepts of chemistry is that of chemical equilibrium reactions. For example, suppose certain compounds $A, B, C, \ldots$ are composed of elements that can be rearranged to form a second set of compounds $P, Q, R, \ldots$ according to the relationship

$$aA + bB + cC + \cdots = pP + qQ + rR + \cdots \tag{6-2}$$

In Eq. (6-2) the lowercase letters $a, b, c, \ldots$ represent the molar proportions of the corresponding compounds. An equilibrium mixture containing a definite proportion of each compound will eventually result, provided two conditions are fulfilled

1. There must be operating some physical mechanism whereby the rearrangements indicated by Eq. (6-2) can take place freely.

Table 6-2 COMPARISON OF THE PROPERTIES OF DEUTERIUM WITH HYDROGEN AND WATER WITH HEAVY WATER

Property	H_2	D_2	H_2O	D_2O
Normal boiling point	20.38°K	23.59°K	100.00°C	101.42°C
Triple point	13.92°K	18.71°K	0.0°C	3.802°C
Latent heat of fusion, cal/mole	28.0	52.3	1435	1522
Density (20°C), g/cm^3	....	....	0.9982	1.1059
Viscosity, millipoises	....	....	10.9	12.6

2. The overall system is closed and therefore not dependent on time. A quasi-equilibrium situation can sometimes result if the overall composition of the system changes slowly compared with the rate at which the rearrangement of Eq. (6-2) takes place.

In the simplest case the equilibrium proportions of the compounds are related by the equation

$$\frac{[P]^p\,[Q]^q\,[R]^r\cdots}{[A]^a\,[B]^b\,[C]^c\cdots} = K \qquad (6\text{-}3)$$

where the square brackets indicate molar concentrations of the compound concerned, and K is the *equilibrium constant*. In many real reactions Eq. (6-3) is only a poor approximation, and methods of physical chemistry have been developed to circumvent this difficulty.

In the case of isotopic exchange reactions, the net rearrangement to be considered is the interchange of two different isotopes of the same element between two different compounds containing that element. This is, therefore, a very special case of the general equilibrium outlined above. It is important to remember that the conditions 1 and 2 above apply equally well to isotopic equilibrium. It is because of their rather simple nature that isotopic equilibrium reactions are so important; for the inherent simplicity makes possible rather precise computation of the equilibrium distribution, a computation which is very difficult or impossible in the general case.

For the general isotopic equilibrium we write

$$aA + bB^* = aA^* + bB \qquad (6\text{-}4)$$

where A and B are two compounds containing the same element, and the asterisk indicates the position of the heavier isotope of that element.

For this reaction the equilibrium constant will be equal to

$$K = \frac{[A^*]^a\,[B]^b}{[A]^a\,[B^*]^b} \qquad (6\text{-}5)$$

Unlike the case of many chemical equilibria, Eq. (6-5) is a rather precise expression of the isotopic equilibrium configuration.

In general, an equilibrium represents a configuration of minimum potential energy. In the case of chemical equilibria the potential energy is the energy contained in the molecules of the compounds. Since the optical spectra of compounds are determined by molecular energy levels, one would expect a relationship between the optical spectra of compounds and the equilibrium constant of a chemical equilibrium. Statistical mechanics is the branch of science which forms a link between these two quantities. It was Urey who pointed out that the difficulties in applying statistical mechanics to the calculation of equilibrium constants were greatly reduced in the case of isotopic exchange equilibria. His work has been extended by J. Bigeleisen and M. G. Meyer (1947), A. P. Tudge and H. G. Thode (1950), and by many others. The result is a relationship between the equilibrium constant and vibrational energy levels, through mathematical functions involving the partition of energy among the possible vibrational energy levels of the molecule. These partition functions can be calculated from the frequencies of the vibration lines of the optical spectra of the compounds. A brief account of the development of the final equation is given in Appendix C. The result is that the equilibrium constant K is determined by the ratios of the partition functions $Q(A^*)/Q(A)$ and $Q(B^*)/Q(B)$, such as

$$K = \frac{Q(A^*)}{Q(A)} \frac{Q(B)}{Q(B^*)} \tag{6-6}$$

where the ratios of the partition functions are of the form

$$\frac{Q(A^*)}{Q(A)} = \prod_i \frac{\sigma_i^* \nu_i^* e^{-h\nu_i^*/kT}(1 - e^{-h\nu_i/kT})}{\sigma_i \nu_i e^{-h\nu_i/kT}(1 - e^{-h\nu_i^*/kT})} \tag{6-7}$$

where σ = symmetry factor
ν = vibrational frequency
T = absolute temperature
h/k = ratio of Planck's constant to Boltzmann's constant

The values of ν_i are obtained from infrared or Raman spectroscopy.

From Eq. (6-6) equilibria can be predicted for isotopic exchange reactions involving all but the very lightest elements, provided that the requisite spectroscopic data are available. Even when such data are lacking, it is possible to predict the general behavior of exchange reactions from the form of the expressions (6-6) and (6-7). It is obvious that monatomic ions have no vibrational energy, and therefore the ratio of the partition functions for isotopic monatomic ions is unity. The population of the various energy levels depends on the temperature at which the exchange reactions are taking place, and therefore the extent of fractionation as determined by the equilibrium constant is always a function of temperature. In cases in which Eq. (6-7) applies and $\sigma_i = \sigma_i^*$, the ratio of the partition functions approaches unity as the temperature becomes very large, and therefore the extent of fractionation decreases with increasing temperature. A third generality is that heavier isotopes tend to concentrate in molecules in which they are present in the highest valence state. For example, in the reaction

$$H_2S^{34} + S^{32}O_4^{--} \rightleftharpoons H_2S^{32} + S^{34}O_4^{--}$$

sulfur-34 is enriched in the sulfate ion. H. G. Thode has found a value of 1.074 at 25°C for the equilibrium constant of this reaction.

Small differences in isotopic composition, such as are important in geology and geophysics, are determined by comparing the sample with a standard. The result is expressed as the *per mil* difference from the standard as obtained from the expression

$$\delta = \left(\frac{R_{\text{sample}}}{R_{\text{standard}}} - 1 \right) \times 1000\%$$

The symbol R refers to the observed abundance ratio of the isotope under study.

6-5 SULFUR

Many examples could be quoted to illustrate the importance of isotopic exchange in geological studies. The work originated by Thode on the fractionation of sulfur isotopes deserves special mention, not only because it was a precise and systematic check between experimentally predicted isotopic equilibrium constants and their measured values, but also because it has thrown light on problems of an amazingly diversified nature. Sulfur is a light element – atomic weight 32.06 – and its most abundant isotopes of masses 32 and 34 have a relatively large proportionate mass

Table 6-3 EQUILIBRIUM CONSTANTS CALCULATED FOR SULFUR-ISOTOPE EXCHANGE REACTIONS

Ion	$\frac{S^{34}O_4^{--}}{S^{32}O_4^{--}}$	$\frac{S^{34}O_2}{S^{32}O_2}$	$\frac{H_2S^{34}}{H_2S^{32}}$	$\frac{PbS^{34}}{PbS^{32}}$	$\frac{S^{34--}}{S^{32--}}$	Temp, °C
$\frac{Q(A^*)}{Q(A)}$	1.101	1.053	1.015	1.010	1.000	0
	1.088	1.045	1.013	1.009	1.000	25
Ion	Equilibrium constant K when equilibrium is with ions in first column at temperatures in last column					
$\frac{S^{34}O_4^{--}}{S^{32}O_4^{--}}$	1.000	1.046	1.085	1.090	1.101	0
	1.000	1.041	1.074	1.078	1.088	25
$\frac{S^{34}O_2}{S^{32}O_2}$		1.000	1.037	1.043	1.053	0
		1.000	1.032	1.036	1.045	25
$\frac{H_2S^{34}}{H_2S^{32}}$			1.000	1.005	1.015	0
			1.000	1.004	1.013	25
$\frac{PbS^{34}}{PbS^{32}}$				1.000	1.010	0
				1.000	1.009	25
$\frac{S^{34--}}{S^{32--}}$					1.000	0
					1.000	25

difference. Therefore large fractionation effects would be expected. In addition, isotopic ratios for sulfur can sometimes indicate the presence and nature of biological activity, as in the occurrence of large masses of native sulfur in some petroliferous anhydrite and salt domes. The study of sulfur isotopes is of particular interest because of the wide occurrence of many sulfur and sulfide minerals and their great economic importance.

That sulfur isotopes can be significantly enriched through equilibrium reactions is indicated by Table 6-3, which shows partition functions and equilibrium constants calculated for sulfur-bearing ions. If, as was originally supposed, isotopes were identical chemically, all the numbers in this table would be identically equal to unity. On the contrary, the equilibrium constants differ significantly from unity in all cases, and in some reactions the difference equals 10 percent. As theory predicts, the equilibrium constants are nearer unity at high temperatures. It is interesting that the ratios of the partition functions for monatomic sulfur ions are unity at all temperatures. This is true for all monatomic ions since such ions have no vibrational energy.

A number of determinations have been made of the S^{32}/S^{34} ratio in sulfur extracted from the troilite phase of meteorites, and each laboratory conducting such tests has concluded that the ratio is constant within the limits of experimental error. For this reason, such sulfur is often chosen as a primary standard for sulfur-isotope measurements (W. U. Ault and M. L. Jensen, 1962). Significant variations occur in other meteorites (I. R. Kaplan and J. R. Hulston, 1966), especially in carbonaceous chondrites. However, the average value seems to be close to the troilite value of 22.22.

The fractionation of sulfur isotopes in nature is controlled largely by the oxidation and reduction processes involved. Presumably, juvenile sulfur in the earth is similar in isotopic composition to meteoritic sulfur. The oxidation of juvenile sulfur usually produces sulfates impoverished in S^{32}. Conversely, the reduction of sulfates (as by bacteria of the species *Desulfovibrio desulfuricans*) results in an enrichment of S^{32}. The amount of isotopic fractionation is closely related to the detailed kinetics of the process. In nature, variations in the isotope ratio S^{32}/S^{34} exist which deviate in both directions from the troilite value by more than 1 percent. For further information see W. T. Holser and I. R. Kaplan (1966), J. S. Lewis and H. R. Krouse (1969), and Y. Kajiwara et al. (1969).

Sulfur-isotope investigations lend themselves to studies of direct economic significance. These include applications to elemental sulfur deposits and sedimentary uranium deposits, as well as sulfide ores (M. L. Jensen, 1962). It seems possible to apply the technique to distinguish syngenetic from epigenetic sulfide-ore deposits. Deposits of hydrothermal or igneous origin show little variation in sulfur-isotope ratios, which usually resemble those ratios in igneous rocks. However, the biological effects associated with syngenetic deposits tend to produce variable sulfur ratios richer in the heavier isotope (M. L. Jensen, 1963; E. DeChow and M. L. Jensen, 1965).

6-6 OXYGEN AND HYDROGEN

Oxygen is an extremely important participant in vital chemical reactions, as well as an important constituent of most minerals and rocks. It not only is involved in the respiration of living animals and in the photosynthesis of living plants but also incorporated into the shells of many animals which may be preserved as fossils. Because of its wide occurrence and the relatively large fractional mass difference between its abundant isotopes, isotopic-abundance measurements of oxygen have been of great value to science.

Experiments show that the ratio of oxygen-18 to oxygen-16 in air (excluding water vapor) is constant to within a few parts per thousand even up to altitudes exceeding 50 km, although oxygen dissolved in sea water is enriched in the heavier isotope, probably as a result of marine biochemical cycles.

Extensive studies of the fractionation of oxygen isotopes in respiration and photosynthesis have been made by M. Dole (1955) and his associates. In general, photosynthesis yields an enrichment of O^{16}, while respiration produces an enrichment in O^{18}. The isotopic fractionation factor corresponding to the respiration of spinach heads is reported to exceed unity by 2.5 percent, while that for human respiration exceeds unity by 1.8 percent.

The isotopic composition of both sea and fresh water has been studied in detail in order to clarify our understanding of some of the mechanisms by which surface waters are transported, and also to aid in the calculation of paleotemperatures, as discussed below. Early researches of this type by S. Epstein and T. Mayeda (1953) and by I. Friedman (1953) have been extended by H. Craig (1963), who studied both oxygen- and hydrogen-isotope ratios in samples of marine and fresh waters, including rain waters, snow, and waters from hot springs, and by A. C. Redfield and I. Friedman (1965). Rain and lake waters appear to be distilled from the ocean under near-equilibrium conditions. The result is that the ratios deuterium/hydrogen and O^{18}/O^{16} are highly correlated, variations in the former isotope ratio being about eight times those in the latter. Water in equatorial regions, where evaporation is rapid, is enriched in the heavier isotopes. Variations in the oxygen isotope ratio are summarized in Fig. 6-6. Melt water from glaciers is isotopically light and is also essentially free of salt. The mixing of this melt water in the oceans results in a general correlation between salinity and isotopic composition.

The large and very regular variation of the isotopic composition of water provides an exceedingly valuable tool for discovering the origins of many surface waters. Brines associated with the oil fields of central North America have been shown not to be trapped sea water, but composed of water derived from local surface waters with the addition of salt and oxygen-isotope equilibration with the local sediments (R. N. Clayton et al., 1966). The hydrogen preserves the isotopic composition of the surface waters, but the exchange with the sedimentary rocks, which are much richer in O^{18}, enriches that isotope in the trapped waters. Similar

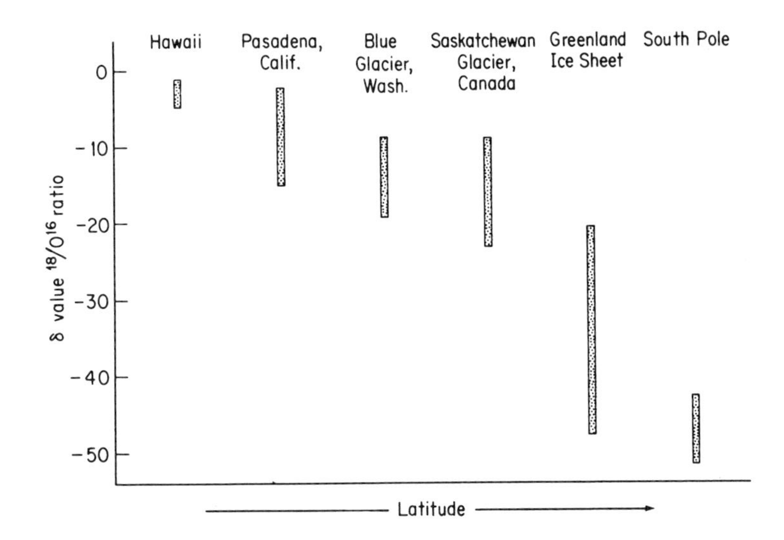

FIGURE 6-6
Plot showing that the O^{18}/O^{16} ratio of precipitation becomes more negative with higher latitudes. (*After R. P. Sharp, 1960.*)

studies show that geothermal waters are composed primarily of recirculated surface waters (H. Craig, 1963).

The isotopic composition of oxygen in glacier ice varies substantially, according to temperature, altitude, and season of precipitation. The previous history of the air masses from which the precipitation falls also affects the O^{18}/O^{16} ratios. Again, although the ratios are modified once the snow is on the ground, they retain general characteristics related to the site of deposition. Thus O^{18}/O^{16} ratios can be used as natural tracers within a glacier system to determine the site and season of accumulation and the subsequent history of the materials.

That the ratio becomes more negative with colder temperatures can be seen within the annual layers of accumulated snow and firn, where the ratios range from most negative in the winter snow to less negative in both the fall and spring snow. This annual curve of ratio variations can be used independently to identify and define the annual layers of snow. C. S. Benson and S. Epstein (1960) have applied this technique in Greenland (Fig. 6-7) and have identified annual layers in cores from a depth of over 400 m. Ratios of O^{18}/O^{16} thus provide a valuable tool for determining rates of accumulation and variations in the climatic environment of large glaciers.

Identification of annual layers in antarctic snow and firn has been made by S. Epstein et al. (1963) using oxygen-isotope ratios. Results from the South Pole, Byrd, Little America 5, and Wilkes satellite stations suggest accumulation rates 20 to 100 percent greater than those determined by surface measurements and pit stratigraphy. Interpretation is complicated, however, by wind action, which may

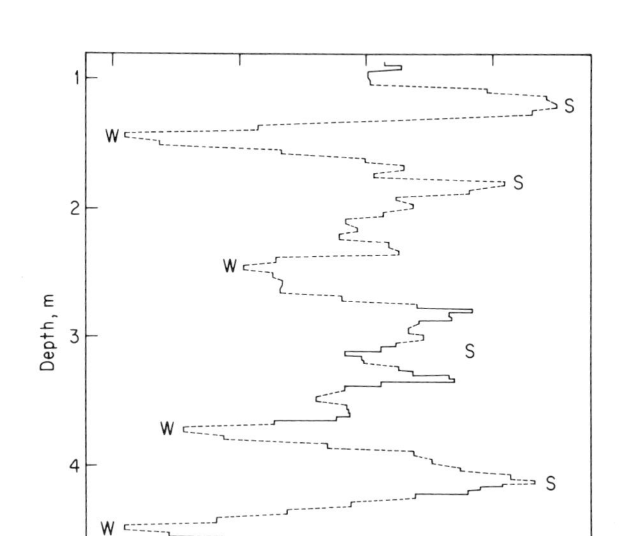

FIGURE 6-7
Annual layering in ice in the Greenland ice sheet. *S* = summer layer; *W* = winter layer. (*After C. S. Benson and S. Epstein, 1960.*)

transport snow a considerable distance and perhaps even produce mechanical separation favoring accumulation of ice particles with higher δ values.

Oxygen isotopes can also be used to study relationships within the ice tongue. Ratios of ice samples collected from the firn edge to the tongue show a general trend toward more negative values downglacier, giving support to deductions concerning flow lines in a valley glacier (Fig. 6-8). The ice toward the tongue should come from the higher part of the accumulation area, so that its ratio should be more negative, as is the case. Analyses of samples taken along profiles across a glacier can also help identify the source of different ice streams composing the glacier.

The calculation of paleotemperatures by measurement of the oxygen-isotope ratios provides a remarkable example of the usefulness of isotope studies. The details of the method are given in a series of papers published since 1951 by H. C. Urey, H. A. Lowenstam, S. Epstein, C. R. McKinney, R. Buchsbaum, and C. Emiliani. It has been pointed out above that equilibrium reactions in general result in a small but definite separation of isotopic molecules and that the equilibrium constant related to this separation is a function of the temperature at which the

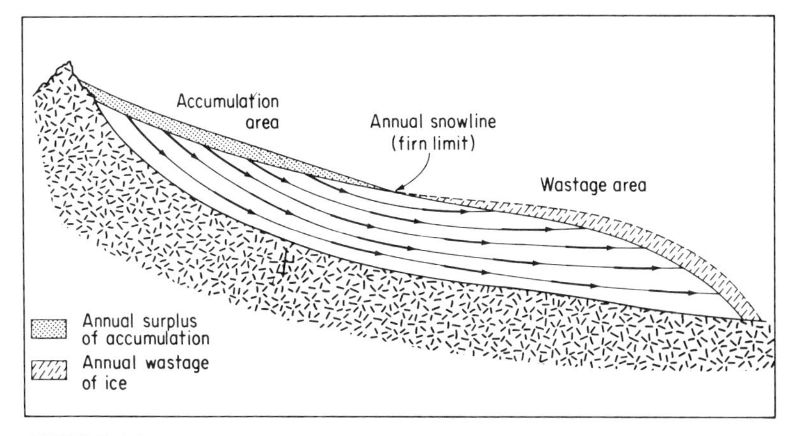

FIGURE 6-8
Longitudinal section through a small valley glacier showing the accumulation-wastage area relationship and the deduced longitudinal flow lines. (*After R. P. Sharp, 1960.*)

equilibrium is attained. The growth of marine shells is slow, and the assumption that it takes place under equilibrium conditions is a reasonable one. The equilibrium between oxygen in water and in carbonate was studied theoretically by J. M. McCrea (1950), who predicted a temperature coefficient of the equilibrium constant of only 0.0176 percent/°C. The measurement of this small change demanded more precise mass spectrometry than had been previously available, and a special mass spectrometer for these measurements was devised and built. The instrument used a bridge method to measure directly the ratio of the abundances of the two isotopes and provided for the rapid comparison of the sample ratio with that of a standard.

In practice, a quadratic relationship is assumed for the variation of O^{18} (enriched relative to the standard) with temperature during shell formation. The constants in this formula are determined by measuring the oxygen-isotope ratios in modern shells grown in water of known salinity at a known temperature. Corrections must be made for any variations in the O^{18} content of the water in which the animal grew. This correction is made possible by noting the way in which the isotope ratios of oxygen vary with the salinity and temperature of the water. In many cases these corrections can be made rather precisely, and paleotemperatures can be determined with a precision approaching 1°C.

The remarkable precision of the method is illustrated by measurements which were made on the growth rings of the shell of a belemnite which grew in the Jurassic period. Samples were taken of the rings on both sides of the center of the fossil, and temperatures calculated from the mass-spectrometer analyses. Figure 6-9 shows a picture of this fossil, and Fig. 6-10 a graph of the isotopic compositions

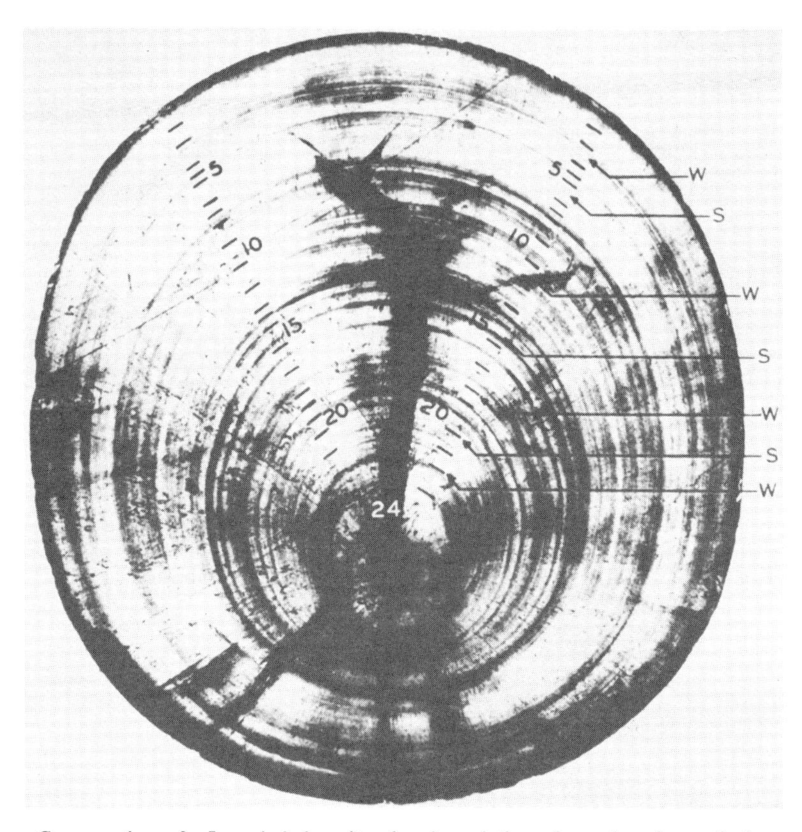

Cross section of a Jurassic belemnite, showing relation of samples of growth rings. *W* and *S* refer to winter and summer regions. (*After H. C. Urey, H. A. Lowenstam, S. Epstein, and C. R. McKinney, 1951.*)

and temperatures deduced. The history of the fossil was deduced by Urey and his coworkers, who report:

> This Jurassic belemnite records three summers and four winters after its youth, which was recorded by too small amounts of carbonate for investigations by our present methods; warmer water in its youth than in its old age, death in the spring and an age of about four years. The maximum seasonal variation in temperature is about 6°C. The mean temperature was 17.6°C.

The principal disadvantage of this method is the fact that the isotopic composition of the water in which the animal grew is in most cases unknown. In principle, this uncertainty has been removed if a second oxygen-bearing compound is incorporated in the shell, and if the dependence on temperature of the oxygen-isotope fractionation of the second compound differs from that for

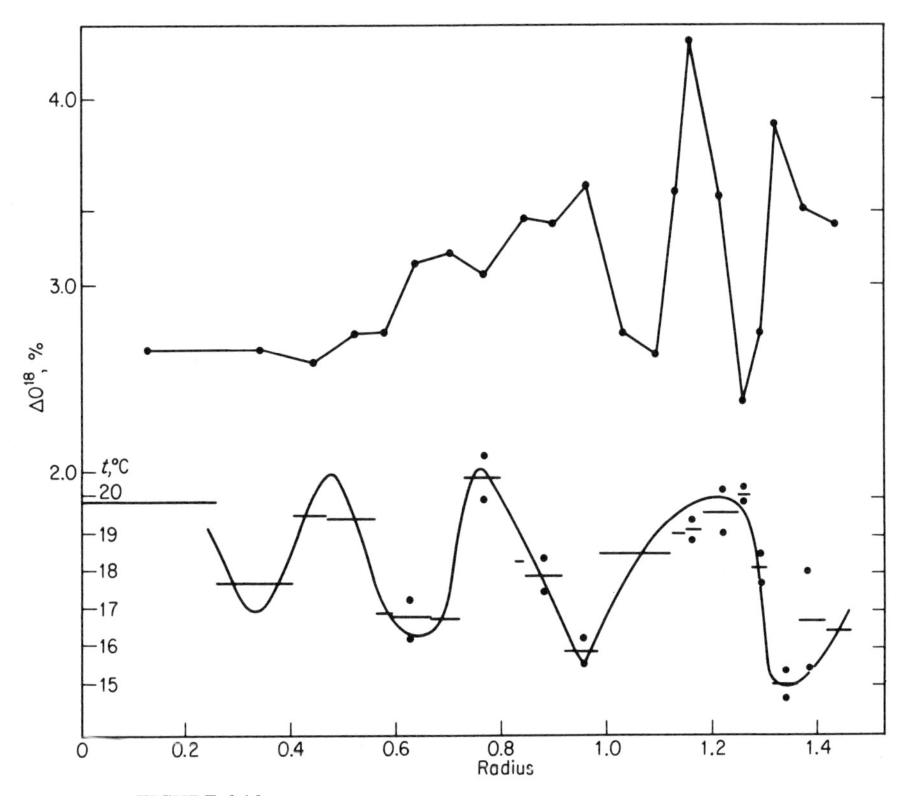

FIGURE 6-10
Variation of temperature and O^{18} content with radius of Jurassic belemnite. (*After H. C. Urey, H. A. Lowenstam, S. Epstein, and C. R. McKinney, 1951.*)

carbonate. In this case two independent quantities can be determined simultaneously, the temperature of equilibration and the isotopic composition of the sea water.

The ideal situation occurs when three oxygen-bearing compounds can be studied experimentally for the same system. If it can be shown that the observed isotopic composition of the oxygen corresponds to the known equilibrium for all possible pairs, the temperature so calculated is almost certain to be the temperature at which oxygen ceased to be exchanged between the compounds. The necessary data about the isotopic equilibria, which require much detailed laboratory study to obtain, has been determined for several important mineral pairs by H. P. Taylor and S. Epstein (1963), D. A. Northrup and R. N. Clayton (1966), J. R. O'Neil and R. N. Clayton (1964), J. R. O'Neil and S. Epstein (1966), and J. R. O'Neil and H. P. Taylor (1966, 1967). Thus the way has been prepared for the determination of the temperatures of formation of a wide variety of rocks and mineral deposits.

6-7 CARBON AND OTHER ELEMENTS

Carbon is another element which has been the subject of extensive mass-spectrometer studies. Part of the interest in the ratio of the abundances of the stable isotopes carbon-13 and carbon-12 stems from the importance of the radioactive isotope carbon-14 (Sec. 5-10). The latter is present in living carbon in an amount determined by its production by cosmic rays in the earth's atmosphere. In dead materials the radioactive carbon slowly decays and provides a time scale which is extremely useful in determining the ages of archeologically important objects and of some very recent geological formations. Since the C^{14} content of living objects should be related to the C^{13}/C^{12} ratio, a knowledge of this ratio for a variety of living materials is important in radiocarbon dating and indirectly for a better understanding of the circulation of the atmosphere.

However, carbon isotopes are interesting to geologists, apart from their usefulness in radiocarbon dating. Like oxygen, carbon is one of the principal constituents of living materials. It is a relatively light element, and the difference between the masses of its stable isotopes is 8 percent. Because of this rather large difference and the complexity of the chemical reactions in which carbon participates, substantial differences in the C^{13}/C^{12} ratio in nature would be expected, and they are observed. The ratio of the abundances of the carbon isotopes in a typical limestone provides a convenient standard for comparison. H. Craig and G. Boato (1955) estimated a total variation of about 4.5 percent for the C^{13}/C^{12} ratio, most substances containing less C^{13} than does limestone. More recent results indicate a larger variation. The *active exchange reservoir* (including the biosphere, atmospheric carbon dioxide, and ocean bicarbonates) is estimated to be about 0.3 percent poorer in C^{13} than the limestone standard. Terrestrial and marine plants are impoverished in C^{13} to the extent of 2.5 and 1.3 percent, respectively. It is believed that the transfer of carbon from the exchange reservoir to sedimentary rocks involves a 0.9 percent depletion of C^{13}. There is evidence that the abundances of the carbon isotopes have remained the same for the past 1000 m yr. Therefore, if the carbon-bearing sediments are considered to be permanently removed from the active reservoir, any juvenile carbon (entering for the first time into the active cycle at the earth's surface) must be 1.2 percent poorer in C^{13} than the limestone standard.

The isotope ratios of carbon extracted from chondritic meteorites have been measured and found to lie well within the terrestrial range. This is also found to be true for most other elements investigated.

In this discussion, the isotope-abundance variations for sulfur, oxygen, and carbon have been chosen to illustrate some of the methods used, largely because the variations can be correlated easily with well-known physical and chemical processes. Other elements, including lithium, boron, nitrogen, silver, calcium, magnesium, copper, silicon, chlorine, potassium, titanium, and uranium, have also been investigated, some thoroughly and others in a cursory manner.

6-8 THE ORIGIN OF THE ELEMENTS

Chapters 1 and 5 both refer to the question of the age and origin of the earth. In those chapters it has been shown that many meteorites shared a gross event and that this event resulted in a major dispersion of uranium/lead ratios in the various meteorites. The time of this event can be calculated to be about 4550 m yr ago. Considerable information is now also available on the uranium/lead ratio in the earth. Using these data, it is found that lead in the earth 4550 m yr ago closely resembled that in the troilite phase of iron meteorites, which is believed to be the earth's primeval lead. From these facts, and the knowledge that meteorites apparently originate within the solar system, there seems to be strong evidence that the earth also participated in the event which we date as the formation of the meteorites.

One cannot help but be curious about the events which preceded the formation of the earth. This is a very difficult problem. The fact that substantial progress has been made in this direction is a tribute to those many scientists who have worked in this field.

It is thought that the elements were formed in stellar interiors, through processes that involved fusion of hydrogen nuclei and the "burning" of helium nuclei to form the elements of low mass, with the possible exception of deuterium, lithium, beryllium, and boron, which seem to have been produced at lower temperatures by high-energy-particle irradiation. These processes are not capable of explaining the existence of elements of higher mass, which require the bombardment of the lighter nuclei by neutrons. The results of such a bombardment vary according to the rate of the process. If the bombardment is slow enough, the elements formed are able to decay between subsequent neutron additions to more stable nuclei (through β-particle emission). This process is the so-called *s* process, where the *s* is used to indicate "slow." If the neutron additions take place very much more rapidly, the so-called *r* processes occur; for such processes all neutron additions take place before β-decays have a chance to occur. The details of these processes have been rather carefully worked out, and in many cases it is clear which elements are produced by *r* processes and which by *s* processes. As an example, only the *r* process is able to produce transuranic elements. A convenient summary of these ideas has been given by W. A. Fowler and W. E. Stephens (1968). This *Resource Letter* also gives a very complete and useful list of references related to this question.

Thus the mechanism by which the elements were formed is understood in some detail. The companion question is, when did this formation occur? It happens that there are also ways to approach the solution of this question – once again through those elements which are radioactive by measurement of their daughter products. A number of elements have been used, the most interesting being the isotope iodine-129. Iodine-129 decays through β emission to xenon-129 with a half-life of 17 m yr. This half-life is very short, and no iodine-129 now exists in nature. However, it can be predicted that this isotope should be produced at the time of formation of the other elements.

It can be expected that the condensation of planetary material would incorporate any of the iodine present, but would be unlikely to incorporate much xenon. Xenon-129 formed within the condensed materials would be trapped and might be observed today. J. H. Reynolds, who has been foremost among those working in this field, has coined the term "xenology" for the study of xenon-isotope abundances in meteorites (Reynolds, 1963). The presence of very small amounts of excess xenon-129 has been established in a number of meteorites. Interpretation of the amounts obtained is rather dependent on the model used for galactic synthesis, but values of a time period of a few tens of millions of years are typical. Such figures suggest that the elements now found in meteorites were formed not long before the condensation of the meteorites themselves. In the case of the two meteorites Bruderheim and Abee, the formation intervals seem to be significantly different.

The occurrence of excess xenon-129 in meteorites is referred to as the *special anomaly*. In addition, there is a *general anomaly* – there are smaller but significant abundance anomalies in many of the isotopes of xenon in meteorites. Interpretation of the general anomaly leads to a number of tantalizing problems. However, most of these are well beyond the scope of this book.

SUGGESTIONS FOR FURTHER READING

DOE, B. R.: "Lead Isotopes," Springer-Verlag New York Inc., New York, 1970.

FAURE, G., and J. L. POWELL: "Strontium Isotope Geology," Springer-Verlag New York Inc., New York, 1972.

FOWLER, W. A., and W. E. STEPHENS: Origin of the Elements, *Resource Letter* OE-1, *Am. J. Phys.,* **36**:289 (1968).

RANKAMA, K.: "Progress in Isotope Geology," John Wiley & Sons, Inc., New York, 1963.

SLAWSON, W. F., and R. D. RUSSELL: Common Lead Isotope Abundances, in H. L. Barnes (ed.), "Geochemistry of Hydrothermal Ore Deposits," Holt, Rinehart and Winston, Inc., New York, 1967.

TAYLOR, HUGH P., Jr.: Isotope Geochemistry of Oxygen and Hydrogen: A Brief Review, *Trans. Am. Geophys. Union,* **47**:287 (1966).

——: Stable Isotopes, *Trans. Am. Geophys. Union,* **48**:686 (1967).

WASSERBURG, G. J.: Geochronology and Isotope Data Bearing on Development of the Continental Crust, in P. M. Hurley (ed.), "Advances in Earth Science," The M.I.T. Press, Cambridge, Mass., 1966.

7

THERMAL HISTORY OF THE EARTH

7-1 INTRODUCTION

Of all the problems connected with the constitution of the earth's interior, the thermal properties are the least well understood, and a more complete understanding of them would go a long way toward solving many of the outstanding problems in geophysics. The temperature at a given point on the earth's surface depends mainly on the radiation from the sun which reaches it and on the angle with the surface at which the radiation arrives. The average solar heat flow reaching the ground on the continents is of the order of 10^{-2} cal/cm^2 sec. Thus the heat flow from the earth's interior, which is of the order of 10^{-6} cal/cm^2 sec (Sec. 7-2), is negligible in comparison and has no influence on the atmospheric temperature and climate. It has been estimated that if one wished to boil some water to make a cup of coffee, using only the heat flow through an area equal to the cup's base, one would have to wait 15 years for the coffee. On the other hand, the total heat flow through the earth's surface is an impressive 30×10^{12} watts, which may be compared with the average volcanic heat output of about 0.1×10^{12} watts (and the solar radiation received on the earth of $90{,}000 \times 10^{12}$ watts).

One of the most remarkable facts about the earth's heat is the extreme slowness with which it travels through the soil and rocks by conduction. At 50 cm below the surface of the ground the daily variations of temperature are hardly felt, seldom producing a change of more than 1°C at that depth, and the effect arrives there from half a day to a day late. A few meters below the surface only the seasonal changes in temperature can be detected, and they arrive months late. At 10 m below the surface the effect of the seasonal changes is about a year late and temperature fluctuations are extremely small. As for the lingering effect of temperatures, the effect of the cold of the last ice age, which ended about 11,000 years ago, is still appreciable at a depth of a few thousand meters. These are but examples of the extraordinary large thermal inertia of the earth resulting from its large size and small thermal diffusivity. The present surface heat flow is thus due essentially to the thermal conditions that prevail in the top few hundreds of kilometers, no heat from below that level having yet reached the surface by conduction. It also follows that the rate of cooling below about 500 km is very slow and may well be compensated by radioactive heating; i.e., although the earth is cooling near the surface, it may be heating up at greater depths.

7-2 HEAT–FLOW MEASUREMENTS

7-2.1 Methods

It is found that, apart from the top few tens of meters of the earth's crust which are subject to seasonal changes, the temperature within the earth steadily rises with increasing depth. In some of the deep oil wells in California and elsewhere, the temperature exceeds the boiling point of water at atmospheric pressure. Temperature gradients at the surface vary considerably even in quiet areas, i.e., areas far removed from any volcanic activity, ranging from less than 10 to more than 50°C/km. This variation is mainly due to differences in the thermal conductivity. However, measurements from many parts of the world have shown that the heat flow, which is the product of the temperature gradient and the thermal conductivity, is reasonably constant, the average value being about 1.5×10^{-6} cal/cm^2 sec. In the rest of this section, all heat-flow measurements will be given in units of 10^{-6} cal/cm^2 sec; the units will be omitted to avoid constant repetition.

On land the heat flow can be measured in boreholes or in mines and tunnels. The temperature gradient in boreholes can be determined by mercury maximum thermometers enclosed in sealed glass envelopes, although they take a considerable time to come to equilibrium. If many measurements are to be made, electrical thermometers are to be preferred. However, these have the disadvantage that they require an electric cable, which is heavy and expensive, as well as a power-driven winch. [For further details see E. C. Bullard (1960), A. D. Misener and A. E. Beck (1960), and A. E. Beck (1965).]

The temperatures measured in a borehole do not necessarily represent the undisturbed temperature of the rocks owing to the disturbance caused by the drilling operations and heat exchange by the drilling fluid. The time for the disturbance to decay depends on the diameter of the hole, the rock type, and the time spent in drilling. E. C. Bullard (1947) has estimated that, once a borehole has been drilled, it takes about twenty times the time taken to drill the borehole for thermal equilibrium to be re-established. For the lower parts of the hole the time is less, and equilibrium may be reached in a few days at the bottom, although it may take months in the upper part. The situation is extremely complex, however. J. C. Jaeger (1961) has found that the controlling factor is the mass velocity of the drilling fluid. For the relatively small water flows used in small-core diamond drilling, the effect on the temperature gradient is negligible except near the top and bottom of the hole, and measurements may safely be made about two days after the cessation of drilling. This is not true for the large flows used in rotary drilling, where the mud temperature is the controlling factor. Conditions may also be greatly affected by the natural circulation of ground water—the temperature gradient is often low in the upper parts of holes penetrating permeable rocks.

The temperature of the rock in a mine or tunnel is likewise disturbed by ventilation, and holes must be drilled into the shaft walls, in some cases upward of 20 m. Another disturbance in the temperature gradient is produced by the topography of the earth's surface. E. C. Bullard (1938) has shown how to allow for this; the correction may reach as much as 30 percent in mountainous areas. Inclined beds or rock domes with a conductivity different from that of the neighboring rocks can also affect the temperature distribution by as much as 20 percent (Bullard, 1954a).

The temperature gradient at sea is measured by forcing a probe into the soft sediments and determining the temperature gradient along the probe. The first probes (Bullard, 1954a; Bullard et al., 1956) were 3 to 5 m long and 2.5 to 4 cm in diameter. The temperature difference between two points, one near the top and the other at the bottom of the probe, is measured by thermocouples or thermistors. The probe is lowered on a steel cable and allowed to remain for a while in the nearly isothermal bottom water to come to temperature equilibrium. The winch is then run out at a speed of about 2 to 3 m/sec until the probe has penetrated into the bottom, where it is left for about 30 to 40 min. The probe is often bent sharply near the top, owing to the sideways pull of the cable on the top of the recorder while the instrument is in the bottom. When the probe penetrates the sediments, it is heated by friction, usually to above the temperature of the surrounding sediments. The recorded temperature thus rises rapidly for 2 or 3 min as the frictional heat is conducted inward to the thermocouples or thermistors, and then falls slowly as the probe comes to the temperature of the surrounding sediments. It is not possible to wait for complete equilibrium, and a correction must be applied (Bullard, 1954a). With a probe 2.5 cm in diameter, the correction is about −10 percent to the last observed temperature.

The measurement of heat flow with such a probe is very laborious, since it is necessary to make a separate lowering of a corer at each station to obtain a sample for the determination of the thermal conductivity. A more convenient instrument combining a corer and a temperature probe has been devised by R. Gerard et al. (1962). In this instrument very small temperature-sensing probes are mounted on outriggers on the penetrating vehicle, which is itself a coring device. A piston corer can penetrate up to 20 m, allowing wider separation of the thermal sensing elements, thus resulting in greater accuracy in the temperature-gradient measurements. The probes housing the thermal elements are only about 3 mm in diameter and soon come to thermal equilibrium with the sediments; because of their small mass, only about 1/2 min is required for them to come within a few percent of their final value. Further details of the techniques in measuring the heat flow through the ocean flow have been given by M. G. Langseth (1965).

The annual variation of temperature in the bottom water of the deep oceans is negligible; in shallow seas, however, it is several degrees and prevents meaningful heat-flow measurements being taken. S. R. Hart and J. S. Steinhart (1965), however, have investigated the feasibility of measuring the heat flow in Lake Superior at depths of over 250 m, using oceanic techniques. The temperature gradients obtained were not linear, as would be expected under steady-state conditions; the nonlinearity is due mainly to the small seasonal variations in temperature of the bottom water. Knowing the annual cycle of water temperature, the temperature gradient may be corrected for climatic variations and meaningful results obtained from measurements in lake bottoms. Heat-flow measurements have now been obtained by J. S. Steinhart et al. (1969) from 145 locations in Lake Superior and provide the most comprehensive picture of local and regional variations for a continental area. They were able to draw a contoured heat-flow map in some detail covering most of the lake, permitting for the first time a regional correlation with other geophysical parameters such as gravity and magnetic field variations. The range of heat flow was surprisingly large, with a variation of almost a factor of 3 between 0.5 and 1.45. Moreover, these variations can occur within very short distances. In the northeastern part of the lake, for example, the heat flow drops from a plateau of about 1.45 to values as low as 0.5 in a distance of only 23 km. More typically, the scale of the heat-flow variations is larger (about 50 to 100 km), and there are generally rather smooth and continuous changes from one anomaly to the next. Some regions show almost no variation in heat flow; the southwestern part of the lake gives values within 10 percent of 1.15 over an area exceeding 15,000 km^2. The overall average of all the measurements was 0.99. R. P. Von Herzen and V. Vacquier (1967) have also made successful heat-flow measurements through the floor of Lake Malawi in Central Africa; they also found a systematic regional variation.

The thermal conductivity of continental rocks may be measured *in situ* or in the laboratory; in either case a troublesome contact resistance between the specimen and temperature-measuring device has to be eliminated. In the laboratory, transient, absolute steady-state, or steady-state comparison methods may be used [see A. D. Misener and A. E. Beck (1960) for a review of these methods]. Laboratory methods suffer from a number of disadvantages: they require additional apparatus and time for the preparation of samples; only fairly sound rock can be used, and cores may not be recoverable from highly sheared zones; they use specimens which may be far too small to be representative of even sound rock and which are generally at pressures only a little above atmospheric pressure instead of at those prevailing *in situ.* The most easily realized and practical of the *in situ* methods involves the use of an electrically heated cylindrical probe. J. H. Blackwell (1954) has developed an approximate theory of radial heat flow which can be applied to a hollow heating probe for use in short holes in a tunnel wall, and J. C. Jaeger (1956) has given an exact theory for use with a solid probe in deep boreholes.

The thermal conductivities of the sediments on the ocean floors can also be determined in the laboratory by steady-state and transient methods. The conductivities thus measured must be corrected to the temperatures and pressures existing on the ocean floor. It is found that the thermal conductivity depends primarily on the moisture content, and only to a small extent on the mineral composition of the solid particles. Conductivity measurements can now be made aboard ship by a needle-probe method developed by R. P. Von Herzen and A. E. Maxwell (1959); the immediate determination prevents any significant dehydration or migration of water within the core which might otherwise introduce errors. *In situ* measurements of the thermal conductivity of ocean-floor sediments have also been made. The technique has been described by J. G. Sclater et al. (1969) and depends on the transient heating of a small probe attached to a slider on the temperature-gradient probe. C. R. B. Lister (1970a) has also obtained *in situ* sediment-conductivity measurements by means of a Bullard type of probe.

J. B. Walsh and E. R. Decker (1966) found that the observed thermal conductivity for dry-rock samples may differ by more than 15 percent from the conductivity for the rocks *in situ,* saturated and under hydrostatic pressure. Error in the apparent conductivity arises because cracks, which are present in most rocks, form effective barriers to the flow of heat when rocks are dry, but have a negligible effect when the rock is saturated or when the rocks are closed by stress. G. Simmons and A. Nur (1968) have compared the physical properties of granites *in situ* with those measured in the laboratory. They found that the presence of open cracks significantly affects many properties, including the thermal conductivity. Many authors have noted that the heat flow beneath the Precambrian shields (Canadian, Australian, Baltic) is about 20 percent lower than the "normal" heat flow beneath continental areas. Simmons and Nur suggest that perhaps the thermal conductivities which are measured in the laboratory and used to calculate the heat flow for rocks for the Precambrian shield areas may be too low by as much as 10 to 15 percent, roughly the difference between apparent heat flow beneath the shields and "normal" heat flow beneath continental areas.

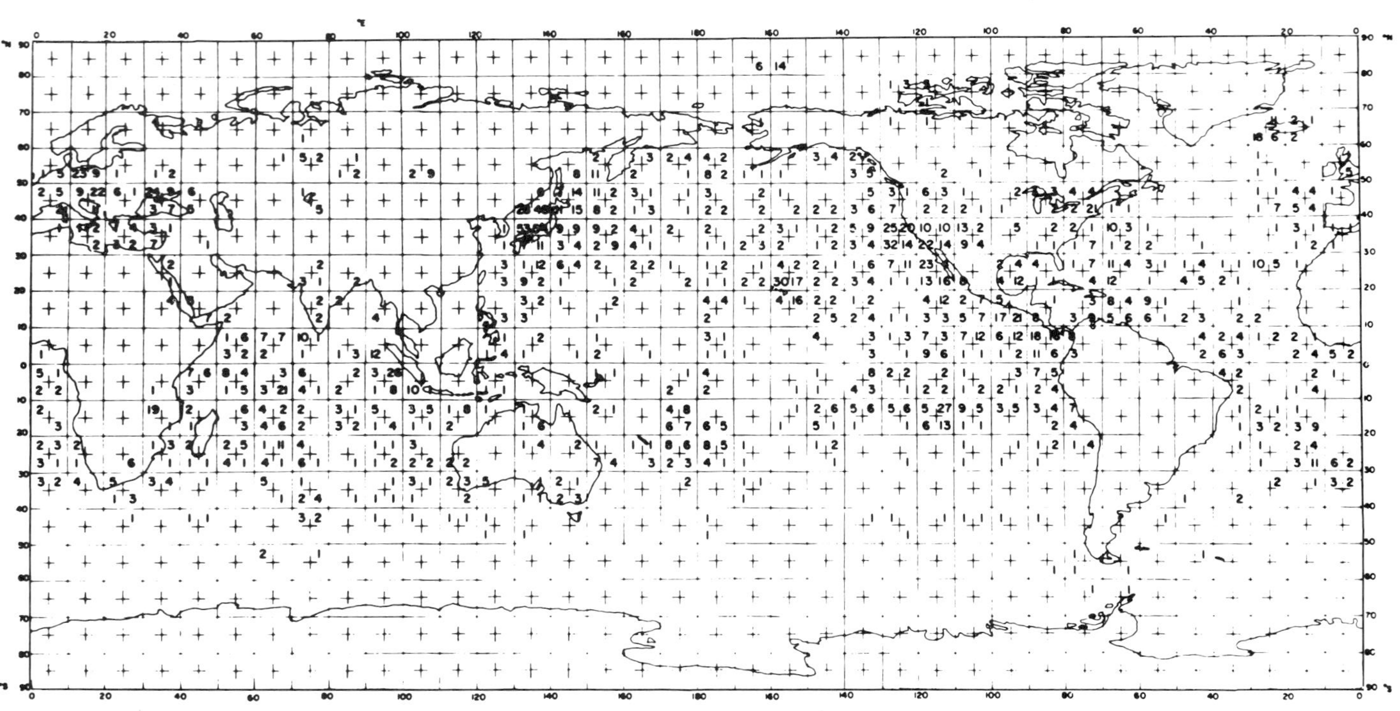

FIGURE 7-1
Distribution of heat-flow stations. The number of stations is given for each 5° (latitude) by 5° (longitude) area. (*After K. Horai and G. Simmons, 1969.*)

7-2.2 Results

Compared with other geophysical measurements, the number of heat-flow determinations is still very small, less than 2000 at the end of 1964, most of these being obtained since 1960. The first comprehensive listing of heat-flow data was made by W. H. K. Lee (1963); it was subsequently revised by W. H. K. Lee and S. Uyeda (1965) and W. H. K. Lee and S. P. Clark, Jr. (1966). Several hundred additional heat-flow values have since been published and have been tabulated by G. Simmons and K. Horai (1968), who hope to keep the list up to date by issuing semiannual summaries. The geographical distribution of heat-flow measurements is very uneven, about 90 percent being obtained in the oceans. K. Horai and G. Simmons (1969) have carried out a spherical harmonic analysis (to the seventh order) of terrestrial heat flow using almost 3000 values. The world distribution of heat-flow stations in $5^\circ \times 5^\circ$ squares is shown in Fig. 7-1, and values of the heat flow (in units of 10^{-7} cal/cm^2 sec) averaged over these $5^\circ \times 5^\circ$ areas are given in Fig. 7-2. Coverage is still poor in the African, South American, and Antarctic continents, and this causes some misleading features in the analysis. The average heat flow for continents (1.65 ± 0.89) is almost identical with that for oceans (1.64 ± 1.11). The average value for shield areas (1.04 ± 0.42) is significantly lower than the world average. The average value for continental lowlands (1.60 ± 0.54) is slightly lower than the world average. In contrast, the average values for the island arcs and marginal seas along the circum-Pacific belt (2.01 ± 0.94) and the oceanic ridge system (1.92 ± 1.64) are considerably higher. These heat-flow values seem to reflect the differences in regional geology in these areas.

The average heat flow through continental orogenic belts decreases with the age of the orogeny to an approximately constant value for the Precambrian shields and platforms (B. G. Polyck and Y. B. Smirnov, 1968). The average heat flow for provinces of the North Pacific also decreases with the age of the province; the mean heat flow through the province younger than 10 m yr is 2.82, whereas the mean heat flow through provinces older than middle Cretaceous is 1.15. J. G. Sclater and J. Francheteau (1970) have presented models of the oceanic and continental crust and upper mantle which can explain the near equality of heat flow through the Precambrian shields and old ocean basins within the framework of plate tectonics.

7-2.3 Oceanic Measurements

A review of oceanic heat-flow studies up to 1964 has been made by R. P. Von Herzen and M. G. Langseth (1966); the same authors have discussed the tectonic implications of oceanic heat-flow data, including the most recent measurements (M. G. Langseth and R. P. Von Herzen, 1970). In the Eastern Pacific Ocean heat-flow measurements range from practically zero to more than 8, the crest of the East Pacific rise being associated with values that are higher than normal (R. P. Von Herzen and S. Uyeda, 1963). The average heat flow through a strip 200

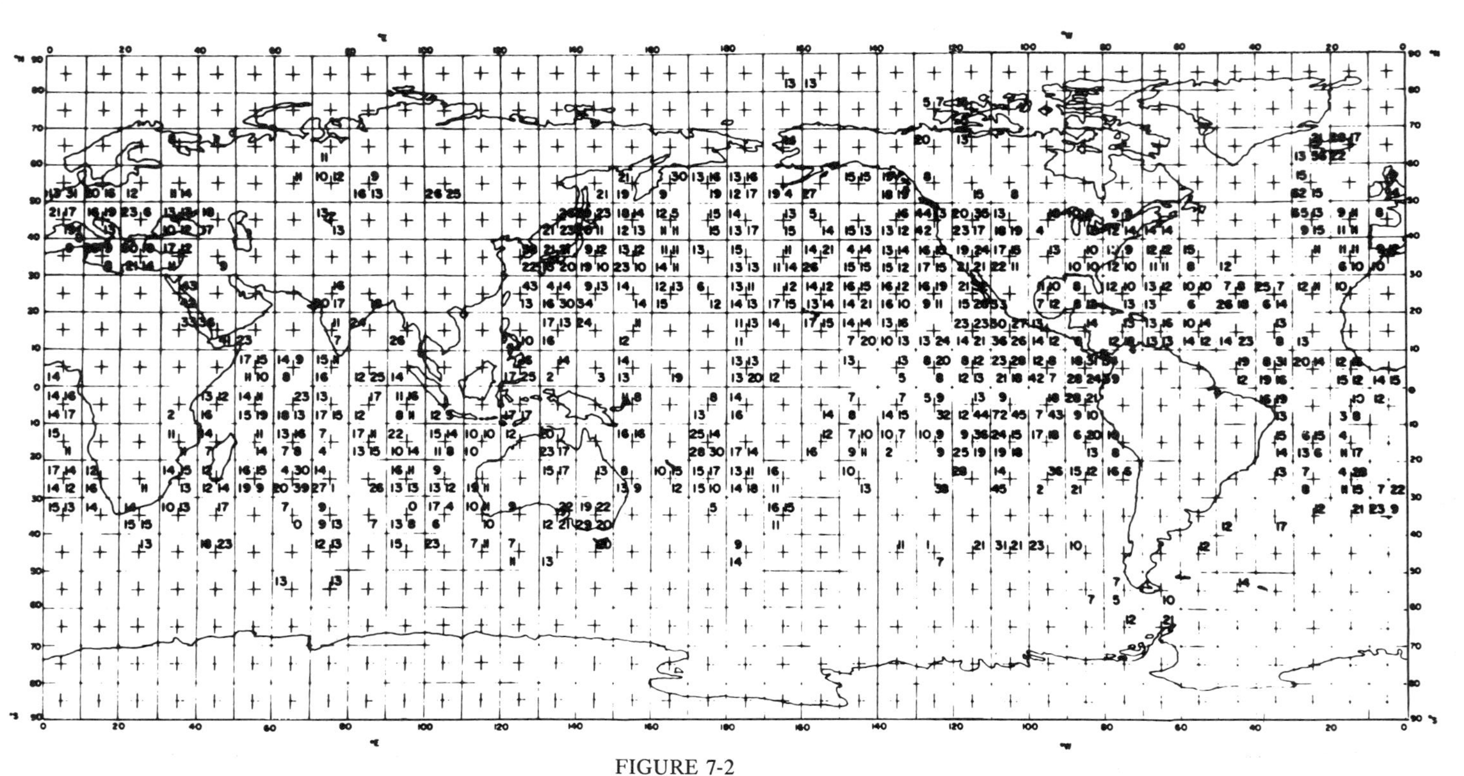

FIGURE 7-2
Distribution of heat flow, in units of 10^{-7} cal/cm^2 sec, averaged for 5° (latitude) by 5° (longitude) areas. (*After K. Horai and G. Simmons, 1969.*)

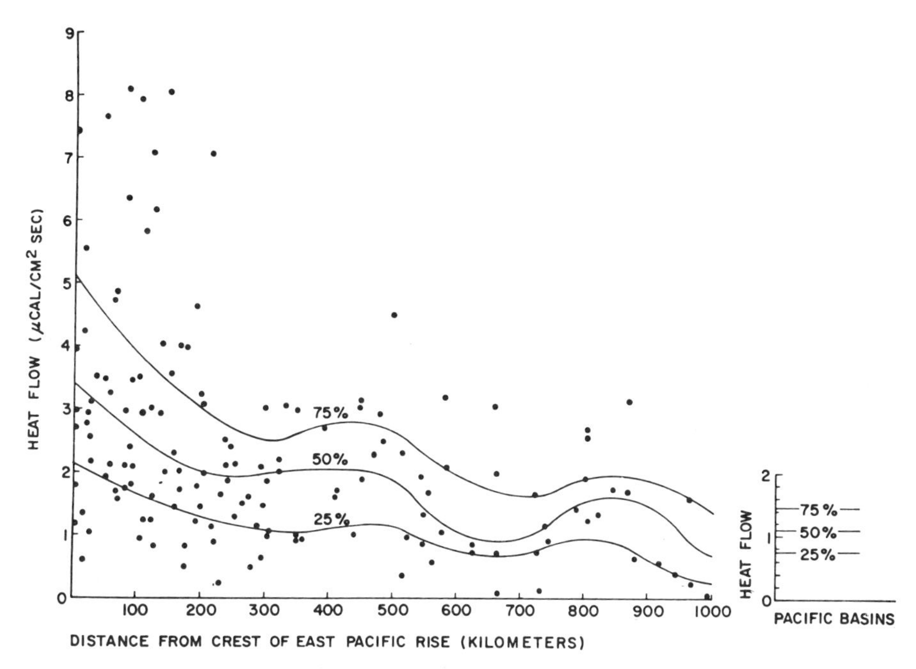

FIGURE 7-3
Heat-flow values versus distance from the crest of the East Pacific rise (50°S to 20°N). 75-, 50- and 25-percentile lines are given for values from the East Pacific rise and the Pacific basins. For example, the 50-percentile line separates the data points into half above and half below it. *(After W. H. K. Lee and S. Uyeda, 1965.)*

to 300 km wide at the crest is about 3, the highest values occurring in two narrow zones approximately parallel, one on either side of the crest (Fig. 7-3). The source of the high heat flow in each of these zones is probably a region of unusually high temperature several tens of kilometers wide located about 10 km beneath the ocean floor. Two regions (approximately 2 to 4 x 10^6 km^2 in area) near the equator, one on either side of the rise, show generally low heat-flow values. The average in one region is 0.68 and in the other 0.65, about one-half the normal oceanic values. Figure 7-4 shows averages for 2° x 2° squares of the heat flow over the equatorial East Pacific, and contours of a fourth-order polynomial fitted to these averages.

M. G. Langseth et al. (1965) obtained 65 new measurements of heat flow in the East Pacific Ocean and confirmed the broad conclusions of Von Herzen and Uyeda. Figure 7-5 shows measurements over the East Pacific rise from both these groups of workers. V. Vacquier et al. (1967) reported 197 new heat-flow measurements in the East Pacific Ocean off the coast of North and Central America. With 367 previous measurements they were able to draw a contour map of the area. While most of the heat-flow values to the west of the rise are around

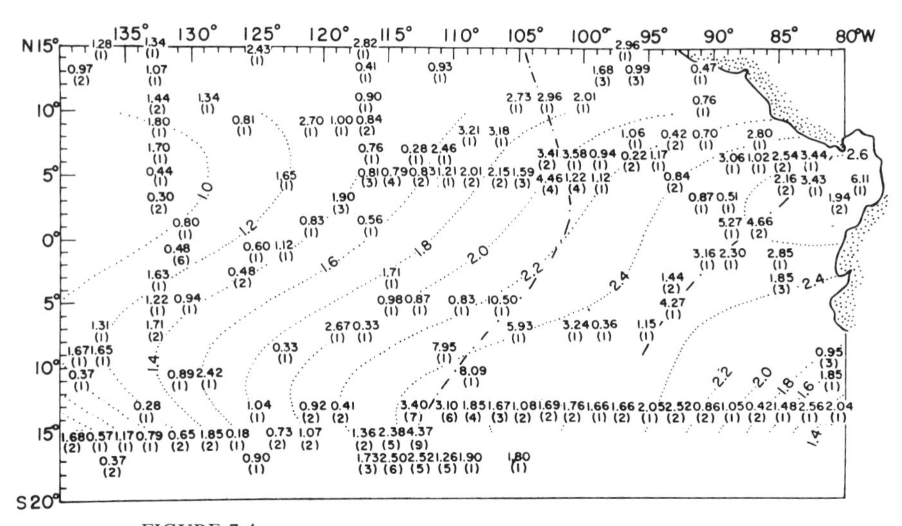

FIGURE 7-4
Averages of heat flow for 2° x 2° squares over the equatorial East Pacific and contours of a fourth-order polynomial fitted to these averages. Approximate axes of the East Pacific rise and the Galapagos rise are shown by dashed lines. (*After M. G. Langseth, Jr., X. Le Pichon, and M. Ewing, 1966.*)

normal, across the rise and to the east of it the values vary widely, widespread areas of high and low heat flows being intermixed. Three regions, corresponding roughly to the three main bathymetric provinces, can be recognized:

1 The area north of 25°N, encompassing the Mendocino and Murray fracture zones and the Mason-Raff magnetic lineations. No correlation was found between the magnetic and heat-flow measurements. There is a sharp change in heat flow where the Gorda ridge intersects the Mendocino fracture zone.

2 The area west of 100°W and south of 25°N, encompassing the East Pacific rise, the Gulf of California, the Clarion and Clipperton fracture zones, the California seamount province, as well as a large section of the Pacific basin.

3 The area east of 100°W, encompassing the Middle America trench; the Cocos, Galapagos, and Carnegie ridges; the Galapagos platform; and the Guatemala basin.

In the second and third regions there are large areas having an average outward flow of heat greater than 3.0. These areas of high heat flow are adjacent to equally large areas of subnormal heat flow and are in striking contrast with the uniformity of heat flow recently discovered in the Northwest Pacific basin (V. Vacquier et al., 1966) that seems to be typical of other ocean basins. Contrary to the conclusions of R. P. Von Herzen and S. Uyeda (1963), V. Vacquier et al. believe that it is

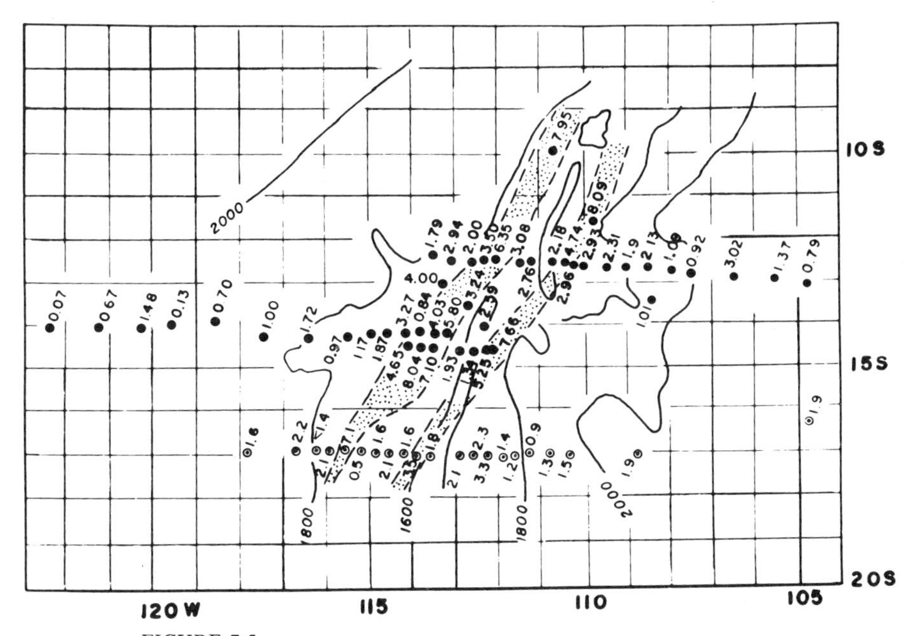

FIGURE 7-5
Heat-flow measurements on the East Pacific rise. (*Solid dots after R. Von Herzen and S. Uyeda, 1963; circled dots after M. G. Langseth, Jr., P. J. Grim, and M. Ewing, 1965.*)

unlikely that the distribution of these areas of high and low heat flow east of the East Pacific rise can be explained by large-scale thermal convection cells in the mantle: Only a mantle deficient in radioactive heat sources can explain the widespread area of low heat flow, and it is difficult to suggest a mechanism that might account for this deficiency.

The contrast in heat-flow values between oceanic ridges and basins is also found in the results from the Indian Ocean (W. H. K. Lee and S. Uyeda, 1965). Values from the ridge are widely scattered but higher than average, especially near the crest; values from the basins are fairly uniform and lower than average. A total of 351 heat-flow measurements were made during the International Indian Ocean Expedition (IIOE). M. G. Langseth and J. P. Taylor (1967) obtained 41 additional measurements and reviewed all published values. The arithmetic mean of the 392 determinations is 1.46 (±1.08). Figure 7-6 shows the average values within $5° \times 5°$ squares, with some measurements from the adjacent continents. The average, 1.46, includes values from the Mid-Indian Ocean Ridge, the Red Sea, the Gulf of Aden, and the Indonesian arc, where values that differ widely from the mean are observed. Excluding such areas, the mean heat flow from the deep, thickly sedimented basin is 1.32 (±0.38). On the basis of recent geophysical observations in the Indian

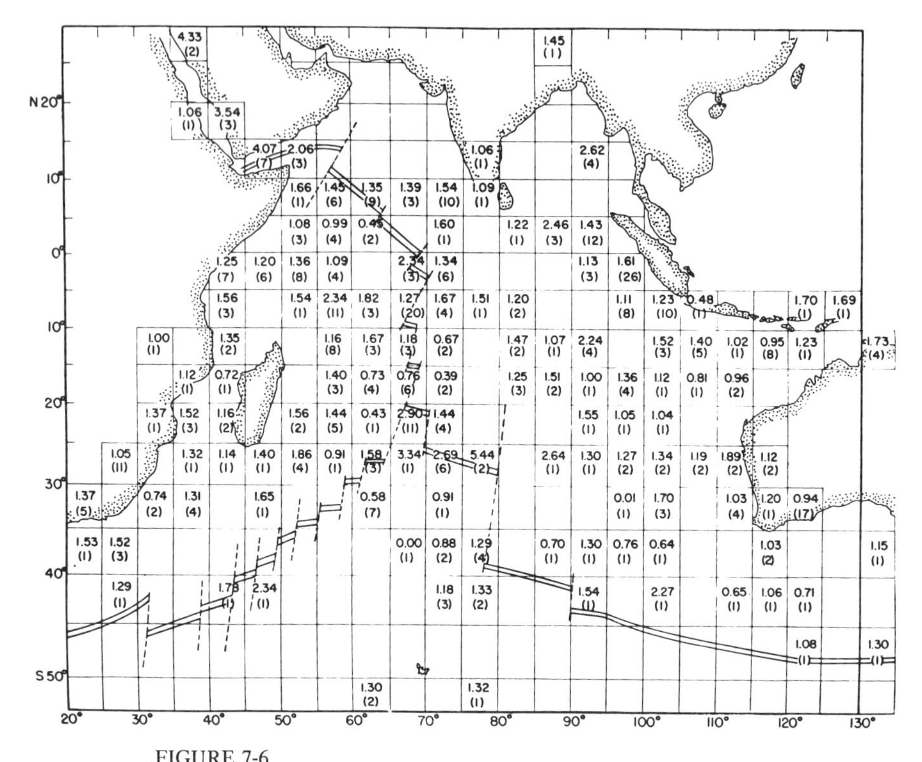

FIGURE 7-6
Average heat flow for 5° x 5° squares over the Indian Ocean. Averages of measurements on adjacent continents are also shown. (*After M. G. Langseth, Jr. and P. T. Taylor, 1967.*)

Ocean, Langseth and Taylor re-examined the trend of the axis of the Mid-Indian Ocean Ridge. The zone of rough topography between the equator and 35°S and longitude 64 to 68°E is now interpreted as a fracture zone that displaces the Carlsberg ridge from the southeast branch of the Mid-Indian Ocean Ridge.

The results from the Atlantic Ocean are more complex. M. G. Langseth, X. Le Pichon, and M. Ewing (1966) have described the broad regional pattern of heat flow in the Atlantic using 179 new measurements, in addition to 197 earlier ones. They found that there was a high degree of symmetry between the heat flow in the western and eastern basins situated at the same latitude. Compared with the world average, the northern basin has a low heat flow of about 1.15, and the equatorial basin a higher heat flow of about 1.50. The average heat flow over the Mid-Atlantic Ridge is within 20 percent of the heat flow over the basins, the higher values being confined to a narrow axial zone. Figure 7-7 shows averages of heat flow for 5° x 5° squares over the North and Equatorial Atlantic and contours of a fourth-order polynomial fitted to these averages. The authors conclude that the

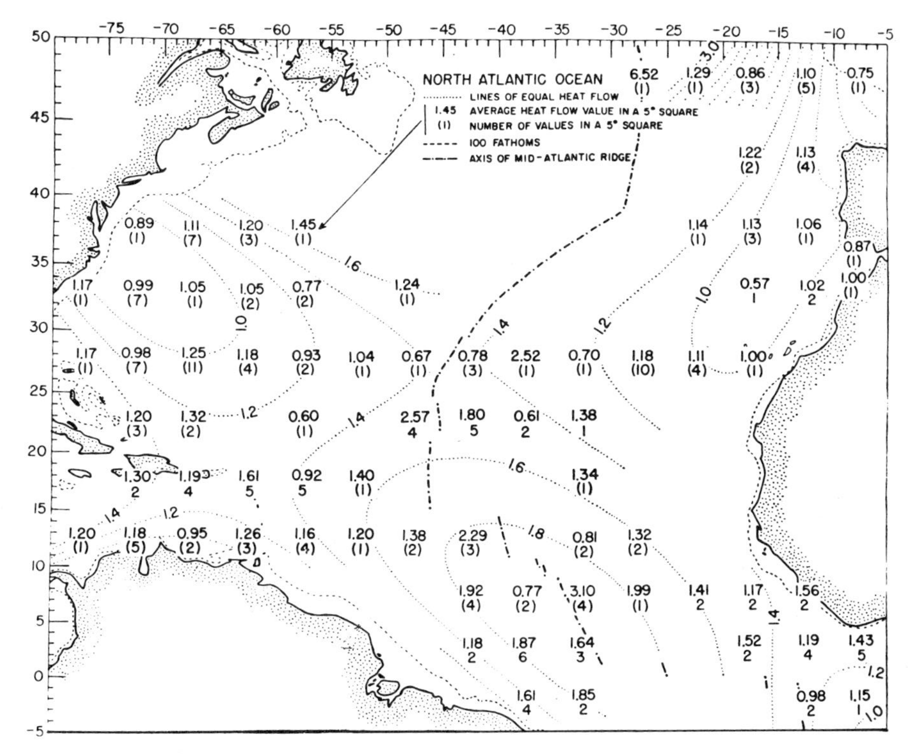

FIGURE 7-7
Averages of heat flow for 5° x 5° squares over the North and Equatorial Atlantic and contours of a fourth-order polynomial fitted to these averages. (*After M. G. Langseth, Jr., X. Le Pichon, and M. Ewing, 1966.*)

absence of a wide heat-flow maximum over the Mid-Atlantic Ridge does not lend support to continuous continental drift of the spreading-floor type during the Cenozoic era. In contrast, the wide maximum heat flow over the East Pacific and Galapagos rise system and the total heat outflow are consistent with a rate of spreading of the ocean floor in the Pacific of the order of 1 cm/year (see Chap. 16 for further discussion). The results, they claim, strengthen other geological and geophysical data which suggest that the Mid-Atlantic Ridge is in an advanced stage of evolution. Measurements across the Reykjanes ridge in the North Atlantic by K. Horai et al. (1970) confirm the coincidence of two peaks of high heat flow with topographic depressions on both sides of the ridge crest as has been found in other oceans. This is strikingly illustrated in Fig. 7-8, which shows a further correlation with anomalies in the earth's magnetic field.

K. Horai and S. Uyeda (1964) found some special features of the heat-flow measurements in and around Japan which may be characteristic of the thermal state

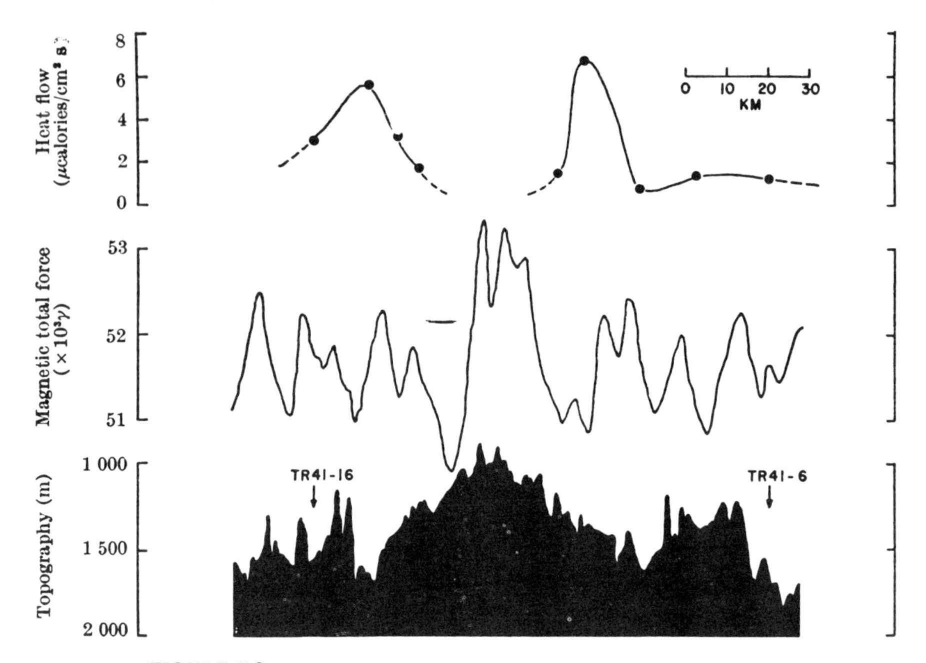

FIGURE 7-8
Profiles of heat flow, topography, and total force of the earth's magnetic field across the crestal area of the Reykjanes ridge near 60°N. (*After K. Horai et al., 1970.*)

in the crust and upper mantle of an island-arc orogenic zone (Fig. 7-9). They found regions of high heat flow on the inner side of the Japanese arc where the value exceeds 2. These high heat-flow regions coincide with those of Cenozoic volcanic activity. A region of low heat flow (<1) was found on the Pacific Ocean side of Northeastern Honshu; this low heat-flow area seems to be bounded in the east by the Japan trench. Apart from these two regions, all land heat-flow measurements lie between 1 and 2; most of these non-high heat-flow measurements come from metamorphic areas where orogenic activity ceased in Mesozoic times.

M. Yasui et al. (1966) made 116 measurements of heat flow in the Japan Sea area. The most characteristic feature is the high values distributed almost all over the area. A weak positive correlation was also found between heat flow and water depth. Contrary to what would be expected, heat flow is higher in the northern deep basin, which has a typically oceanic crust, than in the southern part with a continental crust.

In a later investigation M. Yasui et al. (1968b) reported 38 additional values. Together with the 117 earlier ones, they confirmed that the heat flow in the Sea of Japan is generally very high and uniform, the mean of the 155 measurements being

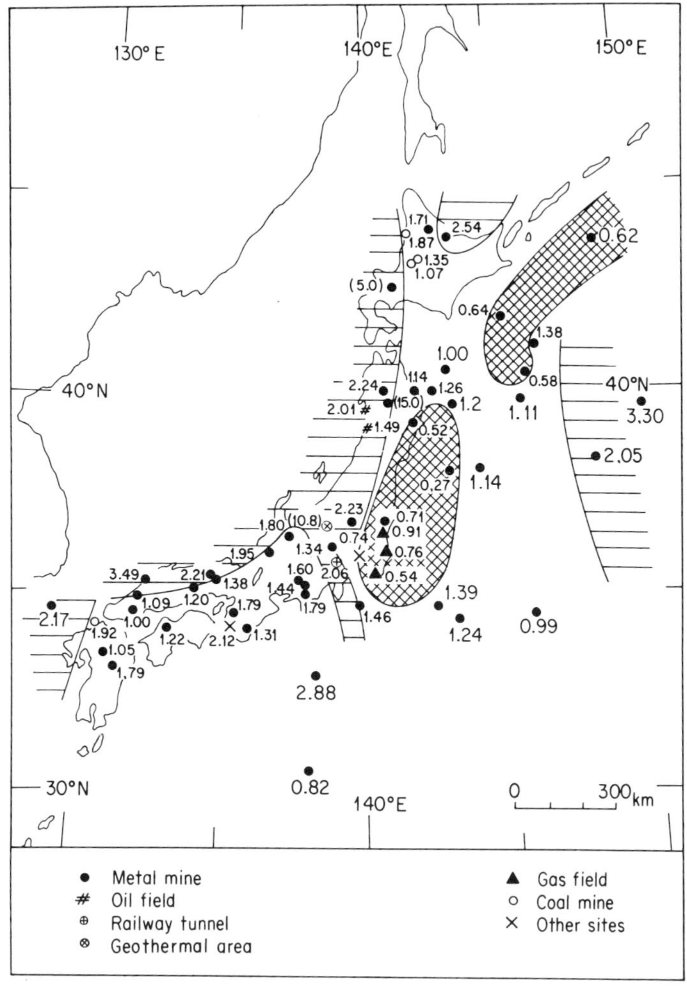

FIGURE 7-9
Distribution of heat flow in and around Japan. (*After K. Horai and S. Uyeda, 1964.*)

2.22 (±0.53). The heat flow is particularly high in the northeastern basin and the westernmost plateau, while an east-west zone crossing the Yamato rise has relatively low heat flow. The cause of the uniformly high heat flow over the entire Sea of Japan is not known, although the authors suggest the possibility of hidden volcanic activity.

M. Yasui et al. (1967, 1968a) also obtained 51 heat-flow measurements in the Okhotsk Sea, another marginal sea of the Northwestern Pacific separated from the

Pacific by a chain of Cenozoic volcanoes. Their object was to determine whether the heat-flow pattern in the Japan Sea, a marginal sea on the continental side of an archipelago, is characteristic of such an area. The 51 heat-flow values ranged from 0.72 to 4.50, with a mean of 2.05 ± 0.70. The Okhotsk Sea as a whole is characterized by a high average value and a belt of relatively low heat flows in the middle part of the sea. The Kuril basin, like the Japan Sea, has a uniformly high heat flow (mean value 2.23 ± 0.26). In the shallow northern region the heat-flow distribution is complicated. The authors show that a prominent elongated anomaly of relatively low heat flow cannot be caused by environmental effects. The anomaly coincides with the zone of deep-focus earthquake epicenters, although it is difficult to account for such a correlation. J. G. Sclater and H. W. Menard (1967) also found normal to subnormal heat flow in the seismically active area of the South Fiji basin, just south of the geothermal area of the north Fiji plateau. Again, K. Nagasaka and T. Kishii (1968) found that heat flow in the Sulu Sea is as high as that in the Japan Sea and the Okhotsk Sea, while that in the Celebes Sea is subnormal – and the Celebes Sea is much nearer to the deep-earthquake zone than the Sulu Sea. This is a further indication that such (low) heat-flow anomalies may be correlated with the downward plunge of mantle convection and the mechanism of deep-focus earthquakes.

V. Vacquier et al. (1966) reported 65 new heat-flow measurements in the Northwestern Pacific off Japan. Figure 7-10 summarizes the heat-flow values in the area as well as those in the Japan Sea, the Sea of Okhotsk, and the Japanese mainland. The most remarkable feature in the area seaward of the island arc is the uniformity of the heat flow; the Northwest Pacific basin, including the Northwest Pacific rise and Emperor seamounts, has uniform and slightly subnormal heat flow, the average being 1.15 ± 0.37. This is in sharp contrast with the generally high and nonuniform heat flow off the shores of North and Central America. Higher heat-flow values were obtained on crossing the Izu-Mariana (Bonin) arc, the average being 1.73 ± 0.93 along a strip about 250 km wide just west of the axis of the arc. This high heat flow may be an extension of the high heat-flow zone of central Honshu.

Measurements of the heat flow in the Canadian Arctic have also been made. A. H. Lachenbruch and B. V. Marshall (1964, 1966, 1968) instituted a continuing program of heat-flow observations from a drifting ice station. The mean heat flow through the Canadian abyssal plain is 1.37. This is significantly higher and more uniform than that in the highlands to the west and north, where the mean value is 1.16. There is a suggestion that a major compositional discontinuity occurs in the crust at the margin of the Canadian basin. Y. A. Liubimova et al. (1969) have also measured the heat flow through the floor of the arctic basin from a drifting ice station. Most of the measurements were made on the slopes of the Lomonosov ridge, with some on a path crossing the ridge near the North Pole. The heat-flow field appeared quite stable. They obtained a mean value of 1.96 ± 0.23, which is about 40 percent higher than the global mean. The maximum value was 2.72 ± 0.2,

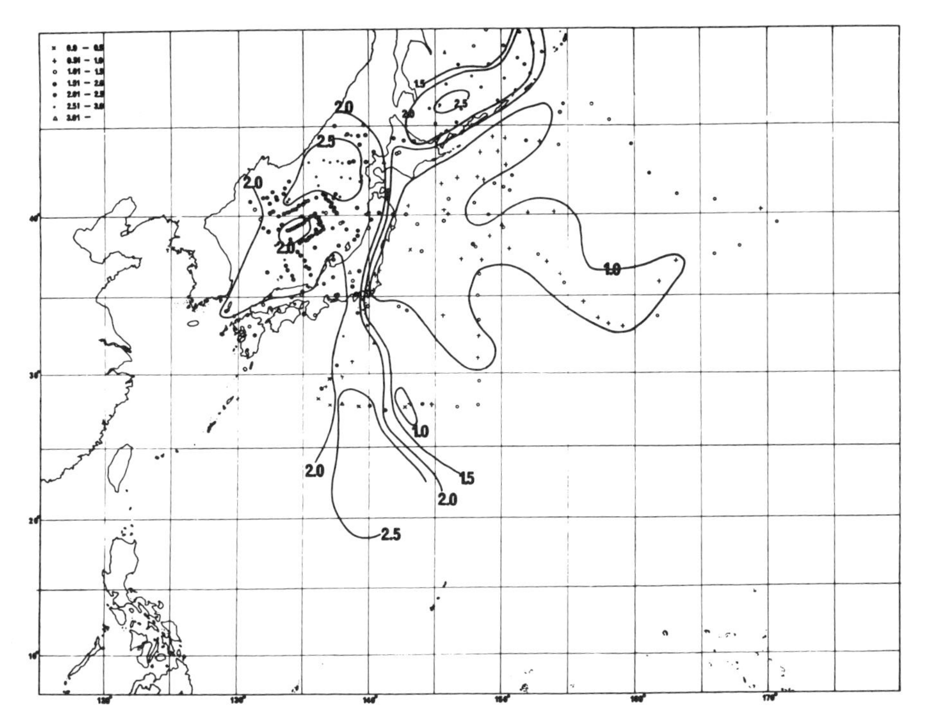

FIGURE 7-10
Summary of heat-flow values in the Northwestern Pacific and related areas. (*After V. Vacquier, S. Uyeda, M. Yasui, J. Sclater, C. Corry, and T. Watanabe, 1966.*)

and the minimum 1.4 ± 0.1. Near the North Pole the heat flow was 2.1 ± 0.1. W. S. B. Paterson and L. K. Law (1966) have also made measurements from sea ice in the general area of southern Prince Patrick Island in the Canadian arctic archipelago, in water depths of between 200 to 600 m. They obtained a value for two stations on the continental shelf of 0.46, while the mean heat flow for five stations in the channels to the east of Mould Bay was 1.46. A. H. Lachenbruch and B. V. Marshall (1969) have given a summary of arctic heat-flow measurements, including some of the difficulties of interpretation of the results of land measurements due to permafrost and the thermal effects of nearby bodies of water.

7-2.4 Measurements on Land

As already mentioned, it is more difficult to obtain heat-flow measurements on land than through the ocean floors, and there are far fewer determinations. No attempt will be made to review the results in detail. Reliable heat-flow measurements were first made in 1939 in South Africa by E. C. Bullard and in

Great Britain by A. E. Benfield; since then measurements have been obtained for North America, Asia, Europe, and Australia. Useful regional summaries have been given by E. A. Lubimova and B. G. Polyak (1969) for Eurasia and by R. K. Verma and H. Narain for India (1968).

Since 1965 a large number of heat-flow provinces have been recognized both on the continents and in the oceans. The boundaries of most major heat-flow provinces seem to correspond to those of major physiographic and tectonic provinces. R. F. Roy et al. (1968a) reported 138 new measurements of heat flow in the United States. In several areas station density was sufficient for preliminary contouring and for correlation with basement geology and radioactivity. The most striking result is the relatively high values of heat flow within the Basin and Range province, a region notable for thermal springs, widespread Tertiary volcanism, and, according to recent geophysical studies, an abnormally thin crust and low seismic velocities in the underlying mantle. All these features seem to be consistent with abnormally high temperatures at the crust-mantle boundary. At the same time, the rapid westward decrease of heat flow near the Nevada-California boundary precludes a deep source for the excess heat (the heat flow falls from 1.9 at Schurz, Nevada, to 1.1 at Gardnerville, Nevada, a distance of 60 km). In New England and New York, with no volcanism more recent than the Triassic, large variations of heat flow appear to be closely correlated with the bedrock radioactivity, with values ranging from 0.8 in the anorthositic rocks of the Adirondacks to 2.0 in the highly radioactive Conway granite. The variations could be accounted for in terms of a relatively thin layer, about 6 km thick, having the radioactivity of the surface bedrock.

F. Birch et al. (1968), in recent studies in New England and New York, found that lateral variations in heat flow could be correlated with the radioactivity of the plutonic rocks in which the measurements of heat flow were made. They found a linear relationship of the form

$$Q = a + bA \tag{7-1}$$

where Q is the heat flux at the surface, and A is the radioactive heat production of the surface rocks. It is assumed that the radioactivity measured at the surface is constant to a depth b, but varies from pluton to pluton. The heat flow a from the lower crust and upper mantle remains constant throughout the region, and the variable radioactivity of the upper crust generates the variable heat flow observed at the surface. A. H. Lachenbruch (1968) also obtained the linear relationship (7-1) between Q and A for heat-flow measurements from the central Sierra Nevada. However, to allow for differential erosion (for which there is much evidence), Eq. (7-1) can be preserved only if the crustal heat sources show an exponential decrease with depth.

In a later paper, R. F. Roy et al. (1968b) used combined radioactivity and heat-flow measurements in plutonic rocks at 38 localities in the United States to define three heat-flow provinces: the eastern United States, the Sierra Nevada, and a zone of high heat flow in the western United States, which includes the Basin and Range province. Using the linear relationship (7-1) between Q and A, they found that in the eastern United States, $b = 7.5$ and $a = 0.79$; in the Sierra Nevada, $b = 10.1$ and $a = 0.40$; and in the Basin and Range province, $b = 9.4$ and $a = 1.4$; where the units for b are kilometers and for a, 10^{-6} cal/cm^2 sec. Data from the Australian shield fall near the line characteristic of the eastern United States, which may have a broad applicability to stable portions of other continents as well. The similarity of all the slopes indicates that most local variations in heat flow are due to sources in the uppermost 7 to 11 km of the earth's crust and that the contribution from the lower crust and upper mantle is quite uniform over large regions. The intercept values can be used to infer the proportion of heat flow from the mantle.

7-2.5 Heat flow and gravity variations

From a study of the orbits of a number of artificial earth satellites, W. H. Guier (1965) has determined the shape of the geoid. The principal anomalies are a high in the Western Pacific, centered on New Guinea, and a low over India and the Indian Ocean. W. H. K. Lee and G. J. F. MacDonald (1963), from a spherical harmonic analysis (to the third order) of the terrestrial heat flow, have shown that these geoid anomalies correlate roughly with the main features of the heat-flow field; where the heat flow is lower than average, the geoid is raised, and vice versa. Such a correlation is in a sense to be expected if the anomalies in both fields are associated with convection in the upper mantle. Lee and MacDonald (1963) also showed, by an order-of-magnitude calculation, that such anomalies are consistent with convection currents having velocities of a few centimeters per year. This work has been followed up by C. Wang (1965), who supports the above correlation, although studies by W. E. Strange and M. A. Khan (1965) indicate no general correlation between free-air gravity anomalies and heat flow except over shield areas where low heat-flow values seem to be associated with negative gravity anomalies. They have shown that the expansion of heat-flow data into spherical harmonics is not meaningful outside the restricted areas where the bulk of the heat-flow data exists. W. M. Kaula (1967) also found no significant correlation on a global scale between heat flow and gravity variations, although there appears to be some evidence of an inverse correlation on a local scale. The question, however, is not completely resolved as yet. R. W. Girdler (1967) has carried out a detailed statistical analysis of all available data, and believes undulations of the geoid are due to temperature differences in the mantle. He found that the heat flow over all regions of negative gravity is significantly higher than the heat flow over all regions of positive gravity, with one notable exception; the region of negative gravity to the west of the Atlantic Ridge has lower heat flow than the world mean.

The results of the more detailed spherical harmonic analysis (up to the seventh order) of the terrestrial heat-flow field by K. Horai and G. Simmons (1969) show no correlation with the earth's gravitational field. The basic nature of these two quantities in the earth is probably quite different. A large part of the mass anomaly that produces undulations in the geoid is probably in the earth's mantle; on the other hand, crustal and upper-mantle inhomogeneities are more likely to be the major cause of the spatial variations in the observed heat flow. This is borne out by M. N. Toksöz et al. (1969), who analyzed the global variations of a number of different geophysical measurements in an attempt to explain the nature of the broad lateral heterogeneities in the earth's mantle. They found that lateral mass anomalies and/or density variations are larger by at least an order of magnitude from that required to explain the satellite data. Surface heat-flow variations are controlled primarily by the shallow structure and tectonic features of the earth, and are uncorrelated with geopotential variations (as determined by satellites). At long wavelengths, gravitational, heat-flow, and seismic travel-time variations are not correlated with topographic elevations.

7-3 TEMPERATURES IN THE PRIMITIVE EARTH AND THE EARTH'S INNER CORE

In any investigation of the thermal properties of the earth, sooner or later the question of its initial temperature and the manner in which it has cooled (or heated) since its birth is bound to arise. This in turn leads to the question of the origin of the earth and to deeper and more far-reaching astrophysical problems, and was the reason why the origin of the solar system was discussed in some detail in Chap. 1.

All thermal histories of the earth are based on either a hot or a cold origin; should a cold origin lead, at some later stage in the evolution of the earth, to a molten state, the subsequent thermal histories, whatever the initial origin, would be the same. A cold origin is at present preferred, and quite low initial temperatures (well below the melting point of the silicate rocks that form the mantle) are generally assumed. Estimates of the time of accretion are of the order of 10^8 years. Possible heat sources that could raise the temperature of the earth during this period of accretion are the radioactive decay of both long-lived and short-lived isotopes, chemical reactions, and the conversion of kinetic energy into thermal energy. Because of the comparatively short time of accretion, the temperature increase due to the radioactive decay of long-lived isotopes could have contributed (G. J. F. MacDonald, 1959). Short-lived radioactive isotopes could have contributed to the initial heat of the earth if the time between the formation of the elements and the aggregation of the earth was short compared with the half-lives of the isotopes. The important short-lived isotopes are U^{236}, Sm^{146}, Pu^{244}, and Cm^{247}, all of which have half-lives sufficiently long to have heated up the earth during the 10^7 to 10^8 years after the initial formation. MacDonald (1959) estimates that if all this heat was retained by the earth, a temperature increase of the order of 2000 to 3000°C may be possible.

The temperature of the material within the aggregating earth will also increase because of adiabatic compression. Although data (particularly on the variation of the coefficient of thermal expansion) are rather uncertain, a rise in temperature from this source of several hundred degrees seems likely. However, the largest source of available energy is the potential energy due to the mutual gravitational attraction of the particles of the dust cloud. This energy, upon aggregation, is either converted into internal energy or radiated away. It is difficult to estimate the total contribution from this source because of the uncertainty of the physical process of accretion. The result depends quite critically on the temperature attained at the surface of the aggregating earth and on the transparency of the surrounding atmosphere to radiation.

Comparatively low surface temperatures (of the order of a few hundred degrees) have been predicted, mainly because the atmosphere of the primitive earth was assumed transparent, so that the thermal energy of the impinging particles was immediately re-radiated into space. A. E. Ringwood (1960) has argued, however, that during these early years, the primitive earth will have a large reducing atmosphere. In the presence of these reducing agents (chiefly carbon and methane), the accreting material will be reduced to metallic alloys, principally of iron, nickel, and silicon. The outer regions of the earth will thus be metal-rich and dense (referred to zero pressure) compared with the interior. Such a state is gravitationally unstable, and convective overturn will follow, leading to a sinking of the metal-rich outer region into the center. This will release further heat due to the energy of gravitational rearrangement. Ringwood believes the whole process is likely to be catastrophic, since the overturn will be accelerated as the initial temperature rises. F. Birch (1965) has estimated the loss of gravitational energy on core formation for the case of simple unmixing of two components whose equations of state are found from the present density distribution. With an approximate allowance for thermal expansion, he finds that the mean energy available for heating is 400 cal/g, equivalent to a mean rise of temperature of about 1600°.

H. C. Urey (1962), on the other hand, has put forward convincing evidence that the earth, accumulating at low temperatures, has at no time in its history become completely molten. The earth has lost most of its hydrogen, helium, and other gaseous materials, and this must have taken place at low temperatures, since otherwise many fairly volatile elements such as Hg, As, Cd, and Zn would have been lost as well, which is not the case. Urey thus argues that any heat must have been lost before this separation, and thus was not available for producing high temperatures in an accumulating earth. Radioactive heating would raise the temperature until viscosity was sufficiently low, and convection would then occur. This would dissipate the heat, and no general and complete melting of the earth is likely to have taken place. It must be confessed that it is not possible at the moment to decide whether or not the earth passed through a completely molten state some time in its history. If the earth was fluid, the mechanism of cooling is reasonably straightforward. The heavy material of the core settled to the center and

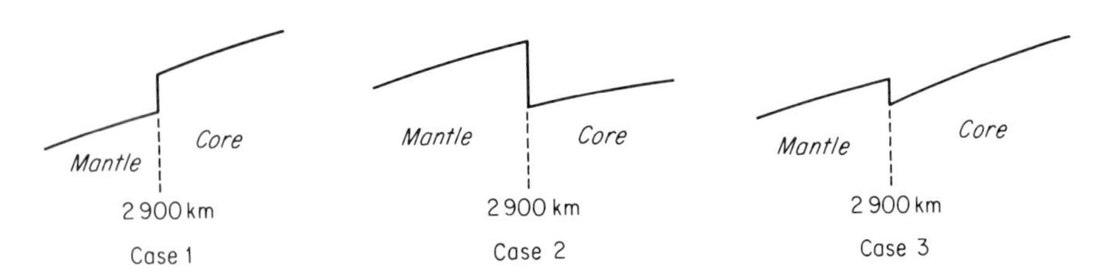

FIGURE 7-11
Possible forms of the melting-point–depth curve in the neighborhood of the core boundary. (*After J. A. Jacobs, 1954.*)

stayed there. In general, liquids as they cool contract and become denser. Thus the matter at the surface of the earth cooled by radiation into space and then sank through the molten liquid below. This would set up irregular convection currents and ensure a continued supply of heat to the surface. Before trying to estimate the temperature distribution within the earth, however, it is instructive to consider first the formation of the innner core.

All evidence indicates that at least part of the earth's core is liquid, while the mantle is solid, and any theory of the thermal history of the earth must satisfy at least these two conditions. It has been suggested, however, that the core contains a solid inner core beginning at a depth of approximately 5000 km (Sec. 2-3). K. E. Bullen (1963) has shown that the rise in velocity of longitudinal waves at this depth can be explained by assuming the inner core to be solid and of the same composition as the rest of the core. F. Birch (1952) has also come to the conclusion that the inner core is most probably crystalline iron and the outer part liquid iron, perhaps alloyed with a small fraction of lighter elements. If that is true, one is faced with the problem of giving a physical explanation of how the earth could have cooled to leave the mantle and inner core solid while the outer part of the core remained liquid. Assuming the core to consist of iron and the mantle of silicates, J. A. Jacobs (1953a) has offered the following explanation of this point.

At the boundary between the silicate mantle and the iron core there must be a discontinuity in the melting-point–depth curve, although the actual temperature must be continuous across the boundary. The form of this discontinuity could, mathematically, take any of the three cases shown in Fig. 7-11. Case 1, in which the melting-point curve in the core is always above that in the mantle, is impossible, for the actual temperature curve must lie below the melting-point curve in the mantle, above the melting-point curve in the core, and yet be continuous across the boundary. Cases 2 and 3 are both possible. In case 2, the melting-point curve in the core never rises above the value of the melting point in the mantle at the core boundary, while in case 3 it exceeds this value for part of the core. Considering first case 3, the melting-point curve will be of the general shape shown in Fig. 7-12.

As the earth cooled from a molten state, the temperature gradient would be essentially adiabatic, there being strong convection currents and rapid cooling at the surface. Solidification would commence at that depth at which the curve

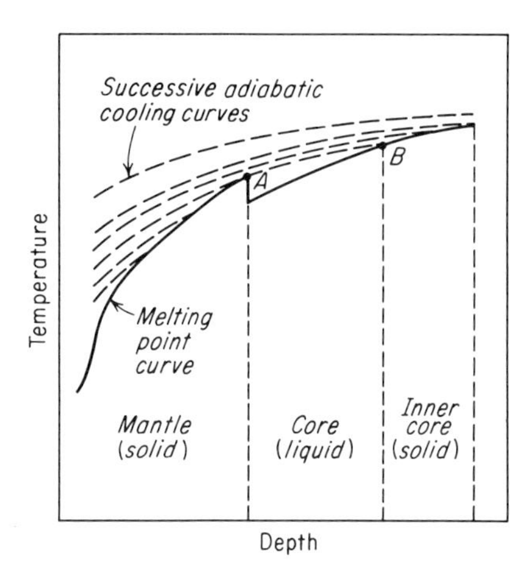

FIGURE 7-12
Melting-point curve and successive adiabats in the earth's interior. (*After J. A. Jacobs, 1953a.*)

representing the adiabatic temperature first intersected the curve representing the melting-point temperature. It is suggested, therefore, that solidification began at the center of the earth, and not at the boundary of the core and mantle as has been supposed. A solid inner core would continue to grow until a curve representing the adiabatic temperature intersected the melting-point curve twice, once at A, the boundary of the core and mantle, and again at B, as shown in the figure. As the earth cooled still further, the mantle would begin to solidify from the bottom upward. The liquid layer between A and B would thus be trapped. The mantle would cool at a relatively rapid rate, leaving this liquid layer essentially at its original temperature, insulated above by a rapidly thickening shell of silicates and below by the already solid (iron) inner core.

In the above argument no specific values of the temperatures are postulated, and the behavior of the adiabatic and melting-point curves need not be known exactly. If they vary qualitatively as shown, however, the above argument does give a physical explanation for the existence of a solid inner core.

It remains to consider the possibility envisaged in case 2 of Fig. 7-11. It follows by reasoning similar to that given above that, in this case as the earth cooled from a molten state, the entire core would be left liquid. Finally, if the earth had a cold origin and never became completely molten, then as the temperature increased with time, either case 2 or case 3 could lead to a liquid outer core with a solid inner core.

7-4 MELTING–POINT AND ADIABATIC–TEMPERATURE GRADIENTS

The argument of the preceding section will now be analyzed a little more fully, and in particular, estimates will be made of the melting-point and adiabatic-temperature gradients. A knowledge of the adiabatic gradient is of interest because it represents the minimum initial gradient that is likely to exist under any theory of the origin of the earth. A. E. Benfield (1950), has shown that even in an initially cold earth with no radioactive matter, compression alone would lead to an adiabatic gradient. The melting point of the material within the earth depends on the pressure, and hence on the depth below the surface. The effect of pressure on the melting point is given by the Clausius-Clapeyron equation

$$\frac{dT_m}{dp} = \frac{T_m}{L}\left(\frac{1}{\rho_1} - \frac{1}{\rho_2}\right) \tag{7-2}$$

where T_m = melting point
L = latent heat of fusion
ρ_1, ρ_2 = densities in liquid and solid states

Assuming hydrostatic equilibrium, the variation of pressure p with depth z is given by

$$\frac{dp}{dz} = g\rho \tag{7-3}$$

From Eqs. (7-2) and (7-3) it follows that in a liquid layer

$$\frac{dT_m}{dz} = \frac{gT_m}{L}\left(1 - \frac{\rho_1}{\rho_2}\right) \tag{7-4}$$

Taking values typical for silicate rocks, for example, T_m = 1300°C, L = 100 cal/g, and $\rho_1/\rho_2 = 0.9$, the gradient is approximately 3°C/km. Again the increase in temperature dT for a reversible adiabatic increase of pressure dp is given by

$$dT = \frac{T\alpha}{\rho c_p}\, dp \tag{7-5}$$

where α = volume coefficient of thermal expansion
c_p = specific heat at constant pressure

From Eqs. (7-3) and (7-5), the adiabatic-temperature gradient is given by

$$\frac{dT}{dz} = \frac{g\alpha T}{c_p} \tag{7-6}$$

which condition is satisfied very closely in a fluid cooling by convection. If the gradient were exceeded, convection currents would increase in strength and redistribute the temperature adiabatically. Conversely, if the gradient became less, convection currents would be damped down by viscosity, cooling at the top would become more rapid, and the gradient would increase again. Taking typical values, namely, $g = 981$ cm/sec^2, $\alpha = 2 \times 10^{-5}$ per degree centigrade, T = 1400°C, and c_p = 0.2 cal/g°C, Eq. (7-6) gives an approximate value of 0.3°C/km for the adiabatic gradient.

Although the numerical estimates obtained from Eqs. (7-4) and (7-6) are only approximate, they do show that the melting-point gradient is greater than the adiabatic, and refinements in the evaluation of these gradients are not likely to change this result. Thus, as the earth cooled, the melting point was first reached at depth, as was assumed in Fig. 7-12. The above argument assumes that there was no internal generation of heat due to radioactivity and that the gradients given by Eqs. (7-4) and (7-6) are constant. This last assumption is almost certainly not valid.

Most of the data on the earth's interior discussed in Chap. 2 have been obtained from an interpretation of seismic data, and it is natural to ask whether seismic evidence can yield any information on the thermal properties of the earth. Actually, considerable headway has been made in this direction by combining seismic data with the theory of the solid state. In particular, R. J. Uffen (1952) has obtained a refinement of the melting-point gradient which does not require a precise knowledge of the chemical or mineralogical constitution of the mantle, while several workers have sought to improve the adiabatic-temperature gradient.

F. E. Simon (1953) has used a semi-empirical equation to estimate the melting point of iron, and hence obtained melting-point–depth curves in the core. E. C. Bullard (1954b) has also used Simon's equation to estimate melting points in both the mantle and the core. A feature of all these estimates is much lower values for the melting point in the core than in the mantle. Experimental work by H. M. Strong (1959) on the fusion of iron up to a pressure of 96,000 atm also leads, by extrapolation, to comparatively low values for the melting point in the core.

J. J. Gilvarry (1956), under certain assumptions, has given a theoretical justification for Simon's equation, and obtained fusion temperatures for both the mantle and the core. Unlike other results, he obtains melting temperatures for the inner core in excess of those in the mantle (1957). On the basis of Gilvarry's curves, the explanation put forward by J. A. Jacobs (1953a) for the formation of a solid inner core as the earth cooled from a completely molten state still holds. However, if the earth had a cold origin and never became completely molten, which seems more likely, a solid inner core and liquid outer core are still possible. A rising temperature curve at depth in the interior (due to compression and radioactivity) could eventually meet the melting-point curve at the core-mantle boundary, and melting would thence progress inward, leaving a solid inner core.

Assuming the existence of a solid inner core, it is possible to estimate the actual temperature at the core-mantle boundary using values that have been obtained for the melting-point and adiabatic-temperature curves for the core. Since the boundary B (Fig. 7-12) of the inner core is the point of transition between the liquid and solid state in the core, the melting point B must be the actual temperature there. Hence, by drawing the adiabat through B, the actual temperature in the outer liquid core can be estimated. Using Simon's value of 3900°K for the melting temperature at B, Jacobs (1954) obtained a value of 3600°K for the actual temperature at A, the core-mantle boundary. Such arguments, however, can be dangerous because it is all too easy to forget the assumptions on which they rest. Estimates of "actual" temperatures in the above case were based, among other things, on the hypothesis that the inner core is

Table 7-1 DECAY CONSTANTS AND HEAT GENERATION OF ABUNDANT RADIOACTIVE ISOTOPES

Isotope	Disintegration energy		Decay constant, 10^{-10} year^{-1}	Half-life, 10^9 years	Heat generation		Percentage of abundance
	Mev/atom	ergs/atom $\times 10^{-6}$			joules/g year	ergs/g sec	
U^{238}	47.4	75.9	1.54	4.51	2.97	0.94	99.27
U^{235}	45.2	72.4	9.71	0.71	18.0	5.7	0.72
Th^{232}	39.8	63.7	0.499	13.9	0.82	0.26	100.0
K^{40}†	0.71	1.14	5.5	1.3	0.94	0.298	0.0119
K^{40}‡	...	...	5.30	1.25	0.92	0.288	0.0119
Rb^{87}	0.044	0.070	0.139	50	6.7×10^{-3}	2.1×10^{-3}	27.8
U	...	...	...	...	3.07	0.97	...
K	...	...	...	...	1.13×10^{-4}	3.55×10^{-5}	...
Rb	...	...	...	...	1.9×10^{-4}	5.8×10^{-5}	...

† Birch (1951).
‡ Aldrich and Wetherill (1958).

solid – it is all too easy to employ a circular argument and use such "actual" temperatures to prove that the inner core is solid!

7-5 HEAT FLOW AND RADIOACTIVITY

Although the heat flow through the earth is very small, the total energy released over the course of geological time is considerable. The cooling of dikes and sills and chemical changes in the near-surface rocks cannot account for anything like the observed heat flow. It was first supposed that the observed heat flow was due to the original heat which remained after the molten earth solidified. On this assumption Kelvin found that the age of the earth, i.e., measured from the time of solidification, was only 30 m yr. This estimate could be considerably increased if the thermal conductivity is assumed to increase with depth, but it is extremely unlikely that such high conductivities as would be necessary to obtain an age compatible with recent estimates do in fact exist. It is practically certain that of the observed heat flow no more than 20 percent can come from the original heat of the earth; the rest must be due to the radioactivity of the rocks.

The naturally radioactive substances of long period which occur in sufficiently high concentrations in rocks to have a decided thermal effect are uranium-238, uranium-235, thorium-232, and potassium-40. One of the main difficulties in tracing the thermal history of the earth is the unknown distribution and concentration of these radioactive materials in rocks. Table 7-1 gives the decay constants of the abundant radioactive isotopes in the earth, together with their rates of heat production, and Table 7-2 gives the average heat production by igneous rocks. The values can at best be regarded as only a general indication of the

Table 7-2 HEAT PRODUCTION BY IGNEOUS ROCKS

Type of rock	Heat produced by U, ergs/g year	Heat produced by Th, ergs/g year	Assumed content of K, 10^{-4} g/g	Heat produced by K, ergs/g year	Total heat production, ergs/g year
Granites	117	84	300	34	235
Acidic	126	109	340	38	273
Intermediate	43	36	263	29	108
Intermediate	81	81	263	29	191
Basalts	25	41	57	6.4	72
Basic lavas	26	28†	49	5.5	59
Hualalai basalt	15	16†	56	6.3	37
Twin Sisters dunite (neutron activation)	0.034	0.036†	0.1	0.01	0.08
Dunites	0.42	0.44†	0.1	0.01	0.87

† Calculated on basis of Th/U = 4.

distribution of radioactive elements, and many more determinations are desirable. For further information see D. M Shaw (1968). It is to be noted that the most recent estimates are considerably less than the earlier ones. The rate of heat production at a time t years ago was $e^{\lambda t}$ times the present rate, where λ is the decay constant. Because of the shorter half-lives of U^{235} and K^{40}, the relative contribution to the heat production by these two isotopes was much greater 4500 m yr ago. The radioactive isotope Rb^{87} is more abundant than K^{40} by a factor of 20, but the specific activity of Rb^{87} is so low that it contributes virtually nothing to the thermal conditions in the earth. It can be seen from Table 7-2 that granite and other acidic rocks are more radioactive than basalts and ultrabasic rocks and that the radioactive elements are concentrated in the crust. If the whole of the crust consisted of granite, the heat generated would be about twice the observed heat flow; even 25 km of granite resting on 10 km of basalt would produce too much. In fact, if the whole earth contained anything like as much radioactivity as is found in crustal rocks, it would be molten at a depth of a few hundred kilometers. This fact is contrary to all seismic evidence, which indicates that it is solid at least to a depth of about 2900 km. Because of the absence in the crust under the oceans of granitic rocks, with their relatively high radioactive content, it might be expected that the heat flow through the ocean floor would be considerably less than that observed through the continents. However, measurements to date do not support this theory; it appears that the average heat flow through the earth is about the samè whether through the continents or through the ocean floors (Sec. 7-2).

The work of E. C. Bullard, A. E. Maxwell, and R. Revelle (1956) has shown that the heat production in the sediments due to biological activity and radioactivity is less than 1 percent of the observed value, and since the oceanic crust can produce at most less than 10 percent of the surface heat flow, the remaining heat must come from the mantle. A distribution of radioactivity through the upper 200 km of the mantle would require a radioactive-heat production much greater than that of ultrabasic rocks or stony meteorites. On this assumption the temperatures at depths of 300 km would be some 300 to 600°C higher under the oceans than at corresponding depths under the continents. On the other hand, a distribution of radioactivity to depths greater than 200 km would lead to temperatures above the melting point unless the heat so generated could be brought to the surface. This could be accomplished by an increase in the thermal conductivity with depth (for which there is some evidence) or by slow convection currents in the mantle. Convection is likely to occur in any fluid in which there exist horizontal temperature gradients or vertical temperature gradients exceeding the adiabatic, since such temperature gradients will cause density gradients and so disturb any pre-existing hydrostatic equilibrium. The mantle is not a fluid in the ordinary sense, yet it is capable of plastic deformation. The main difference between the rheological properties of the mantle and those of an ordinary fluid are that the former has a yield point and a very high effective viscosity (Chaps. 10 and 11). These properties need not preclude the possibility of convection in systems which are sufficiently large; thermal stresses that would not induce convection in a block of stone can do so if the dimensions of the block are of the order of 10^8 cm. A. R. McBirney (1963) believes, on the other hand, that the pattern of heat-flow measurement across the oceanic rises is better explained as the result of refraction of the regional heat flow by local conductivity anomalies in the crust and upper mantle.

7-6 THE THERMAL HISTORY OF THE EARTH

The classical approach to the earth's thermal history is to formulate it as an initial boundary-value problem with calculations based on the theory of heat conduction in a solid. For the solution to be acceptable, it must give a present temperature distribution that is within the limits deduced from other geophysical evidence. It must also give a present surface heat flow comparable with that which is observed.

A. Holmes (1915) was the first to calculate the radial internal temperature distribution, assuming constant density, specific heat, thermal conductivity, and rate of radioactive heat production. He showed that about three-fourths of the surface heat flow is radiogenic, the remainder coming from the original heat content. L. B. Slichter (1941) made extensive studies of steady temperature distributions for several earth models, and J. A. Jacobs and D. W. Allan (1954, 1956), using digital computers, carried out detailed calculations of an analytical solution of the heat-conduction equation, again assuming constant specific heat and conductivity but taking into account the previously neglected time dependence of radioactive heat production. The importance of radiative heat transfer within the earth was first demonstrated by S. P. Clark (1956, 1957), and subsequent calculations on the earth's thermal history by E. A. Lubimova (1958) and G. J. F. MacDonald (1959) were dominated by such considerations. Recent reviews of the earth's thermal history have been given by W. H. K. Lee (1967) and E. A. Lubimova (1967).

Two main processes account for heat conduction in solids. Below 1000°K, energy transfer is mainly via thermoelastic waves, i.e., phonon or lattice conduction. Lattice conduction may be viewed as the propagation of anharmonic lattice waves through a continuum, or as the interaction between quanta of vibrational thermal energy. Above 1000°K, radiative heat transfer, or photon conduction, begins to dominate. Photon conduction is transmission or absorption and re-radiation of electromagnetic energy. Above the Debye temperature (which is about 500°K for the mantle) the lattice conductivity of most materials decreases with increasing temperature as $1/T^m$, where m is between 1 and 2. The effect of increasing pressure is to increase the lattice conductivity, for example, by about 2 percent on compressed samples of gabbro to 700 bars. For the earth, pressure and temperature effects on lattice conductivity seem to counterbalance each other. On the other hand, photon conduction depends on T^3, and this plays the dominant role at higher temperatures in the earth's mantle. Recently, some doubt has been cast on the importance of radiative heat transfer in the mantle. Y. Fukao (1969) found that the sum of the lattice conductivity and photon conduction of olivine is approximately constant (≈ 0.012 cal/cm sec deg) in the temperature range from 300 to 1300°K. This is very different from the simple T^3 dependence which assumes constant (i.e., temperature-independent) absorption.

G. J. F. MacDonald was the first to apply finite differences to solve numerically the heat-conduction equation, and was able to construct more complicated models of the earth's thermal history (MacDonald 1959, 1963, 1964); more recently R. T. Reynolds et al. (1966) have modified MacDonald's calculations to include the effects of melting.

Although calculations based on the theory of heat conduction are straightforward, the results are not necessarily applicable to the earth. The data required are poorly known, and conduction does not describe all the processes of heat and mass transfer inside the earth. Large-scale convection is extremely efficient in transporting heat, and such convective heat transfer will dominate thermal lattice and radiative heat conduction even for a small velocity of the order of 10^{-2} cm/year. Both G. J. F. MacDonald (1963) and L. Knopoff (1964) have presented evidence against large-scale convection in the mantle, but since their arguments depend quite critically on the assumed rheological behavior of the earth, they cannot be regarded as conclusive.

Current theories on the origin of the earth suggest an initial uniform distribution of radioactive elements. However, geochemical data indicate that radioactive elements are now concentrated toward the earth's surface. If such fractionation were completed during the earth's early history, thermal calculations based on an estimate of the present distribution of radioactivity may be justified. It would appear, however, that conditions in the past were unfavorable for an early fractionation of radioactive elements. Moreover, early fractionation is difficult to reconcile with the geological evidence which suggests continuous outgassing and differentiation of the mantle.

An investigation of the thermal history of the earth taking into account convection and fractionation of radioactive elements is, however, a formidable undertaking. The difficulties are twofold: mathematical and physical. A mathematical treatment of the problem entails formulation and solution of the complete field equations of a multicomponent, multiphase, and radioactive continuum of varying properties. The physical difficulties arise mainly from the lack of understanding of the earth's rheological behavior and fractionation processes. The most recent and detailed discussion of this problem is due to W. H. K. Lee (1967, 1968).

The most important conclusion of Lee's work is that the earth has been extensively subjected to selective fusion. Following A. E. Ringwood's (1966b) ideas on the origin of the earth, which postulate that the earth was formed from initially cold and unsorted conglomerations of cosmic dust, Lee assumes that the earth's initial temperature was similar to the melting point of iron. An important consequence is that the earth's surface heat loss has been fairly constant throughout geological time. Because radiogenic heat production increases exponentially backward with time, the constancy of surface heat flow requires that the radioactive elements were deeply buried during the early history, and have since been migrating toward the surface due to partial melting. To allow for partial melting, Lee developed mathematical techniques to treat the earth's thermal history beyond simple heat-conduction theory, taking into account latent heat, convection, and fractionation of radioactive elements. Latent heat tends to "lock" temperatures at the melting point, and convection depends critically on the earth's rheological properties. Lee has shown that large-scale convection within the earth is unlikely, and that heat transfer by small-scale penetrative convection is unimportant. Such convection is of great importance, however, as a means of moving the radiogenic heat sources upward.

Lee has constructed a number of spherically symmetric models of the mantle, covering the possible range of thermal data and initial conditions. Two sets of calculations for each model were carried out to examine the effects of upward migration of radioactive elements. In each case, latent heats of melting for a multicomponent chemical system were taken into account. His results indicate that all models with stationary heat sources are unacceptable. To yield the present observed surface heat flow and an acceptable internal temperature distribution, a combination of special and *ad hoc* assumptions must be made regarding the radioactive elements, the initial temperature, and radiative transfer. On the other hand, several models with nonstationary heat sources give acceptable results. A chondritic earth model is unacceptable because it produces too much heat, in addition to other geochemical objections. Many difficulties can be resolved if we adopt A. E. Ringwood's (1966b) theory that the primordial materials forming the earth had a composition similar to Type I carbonaceous chondrites. Lee thus used an Orgueil and a modified Orgueil model, but found that they produced present surface heat flows that were about 30 percent too high. This implies that either the earth has less radioactivity than the Orgueil or modified Orgueil models or that the initial temperature was lower than the fusion curve of iron. The Wasserburg model or a lower initial temperature model gives the required surface heat flow. G. J. Wasserburg et al. (1964) proposed the adoption of the terrestrial ratio $K/U = 1 \times 10^4$ rather than the value 8×10^4 found in chondrites. Since the earth probably lost a considerable amount of volatiles (including K, and perhaps also Th and U) during the final stages of accretion, Lee favors the Wasserburg model of radioactivity. The heat production of various radioactive earth models throughout geological time is shown in Fig. 7-13. Figure 7-14 shows the present temperature distribution in the mantle for a nonfractionated model (1*A*) and a fractionated model (1*B*). Calculations for different times in the earth's history show that for the nonfractionated model 1*A*, temperatures rose rapidly and approached the melting

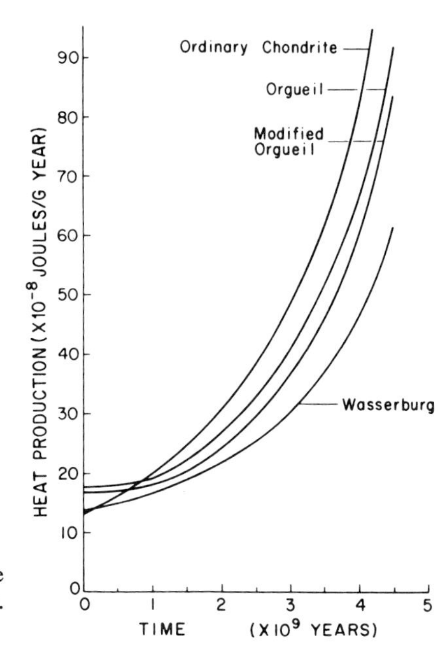

FIGURE 7-13
Heat production by various radioactive earth models throughout geological time. (*After W. H. K. Lee, 1967.*)

point of dunite after 1500 m yr. After 4500 m yr the temperature of most of the mantle is at the melting point of dunite. This model must thus be rejected, not only because it implies that the mantle is largely molten, but also because it yields a surface heat flow about one-half of that observed. When fractionation of radioactive elements is allowed (model 1*B*), the rapid rise of temperature is halted, because the upward migration of the radioactive elements has depleted the heat sources in the lower mantle and increased the surface heat loss. In this model the temperature is below the melting point of dunite at all times. The present temperature distribution for fractionated models of different total radioactivities is shown in Fig. 7-15.

The acceptable fractionated models confirm many important deductions: (1) The surface heat flow is more or less constant throughout geological time; (2) the earth has never been completely molten, but has always been near or within the range of the melting temperature of basalt and dunite; (3) the internal temperature has changed little from the initial temperature, especially during the last 3000 m yr; and (4) "magma bubbles" might have brought radioactive elements up toward the surface continuously, and thus have prevented the deep interior from completely melting. Ignorance of the actual rheological conditions in the earth does not have a highly critical effect on this process, since changes in viscosity will be compensated for by changes in the bubble size over a considerable range of values.

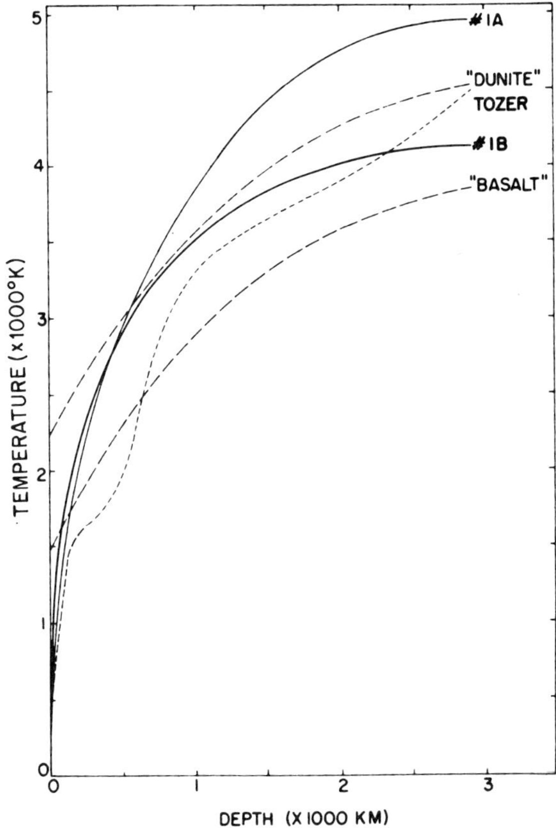

FIGURE 7-14
Present temperature distribution within the mantle for a nonfractionated model (1*A*), and for a fractionated model (1*B*). Dashed curves are melting temperatures for dunite and basalt. The short-dashed curve is the present temperature estimated by Tozer. (*After W. H. K. Lee, 1967.*)

Lee concludes that the slow upward migration of radioactive elements, rather than radiative heat transfer, plays the dominant role in the earth's thermal history, and rejects the idea that the fractionation of radioactive elements took place early in the earth's history. He argues instead for a continuous differentiation of the mantle throughout geological time; his results also support the hypothesis that the earth's atmosphere, hydrosphere, and crust have been accumulated by degassing and partial melting of the mantle. T. C. Hanks and D. L. Anderson (1969) have investigated the early thermal history of the earth using a different approach. Most theories of the earth's main magnetic field ascribe its origin to motions in the fluid core of the earth (Sec. 8-3). They thus examine the thermal history of the earth

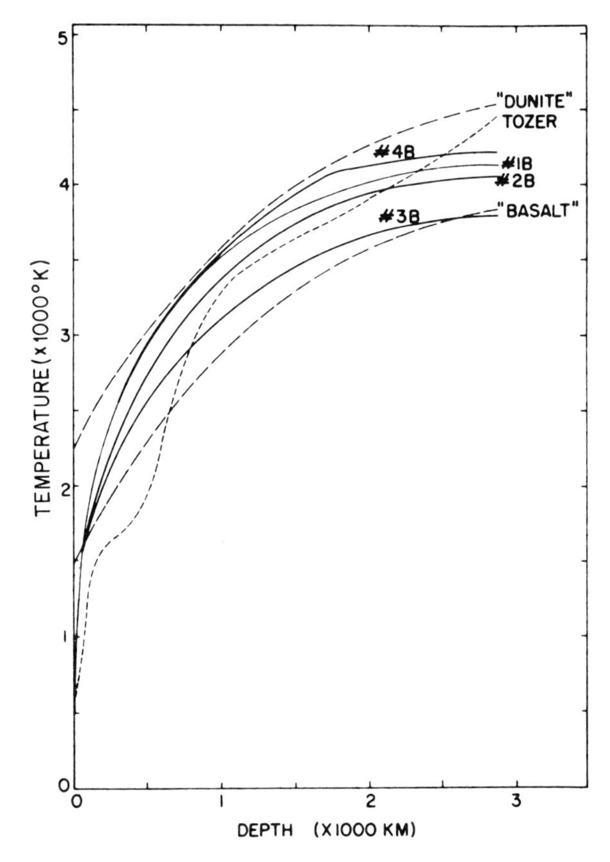

FIGURE 7-15
Present temperature distribution within the mantle for fractionated models of different total radioactivities: Orgueil (1*B*), modified Orgueil (2*B*), Wasserburg (3*B*), and ordinary chondrite (4*B*). Dashed curves are melting temperatures for dunite and basalt. The short-dashed curve is the present temperature estimated by Tozer. (*After W. H. K. Lee, 1967.*)

with the additional constraint that core formation must occur before the emplacement of the oldest known rock possessing remanent magnetism. The thermal consequences of core formation imply that the event must have antedated the oldest known surface rock (3400 m yr). To fulfill this constraint, all models of the earth considered must accrete in a period less than ½ million years (4500 m yr ago); it is most probable that the accretion period was less than 200,000 years and that large-scale differentiation of the earth upon accretion took place. This would imply a hot-origin hypothesis. One consequence is an early enrichment of radioactive elements in the outer regions of the earth; the temperatures in the earth's deep interior thus reflect residual accretional energy (including that of core formation) and are not primarily a result of radioactive heating.

SUGGESTION FOR FURTHER READING

Terrestrial Heat-flow, American Geophysical Union, Geophysics Monograph Series, No. 8, 1965.

8

GEOMAGNETISM

8-1 GENERAL FEATURES OF THE EARTH'S MAGNETIC FIELD

The existence of the earth's magnetic field was appreciated and utilized in the mariner's compass long before the field's origin within the earth was suspected. In 1180, Alexander Neckam referred to the directional property of a magnetized needle, although it was not until 1600 that the true nature of the magnetic field observed at the earth's surface was revealed by the experimental work of William Gilbert, whose famous treatise "De magnete" has been described as the first modern scientific work. The intensity and direction of magnetization not only vary from place to place, but also show a time variation, the *secular variation*. Over a period of a hundred years or so this variation may be considerable. Even today one can only speculate upon the origin and secular variation of the earth's magnetic field, which still rank among the most important unsolved problems in geophysics.

In a magnetic compass the needle is weighted so that it will swing in a horizontal plane, and its deviation from geographical, or true, north is called the *declination* D, or by mariners the *variation*. D is reckoned positive or negative according as the deviation is east or west of geographical north. The vertical plane through the magnetic force $\mathbf{F}$ (or its horizontal component H) is called the local *magnetic meridian*. Thus the declination D at any point P is the angle between the magnetic meridian and the geographical meridian through P. A magnet perfectly balanced about a horizontal axis at right angles to the magnetic meridian, so that it can swing freely in this plane, is called a *dip needle.*[1] Over most of the Northern Hemisphere the north-seeking end of the needle will dip downward, the angle it makes with the horizontal being called the *magnetic dip*, or *inclination* I. Over most of the Southern Hemisphere, the north-seeking end of the needle points upward, and the inclination I is considered negative. The curve along which I equals zero is called the *magnetic equator*. At points on the earth's surface where the horizontal component of the earth's magnetic field vanishes, the dip needle will rest with its axis vertical. Such points are called *dip poles*; two principal poles of this kind are situated near the north and south geographical poles. Their positions at present are approximately 75½°N, 101°W, and 66½°S, 140⅓°E. They are not diametrically opposite, each being over 2500 km from the point antipodal to the other, and they must not be confused with the geomagnetic poles, which will be defined later. There are also regions of local magnetic disturbances, many of which are caused by ore deposits. Some of these disturbances are so great that they give rise to local dip poles and intensities three times the normal value in those regions. Striking examples are the anomalies at Kursk, south of Moscow, and at Berggiesshubel, in Germany. The great anomaly in South Africa resulting from the Pilansberg system of Precambrian volcanic dikes shows strong magnetization in a direction opposite to the present-day magnetic field.

The intensity of the magnetic force at any point is denoted by $\mathbf{F}$. The horizontal component of $\mathbf{F}$ is called H and is always considered positive, whatever its direction. The vertical component of $\mathbf{F}$ is called Z and is reckoned positive if downward. Thus Z has the same sign as I. H may be further resolved into two components X and Y. X is the component along the geographical meridian and is reckoned positive if northward; Y is the component transverse to the geographical meridian and is reckoned positive if eastward. The magnetic elements F, H, X, Y, Z, D, I are connected with one another by the following relations (Fig. 8-1):

$$X = H \cos D \qquad Y = H \sin D$$

so that

$$\tan D = \frac{Y}{X} \qquad \text{and} \qquad H^2 = X^2 + Y^2$$

Also

$$H = F \cos I \qquad Z = F \sin I$$

so that

$$\tan I = \frac{Z}{H} \qquad \text{and} \qquad F^2 = H^2 + Z^2 = X^2 + Y^2 + Z^2$$

[1] The *dip needle* used in geophysical prospecting is weighted so that it rests nearly horizontally.

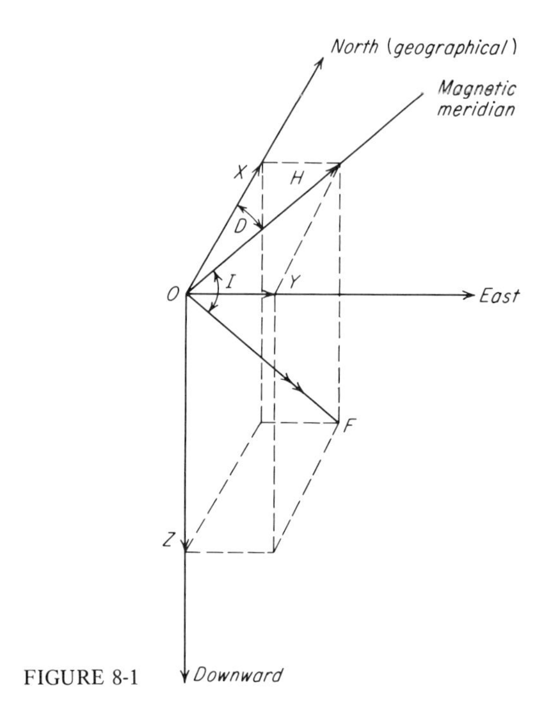

FIGURE 8-1

The unit of force used in geomagnetism is the gamma (γ), which is defined as 10^{-5} gauss (Γ).[1] The total intensity F of the earth's magnetic field reaches a maximum near the magnetic poles: It obtains a value of just over 0.6Γ near the north dip pole and a value of just over 0.7Γ near the south dip pole. Its minimum value, about 0.25Γ, is obtained near the Tropic of Capricorn off the west coast of South America. In some areas the value of F may reach 3Γ or even more, owing entirely to local concentrations of magnetic ores.

The variation of the magnetic field over the earth's surface is best illustrated by isomagnetic charts, i.e., maps on which lines have been drawn through all points at which a given magnetic element has the same value. Contours of equal intensity in any of the elements X, Y, Z, H, or F are called isodynamics. Figures 8-2 and 8-3 are world maps showing contours of equal declination (isogonics) and equal inclination (isoclinics). Figure 8-4 is a simplified isogonic chart of the north polar regions, showing that the isogonics converge at both the magnetic and geographical

[1] Strictly speaking, the unit of magnetic field is the oersted, the gauss being reserved for magnetic induction, and some geophysicists define the gamma in terms of the oersted. However, the distinction is somewhat pedantic in geophysical applications since the permeability of air is virtually 1 in cgs units. We have followed most of the geophysical literature by retaining the term gauss in both contexts.

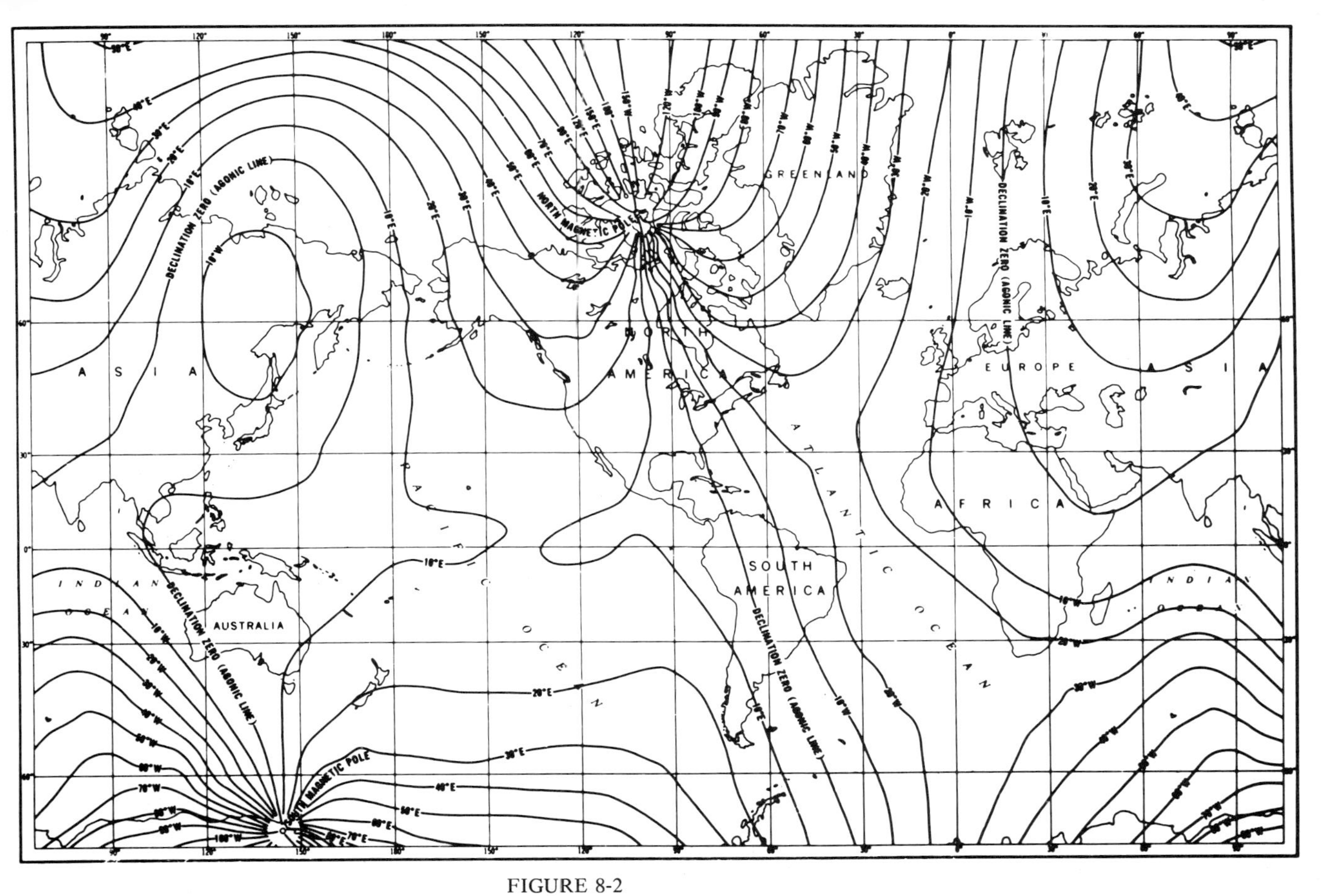

FIGURE 8-2
World map showing contours of equal declination (isogonics) for 1955. Mercator projection. (*After J. H. Nelson, L. Hurwitz, and D. G. Knapp, 1962.*)

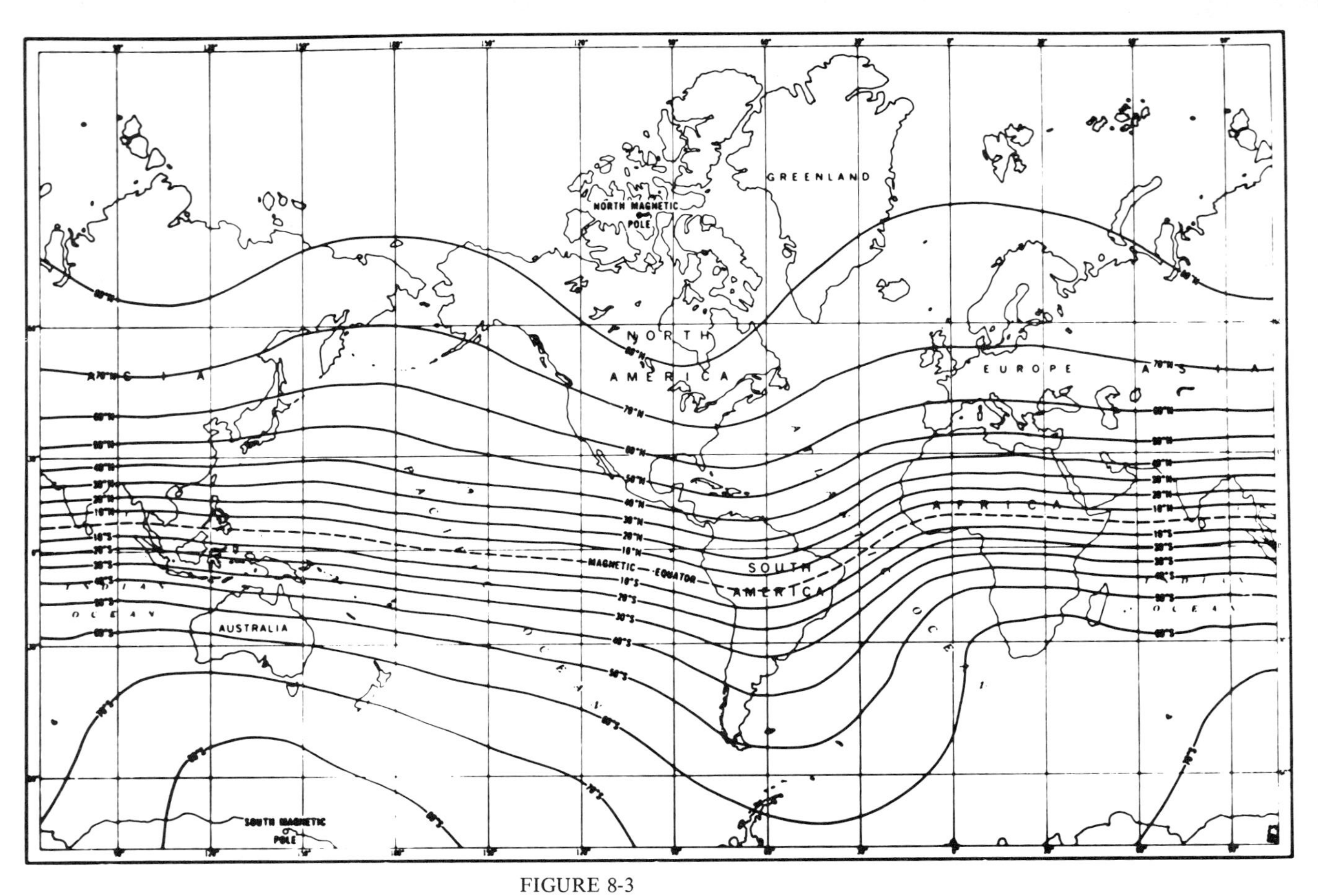

FIGURE 8-3
World map showing contours of equal inclination (isoclinics) for 1955. Mercator projection. (*After J. H. Nelson, L. Hurwitz, and D. G. Knapp, 1962.*)

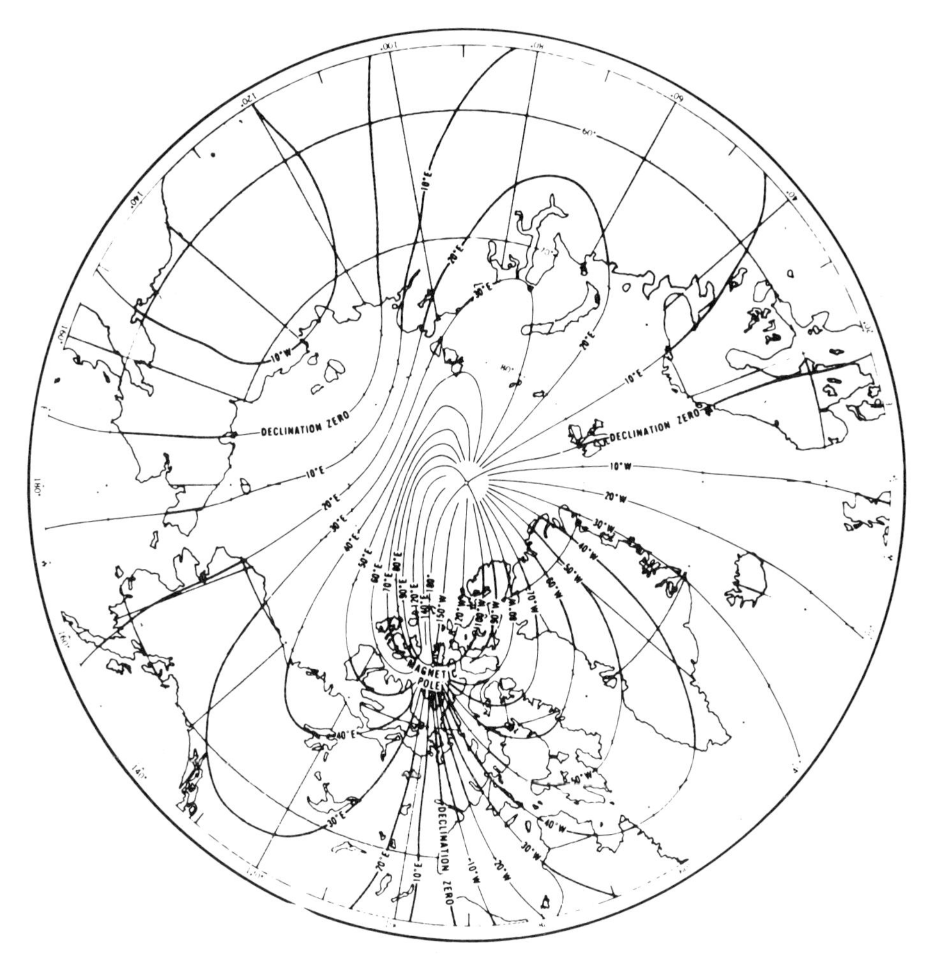

FIGURE 8-4
Simplified isogonic chart of the north polar regions for 1955. The isogonic lines converge at the magnetic poles as well as at the geographic poles. (*After J. H. Nelson, L. Hurwitz, and D. G. Knapp, 1962.*)

poles. It is remarkable that a phenomenon (the earth's magnetic field) whose origin lies within the earth should show so little relation to the broad features of geography and geology. The isomagnetics cross from continents to oceans without disturbance and show no obvious relation to the great belts of folding or to the pattern of submarine ridges. In this respect the magnetic field is in striking contrast to the earth's gravitational field and to the distribution of earthquake epicenters, both of which are closely related to the major features of the earth's surface.

It has already been mentioned that the earth's magnetic field is not constant

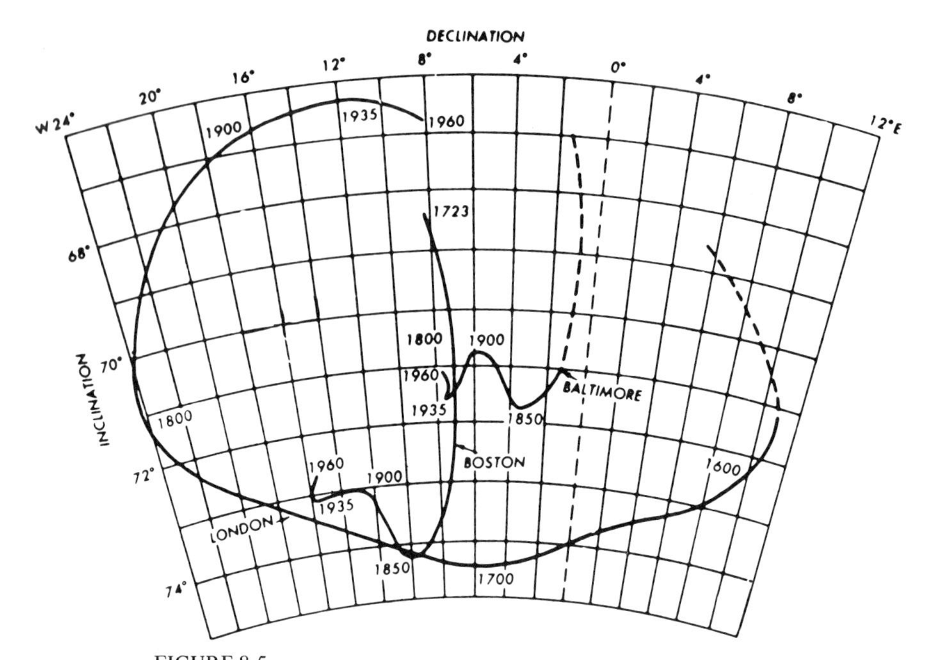

FIGURE 8-5
Secular change of declination and inclination at London, Boston, and Baltimore. (*After J. H. Nelson, L. Hurwitz, and D. G. Knapp, 1962.*)

but is continually changing. There are two distinct types of change: transient fluctuations and long-term secular changes. Transient variations arise from causes outside the earth, and produce no large or enduring changes in the earth's field. They are discussed in the next chapter. The secular change, on the other hand, is due to causes within the earth, and over a long period of time, the net effect may be considerable.

In 1635 H. Gellibrand first discovered that the declination changed with time. He based his conclusions on the observations given in Table 8-1, which were made in London. Such changes in the magnetic field with time are observed in all magnetic elements. If successive annual mean values of a magnetic element are

Table 8-1

Date	Observer	Declination
Oct. 16, 1580	William Burrow	11.3° E
June 13, 1622	Edmund Gunter	6.0° E
June 16, 1634	Henry Gellibrand	4.1° E

obtained for a particular station, it is found that the changes are in the same sense over a long period of time, although the rate of change is not usually constant. Over a period of a hundred years or so this change may be considerable. Thus the H component at Cape Town has decreased by 21 percent in the hundred years following the first observations in 1843. Figure 8-5 shows the changes in declination and inclination at London, Boston, and Baltimore. A compass needle at London was $11\frac{1}{2}$°E of true north in 1580 and $24\frac{1}{4}$°W of true north in 1819, a change of almost 36° in 240 years.

The curves suggest that there might be a cyclic variation, the magnetic pole precessing around the geographical pole (at London the period of such a precession would be about 480 years). However, the variations at other stations indicate different periods, and it is extremely doubtful that there is any significant periodicity in the variation. The secular change appears to be a regional rather than a planetary phenomenon. Isopors, i.e., lines of equal secular change in an element, appear to form sets of ovals centering on points of local maximum change, called isoporic foci. Figures 8-6 and 8-7 show the secular change in the vertical component Z for 1922.5 and 1942.5. It is clear that considerable changes take place in the general distribution of the isopors even within 20 years. Note the center of rapid secular change in South Africa, which appears to have arisen quite suddenly just before the beginning of the present century. The secular variation is anomalously large and complicated over and around the Antarctic. In an area about 1000 km in linear extent between South Africa and Antarctica, the secular change in the total intensity is at present -220γ/year, which is about 18 times as large as the average rate. The secular change is particularly remarkable in the area of East Antarctica, where at present there are two isoporic foci, one positive and the other negative, the intensities being about $+200\gamma$/year and -100γ/year.

The secular variation also appears to be markedly smaller in the Pacific hemisphere (between 120°E and 80°W) than over the rest of the earth's surface. If this is a permanent feature of the magnetic field, it must indicate systematic differences in physical properties at depth within the mantle, since the field cannot be affected by differences between oceanic and continental crustal structure. This would have serious consequences, since most of the models of the earth's interior are based on the assumption of a spherically symmetrical earth.

The variations of the earth's field are recorded at magnetic observatories all over the world on variometers, which are instruments designed to give not absolute values of the magnetic elements but only the variations. A continuous photographic record of these variations is obtained, the traces being called magnetograms. An absolute instrument is also installed at an observatory in order to calibrate the *base lines* from which the recordings of the magnetograms are measured. Apart from these fixed observatories, portable absolute instruments are available for carrying out magnetic surveys both on land and at sea.

Much effort was expended early in this century to extend magnetic measurements over the oceans and unexplored regions of the world in an attempt to obtain

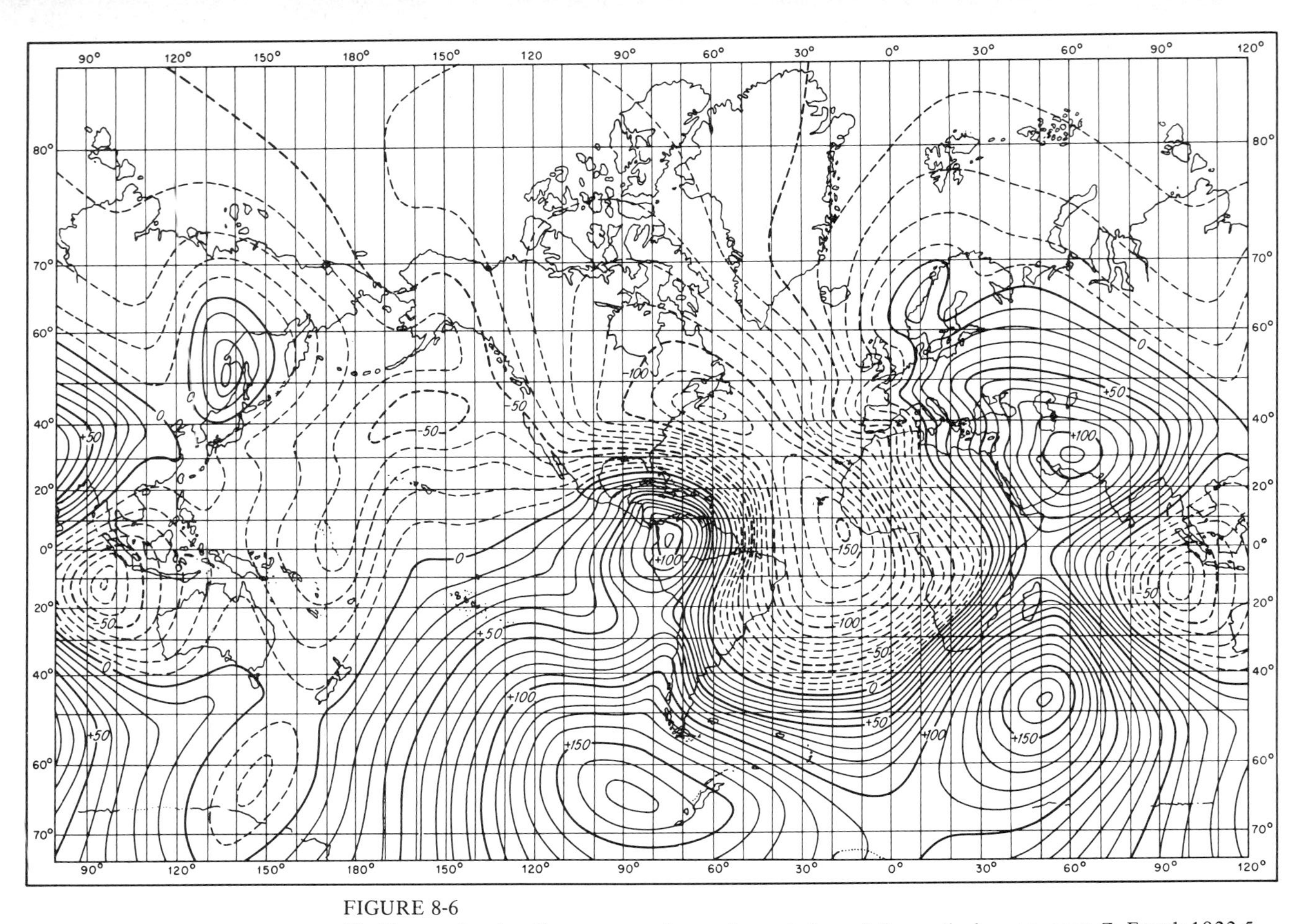

FIGURE 8-6
World map showing the geomagnetic secular variation of the vertical component Z. Epoch 1922.5. (*After E. H. Vestine, L. Laporte, C. Cooper, I. Lange, and W. C. Hendrix, 1947.*)

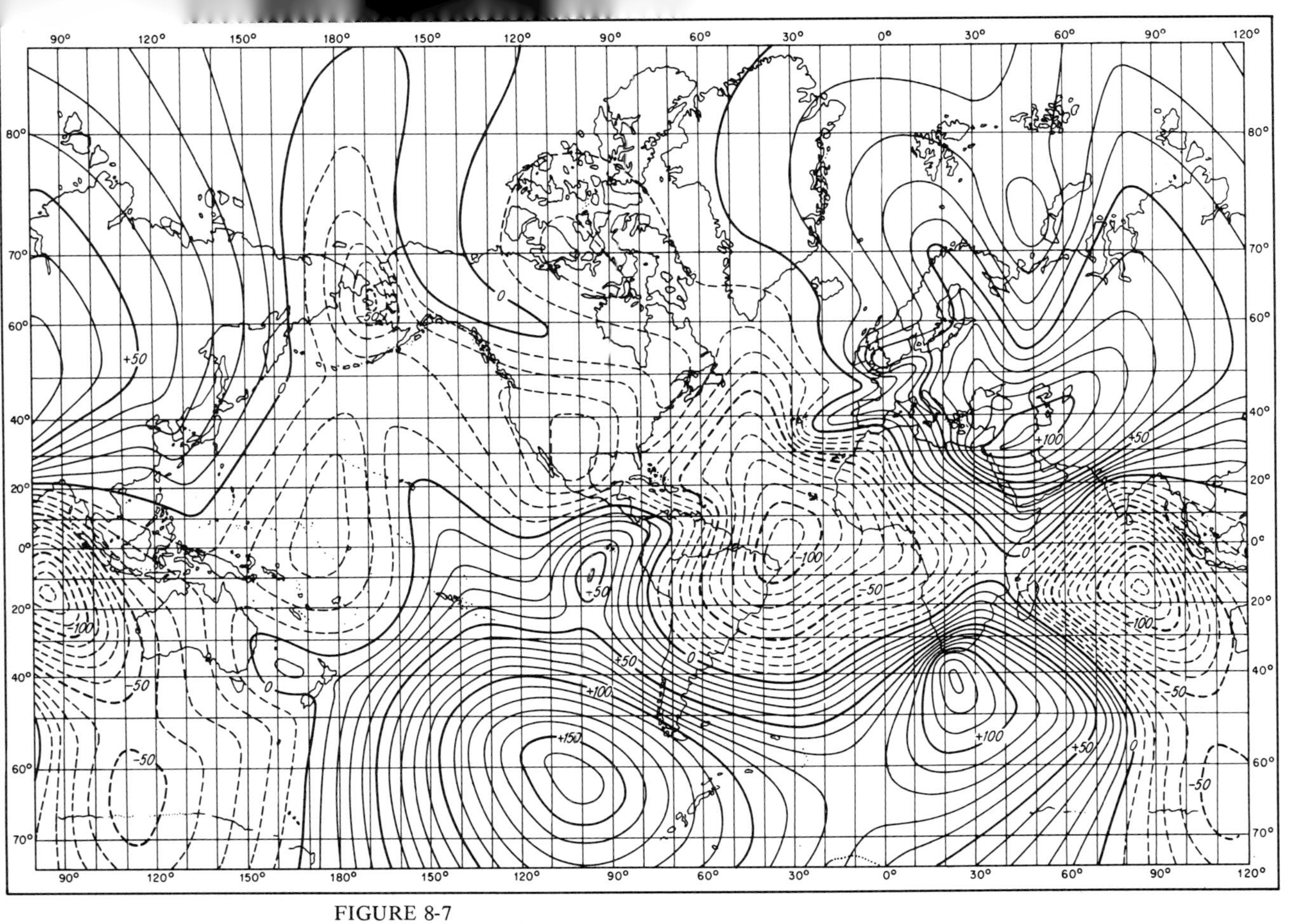

FIGURE 8-7
World map showing the geomagnetic secular variation of the vertical component Z. Epoch 1942.5. *(After E. H. Vestine, L. Laporte, C. Cooper, I. Lange, and W. C. Hendrix, 1947.)*

a more precise description of the earth's field and secular variation. Much of the initiative came from the Carnegie Institute of Washington, whose nonmagnetic ship the *Carnegie* was unfortunately destroyed by fire in Apia harbor, Samoa, in 1929. Since then few other cruises by nonmagnetic ships have been made until the Soviet Union fitted out their nonmagnetic ship *Zarya* during the International Geophysical Year. At present this is the only nonmagnetic ship that can make vector measurements of the magnetic field. Many oceanographic research vessels, however, use a magnetometer mounted in a fish and towed sufficiently far behind the ship so that the magnetic field of the ship does not influence the readings. Only total field strength F is generally obtained in this way, but such measurements have proved of great importance in tectonics (Sec. 14-3). For some time now measurements of the geomagnetic field have also been made by aircraft and have proved a valuable tool in geophysical exploration for mineral deposits. In these surveys the detail is too fine for mapping purposes because the flights have been carried out at low altitudes in a search for local anomalies. The first airborne surveys measured only the total field, but it has now proved possible to install three-component magnetometers which will measure F, D, and I. Canadian scientists have surveyed much of Canada from the air and have also made crossings of the Atlantic and Pacific Oceans. A major program of airborne surveys (Project Magnet) is at present being carried out by the U.S. Hydrographic Office. Although these surveys cannot compete with ground stations in accuracy, much valuable information can be obtained from parts of the world whose inaccessibility would make ground surveys extremely difficult. Because of the irregularity of the changes in the earth's field which usually cannot be predicted very far in advance, magnetic surveys must be repeated continually. Depending on the rapidity of the variation, a survey should be repeated every 15 to 30 years.

Determinations of the main geomagnetic field have been carried out for more than a century. One of the most detailed is that of E. H. Vestine and his colleagues (1947), who carried out an exhaustive analysis of both the main field and its secular variation between the years 1912 and 1942; in particular, they published detailed charts and tables for the four epochs 1912.5, 1922.5, 1932.5, and 1942.5. There is still a poor distribution and uneven quality of the raw data, although the situation has improved through the use of aircraft and satellites (see, for example, L. O. Tyurmina and T. N. Cherevko, 1967). There have been a number of new analyses in recent years, e.g., that of J. C. Cain et al. (1965), who evaluated the main field during the interval 1940 to 1962; B. R. Leaton et al. (1965), who estimated the geomagnetic field and secular change for the epoch 1965.0; and L. Hurwitz et al. (1966), which was used as the basis for the 1965 United States World Magnetic Charts as published by the U.S. Naval Hydrographic Office. The techniques for such analyses have generally included, first, drawing magnetic charts from the data, and then performing a spherical harmonic analysis on equally spaced grid points from the charts. D. C. Jensen and J. C. Cain (1962) suggested, however, using the raw data for the spherical harmonic analysis rather than first mapping it and using estimated values at chosen grid points. Cain et al. (1965) found this method to be more accurate. For the levels of accuracy now obtainable, it is also necessary to allow for the oblateness of the earth.

A World Magnetic Survey (WMS) was proposed at the Toronto meeting of the International Union of Geodesy and Geophysics in 1957. The need then came from a lack of adequate data for studying the origin of the main field and the secular variation and for preparing more accurate magnetic charts for navigational purposes. Since 1957 the advent of satellites has made the proposal feasible (by permitting global coverage on a time scale short compared with the time in which secular variations could affect the results), and at the same time has given additional reasons for its need. In space science an accurate knowledge of the earth's main field is needed for many purposes, such as (1) the determination of the motions and distributions of trapped particles and the trajectories of solar particles and cosmic rays as they approach the earth; (2) a description of the detailed geometry of the magnetic field lines for field-dependent radio transmissions, locating conjugate points, etc.; and (3) the provision of an accurate reference for satellite magnetic field experiments designed to determine the sources of magnetic disturbance. Thus the objective of the WMS is, first, a three-dimensional description of the field (latitude, longitude, and altitude), and second, a four-dimensional description in which the secular changes are sufficiently well known to write the coefficients as a function of time. Further details are given by J. P. Heppner (1963).

In 1967, J. C. Cain et al. submitted to the World Magnetic Survey Board a standard geomagnetic reference field. This model, which supersedes their earlier analysis (J. C. Cain, 1966; S. J. Hendricks and J. C. Cain, 1966), is described by a series of 120 spherical harmonic coefficients and their first and second time derivatives from an epoch 1960.0. It was derived from a sample of all magnetic survey data available from 1900 to 1964, together with a global distribution of preliminary total field observations from the OGO 2 spacecraft for epoch 1965.8.

8-2 FIELD OF A UNIFORMLY MAGNETIZED SPHERE

Using potential theory that had been developed by Laplace and Poisson, Gauss showed that the field of a uniformly magnetized sphere is an excellent first approximation to the earth's magnetic field. He also showed how to determine the field of this type which most closely fits the observed data at the earth's surface. Gauss further analyzed the irregular part of the earth's field, i.e., the difference between the actual field and that due to a uniformly magnetized sphere. With the data he had at his disposal he concluded that both the regular and irregular parts of the earth's field were of internal origin. This was nearly 2½ centuries after William Gilbert had come to the same conclusion as a result of experimental studies. Gilbert had explored the variation in direction of magnetic force over the surface of a piece of the naturally magnetized mineral lodestone which he had cut in the shape of a

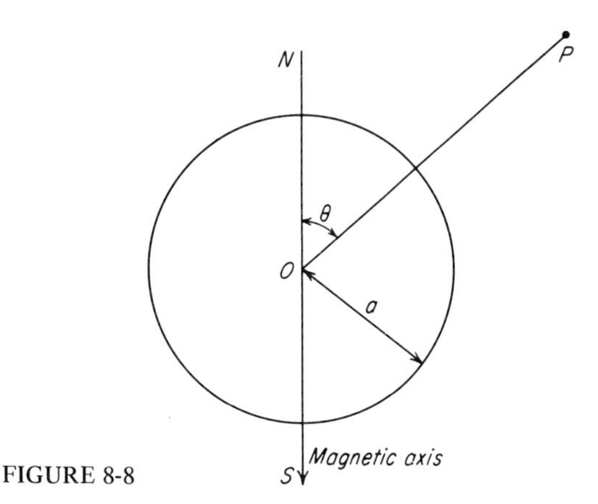

FIGURE 8-8

sphere. He found that the distribution of dip was in agreement with what was then known about the earth's field, and he concluded that the earth behaved as a large magnet, its magnetic field being due to causes from within the earth, and not from any external agency, as was supposed at that time.

Since the north-seeking end of a needle is attracted toward the northern regions of the earth, those regions must have opposite polarity. Thus consider the field of a uniformly magnetized sphere whose magnetic axis runs north-south, and let P be any external point distant r from the center O, and θ the angle NOP; that is θ is the magnetic colatitude (Fig. 8-8). If the intensity of magnetization is J and the radius of the sphere is a, then the magnetic moment $M = \frac{4}{3}\pi a^3 J$. The magnetic potential V at P is given by

$$V = \frac{-M \cos \theta}{r^2} \tag{8-1}$$

The inward radial component of force corresponding to the magnetic component Z is given by

$$Z = \frac{\partial V}{\partial r} = \frac{2M \cos \theta}{r^3} \tag{8-2}$$

The component at right angles to OP in the direction of decreasing θ, corresponding to the magnetic component H, is given by

$$H = \frac{1}{r}\frac{\partial V}{\partial \theta} = \frac{M \sin \theta}{r^3} \tag{8-3}$$

Hence the angle of dip I is given by

$$\tan I = \frac{Z}{H} = 2 \cot \theta \tag{8-4}$$

and the total intensity F by

$$F = \sqrt{H^2 + Z^2} = \frac{M}{r^3}\sqrt{1 + 3\cos^2\theta} \tag{8-5}$$

The equations of the lines of force can easily be obtained from Eqs. (8-2) and (8-3) (see Appendix C).

The maximum value of Z on the surface of the sphere ($r = a$) is $Z_0 = 2M/a^3$ and occurs at the poles. The maximum value of H on the surface is $H_0 = M/a^3$ and occurs at the magnetic equator. Thus $Z_0 = 2H_0$, a relationship which is approximately true for the earth, since $Z_0 \approx 0.6\Gamma$ and $H_0 \approx 0.3\Gamma$. The regular field which best approximates the earth's field can be obtained by a spherical harmonic analysis. This regular field has its geomagnetic axis joining the points 79°N, 70°W and 79°S, 110°E and is inclined at 11° to the earth's geographical axis. If the regular field were the total field, the dip poles and geomagnetic poles would be the same. The value of H_0 for the regular part of the earth's field (1945) was found to be 0.312Γ. This corresponds to a magnetic moment $M = H_0 a^3 = 8.06 \times 10^{25}$. If the regular field were due to uniform magnetization throughout, the intensity of magnetization

$$J = \frac{3M}{4\pi a^3} = \frac{3H_0}{4\pi}$$

would have the value 0.075. The surface layers of the earth are in general not magnetized to anything like this extent.

Measurements of H_0 have been made for over a century and indicate a decrease in H_0, and hence in the earth's magnetic moment, of about 4 percent over this period. Compared with other geophysical and geological phenomena, this secular change is extremely rapid. The cause of this secular variation and the origin of the earth's magnetic field as a whole will be discussed in the next section. It has already been mentioned that both the regular and irregular parts of the earth's field have an internal origin, derivable from potential functions. The potential of the total field can be expanded in spherical harmonics, the coefficients being determined from worldwide surface measurements. The first three coefficients represent a dipole, the regular field already discussed. The next five coefficients represent quadrupole terms. It can be shown that these eight coefficients of the expansion are equivalent to a single dipole displaced from the center of the earth plus two remaining quadrupole terms. The center of this eccentric dipole is about 300 km from the center of the earth toward Indonesia. A nonpotential field would imply the existence of electric currents flowing across the earth's surface. They could be detected by the evaluation of line integrals of the horizontal magnetic force around closed circuits on the earth's surface, the value of such an integral being $4\pi i$, where i is the total electric current flowing through the circuit. Such line integrals as have been evaluated do not vanish exactly. The most thorough analysis of the earth's surface magnetic field has been carried out by E. H. Vestine and

J. Lange for the epoch 1945, and their results indicate earth-air currents of the order of 10^{-12} amp/cm^2. However, the results have a completely random distribution, which indicates that they are probably based on incomplete and imperfect magnetic data.

8-3 THE ORIGIN OF THE EARTH'S MAGNETIC FIELD

A spherical harmonic analysis of the magnetic field observed at the surface of the earth shows that its source is predominantly internal. Superimposed on this field, however, is a rapidly varying external field giving rise to transient fluctuations. Unlike the secular variation, which is also of internal origin, these transient fluctuations produce no large or enduring changes in the earth's field. They are mostly due to solar effects which disturb the ionosphere and give rise to a number of related upper-atmospheric phenomena such as magnetic storms and aurora. Such events are discussed in the next chapter.

There has been much speculation as to the cause of the earth's main field, and no completely satisfactory explanation has as yet been given. All possible ways in which a magnetic field can be produced will be examined to see whether any are applicable to the case of the earth's field.

Consider first the possibility of permanent magnetization. The temperature gradient in the crust is approximately 30°C/km, so that at a depth of about 25 km a temperature of the order of the Curie point for iron, namely 750°C, is reached. Thus, unless the Curie point increases with increasing pressure, all ferromagnetic substances will have lost their magnetic properties at greater depths. Experimental work has been carried out at moderate pressures and indicates that at 3600 atm (which is equivalent to a depth of about 13 km in the earth), the change in the Curie point is not more than 10°C. Hence, to account for the earth's magnetic moment, an intensity of magnetization in the earth's crust of about 6Γ is necessary, which is quite impossible. Permanent magnetization also fails to account for other features of the earth's magnetic field, such as the close proximity of the magnetic and geographical poles and the high rate of change and westward drift of the nondipole field. The magnetization of rocks depends on the amount and nature of the iron oxide minerals contained in them, and for most rocks the intensity is less than $10^{-2}\Gamma$. S. A. Deel (1945) first pointed out the absence of anomalies in the geomagnetic field on a scale intermediate between local ones (of crustal origin) and those of some thousands of kilometers in extent as are observed on world maps. Presumably, therefore, there are no appreciable magnetic sources between the Curie point isotherm and the earth's core.

Consider now the possibility that the earth's magnetic field arises from the rotation of the earth. In a ferromagnetic body the magnetic moments of the atoms are associated with their angular momenta. For a rotating body gyroscopic effects cause a partial lining up of the elementary magnetic moments along the axis of rotation. Such a body is uniformly magnetized with an intensity proportional only to its angular velocity and independent of the size of the body. The effect can be observed in the laboratory at high speeds, but the angular velocity of the earth is so small that the magnetization arising from such causes is negligible, being less than about $10^{-6}\gamma$.

If the earth is assumed to have either a surface or a volume electric charge, a magnetic field would be produced by its rotation. However, the field produced in this way would appear different to an observer on the earth and to one in space not sharing the earth's rotation. If this difficulty is removed by considering the earth electrostatically neutral – a negative surface charge and a positive volume charge leading to a net dipole moment – electric fields of the order of 10^8 volts/cm are required to account for the earth's magnetic moment. Such potential gradients are many orders of magnitude greater than the dielectric strength of most materials.

Some workers have postulated changes in the fundamental laws of physics which would manifest themselves only in rotating bodies of cosmic size, but they have not met with much success. The latest attempt was made by Lord Blackett in 1947, who suggested that the dipole moment of a massive rotating body is proportional to its angular momentum. This was prompted by the discovery by H. W. Babcock (1947) that the star 78 Virginis possessed a magnetic field.

If such a fundamental property did exist, it should be possible to detect it by measuring the radial variation of the geomagnetic field in mines. Blackett's theory would suggest that the cause of the earth's field is distributed throughout the bulk of the earth, so that, on descending a mine, the horizontal field should decrease with depth. On the other hand, in nonmagnetic strata the field (supposed to originate in deeper levels) should increase downward as $1/r^3$. No measurements have of course been made at any appreciable depth within the earth, but observations by S. K. Runcorn et al. (1951) in coal mines in England gave no support to Blackett's theory. The discovery of the apparent reversals of stellar magnetic fields and of the sun's and the strong possibility of reversals of the earth's magnetic field (Sec 8.9) have thrown further doubt on the hypothesis.

During the Danish *Galathea* expedition (1950–1952), measurements were made in the Pacific Ocean of the vertical gradient of the geomagnetic field. Although the existence of large local anomalies over a rugged ocean bottom made some measurements unreliable, most results indicated a distinct increase in H and Z with increasing depth, with not a single decrease. A further consequence of Blackett's fundamental hypothesis is that a small magnetic field should be produced by a dense body rotating with the earth. In a long and detailed paper, Blackett (1952) describes the results of a "negative experiment" in which a sensitive astatic magnetometer failed to detect a field of the order of magnitude predicted near dense bodies at rest in the laboratory. Finally, any rotational theory can explain only the component of the magnetic moment along the axis of rotation. Further arguments are necessary to account for the transverse component, which is about one-fifth of the axial. We are thus finally led to the possibility that electric currents flow in the earth's interior and set up a magnetic field by induction.

8-4 THE DYNAMO THEORY OF THE EARTH'S MAGNETIC FIELD

Electric currents may have been initiated by chemical irregularities which separated charges and thus set up a battery action, generating weak currents. Paleomagnetic measurements have shown that the earth's main field has existed throughout geological time and that its strength has never differed widely from its present value. H. Lamb showed in 1883 that electric currents generated in a sphere of radius R and electrical conductivity σ and left to decay freely would be reduced by electrical dissipation by Joule heating to e^{-1} of their initial strength in a time not longer than $4\sigma\mu R^2/\pi$. This time is of the order of 10^5 years, whereas the age of the earth is more than 4×10^9 years, so that the geomagnetic field cannot be a relic of the past, and a mechanism must be found for generating and maintaining electric currents to sustain the earth's continuing magnetic field. A process that could accomplish this is the familiar action of the dynamo. The dynamo theory of the earth's magnetic field was due originally to Sir Joseph Larmor, who in 1919 suggested that the magnetic field of the sun might be maintained by a mechanism analogous to that of a self-exciting dynamo. The earth's core is a good conductor of electricity and a fluid in which motions can take place; i.e., it permits both mechanical motion and the flow of electric current, and the interaction of these could generate a self-sustaining magnetic field. The secular variation also lends support to a dynamo theory, the variations and changes in the earth's magnetic field reflecting eddies and changing patterns in the motions in the core.

It has not proved possible to demonstrate the existence of such a dynamo action in the laboratory. If a bowl of mercury some 30 cm in diameter is heated from below, thermal convection in the mercury will be set up, but no electric currents or magnetism can be detected in the bowl. Such a model experiment fails because electrical and mechanical processes do not scale down in the same way. An electric current in the bowl of mercury would have a decay time of about 1/100 sec. The decay time, however, increases as the square of the diameter of the bowl, and an electric current in the earth's core would last for about 10,000 years before it decayed. This time is more than sufficient for the current and its associated magnetic field to be altered and amplified by motions in the fluid, however slow. The dynamo theory suggests that the magnetic field is ultimately produced and maintained by an induction process, the magnetic energy being drawn from the kinetic energy of the fluid motions in the core. A group of particles moving at different speeds in the fluid may pull laterally on some magnetic lines of force, thus stretching them. In this process of stretching they will gain energy – energy which is taken from the mechanical energy of the moving particles.

Even after the existence of energy sources sufficient to maintain the field has been established, there remains the outstanding problem of sign, i.e., it must also be shown that the inductive reaction to an initial field is regenerative and not degenerative. The problem is extremely complex, involving both hydrodynamic and electromagnetic considerations, and a complete solution may not be possible. The question that has to be answered is this: Do there exist motions of a simply connected, symmetrical, homogeneous, and isotropic fluid body which will cause the body to act as a self-exciting dynamo and produce a magnetic field in the absence of any sustaining field from an external source? In an engineering dynamo, the coil has the symmetry of a clock face in which the two directions of rotation are not equivalent; it is this very feature which causes the current to flow in the coil in such a direction that it produces a field which reinforces the initial field. A simple body such as a sphere does not have this property; any asymmetry can exist only in the motions. This is the crux of the problem – whether asymmetry of motion is sufficient for dynamo action or whether asymmetry of structure is necessary as well.

Other possible causes of electromotive forces deep within the earth are thermoelectric and chemical. A likely place for either would be a major contact between dissimilar materials or between the same substances under markedly different physical conditions. The suggestion that thermoelectric currents circulate within the earth was first put forward by W. M. Elsasser in 1939. Thermoelectric electromotive forces are generated whenever two materials with different electrical properties are in contact at points which are at different temperatures. Elsasser proposed that thermoelectric electromotive forces are due to inhomogeneities in the core material which are created and continuously regenerated by turbulent fluid motion. Only a small fraction of the current generated by such a mechanism could be responsible for the surface magnetic field, the greater part producing only a contained field.[1] S. K. Runcorn (1954) suggested that thermoelectric currents are generated at the core-mantle boundary where there is a contact between two materials with different electrical properties. Temperature differences between different parts of the core-mantle boundary could be due to the eccentricity of the earth (producing a temperature contrast between the poles and the equator) or thermal convection in the core. All thermoelectric currents generated by Runcorn's hypothesis would produce contained fields, and further recourse has to be made to inductive interaction between these fields and the fluid flow in the core. Since the thermoelectric power of materials under the conditions prevailing in the core is unknown, it is extremely difficult to make a quantitative assessment of a

[1] That is, a field with lines of force not cutting the earth's surface. An electric current flow restricted to meridian planes, for example, would produce a purely zonal magnetic field, which would be a contained field.

thermoelectric theory. It seems that rather extreme assumptions are necessary to make any theory satisfactory – either an extreme geometry or extreme and implausible values of some of the physical properties of the material in the core and the lower mantle. It also appears that the convective heat flow demanded by the theory is excessive, and it is not at all certain that the required temperature differences can be realized.

8-5 THE SECULAR VARIATION AND WESTWARD DRIFT

All evidence indicates that the source of the secular variation, like that of the main field, lies within the earth's core, originating most probably in the top layers some 50 to 100 km thick. Variations occurring at great depths would not be observed, owing to the screening effect of the superimposed layers of electrically conducting fluid, while it is hardly likely that variations having the rapidity of the secular variation can take place in the solid mantle. E. H. Vestine and his colleagues in a series of papers (A. B. Kahle et al., 1967a,b; R. H. Ball et al., 1968) have inferred surface fluid motions of the earth's core from a knowledge of the magnetic field and its secular change observed at the earth's surface. The velocity patterns for epoch 1960 show an upflow in southern Africa and a downflow in the Pacific, and four horizontal rotational cells, corresponding roughly to geostrophic flow around the vertical flow. Comparison of the velocity patterns at epochs 1855, 1912, 1933, and 1960 shows the African upflow to be a strikingly persistent feature.

As early as 1692, E. Halley noticed that certain nonaxial features of the geomagnetic field drifted westward (at a rate of about 0.5 deg/year), and this result has since been confirmed by many workers. The exhaustive analyses of E. H. Vestine and his colleagues (1947) and the work of C. Gaibar-Puertas (1953) indicate a clear tendency for the isoporic foci to drift westward. E. C. Bullard et al. (1950) carried out a statistical analysis of Vestine's charts and found the mean westward drift between 1912.5 and 1942.5 of the nondipole field to be 0.18 deg/year and of the secular variation field 0.32 deg/year.

The westward drift of the geomagnetic field has been interpreted to imply that the outer core is rotating more slowly than the mantle. E. C. Bullard et al. (1950) related the westward drift to the effect of Coriolis forces on the motions in the core. They suggested that a differential angular velocity would be set up between the outer and inner core, the interchange of fluid across the outer core due to convection causing material with a smaller transverse velocity to rise to the top of the core. W. Munk and R. Revelle suggested later (1952) that the westward drift may be related to the irregular fluctuations in the length of the day. Calculations by S. K. Runcorn (1954) show that this suggestion is reasonable, provided a substantial part of the outer core is involved in the fluctuations. It is implied that there exists an adequate couple to produce the observed change in angular velocity; this could be caused by viscous or electromagnetic coupling. However, as Bullard et al. (1950)

pointed out, viscous coupling would lead to an eastward drift. Since the angular velocity of the mantle diminishes because of tidal friction, then, at any moment, the core will rotate slightly faster than the mantle and eastward with respect to it. This difficulty can be overcome by electromagnetic coupling. If there is an interchange of matter between the inner and outer parts of the core, the inner portion rotates somewhat faster and the outer somewhat slower than the core as a whole. The mantle is electromagnetically coupled to the entire core, and the outer part moves westward with respect to it. The magnitude of the electromagnetic torque is difficult to estimate but is possibly sufficient, provided the most favorable of the uncertain values of the parameters are chosen. W. M. Elsasser and H. Takeuchi (1955) have shown that fluctuations of the order of a few tenths of a gauss in the toroidal components of the core's field are sufficient to explain the observed irregularities of rotation. M. G. Rochester (1960) has investigated time-dependent perturbations of the mantle-core coupling and shown that reasonable fluctuations of the fields at the core-mantle boundary are in fact capable of explaining changes in the length of the day of the order of 1 msec in 10 years.

8-6 DYNAMO MODELS

There are two major problems to be solved in a dynamo theory of the earth's main field. First, it is necessary to show that some pattern of hydrodynamic flow exists which can produce a (predominantly) axial dipole field. Second, it must be shown that this flow exists in the earth's core.

Sir Joseph Larmor (1919) originally suggested that through inductive interaction with a small inducing field parallel to the magnetic axis, steady meridional circulation of matter might produce zonal electric currents which amplify the inducing field. The effect of such a motion would be to stretch the lines of force within the core, the field outside remaining unchanged. T. G. Cowling (1934) showed, however, that steady motion confined to meridian planes cannot amplify a field of the required type. W. M. Elsasser (1947) further showed that purely zonal motion parallel to latitude circles cannot amplify an inducing dipole field. Cowling's result has been extended by G. E. Backus and S. Chandrasekhar (1956), and it appears that homogeneous dynamos must possess a low degree of symmetry. If convection alone were active, the fluid particles would describe closed curves confined to planes, and no dynamo would result. If the system rotates, the paths of the particles are twisted into three-dimensional shapes by the action of the Coriolis force.

There are in fact a number of "nondynamo" theorems that prohibit particular types of motion in a sphere from acting as dynamos. These exclude most kinds of symmetry. Thus an axially symmetric field cannot be produced by a dynamo, although G. Roberts has shown (1971) that a symmetric motion can maintain an asymmetric field. Proofs that particular motions act as dynamos are few [see, for example, those of G. Backus (1958) and H. Herzenberg (1958), which are discussed later]. More recently S. Childress (1969) and G. Roberts (1970, 1971) have investigated dynamos in infinite bodies of fluids. In particular, Roberts has shown that "almost all" motions spatially periodic in three dimensions act as dynamos, and that these spatially periodic dynamos can be wrapped around an axis and enclosed in a sphere. These dynamos are remarkable in that, although the motion extends to infinity, the field does not. Also the field, unlike the motion, is not spatially periodic.

In their original work on the dynamo problem W. M. Elsasser (1946, 1947) and E. C. Bullard (1949a) separated the electromagnetic and hydrodynamic problems and attempted to solve the former only; i.e., they assumed a particular motion in the earth's core, together with a magnetic field, and calculated the electromagnetic interaction occurring within such a system. They drew up tables showing what interactions are possible between different types of fluid motions and a given field. It appears that, in order to produce the poloidal field outside the earth, a much more powerful toroidal field must exist in the earth's core.

In 1954 E. C. Bullard and H. Gellman, after a considerable amount of computation, appeared to have found one particular set of motions in the core that could set up a self-exciting dynamo. (They could not, of course, show that these motions do in fact exist.) Some doubt was cast, however, on the convergence of their solutions, which included harmonics up to degree and order 4. R. D. Gibson and P. H. Roberts showed later (1969) that the solution including harmonics up to degree and order 5 was very different. It may be that the velocity field chosen by Bullard and Gellman was approximately symmetric about an axis and had planes of symmetry. F. E. M. Lilley (1970) appears to have obtained better convergence by introducing asymmetry.

In 1958 G. Backus and H. Herzenberg, working independently, each showed that it was possible to postulate a pattern of motions in a sphere filled with a conducting fluid in such a way that the arrangement acts as a dynamo producing a magnetic field outside the conductor. In each case the motions were physically very improbable; however, rigorous mathematical solutions were obtained, as was not the case with Bullard and Gellman's numerical solution. The motions obtained by Backus all involved periods when the fluid is at rest. He needs the periods of rest to ensure that other fields generated by induction will not develop in such a fashion that they eventually destroy the whole process. The model of the core obtained by Herzenberg consists of two spheres (which may be pictured as two eddies), each of which rotates as a rigid body at a constant angular velocity about a fixed axis. About a half of all possible configurations can act as dynamos if the velocities, positions, and radii of the rotating spheres are suitably adjusted. The essential point of Herzenberg's dynamo is that the axially symmetric component of the magnetic field of one of the spheres is twisted by the rotation resulting in a toroidal field which is strong enough to give rise to a magnetic field in the other sphere. The axial component of this field is twisted as well and fed to the first sphere. If the rotation of the spheres is sufficiently rapid, a steady state may be reached.

F. J. Lowes and I. Wilkinson (1963) have built a working model of what is effectively a homogeneous self-maintaining dynamo based on Herzenberg's theory. For mechanical convenience they used, instead of spheres, two cylinders placed side by side with their axes at right angles so that the induced field of each is directed along the axis of the other. If the directions of rotation are appropriate, any applied field along an axis of rotation will lead, after two stages of induction, to a parallel induced field. If the velocities are large enough, the induced field will be larger than the applied field, which is no longer needed, i.e., the system would be self-sustaining. T. Rikitake and Y. Hagiwara (1966) have studied the stability of a Herzenberg dynamo and concluded that such a dynamo is unstable for a small disturbance applied to its steady state. No direct application is possible, however, to the case of the earth's dynamo since numerical integrations could only be performed for parameters very different from those in the earth's core.

Finally, it must be pointed out that solutions in which Maxwell's equations are solved for specified velocities are of limited geophysical interest since there is no guarantee that there are forces in the earth's core that can sustain them. In a dynamical theory the velocities would be calculated from assumed forces – almost nothing has been done on this problem. There have been a number of suggested causes of fluid motions in the earth's core, although only two, thermal convection and precession, seem at all possible. It is extremely difficult to assess quantitatively the effects of these two processes. Thermal convection will occur in the core if the transport of heat radially outward exceeds the heat transport by thermal conduction alone. The material of the liquid (outer) core should be very nearly in a state of chemical equilibrium, uniformity being maintained by the mixing action of the convective motion. Thus it is unlikely that any motion is due to variations in the physical properties of the liquid outer core; rather it must be determined by boundary conditions. Either the heat supplied by the inner core to the outer liquid core exceeds the amount that can be carried away from there by conduction alone, or the heat flow in the mantle adjacent to the outer core exceeds the purely conductive heat transport in the core itself (which is extremely unlikely). J. A. Jacobs (1953b) has estimated that the radioactive content of the inner core need be less than 1 percent of that in the crust for the heat flow at the boundary of the inner core to exceed that which can be carried away by conduction alone and hence set up convection. But even this small amount presents difficulties, since there are geochemical problems in having any radioactivity in the inner core, and neutron-activation analyses have failed to find any in iron meteorites (G. W. Reed and A. Turkevitch, 1956; G. L. Bate et al., 1958).

Discussions of thermal convection are not too satisfactory, since we have no means of estimating with any real certainty the total heat generated in the core, and our estimates of such physical parameters as the thermal conductivity of the core may well be out by a factor of 5 or more. The inhibiting effects of viscosity, rotation, and magnetic fields on convection are also difficult to assess.

The rotational axis of the earth precesses about a 24° cone with a period of 27,000 years. Nutation is a fluctuation of smaller amplitude and period superposed on this precession. If the core fails to precess with the mantle, strong fluid motions might be induced which would give rise to much stronger magnetic fields than are observed. It was for this reason that E. C. Bullard (1949a) suggested that precession (and nutation) do not have an appreciable effect on core motions. The mathematical difficulties of a complete discussion of the problem are very great, however, and the conclusions reached are rather unsatisfactory. Precession cannot definitely be ruled out as a major influence on fluid motions in the core. In fact W. V. R. Malkus (1968) has carried out experiments in the laboratory that lead him to believe that precessional torques acting on the earth can sustain a turbulent hydromagnetic flow in the liquid core which would drive the earth's dynamo.

8-7 MAGNETOHYDRODYNAMICS AND THE EARTH'S CORE

In the absence of a magnetic field, an ordinary fluid can transmit only compressional waves which travel with the same speed in all directions. When a magnetic field is present, if the fluid is a good conductor of electricity (and not too viscous), it can then transmit shear waves which travel along lines of magnetic force with a speed that depends on the strength of the magnetic field. Compressional waves are also affected by the presence of a magnetic field; the additional elasticity of the medium due to the magnetic lines of force increases their propagation speed by an amount that depends on the component of the magnetic field vector perpendicular to the direction of propagation.

In the hydrodynamic processes in the earth's liquid core that are responsible for the main geomagnetic field and its secular variation, R. Hide (1966) has shown that "magnetohydrodynamic" shear waves may play an important role. On the other hand, any shear wave in the core excited by a seismic wave impinging on the boundary of the core will suffer complete attenuation within a short distance of the point of incidence.

Hide (1966) has investigated small-amplitude hydromagnetic oscillations of a rotating spherical shell of incompressible fluid about a mean state characterized by a uniform (mainly toroidal) magnetic field. In the absence of rotation, the oscillations correspond to the superposition of ordinary nondispersive hydromagnetic waves propagating at the Alfvén speed in both directions along the magnetic lines of force. The effect of rotation is to reduce the phase speed of waves propagating in one of these directions and to increase the phase speed of waves moving in the other direction. Planetary-scale oscillations of the core are so strongly influenced by the earth's rotation that both types of wave are highly dispersive. The periods of the slow waves are decades, or even centuries, if the strength of the toroidal magnetic field in the core lies between 50Γ and 200Γ. Those of the fast waves are of the order of days. Thus the slow waves have oscillation periods and dispersion times comparable with the time scale of the secular variation. On the other hand, the electrical conductivity of the mantle, though weak, is sufficient to suppress from the magnetic record at the earth's surface any manifestation of magnetic variations in the core on the time scale of the fast waves. Any magnetic signals generated inside the core and introduced at the core-mantle boundary with periods less than approximately four years are effectively cut off from observation by the conducting lower mantle. The slow waves should move westward relative to the core material, and Hide suggests that the westward drift of the geomagnetic field relative to the earth's surface is a manifestation of free hydromagnetic oscillations. However, until we know much more about the spectrum of motions in the earth's core, and the theory of hydromagnetic turbulence has advanced beyond its present rather rudimentary state, any proposed mechanism for exciting free hydromagnetic oscillations must be speculative.

8-8 PALEOMAGNETISM

Since the late 1940's, paleomagnetism has become one of the most widely studied subjects in geophysics. As a result, much valuable information has been obtained in many different fields, and new light has been shed on such controversial issues as polar wandering and continental drift. However, although a considerable amount of research has been undertaken, definite answers cannot as yet be given to many of the questions. The classic early work is that of R. Chevallier (1925), who showed that the remanent magnetizations of several lava flows on Mt. Etna were parallel to the earth's magnetic field measured at nearby observatories at the time the flow erupted. The recent intense interest in paleomagnetism goes back to the work of E. A. Johnson, T. Murphy, and O. W. Torreson (1948), who examined the remanent magnetism of a series of varved clays in New England. These clays were laid down in stream beds and contain small quantities of detrital magnetic material in the form of fine grains which were oriented in the direction of the earth's field at the time of deposition. By measuring changes in the direction and intensity of magnetization with depth over a continuous vertical sequence, it was possible to trace the history of the earth's magnetic field as far back as 15,000 B.C. The results indicate that, during the whole of this time, the field has undergone fluctuations in direction about geographical north (Fig. 8-9). More recent investigations confirm this view, and it seems that for at least several million years the earth's main field has corresponded on the average to that which would be produced by an axial magnetic dipole situated at the center; in addition, there has been a continuous secular variation about the mean direction with a time period of from 500 to 1000 years (cf. Fig. 8-5). Striking statistical evidence that it was axisymmetrical in Mesozoic times (5 to 50 m yr ago) has been obtained by O. W. Torreson et al.

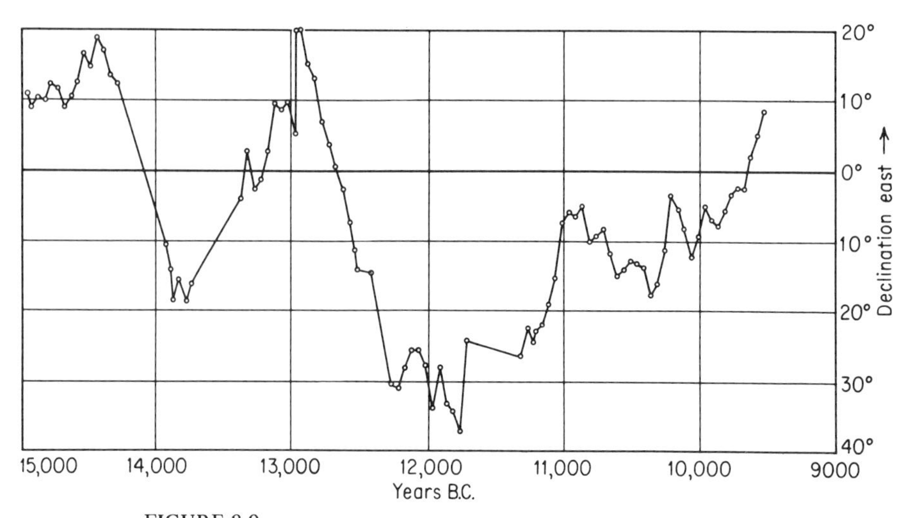

FIGURE 8-9
Declination of the earth's magnetic field in New England, 15,000–9000 B.C. (*After E. A. Johnson, T. Murphy, and O. W. Torreson, 1948.*)

(1949). Figure 8-10 shows how the mean declination of flat-lying sedimentary rocks of Mesozoic age gives a better fit to an axial dipole than to the present dipole.

Most rocks owe their magnetism to the various oxides of iron which they contain. However, the manner in which igneous and sedimentary rocks acquire their permanent magnetism is very different. Igneous rocks have been injected into pre-existing rock and extruded at the surface in a liquid state and then subsequently cooled and solidified. When lava cools and freezes following a volcanic outburst, it will acquire a permanent magnetization dependent on the orientation and strength of the geomagnetic field at that time. This magnetization may remain practically constant because of the small capacity for magnetization in the earth's field after freezing, and is much larger than would be acquired in the present geomagnetic field at 20°C. It has been established experimentally that the permanent magnetization acquired by an igneous rock which has been heated and cooled through the Curie point in magnetic fields of the order of the earth's field (this permanent magnetization is called thermoremanent magnetization, TRM) is considerably greater than the permanent magnetization acquired by exposing it to the same field at room temperature. It therefore seems probable that the present direction of magnetization of an igneous rock as found in the field is the same as that of the local magnetic field at the time of solidification, allowing for any subsequent tectonic movements.

Sedimentary rocks may be magnetized by the orientation in the earth's field of small grains of magnetic material such as magnetite when deposited in shallow

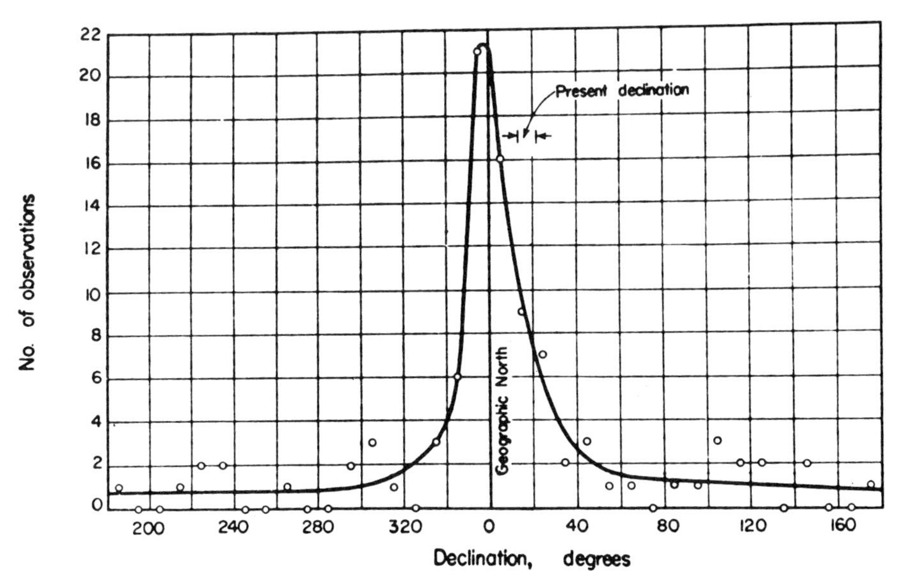

FIGURE 8-10
Frequency distribution of declination measurements on Mesozoic rock samples. (*After O. W. Torreson, T. Murphy, and J. W. Graham, 1949.*)

rivers, lakes, and seas. The magnetic grains tend to align themselves along the direction of the earth's field while the sediment is still wet and unconsolidated. Later the material may harden because of compression, and the magnetic grains may get locked in their original positions. In addition, many sedimentary rocks undergo marked physical and chemical changes while they are being consolidated by compression and perhaps subjected to heating, and their magnetization may be altered by these processes. Chemical magnetization is probably the main origin of the magnetization of the well-known red sandstones. One of the principal uncertainties in the interpretation of paleomagnetic data is that of deciding how much the rocks have changed physically and chemically since they were laid down.

Compared with the large number of ancient field directions which go back to Precambrian times, comparatively few accurate determinations have been made of the intensity of the geomagnetic field in the past. The pioneering work in this field was done by E. Thellier (1938) and J. G. Koenigsberger (1938). In a review article E. and O. Thellier (1959) describe the difficulties of the problem and the reasons for such a small number of intensity measurements. More recently P. J. Smith (1967) has reviewed all paleomagnetic intensity measurements, both from historic (archeomagnetic) and geological specimens. It appears that the geomagnetic dipole is not constant within any given polarity, but fluctuates in strength, possibly with a period of the order of 10^4 years. Because of these fluctuations and nondipole

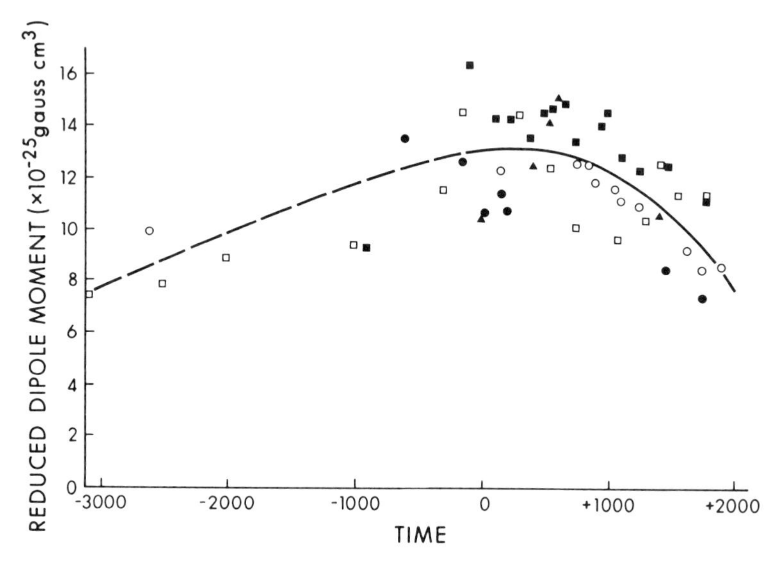

FIGURE 8-11
Historic and archeological reduced dipole moments plotted against time. Dipole moments have been calculated using present geomagnetic latitudes. Times A.D. are expressed as positive numbers; times B.C. as negative numbers. The different symbols represent the results of different investigators. (*After P. J. Smith, 1967.*)

variations, it is impossible to draw conclusions regarding the mean strength of the geomagnetic dipole at any given time if only a few rock samples are measured.

During the past 2000 years, the geomagnetic dipole moment has apparently decreased by about one-third of its peak value to the present directly observed value of 8.0×10^{25} gauss/cm^3. Prior to the last two thousand years the dipole moment was increasing (Fig. 8-11). The scatter of the points is large, and in order to show the main trend, the results have been averaged in groups of 10. It appears that the curve in Fig. 8-11 is due neither to secular variation of the nondipole field nor to wobble of the main dipole. Over the past 150 years the geomagnetic dipole moment has been decreasing at the rate of about 7 percent per century, in good agreement with the value of 5 percent per century deduced from direct field measurements.

Field intensities in the geological past are given in Fig. 8-12, which shows that for at least the past 400 m yr the mean geomagnetic dipole moment was smaller than the earth's present dipole moment. However, in view of the paucity of data before the Tertiary, it would be unwise to consider this increase of the mean geomagnetic dipole moment with time as much more than a possibility.

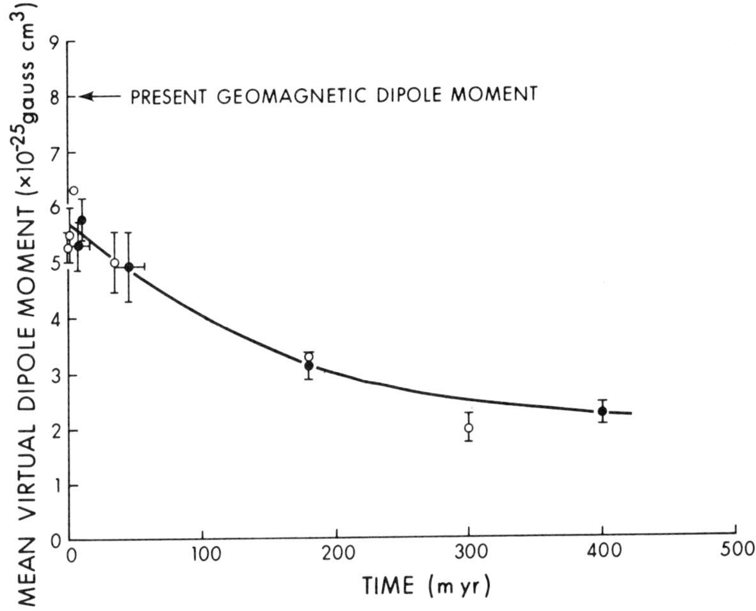

FIGURE 8-12
Mean geological virtual dipole moments (VDM) plotted against time (m yr), i.e., equivalent dipole moments which would have produced the measured intensity at the magnetic paleolatitude of the samples. • signifies accurate (Class 1) values. ○ signifies approximate (Class 2) values. Error bars in VDM represent standard errors of the means; error bars in time represent the range of potassium-argon ages where appropriate. (*After P. J. Smith, 1967.*)

8-9 FIELD REVERSALS

One of the most interesting results of paleomagnetic studies is that many igneous rocks show a permanent magnetization approximately opposite in direction to the present field. The reason for this reverse magnetization has been the subject of much speculation, and it has not yet been settled. Reverse magnetization was first discovered in 1906 by B. Brunhes in a lava from the Massif Central mountain range in France. Since then examples have been found in almost every part of the world, including Spitsbergen, Greenland, Europe, Australia, Japan, and South Africa, where H. Gelletich (1937) discovered that the huge Pilansberg dike system was reversely magnetized. After the last war, the problem was investigated in detail by J. M. Bruckshaw and E. I. Robertson (1949) on dikes in northern England, by A. Roche (1950a, b, 1951, 1953) on Tertiary lava flows in central France, by J. Hospers (1951, 1953, 1954) on Tertiary lava flows in Iceland, and by C. D. Campbell and S. K. Runcorn (1956) on the late Tertiary lavas of the Columbia River basalts. In each case it was found that about one-half of the flows were normally magnetized and one-half reversely. There were no intermediate cases. Neither Roche nor Hospers could detect any difference by laboratory experiments between the magnetic properties of the normal and reversed rocks, and both concluded that the earth's field reversed its polarity several times during the Tertiary period – probably about every half million years – the duration of the reversal process being of the order of 10,000 years. More recently an extensive paleomagnetic survey has been made of eastern Iceland by P. Dagley et al. (1967), where some 900 separate lava flows lying on top of each other were sampled. The direction of magnetization of more than 2000 samples representative of individual lava flows was determined covering a time interval of 20 m yr. At least 61 polarity zones, or 60 complete changes of polarity, were found, giving an average rate of at least three inversions per million years.

There is no a priori reason why the earth's field should have a particular polarity, and in the light of modern views on its origin, there is no fundamental reason why its polarity should not change. There have been many cases where reversely magnetized lava flows crossed sedimentary layers. Where the sediments have been baked by the heat of the cooling lava flow, they were also found to be strongly magnetized in the same reverse direction as the flow. In fact, in all reported cases the direction of magnetization of the baked sediment is the same as that of the dike or lava which heated it, whether normal or reversed. It seems improbable that the adjacent rocks, as well as the lavas themselves, would possess a self-reversal property, and such results seem difficult to interpret in any other way than by a reversal of the earth's field.

Before such an explanation is accepted, however, it must be asked whether any physical or chemical processes exist whereby a material could acquire a magnetization opposite in direction to that of the ambient field. J. W. Graham (1949) found some sedimentary rocks of Silurian age which were reversely magnetized. He was able to identify the precise geological horizon over a distance of several hundred miles by the presence of a rare fossil which existed only during a short geological period. He found that some parts of the horizon were normally magnetized and some reversely, and argued that this could not be accounted for by a reversal of the earth's field which would affect all contemporaneous strata alike. (However, if the time scale of reversals in Silurian times was as short as Hospers found for Tertiary rocks, Graham's fossil might well have survived at least one reversal.) Graham thus wrote to Professor L. Néel of Grenoble and asked him if he could think of any process by which a rock could become magnetized in a direction opposite to that of the ambient field. Néel (1951, 1955) came up with not one but four possible mechanisms, and within two years two of these four mechanisms were verified, one by T. Nagata for a dacite pumice from Haruna in Japan, and one by E. W. Gorter for a synthetic substance in the laboratory.

The first and third of these mechanisms involve only reversible physical changes, while the second and fourth involve in addition irreversible physical and/or

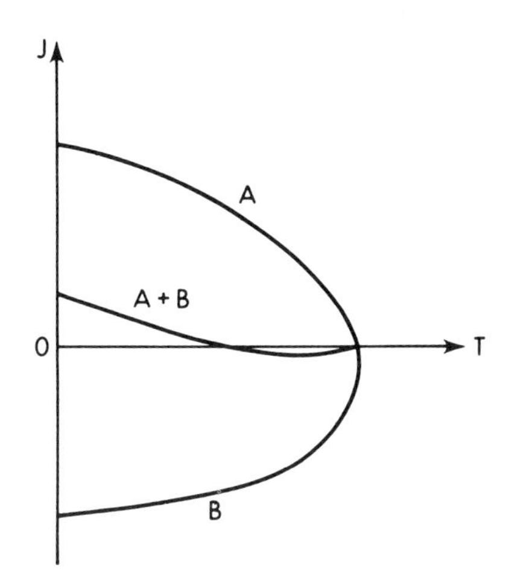

FIGURE 8-13

chemical changes. In his first mechanism Néel imagined a crystalline substance with two sublattices A and B, with the magnetic moments of all the magnetic atoms in lattice B oppositely directed to those of lattice A. If the spontaneous magnetization of the two sets of atoms J_A and J_B varied differently with temperature, Néel suggested that the resultant magnetization of the whole, $J_A + J_B$, could reverse with change in temperature (Fig. 8-13). E. W. Gorter and J. A. Schulkes two years later synthesized a range of substances with the properties predicted by Néel, although no naturally occurring rock has been found which behaves in this manner.

Néel's second mechanism is a modification of the first, in which $J_A > J_B$ at all temperatures, so that no reversal would take place. However, Néel suggested that subsequent to the formation of such a substance, chemical or physical changes might occur which would lead to the demagnetization of lattice A, leaving the reverse magnetization of lattice B predominant. No evidence of such a possibility actually occurring in nature has yet been found.

For his third mechanism, Néel considered a substance containing a mixture of two different types of grains, A and B, one with a high Curie point T_A and a low intensity of magnetization J_A, and the other with a low Curie point T_B and a high magnetization J_B (Fig. 8-14). When such a substance cools from a high temperature, substance A, because of its higher Curie point, becomes magnetized first in the direction of the ambient field. When the temperature falls below T_B, substance B becomes magnetic but will be subject to the dual influence of the ambient field and of the field due to the grains of substance A. Néel suggested that under suitable geometrical conditions, the resultant direction of magnetization of B can be on the average opposite to that of the ambient field. At room temperature the greater value of J_B makes the resultant magnetization of the whole in the opposite direction to the ambient field. In 1951 T. Nagata found a dacite pumice

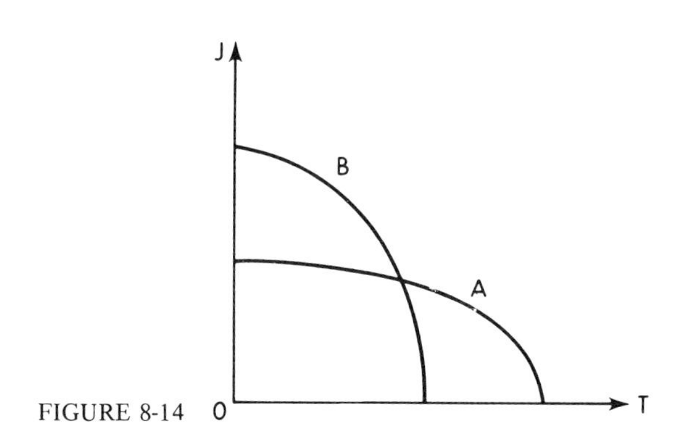

FIGURE 8-14

from Haruna in Japan which was reversely magnetized in the field and which behaved in the laboratory in the way which Néel had predicted. However, the great majority of igneous rocks which show reversal in the field do not show this reversal property in the laboratory.

Néel's fourth mechanism, like his second, involves the possibility of subsequent demagnetization by physical or chemical changes. Thus reverse magnetization might be possible later in time, even though initially the intensity of the *B* component was not large enough. Since discrete magnetized grains free to rotate must align themselves along and not against the field, only the second and fourth of Néel's mechanisms could apply to sedimentary rocks; igneous rocks could, in theory, become reversely magnetized by any of them.

Reversals occurred during the Precambrian, and have been observed in all subsequent periods except the Permian. There is no evidence that periods of either polarity are systematically of longer or shorter duration. P. Dagley et al. (1967), in their monumental work on Icelandic lavas, found 73 lava flows whose directions of magnetization were anomalous, and of these at least 55 were unambiguously intermediate—magnetized neither in the normal nor the opposite direction. This was about 5 percent of the lava flows examined, and indicates that reversal does not simply involve a change in the sense of the dipole. These intermediate directions could be caused either by the continuous change of the dipole axis (with possibly also a change in strength) or by the reduction in strength of the dipole without change in orientation, thus allowing the nondipole part of the field to make a proportionally larger contribution, and so give the anomalous directions. N. D. Watkins (1963) has examined the behavior of the geomagnetic field during the Miocene from the lavas of the Columbia River Plateau in southeastern Oregon. In one section a gradual transition between normal and reversely magnetized flows was found across a succession of six flows which may represent intermediate directions of the geomagnetic field during the process of reversal. Since this reversal is likely to be accompanied by a decrease in intensity of the main dipole, nondipole sources could be expected to control the ambient field during this period. Hence a more rapid variation of magnetic direction would occur, resulting in a scatter of measured paleomagnetic directions. P. J. Smith (1967) has examined the limited data from transition zones and has come to the conclusion that during a field reversal the dipole moment reduces to zero although the nondipole field remains.

An extremely interesting finding is that almost all rocks of Permian age that have been studied have reversed polarity. If the field-reversal hypothesis is incorrect, it follows that mineral assemblages necessary for self-reversal are abundant in Carboniferous and Triassic rocks (both these periods have many reversals) but are all but absent in Permian rocks. Such a conclusion is very difficult to believe; it is far more plausible to assume that the field did not alternate during the Permian.

E. Asami (1954a,b) has examined some early Pleistocene lavas at Cape Kawajiri, Japan. Several hundred specimens were taken from closely spaced sites along the coastline. Along some stretches of the coast all the magnetization was normal; in other stretches it was reversed; and on some stretches normal and reversed were found close together. Such results show that one must be cautious about interpreting all reversals as due to a field reversal, and the problem of deciding which reversed rocks indicate a reversal of the field may in some cases be extremely difficult. To prove that a reversed-rock sample has been magnetized by a reversal of the earth's field, it is necessary to show that it cannot have been reversed by any physiochemical process. This is a virtually impossible task, for physical changes may have occurred since the initial magnetization or may occur during certain laboratory tests. More definite results can come only from the correlation of data from rocks of varying types at different sites and by statistical analyses of the relation between the polarity and other chemical and physical properties of the rock sample. If the dipole field of the earth has reversed, it is most probably a result of physical processes occurring in the core of the earth, and should thus be quite uncorrelated with physical processes associated with the crust and outer mantle or the atmosphere, such as orogenic and volcanic activity or climatic changes. A number of workers have reported chemical differences between normally and reversely magnetized lava sequences from various parts of the world (see, for example, R. L. Wilson and N. D. Watkins, 1967). Reversely magnetized lavas appear to be more highly oxidized than normal ones; no differences in the distribution of other elements involved in the magnetic minerals have been observed. The significance of this correlation is not yet clear; it is hard to imagine what physical connection can exist between the polarity of the earth's magnetic field and the state of oxidation of lavas which became magnetized in that field.

If the origin of reversals is one of the instantaneous self-reversal mechanisms (such as that of the Haruna dacite), normally and reversely magnetized rocks should be randomly distributed throughout a group of rocks of different ages. If reversals are due to one of the time-dependent self-reversal mechanisms, reversals should be increasingly abundant in older rocks. If, on the other hand, reversals are due to geomagnetic field reversals, normal and reversely magnetized groups of rocks should be exactly the same age over the entire earth; and unless it so happened that the earth's field suffered more reversals in the past, the proportion of reversed magnetizations should not be greater among older rocks. Although there can be no doubt that self-reversal occurs in some rocks, the stratigraphic distribution of normally and reversely magnetized rocks strongly supports the field-reversal hypothesis.

Four major normal and reversed sequences during the past 3.6 m yr have been found. These major groupings have been called geomagnetic polarity epochs, and they have been named by A. Cox et al. (1964) after people who have made significant contributions to geomagnetism. Superimposed on these polarity epochs are brief fluctuations in magnetic polarity with a duration that is an order of magnitude shorter. These have been called polarity events and have been named after the localities where they were first recognized (Fig. 8-15). The best statistical estimates of the ages of the boundaries between geomagnetic polarity epochs are Gilbert-Gauss boundary, 3.36 m yr; Gauss-Matuyama boundary, 2.5 m yr; and Matuyama-Brunhes boundary, 0.70 m yr. The duration of polarity events is estimated to vary from 0.70 to 0.16 m yr, and the best estimate of the time required for the earth's field to undergo a complete change in polarity is 4600 years. There has been much speculation on the possibility of short intervals of reversed polarity within the Brunhes epoch of normal polarity. The first report of a possible reversal (the Laschamp event) was made by N. Bonhommet and J. Babkine (1967), who found reversed magnetizations in two different formations of the Chaine des Puys (Auvergne, France) volcanic chain of Quaternary rocks. Age determinations by the C^{14} method of the lava flow of the Puy de Laschamp placed the event between 7000 to 9000 years ago, while K-Ar ages indicate an upper limit of 20,000 years for the end of the event. More recently J. D. Smith and J. H. Foster (1969) have examined the magnetic record of seven cores of deep-sea sediments and established the existence of a short period of reversed polarity (the Blake event) in the upper part of the Brunhes epoch. The reversed zone in the cores correlates well with paleontological boundaries and is estimated to have existed between 108,000 and 114,000 years ago (±10 percent).

R. Hide (1966) has suggested that changes in the radial velocity of the fluid motions in the earth's core might in some cases be impressed from outside. Horizontal temperature variations of only a few degrees and "topographical features" (bumps) only a few kilometers high at the core-mantle interface should produce quite marked effects on core motions, and could in principle bring about reversals. Gradual changes in the radius of the core and in the potency of the mechanism that drives core motions should also produce occasional reversals. Hide thus suggests that two general types of reversal should occur: "forced" reversals due to changes impressed from outside the core, and "free" reversals that would arise even in the absence of impressed changes. Each type of reversal would be characterized by its own range of time scale. Major geological events are associated with large-scale motions in the mantle. If these motions penetrate to a sufficient

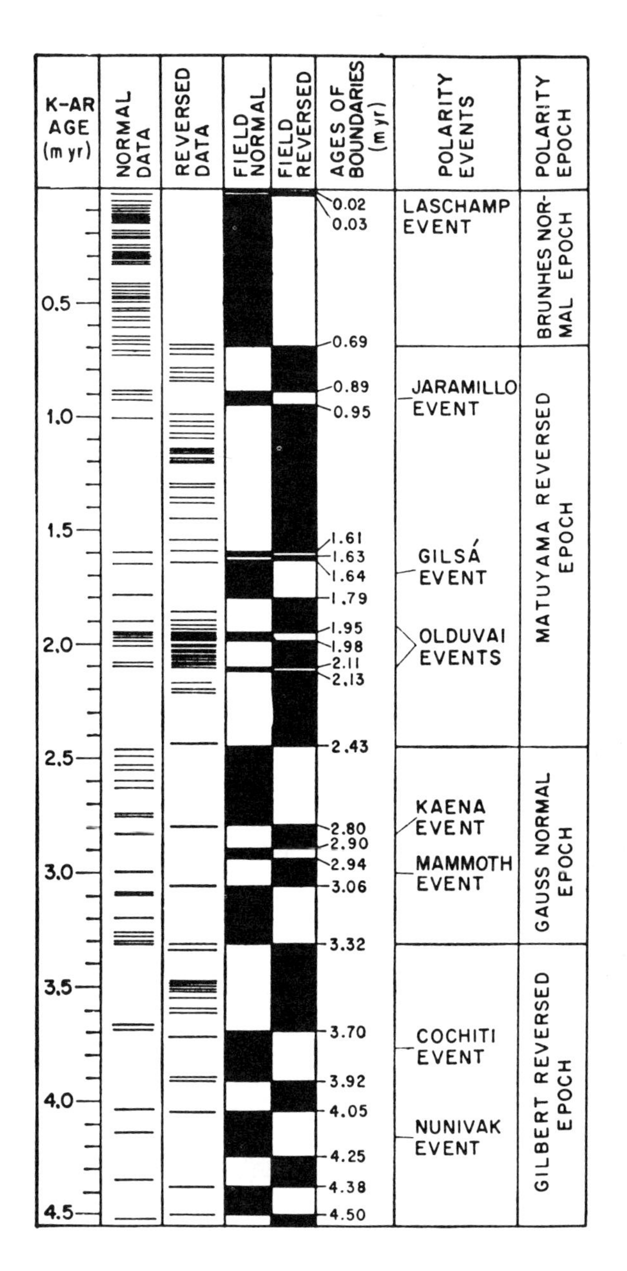

FIGURE 8-15
Time scale for geomagnetic reversals. Each short horizontal line shows the age as determined by potassium-argon dating and the magnetic polarity (normal or reversed) of one volcanic cooling unit. Normal polarity intervals are shown by the solid portions of the "field normal" column, and reversed polarity intervals by the solid portions of the "field reversed" column. The duration of events is based in part on paleomagnetic data from sediments and magnetic profiles. (*After A. Cox, 1969b.*)

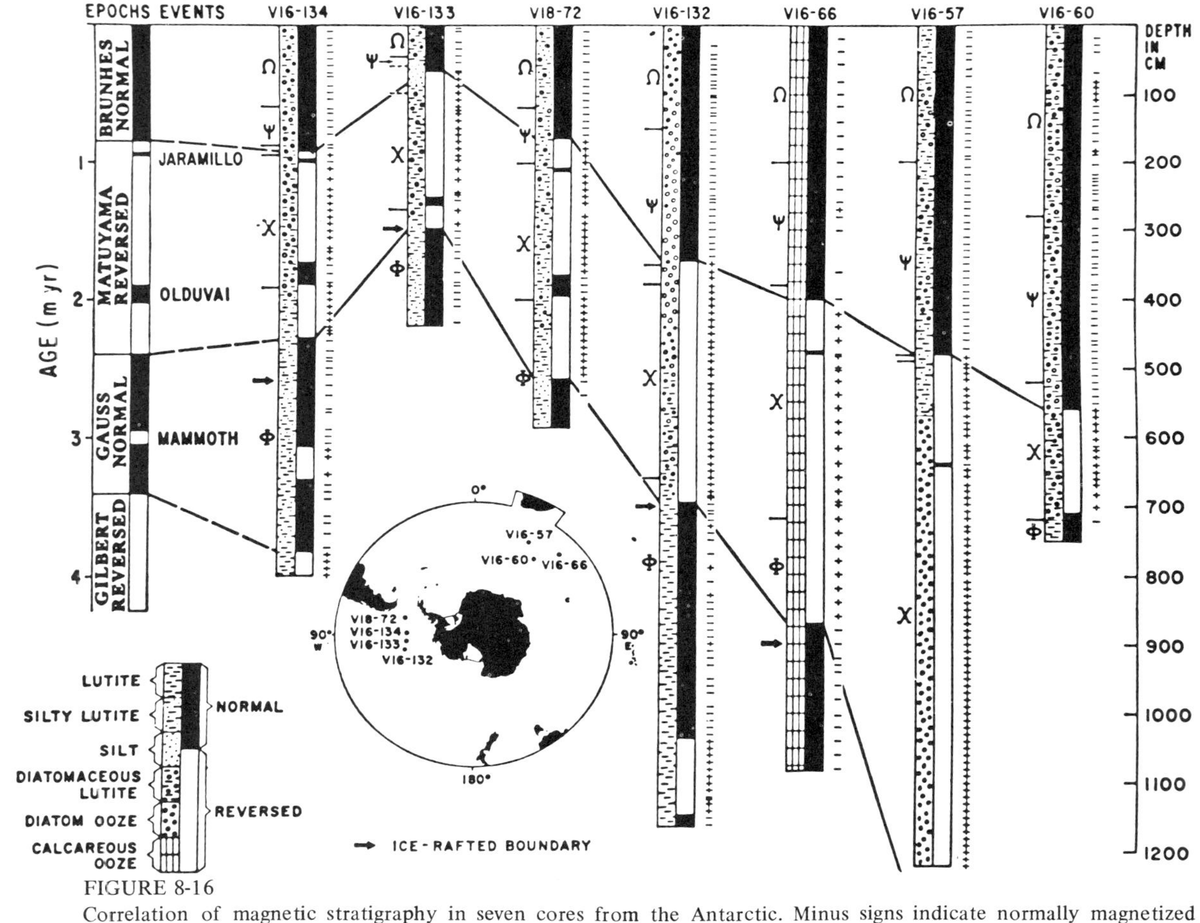

FIGURE 8-16
Correlation of magnetic stratigraphy in seven cores from the Antarctic. Minus signs indicate normally magnetized specimens; plus signs, reversely magnetized. Greek letters denote faunal zones. Inset: source of cores. (*After N. D. Opdyke, B. Glass, J. D. Hays, and J. Foster, 1966.*)

depth to produce horizontal variations in the physical conditions that prevail at the core-mantle interface, "forced" reversals should be strongly correlated with other worldwide geological phenomena. "Free" reversals, however, should show no such correlations.

That the foregoing classification of reversals might be a useful one receives some support from observations. Thus there are "paleomagnetic intervals" in which almost all rocks are reversely magnetized, such as the period from the upper Carboniferous to the upper Permian, a time span of 40 to 50 m yr, some 300 m yr ago (E. Irving, 1964a). The "events" and "epochs" found by A. Cox et al. (1964) in their paleomagnetic study of carefully dated rocks up to 4 m yr old occur on time scales of the order of 100,000 and a million years, respectively. These much shorter "intervals" are probably caused by "free reversals."

Geologists are now applying the reversal time scale to establish age relationships among rocks that would be difficult to date by any other means. A particularly important application is the determination of the ages of deep-sea sediments. N. D. Opdyke et al. (1966) have found a polarity record in cores going back to the Gilbert epoch (3.6 m yr), in which the pattern of reversals is remarkably similar to the pattern of reversals obtained from igneous rocks on land (Fig. 8-16). Thus polarity studies can provide a method for determining rates of sedimentation and for establishing worldwide correlations among different deep-sea sediments. It is very striking that the long sequence of 60 changes of polarity identified in Icelandic lavas by P. Dagley et al. (1967) agrees both in character and approximately in chronology with the sequence of magnetic reversals which appears to be fossilized in the igneous rocks spreading outward from the ocean ridges. A particularly interesting feature is that 32 reversals of the earth's magnetic field ago, they found a long period of normal magnetization which coincides with one already reported from studies of the ocean floor near the East Pacific rise.

8-10 POLAR WANDERING

If it is assumed that the geomagnetic field at the earth's surface averaged over several thousands of years can be represented by a geocentric dipole with its axis along the axis of rotation, it is possible, by measuring the present direction of magnetization of a suite of rocks, to deduce the position of the earth's rotational axis relative to the location of the rocks at the time when they were laid down. The measured declination will give the azimuth of the land mass at the time the rock became magnetized, and the measured inclination will give the geographical latitude [see Eq. (8-4)]. The accuracy of the interpretation of any results will depend on the accuracy with which the rocks can be dated and on the reliability of the magnetic measurements, which depend on the stability of the magnetization. Figure 8-17 shows the polar-movement paths for several continental areas; an interpretation of these and other paleomagnetic results involves a consideration of the

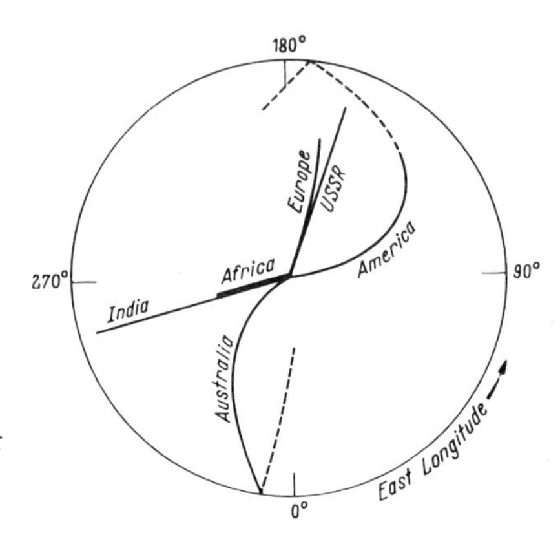

FIGURE 8-17
Schematic polar-movement paths for several continental areas. (*After R. L. Wilson, 1965.*)

hypotheses of polar wandering and continental drift, which will be discussed in detail in the later chapters of this book.

It must be realized that a movement of the entire earth relative to the poles of rotation can be achieved by internal effects; it is only the vector of angular momentum that has to remain fixed in space. T. Gold (1955) pointed out that a beetle weighing 1 g could overturn a stationary sphere resembling the earth by walking 10^{27} times around it; for a rotating sphere, the beetle can cause any desired angle of movement of the poles relative to the sphere by merely sitting in the right place. The actual earth, however, possesses greater stability than this by virtue of its shape. It is spinning around a principal axis of inertia of greater moment than the other two, and any departure from this condition implies an increase of energy if the angular momentum is to remain the same. If the earth were a perfectly rigid body so that it had complete stability of shape, it would maintain indefinitely the same axis of rotation within close limits despite geological or other events occurring on the surface.

Information on the earth's rigidity to forces applied over a long period can be obtained from the "free," or "Eulerian," nutation, whereby the axis of rotation moves by a small angle, of the order of ¼″ relative to the earth with a period of about 14 months. (See also Sec. 4-2.) This has the effect of distorting the earth as a whole and thereby moving the axis of symmetry of the spheroid. Any plastic flow of the body of the earth in response to the forces resulting from this slightly misaligned spin would show itself in a damping of the motion. In fact, the motion is found to be very significantly damped, decaying in the absence of disturbances to $1/e$ of its initial value in 10 periods or less. This damping was at first attributed to the liquid core of the earth, but it has now been shown that it cannot be due to viscosity there. The source of the dissipation must thus be in the mantle and due to plastic flow. Plastic flow is also the most common nonelastic behavior of solids when they are heated to temperatures not far below their melting point. If the damping is attributed to plastic flow, the approximate magnitude of the effect can be calculated and an estimate can be made of the influence this amount of plasticity would have on polar wandering. Hence, if the nature and magnitude of a disturbance such as a redistribution of mass on the earth's surface were known, the rate and extent of polar wandering which it would cause can be estimated. T. Gold (1955, 1956) has estimated that, if a continent the size of South America were raised by 3 m in middle latitudes, the earth would rotate through 90° in about 1 million years. W. Munk has shown that there is a large diminution of the effects if isostatic compensation is maintained, but that nevertheless the effects remain geologically significant.

A movement of the poles on the earth would cause a certain amount of redistribution of mass. The largest effect would be the change in polar glaciation, and hence in sea level. This could work in a sense either to help or to hinder polar wandering. Gold (1955) has shown that it is possible to have a "pole trap," where the induced changes of mass would counteract the other disturbances and hold the axis until either a sufficiently large change occurred elsewhere or the distribution of land and water in the polar zone giving rise to the pole trap was substantially altered. If any such mechanism has operated, one would expect polar wandering during geological time to have occurred in a series of jerks rather than a steady movement. The paleomagnetic and paleoclimatic evidence leaves little doubt that the position of the pole relative to the land masses has changed throughout geological time. The foregoing discussion also indicates the probability of polar wandering. P. Goldreich and A. Toomre (1969) have shown that large *relative* displacements of material need not be involved to account for large shifts in the location of the pole; large angular displacements of the earth's rotation axis relative to the entire mantle have occurred on a geological time scale, owing to the gradual redistribution (or decay or manufacture) of density inhomogeneities within the earth by the same convective processes that are responsible for continental drift. The question as to whether polar wandering as well as continental drift is necessary to explain the paleomagnetic results is discussed in Sec. 14-7.

SUGGESTIONS FOR FURTHER READING

S. Chapman and J. Bartels, "Geomagnetism," Oxford University Press (1940), has been a standard text on geomagnetism for many years. S. Chapman has also published a small book in Methuen's Monographs on Physical Subjects entitled "The Earth's Magnetism" (1951). More recently S. Matsushita and W. H. Campbell have edited a two-volume set entitled "Physics of Geomagnetic Phenomena," Academic Press (1967). Additional information on the earth's main field is given by:

HIDE, R.: "The Hydrodynamics of the Earth's Core," Physics and Chemistry of the Earth Series, vol. 1, pp. 94–137, Pergamon Press, New York, 1956.

—— and P. H. ROBERTS: "The Origin of the Main Geomagnetic Field," Physics and Chemistry of the Earth Series, vol. 4, pp. 27–98, Pergamon Press, New York, 1961.

JACOBS, J. A.: "The Earth's Core and Geomagnetism," Pergamon Press, New York, 1963.

In paleomagnetism, apart from the excellent review articles by A. Cox and R. R. Doell (1960) and R. R. Doell and A. Cox (1961), further information is given by:

BLACKETT, P. M. S.: "Lectures on Rock Magnetism," Weizmann Science Press of Israel, 1956.
IRVING, E.: "Paleomagnetism," John Wiley & Sons, Inc., New York, 1964.
NAGATA, T.: "Rock Magnetism," 2d ed., Maruzen Co., Ltd., Tokyo, 1961.
STRANGWAY, D. W.: "The History of the Earth's Magnetic Field," McGraw-Hill Book Company, New York, 1970.

9

PHYSICS OF THE UPPER ATMOSPHERE

9-1 TRANSIENT MAGNETIC VARIATIONS

In its broadest sense, geophysics ranges from the physics of the upper atmosphere to the physics of the earth's deep interior. Thus meteorology, hydrology, and oceanography are all parts of geophysics, although it has not been possible to include an account of those disciplines in this book. However, the physics of the upper atmosphere, or *aeronomy* as it is called, is intimately connected with geomagnetism, and an account of some of the more important phenomena in that field is given in this chapter.

The continuous magnetic records of any observatory show that on some days all three elements exhibit smooth and regular variations, while on other days they are disturbed and show irregular fluctuations. Days of the first kind are called quiet days, and days of the second, disturbed days. By an international scheme which is in operation, at each observatory a figure K between 0 and 9 is assigned to describe the magnetic conditions for each period of three Greenwich hours 0–3, 3–6, etc.

K indices are a measure, for an interval of 3 hours, of the intensity of magnetic disturbance as shown on the magnetograms of an observatory. Thus they incorporate also any local effects such as the systematic diurnal variations in geomagnetic activity. There is therefore a need for an abstract of the individual K

indices to express worldwide features of geomagnetic disturbances over a 3-hour period. An average of all individual K indices would not be satisfactory, owing to the inadequate geographical distribution of magnetic observatories. Thus a new index K_p has been designed to measure "planetary" variations in magnetic activity. It is based on "standardized" indices which have been freed as far as possible from local features. K_p indices are given to thirds as follows: The intensity interval 1.5 to 2.5, for example, is divided equally into three thirds designated as 2–, 2o, and 2+. This provides 28 grades of K_p from 0o, 0+, 1–, 1o, 1+, . . . , 8+, 9–, 9o.

The definition of K_p was chosen so that the whole range of geomagnetic activity from the quietest conditions to the most intense storm could be expressed by a single digit and an affix. This was achieved by a quasi-logarithmic relation between the amplitudes of disturbance in the 3-hour interval and K_p. To obtain a linear scale, K_p may be converted into a 3-hour equivalent planetary amplitude a_p, by means of Table 9-1. At a standard station in about 50° geomagnetic latitude, a_p may be thought of as the range of the most disturbed of the three field components expressed in the unit 2γ; for example, the range in a 3-hour interval with K_p = 4+ is 2 x 32, that is, 64γ. The average of the eight a_p values for a day is called A_p.

An old measure for the activity of a Greenwich day is the international character figure C_i ranging from 0.0 (very quiet) to 2.0 (intense storm). It is possible to derive from the A_p value a similar measure (called the daily planetary character figure C_p) which varies between 0.0 and 2.4.

It is found that, in general, day-to-day changes in the intensity of any disturbance follow a similar pattern over a wide area; similarly, quiet conditions are usually widespread. Most days show some magnetic disturbance, but except in periods of very violent activity, it is found that the disturbance D is superimposed on a regular daily variation. This is called the solar daily variation S. There is also a regular daily variation L which depends on lunar time. L is of much smaller magnitude than either S or D and cannot usually be recognized at sight on a magnetogram as S and D can. S is seen in its pure form on quiet days, when it is denoted by S_q.

Since S, D, and L, unlike the secular variation, produce no long or enduring changes in the earth's field, they are called transient magnetic variations. It is found that each magnetic element is affected in a characteristic way by each of the three types of variation S, L, and D. For a detailed study of transient magnetic phenomena, results must be obtained from a number of observatories widely

Table 9-1

K_p	= 0o	0+	1–	1o	1+	2–	2o	2+	3–	3o	3+	4–	4o	4+
a_p	= 0	2	3	4	5	6	7	9	12	15	18	22	27	32

K_p	= 5–	5o	5+	6–	6o	6+	7–	7o	7+	8–	8o	8+	9–	9o
a_p	= 39	48	56	67	80	94	111	132	154	179	207	236	300	400

distributed geographically. The type and range of variation vary throughout the year, showing a seasonal change, and the range and incidence of D also vary from year to year. At most observatories, mean values of the elements are measured from the magnetograms for hourly intervals. The mean of the hourly values for each day (i.e., the daily mean) and the mean of the hourly values for each hour for all the days of each month (i.e., the mean daily variation for the month) can then be evaluated. The mean of the daily mean values for the month is called the monthly mean value. When the monthly mean value is subtracted from the corresponding sequence of hourly means, i.e., from the daily variation, a sequence of hourly departures from the mean is obtained; it is called the mean daily inequality for the month.

9-2 QUIET–DAY SOLAR DAILY VARIATION S_q

The mean daily variation obtained from the five quietest days of the month is denoted by S_q. At a given epoch, S_q depends mainly on latitude and local time and not also on longitude; i.e., it is virtually the same at all stations around any circle of latitude at corresponding local times. Figures 9-1 to 9-4 show the quiet-day daily variation S_q in H, Z, D, and I at different seasons for four well-distributed observatories in the United States. These curves are based on days selected because of their freedom from storms, and represent the average of a large number of days, so that minor irregularities are smoothed out. The actual curves for any observatory undergo seemingly fortuitous changes from day to day (both in amplitude and shape), so that average curves are hardly ever realized. For stations in southern latitudes, the curves for H are approximately the same as those obtained at stations in corresponding northern latitudes. The curves for Z and D reverse on crossing the equator. An outstanding feature of the daily variation is the equatorial enhancement of the daily range of the horizontal intensity. At the center of a narrow belt, no more than 2 or 3 degrees of latitude in width approximately centered on the magnetic dip equator, the daily range of the horizontal intensity is nearly three times as large as the range at points no more than 500 km to the north or south.

An alternative representation of S_q at any station can be given by means of a vector diagram. If $\mathbf{O'P}$ represents the total magnetic force $\mathbf{F}$, then as $\mathbf{F}$ varies, the end P of the vector $\mathbf{O'P}$ will move in space (O' being a fixed origin). If the mean position of P is 0, then $\mathbf{OP}$ represents the difference between $\mathbf{F}$ at any instant and its mean value. For the S_q variation, P will describe a closed curve which will in general not be a plane curve. The horizontal projection can be obtained from the variations in H and D alone, while the variations in H and Z alone give the projection on the magnetic meridian plane. Figure 9-5 illustrates the S_q horizontal vector diagram for Greenwich. Points corresponding to exact hours of local solar time are marked on the curve, so that it is possible also to estimate the rate of variation. The figure shows that the S_q variation is greater and more rapid during the hours of daylight than of darkness, suggesting that the sun is the ultimate cause

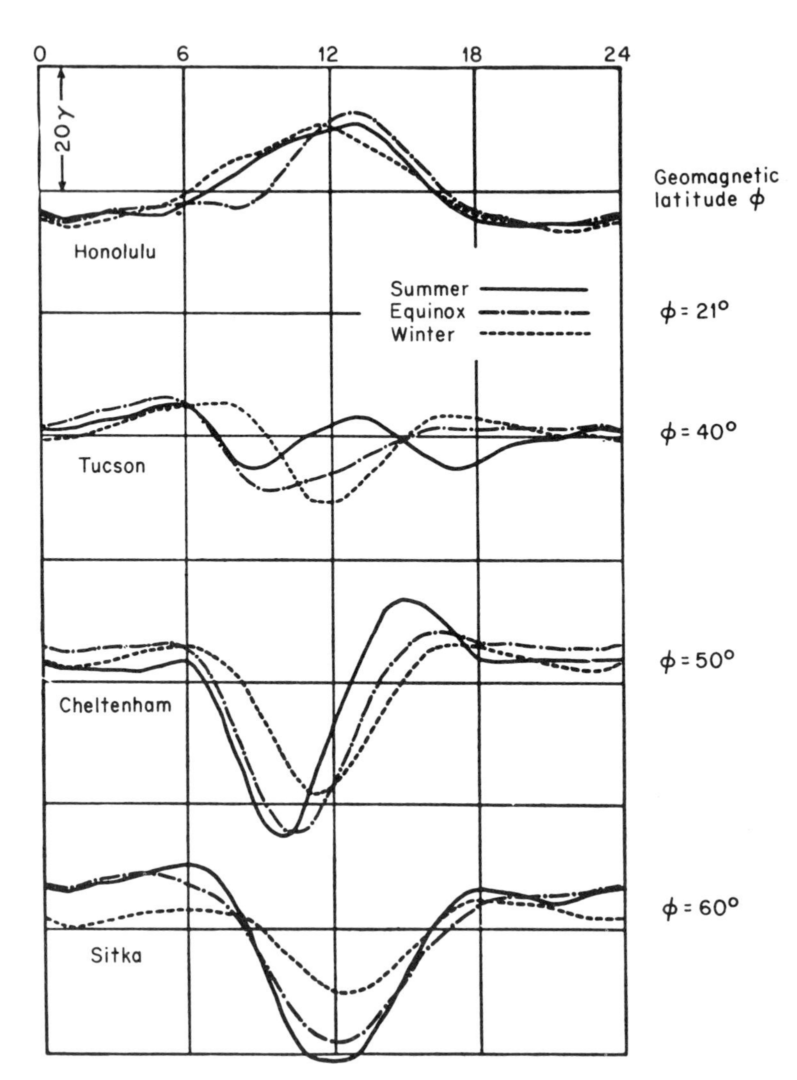

FIGURE 9-1
Daily variation of horizontal intensity, H. In this figure and in Figs. 9-2, 9-3, and 9-4, each curve shows the average variation for a large number of magnetically quiet days, for the observatory, season, and magnetic element specified. The time is local mean time. The geometric latitudes of the four observatories are Honolulu, 21°; Tucson, 40°; Cheltenham, 50° Sitka, 60°. (*After J. H. Nelson, L. Hurwitz, and D. G. Knapp, 1962.*)

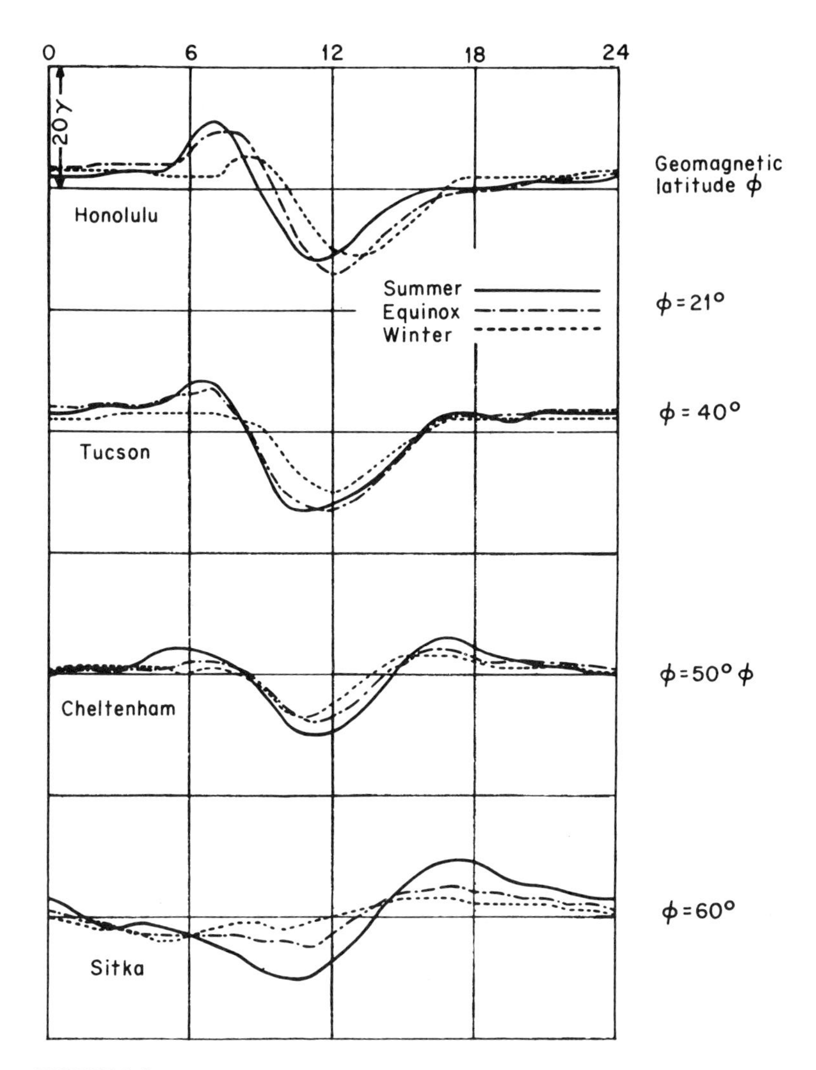

FIGURE 9-2
Daily variation of vertical intensity, Z.

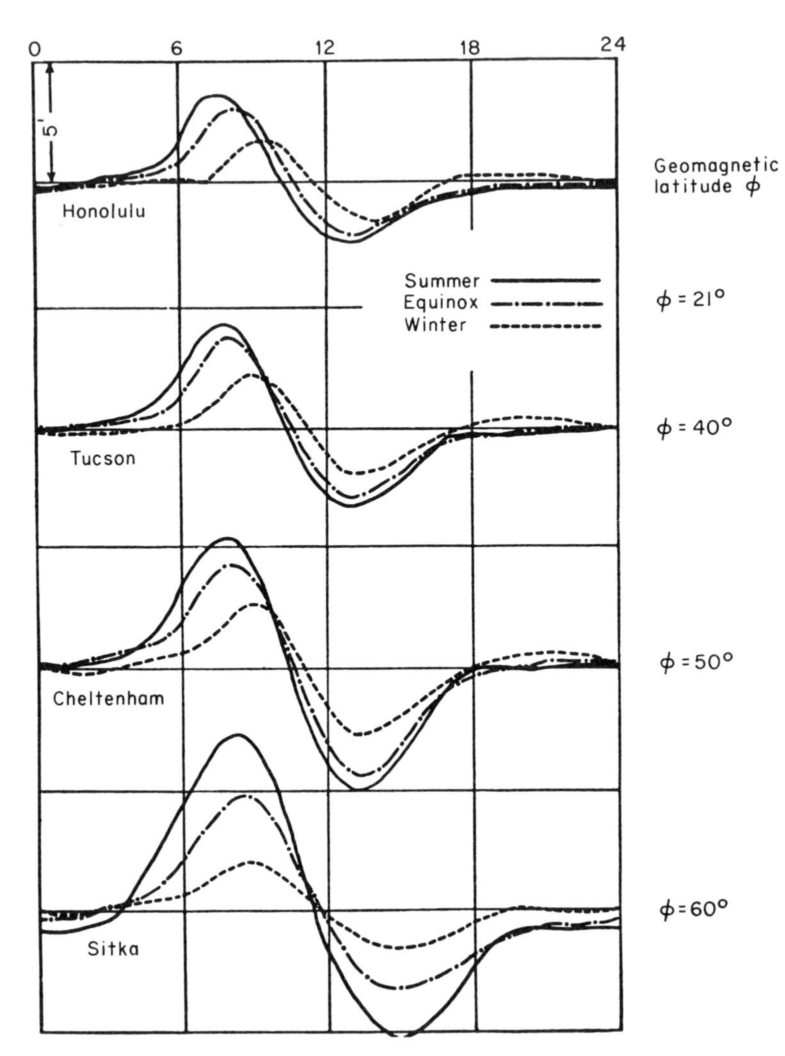

FIGURE 9-3
Daily variation of magnetic declination, *D*.

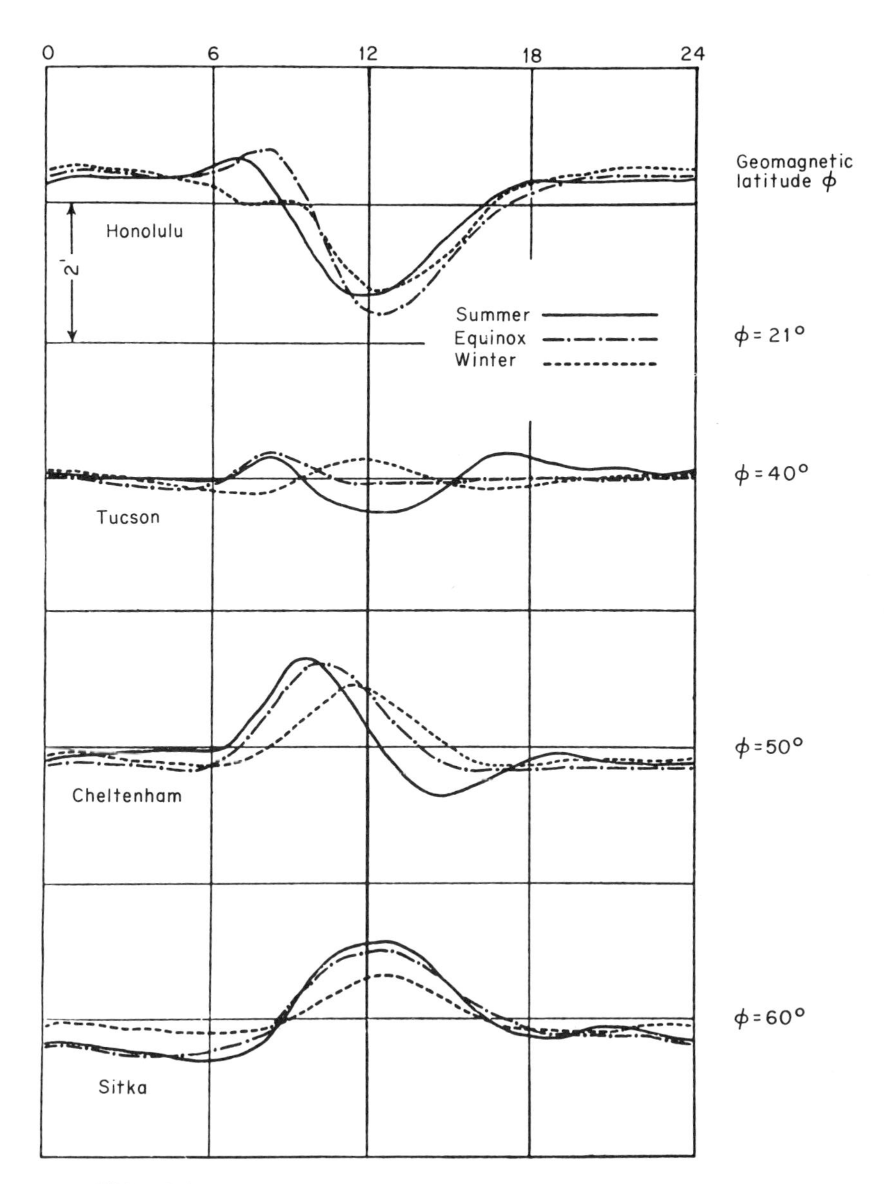

FIGURE 9-4
Daily variation of magnetic inclination, I.

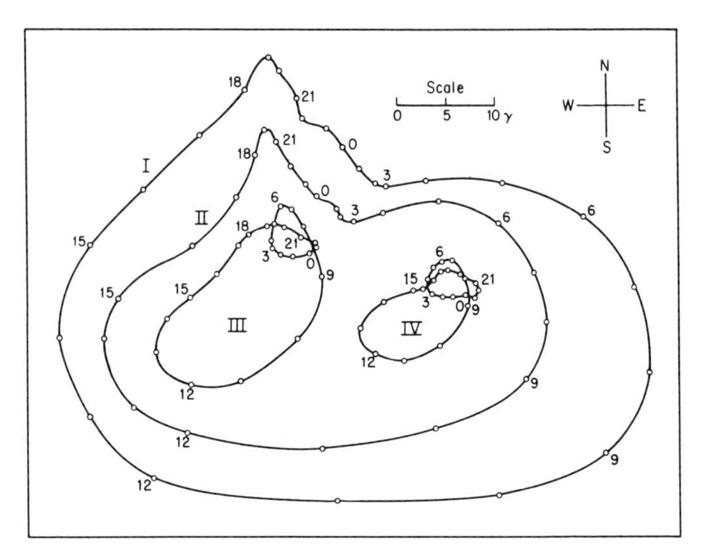

FIGURE 9-5
Quiet-day vector diagrams of the daily variation of magnetic force in the horizontal plane at Greenwich, 1889–1914. I: June, sunspot maximum years. II: June, sunspot minimum years. III: December, sunspot maximum years. IV: December, sunspot minimum years. (*After S. Chapman and J. Bartels, 1940.*)

of S_q. This is also shown in the figure by the contrast between the summer and winter vector diagrams. This feature is also present in the diagrams for other stations, the amplitude of S_q being greater at the stations which are more exposed to the sun. There is one further significant fact which indicates the close connection between S_q and the sun. The amplitude of S_q is from 50 to 100 percent greater in years of sunspot maximum than in years of sunspot minimum. It is also found that magnetic disturbances are more frequent and generally more intense in years of sunspot maximum, showing a periodicity of approximately 11 years.

A spherical harmonic analysis of the S_q field shows that it is not produced wholly inside the earth (as is substantially the case for the earth's main field) nor wholly above it. The major (external) part of the S_q field, written $S_q{}^e$, has its origin above the earth, while the minor (internal) part, written $S_q{}^i$, has its source inside the earth. The S_q variations due to the internal and external parts of the field reinforce one another so far as their horizontal components are concerned, although their vertical components are opposed. A varying magnetic field surrounding the earth must induce electromotive forces within the earth, causing electric earth currents. These in turn will produce a varying magnetic field, both inside and outside the earth, so that it is natural to suppose that the $S_q{}^e$ field produces the $S_q{}^i$ field by electromagnetic induction. It is not possible to establish this conclusively because the distribution of electrical conductivity σ throughout the earth is not known. However, calculations for a model earth with a reasonable distribution of σ lend considerable support to the hypothesis, which is now generally accepted. B. N. Lahiri and A. T. Price (1938) have shown that to account for the observed S_q variation, the electrical conductivity must be very small, down to a depth of about 600 km, when it rises rapidly, reaching a value of at least 10^{-2} $\text{ohm}^{-1}\ \text{cm}^{-1}$ at a depth of 900 km. No information can be obtained by this method for greater depths since the induced currents do not penetrate appreciably farther than this. For further information on the electrical conductivity of the earth, see D. C. Tozer (1959), A. T. Price (1967, 1970), and R. J. Banks (1969).

The air above the earth's surface is virtually nonmagnetic and nonconducting up to a height of about 70 km, above which there are two main ionized layers, the E layer at about 100 km and the F layer at about 200 to 300 km (Sec. 9-5). The ionization in these layers is renewed daily by the sun, decreasing during the afternoon and night. Since the S_q field is more intense over the sunlit than over the dark side of the earth, and since the $S_q{}^e$ field must be produced in some ionized region, it is reasonable to suppose that it has its origin in a thin spherical layer concentric with the earth. From a spherical harmonic analysis of the S_q field, it is possible to calculate the strength and direction of a current system flowing in such a layer which could produce the $S_q{}^e$ field. Such overhead current systems are illustrated in Fig. 9-6. It can be seen that at the equinoxes the currents flow in four main circuits, two north and two south of the equator; at the solstices the circuits in the summer hemisphere are intensified and extend across the equator into the winter hemisphere. The two circuits on either side of the equator are approximately situated one over the sunlit and one over the night hemisphere, the former being far more intense. The total current flow in the day circuit is approximately 62,000 amp at the equinoxes and 89,000 amp in summer. A tremendous amount of additional data have been obtained during the IGY and IQSY, and several quite different methods have been developed for carrying out a worldwide analysis of the S_q field and its corresponding ionospheric current system. These methods have been reviewed in detail by A. T. Price (1969), who also discusses the differences in the results obtained.

In 1882, Balfour Stewart suggested that the electric currents are due to electromotive forces induced by the daily periodic motion of the air in the presence of the earth's main magnetic field. This theory, known as the *dynamo theory*, was first investigated in detail by A. Schuster, in 1908, and later by S. Chapman, in 1914 and 1919, although it was not until 1924 that E. V. Appleton demonstrated the existence of the highly conducting E layer. Earlier, in 1901, G. Marconi had transmitted radio signals across the Atlantic around the curvature of the earth, and A. E. Kennelly and O. Heaviside independently revived the idea of an electrically conducting region high in the atmosphere which would reflect these radio waves back to the ground and thus prevent them from being lost to outer space.

A very important contribution to our knowledge and understanding of the sources of the S_q variation has been made in recent years by the detection and

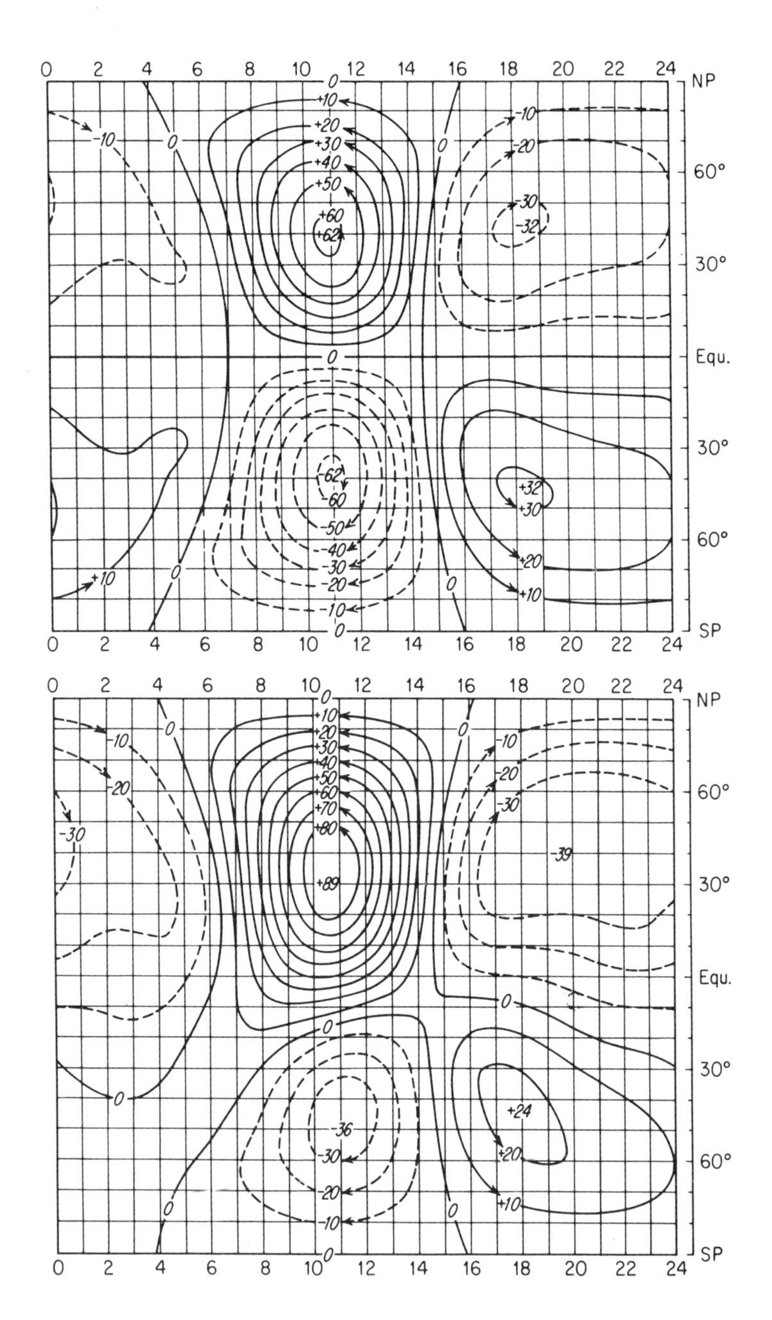

FIGURE 9-6
Atmospheric systems of electric currents which could produce the solar daily magnetic variation. Above, at the equinoxes; below, in June. (*After J. Bartels*).

evaluation of ionospheric currents from magnetic field profiles obtained from magnetometers flown in rockets. Ionospheric currents were first detected by rocket-borne magnetometers flown in the region of the magnetic equator (S. F. Singer et al., 1951; L. H. Cahill, 1959). Since then there have been a number of other successful rocket observations. T. N. Davis et al. (1967) investigated the equatorial electrojet using rocket-borne rubidium vapor magnetometers launched near the coast of Peru. The lower boundary of the equatorial electrojet was found to be near an altitude of 87 km. A steep vertical gradient in current density appeared near an altitude of 100 km, and the maximum current density, 10 amp/km^2, was measured at an altitude of 107 km directly above the measured location of the magnetic equator. The contour for the half maximum value of current density crosses the equator at altitudes of 102 and 114 km and extends to 300 km north of the equator.

P. A. Cloutier and R. C. Haymes (1968) measured the magnitude and direction of the S_q current system with cesium vapor magnetometers flown aboard sounding rockets from Wallops Island, Virginia. Deviations of the magnitude of the geomagnetic field from the value expected from a spherical harmonic expansion indicated ionospheric currents. Surface magnetograms showed no corresponding magnetic fluctuations at the times the deviations were observed, indicating that the rockets had penetrated localized field sources. The current sheets were thinner, at a lower altitude, and more intense than those measured on earlier flights.

9-3 ATMOSPHERIC TIDES

The behavior of ocean tides has been well understood since the nineteenth century—their cause is gravity. A point on the surface of the earth which is nearer the sun than the center of the earth is subject to a slightly greater inward gravitational and slightly less outward centrifugal force (arising from the earth's orbit around the sun). Conversely, a point on the far side of the center of the earth is subject to a little less gravitational force and a little more centrifugal force. The net result is a force which tends to push the surface material toward points directly in line with the sun. In a sense the tides are permanent, being fixed on a line from the sun to the center of the earth. Because of the earth's spin, the water at any point on its surface surges up and down with a 12-hour period. The same arguments apply to the earth-moon system, but because of the proximity of the moon, its effect is 2.2 times as strong as that of the sun. Also, since the period of the moon's orbit around the earth is 24h 51m, the lunar tide has a period of 12h 25.5m.

The same forces act on the atmosphere and must also produce tides in it. However, the identification and determination of the lunar atmospheric tide by recording the barometric pressure at ground level are very difficult, even in the tropics. It is so small that it is overlaid by other effects, including errors of measurement, and highly sophisticated methods of analysis of data accumulated over very long time intervals must be employed. Strangely enough, the solar influence is much easier to distinguish, which is exactly opposite to the case of oceanic tides. Laplace suggested early in the nineteenth century that the solar atmospheric tide might be caused by thermal rather than gravitational action of the sun on the atmosphere. However, if this were the case, the main heating effects should vary diurnally, not semidiurnally, and it is necessary to explain why the amplitude of the 12h pressure wave is greater than that of the 24h wave. A way out of this difficulty was indicated by Kelvin in 1882. He suggested that the atmosphere may have a free period of oscillation very close to 12 hours, so that any solar, semidiurnal disturbing force would produce a much greater response than a disturbance of any other period; if the atmosphere does possess a free period close to 12 hours, there is no need to invoke solar heating at all. A great amount of theoretical work has been carried out since Kelvin's time to determine the free oscillations of the terrestrial atmosphere and its response to both gravitational and thermal excitation. It has been shown that the free oscillations of the atmosphere are the same as those of an incompressible and homogeneous ocean of "equivalent depth" h, which depends on the vertical temperature distribution of the atmosphere. G. I. Taylor (1936) showed that in general there is an infinite number of equivalent depths for a given atmospheric temperature distribution, and that h is related to the velocity V of long atmospheric waves. Such waves were observed during the eruption of the volcano Krakatoa in 1883, and gave a value for h of approximately 10 km. On the other hand, for a free oscillation with a period of 12 hours, theory indicates an equivalent depth h of 7.8 km.

Before the advent of rocket data, information about the temperature distribution above 30 km came mainly from observations of the anomalous propagation of sound. These observations indicated temperatures around 50 km to be well above those at ground level. C. L. Pekeris (1937), assuming a temperature distribution with a maximum of 350°K at 60 km, showed that two values are obtained for $H = 10$ km, as suggested by the Krakatoa waves, and 7.8 km, very close to the value required for strong resonance magnification. Since then, however, rocket data have given a much more reliable temperature distribution of the upper atmosphere, with temperatures considerably lower than 350°K at altitudes of 50 to 60 km. With these lower temperatures, the value $h = 10$ km is still obtained, but not the value $h = 7.8$ km, and the resonance magnification for the solar semidiurnal oscillation is quite small. Thus gravitational excitation cannot contribute appreciably to the observed magnitude of the solar semidiurnal pressure oscillations, and it became necessary to examine Laplace's suggestion of thermal excitation in more detail.

Only a small amount of the incoming solar-radiation energy is absorbed in the atmosphere before reaching the ground. Thus the heating of the atmosphere proceeds mainly from the ground upward by turbulence and long-wave radiation. These processes are effective only through a very limited height range, and hence the resulting temperature oscillation is negligible at an altitude of a few hundred meters. H. K. Sen and M. L. White (1955) and M. Siebert (1961) pointed out that although the amount of incoming solar energy absorbed directly in the atmosphere is small, it must cause a daily temperature variation in the atmosphere which makes a very significant contribution to atmospheric oscillations, since it affects the whole atmosphere. Furthermore, at higher layers in the atmosphere, the presence of ozone becomes important. Because of its high absorptive power in certain parts of the ultraviolet region, the ozone that is present heats the upper atmosphere considerably between about 30 to 50 km. The ozone also produces pronounced daily temperature variations in this whole layer which must contribute to the diurnal oscillations and their higher harmonics. S. T. Butler and K. A. Small (1963) found that by far the largest part of the semidiurnal pressure oscillation is due to the temperature oscillation in the ozone layer. It is generally agreed now that the diurnal tide in the atmosphere is thermally driven and that the main sources of the thermal drive are the direct insolation absorption by water vapor in the troposphere and by ozone in the stratosphere-mesosphere. However, it was not until 1967 that R. S. Lindzen solved Laplace's tidal equations and obtained the theoretical response of the atmosphere's winds, temperatures, etc., to these known drives. The calculated winds are in reasonable agreement with observations of the diurnal wind oscillation from the troposphere to the upper mesosphere and are in striking agreement with both the magnitude and complicated structure of winds in the mesosphere observed at single times. Lindzen also showed that the small amplitude of the diurnal surface-pressure oscillation results from most of the thermal drive being used to activate trapped modes which do not propagate to the ground. In a later paper, Lindzen (1968) showed that the semidiurnal surface-pressure oscillation is approximately the same for widely differing basic temperature profiles. This is in contrast to the results obtained by the old resonance theories (C. L. Perkeris, 1937), and these have finally been abandoned now. Further work on atmospheric dynamo action has been reported by J. D. Tarpley (1970a,b).

Direct evidence of tides in the high atmosphere has been obtained through observations of the drift of meteor trails. When a meteor penetrates the atmosphere down to a level of about 80 to 100 km, it produces an ionized trail during its disintegration as it collides with the air molecules. This trail drifts along with the air. From observations of the reflection of radio signals from these meteor trails, it is possible to compute the wind velocity. The data show that the amplitudes of both the diurnal and semidiurnal solar variations of the wind components are about 100 times greater in the high atmosphere than at the ground.

In the atmosphere-dynamo theory of the S_q variations, it is argued that the tidal motion of the air in the ionosphere in a horizontal direction across the lines of force of the earth's magnetic field induces electric currents. The field coils of the dynamo correspond to the earth's field, and the moving armature to the ionized air in tidal motion. In the ionosphere the horizontal tidal flow across the vertical component of the earth's magnetic field gives rise to an emf which tends to drive the current in a horizontal direction perpendicular to the tidal flow. To calculate what currents these emfs will cause requires a knowledge of the electrical conductivity of the atmosphere at any time and place. This in turn depends on the electron and ion concentrations and on the magnetic field, the latter dependence being a great complication since it means that the conductivity depends on the direction of current flow as well as on the electron and ion concentrations. A further complication arises from electrical polarization effects. The currents due to the dynamo emfs alone do not flow in closed circuits, so that electric charge of one sign builds up at one end of the path, and of opposite sign at the other. These charge accumulations in turn produce electric fields which influence the current flow. When a steady state has been reached, these additional polarization emfs will be just such that, when combined with the dynamo emfs, they give rise to currents which flow only in closed circuits.

9-4 MAGNETIC STORMS

It is found that the intensity of magnetic disturbances increases from low to high latitudes up to about magnetic latitude 65°, the latitude of the auroral zones, i.e., the zones where auroras are most frequent. Within these zones, the intensity, although considerable, decreases slightly toward the magnetic poles. Magnetograms are seldom completely undisturbed in high latitudes. Although disturbances of low intensity may be confined to a small area of the earth's surface, large disturbances, called magnetic storms, are worldwide phenomena. Intense magnetic storms usually commence suddenly at almost the same instant (less than ½ min) all over the earth. Some storms do not begin so abruptly, but their commencement can usually be fixed within an hour.

The records of individual storms differ greatly among themselves, but the changes in the field at the earth's surface may be analyzed into three main parts:

1 A part depending on time measured from the commencement of the storm. This is known as the storm-time variation D_{st}.

2 A daily variation in addition to that present on quiet days and much greater in intensity and markedly different in character. This is known as the disturbance daily variation S_D. M. Sugiura and S. Chapman (1960) found that both the amplitude and the time of the maximum amplitude of the disturbance daily variation change during the course of a storm. They suggested that the disturbance daily variation at a certain period during a storm or during individual storms be called D_s, and that S_D be considered as an average of D_s.

3 An irregular part most marked in high latitudes.

In middle and lower latitudes, the storm-time part of the horizontal intensity H rises to a maximum within an hour or two of the commencement and remains above its initial value for a period of 2 to 6 hours. This is called the initial phase. H

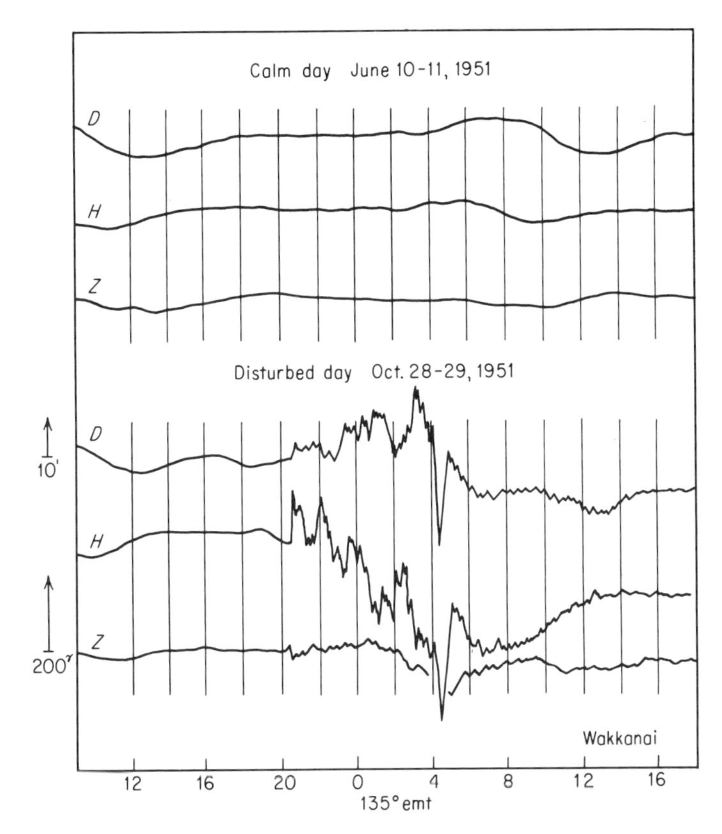

FIGURE 9-7
Magnetograms from Wakkanai, Japan.

then decreases, attaining after several hours a minimum which is much more below the initial undisturbed value than the maximum was above it. This is called the main phase and is followed by a gradual recovery which may last for several days. The main phase generally lasts from 12 to 14 hours and tends to be noisy. Often large positive and negative excursions with amplitudes of the order of hundreds of gammas and periods of about ½ hour occur. The greater the storm, the more rapid is the development of these phases. Except in high latitudes, the declination shows little or no storm-time change, while Z shows changes much smaller than those in H and opposite in sign; i.e., when H decreases, Z increases, and vice versa. Again H has the same direction in both Northern and Southern Hemispheres, whereas Z has opposite directions north and south of the magnetic equator. Moreover, the range of D_{st} in H decreases with increasing latitude; that is, D_{st} changes in H are greatest at the equator, where those in Z are zero. Figure 9-7 illustrates the behavior of D,

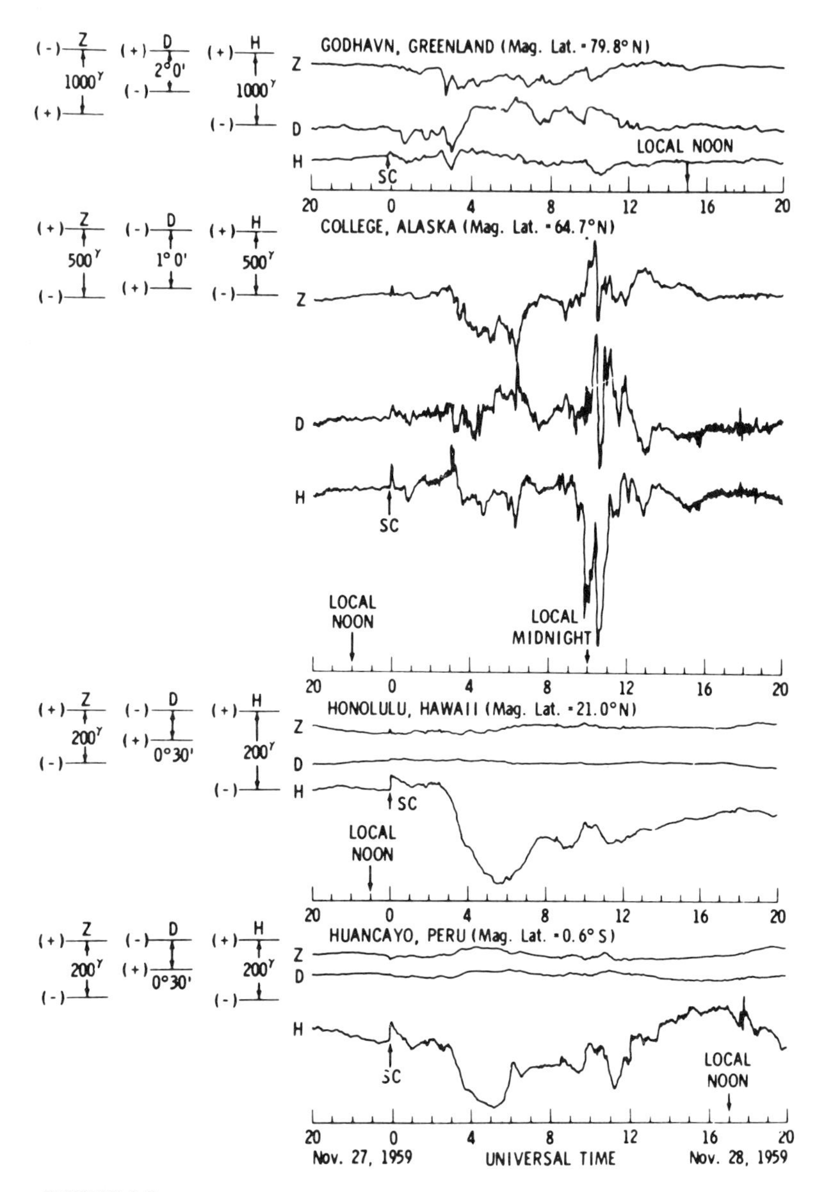

FIGURE 9-8
Example of a magnetic storm of moderate intensity at different latitudes. (*After M. Sugiura and J. P. Heppner, 1965.*)

H, and Z on a quiet day and during a magnetic storm at Wakkanai, Japan. Figure 9-8 shows the striking difference in the storm-time variations in low latitudes (Honolulu) and those in the auroral zone (College, Alaska). On crossing the auroral zone toward the magnetic pole, the characteristics undergo a further change to another type peculiar to the polar caps (see Godhavn, Fig. 9-8). The disturbance daily variation S_D, although depending on local time, is not of constant intensity, but decreases from the first to the second day of a storm. Its variation is also very different from that of S_q, both in times of maximum and minimum and in relation to latitude; moreover, the S_D curves show no greater intensity of change during the sunlit hours than during the hours of darkness.

In latitudes greater than about 60° the disturbance field is far more complex. It is more intense and more variable both in time and in geographical distribution. One of the most striking features of the character of the D field is the increasing predominance of the S_D part of the field as compared with the D_{st} part as the auroral zone is approached from low latitudes, although in tropical latitudes the range of the storm-time variation exceeds that of S_D.

Figure 9-9 shows that the S_D variation in Z has a constant phase from the equator to as far north as Sitka in magnetic latitude 60°N, with a morning minimum and an evening maximum, the amplitude increasing greatly with latitude. The curves for polar stations are given in Fig. 9-10. These curves, with the exception of Novaya Zemlya, also show a very large range, but their phases are all opposite to that for Bossekop and other stations of lower latitude. The results for Novaya Zemlya are transitional between the two sets of curves. This reversal of phase occurs within a narrow belt near the auroral zone. The S_D variation in the horizontal plane also shows a striking change of type on crossing into the auroral zone, as is illustrated by the horizontal-force vector diagram for S_D (Fig. 9-11). Thus at Sitka the figure is still roughly oval in form, elongated in the direction transverse to the magnetic meridian, but for stations near the auroral zone, like Sodankyla and Bossekop, the diagram is very narrow in that direction and much elongated in the direction normal to the zone. On passing well inside the zone, the vector diagram becomes oval again (cf. Kingua-Fjord and Cape Evans).

Harmonic analysis of the D field indicates that its origin, like that of S_q, is partly external and partly internal. The internal part D^i can be accounted for by electric currents induced within the earth by the external part D^e. The origin of the D^e field is not known with any certainty. It is possible, however, to determine electric current systems which, independent of any theory of their origin, should flow in the ionosphere to produce the observed S_D and D_{st} variations. Figure 9-12 shows the idealized electric current systems for the S_D, D_{st}, and $S_D + D_{st}$ variations as deduced by S. Chapman from 40 magnetic storms of moderate intensity.

The $S_D + D_{st}$ current system is completely unlike that for S_q both in its great intensity along and within the auroral zones and in its lack of any marked difference of intensity between the sunlit and the dark hemispheres. It must be

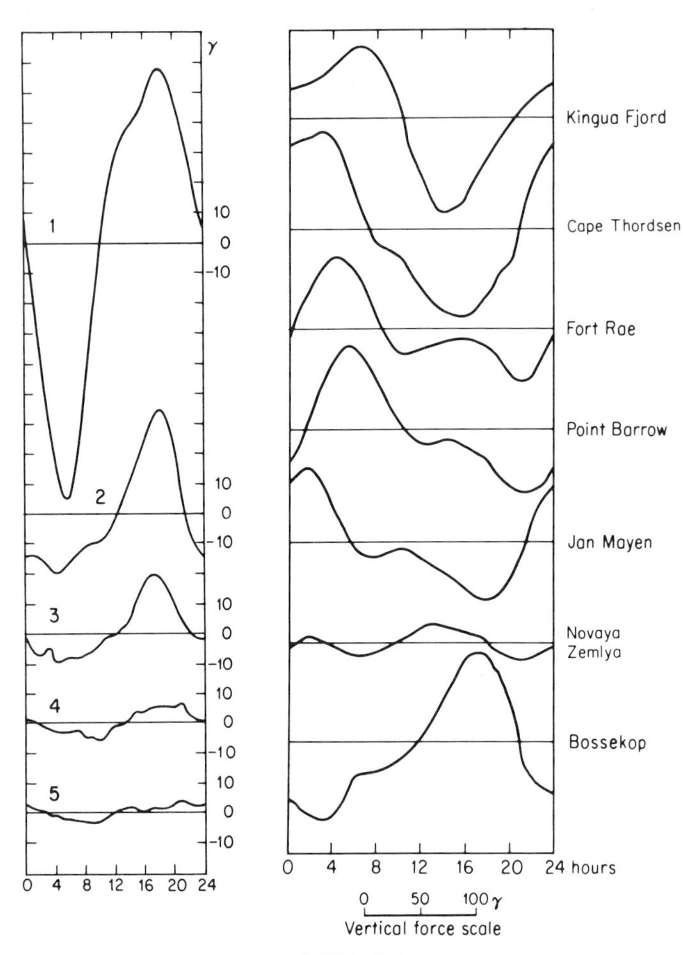

FIGURE 9-9
S_D variation in Z for the five groups of observatories. (1) Sitka (60°); (2) Pavlovsk (56°); (3) Pola, Potsdam, Greenwich (51°); (4) Zikawei, San Fernando, Cheltenham, Baldwin (40°); (5) Batavia, Puerto Rico, Honolulu (23°). Figures in parentheses are the magnetic latitudes, or mean magnetic latitude in the case of a group of observatories. (*After S. Chapman and J. Bartels, 1940.*)

FIGURE 9-10
Annual mean daily variation of vertical force in polar regions. (*After S. Chapman and J. Bartels, 1940.*)

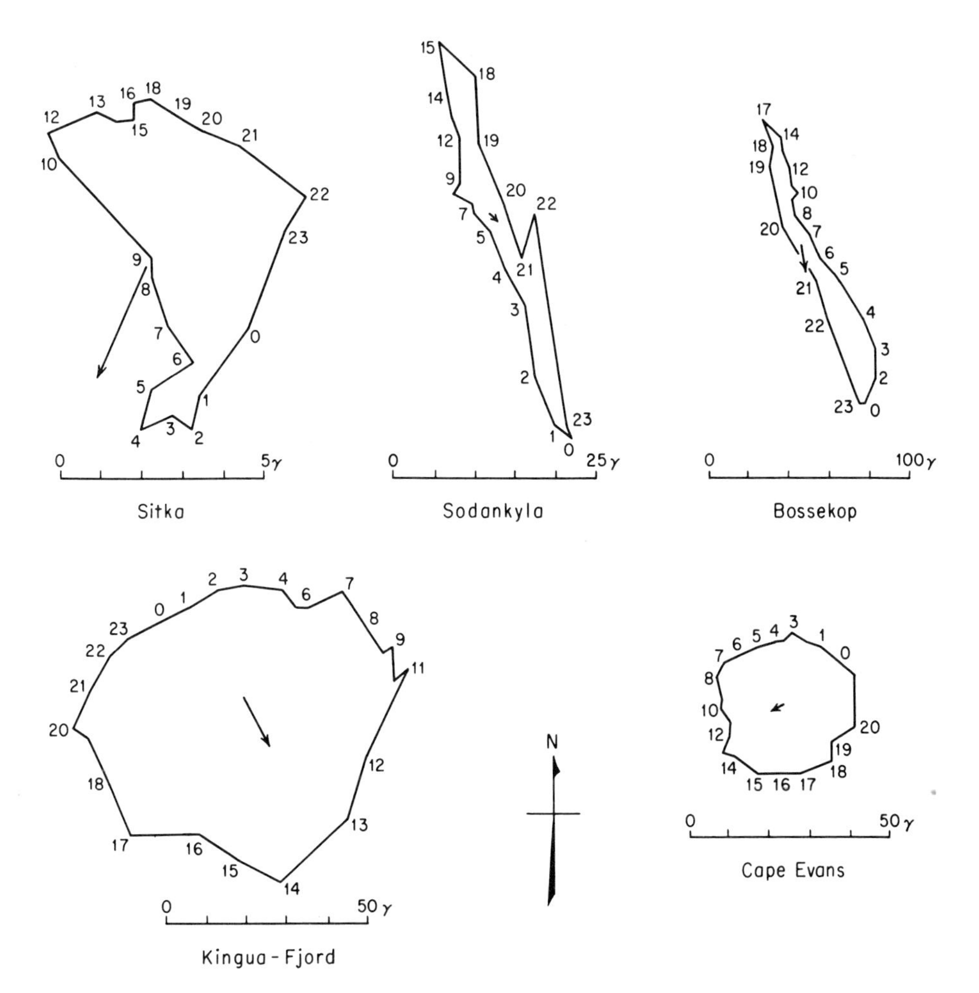

FIGURE 9-11
Horizontal-force vector diagrams in high latitudes, N and S. (*After S. Chapman and J. Bartels, 1940.*)

appreciated that such current systems are no more than a convenient mathematical model whereby the observed magnetic variation can be described. The D_{st} field can also be ascribed to a ring current around the equator at a distance of several thousand kilometers.

In addition to large-scale geomagnetic variations, there are disturbances of much shorter duration, such as a bay which is a departure from an otherwise undisturbed record in the form of a V, or bay of the sea. Abrupt impulsive changes (sudden impulses) may also occur, while variations with periods roughly from 0.1

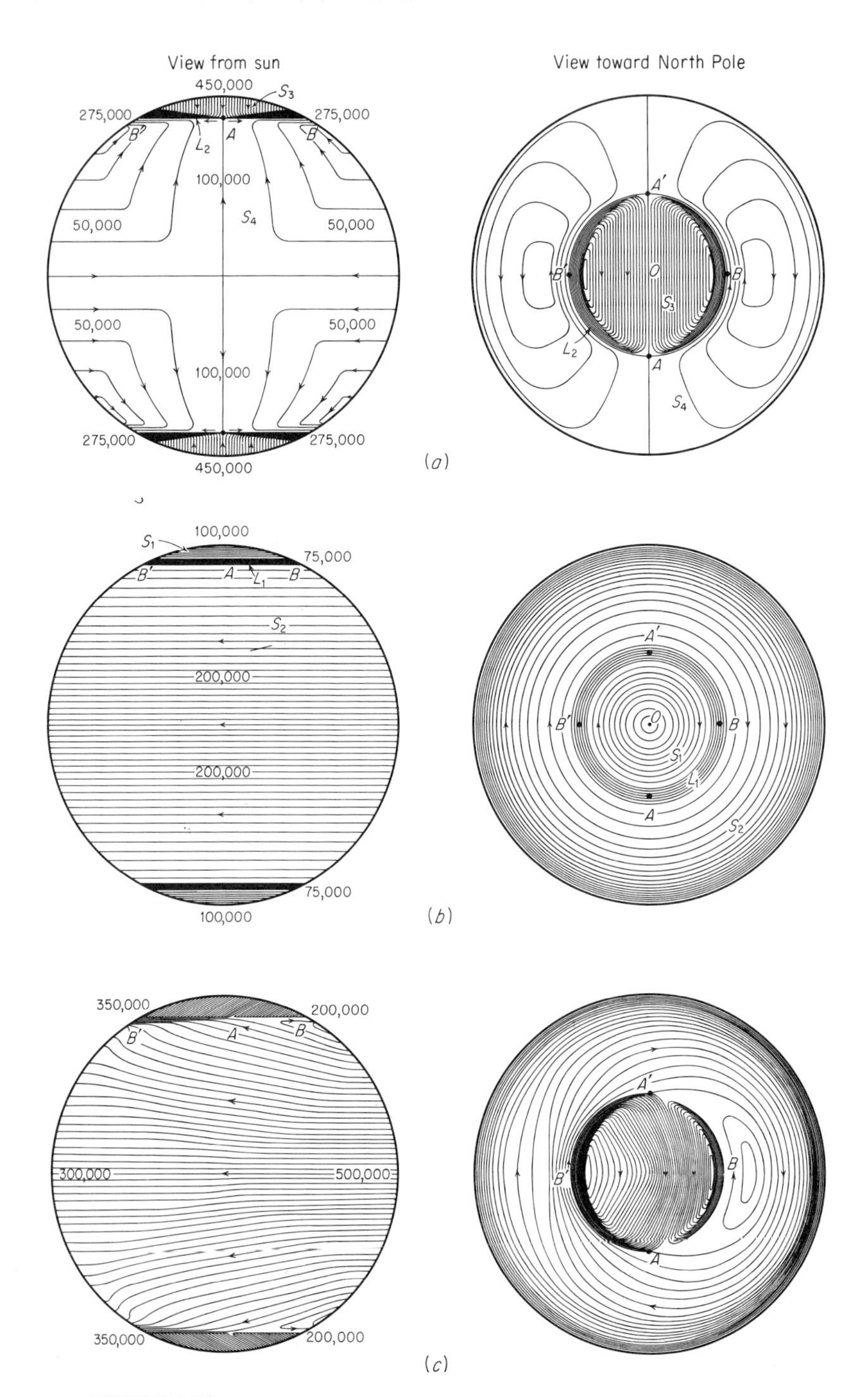

FIGURE 9-12
Idealized electric current systems which could produce (*a*) S_D variation, (*b*) D_{st} variation, and (*c*) $S_D + D_{st}$ variation. (*After S. Chapman and J. Bartels, 1940.*)

sec to 10 min are grouped together and called geomagnetic micropulsations. Information on these rather specialized phenomena can be found in recent review articles (see, for example, V. A. Troitskaya, 1964, 1967; W. H. Campbell, 1967; and J. A. Jacobs, 1970).

9-5 THE PHYSICAL PROPERTIES OF THE UPPER ATMOSPHERE

The atmosphere is divided into a number of regions based on its thermal structure. The lowest region, the troposphere, extends up to an altitude of about 10 km over the poles and 16 km above the equator. This is essentially the region of meteorological interest. The upper boundary of the troposphere is the tropopause, and the region above it is called the stratosphere. Its upper boundary, the stratopause, has been variously defined; we take it to be the temperature maximum which occurs around 50 km. Above the stratopause is the mesosphere, which extends up to the temperature minimum near 80 km. The mesosphere is often taken to be the broad region around the temperature maximum rather than just the upper half of it. Above the mesopause, which is the upper boundary of the mesosphere, is the thermosphere, in which the temperature rises rapidly up to about 200 km and less rapidly above that height, becoming essentially isothermal above 400 km. The exosphere is defined as the upper portion of the thermosphere, where the atmospheric gases are so rarefied that collisions can generally be neglected. The base of the exosphere varies between altitudes of about 350 and 700 km during the sunspot cycle. Figure 9-13 gives the vertical distribution of temperature at two extremes of the solar cycle, and shows that great changes in atmospheric structure occur during the solar cycle.

The atmosphere is heated nonuniformly by the sun, causing its physical properties to change both in space and time. The fundamental relationship between pressure and height is given by

$$dp = -\rho g \, dh \tag{9-1}$$

where p = pressure
ρ = density
g = acceleration of gravity at altitude h

This can be written in the form

$$\frac{dp}{p} = \frac{-dh}{H} \tag{9-2}$$

where $H = kT/mg$ = scale height (9-3)
m = average particle mass
T = temperature at height h
k = Boltzmann's constant

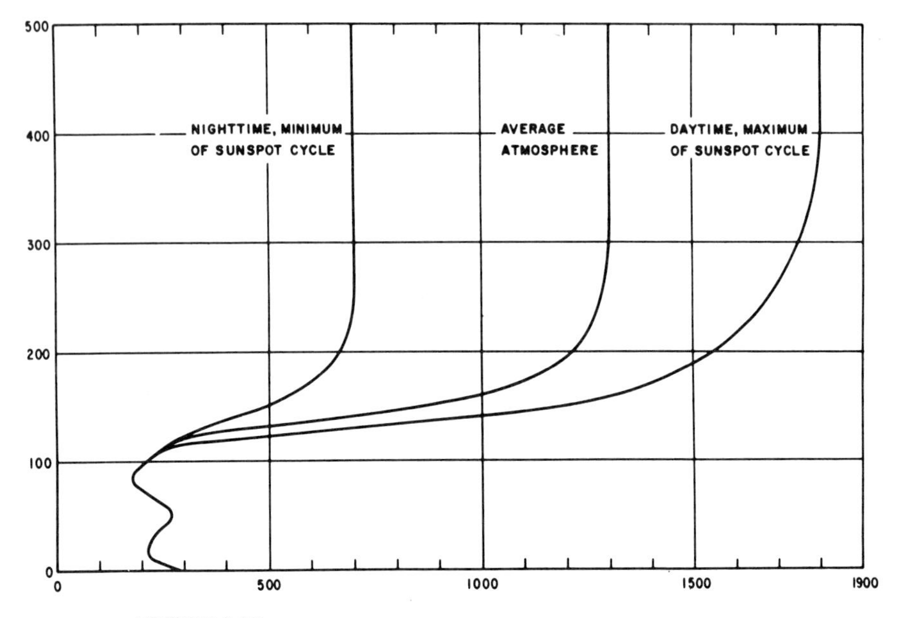

FIGURE 9-13
Atmospheric temperature distributions typical of daytime conditions near the maximum of the sunspot cycle, of nighttime conditions near the minimum of the sunspot cycle, and of an average, in-between situation. (*After F. S. Johnson, 1965.*)

On integrating Eq. (9-2), we obtain

$$\frac{p}{p_0} = \exp\left(-\int_o^h \frac{dh}{H}\right) \tag{9-4}$$

where p_0 is the pressure at the level $h = 0$. Thus the pressure ratio can be determined for any two levels between which the distribution of temperature and mean particle mass is known. If the temperature and average particle mass are constant, then neglecting the variation of g with altitude, Eq. (9-4) reduces to

$$\frac{p}{p_0} = \exp\frac{-h}{H} = \exp\frac{-mgh}{kT} \tag{9-5}$$

Further,

$$\frac{\rho}{\rho_0} = \frac{n}{n_0} = \frac{p}{p_0}$$

where ρ and n are the atmospheric density and particle concentration at height h, and ρ_0 and n_0 the corresponding quantities at the reference level $h = 0$. The variation of g with height may be taken into account by using the relationship

$$g = \frac{g_0 R_0{}^2}{(R_0 + h)^2} \tag{9-6}$$

where R_0 is the distance from the center of the earth to the reference level, and g_0 is the acceleration of gravity at that level.

Atmospheric temperature and pressure are measured directly in the troposphere by using instruments carried above by balloons, and the height of the observation is then determined from Eq. (9-1). However, above the troposphere it becomes difficult to measure temperature directly because the temperature-measuring elements come into radiative equilibrium with their distant surroundings rather than into conductive equilibrium with their immediate surroundings. At such altitudes pressure is measured as a function of altitude, and temperature is determined from sound-velocity experiments using grenades carried aloft in rockets. Above heights of about 100 km the pressure becomes difficult to measure and sound does not propagate well, so that density is the quantity most commonly observed as a function of altitude. These measurements are generally made either with vacuum gages or by observing the rate of orbital decay of satellites.

The atmosphere consists of a mixture of nitrogen and oxygen molecules up to about 120 km, a layer consisting of predominantly atomic oxygen between that height and 1000 km, a layer of helium from 1000 to 2500 km, and a hydrogen atmosphere extending out into interplanetary space above 2500 km. One of the most surprising results derived from satellite drag data is that there is a strong diurnal change of density and temperature above an altitude of about 200 km. The density increases shortly after sunrise, reaches a maximum in the afternoon (at about 1400 LMT), and then decreases rapidly until sunset. During the night it decreases further but at a slower rate. The amplitude of these density variations increases with altitude. At 200 km the variation is only a few percent, but at 700 km it is a factor of 10. The upper atmosphere is extremely sensitive to solar control, undergoing variations in density by as much as a factor of 100 in extreme cases and experiencing temperature changes of hundreds of degrees.

The radiation from the sun contains sufficient energy at short wavelengths to cause appreciable photoionization of the earth's atmosphere at high altitudes. The recombination of ions and electrons which are formed in this way proceeds slowly at the low gas densities which exist at high altitudes, so that fairly high concentrations of electrons persist even throughout the night. During the daytime several distinct ionospheric layers (D, E, F_1, and F_2) can be recognized, although the separation between them is not as distinct as was originally believed. Large diurnal effects also occur, particularly in the lower ionosphere; the F_1 and F_2 layers join at night, and the D layer virtually disappears. Figure 9-14*a* and *b* shows a typical daytime and nighttime electron distribution as a function of altitude. The electron density varies with the 11-year solar cycle and is generally higher during

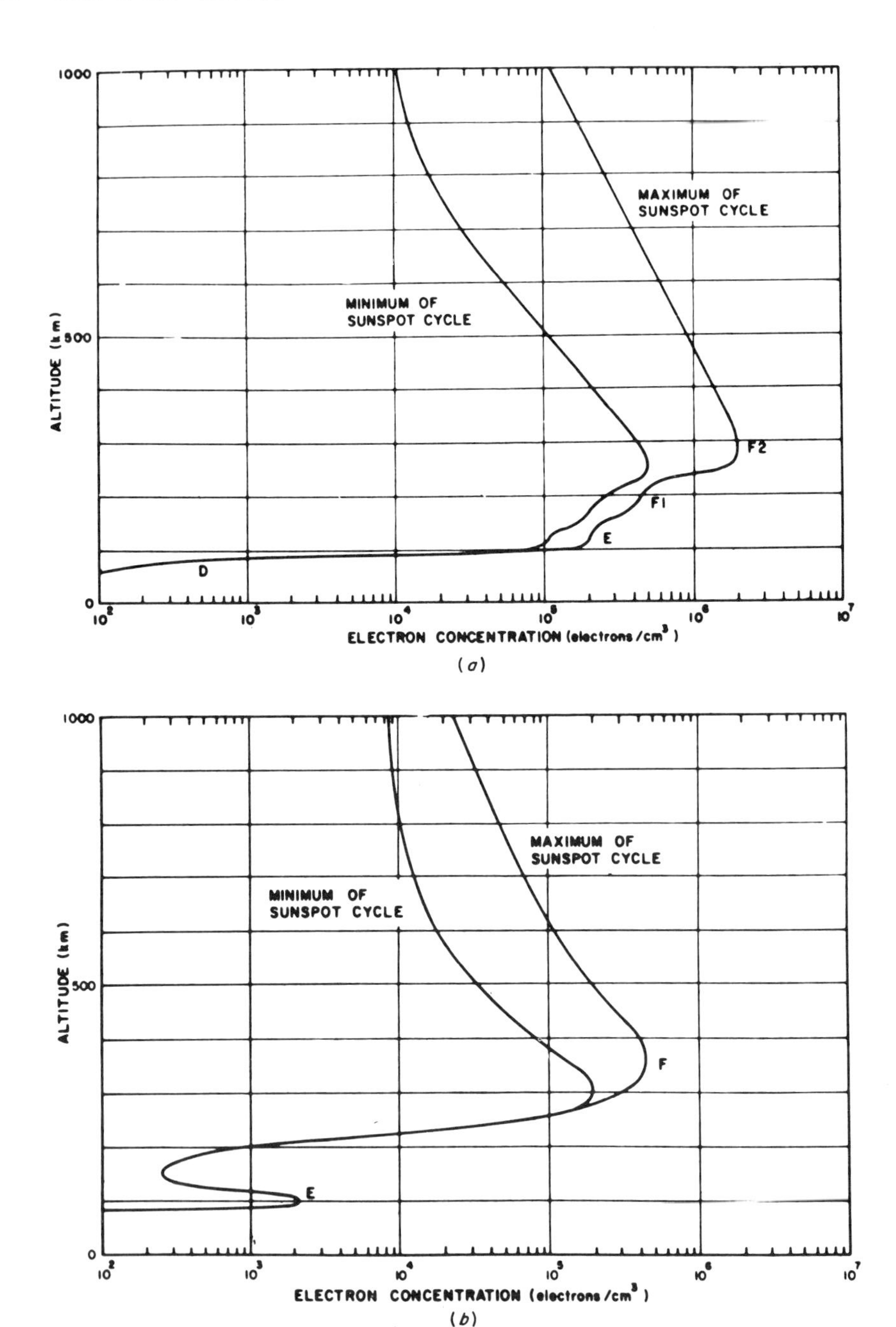

FIGURE 9-14
Normal electron distributions at the extremes of the sunspot cycle. (*a*) Daytime, (*b*) nighttime. (*After W. B. Hanson, 1965.*)

sunspot maximum and lower during sunspot minimum. Above its maximum in the F_2 layer, the electron density decreases out to several earth radii where the earth's magnetic field is confined by the solar wind.

The D layer extends from about 60 to 85 km. It is thought that the ionizing radiation is hydrogen Lyman alpha radiation and that the primary ions formed are NO^+. The maximum electron density occurs at about 80 km and is of the order of 10^3 electrons per cubic centimeter. The electrons in the D layer essentially disappear at night, though the actual physical processes involved are not well understood. Because of the high gas densities in the D layer, the electron collision frequencies are high there and the layer acts as a strong absorber of electromagnetic energy. At night when the electron density in the D layer is negligible, the attenuation of radio signals is much less than during the day. The range from about 85 to 140 km is called the E layer. It is generally believed that soft solar x-rays are mainly responsible for the photoionization which occurs in this range. The main ionic species present are the diatomic ions $O_2{}^+$ and NO^+. NO^+ is the predominant ion, except near 100 km during the daytime when the two ion concentrations are comparable. The electron density in the E layer is of the order of 10^5 electrons per cubic centimeter at noon during sunspot minimum and increases by about 50 percent during sunspot maximum. The electron concentration is highest around local noon, falling off rather symmetrically with time on either side. A rather common perturbation of the electron density in the E layer occurs around 100 km and consists of a thin layer, where the electron concentration may be as much as twice as high as the normal concentration above and below the layer. The layers are only a few kilometers thick and are called sporadic E. The boundaries of the F_1 layer are not clearly defined; it may be taken rather arbitrarily as extending from 140 to 200 km. The predominant ions are NO^+ and $O_2{}^+$ at the lower-altitude boundary, with a gradual transition to the upper boundary, where O^+ is the principal ion. The electron density is typically of the order of 2.5×10^5 electrons per cubic centimeter at noon during sunspot minimum and 4×10^5 electrons per cubic centimeter at noon during sunspot maximum. Large diurnal variations occur in the F_1 layer, and the layer is not present at night. In the F_2 layer the principal ions present are O^+ and N^+, with O^+ greatly predominant. The limits of the F_2 layer vary considerably but have been taken to be from 200 to about 1000 or 2000 km, the lower limit being determined by the distribution of the electron concentration, and the upper limit by the change in ion composition. Sometimes the F region shows a diffuse character which is attributed to clouds of ions having concentrations different from the surroundings. This condition is called spread F and occurs mainly at night. At greater altitudes the predominant ion constituent changes from atomic oxygen ions to protons. F. S. Johnson (1960) suggested that this region be called the protonosphere to distinguish it from the lower ionized layer containing heavier ions, which is normally referred to as the ionosphere. The protonosphere is the medium responsible for the propagation of radio whistlers. These low-frequency (1 to 30 kHz) signals are generated by lightning strokes and follow paths along the earth's magnetic field lines from one hemisphere to the other (L. R. O. Storey, 1953). They propagate at very large distances from the earth and at present provide the main experimental evidence on electron densities at great heights.

9-6 AURORAS AND AIRGLOW

Another phenomenon which is intimately connected with magnetic storms is the aurora. There are two kinds of atmospheric luminosity: (1) that caused by light coming from outside the atmosphere and modified by the air or its dust or cloud particles and (2) that caused by light generated within the atmosphere. The first class includes rainbows, halos, and the beautiful colors of twilight and dawn; the second includes lightning, meteors, auroras, and airglow.

Lightning is due to strong electric discharges associated with thunderclouds in the lowest layer of the atmosphere (the troposphere). *Meteor light* is caused by the entry at high speed of small or large pieces of solid matter from outside the atmosphere. *Airglow* is due to reactions in the high atmosphere, energized by the sunlight absorbed during the daytime. The sky over the earth is suffused day and night with a faint glow which is invisible to the eye; not only is it too faint, but its strongest radiations lie outside the visible band of the spectrum. Nightglow is faintest at the zenith and grows in intensity down the sky, reaching a maximum about 10° above the horizon. The height of airglow has been estimated to be between 100 and 200 km. Most of the visible light comes from certain forbidden lines in the spectrum of atomic oxygen. The intensity radiated in the infrared is much greater and is due to certain transitions in the hydroxyl molecule (OH). If the visible light were of the same intensity, there would be perpetual twilight under clear conditions. One of the most remarkable features of the airglow is the appearance of the yellow sodium line at quite high altitudes. There is no evidence of any radiation from other metallic atoms, and there is no really satisfactory explanation of the presence of sodium.

Auroras are due to the entry into the atmosphere of streams of positively charged atoms, mainly of hydrogen, together with electrons. They come from the sun with speeds far exceeding those of meteors and cause very beautiful and spectacular luminous displays, particularly in the polar regions. Auroral displays have a great variety of forms, and an international classification has been adopted. One of these displays is illustrated in Fig. 9-15. The height of auroras is about 100 km above the earth's surface, and the zones of greatest frequency are belts about 23° from the magnetic poles. In the Northern Hemisphere this runs through Alaska between Point Barrow and Fairbanks, across Canada and the southern tip of Greenland, over the northern edge of Norway, and off the north coast of Russia and Siberia (Fig. 9-16).

Although magnetic storms are not always accompanied by auroras, all auroral displays occur at times when the magnetic field is disturbed, and there must be some

FIGURE 9-15
Auroral display at Fort Churchill, Manitoba, Jan. 30, 1956. *(Courtesy of R. Montalbetti.)*

correlation between the two phenomena. Apart from visual observations, auroral information is now obtained by the "all-sky" camera and the spectrograph. The all-sky camera is a 16-mm motion-picture camera pointing downward onto a convex mirror which gives an image of the entire sky. Spectrographic analysis of the light from auroras can give information about the amount of energy radiated, the kinds of atoms and molecules present in the atmosphere, their temperature, and something about their mode of excitation. The auroral spectrum is rich in lines and bands. The green color arises from the atomic oxygen line (5577 A), the red colors are due either to the red line of atomic oxygen (6300 A) or to certain bands of N_2. A strong feature of the spectrum arises from transitions, not in the neutral N_2 molecule, but in the ion N_2^+. There are also present certain lines from atomic nitrogen, as well as other N_2 bonds and atomic oxygen lines. The particles that contribute most to auroral emission have been found to be electrons with energies below 25 kev, although electrons with energies as high as 100 kev are known to be associated with auroras. Most of the auroral light seems to be produced by electrons near 10 kev; certain auroras are excited by both electrons and protons, and others

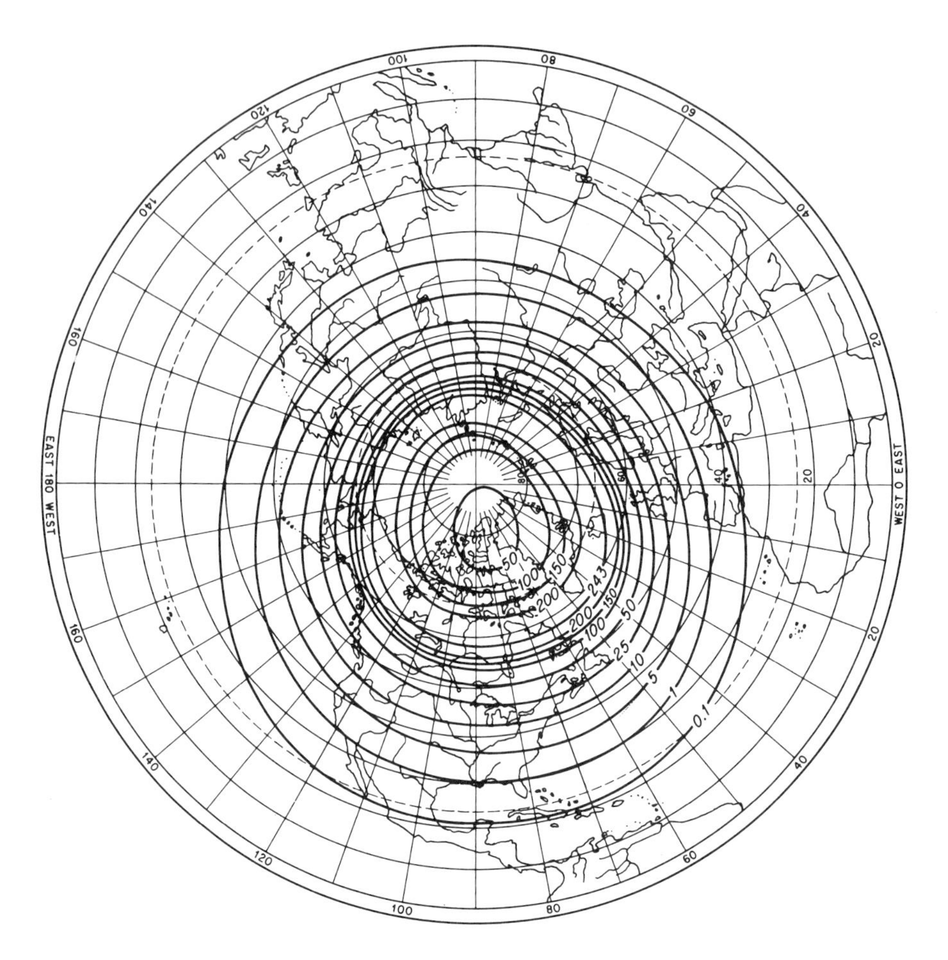

FIGURE 9-16
Map showing the frequency distribution of auroras in the Northern Hemisphere. The figures represent the mean number of days with auroras in a year. (*After E. H. Vestine, 1944.*)

predominantly by either electrons or protons. The narrowness of some auroral arcs is consistent with electrons being the active particles; the cyclotron radius would be too large if protons were the cause.

There is really no satisfactory explanation of the source of auroras. Satellite experiments have, however, eliminated two seemingly acceptable theories. C. Störmer (1955) explained auroras as due to solar particles that entered the earth's magnetic field directly near the auroral zones. Both Russian (K. I. Gringauz et al., 1961) and U.S. (J. A. Van Allen and L. A. Frank, 1962) deep-space probes have shown that sufficient fluxes of such energetic solar particles do not exist in the interplanetary medium away from the earth. The detection by early low-altitude satellites of large fluxes of electrons with sufficient particle energies to contribute to the production of auroras led to the hope that auroras were caused by the dumping of trapped electrons from the Van Allen radiation belts. However, measurements by both Russian (K. I. Gringauz et al., 1961) and U.S. (B. J. O'Brien et al., 1962) scientists have shown that such a source is insufficient; a flux tube of the Van Allen belt would be drained in a matter of seconds by a strong aurora, which itself could last for hours. In fact, the results from the Injun satellite (B. J. O'Brien, 1962) indicated an *increase* in the trapped-particle population when large fluxes of electrons entered the atmosphere.

9-7 THE MAGNETOSPHERE

In recent years transient variations in the geomagnetic field have been considered from the viewpoint of hydromagnetic waves generated by interactions between the geomagnetic field and solar plasma, i.e. the ionized gas moving out from the sun. This gas, which flows radially outward, has been called the *solar wind*. The best direct evidence for the existence of a continuous solar wind comes from the observations of comet tails by L. Biermann (1953). Impacts between the solar wind and gaseous particles in the tail of a comet cause the latter always to point away from the sun. Ejection of particles from the sun to form a "wind" is probably the result of an acceleration process involving changing magnetic or electric fields; i.e., the sun acts as a giant particle accelerator.

In a series of papers beginning in 1957 on the investigation of the expansion of the solar corona into interplanetary space, E. N. Parker developed a magnetohydrodynamic theory of the solar wind (see, for example, E. N. Parker, 1963). An essentially static situation had been considered for many years to represent the steady-state interplanetary medium into which transient tongues of plasma were ejected following solar activity, thereby leading to magnetic storms. Parker showed that the only reasonable model of the interplanetary medium utilizing all available information on coronal temperatures and densities was of necessity hydrodynamic and, most important, supersonic. The theory predicted that velocities of 400 to 1000 km/sec would be observed at the orbit of the earth. Since then the investigation of interplanetary space by artificial satellites and space probes has confirmed the model of the solar wind as developed by Parker. Satellites have shown that the interplanetary medium in the vicinity of the earth is not just empty space, but instead is filled with a highly tenuous plasma which is being continuously blown radially out from the sun at speeds averaging 300 to 500 km/sec. The wind is very "gusty," however, showing fluctuations in energy, energy spread, and density in times of the order of hours. The solar plasma consists primarily of ionized hydrogen (protons and electrons) and is electrically neutral. The density is of the order of 10 ions per cubic centimeter. Imbedded in the solar wind is an interplanetary magnetic field whose strength is of the order 5γ during quiet solar periods but increases to many times this value during periods of high solar activity. Its energy density is much smaller (approximately 1 percent of that of the solar wind), and it is thus carried along by the solar wind.

Plasmas and magnetic fields tend to confine one another. If a streaming plasma encounters a magnetic object such as a magnetized sphere, the plasma will confine the magnetic field to a limited region around the object. The object in turn will tend to exclude the plasma, thus creating a hole or cavity. The size of the cavity is determined by the energy density of the streaming plasma and the degree of magnetization of the object. In addition, if the velocity of the plasma is great enough to be highly supersonic, i.e., if the velocity is much greater than the Alfvén velocity in that medium, a detached shock wave may be produced in a region ahead of the cavity boundary. This is analogous to the formation of the detached shock front of an aerodynamic object traveling at hypersonic speed (above about a Mach number of 5). The analogy, however, is by no means perfect. In aerodynamics the shock wave results from collisions of particles and is about one mean free path thick. In the solar wind a Coulomb collision mean free path (approximately 10^{14} cm) is so large that collisions play no part in the observed shock wave. This collisionless shock is produced by the action of the magnetic field, and the characteristic dimension is the cyclotron radius, not the mean free path. A 1-kev proton in the interplanetary magnetic field of 5γ at IAU has a cyclotron radius of about 1000 km.

S. Chapman and V. C. A. Ferraro (1931, 1932, 1933) had predicted the confinement of the earth's magnetic field inside an elongated cavity during magnetic storms. The region inside the cavity is called the *magnetosphere*, and the boundary the *magnetopause*. The satellites Explorer 12 and Explorer 18 verified the existence of the magnetopause and a bow shock wave. [See Fig. 9-17, where the magnetopause is encountered at 13.6 earth radii (R_e), as indicated by the sharp change in magnitude and direction of the field. At $20R_e$ the satellite passed through the bow shock wave, the field suddenly becoming steady at about 5γ in a relatively constant direction.] The region between the magnetopause and the shock wave is referred to as the *magnetosheath*, or *transition region*. Outside this transition region, i.e., beyond the shock wave, conditions are characteristic of the interplanetary medium, and the presence of the magnetized earth has little or no effect.

The dimensions of the cavity depend on the intensity of the solar wind, although large changes in the solar-wind intensity produce comparatively small changes in the size of the cavity. The distance from the center of the earth to the magnetopause in the solar direction is about $10R_e$, although distances less than $8R_e$ and greater than $13R_e$ have occasionally been observed. The shock wave is located several R_e beyond this distance. At 90° to the solar direction both the magnetopause and shock wave are observed to flare out to distances about 30 to 50 percent greater than in the subsolar direction. In the antisolar direction the cavity extends out to very large distances, very likely as far as the moon ($\approx 60R_e$) or farther. No definite closure of the magnetospheric tail has yet been observed by satellites. A schematic diagram of the geomagnetic field in the noon-midnight meridian plane is shown in Fig. 9-18.

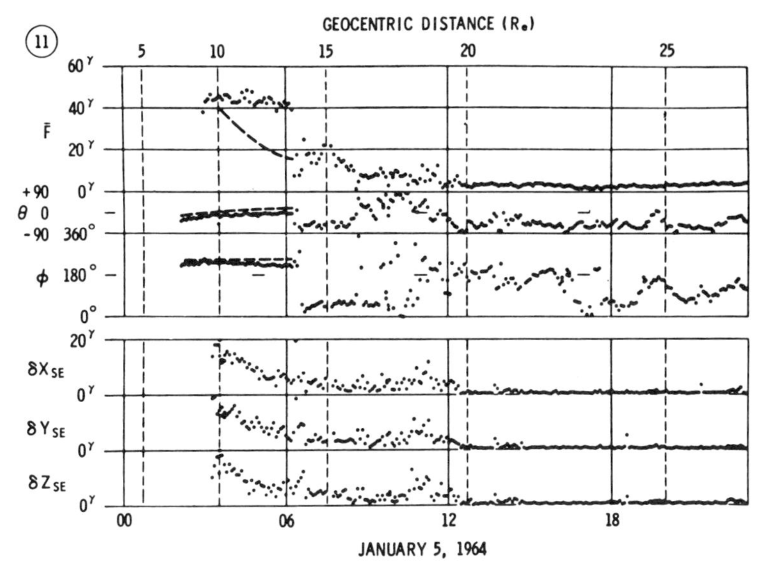

FIGURE 9-17
Magnetic field data from orbit 11 of Explorer 18. The lower curves give the variance in the field. The magnetopause is at $13.6R_e$. The second transition at $20R_e$ to an ordered field outside is the location of the bow shock wave. (*After N. F. Ness, C. S. Scearce, and J. B. Seek, 1964.*)

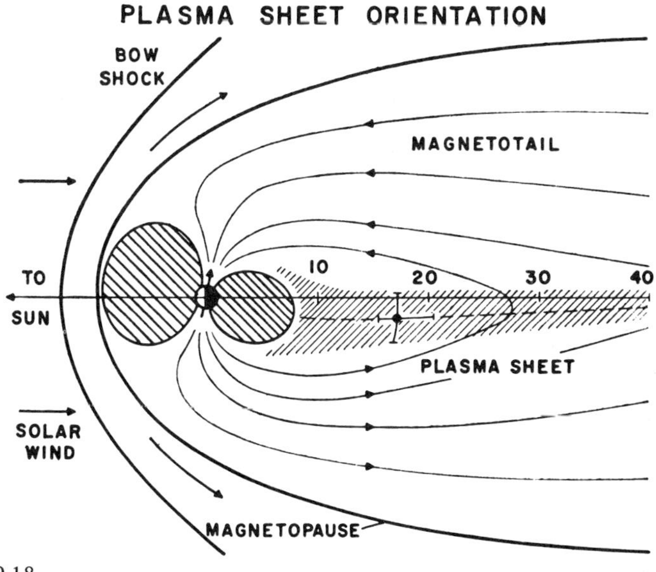

FIGURE 9-18
Approximate configuration of the magnetosphere in the solar magnetospheric noon-midnight meridional plane for a 12° tilt of the magnetic dipole axis. (*After S. J. Bame et al., 1967.*)

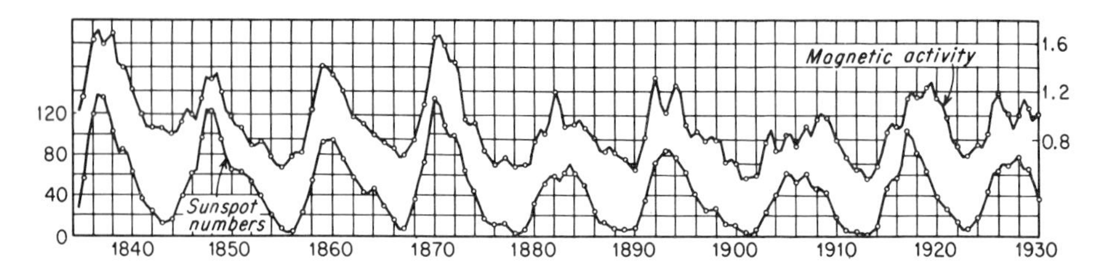

FIGURE 9-19
Variation of the annual mean sunspot number (lower curve) and the annual mean magnetic activity (upper curve) between 1835 and 1930. (*After S. Chapman and J. Bartels, 1940.*)

9-8 THEORIES OF MAGNETIC STORMS AND AURORAS

Magnetic storms are closely related to solar phenomena. This is strikingly revealed in Fig. 9-19, which shows the close parallel between the curves of relative sunspot number and magnetic activity over a number of years. Another close relationship with the sun is the recurrence tendency of magnetic storms, i.e., the tendency for magnetic storms to recur after a period of one or more solar rotations (27 days), although this tendency is characteristic of the smaller storms only. Again there is a time lag between notable magnetic activity on the earth and notable solar activity, about 1 day for great magnetic storms and 2 to 3 days for smaller disturbances.

The close connection between solar and magnetic activity and auroras suggests that the cause of the terrestrial phenomena may lie in disturbed areas of the sun's surface. The time lag of about 1 to 3 days between the central-meridian passage of disturbed areas on the sun and the onset of a magnetic storm shows that the storm cannot be caused by ultraviolet light or other electromagnetic radiation, but rather that magnetic storms and auroras are caused by the emission of a stream of charged corpuscles from the sun. Further evidence has been obtained by A. B. Meinel, who showed that hydrogen atoms traveling at a speed of several thousand kilometers per second enter the atmosphere during auroral displays. This is comparable with a speed of 1000 km/sec inferred from the time lag of about a day between notable solar activity and associated magnetic disturbances.

Since the rotation of the sun about its axis and the revolution of the earth about the sun are in the same sense, a solar stream of corpuscles will overtake the earth in its orbit. It will resemble the curved stream of water issuing from a rotating garden sprinkler, although the motion of an individual particle will be very nearly radially outward from the sun (Fig. 9-20). In 1896 K. Birkeland showed in the laboratory that when a stream of cathode rays (electrons) is projected toward a magnetized sphere, the rays are deflected toward the poles by the magnetic field. The mathematical theory of the motion of a stream of charged particles (of the same sign) ejected from the sun and entering the earth's magnetic field has been developed by C. Störmer (1955), who was able to account for many features of the

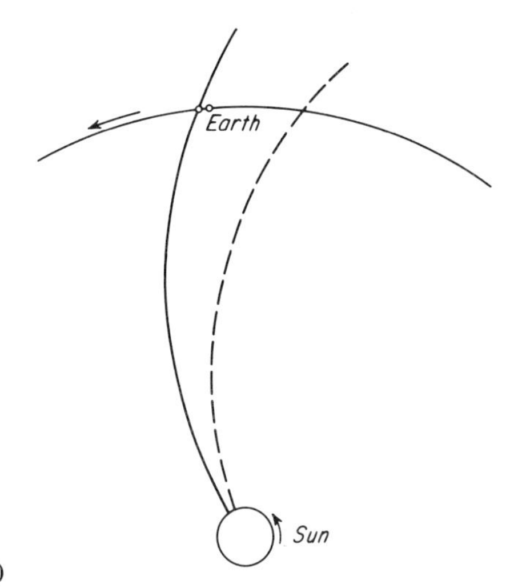

FIGURE 9-20
(*After S. Chapman and J. Bartels, 1940.*)

auroras. There is one fatal objection to this theory, however. To produce the auroras and magnetic storms, the density of charged particles must be high. If the stream contains charged particles of one sign only, they would be dispersed by mutual electrostatic repulsion long before the stream reached the earth. To overcome this difficulty, F. A. Lindemann suggested that the solar corpuscular stream might be electrostatically neutral but ionized, and S. Chapman and V. C. A. Ferraro (1931, 1932, 1933) have developed a theory based on this assumption.

Many theories of magnetic storms and auroras have been proposed to explain the observed characteristics. As Ferraro himself has remarked, "The history of magnetic storms and auroras is strewn with the wrecks of discarded theories." The work of Chapman and Ferraro has dominated thinking during the last few decades. In their theory a neutral stream of ionized particles is ejected from the sun, the sudden commencement of a magnetic storm being interpreted as the compression of the geomagnetic field due to the approach of the corpuscular stream toward the earth. As the stream advances, the flow of charged particles under the influence of the earth's magnetic field forms a curved hollow in the vicinity of the earth. The development of the D_{st} field in the main phase of magnetic storms is thus explained by the formation of an equatorial electric ring current. The Chapman-Ferraro theory was extended by D. F. Martyn (1951), but although their ideas can explain many features of magnetic storms and related phenomena, they cannot give a quantitative explanation for them. Their failure was due to the fact that they did not take into account the then unknown existence of the solar wind and plasma that surround the earth.

When a transient plasma stream arrives at the boundary of the magnetosphere with a speed substantially greater than that of the steady solar wind, the outer region of the magnetosphere must experience a strong impact. A direct effect of the impact is the sudden compression of the magnetic field and the plasma in that region. This compression is transmitted inward as a hydromagnetic perturbation which, on arrival at the earth, is observed as a sudden increase in the magnetic field, which we call a sudden commencement. The compressional effect is observed in its simplest form in low and moderate latitudes. In higher latitudes the consequence of the impact is less straightforward. The equatorial portion of magnetic field lines originating from high latitudes is first distorted inward by the impulsive force. The distortion is then transmitted along the field lines toward the earth in high latitudes as Alfvén waves.

In low latitudes the increased level of H caused by the sudden commencement is maintained for a few hours until the large decrease of the main phase begins. In a relatively short time after the first contact of the front of the solar stream with the sunward surface of the magnetospheric boundary, the magnetosphere is completely immersed in the storm plasma stream, and hence will be in a more compressed state than it was prior to its arrival. This is the reason for the increased level of H in the initial phase. The new equilibrium between the intensified solar wind and the earth's magnetic field is maintained until the solar plasma cloud has completely passed the earth and the solar-wind pressure returned to its normal level.

Analyses of storm records from many observatories around the world show that the worldwide decrease in H in the main phase is due to a uniform magnetic field. Such a field could be produced by a ring-shaped westward current encircling the earth. S. Chapman and V. C. A. Ferraro (1931, 1932, 1933) were the first to investigate in detail the formation, equilibrium, and stability of a ring current. At the time they presented their theory, the existence of a plasma around the earth, extending to distances of several tens of thousands of kilometers, was not known, nor was it known then that there exists a steady plasma flow out from the sun. Charged particles trapped in the earth's magnetic field drift longitudinally, the direction of the drift being eastward for electrons and westward for positive ions. The net current is therefore westward. A charge gyrating about a magnetic field line constitutes a small current loop which is equivalent to a small magnetic dipole. This dipole is so oriented as to reduce the magnetic field. Thus the particle has a diamagnetic effect. L. A. Frank (1967) has obtained direct evidence for a ring current at a distance of about $3.5R_e$ by measuring, on the satellite OGO 3, the differential energy spectra of protons and electrons in the outer Van Allen radiation belt in the energy range from ≈ 200 ev to 50 kev. The total energy of these low-energy protons and electrons within the earth's magnetosphere is sufficient to account for the reduction of the geomagnetic field $D_{st}(H)$ observed at the earth's surface at low and middle latitudes during the main phase of magnetic storms. The slow decay of the ring current is probably due to a charge exchange process between the protons in the ring current and the ambient slow neutral hydrogen in the magnetosphere.

9-9 COSMIC RAYS

Primary cosmic rays are submicroscopic particles that travel in space outside the earth's atmosphere at speeds nearly equal to the speed of light. The atmosphere is being continually bombarded by high-energy particles. They are deflected by the earth's magnetic field and collide with nitrogen and oxygen nuclei, giving rise to secondary cosmic rays. The main portion of the radiation is isotropic and time-independent. There are, however, small variations with direction, latitude, time of day, and season of the year. These are mainly due to the influence of the earth's magnetic field and the atmosphere; they will be discussed later. The primary radiation is composed approximately of 94 percent protons, 5 percent α particles, and 1 percent heavier nuclei. Only recently (1961) have electrons been identified in the primary radiation. The electron/proton ratio, however, is only about 1 percent; i.e., the electron flux is not very large.

There has been much speculation on the origin of cosmic rays. It has been suggested that they were created by some sudden catastrophic process or by a tremendous explosion that gave birth to the universe. Such an origin does not seem very likely, and it now appears that they are being produced continually somewhere in the system of stars that forms our galaxy; perhaps, as Alfvén has suggested, they are confined by magnetic fields to the interior of galaxies. It has been shown that cosmic rays cannot be generated by electrostatic fields within the galaxy; it would be necessary to assume exceedingly high potential differences, and since regions of high conductivity occur quite frequently within the galaxy, it would not be easy to maintain such electrostatic fields. It is necessary to consider the acceleration of cosmic rays by magnetic fields which would themselves have to be in motion, since charged particles cannot gain energy in a stationary magnetic field. Although variable magnetic fields will produce electric fields which will impart a motion to the conducting matter as a whole, the magnetic fields will not give rise to strong acceleration of individual ions which could be identified with the primary particles of cosmic radiation. E. Fermi (1949) has shown, however, that repeated accelerations can give rise to a spectacular increase of energy.

It is possible that magnetic fields near the earth's orbit prevent some of the particles with smaller energies from reaching our planet. It is also possible that cosmic-ray particles cannot escape from the neighborhood in which they are created unless they possess a minimum amount of energy. It has been suggested that the bulk of the radiation may come from double stars whose two partners possess strong magnetic moments; from variable stars whose exceedingly strong magnetic fields reverse direction with periods of a few days; or from supernovas. (See, for example, V. L. Ginzburg and S. I. Syrovatskii, 1964.) A. Unsöld has suggested that radio stars produce high-energy particles, similar to those emitted by the sun when it is in a disturbed state, and that the main component of cosmic radiation comes from radio stars in the center of our own galaxy. More recently G. R. Burbidge and F. Hoyle (1964) suggested that cosmic radiation pervades the

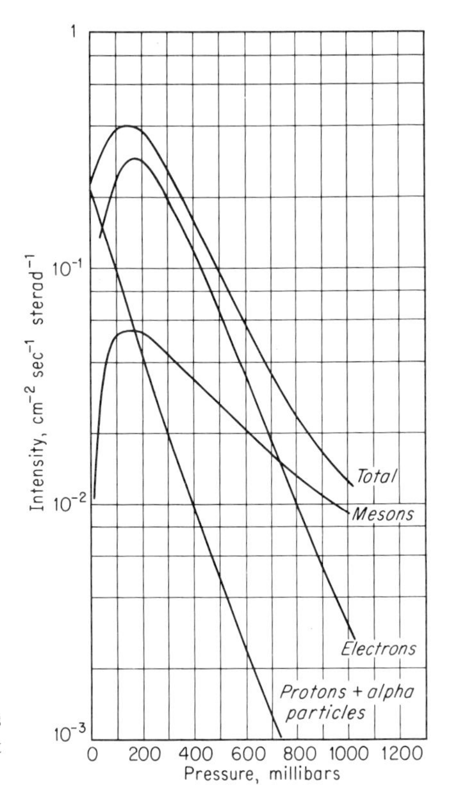

FIGURE 9-21
Height-intensity curves for the various components of cosmic radiation at 50° N. (*After G. Puppi.*)

universe, or at least a local supercluster of galaxies, at about the level observed near the solar system, and possibly originates in quasi-stellar objects.

As cosmic rays approach the earth they become deflected by the earth's magnetic field and their intensity exhibits a number of variations. The intensity increases with height above the earth's surface, reaching a maximum appreciably below the top of the atmosphere, the actual distribution depending on latitude (Fig. 9-21). The intensity is very small at the top of the atmosphere, indicating that most of the phenomena observed are secondary effects produced within the atmosphere. These secondary cosmic rays are produced by interactions between the primaries and nuclei in the atmosphere and by subsequent nuclear interactions or radioactive decay of the first products. Cosmic rays are often divided roughly into two classes by the experimental distinction of whether or not they penetrate through 10 to 15 cm of lead. The nonpenetrating, or soft, component includes almost all the photons, electrons, and positrons, while the hard component includes most of the μ mesons. One of the outstanding features of the hard component is its extraordinary penetrating power; μ mesons can be detected in deep mines and at great depths under water.

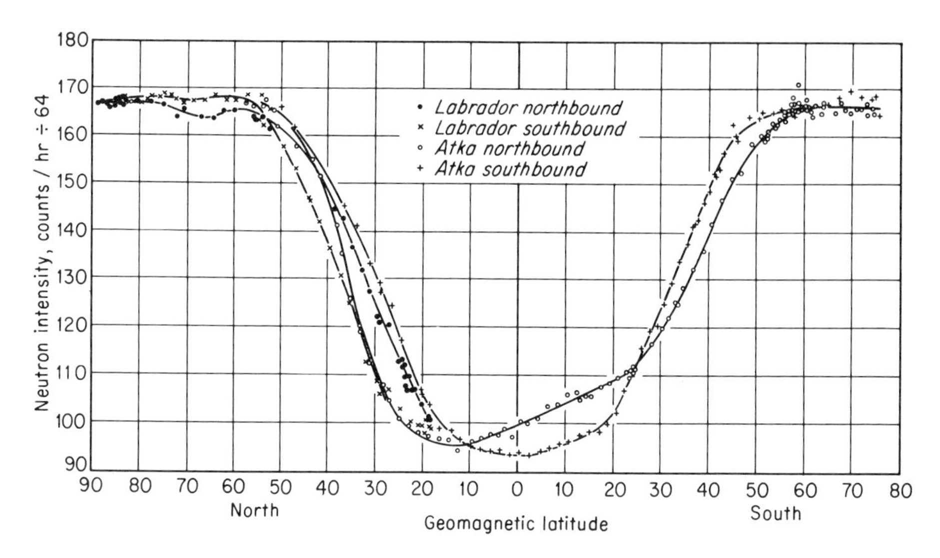

FIGURE 9-22
Variation of cosmic-ray nucleon component (neutron intensity) with latitude, using geomagnetic coordinates based on the 1945 epoch geomagnetic survey. Sea level, 1954–1955. (*After D. C. Rose, K. B. Fenton, J. Katzman and J. A. Simpson, 1956.*)

Figure 9-22 shows the variation in intensity with geomagnetic latitude. The variation is small at high latitudes, the graph exhibiting a knee at about 35 to 40°, and increases with altitude. There are also small seasonal changes which are more apparent at high latitudes and opposite in the two hemispheres. The maximum intensity occurs during the colder part of the year. The intensity also exhibits a diurnal change, with a maximum at about 1400h and a minimum shortly after 2400h, both local mean time. This variation is, however, small and appears to be independent of latitude and altitude. Erratic changes may also occur during magnetic storms.

Since at large distances from the earth local anomalies in the magnetic field become negligible, the possibility arises that the orientation and center of the equivalent dipole moment of the earth's magnetic field may be determined by cosmic-ray measurements. In other words, cosmic rays may be used as "probes" to explore the otherwise inaccessible magnetic field around the earth. Since for any longitude the cosmic-ray intensity reaches a minimum at the effective geomagnetic equator of this outer field, a series of determinations of the minimum intensity at several longitudes defines its effective equatorial plane. The first measurements of the neutron intensity from the nucleonic component were made aboard the *U.S.S. Atka* which crossed the equator at 30°W and 100°W on its 1954–1955 Antarctic expedition. The meson intensity from the vertical was also measured with a

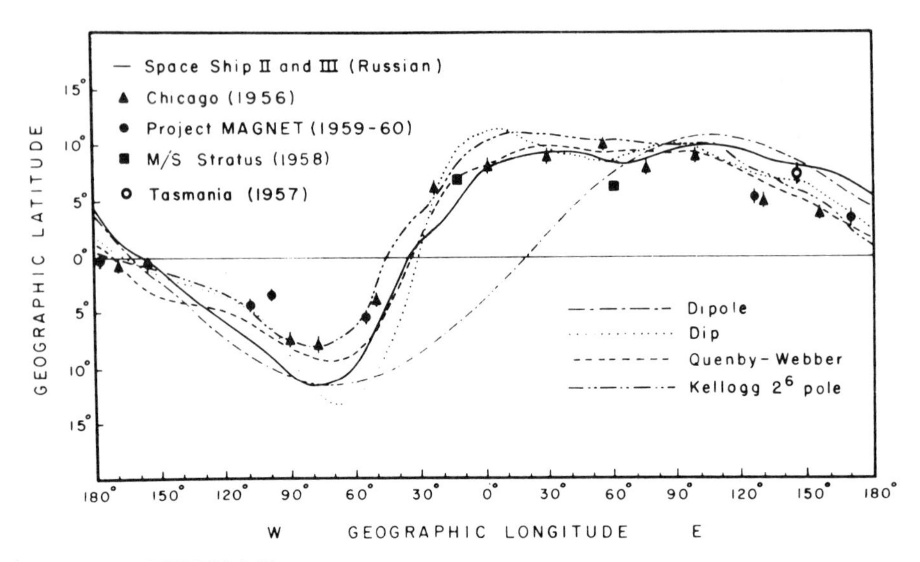

FIGURE 9-23
Determination of the location of the cosmic-ray equator by Project Magnet and other surveys. (*After H. Coxell, M. A. Pomerantz, and S. P. Agarwal, 1966.*)

threefold vertical counter telescope, and the positions of the minima agreed with those obtained by nucleonic-component measurements. Further measurements were made aboard the *U.S.S. Arneb*, during a second Antarctic expedition in 1955-1956, which crossed the equator at 100°W and 65°E. These early results showed large discrepancies between the effective geomagnetic equator derived from cosmic-ray measurements and the geocentric dipole equator, and this was confirmed by additional measurements by many other workers. The magnetic dip equator, on the other hand, fits the data fairly well. It has now been established (see, for example, J. J. Quenby and W. R. Webber, 1959) that in order to account satisfactorily for the distribution of cosmic-ray intensity over the surface of the earth, the effect of the nondipole part of the earth's internal field must be considered. Observations of the intensity of the nucleonic component have more recently been obtained with an airborne neutron monitor (U.S. Naval Oceanographic Office, Project Magnet 1) in a worldwide survey extending from the equator to the north and south geomagnetic poles. Figure 9-23 shows the position of the cosmic-ray equator determined by Project Magnet 1; earlier experimental determinations and a theoretical prediction by P. J. Kellogg (1960) are included for comparison. Kellogg used the 48-parameter expression for the earth's magnetic field as given by H. F. Finch and B. R. Leaton for the epoch 1955. The observations obtained by Project Magnet 1 are in good agreement with Kellogg's theoretical curve, which also coincides with the cosmic-ray equator defined by Quenby and

Wenk (1962). The location of the cosmic-ray equator has now been determined by many different experimenters at discrete longitudes. Also, with the improvements in high-speed digital computers in recent years, it has been possible to compute theoretical cosmic-ray equators more accurately. A comparison of all the theoretical and experimental determinations has been presented recently by M. A. Shea (1969).

SUGGESTIONS FOR FURTHER READING

S. Chapman and J. Bartels, "Geomagnetism," Oxford University Press (1940), has been a standard text on all aspects of geomagnetism for many years. S. Chapman has also published a small book in Methuen's Monographs on Physical Subjects entitled "The Earth's Magnetism" (1951). "The Upper Atmosphere," by S. K. Mitra (Asiatic Society Monograph Series, vol. 5, 1952), is another older but classic work. More recently S. Matsushita and W. H. Campbell have edited a two-volume set entitled "Physics of Geomagnetic Phenomena," Academic Press (1967). In recent years there have been many excellent books written on all branches of upper atmospheric physics, some of a review nature and some on specialized topics. It is impossible to give a comprehensive list, but some are as follows:

ALFVÉN, H., and C. G. FÄLTHAMMAR: "Cosmical Electrodynamics," 2d ed., Oxford University Press, Fair Lawn, N.J., 1963.

CHAMBERLAIN, J. W.: "Physics of the Aurora and Airglow," International Geophysics Series, vol. 2, Academic Press, Inc., New York, 1961.

DAVIES, K.: Ionospheric Radio Propagation, U.S. Dept. of Commerce, National Bureau of Standards Monograph 80, 1965.

DeWITT, C., J. HIEBLOT, and A. LEBEAU (eds.), "Geophysics and the Earth's Environment," Gordon and Breach, Science Publishers, Inc., New York, 1963.

DUNGEY, J. H.: "Cosmic Electrodynamics," Cambridge University Press, New York, 1958.

HESS, W. N. (ed.): "Introduction to Space Science," Gordon and Breach, Science Publishers, Inc., New York, 1965.

KING, J. W., and W. S. NEWMAN (eds.), "Solar-Terrestrial Physics," Academic Press, Inc., New York, 1967.

RATCLIFFE, J. A.: "Physics of the Upper Atmosphere," Academic Press, Inc., New York, 1960.

"Satellite Environment Handbook" (F. S. JOHNSON, ed.), Stanford University Press, Stanford, Calif., 1961.

Space Research, *Proc. Int. Space Sci. Symp.*, North-Holland Publishing Company, Amsterdam, published annually since 1960.

10

THE STRUCTURE AND COMPOSITION OF THE EARTH'S MANTLE AND CORE

10-1 INTRODUCTION

In a series of papers written between 1940 and 1942, K. E. Bullen divided the earth into a number of regions based on velocity-depth curves (see, for example, Bullen, 1963a). His nomenclature (Table 2-2) has since been widely used and, in spite of uncertainties in the boundaries between the different regions, continues to serve as a useful basis in discussing the earth's interior. The upper mantle consists of layer B, extending from the base of the crust (layer A) to a depth of about 400 km, and layer C, which is a transition zone between depths of about 400 and 1000 km. The lower mantle, below a depth of about 1000 km, is called D. Bullen later subdivided D into D' and D'', the region D'' being the bottom 200 km. The core has been divided into an outer (fluid) core E, a transition region F, and an inner (solid?) core G. More recent work (Sec. 10-10) indicates additional fine structure in the core, and further subdivisions have been proposed. The composition of the mantle and the core and the nature of some of the boundaries between the different regions will be discussed in the following sections. A recent review of the structure and composition of the mantle has been given by D. L. Anderson (1966a).

10-2 THE NATURE OF THE MOHOROVIČIĆ DISCONTINUITY

The crust and mantle are separated by a seismically determined boundary known as the Mohorovičić discontinuity, or "Moho" (Sec. 2-4). The depth to the Moho is usually about 30 to 50 km below the continents, but in general is not much more than 5 km beneath the ocean floors. At the Moho there is a discontinuous jump in the velocity of both P and S waves. It is not certain how "discontinuous" the velocity function is, but the change is thought to take place within a few kilometers at the most. There are two possibilities at the Moho: Either it represents a chemical boundary separating solid phases having different chemical compositions or it is a phase boundary, separating two solid phases.

Seismic velocities in the crust are in the range of experimental velocities at kilobar pressures for common rocks such as granite and gabbro; below the Moho, seismic velocities are fairly consistent with those found experimentally for relatively rare dense rocks such as dunite, peridotite, and eclogite. The generally accepted interpretation of the Moho is that it is a chemical discontinuity, separating crustal rocks characterized by a high feldspar content from underlying dunite or peridotite, essentially magnesium-iron olivines and pyroxenes. Two reactions have been suggested, however, as possible phase changes at the Moho, the gabbro-eclogite reaction and the olivine-serpentine reaction.

In 1959, H. H. Hess questioned the supposition that the oceans had a basaltic crust and proposed that the main oceanic crustal layer was more probably serpentine. Serpentine, which may be formed by hydration of olivine, decomposes above about 500°C, this temperature being nearly independent of water pressure. Hess suggested that the position of the suboceanic discontinuity might be determined by the 500° isotherm of an earlier time, "fossilized" at the lower present temperature of 150 to 200°C (see Hess, 1962, for more details). Although this is difficult to reconcile with the dynamical effects supposed to follow from subsequent hydration and dehydration, 500°C is consistent with estimates of temperature at the base of the normal continental crust, and serpentinization of peridotite can reduce the seismic velocities to values in the range of those in the lower crust. More recently experimental studies have been carried out on the deformation of serpentinite by C. B. Raleigh and M. S. Paterson (1965), who also discussed the tectonic implications of their results. A detailed study has also been carried out on a 300-m serpentine core drilled near Mayaguez, Puerto Rico (C. A. Burk, 1964).

In more recent years the gabbro-eclogite reaction has attracted attention again. Gabbro and eclogite, rocks with sharply contrasting mineralogy, have essentially identical composition. The mean density of eclogite is, however, about 10 percent greater than that of gabbro, a difference which is of the same order of magnitude as that indicated by seismic evidence to exist at the Moho. It had been suggested as long ago as 1912 by L. L. Fermor that the Moho was a phase change from gabbro to eclogite rather than a change in chemical composition. The phase

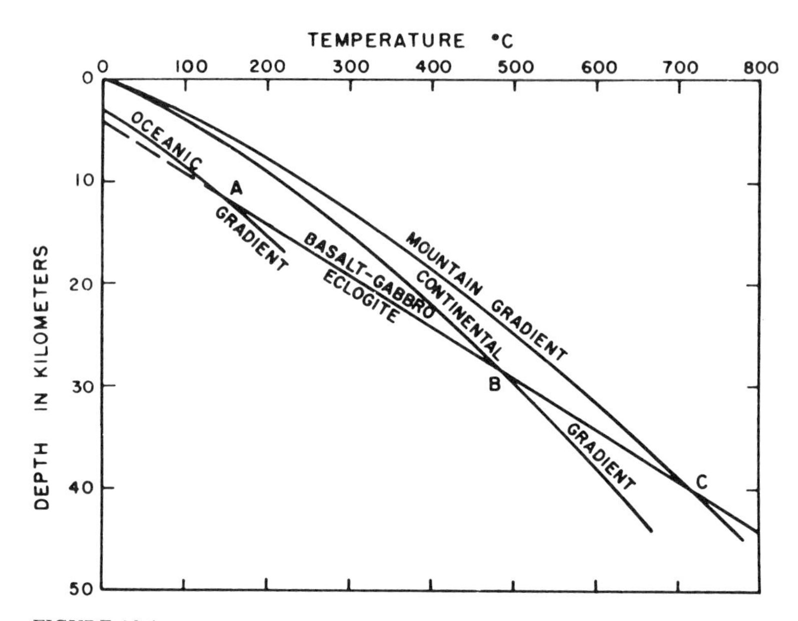

FIGURE 10-1
Postulated temperature gradients under mountain ranges, continental areas, and oceanic regions. (*After G. C. Kennedy, 1959.*)

change would take place at a different pressure for a different temperature, so that if the earth's temperature increases with depth a little more rapidly under mountain ranges than under continents, the transition will take place at a greater depth (Fig. 10-1, curve *C*); likewise, if the earth's temperature increases with depth a little more slowly under the oceans than under the mountains and continents, the transition is at a shallower depth (Fig. 10-1, curve *A*). Thus this single transition can explain the varying depths to the Moho under oceans, mountains, and continents. G. C. Kennedy (1959) suggested that such a phase change can explain a number of problems connected with the origin of the surface features of the earth:

1 The uplift of large continental areas, once at sea level, to form high plateaus is explained as a consequence of a rise in temperature of the rocks near the Moho by a few tens of degrees. If this happened, the phase change would migrate downward to much greater depths. The dense rock below the Moho would become light rock, and the change in volume would raise the continents to higher levels.

2 The lifetime of continents and mountain ranges is much greater than rates of erosion would suggest. This may be explained as the result of a decrease in pressure at the Moho below the mountains as the tops are eroded away. Dense rock at the Moho would then be converted to light rock; i.e., lighter roots beneath the mountains would be recreated to keep them at high elevations.

3 Sediments of low density filling troughs along the margins of continents are apparently able to subside into this higher density substratum. In most cases the sediments, after accumulating for perhaps 100 m yr and reaching a thickness of as much as 30,000 m, may be slowly folded and uplifted to form mountain ranges which may initially stand as high as 7000 m. The first effect of the deposition of sediments in troughs would be to increase the pressure at the base of the trough, with little change in temperature. Thus the discontinuity would migrate toward the surface. The trough would sink, not only because of the added load of sediments at the surface, but also because light rock would be converted into denser rock at the Moho below the trough, with a consequent decrease in volume of the material below. The new sediments filling the trough are of low thermal conductivity and possibly richer in radioactive material than the surrounding rock. Thus, given sufficient time, the temperature would slowly rise at the bottom of the trough, and although the Moho would first migrate upward, it would ultimately migrate downward, due to the rise in temperature. Thus troughs might sink for a considerable time and then be uplifted to form mountain ranges.

Although attractive in many ways, there are a number of difficulties to the above hypothesis. The principal minerals of gabbro are plagioclase feldspar and pyroxene, usually augite or hypersthene, while those of eclogite are typically magnesium-rich garnet and soda-rich pyroxene (omphacite). The relation between these rocks is not one of simple polymorphism; gabbro must recrystallize completely to form eclogite, no single mineral transforming without chemical change, even though the bulk composition remains fixed. Thus, if the Moho represents a change in crystal structure, it would not occur at a definite depth, but should extend over a range of pressures. Although seismic evidence indicates that the discontinuity may possibly be spread over some kilometers under the continents, it is extremely difficult to interpret the seismic data in this way under the oceans where the Moho is at a very shallow depth.

E. C. Bullard and D. T. Griggs (1961) have estimated pressure-temperature relations in the earth and found that the oceanic and continental transitions cannot both result in the high-pressure form lying beneath the Moho. They have shown that the temperature-depth curve under the oceans as drawn in Fig. 10-1 is incorrect, the curve actually crossing the transition from below to above and not from above to below; this result does not depend critically on the data used to construct the temperature-depth curves. They have also shown that variations in depth of the Moho from place to place are much less than would be expected from variations in heat flow. An extremely large increase in crustal thickness under the oceans is indicated if the surface heat flow is increased above average values. Similar results have been reached by G. W. Wetherill (1961) as a result of much more detailed calculations. He concluded that the same phase transition can produce both the oceanic and continental Moho only if the phase transition has very special properties and unreasonable crustal models are used.

P. J. Wyllie (1963) has pointed out that the two hypotheses of a chemical discontinuity and a phase change need not be mutually exclusive. He suggested that there may be two discontinuities beneath the crust, a chemical discontinuity of global extent between material of basaltic composition at the base of the crust and material with the composition of feldspathic peridotite in the upper mantle. The depth to this discontinuity in oceanic areas differs from that in continental areas. In addition to this chemical discontinuity, there may also exist (at the appropriate depth, depending on the geothermal gradient) another discontinuity marking the gabbro-eclogite phase transition. Whichever discontinuity is nearer the earth's surface would be detected as the Moho by seismic measurements. It would probably be difficult by such methods to detect a deeper transition, although refinements in techniques may ultimately determine whether a second discontinuity exists below the Moho. In this respect it is interesting to note that magnetotelluric measurements in Massachusetts indicate a rapid change of resistivity at a depth of about 70 km, which is deeper than the Moho in that region.

W. B. Joyner (1967), like P. J. Wyllie, believes that the hypothesis that the Moho is a phase change is not applicable on a worldwide basis, applying only to parts of the continents, and offers the best explanation for the occurrence of great thicknesses of shallow-water sedimentary deposits. Joyner carried out a numerical investigation of the history of a sedimentary basin based on the assumption that such a phase change does occur. Subsidence due to isostasy alone is inadequate to explain the observed thickness of shallow-water sediments; the phase-change hypothesis, on the other hand, provides a mechanism that can account for the deposition of substantial thicknesses of sediments in relatively shallow basins. As far as the consequences of the phase-change hypothesis for subsidence and uplift are concerned, it does not matter whether or not the phase change is identified with the seismically defined Moho. What does matter is whether or not a phase-transformation zone occurs within material of such bulk composition that significant volume changes accompany the change in phase.

A very thorough and detailed experimental investigation of the gabbro-eclogite transformation in several basalts has recently been carried out by D. H. Green and A. E. Ringwood (1967). They concluded (A. E. Ringwood and D. H. Green, 1966) that the hypothesis that the continental Moho is caused by an isochemical transformation from gabbro to eclogite must be rejected on a number of grounds.

Figure 10-2 shows, in a simplified form, the phase assemblages which were found in the typical basalts studied at 1100°C as a function of pressure. For each basalt there are clearly three principal mineral stability fields corresponding closely with naturally observed mineral assemblages. The low-pressure assemblage is that of

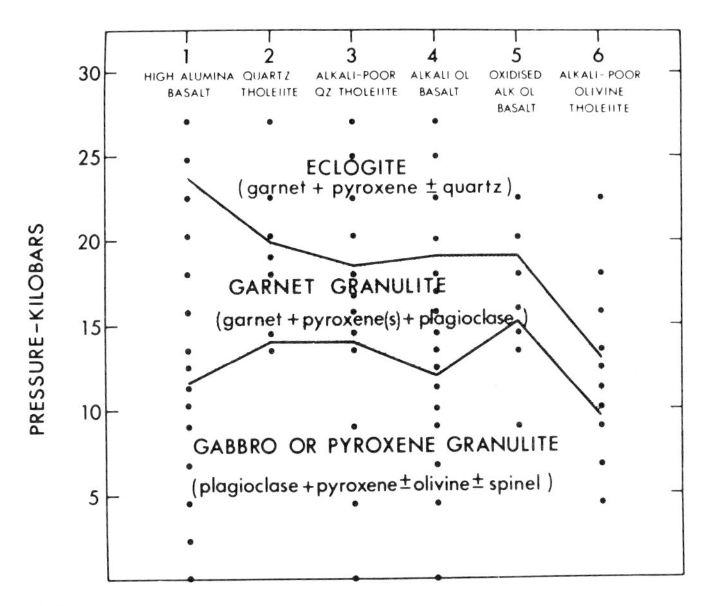

FIGURE 10-2
Principal mineral assemblages displayed by several basaltic compositions as a function of pressure at 1100°C. Solid circles denote experimental runs. (*After A. E. Ringwood and D. H. Green, 1966.*)

gabbro or pyroxene granulite. In all the basalt compositions studied, the transformation from gabbro or pyroxene granulite to eclogite proceeds through an intermediate mineral assemblage characterized by coexisting garnet, pyroxene(s), and plagioclase. This possesses an extensive stability field varying from 3.5 to 12 kbar in width. Figure 10-2 also shows that rather modest changes in chemical composition cause large changes in the pressures and widths of the gabbro-eclogite transformation.

Figure 10-3 shows the extrapolated stability fields of eclogite, garnet granulite, and pyroxene granulite or gabbro for a basalt of quartz tholeiite composition. The broken lines *AB* and *CD* are the experimental boundaries. Extrapolation is based upon the average of the gradients *AB* and *CD* with the assumption that the width of the garnet granulite zone is proportional to absolute temperature. The mean gradient thus obtained is 21 bars/°C. It can be seen from Fig. 10-3 that the temperature on the garnet granulite–eclogite boundary at a pressure corresponding to the base of the normal continental crust (10 kbar) is 670°C. If the temperature at the base of the crust is lower than this, eclogite would be the stable form of a basalt of this composition throughout the crust. In stable continental regions of normal crustal thickness and characterized by heat flows between 0.8 and 1.5 x 10^{-6} cal/cm^2 sec, the temperature at the base of the crust is usually less than 670°C on most

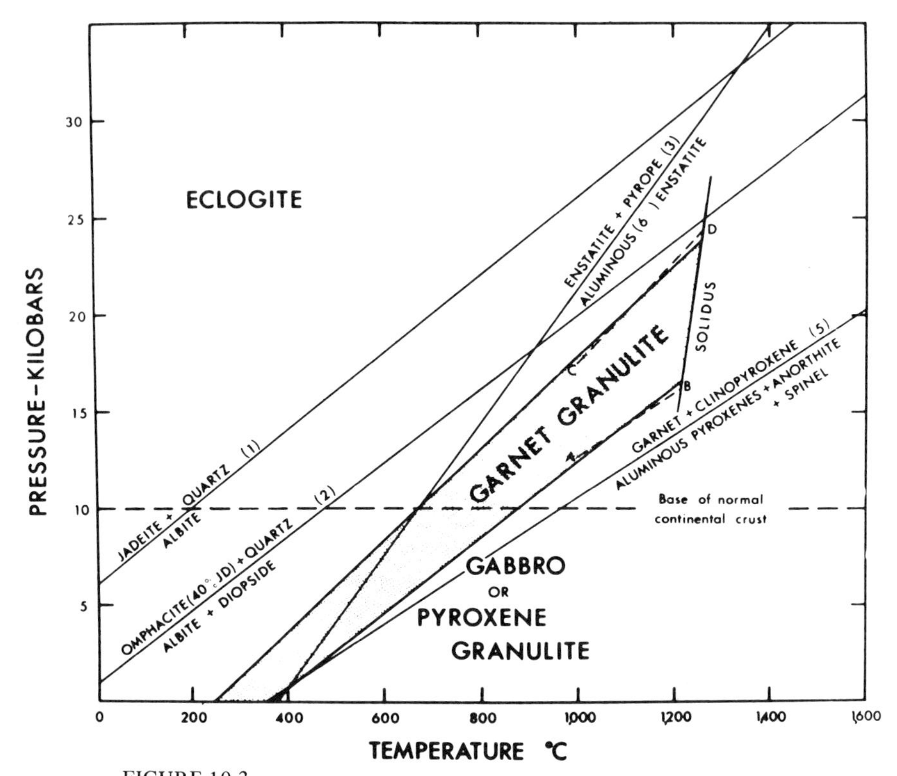

FIGURE 10-3
Extrapolated stability fields of eclogite, garnet granulite, and pyroxene granulite or gabbro for a basalt of quartz tholeiite composition. The broken lines *AB* and *CD* are the experimental boundaries. (*After A. E. Ringwood and D. H. Green, 1966.*)

reasonable assumptions of radioactivity distribution. This conclusion is practically certain for Precambrian shields characterized by mean heat flows of 1.0×10^{-6} cal/cm^2 sec; the temperature at the base of the crust in these regions is probably less than 450°C. It thus appears that eclogite is the stable modification of quartz tholeiite throughout very large regions of normal continental crust. This conclusion was unexpected and, if correct, would have profound tectonic consequences. The only escape from this conclusion would be to introduce a much larger curvature into the extrapolated phase boundaries than appears plausible. The foregoing experimental observations, when considered as a whole, strongly suggest that eclogite is the stable modification of most rocks of basaltic composition under dry conditions within large regions of continental crust.

In addition to the difficulty described above, A. E. Ringwood and D. H. Green (1966) showed that there are a number of other objections to the hypothesis that the Moho in normal continental regions is caused by an isochemical phase transformation from gabbro or basalt to eclogite. These are:

1. Experimental evidence on the effect of temperature on the pressure required for the gabbro-eclogite transformation cannot be reconciled with the rather small differences in crustal thickness in normal continental areas characterized by widely differing surface heat flows and, by inference, widely different temperatures at the base of the crust.
2. In most rocks of basaltic composition, the transformation from gabbro to eclogite occurs over a broad pressure interval, and the rate of change of seismic velocity is approximately uniform across the interval. Furthermore, the effective breadth of the transformation in the earth would be greatly expanded, owing to the tendency of geotherms to cross phase boundaries at low angles. The large effective width of the transformation in the earth makes it impossible to explain the Moho, which requires a substantial velocity increase within a depth of approximately 5 km.
3. Small changes in basaltic chemical composition have a large effect upon the pressure required for a gabbro-eclogite transformation. On a small scale, this would cause further smearing of the transformation zone if the Moho were caused by a phase change. On a large scale it would lead to improbably large fluctuations in the thickness of the crust.
4. The average density of eclogites is 3.5 g/cm^3, whereas the density of the upper mantle is generally believed to lie between 3.3 and 3.4 g/cm^3.
5. Most current advocates of the phase-change model argue that the eclogite layer (density 3.5 g/cm^3) immediately below the continental crust passes downward into peridotite (density 3.3 g/cm^3). Such a configuration would possess a high degree of gravitational instability and is inherently improbable.

Although the gabbro-eclogite transformation is not believed to play a significant role in the structure of stable continental regions, it may be of major importance in tectonically active areas where the Moho cannot be clearly recognized, e.g., regions of recent orogenesis, continental margins, island arcs, and mid-oceanic ridges. A. E. Ringwood and D. H. Green (1966) suggest that the basalt-eclogite transformation may provide a tectonic engine of great orogenic significance. Large volumes of basalt, when extruded and intruded at or near the earth's surface, may, on cooling, become transformed to eclogite under suitable circumstances. Because of the high density of eclogite, such large-scale transformations would generate gravitational instability. Large blocks of eclogite would sink through the crust, dragging it down initially into a geosyncline, and later causing extensive deformation (folding). Because the density of eclogite is greater than that of the ultramafic rocks which make up most of the mantle, blocks of eclogite would sink deep into the mantle and may undergo partial fusion, leading to the generation of andesitic and granodioritic magmas which rise upward and intrude the folded geosyncline.

10-3 THE UPPER MANTLE (LAYER *B*)

Static high-pressure equipment using large presses can reach a pressure of only about 100 kbar, which is equivalent to a depth of about 300 km in the earth. Measurements of the elastic properties of rocks have been made up to only about 15 kbar, which corresponds to a depth not much below the crust. Shock-wave data on rocks, on the other hand, can be compared fairly directly with data available from seismology (Sec. 10-4), and rapid progress is now being made to infer the composition and physical state of the material at all depths in the earth's interior.

The properties of the mantle are governed to an important degree by the response of the material to changes in pressure and temperature. Compositional variation affects the details of the overall picture, but the main features are independent of composition within the limits set by the cosmic abundances of the elements. The pressure distribution of the mantle is reasonably well established, but temperatures are much less certain. At depths less than about 200 km the effect of temperature in changing the properties of the mantle is comparable with the effect of pressure because of the high thermal gradient, and for certain properties, temperature may be the controlling factor. This is not the case for greater depths, and throughout most of the mantle, pressure is the dominant parameter affecting the properties of the material.

In a detailed discussion of petrological models for the upper mantle, S. P. Clark and A. E. Ringwood (1964) considered two models. In one the upper mantle was assumed to be composed of ultrabasic rock, which Ringwood (1962a) called pyrolite, and in the other, to be basic in composition and in the eclogite facies. In that case the Moho is a phase change. This hypothesis runs into severe difficulties, which have already been discussed (Sec. 10-2). Other difficulties are the high density implied by the eclogite model and the problem of accounting for the seismic structure of the upper mantle. Pyrolite is equivalent to a mixture of approximately 3 parts of dunite to 1 part of basalt. Because of the very low content of radioactive elements in dunite, the radioactivity of pyrolite is about one-quarter that of basalt. In their ultrabasic model the mean chemical composition of the upper mantle and crust over any extensive area of the earth is, to a first approximation, the same. Fractional melting of this primitive material (pyrolite) provides the source of basaltic magma and leaves residual dunite or peridotite. Continents are considered to have segregated essentially vertically owing to differentiation by fractional melting of the primitive pyrolite. As a result the upper mantle is chemically zoned. Immediately below the Moho the mantle consists predominantly of dunite or peridotite, with minor residual pockets of eclogite. This zone passes downward, perhaps to a depth of 200 km or more beneath Precambrian

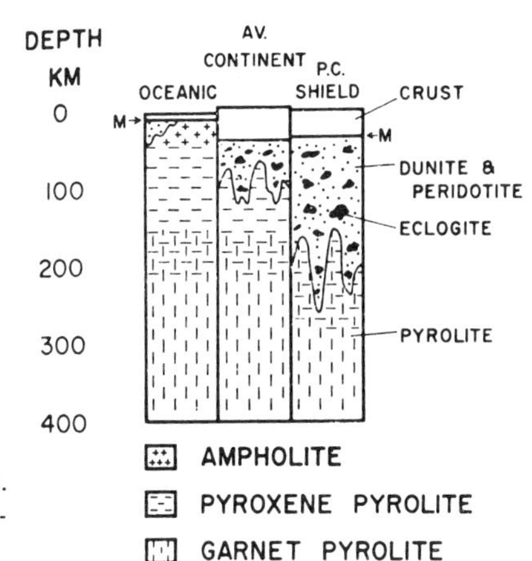

FIGURE 10-4
Petrological model for the upper mantle. (*After S. P. Clark, Jr., and A. E. Ringwood, 1964.*)

shields, into the primitive pyrolite. Under the oceans the pyrolite may extend right up to the Moho, or alternatively, a thin zone of dunite-peridotite (perhaps 25 km thick) may occur between the crust and the pyrolite zone. In "normal" continental areas the thickness of the residual peridotite zone may be intermediate between that beneath the oceans and that beneath Precambrian shield areas. Figure 10-4 illustrates schematically Clark and Ringwood's model of the upper mantle. Laboratory and field data (A. E. Ringwood 1962a,b; D. H. Green and A. E. Ringwood, 1963) have shown that material of pyrolite composition can crystallize in one of four distinct mineral assemblages, according to conditions of pressure and temperature. The two principal mineral assemblages are olivine + aluminous pyroxenes ± spinel (pyroxene pyrolite) and olivine + low-alumina pyroxenes + garnet (garnet pyrolite). Because of the intrinsically lower seismic velocity of the pyroxene pyrolite compared with the overlying dunite peridotite and the underlying garnet pyrolite, it would constitute a low-velocity (LV) zone.

It is not known for certain what is the cause of the LV zone at a depth of about 100 km. It may, as suggested above, represent a different mineral assemblage than the adjacent regions of the mantle, or the material in this zone may be partially molten; it may also be caused by a thermal gradient so large that the effects of pressure are canceled out. There is increasing evidence (see, for example, I. B. Lambert and P. J. Wyllie, 1970) that the LV zone may be due to incipient melting of mantle eclogite and peridotite in the presence of traces of water. The LV zone is also present in oceanic areas, but is virtually absent in stable shield areas. The LV zone is terminated fairly abruptly at a depth of about 150 km, indicating a sudden change in the physical state or the composition of the material at that depth. The magnitude and abruptness of the velocity change favor a compositional change. Perhaps the lighter fraction of the mantle, which also has a lower melting point, has migrated upward, leaving behind a refractory residue which not only has

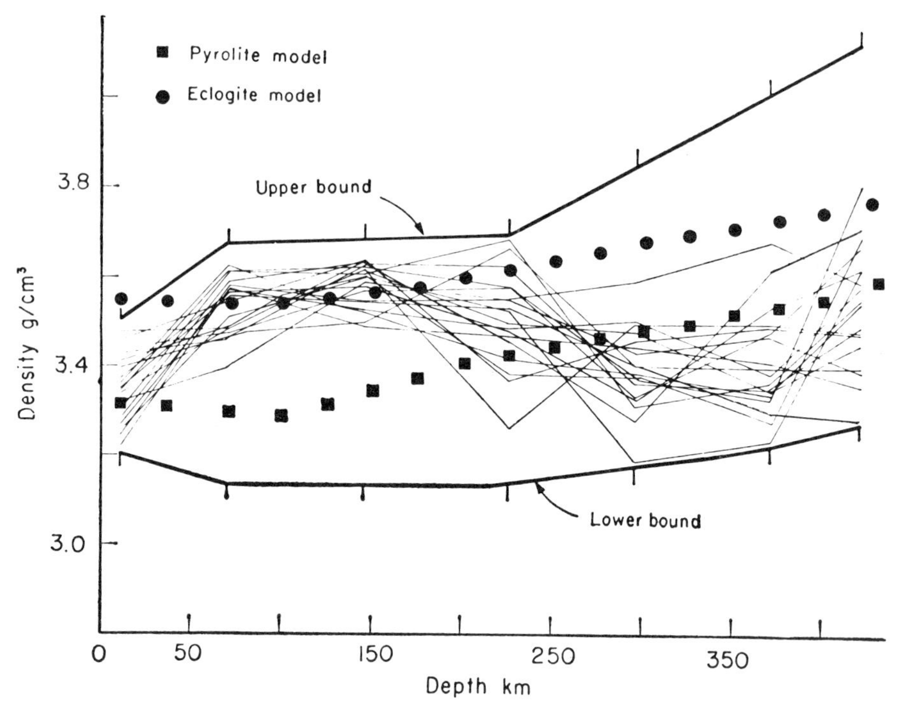

FIGURE 10-5
Successful density models for the suboceanic upper mantle. Bounds define the range permitted in the Monte Carlo selection. Points show density values according to S. P. Clark, Jr., and A. E. Ringwood (1964) for pyrolite and eclogite mantles. (*After F. Press, 1969.*)

higher velocities, but is further from its melting point. The LV zone may represent a great reservoir of magma held in a solid matrix as water is held in a sponge. Since molten rock is enriched in radioactivity, a partially molten zone is self-perpetuating. The conductivity of rock is so low that internally generated heat is effectively held in the earth unless the molten rock is allowed to escape to the surface or to shallow depths.

F. Press (1968a,b, 1970b) used a Monte Carlo method to obtain the density distribution of a number of earth models (Sec. 2-5). In a later paper (1969), using additional data, including both Rayleigh and Love wave phase velocities, he obtained a new determination of the density in the suboceanic mantle. Figure 10-5 shows the density distributions to a depth of 400 km for 18 successful models. At the very top of the mantle the densities occupy the entire range permitted in the Monte Carlo selections, indicating that the data are insufficient to constrain the models to narrow bounds. However, in the vicinity of 100 km, the densities fall in the narrow band between 3.5 and 3.6 g/cm^3, which lies in the upper part of the permissible range. In the depth range 250 to 400 km, control of density deteriorates again, with the models filling more than half the permissible range. Also plotted in Fig. 10-5 are densities computed by S. P. Clark and A. E. Ringwood (1964) for petrologic models of the mantle composed of pyrolite and eclogite. Their eclogite model alone is consistent with Press's results between 80 and 150 km. Either model is acceptable above this region, and the pyrolite model is weakly favored in the region near 300 km. All successful models show an LV zone for shear waves centered at depths between 150 and 250 km.

Recent data in support of sea-floor spreading and continental drift imply that the suboceanic mantle-crust system consists of a lithosphere about 100 km thick which behaves mechanically like a rigid plate. (X. Le Pichon, 1968; B. Isacks et al., 1968. See also Sec. 14-10, where these concepts are discussed in detail.) It is underlain by the asthenosphere, which is associated with the LV zone and presumably is a region of low strength. These properties of the asthenosphere probably result from partial melting forming the basaltic magma and peridotite or dunite residue. The lithospheric plate is produced beneath the ocean ridges from which it spreads away, cooling in the process.

Press's results uniquely associate high densities of 3.5 to 3.6 g/cm^3 with at least the lower half of the lithosphere and suggest reduced densities (3.3 to 3.5 g/cm^3) in the asthenosphere. The high density for the lower part of the lithosphere indicates that it is predominantly of eclogitic composition. Press therefore proposes that fractionation of eclogite is a key element in the synthesis of the lithosphere: The mechanism involves the basalt-eclogite phase transition (see also Sec. 10-2). A shell of eclogite around the earth has been proposed by many authors. A. E. Ringwood and D. H. Green (1966) proposed that the transformation of small pockets of basalt to eclogite in the crustal segment of the lithospheric plate drags the crust down near the continental margins, island arcs, or both. They did not envisage, however, the large-scale transformation to eclogite proposed by Press.

10-4 SHOCK–WAVE STUDIES

Studies of wave-propagation measurements in rocks show that V_p depends principally upon density and mean atomic weight. However, most common rocks have mean atomic weights close to 21 or 22, regardless of composition, and rocks or minerals of very different composition may have the same densities and seismic velocities. Thus it is not easy to infer chemical composition from seismic data, and laboratory experiments at the conditions of pressure and temperature that exist deep within the earth are highly desirable. The pioneering experimental work of P. W. Bridgman up to pressures of 10^5 atm corresponds to a depth of only 300 km within the earth. However, in the last few years, dynamic determinations of the compressibility of minerals and rocks have been made by a number of workers up to pressures in excess of 5×10^6 atm, which is greater than the pressure at the center of the earth. These high pressures are created for very short time intervals behind the front of a strong shock wave set up by an explosive charge, and are an order of magnitude greater than pressures which can be obtained by static methods. M. H. Rice et al. (1958) determined the equation of state for a number of materials up to pressures of 500 kbar. The agreement between the results and the static measurements of Bridgman strengthened the soundness of the basic assumptions upon which their interpretation rested. To interpret the data, the equation of state as determined from the shock-wave data, which is neither adiabatic nor isothermal, must be reduced to a reference temperature. Temperatures in the shock front are not generally known, and additional measurements or assumptions must be made to reduce the pressure-density data to those at absolute zero. It should be mentioned that equations of state of materials derived from shock-wave experoments have been regarded with some skepticism in the ultrahigh-pressure range by L. Knopoff and J. N. Shapiro (1969).

High-pressure research has been directed primarily to a study of the elements, although a number of studies have been carried out on materials likely to exist in the earth. It is impossible to review in detail all this extremely important work, and the reader is referred to two excellent review articles by F. Birch (1963) and T. J. Ahrens et al. (1969a). More detailed information can be found in the following references; the list is not intended to be comprehensive but rather to indicate some of the areas investigated and the problems encountered: T. J. Ahrens et al. (1964, 1969b); L. V. Al'tshuler et al. (1958a,b, 1960, 1962, 1965); D. L. Anderson and H. Kanamori (1968); A. A. Bakanova et al. (1965); R. G. McQueen et al. (1960, 1963, 1964, 1966, 1967); J. N. Shapiro and L. Knopoff (1969); H. Takeuchi and H. Kanamori (1966); R. F. Trunin et al. (1965); J. Wackerle (1962); J. M. Walsh et al. (1955, 1957); C. Y. Wang (1967, 1968, 1969).

Shock-wave data are at present the only source of information on the compressibility and polymorphism of silicates and oxides at pressures in excess of 300 kbar. These data complement the lower-pressure ultrasonic and x-ray diffraction data and the relatively low-pressure, high-temperature phase-equilibria studies on silicates and analog components. It is now possible to make direct comparisons of seismic data with the density and compressibility of a variety of materials tested with shock-wave techniques.

Numerical calculations have been carried out that determine the equation of state of the materials which constitute the mantle and core of the earth; comparison of these equations of state with those obtained experimentally indicate the type of materials that may exist in various regions of the earth. The general results are in close agreement with the conclusions reached earlier by F. Birch and others, viz., that the mantle is made up of high density oxides of silicon, iron, and magnesium, and that the core is mostly iron alloyed with some light elements. R. G. McQueen et al. (1967) found that a high-pressure modification of an olivine rock with a mean atomic weight of about 21.7 and an initial density of about

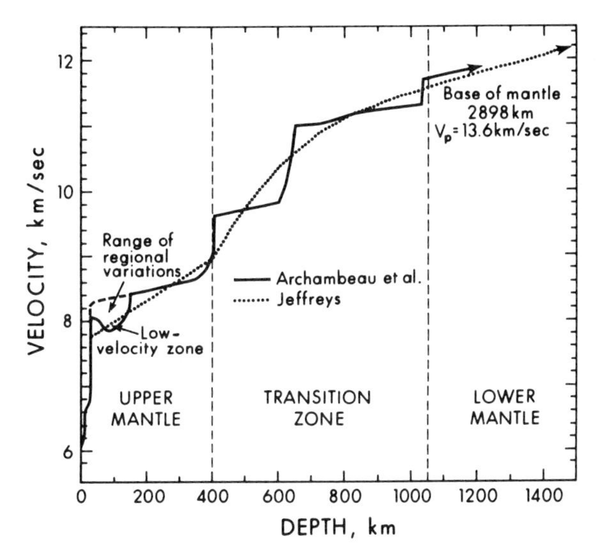

FIGURE 10-6
Distribution of seismic *P*-wave velocities in the outer 1200 km of the earth. [(*According to H. Jeffreys (1939) and C. B. Archambeau, E. A. Flinn, and D. G. Lambert (1969).*]

3.36 g/cm^3 satisfies the pressure-density and seismic requirements for the inner mantle. A. S. Balchan and G. R. Cowan (1966) measured the compressibility of iron-silicon alloys of 4.0 and 19.8 wt % silicon to pressures of 2.7 Mbar. Their results are consistent with an outer core containing 14 to 20 wt % silicon in iron, as had been proposed by A. E. Ringwood (1959) and by G. J. F. MacDonald and L. Knopoff (1958). In addition, S. B. Kormer and A. I. Funtikov (1965) analyzed shock-wave data on commercial ferrosilicon containing 17.4 percent silicon and 1.1 percent carbon in iron and again obtained better agreement with the properties of the outer core than pure iron.

10-5 THE TRANSITION LAYER *C*

Surface-wave studies (see, for example, D. L. Anderson, 1966a, 1967a,b) have shown that the abnormally high velocity gradients in the transition layer are concentrated in two relatively narrow zones 50 to 100 km thick instead of being spread out uniformly over some 600 km as was previously thought. Later studies, using travel times and the apparent velocities of body waves, have verified the presence of these two transition layers (Fig. 10-6). In most recent models of the upper mantle there is a very rapid increase in velocity between about 100 and 150 to 175 km and major discontinuities starting at 320 to 365 and 620 to 640 km. The velocity gradients of the adjacent sections of mantle are appropriate for normal compression, including that section of the mantle between 440 to 620 km which lies in the middle of what has been designated as the transition layer. It has been shown that the locations of these transition layers, their general shape and their thicknesses, are consistent with, first, the transformation of magnesium-rich olivine to a spinel structure, and then a further collapse to a material having approximately the properties of the component oxides.

F. Birch (1939, 1952), using an equation of state based upon finite-strain theory, investigated the rate at which seismic velocities should increase with depth in a homogeneous medium with an arbitrary temperature gradient. He concluded that the properties of the upper mantle were consistent with the layer being composed of familiar minerals such as olivines, pyroxenes, and garnets. However, the rate of increase of velocity with depth in layer C was too great for it to be homogeneous, and Birch proposed that phase changes leading to close packing occurred in this layer. Below 1000 km the elastic properties of the mantle were consistent with those of close-packed oxide phases which remained unchanged to the core boundary. Although Birch's arguments, particularly those relating to phase transformations, met with some opposition, experimental investigations have since confirmed them.

The difficulty about testing Birch's hypothesis experimentally was that, until 1963, available static high-pressure-temperature (p,T) apparatus was incapable of reproducing the (p,T) conditions in the earth at depths greater than about 300 km. Thus, prior to 1963, it was necessary to use *indirect* experimental methods based on thermodynamics, comparative crystal chemistry, and in particular upon the study of germanate isotypes of silicates. This latter technique is due to the fortunate circumstance that germanates were often found to display the same kinds of phase transformations as the corresponding silicates, but at much lower pressures. A. E. Ringwood (1969) has given an excellent summary with a very complete bibliography of the results of this indirect phase of investigations which covered the period 1956 to 1963 and in which he himself played so great a part. All the germanate olivines and pyroxenes which were studied in the pressure range 0 to 90 kbar were found to be unstable at high pressures and transformed to dense phases, suggesting that the corresponding silicates would transform similarly at higher pressures. During this period Ringwood found that the silicate olivines Fe_2SiO_4, Ni_2SiO_4, and Co_2SiO_4 could be transformed to spinel structures at 20 to 70 kbar and that the spinels were about 10 percent denser than the corresponding olivines. A high-pressure form of silica (SiO_2), having the rutile structure in which silicon was in octahedral coordination with a density of about 4.3 g/cm^3, was synthesized by S. M. Stishov and S. V. Popova (1961) at a pressure corresponding to a depth of about 100 km and has since been identified in the crushed zone in the Barringer meteorite crater, Arizona. This discovery suggested that the pyroxene family might transform to new structures such as ilmenite, characterized by octahedral coordination of Si^4.

In 1966 an apparatus capable of developing pressures above 200 kbar (corresponding to a depth of 600 km) simultaneously with high temperatures was developed by A. E. Ringwood and A. Major (1966, 1968) in Australia. Other laboratories (in particular in Japan) have also now developed such techniques. Many new phase transformations of geophysical interest have been discovered both in silicates and in germanates—an excellent review has been given by Ringwood (1969). The rapid increase of seismic velocity around 350 to 450 km seems to be caused mainly by the transformation of pyroxenes into a new type of garnet structure and the transformation of olivines to the spinel (or related) structure. Our knowledge of the constitution of the mantle below 600 km rests mainly upon the interpretation of phase transformations in germanate analog systems and on shock-wave studies. These suggest that around 600 to 700 km, garnets and spinels transform to new phases possessing ilmenite, perovskite, and strontium plumbate structures with densities and elastic properties resembling those of isochemical mixed oxides.

10-6 THE LOWER MANTLE *D*

The net effect of the phase changes in the transition zone is to transform familiar silicate structures into close-packed structures similar to those of the dense oxides. An essential feature is the change in coordination of silicon from fourfold to sixfold. Once closest packing is attained, further transitions involving changes in crystal structure are not possible, and the material remains homogeneous over a wide pressure range.

The constitution of the lower mantle has been discussed in detail by F. Birch (1952, 1964). He concluded that the seismic-velocity distribution is consistent with compression of a uniform layer, but the properties of this layer are not those of any known silicate. They resemble those possessed by relatively close-packed oxides such as corundum, periclase, rutile, and spinel. There is no evidence suggesting further phase changes or significant changes in chemical composition. Since velocity and density data provide at most two useful properties of the material in the lower mantle, only three compositional variables can be fixed. Consideration of meteorite and solar abundances indicates that the dominant oxides in the earth are SiO_2, MgO, and FeO, and thus these are the ones whose ratios should be adjusted. The abundance figures allow small amounts of other oxides such as Al_2O_3, CaO, and Na_2O, but they will not have a large effect. Taking the phases in the lower mantle to be an oxide solid solution with the periclase structure and a metasilicate solid solution with the ilmenite structure, S. P. Clark and A. E. Ringwood (1967) have shown that a molecular ratio of MgO/SiO_2 of 1.5 and of $FeO/(FeO + MgO)$ of about 0.1 provide a good fit to the density and velocity data. These ratios are also very plausible on geochemical grounds.

M. A. Chinnery and M. N. Toksöz (1967) and M. N. Toksöz et al. (1967) have determined *P*-wave velocities in the lower mantle from $dt/d\Delta$ measurements using the large-aperture seismic array in Montana and travel times from the LONGSHOT nuclear explosion. Previous models for the *P*-wave velocity profile were based primarily on observations of travel times and amplitudes from earthquakes, with the inherent lack of accurate origin times and epicenters and the variability of measured amplitudes. They found that the velocity structure showed anomalous gradients, or "discontinuities," at depths of about 1200 and 1900 km. It is not possible to say whether these discontinuities are global since the data used came from one small part of the earth, and there is evidence that lateral inhomogeneities persist to considerable depth. Discontinuities in the upper mantle at depths of about 300 and 700 km were detected earlier and explained as possible phase changes (Sec. 10-5). The discovery of discontinuities at depths of about 1200 and 1900 km indicates regions where there are either phase or composition changes or both and, contrary to earlier beliefs, point to departures from homogeneity in the lower mantle.

10-7 THE VISCOSITY OF THE MANTLE

Temperatures in the mantle most closely approach the melting point in the low-velocity (LV) zone at a depth of about 100 to 150 km. At greater depths the melting-point curve and the actual temperature diverge again. This behavior is of the greatest importance to such rheological properties of the mantle as viscosity, plasticity, creep, and strength. At the very low strain rates which are encountered in tectonic deformation, these properties depend in an approximately exponential manner upon the difference between the actual temperature and the melting temperature. Thus the viscosity and strength in the LV zone may be lower than in the regions above and below by many orders of magnitude, and the LV zone is likely to be the site of flow and deformation processes of tectonic importance. Between depths of about 50 and 250 km the mantle below the Precambrian shield is, on the average, 200°C cooler than it is under the oceans; these horizontal temperature differences would result in a wide variation in rheological properties. Thus the mean viscosity and strength of the upper mantle beneath shields will be far higher—probably by orders of magnitude—than under oceans. On both the models of S. P. Clark and A. E. Ringwood (1964) the density decreases with depth beneath the oceans. [This is also true for all F. Press's (1969) oceanic models; see Fig. 10-5.] Beneath the shields the density either remains constant or increases slightly, whereas normal continental regions are intermediate between these extremes. Hence the degree of instability against any convective overturn would appear to be greatest in the ocean basins and least beneath the shields. This also would contribute to the tectonic stability of the shields as compared with the younger continental areas and oceanic regions.

H. Takeuchi (1963) has pointed out that much of the discrepancy between values of the viscosity of the upper mantle derived from the postglacial uplift of

Fennoscandia and the limits imposed by the persistence of long-wavelength gravity anomalies might be resolved by assuming that flow takes place within a relatively thin viscous channel in the upper mantle. This result was confirmed in a later paper by H. Takeuchi and Y. Hasegawa (1965) using additional data for the uplift of the Pleistocene Lake Bonneville and the present ellipticity of the earth. R. K. McConnell (1963) agrees in principle with Takeuchi, although not in his estimate of the thickness of the mobile layer. McConnell finds that the model which best fits the data has an elastic surface layer and a low-viscosity channel in the upper mantle below which there is a considerable, but not infinite, increase in viscosity.

The damping of seismic waves with distance and the decay with time of the earth's free oscillations can be used to determine how much the earth departs from a perfectly elastic body. This seismic measure of anelasticity is expressed by a dimensionless quantity Q, the reciprocal of Q being essentially a measure of the fraction of the elastic energy that is dissipated per cycle. Q is roughly independent of frequency for homogeneous materials, and the observed frequency dependence of Q for long-period surface waves and the earth's free oscillations may be attributed to the variation of Q with depth in the earth. It has been found that Q in the crust is about 450; it falls to about 60 to 130 in the LV zone, depending on whether S or P waves are involved, and begins to rise rapidly at about 400 km. The mean Q below about 600 km is about 2200, and Q is at least 5000 at the base of the mantle (R. L. Kovach and D. L. Anderson, 1964). For comparison, crustal rocks have Q in the range 100 to 200 at room temperature and pressure. Q thus varies by orders of magnitude throughout the mantle, the main features being the existence of a low Q (i.e., extremely dissipative zone) in the upper mantle–in the general region of the LV layer–and the rapid increase in Q below about 400 km, where the attenuation of seismic waves decreases markedly. R. K. McConnell (1963) found a similar trend for the variation of viscosity with depth in the upper mantle. This is illustrated in Fig. 10-7, which shows McConnell's viscosity profile and two Q profiles determined from seismic shear waves. D. L. Anderson (1966b) pointed out that the ratio of viscosity η to Q is roughly constant, at least in the upper mantle, where

$$\frac{\eta}{Q} \approx 0.4 \times 10^{20} \qquad (10\text{-}1)$$

and η is measured in poises.

Assuming that this relation is valid for the rest of the earth, Anderson estimated viscosities at greater depths, using average values of Q in shear for the mantle obtained by Kovach and himself (1964). These values are shown in Table 10-1. The viscosities are all much lower than the 10^{26} estimated for the mantle by G. J. F. MacDonald (1963) from the nonequilibrium shape of the earth, based on the assumption that the viscosity is constant throughout the mantle. D. P. McKenzie (1966) attempted to justify a stress-independent viscosity and applied the Navier-Stokes equation to creep within the mantle. He also obtained a very high

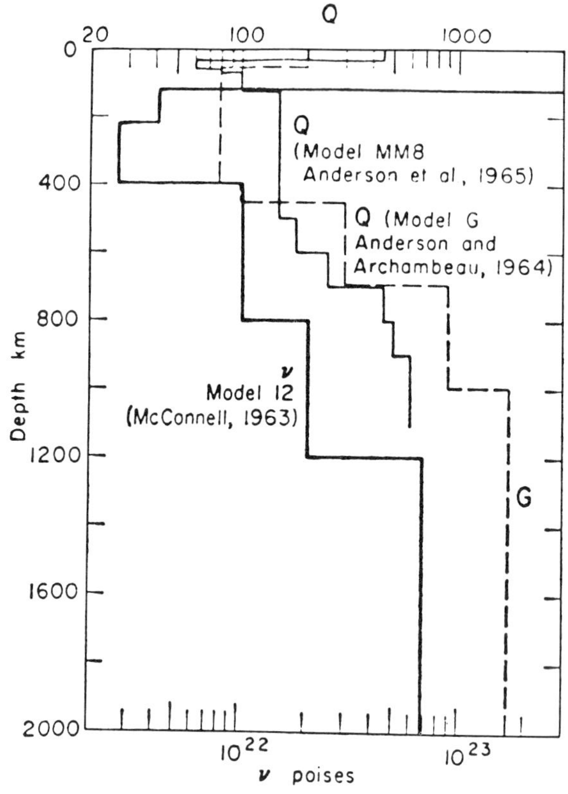

FIGURE 10-7
Variation of seismic anelasticity Q and viscosity η as a function of depth in the earth. (*After D. L. Anderson, 1966b.*)

(6×10^{26}) value for the viscosity of the lower mantle. All evidence indicates, however, that large-scale deformations of the earth involve the upper more than the lower mantle, and that a viscous-layer model as proposed by Anderson is more realistic.

Table 10-1 VALUES OF Q AND η FOR THE EARTH'S MANTLE

	Q	η, poises
Whole mantle	600	2.4×10^{22}
Upper 600 km of mantle	200	8×10^{21}
Lower 2300 km of mantle	2200	10^{23}
Base of mantle	5000	2×10^{23}

SOURCE: D. L. Anderson, 1966.

R. K. McConnell (1968) has carried out a harmonic analysis of the present level of former shoreline features in southeast Fennoscandia (after allowing for the effects of sea-level fluctuations) and found a very rapid increase in relaxation times at long wavelengths under surface loads. Interpreted in terms of elastic and viscous models, his results suggest a rigid upper mantle and crust down to about 120 km. Below this depth, material deforms by creep with viscosities of the order of 10^{21} in a zone between depths of about 100 and 300 km and with viscosities as high as 10^{25} below 800 km. Such a strong viscosity contrast between the upper and lower mantle would have important geophysical consequences; in particular, thermal convection in the lower mantle would be extremely unlikely. On the other hand, R. H. Dicke (1969) has obtained a value of 10^{22} for the viscosity of the deep mantle; a value as low as this would permit the convective transport of heat from the lower mantle and core. Dicke obtained this estimate of the viscosity from a re-examination of the average acceleration of the earth over the past 3000 years based on a new analysis of ancient eclipses.

10-8 THE EARTH'S CORE

The existence of a liquid (outer) core having a density several times that of silicate rocks is now generally accepted. The situation is strikingly similar to that observed in the smelting of iron in blast furnaces. The iron, reduced to the metallic state, sinks to the bottom, forming a single, dense, liquid phase, while the residual silicates float to the surface as slag. A similar process may have occurred within the earth, where the elements iron, silicon, and oxygen are extremely abundant. Part of the iron combined to form iron (or more probably iron-magnesium) silicates, while the excess remained in the metallic state. It is probable that the silicate which formed was an ultrabasic rock like olivine, an iron-magnesium silicate which is the chief constituent of the surface rocks dunite and peridotite. Free iron and the iron silicates form two immiscible phases, and the heavier iron phase settled to the center under the action of gravity. If the earth were ever completely molten, solidification of the mantle would first have taken place at the mantle-core boundary, the lower-melting iron core being soon encased by an insulating sheath of silicate rocks and remaining liquid ever since (Sec. 7-3). This model of the earth was proposed as early as 1873 by J. D. Dana, who had discovered the presence of silicate and metallic nickel-iron phases in meteorites.

There has been much speculation on the origin and evolution of the earth's core, which is probably bound up with the origin of the earth itself. It is a common assumption in most theories of the origin of the earth that the protoearth was homogeneous and that the present differentiation into a core, mantle, and crust occurred late. However, as A. E. Ringwood (1966a) has pointed out, there are a number of difficulties with such a model, and there have been some qualitative considerations in recent years of a nonhomogeneous accretion of the earth and planets. A highly significant development in the study of interstellar matter is the very recent discovery of evidence for the presence in this material of grains of refractory silicates (N. J. Woolf and E. P. Ney, 1969; R. C. Gilman, 1969; R. F. Knacke et al., 1969; W. A. Stein and F. C. Gillett, 1969). Similarities of composition have been inferred between these grains and both meteoritic material and terrestrial minerals (olivine in particular). If the interpretation of the evidence is valid, it implies that silicate grains are a significant component of interstellar dust.

E. Orowan (1969a) pointed out that iron is plastic-ductile, even at low temperatures, provided that it does not contain far more carbon than is found in meteorites. If it is assumed that the planets have agglomerated from solid particles condensed from a gaseous atmosphere around the sun, metallic particles would be expected to stick together when they collide because they can absorb kinetic energy by plastic deformation. They can therefore unite by "cold-welding" or by hot-welding. Silicates, on the other hand, are brittle and break up in a collision, except within a narrow temperature range near their melting point. The agglomeration of the planets may thus start with metallic particles. When the body, built up chiefly from heavy metal particles, is sufficiently large, it can easily collect nonmetallic particles by imbedding them in ductile metal, and later by its gravitational attraction on the fragments resulting from collisions. Orowan thus suggests that planets may arise cold in this way with a metal core already partially differentiated, and that subsequent melting can produce a sharp boundary between core and mantle.

K. K. Turekian and S. P. Clark (1969) have also considered a model of the planets stratified initially due to the inhomogeneous accumulation of the elements. As the primitive solar nebula cooled, elements and compounds would condense in the order of increasing vapor pressure. Assuming a pressure between 10^{-3} and 1 atm, J. W. Larimer (1967) calculated that the order of condensation would be iron and nickel; magnesium and iron silicates; alkali silicates; metals such as Ag, Ga, Cu; iron sulfide; and finally metals such as Hg, Tl, Pb, In, and Bi. This order of condensation is, grossly, that inferred in the earth. Such stratification is usually attributed to the settling of the densest material to the center. High density, however, is associated with low volatility, enabling planets to accrete in a manner that is automatically gravitationally stable. Turekian and Clark thus suggest that the earth's core formed by accumulation of the condensed iron nickel in the vicinity of its orbit, and then became the nucleus upon which the silicate mantle was deposited. The last accumulates would be FeS, Fe_3O_4, the volatile trace elements, organic compounds, hydrated silicates, and rare gases.

It has been shown (e.g., W. M. Latimer, 1950, and H. C. Urey, 1952b) that all the common metals in a gas and dust cloud of solar composition would occur in the form of *oxides* at temperatures below 300°K under equilibrium conditions. This conclusion is of the utmost importance in the case of iron. Since *metallic* iron is an important constituent in meteorites and of the earth, it appears that accretion of the primitive dust in the large bodies was accompanied or preceded by chemical reduction of iron and nickel oxides to the metallic state. A. E. Ringwood (1960) proposed that the earth formed directly by accretion from the primitive oxidized dust in the solar nebula, and that reduction to metal, loss of volatiles, melting, and differentiation occurred simultaneously and as a direct result of the primary accretion process. During the later stages of accretion, when the melting point of the surface was exceeded, metallic iron segregated into masses large enough to flow directly into the core. Gravitational energy liberated during the formation of the core would also contribute heat sufficient to result in complete melting throughout. On this model, segregation of the core occurred as a continuous process during the primary accretion. A catastrophic version of core formation is also a possibility. According to Ringwood's model the earth developed in a state which is grossly out of chemical equilibrium. The deep interior is initially highly oxidized and rich in volatiles, whereas the outer regions are progressively more reduced and poor in volatile components. After melting near the surface, the metal phase consisting of an iron-nickel-silicon alloy collects into bodies which are large enough to sink into the core.

W. M. Elsasser (1963) has considered in some detail the early history of the earth. He starts from a model of an original earth accreted cold with the material uniformly distributed. The main feature of such an earth model is that the melting-point curve of the silicates rises much more steeply with depth than the actual temperature. This implies that the viscosity of the silicates should increase appreciably with increasing depth. As the original earth is heated by radioactivity, the outer layers are then the first to become soft enough to permit iron to sink toward the center; farther down, the fall of iron is slowed down by increased viscosity. It then forms a coherent layer, which, however, is gravitationally unstable and results in the formation of quite large drops. The latter fall rapidly to the center, giving rise to a protocore. Elsasser estimates that the formation of a protocore of a somewhat, but not much, smaller size than the present core probably took no more than several hundred million years.

The fall of iron is controlled by the elastoviscous properties of the silicate matrix. Below the melting point, the viscosity will increase exponentially with decreasing temperature. Thus, instead of a uniform rain of iron toward the center, we have a far more complicated process in which the fall of iron is largely controlled by the variation in temperature at any given depth. As iron fell through the almost fluid silicate material, the general flow pattern of the latter would at first be nearly streamline, but since the viscosity of the mantle increases with depth, the inner portion would not have time to flow but would be pushed outward by the falling drop of iron. Elsasser suggests that the composition of the mantle might thus become asymmetrical, the material displaced to the antipodal point containing iron, while that above the falling drop lost most of its iron. Further differentiation of the iron-free material would occur later, sial rising to the surface, forming a primeval continent antipodal to an ocean above a mantle region relatively rich in iron. Thus the possible presence of a layer of iron near the base of the mantle under the Pacific could be a relic of a primeval asymmetrical distribution of iron.

D. C. Tozer (1965a) has reconsidered the question of the kinetics of core formation and concluded that the simple theory of falling iron masses in silicate material is untenable. He suggests as an alternative a mechanism based on the flow of iron along channels in the silicate phase; the acceptability of this theory depends quite critically on whether iron is able to flow over distances of the order of a kilometer under such conditions. Tozer concludes that in any case core formation is proceeding today much more slowly than in the past, and that it was virtually complete very early in the earth's history.

In 1949 W. H. Ramsey suggested that the discontinuity at the core-mantle boundary might be due to a high-pressure phase change. He thus visualized a chemically homogeneous composition for the earth (below the crustal layers) which he identified as olivine; at the core-mantle boundary there would be a transition from a molecular to a metallic phase. Originally, Ramsey put forward this theory to account for the differences in the observed densities of the terrestrial planets, which he assumed to have a common primitive composition. More recent astronomical data, however, have revised the older figures and make Ramsey's hypothesis improbable. At the time of Ramsey's suggestion P. W. Bridgman had accumulated a large amount of experimental data on the densities and compressibilities of materials up to 10^5 atm. On the other hand, theoretical data at pressures of 10^7 atm and above may be obtained using a quantum-statistical method, based on a Thomas-Fermi-Dirac model for the electrons surrounding the nucleus. In this model the electronic shells of the atoms are pressed together and lose their individual structure. Many investigators have attempted to obtain information relevant to conditions in the earth's interior by interpolating in this pressure range. Their results do not support Ramsey's hypothesis, but suggest that the inner core is of nickel-iron composition, although the outer core may possibly consist of a modification of ultrabasic rock. Finally, a phase change in a multicomponent system almost certainly would be spread over a range of pressures, whereas the core-mantle boundary presents a very sharp discontinuity.

Studies of the physical properties of the earth's core by F. Birch (1952) have indicated that it is about 10 percent less dense than nickel iron and that its seismic velocity is substantially greater than that of nickel iron under comparable (p,T) conditions. These conclusions have been confirmed by recent investigations on the densities and seismic velocities of metals at pressures equal to those in the core using shock-wave techniques (Fig. 10-8; F. Birch, 1963, 1968). Unfortunately shock-wave data supply only one elastic constant, the compressibility, and the effects of temperature can be treated only approximately. Silicates are structurally complicated, and the interatomic forces between constituent ions are not as clear-cut as in simple ionic crystals, so that no one has yet carried out a complete lattice-dynamical or quantum-mechanical calculation for any common rock-forming mineral. Even pressures in the core are too low for the applicability of simplified statistical treatments such as the Thomas-Fermi-Dirac equation of state.

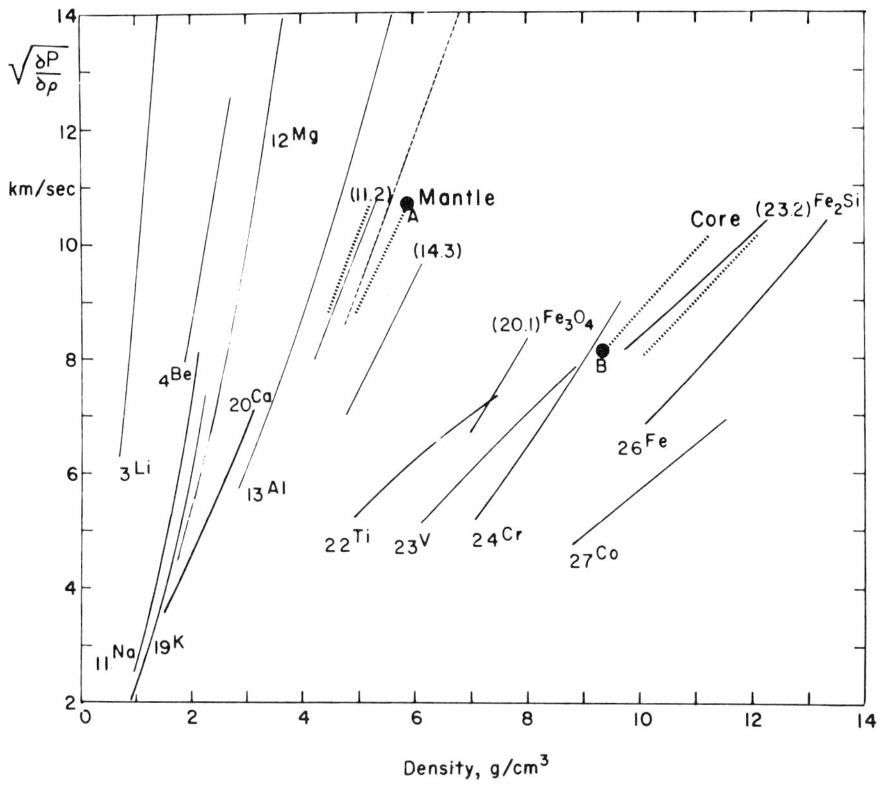

FIGURE 10-8
Hydrodynamic velocity, in kilometers per second, versus density, in grams per cubic centimeter, for several metals and rocks, and the corresponding quantities for mantle and core (limits shown as dotted lines). Atomic numbers or representative atomic numbers (in parentheses) are attached to each curve, except the "olivinite I" of R. F. Trunin et al. (1965), shown as a broken line.

Figure 10-8 is a graph of the hydrodynamic sound velocity (which is related to the seismic velocities) against density for several metals and rocks. The limits of the corresponding quantities for the mantle and core are shown as dotted lines. No adjustment of the data will allow a core of light metals or their oxygen compounds, nor can the mantle be made of heavy metals. Transformation of light compounds to a metallic state may take place in the earth, but the density of the core requires a metal of the transition group, and only iron is sufficiently abundant. The properties of iron are close to those required, and can be adjusted with small amounts of light alloying elements. It appears, therefore, that the earth's core contains a substantial amount of an element with a low density which can also increase the elastic ratio and seismic velocity of iron. Limitations upon possible choices are that the element must be reasonably abundant and miscible with liquid iron and possess chemical properties which will allow it to enter the core. A. E. Ringwood (1966b) has shown that silicon is the most likely extra component of the earth's core. At the mantle-core boundary magnesium oxide could be soluble to an extent of about 10 percent in liquid iron. Such solubility considerations put stringent constraints on the chemical constitution of the earth; B. J. Alder (1966) has shown that neither SiO_2 nor Mg_2SiO_4 is sufficiently soluble to lower the density of the liquid core significantly below that of pure iron.

Many calculations have been carried out in an attempt to estimate the loss of gravitational energy and the consequent rise of temperature associated with core formation. F. Birch (1965) has simplified the problem by assuming that the undifferentiated earth consists of a homogeneous mixture of the materials of the present core and mantle. Only simple unmixing was considered, the total mass remaining unchanged, so that his calculations do not include an allowance for chemical reactions or loss of volatile components. Without allowance for thermal expansion, Birch found that the mean energy available for heating is 600 cal/g; with an approximate allowance for thermal expansion this is reduced to 400 cal/g, which is equivalent to a mean rise of temperature of about 1600°. This suggests that core formation within an originally homogeneous earth would be the most important happening in its thermal history after accretion; it may in fact represent the 4500 m yr age of the earth.

Although we have no direct knowledge of the time of origin of the earth's molten metallic core, nevertheless, if it is assumed that the earth's magnetic field results from motions in a predominantly fluid iron core, a time constraint can be placed on core formation. It must have antedated the oldest known rock possessing remanent magnetism. The age of some rocks in Africa possessing remanent magnetism is 2500 to 2700 m yr. Thus, if the earth is 4500 m yr old, core formation must have occurred within the first 2000 m yr of the earth's history. T. C. Hanks and D. L. Anderson (1969) have carried out a series of calculations on the early thermal history of the earth, taking into account this boundary condition. The results of their calculations are that large-scale differentiation within the earth leading to core formation must almost certainly have taken place during an accretion period lasting 500,000 years or less. In other words, the origin of the earth and its core would have taken place practically simultaneously.

The possibility that the growth of the earth's core has been gradual seems to have been first proposed by H. C. Urey (1952b), who arrived at this hypothesis on geochemical grounds. He argued that the abundance of volatile elements in the earth's crust was greater than would be expected if the earth had ever been molten. Because it is then inferred that the earth was formed cold by a process of accretion, the formation of the core, presumed to be chemically distinct from the mantle, presented a problem not before recognized. Urey's solution was that the separation of iron toward the center could have taken place only gradually, and he raised the question as to whether the growth was yet complete.

S. K. Runcorn (1962a,b) suggested that the core has continued to grow throughout geological time and has developed a theory of continental drift based on its growth. F. A. Vening Meinesz (1952) had argued that the positions of the continents today could result from a large-scale, regular pattern of convective motions in the mantle, continental material tending to congregate at places where the currents are descending. Using S. Chandrasekhar's results (1961) on the convection of a fluid contained in a spherical shell, Runcorn showed that the present radius of the core is only just greater than the value at which the fourth-harmonic convection pattern in the mantle should become less readily excited than the fifth. On this theory the relatively recent time (geologically speaking) at which continental drift took place is at once understandable. The theory also predicts other epochs of large continental displacements when the core radius reached successively the values at which the first-degree convection pattern in the mantle, initially present when the core was only beginning to form, gave place to the second, and the second to the third, etc. These other epochs of continental drift occurred early in the early Precambrian. (See also Sec. 17-4.)

During the transitions from one harmonic to another the continents will be under much stress as the convection pattern changes to a new form. In addition to continental displacements, one would expect during these transition periods a series of orogenies. Radioactive age determinations have mainly been carried out on igneous rocks which come from depth or metamorphic rocks formed in the deeper parts of the crust as the result of orogenic forces (G. Gastil, 1960). It is therefore of interest that the ages obtained for such rocks have been found to be grouped in broad peaks around the dates 2600, 1800, 1000, and 300 m yr ago. The most recent peak covers much of the geological record since the middle Paleozoic, and this is associated with continental drift and the transition from a fourth- to a fifth-degree convection pattern in the mantle. Runcorn suggests that it is natural to identify the 1000 m yr peak with the transition from the third- to the fourth-degree convection pattern, the 1800 m yr peak with the transition from the second to the third, and the 2600 m yr peak with the transition from the first to the second. In this way the dates at which the core radius reached these critical values can be determined. See Fig. 10-9, which also shows that the core started its growth a little over 3000 m yr ago.

The basic question at issue is how long it took the earth's core to form and when this event took place. The total evidence available seems to indicate that the event was comparatively rapid and took place very early in, or simultaneous with, the formation of the earth itself. It would thus seem that Runcorn's theory is not tenable.

10-9 BULLEN'S COMPRESSIBILITY–PRESSURE HYPOTHESIS

From the results of his earth model *A*, K. E. Bullen found that there was no noticeable difference in the incompressibility gradient dk/dp between the base of the mantle and the top of the core. Moreover, there was only a 5 percent difference

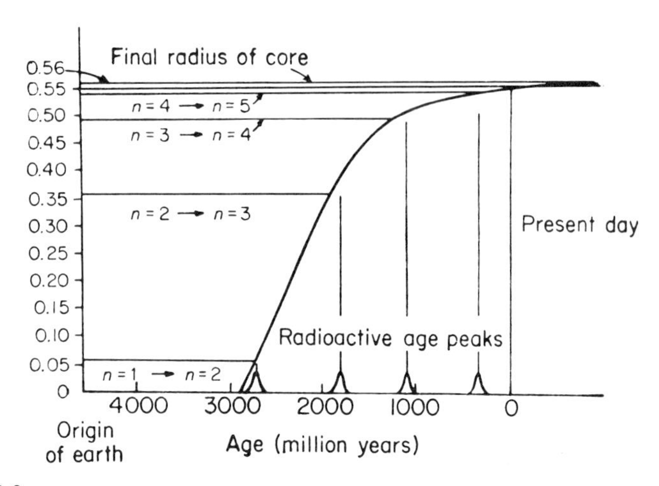

FIGURE 10-9
Growth of the earth's core compared with radioactive-age-determination peaks. (*After S. K. Runcorn, 1962b.*)

in the value of k across the core-mantle boundary. These features are in marked contrast to the large changes in the density and rigidity at the boundary. The change in k is a diminution from the mantle to the core. However, interpolation between experimental data at 10^5 atm and theoretical studies at 10^7 atm indicates that in the transition from the mantle to the core, a slight increase in k could be expected for materials likely to occur in the earth's interior. Because of the smallness in the change in k across the core-mantle boundary, and because this change is opposite in sign to that predicted by such an interpolation, Bullen (1949, 1950) proposed another earth model B, in which he assumed that k and dk/dp are smoothly varying functions throughout the earth below a depth of about 1000 km. This hypothesis, called the compressibility-pressure (k, p) hypothesis, implies that at high pressures the compressibility of a substance is independent of its chemical composition. More recent work in theoretical physics indicates that the hypothesis as stated is a little too general and that there is some small variation of k with atomic number at high pressures. Bullen has also obtained a quadratic representation of k in terms of p for the materials of the earth below 1000 km. His latest (1968) expression is

$$k = 2.34 + 3.00p + 0.10p^2$$

where k and p are measured in 10^{12} dynes/cm^2.

On the basis of his hypothesis, Bullen found that there must be a concentration of more dense material near the base of the mantle (region D''). This material could be a mixture of metallic iron with silicates near the core boundary or an iron-sulfide phase at the base of the mantle.

If the entire core is liquid, so that $V_s = 0$, then, from Eq. (2-3), V_p is given by

$$V_p^2 = \frac{k}{\rho} \qquad (10\text{-}2)$$

Jeffreys' velocity distribution shows a discontinuous jump across the boundary between *F* and *G*, and to accommodate this, *k* would have to increase by 32 percent, excluding the highly improbable case that the density decreases with depth. Assuming Gutenberg's velocity-depth curves, the effective increase in *k* would be 23 percent. On the other hand, as Bullen first pointed out in 1946, if the inner core *G* is solid and thus capable of transmitting *S* waves, Eq. (10-2) is replaced by Eq. (2-3) in *G*, and the increase in V_p can be accounted for without violating his (k,p) hypothesis. Thus both Jeffreys' and Gutenberg's interpretations of the velocity-depth curves indicate that the inner core has significant rigidity comparable with that in the lower mantle.

K. E. Bullen and R. A. W. Haddon (1969) have constructed a series of earth models in which the excess Δk of the incompressibility *k* at the top of the earth's core over the value of *k* at the bottom of the mantle was given different assigned values. Their calculations indicate that, unless the assumed seismic velocities and density gradients are more seriously in error than expected, $\Delta k/k$ does not exceed 2 percent. The most probable value of $\Delta k/k$ is insignificantly different from zero. However, H. Takeuchi and H. Kanamori (1966) have calculated an equation of state for materials of the earth using shock-wave data, and find that the density of the earth's core is about 1 to 1.5 g/cm^3 less than the density of iron at corresponding pressures and temperatures. The variations in the values of the incompressibility of the metals studied are too large (even at 4 Mbar) to support Bullen's incompressibility-pressure hypothesis. They find it difficult, however, to avoid the conclusion that the inner core is solid.

K. E. Bullen and R. A. W. Haddon (1967) have since revised Bullen's original model *B* and constructed a series of models based on the (k,p) hypothesis. The revision was undertaken to take into account the latest values of the moment of inertia of the earth, seismic *P* velocities in the core, and data from the free oscillations of the earth. In a later paper (1969), Haddon and Bullen constructed a series of new earth models incorporating free-earth oscillation data. They found that *k* is practically continuous between mantle and core and that the different approaches to the structure of the deep interior of the earth made in the past by Bullen through models of *A* and *B* type are now converging. (See also Sec. 2-5.)

10-10 CHEMICAL INHOMOGENEITY IN THE EARTH

K. E. Bullen (1963b) has shown (Appendix E) that, allowing for changes in chemical composition and possible phase changes, the density gradient within the earth is given by

$$\frac{d\rho}{dr} = \frac{-\eta g \rho}{\phi} \qquad (10\text{-}3)$$

where $$\phi = \frac{k}{\rho} = V_p{}^2 - \tfrac{4}{3} V_s{}^2 \tag{10-4}$$

and $$\eta = \frac{dk}{dp} + g^{-1} \frac{d\phi}{dr} \tag{10-5}$$

ρ, k, p, V_p, V_s, and g are, respectively, the density, adiabatic incompressibility, pressure, velocity of P and S waves, and gravitational attraction at a distance r from the center of the earth. When $\eta = 1$, Eq. (10-3) reduces to the Adams-Williamson equation (2-6), so that η is a measure of the departure from chemical homogeneity; it is the ratio of the actual density gradient to the gradient that would obtain if the composition were uniform (chemical inhomogeneity includes both changes in chemical composition and phase changes caused by pressure).

From Eq. (10-5) it can be seen that η depends on dk/dp, g, and $d\phi/dr$. On Bullen's compressibility-pressure hypothesis (Sec. 10-9), dk/dp is slowly varying and lies between about 3 and 6 throughout most of the earth's deep interior. Values of g within the earth have already been obtained (Sec. 2-6), while values of $d\phi/dr$ are immediately derivable from the P and S velocity distributions. Thus it is possible from Eq. (10-5) to estimate the degree of departure from chemical homogeneity in any given region. In a later paper, Bullen (1965a) has refined Eq. (10-5); in particular, he has investigated the implications of the variation of k with composition and of the deviation of dk/dp from $(\partial k/\partial p)_{\text{const composition}}$.

In the lower 200 km of the mantle (region D''), the seismic velocity distributions of both Jeffreys and Gutenberg indicate that $d\phi/dr \approx 0$, so that $\eta \approx dk/dp$. Bullen's model A gives $dk/dp \approx 3$ in D'', indicating that the lower 100 to 200 km of the mantle is inhomogeneous. The inhomogeneity is not too severe, however; with $\eta = 3$ it contributes only an extra 0.2 g/cm^3 density increase through D''.

In the transition region F between the outer and inner core, the Jeffreys velocity distribution (characterized by a large negative P-velocity gradient, that is, $d\phi/dr \gg 0$) leads to a value of η of 38, entailing a density increase of the order of 3 g/cm^3 through F. On the other hand, the Gutenberg velocity distribution gives large negative values of $d\phi/dr$, so that η is significantly less than unity (actually

Table 10-2 ASSUMED SEISMIC DATA AND LAYERING IN THE CORE

Layer	Range of r, km	V_p, km/sec
E'	3470–1810	According to H. Jeffreys (1939)
E''	1810–1660	10.03
F	1660–1210	10.31
G	1210–0	11.23

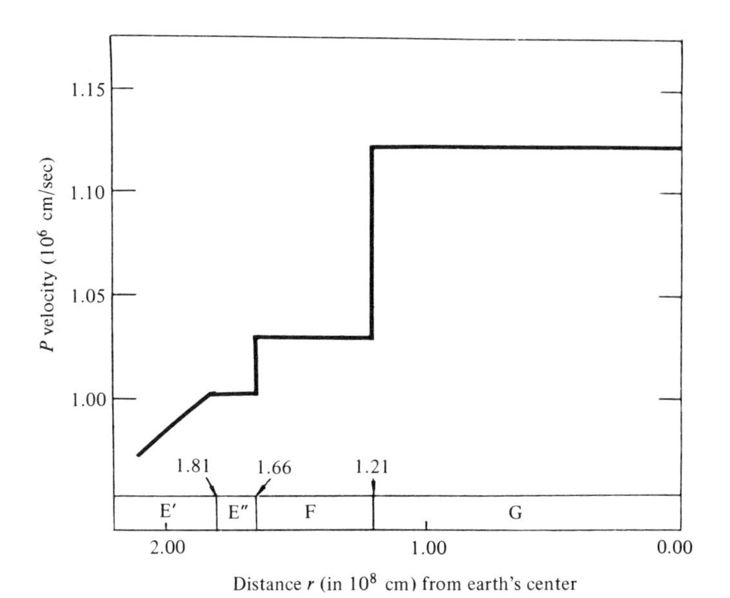

FIGURE 10-10
P velocities and the regions E', E'', F, and G of the lower core. (*After K. E. Bullen, 1965b.*)

negative), implying an unstable distribution of mass. It would appear that seismic velocity gradients much in excess of those in regions D and E cannot exist in the earth's deep interior. An infinite gradient (i.e., a velocity discontinuity), on the other hand, is not impossible, since then the range of depth of any instability would be zero.

In B. A. Bolt's (1962, 1964) revision of Jeffreys' distribution of the velocity of P waves in the deep interior of the earth, the core is divided into four distinct regions, E', E'', F, and G. The velocity distribution down to the bottom of E' is the same as that of Jeffreys for corresponding depths inside E. There are discontinuous jumps in V_p at the E''–F and F–G boundaries, and V_p is constant in E'', F, and G (Table 10-2 and Fig. 10-10).

F. Birch (1963), using shock-wave data at pressures of the order of 10^6 atm, inferred that the density ρ_0 at the center of the earth does not exceed 13 g/cm^3; Bullen (1965b) has investigated the consequences of this limiting value using Bolt's seismic P-velocity distribution in the core. From Eqs. (2-2) and (2-3) it follows that

$$\rho V_p{}^2 = k + \tfrac{4}{3}\mu \qquad (10\text{-}6)$$

Since ρ is not likely to decrease with depth in the core, Bullen's (k,p) hypothesis implies, through Eq. (10-6), that departures from smooth variations of V_p with r

are accompanied by similar departures in the variation of μ rather than of k. Bullen has shown that it is impossible for ρ_0 to be as low as 13 unless there is substantial rigidity in both regions F and G. In addition, it is essential that $d\mu/dr > 0$ over a significant range of depth in the lower core; i.e., the rigidity must *decrease* with increase of depth. Bullen found that ρ_0 could be as low as 12.6 g/cm^3 if suitable assumptions (compatible with the seismic data) are made on the variations of k and μ in the lower core. The value of ρ_0 could actually be reduced to 12.3 g/cm^3 if E'' were chemically homogeneous, which is, however, rather improbable. If F and G are both fluid (i.e., complete absence of rigidity), ρ_0 must be at least 14.7 g/cm^3. This value is sufficiently in excess of 13 g/cm^3 to give additional support to the conclusion that the inner core is solid.

SUGGESTIONS FOR FURTHER READING

GASKELL, T. F. (ed.): "The Earth's Mantle," Academic Press, Inc., New York, 1967.
HURLEY, P. M. (ed.): "Advances in Earth Science," The M.I.T. Press, Cambridge, Mass., 1966.
HART, P. J. (ed.): The Earth's Crust and Upper Mantle, American Geophysical Union, Geophysical Monograph Series, No. 13, 1969.

11

FAULTING, FOLDING, FLOW, AND MOUNTAIN BUILDING

11-1 INTRODUCTION

In deformation, geologists distinguish between flow, of which folding is an important example, and fracturing, which produces joints and faults. Most small fractures are *joints*, or *tension cracks*, which are partings without appreciable relative displacements between the two sides. Less numerous, and usually of greater extent, are *faults*, which are shear failures in which the two sides have been offset. There are two types of faulting, that in which faults terminate by dying out and crust is conserved, and that in which faults end abruptly by changing into spreading zones, where crust is created, and *subduction zones*, where it is reabsorbed back into the mantle (D. A. White et al., 1970).

11-2 FAULTING WHERE CRUST IS CONSERVED: NORMAL, TRANSCURRENT, AND THRUST FAULTS

Observation has shown that during earthquakes the two sides of many faults near the surface have undergone sudden relative displacements. Relative movements exceeding 10 m have been observed during single earthquakes, and repeated

FIGURE 11-1
Oblique aerial photograph of the McDonald Lake fault, a great strike-slip fault along the south shore of Great Slave Lake, Northwest Territories, Canada. The rocks on the two sides are different, so that either the offset is very great or the fault marks the sheared junction between two plates of the lithosphere. (*Department of Energy, Mines and Resources, National Air Photo Library, Ottawa, Canada.*)

movements in the same direction over long periods of time appear to have produced hundreds of kilometers of horizontal displacement along some faults. Large active dislocations, like the San Andreas fault in California, are conspicuous, and so are some inactive, older faults like that illustrated in Fig. 11-1.

During the past century and a half, C. A. Coulomb, O. Mohr, C. L. M. Navier, and E. M. Anderson (1951) developed a theory of faulting. In Sec. 2-1 it was shown that the stresses acting on a body can be represented by three normal stresses (the *principal stresses*) acting along three directions (the *principal axes*) which are at right angles to one another (Fig. 11-2). The maximum stress acting on a body lies along one of these directions, and the minimum stress along another. The three planes, each containing two of the principal stresses and hence normal to the third, are *planes of no shear*. In any body subject to nonhydrostatic stress (i.e., one in which the principal stresses are not all equal) there are two planes parallel to which shear is a maximum. Each of these planes contains the direction of the intermediate

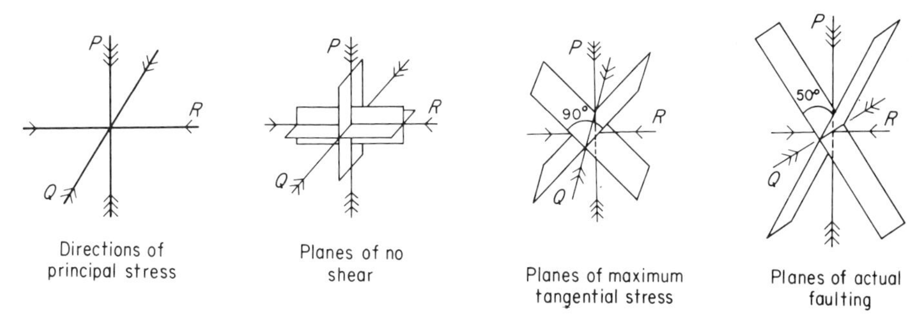

FIGURE 11-2
The principal stresses, $P > Q > R$, act in orthogonal directions. Any two define a plane of no shear, and the bisectors of the angle between P and R are planes of maximum tangential stress. Faulting occurs on planes making a smaller angle, about 25°, with P. In this figure P is vertical and the faults are normal faults. (*After E. M. Anderson, 1951.*)

principal stress and bisects one of the angles between the other two principal axes. One might expect faulting to occur parallel to these *planes of maximum shear stress*, but this is not so. Because of internal friction, actual *fault planes* make smaller angles of 20 to 30° with the direction of maximum stress.

The earth's surface is a plane of no shear, and thus controls the orientation of stresses, so that of the principal stresses one is always nearly vertical and two are horizontal. There are three cases, depending upon whether the direction of greatest, intermediate, or smallest principal stress is vertical (Fig. 11-3). For these three types, Table 11-1 gives the properties predicted by theory and observed in the field (E. M. Anderson, 1951).

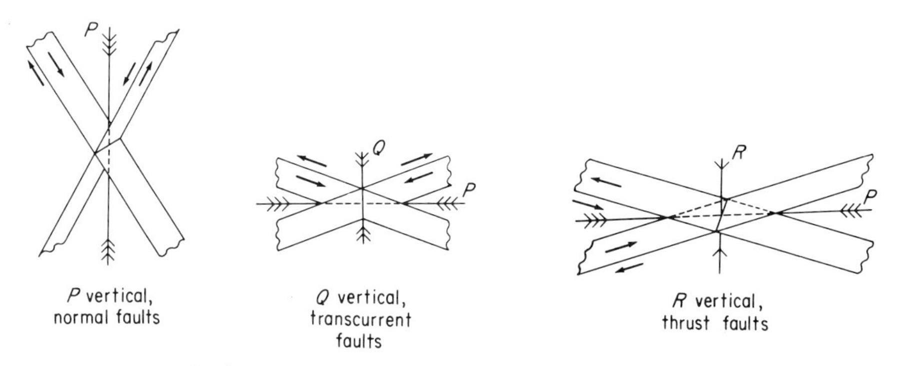

FIGURE 11-3
The orientation of the principal axes (one of which must be vertical) determines the type of faulting. (*After E. M. Anderson, 1951.*)

Table 11-1 CHARACTERISTICS OF THE THREE CLASSES OF FAULT OBSERVED ON LAND

Property	Normal	Transcurrent	Thrust
Principal axis is vertical	Major axis	Intermediate axis	Minor axis
Nature of dip	65° and straight	90° and variable	25°
Strike	Wavy	Straight	Very wavy
Direction of movement	Down dip	Horizontal	Up dip
Connection with mountain-building forces	None; gravity greatest force	Yes; faults at 25° to forces	Yes; faults normal to forces
Nature of fracture	Brecciated	Indeterminate	Sheared
Intrusives along fault plane	Dikes possible	Dikes improbable	Intrusives unlikely

SOURCE: E. M. Anderson, 1951.

11-2.1 Normal Faults

Normal faults occur when the direction of greatest principal stress is vertical and there is relief of pressure in all horizontal directions. The fault planes dip downward under the downthrust blocks at angles rather steeper than 45°. At any depth the vertical pressure is the hydrostatic pressure, and by definition this is greater than that in any other direction. The pressures across the fault planes are therefore less than hydrostatic, and dikes may be intruded along the fault planes. To build folded mountains, horizontal pressures must exceed the vertical force of gravity, and thus normal faults are not to be expected as primary features in orogeny.

11-2.2 Reverse, or Thrust, Faults

Reverse, or thrust, faults occur when the direction of least stress is normal to the surface and the pressure in all horizontal directions is greater than the hydrostatic pressure. The dip of these fault planes is shallower than 45°, and one faulted slab rides over another so that the surface is shortened by thickening. Because the vertical, or hydrostatic, pressure is less than the pressures across the fault planes, intrusions do not tend to follow these planes, but if water is trapped under high pressure in porous rocks, it may lubricate and help movement on thrust planes (W. W. Rubey and M. K. Hubbert, 1959).

11-2.3 Transcurrent Faults

Transcurrent faults occur when the direction of intermediate stress is normal to the surface and the directions of both the largest and smallest principal stresses are parallel to the surface. In this case the fault planes will be approximately vertical and the two sides of faults will be horizontally displaced relative to one another. Two directions of displacement are possible, which may be distinguished in the following manner. If an observer on approaching a fault finds that the farther side has been moved to the right, the fault is called a *right-handed*, or *dextral*, fault. If it has moved to the left, the fault is a *left-handed*, or *sinistral*, fault. The direction in which the fault is approached is immaterial. Successive displacements during earthquakes are in the same direction as the offset and tend to increase it.

Faults tend to be straight in the direction of movement; otherwise, overlaps and open spaces would develop. But in the direction normal to the motion there is no reason for such regularity. This can explain the observed shapes of fault planes and, specifically, why the strike of normal and thrust faults is usually wavy while that of transcurrent faults is straight.

Although E. M. Anderson's theory suggests that transcurrent faults should continue indefinitely, in real faults the motions and displacements are gradually dissipated and the faults die out. This often produces many small branch faults, a process called *horsetailing*. Observations on small transcurrent faults show that their length is many times the maximum offset across them. Compression in the direction of forward motion tends to produce uplift on the advancing side of transcurrent faults.

H. O. Seigel (1950) has examined faulting in anisotropic and faulted rocks and concluded that (1) in such cases only one of the two possible conjugate directions of faulting is likely to develop, (2) failure tends to recur along old faults because they are directions of weakness, and (3) a younger fault intersecting an older fault may follow it for a way before crossing over and reverting to its former direction. This is called *trailing*. Note that the older fault appears to offset the younger.

The terms *wrench faults* and *strike-slip faults* have been used synonymously for transcurrent faults, but now that another class of faults (*transform faults*), also with horizontal displacements, has been recognized (see next section), it is convenient to use the term strike-slip fault to refer to both transcurrent and transform faults collectively or to faults which have not yet been classified.

11-3 FAULTING WHERE CRUST IS NOT CONSERVED: TRANSFORM FAULTS

Some geologists have long considered that parts of the earth's surface have moved large distances relative to other parts. At first they tended to emphasize flow, although they recognized some faults. Thus A. Wegener (1929) referred to the displacement of Greenland relative to the Canadian arctic islands, and E. Argand (1924) and others recognized the existence of great thrust faults in the Alps.

Clearer recognition of great displacements began when W. Q. Kennedy (1946) published evidence that the two sides of the Great Glen fault across Scotland had

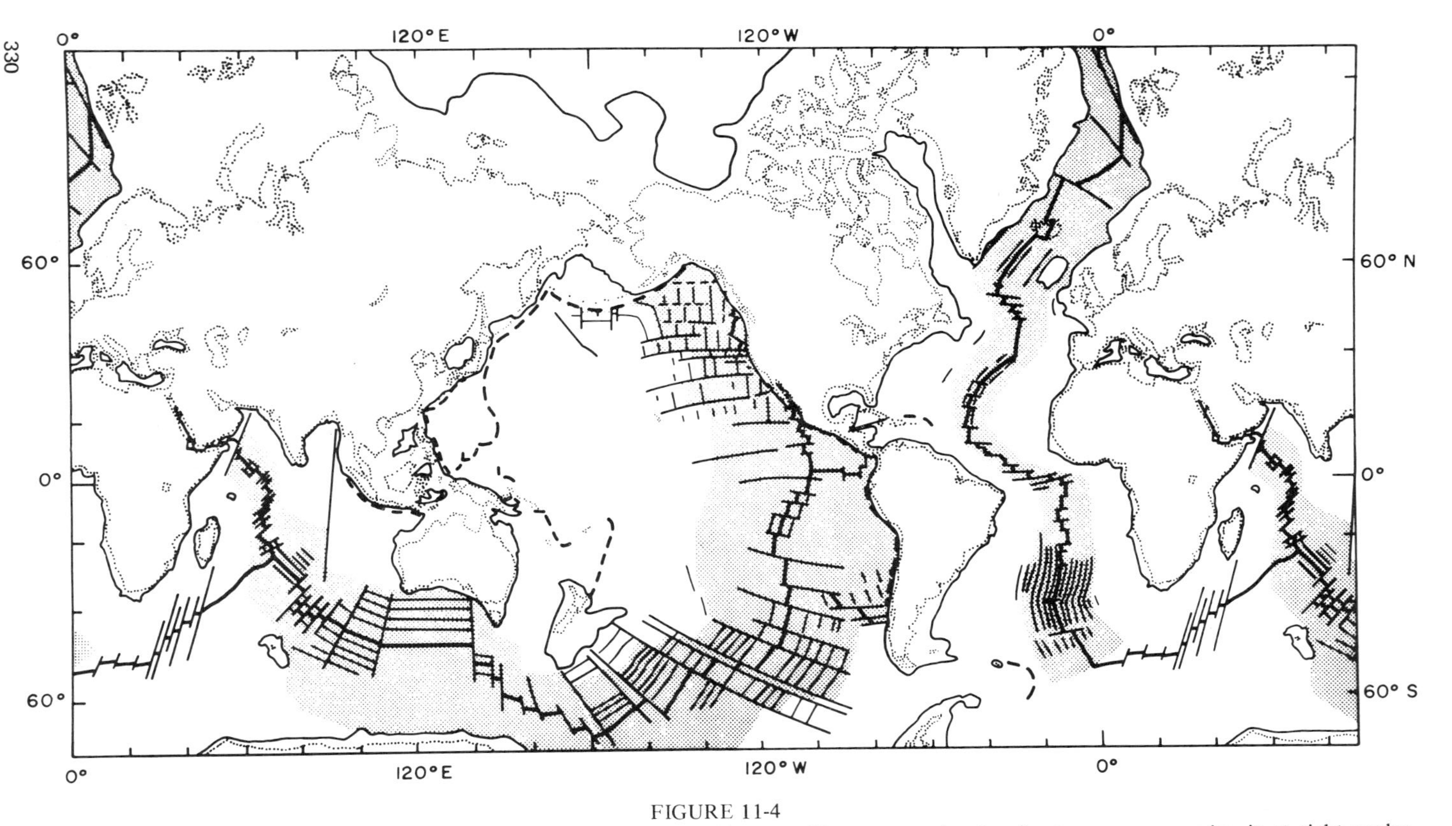

FIGURE 11-4
The mid-ocean ridge system showing fracture zones crossing it at right angles, growth lines parallel to ridges, area of growth during Tertiary (shaded). (*After F. J. Vine, 1970.*)

been offset by 107 km. Soon afterward, H. W. Wellman (1953, 1955) in a series of papers recognized the existence of, and the great displacement along, the Alpine fault in New Zealand. In 1953 M. L. Hill and J. W. Dibblee, Jr., suggested a right-handed displacement along the San Andreas fault of at least 560 km. In 1952 H. W. Menard and R. S. Dietz recognized and named great *fracture zones*, some of which extend for thousands of kilometers across the ocean floors. Later V. Vacquier (1959) and his colleagues showed that these zones offset patterns of magnetic anomalies by hundreds of kilometers (Fig. 14-3).

Scores of faults with great offsets have now been recognized, and H. W. Wellman (1969) and J. T. Wilson (1969) have listed some of them, which are also shown in Fig. 11-4.

According to E. M. Anderson's theory of faulting and other prevailing tectonic theories, such large offsets seemed to be impossible, and many disputed these claims. The problem is difficult because most of these large faults are partly or wholly hidden beneath the sea.

L. V. de Sitter (1964) has drawn attention to another group of puzzling faults called *great fundamental faults.* Although these may be traced for great distances, the rocks on opposite sides of them bear no recognizable relationships (Fig. 11-1). A good example is the fault across North Africa from Agadir, Morocco, to the Gulf of Gabes in Tunis. *Tear faults* which cross thrust sheets and separate parts that have moved differentially are another class of enigmatic faults not explained by Anderson (J. G. Dennis, 1967).

In 1960 H. H. Hess (1962) made a notable advance when he clearly formulated the concept of sea-floor spreading. According to this hypothesis material wells up from the mantle to generate new sea floor along mid-ocean ridges (Sec. 14-9). The lithosphere spreads and is carried away until it is reabsorbed back into the mantle under ocean trenches or young mountains. This contradicts an assumption implicit in Anderson's theory of faulting that the surface of the earth is conserved. If the sea floors do spread, other types of fault can exist which may have large displacements and end abruptly by being transformed into either spreading ridges or closing trenches and mountains. The terminus of a shear is a spreading ridge or a closing trench, and the junction is a *half shear* (Fig. 11-5). Ridges are symmetrical, but trenches are frequently curved, so that two alternatives are possible, and all may be either right- or left-handed; thus there can be six half shears. (To these may be added the change from spreading to closing by rotation without shear about a pole to make seven varieties of *transform*.)

The three right-handed half shears can combine in pairs to form six different *dextral transform faults*, which are illustrated and named in Fig. 11-6, and the left-handed half shears can form six *sinistral transform faults.* The changes produced by movement are shown in Fig. 11-7, and Table 11-2 gives some of the peculiar features of transform faults which distinguish them from transcurrent faults. Folding pieces of paper as shown by one example in Fig. 11-8 will help to visualize the nature and behavior of transform faults.

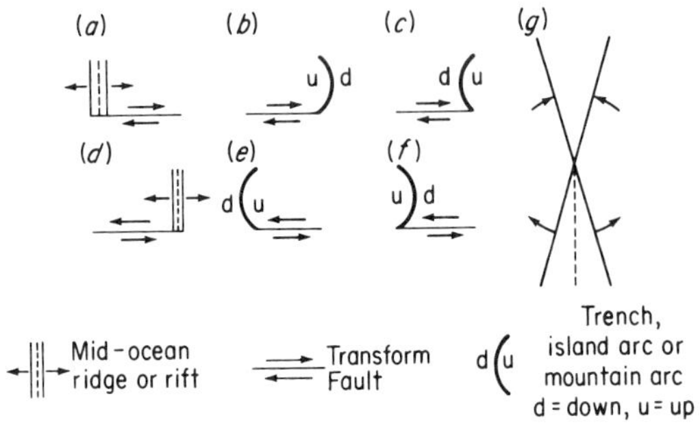

FIGURE 11-5
The seven possible transforms. One is a rotation which changes a mid-ocean ridge into a compressional feature, and three are right-handed half shears in which a shear failure is transformed into a mid-ocean ridge or into an arcuate trench or young mountain, oriented in one of two ways, depending upon the direction in which the overriding convex side faces.

The differences between transcurrent and transform faults listed in Table 11-2 enabled L. R. Sykes (1967, 1968), D. G. Tobin and L. R. Sykes (1968), and B. A. Bolt et al. (1968) to show from the study of recent earthquakes that many important faults are transform. This discovery supported the concept of sea-floor spreading and paved the way for plate tectonics by recognizing that transform faults, mid-ocean ridges, trenches, and young mountains form an integrated system about the world, that "these features are not isolated, that few come to dead ends, but that they are connected into a continuous network of mobile belts about the

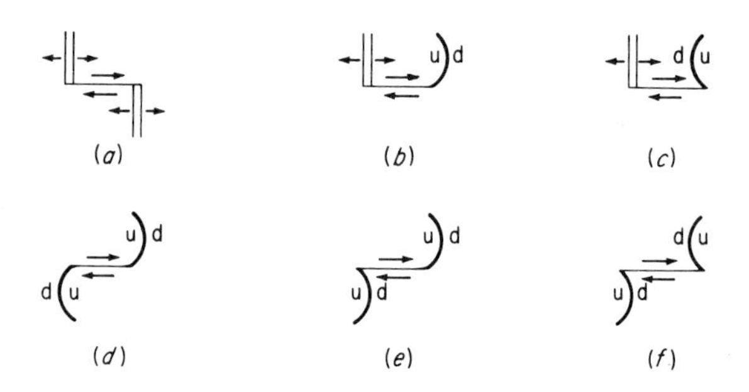

FIGURE 11-6
The six possible types of right-handed transform faults. (*a*) Ridge to ridge; (*b*) ridge to concave arc; (*c*) ridge to convex arc; (*d*) concave arc to concave arc; (*e*) concave arc to convex arc; (*f*) convex arc to convex arc. Note that the direction of motion in type *a* is the reverse of that required to offset the ridge. See Fig. 11-5.

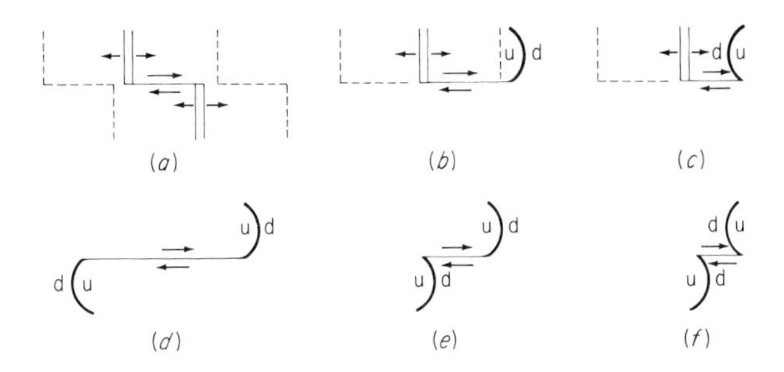

FIGURE 11-7
The appearance of the six types of dextral transform faults shown in Fig. 11-6 after a period of growth. Traces of former positions now inactive, but still expressed in the topography, are shown by dashed lines.

earth which divide the surface into several large rigid plates. Any feature at its apparent termination may be transformed into another feature" (J. T. Wilson, 1965b).

Fundamentally, it was the work of seismologists in plotting the distribution of earthquakes that led to the realization that although the earth's surface is mobile, major movements are confined to narrow zones between large plates. This led to the concept of global plate tectonics (Sec. 14-10) and provided a compromise between the views of those like H. Jeffreys (1970) who had been impressed by the rigidity of the earth's surface features and those like S. W. Carey (1953, 1958) who had noted the evidence of folding and flow in rocks.

D. P. McKenzie and W. J. Morgan (1969) realized that, besides junctions of pairs of transform faults, trenches, and mid-ocean ridges, triple points exist where three plates and three boundaries meet. They discussed the geometry of triple points assuming that (1) ridges are structures which produce lithosphere symmetrically on both sides and lie at right angles to the relative-velocity vector between two

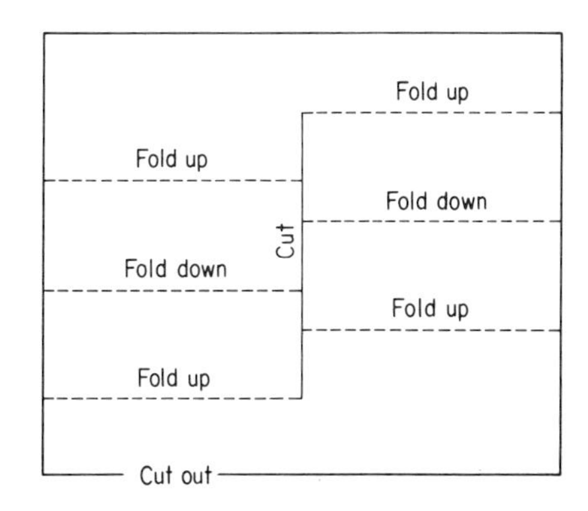

FIGURE 11-8
Method of making a paper model by which to examine the properties of a ridge-ridge transform fault.

Table 11-2 PROPERTIES OF TRANSCURRENT AND SIX TRANSFORM FAULTS

Name	Termination and extent of fault	Extent of earthquakes along fault	Change in the apparent offset with motion	Relationship between directions of apparent offset and motion
Transcurrent	Indefinite	Whole length	Increase	Same
Ridge-ridge transform	Limited and definite	Confined to offset	No change	Reverse
Ridge-convex-arc transform	Limited and definite	Confined to offset	Decrease probable	
Ridge-concave-arc transform	Limited and definite	Confined to offset	Increase	
Convex-arc-convex-arc transform	Limited and definite	Confined to offset	Decrease	Reverse
Convex-arc-concave-arc transform	Limited and definite	Confined to offset	No change necessary	
Concave-arc-concave-arc transform	Limited and definite	Confined to offset	Increase	Same

plates which are separating; (2) transform faults are pure shears parallel to the velocity vector between pairs of moving plates; and (3) trenches are structures which consume lithosphere from one side only and lie at any angle to the velocity vector between the converging plates. They drew trenches as straight lines, although many are arcuate. They found sixteen possible cases, of which two are unstable or transient. At least six are represented by existing examples, as illustrated in Fig. 11-9.

11-4 RHEOLOGY

Faulting implies brittle behavior, but under some conditions rocks do not fracture, but flow. This is generally true in the deep and hot interior of the earth, although some rocks, notably ice and rock salt, can flow on the earth's surface in those especially cold or dry regions where they are exposed. Many other rocks can be deformed slowly at moderate depths and pressure such as prevail in the heart of mountains (S. W. Carey, 1953, 1962; E. S. O'Driscoll, 1964).

Rheology is the study of the flow and deformation of matter, and a fundamental problem in the mechanics of the earth is to determine the proper rheological conditions at any point.

In Chap. 2 it was shown that the state of stress at any point of a body can be described by an array of nine entities p_{ij} $(i,j = x,y,z)$ and that the strain likewise can be represented by another array ϵ_{ij} $(i,j = x,y,z)$. Such arrays are examples of tensors and are called the stress and strain tensors p and ϵ. The rheological behavior of a material will be determined by an equation between p and ϵ called the rheological equation of state. Since the behavior of some materials depends on the rate of strain and the rate at which the stresses are produced, the rheological equation may also include the time derivatives of p and ϵ.

The simplest rheological equation is $\epsilon = 0$, which implies that there is no strain, whatever stresses may be applied. This is the case for a *rigid body* and is the simplest ideal body. Actually, all rheological equations that can be devised describe the behavior of ideal bodies which may form increasingly better approximations to reality, yet nevertheless do not exist in nature. The next two ideal bodies are the *Hooke solid* and the *Newtonian liquid*. For a perfectly elastic body below the elastic limit (a Hooke solid), the relation between shear stress and shear strain is

$$p_{ij} = 2\mu\epsilon_{ij} \qquad i \neq j \qquad (11\text{-}1)$$

where μ is the rigidity, while for a viscous (Newtonian) liquid

$$p_{ij} = 2\eta\dot{\epsilon}_{ij} \qquad i \neq j \qquad (11\text{-}2)$$

where η is the viscosity. The theory of these two ideal bodies has been very highly developed mathematically but cannot explain all the phenomena of actual bodies, especially when the stresses are applied over any length of time.

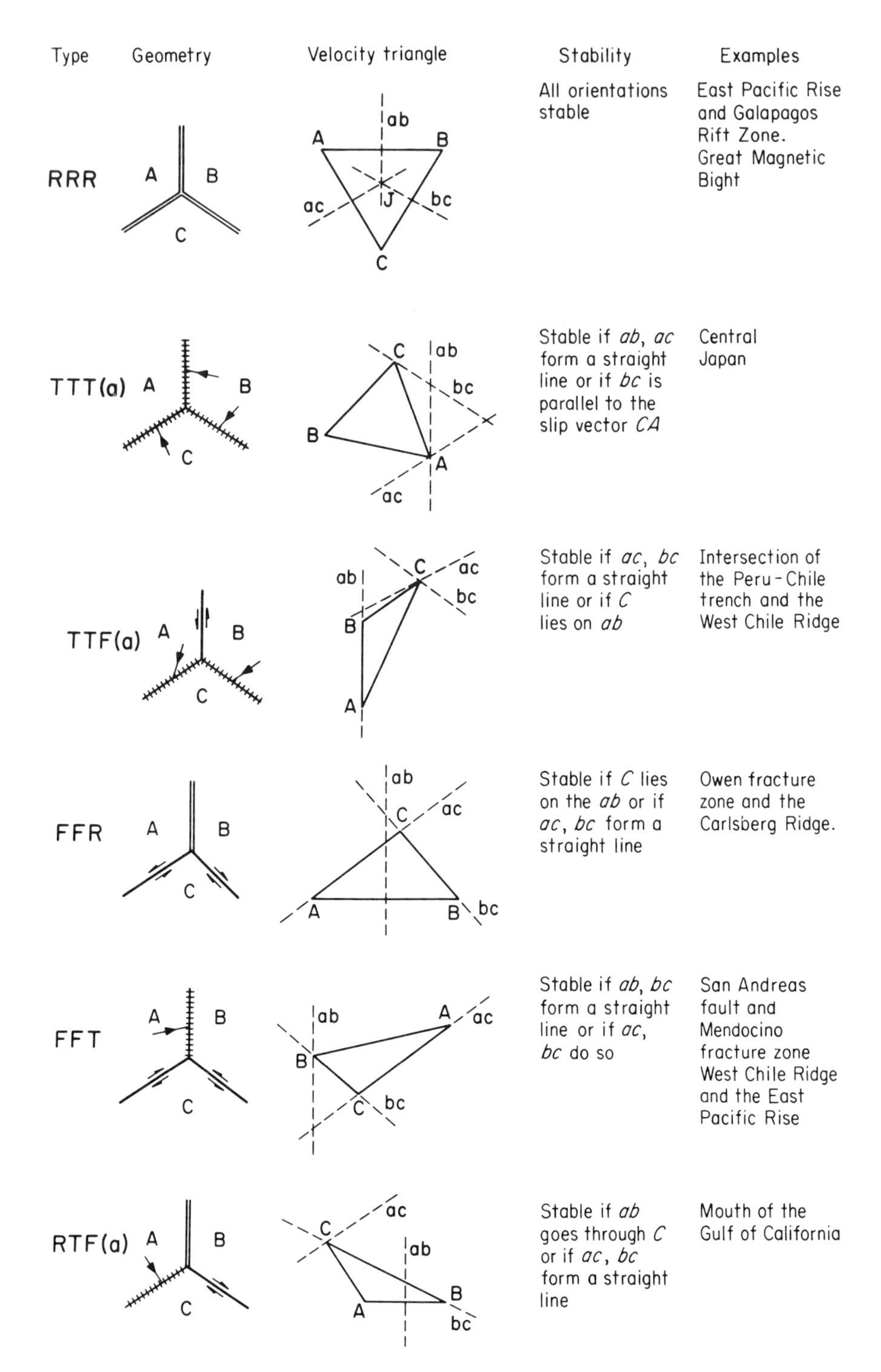

Type
Geometry
Velocity triangle
Stability
Examples
RRR
A B C
ab ac bc J
All orientations stable
East Pacific Rise and Galapagos Rift Zone. Great Magnetic Bight
TTT(a)
A B C
ab bc ac
Stable if *ab*, *ac* form a straight line or if *bc* is parallel to the slip vector *CA*
Central Japan
TTF(a)
A B C
ab ac bc
Stable if *ac*, *bc* form a straight line or if *C* lies on *ab*
Intersection of the Peru-Chile trench and the West Chile Ridge
FFR
A B C
ab ac bc
Stable if *C* lies on the *ab* or if *ac*, *bc* form a straight line
Owen fracture zone and the Carlsberg Ridge.
FFT
A B C
ab ac bc
Stable if *ab*, *bc* form a straight line or if *ac*, *bc* do so
San Andreas fault and Mendocino fracture zone
West Chile Ridge and the East Pacific Rise
RTF(a)
A B C
ac ab bc
Stable if *ab* goes through *C* or if *ac*, *bc* form a straight line
Mouth of the Gulf of California

FIGURE 11-9
Some properties of identified kinds of triple junctions between lithospheric plates. In theory 10 other types could exist. Transform faults are shown as single lines, ridges as double lines, and trenches as hatched lines. In the velocity triangles dashed lines *ab*, *bc*, and *ac* join points the vector velocities of which leave the ◀ geometry of *AB*, *BC*, and *AC*, respectively, unchanged. The relevant junctions are stable only if *ab*, *bc*, and *ac* meet at a point. This condition is always satisfied by *RRR*; in other cases the velocity triangles are drawn to demonstrate instability. The letter *a* after *TTT*, etc., indicates that another form of *TTT* is possible, in this case with each plate rotating in the same direction. (*After D. P. McKenzie and W. J. Morgan, 1969.*)

Some materials behave as a Hooke solid when the stresses are applied for a short time, but over longer time intervals flow like a liquid, exhibiting the phenomenon of *creep*. A possible rheological equation for such a body is a combination of the rheological equations of the Hooke solid and the Newtonian liquid. Omitting the indices $i,j(i \neq j)$, Eqs. (11-1) and (11-2) may be combined to give

$$\dot{\epsilon} = \frac{\dot{p}}{2\mu_M} + \frac{p}{2\eta_M} \tag{11-3}$$

where μ_M and η_M are constants of the body, known as Maxwell constants. A body which obeys Eq. (11-3) is known as a *Maxwell liquid*, after Maxwell, who first postulated such a relation in 1868. If $\mu_M \to \infty$, Eq. (11-3) reduces to Eq. (11-2), i.e., a viscous fluid with viscosity η_M; if $\eta_M \to \infty$, one obtains an elastic solid with rigidity μ_M. A Maxwell liquid also exhibits what is known as *stress relaxation*; if the deformation is kept constant ($\dot{\epsilon} = 0$), the stress diminishes exponentially, reaching a value of $1/e$ of the initial stress after a time η_M/μ_M, known as the relaxation time. If a constant stress is applied ($\dot{p} = 0$), deformation occurs at a constant rate; i.e., the material exhibits creep.

Equations (11-1) and (11-2) could have been combined in another way. A body whose rheological equation is

$$p = 2\mu_K \epsilon + 2\eta_K \dot{\epsilon} \tag{11-4}$$

is called a *Kelvin solid*. A Kelvin solid becomes a Hooke solid if $\eta_K = 0$, and a Newtonian liquid if $\mu_K = 0$. If the strain is constant ($\dot{\epsilon} = 0$), the stress is constant, but if the stress is removed, the strain does not vanish at once as in a Hooke solid; i.e., a Kelvin solid exhibits a delayed elastic effect, or elastic aftereffect. The strain-time curve (for a given stress) for a Kelvin body is illustrated in Fig. 11-10.

Many more ideal bodies can be obtained by introducing a yield stress k. The simplest is a perfectly plastic, or *Saint Venant*, body. For stresses below the yield stress, the body behaves as a Hooke solid; when the yield stress is reached, the body becomes plastic; i.e., under the same stress, the body is continuously deformed; it

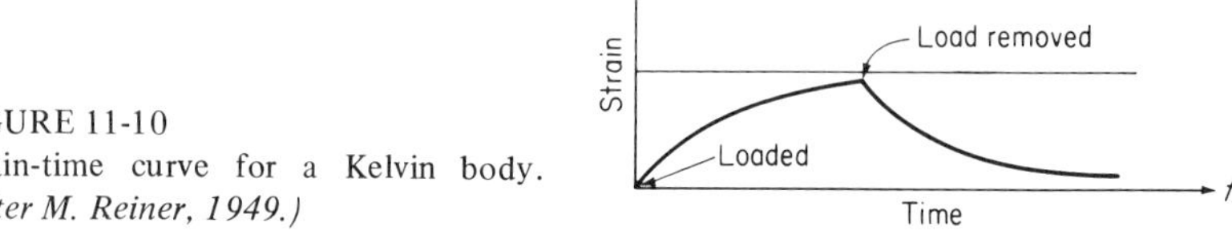

FIGURE 11-10
Strain-time curve for a Kelvin body. (*After M. Reiner, 1949.*)

flows not as a viscous liquid, but plastically. One further ideal body of particular geophysical interest is the *Bingham body*, characterized by the equation

$$p = k + 2\eta\dot{\epsilon} \qquad (11\text{-}5)$$

where k is the yield stress. This reduces to a Saint Venant body when $\eta = 0$.

The above rheological equations are all linear; they lead to exponential creep. However, the creep behavior of various materials, particularly rock, is often not exponential but logarithmic (P. Morlier, 1966; R. C. Parsons and D. G. F. Hedley, 1966). To describe this characteristic, A. E. Scheidegger (1970a) proposed the following rheological equation:

$$\dot{\sigma} = 2\eta\ddot{\epsilon} + \beta(\dot{\epsilon} - C)^2 \qquad (11\text{-}6)$$

where η is the viscosity, and β a creep factor.

For a constant stress σ_0, the strain is a logarithmic function of time; integration of Eq. (11-6) yields

$$\epsilon = A + \frac{2\eta}{\beta}\ln(1 + Bt) + Ct \qquad (11\text{-}7)$$

A special case of Eq. (11-7) is Lomnitz's (1956) law:

$$\epsilon = \frac{\sigma_0}{\mu}\,[1 + q\ln(1 + BT)] \qquad (11\text{-}8)$$

where $A = \sigma_0/\mu$, $2\eta/\beta = q\sigma_0/\mu$, and $C = 0$. μ is the (static) rigidity of the material.

More complex rheological equations could be set up to describe more sophisticated ideal bodies, but with little advantage. The main difficulty from a geophysical point of view is to determine the proper rheological conditions that exist in different parts of the earth. Apart from the crust, it is not easy to reproduce in the laboratory the temperatures and pressures that prevail within the earth, so that one can only speculate on the behavior of the materials below the crustal layers. Moreover, the time element involved is usually so great that it cannot be incorporated into any experimental work. It is as well to realize that such terms as "rigid" and "fluid" have a meaning only when the time interval over which the stresses are applied is specified. The rheological behavior of the earth may well be different for stresses applied over different time intervals.

11-5 DYNAMICS OF FOLDING

There is ample evidence that materials at the surface of the earth yield inelastically under the stresses to which they are subjected. In addition to large-scale faulting which we observe at the surface and detect at depths as earthquakes, crustal rocks are highly deformed in a manner which could result only from plastic flow. Highly folded and contorted sedimentary beds and the formation of entire mountain ranges by the yielding of surface rocks show this clearly.

Under the conditions and lengths of time with which we are ordinarily familiar, most rocks behave as very brittle solids. Therefore the plastic deformation of rocks must take place over lengths of time and often at temperatures and pressures very different from those of everyday life. Little is known about the theory of failure of materials even under laboratory conditions, and the practical applications of the theory to geophysics are of limited use. While the temperatures and pressures involved in near-surface phenomena can be reproduced in the laboratory, the enormous intervals of time involved preclude the possibility of directly observing the plastic deformation of rocks in nature. To obtain better visualization of the nature of the deformation of rocks, many model experiments have been carried out; some of them are very successful in reproducing surface phenomena. Although the successful construction of a model may clarify the events involved in a geological process, it does not generally help in understanding the nature of the processes involved.

The construction of a useful model involves much more than a simple scaling down of the dimensions of a geological area. M. K. Hubbert (1937, 1951) has formulated the known principles of model studies, with regard to their application to geological problems. Where only mechanical phenomena are to be considered, the scale of a model is related to the original by three arbitrarily chosen parameters, for which mass, length, and time are most commonly used. Then the mechanical properties and mass distribution of the model must be so selected that the effects of internal body forces and external forces will result in the same deformations as their counterparts would produce in the original. If the ratios of mass, length, and time in the model to mass, length, and time in the original are given by μ, λ, and τ, respectively, the ratio of all other physical properties can be expressed in terms of them. In problems of a geological nature, forces due to gravity are of prime importance. The proportionality of gravitational forces in the model and in the original requires a point-by-point proportionality of density in the model and original. Furthermore, the acceleration due to gravity is usually the same in the model as in the original, and thus the product $\lambda\tau^{-2} = 1$; that is, $\tau = \lambda^{1/2}$. As a consequence of this relationship, if the time ratio is chosen suitably, the length ratio usually renders the model microscopic in size. Conversely, the time scale imposed by a convenient choice of linear dimensions is usually far too great. Moreover, in many cases the restrictions imposed by the correct scaling ratios require the model to be made of materials having unrealizable physical properties.

Applying dimensional analysis to the folding of a typical granite body, suppose the linear dimensions of the body are reduced by a factor of 2×10^5 (that is, $\lambda = 5 \times 10^{-6}$). For practical purposes the gravitational acceleration must be the same for the model and the original. Therefore $\tau = \lambda^{1/2} = (5 \times 10^{-6})^{1/2} = 2.24 \times 10^{-3}$. If the model is made of a material half as dense as granite, then $\mu = \frac{1}{2}\lambda^3 = 6.25 \times 10^{-17}$. Using these values, the ratio of the strength in the model to that of the original is $\mu\lambda^{-1}\tau^{-2} = 2.5 \times 10^{-6}$. If the strength of granite is taken as 2×10^9 dynes/cm^2, that of the model is 5×10^3 dynes/cm^2. A cube of such material of density 1.5 g/cm^3, larger than 3.3 cm to a side, could not support its own weight. Thus such a model would have to be made of a material having very small strength. Even in this model the time scales involved would be very inconvenient for laboratory studies, more than 2 years being required in the model experiment to reproduce changes taking 1000 years at the earth's surface.

Although most geological problems cannot be scaled down to laboratory experiments, the results of such experiments as can be performed lend general support to the supposition that the apparently brittle and rigid nature of surface rocks is not inconsistent with rock flow over great distances and over long intervals of time to give rise to folded rocks and mountains. Such experiments have been performed by V. V. Beloussov (1961), and by D. T. Griggs and J. W. Handin (1960), among others.

H. Ramberg (1967, 1972) has carried out a number of model experiments of tectonic phenomena which are controlled by gravity. He used large-capacity centrifuges where the centrifugal force plays the same role in the models as the force of gravity does in geological structures. Since the centrifugal force per unit mass may be made several thousand times greater than the gravitational force per unit mass, model materials can be used that are several thousand times stronger, and correspondingly more viscous, than materials in noncentrifuged models of the same size.

11-6 THE PHYSICS OF OROGENESIS

A. E. Scheidegger (1957) has classified stresses according to whether their duration is short, intermediate, or long. Stresses are considered short if they have a typical duration of the order of 3 sec; intermediate, if their typical duration is of the order of 3 years; and long, if their typical duration is 100 m yr. The upper limit of stresses of short duration is about 4 hours. In this range the material of the crust and mantle behaves as an elastic solid with a rigidity of about 2×10^{12} dynes/cm^2 and a Young's modulus of about 5×10^{12} dynes/cm^2. If the elastic limit is exceeded, the material undergoes brittle fracture. This information is obtained from laboratory experiments and the passage of earthquake waves.

The time range for stresses of intermediate duration is from about 4 hours to 15,000 years. Information on this time range comes from a number of sources:

observations of rock behavior in the laboratory, seismic aftershock sequences, earth tides, and the decay of the Chandler wobble. Originally, A. E. Scheidegger (1957) tried to explain the response of the earth to seismic aftershock sequences and the damping of the earth's free nutation (the Chandler wobble) in terms of a Kelvin model. However, new information obtained over the last ten years shows that the Benioff strain-rebound model of aftershocks on which he had based his calculations is no longer valid. G. Ranalli and A. E. Scheidegger (1969) have since carried out a detailed analysis of a number of aftershock sequences and showed that aftershocks are a discontinuous manifestation of overall plastic creep, the earth adjusting itself to a redistribution of stresses by logarithmic creep. Logarithmic rather than exponential creep is also in agreement with the results of laboratory experiments on rock creep. Although the mechanism of energy dissipation in earth tides and the damping of the Chandler wobble are not yet well understood, a rheological model of the earth in the intermediate time range based on logarithmic creep is possible, and does not contradict any of the observational evidence to date.

12

THEORIES OF THE EARTH'S BEHAVIOR

12-1 INTRODUCTION

While the first part of this book has dealt with general physical, geological, and chemical properties of the whole earth, the remainder considers the evolution and behavior of the present crust and upper mantle. V. V. Beloussov (1962), A. E. Scheidegger (1963), A. Holmes (1965), H. Takeuchi et al. (1970), and J. Verhoogen et al. (1970) are among those who have reviewed the subject.

It is important to realize that three different groups of men have studied the earth. These are miners, travelers, and surveyors; and from their early endeavors have descended three branches of earth science: geology, geophysics, and geodesy. All these subjects are old, but until very recently they developed separately and remained distinct, for they employed different methods to observe the earth from different points of view.

The development of earth science may be considered in four stages, with boundaries at about A.D. 1800, when modern concepts of geology emerged; 1900, when radioactivity was discovered; and 1950, when the introduction of instruments which could collect physical data quickly enabled geophysical theories, long known, to be applied to the interpretation of geological observations. Very quickly this has led to the concept of plate tectonics, and perhaps the start of a clearer understanding of the earth's behavior.

12-2 EARLY HISTORY AND THE RECOGNITION OF THE ROLE OF VERTICAL MOVEMENTS

F. D. Adams (1954) and S. J. Gould (1965) have given accounts of the early history of geology. For centuries, ideas about the earth were chiefly based upon the practical lore of miners, upon an account of volcanic eruptions and rock specimens written by Pliny the Elder (A.D. 23-79), and upon a belief in the *cataclysmic theory* of creation of the earth, which an attempt to interpret the Bible literally had dated as occurring in 4004 B.C.

With the Renaissance, observations on the earth's behavior began to replace speculation, revealing the importance of vertical movements. Thus Agricola (1490-1555), besides beginning the proper classification of minerals, appreciated that the land surface is eroded and lowered by rivers, and Leonardo da Vinci (1452-1519) recognized the need for uplift and renewal. Steno (1631-1686) described the advance and retreat of the sea over Italy, and Lomonosov (1711-1765) proposed that vertical motions constitute the fundamental process of tectonics. The latter distinguished between relatively rapid movements in mountain building and slower, gentle warpings like those which have long made the vast Russian and west Siberian plains basins of deposition.

Meanwhile, William Gilbert (1540-1603) and Isaac Newton (1642-1727) established the general principles of the behavior of the earth's magnetic and gravitational fields, and slowly instruments were devised to measure these forces. The disastrous Lisbon earthquake of 1755 and the terrible eruption in Laki, Iceland, in 1783 further stimulated interest in geophysical subjects.

12-3 THE THEORY OF UNIFORMITARIANISM

The start of the Industrial Revolution led to the opening of new mines, quarries, and canals which provided good opportunities for observation.

Taking advantage of these opportunities, J. E. Guettard (1715-1786) became perhaps the first true geologist, for he produced the first mineralogical maps, recognized the volcanic nature of old lavas, and grasped the importance of erosion in shaping landscapes. He and James Hutton (1726-1797) opposed the prevailing notion that the earth had been shaped by cataclysms. Hutton advocated "no action to be admitted except those of which we know the principle" and proposed that the earth was really very old and that the processes operating today are similar to those which have operated in the past. This is the *theory of uniformitarianism.* Although it is only partly correct, the emphasis on slow evolution, on the length of geological time, and on the need for an empirical approach was a great contribution.

For a time A. Werner (1750-1817) maintained that all rocks had crystallized out of a primitive ocean. This led to confusion until Hutton distinguished sedimentary rocks which had been laid down in succession according to the *law of superposition* from igneous rocks which had been injected later to disrupt the orderly succession. About the same time Baron G. Cuvier (1769-1832) and Chevalier de Lamarck (1744-1829) began to relate diagnostic fossils to particular strata, and by the close of the eighteenth century William Smith (1769-1839) had appreciated that across England the Jurassic strata succeed one another like rows of shingles on a roof, each characterized by its own distinctive index fossils.

The next few decades saw the identification of the geological systems and the establishment of a time scale for later geological ages (Sec. 5-3). In 1830 C. Lyell summed up the new discoveries by writing "The Principles of Geology." Many of his concepts are still valid today.

12-4 THE CONTRACTION HYPOTHESIS

During the second half of the nineteenth century vague ideas about uplift were replaced by the *contraction theory* first suggested by Isaac Newton. He realized that the earth is emitting heat and, ignorant of the existence of radioactivity, supposed that it must be cooling. In 1829 Elie de Beaumont proposed that thermal contraction was the cause of the conspicuous folding and thrusting of the rocks he studied in the Alps. In 1887 C. Davison and G. H. Darwin treated the hypothesis mathematically and provided a basis for later work by H. Jeffreys (1970), who estimated that during all geological time the total surface compression along any great circle has been of the order of 200 km. He considered that this was sufficient to have uplifted all past and present mountains. Many, including W. H. Bucher (1933), J. T. Wilson (1954), and A. E. Scheidegger (1963), attempted to explain the growth of mountains and the formation and shape of island arcs under these conditions, but their explanations were vague and imprecise and the theory must be regarded as unsatisfactory.

12-5 THEORIES OF UNDATION, PULSATION, AND VERTICAL TECTONICS

The discovery of radioactivity reduced the likelihood of thermal contraction, and earth scientists looked for other possible causes of tectonic movements.

In 1909 J. Joly proposed that radioactive heating causes periodic melting within the earth, and A. W. Grabau (1936) and J. H. F. Umbgrove (1947) extended this idea and advanced related theories according to which tension and compression have alternated in worldwide cycles. In 1930 E. Haarmann, from physiographic studies, proposed that oscillatory vertical movements of the crust are the primary cause of mountain building and that folding is a secondary phenomenon. R. W. van Bemmelen (1933, 1968) developed this hypothesis and named it the *undation theory*. V. V. Beloussov (1962, p. 356) has stated that this was "the beginning of the lines of development of general ideas that now are the core of geotectonics."

He, with many other Soviet scientists, believes that periodically local melting and differentiation cause parts of the mantle to rise into the crust, producing uplift and metamorphism, followed by cooling, increase in density, and sinking (V. V. Beloussov and I. P. Kosminskaya, 1968). It is a mistake, however, to believe that all Soviet scientists agree with these ideas (A. V. Peyve, 1969).

Modern advocates of the undation theory face certain difficulties. Although heating can produce melting and uplift, it is not clear why this process should be cyclical, and A. E. Scheidegger (1963) has shown that the theory is untenable on rheological grounds.

12-6 THE EXPANSION HYPOTHESIS

The discovery of radioactivity opened the possibility that after a cold accretionary origin the earth has been heating up and expanding (R. Dearnley, 1966; P. Jordan, 1969). E. A. Lubimova (1967, p. 310) examined the question and concluded that an "expanding earth is a natural consequence of thermal evolution of a planet having the chemical composition of chondritic meteorites." She considers the present annual rate of increase of the earth's radius to be 0.07 mm/year, with a total increase during geological time of 100 to 200 km.

Once the idea of expansion due to radioactive heating became current, others espoused expansion for different reasons. Thus P. M. Dirac (1938) suggested that the value of G (the gravitational constant) might be decreasing, and J. Peebles and R. H. Dicke (1962) calculated that this could produce an increase of 0.05 mm/year in the radius. L. Egyed and L. Stegena (1958) suggested a similar rate from elastic rebound and a larger one due to phase changes in the earth's interior.

S. W. Carey (1958) and B. C. Heezen (1960) proposed a much greater expansion involving doubling the earth's radius since the Triassic. This could have opened the existing ocean basins, but does nothing to explain earlier geological history. Another objection is that whereas three rifts strike nearly north and south through the main oceans, only one strikes east and west to encircle the earth in the southern oceans, so that expansion might be expected to change the shape of the earth (F. J. Vine and H. H. Hess, 1970).

Many have raised objections on physical grounds. If expansion has been a universal phenomenon, the dimensions and properties of the sun and solar system and the amount of solar radiation received on earth should all have changed, for which there is no evidence. If the earth alone expanded, no one has yet explained why this should have suddenly happened late in geological time or what materials could have existed of sufficiently high density to have constituted the interior of the earth before expansion. A. E. Beck (1969) from energy considerations and F. Birch (1968) from our knowledge of the behavior of matter at high pressures both consider an increase in radius of more than 100 km during geological time to be unlikely. It is very doubtful if this is adequate to have built all mountains.

12-7 EARLY THEORIES OF CONVECTION CURRENTS

Volcanic eruptions and the observation by miners that the earth's temperature increases rapidly with depth early led to the idea that beneath a surface "crust" the earth might be liquid. In 1839 W. Hopkins suggested that there might be subcrustal convection currents, which O. Fisher (1889, pp. 376-378) described thus:

> The ascending currents are situated beneath the oceanic areas and the descending currents beneath the land. Consequently the liquid magma must flow from the oceanic towards the continental areas, and must acquire a more or less horizontal motion . . . to produce compression along their common boundary. . . .
>
> Where the currents ascend beneath the ocean they would give rise to a tensile stress, the correlative of the compression of the land. Fissures would thus be produced, which would open volcanic vents, and, when filled with solidified lava, become dykes of igneous rock in the suboceanic crust. . . . We recognize two principal types of volcanic regions, coastline and oceanic. We believe the former to be connected with the agencies which have raised the continents which they skirt. . . . The oceanic volcanoes on the other hand appear unconnected with compressive action, for the oceanic islands consist almost all of them of volcanic rocks. . . .
>
> They occupy a medial position with respect to the coastlines, being in the Atlantic rudely parallel to the opposite shores.

The discovery of radioactivity strengthened the case for thermal convection currents in the earth's mantle, and many have considered their possible nature and distribution.

J. W. S. Rayleigh (1916) stated the conditions of instability and concluded that convection was to be expected in the mantle if its viscosity was less than about 10^{26} poises. N. A. Haskell (1937), later supported by W. R. Farrand (1962) and R. K. McConnell (1965), showed that a viscosity as small as 10^{22} poises could explain the postglacial uplift of Fennoscandia, thus demonstrating the probability of convection. H. Jeffreys (1970) showed that the earth's rotation should not affect the pattern of currents. Many authors have discussed possible convection patterns, including O. Ampferer (1906), A. Holmes (1931), C. L. Pekeris (1935), S. Chandrasekhar (1952), and F. A. Vening Meinesz (1948b, 1964). D. T. Griggs (1939) built a mechanical model, but no one produced any results which fitted the surface patterns, S. K. Runcorn (1962c, 1965) revived interest in the problem, but it now seems that all these early attempts may have been wrong in assuming that convection affected the whole mantle. It may be that they are confined to the asthenosphere (Sec. 17-5). Some idea of the extreme rigidity even of the asthenosphere is indicated by N. Kumagai and H. Ito's (1970) measurement of a viscosity of 10^{20} poises for granite. They cut long bars of granite and left them outside, supported at either end, measuring the sag at the center over a period of some years.

12-8 THE PROBLEM OF ISLAND ARCS

A glance at a globe will show that there are many more and better circular arcs on the earth's surface than might be expected from chance and that many are about the same size (Fig. 12-1). A. E. Scheidegger (1963) has reviewed numerous attempts to explain this phenomenon, but under the old theory of fixed continents no one could suggest any satisfactory theory for these conspicuous and regular features.

During the last century E. Suess (1904-1924) and other geologists showed that all island arcs and young folded mountains lie along one of two comparatively narrow belts about the earth. Most of the world's 1050 active volcanoes, including nearly all emitting andesitic lava, lie on the same belts. The worldwide study of earthquakes, which was begun about 1900, has shown that nearly all the world's deep, and most of its shallow earthquake foci lie beneath the same belts (M. Barazangi and J. Dorman, 1969; J. H. Latter, 1969).

Each of these active belts extends more than halfway around the earth, very roughly along a great circle. One belt, which will be referred to as the Eurasian-Melanesian belt, includes the Atlas, Alpine, Turkish, Iranian, Himalayan, and Burmese mountains and the island chains of Indonesia, New Guinea, the Solomons, New Hebrides, and New Zealand. The other, called the East Asian-Cordilleran belt, surrounds the Pacific Ocean clockwise from Indonesia to Antarctica. The two belts meet orthogonally in a T-shaped junction at Celebes Island in Indonesia.

Along these active belts there is a wide variety of volcanic, plutonic, and sedimentary ranges, of single and double island chains, and of plateaus. These elements are so varied and complex that they defied simple classification until geophysical observations showed that they had common deep-seated features. F. A. Vening Meinesz (1929 and 1941) discovered that a large negative gravity anomaly follows the inner side of trenches. H. Benioff (1949) showed that beneath arcs, the foci of intermediate and deep earthquakes form conical zones, since named *Benioff zones*.

Figure 12-2 and Table 12-1 illustrate these relations. The name *primary* is used to refer to arcs and mountains with these deep-seated features. It is useful to distinguish them from other arcs and mountains, which may appear superficially to be rather similar but which have no roots and are called *secondary*.

The active primary elements show some or all of the following diagnostic characteristics: (1) they are underlain by all the world's deep earthquakes and most of its shallow ones; (2) they have associated with them those of the world's active volcanoes which give andesitic lavas and also the young plutonic rocks of similar composition; (3) they are followed by strips of large negative gravity anomalies; (4) they are accompanied by the world's deepest ocean trenches; (5) they rest upon no visible basement of older gneissic rocks; (6) their sedimentary rocks are of greywacke or eugeosynclinal facies. Many of the elements have two other

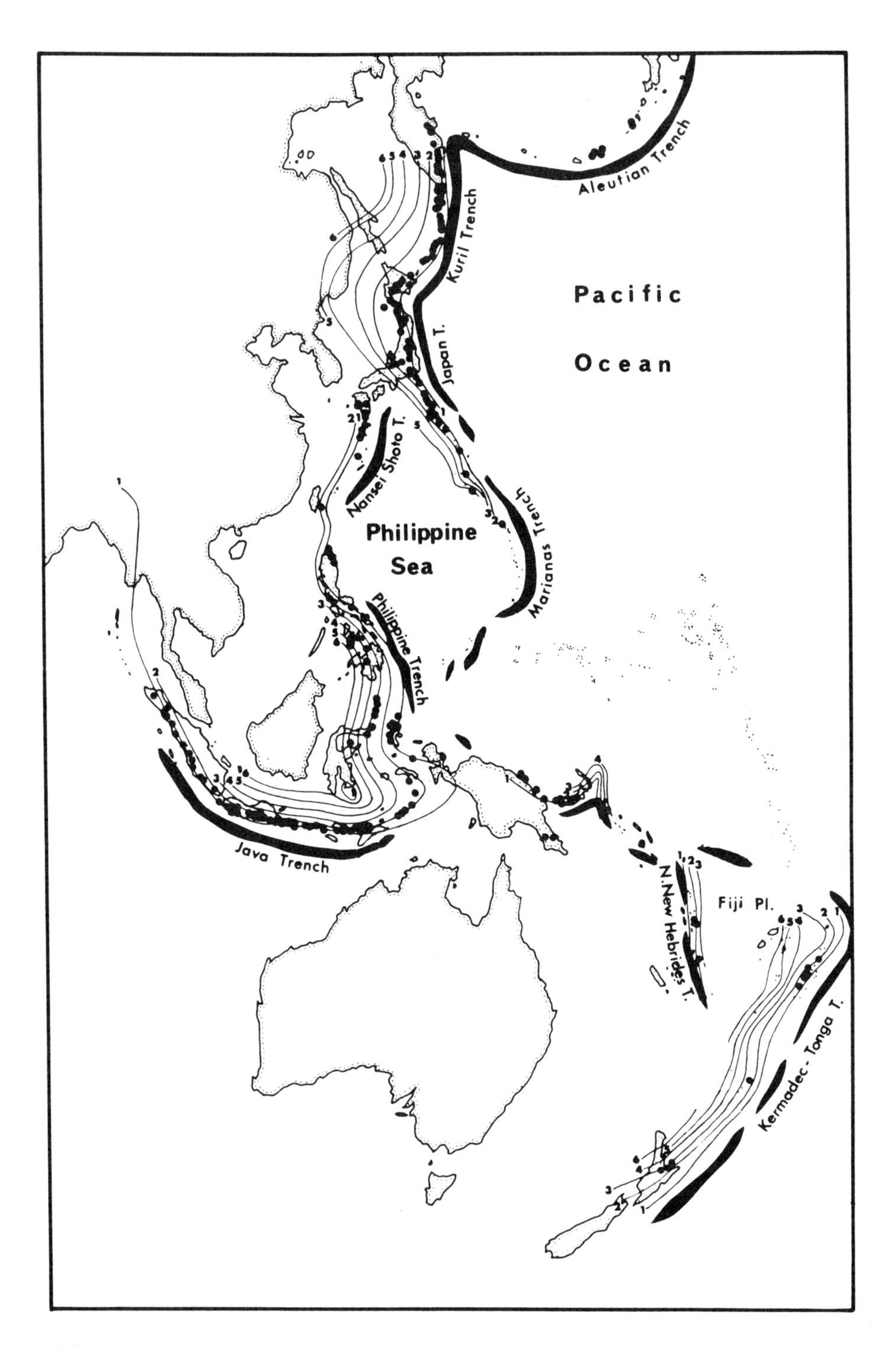
Aleutian Trench
Kuril Trench
Japan T.
Pacific
Ocean
Nansei Shoto T.
Philippine
Sea
Marianas Trench
Philippine Trench
Java Trench
N. New Hebrides T.
Fiji Pl.
Kermadec-Tonga T.

FIGURE 12-1
◀ Island arcs of Western Pacific showing active volcanoes as black dots, trenches in dark tone, and Benioff zones of deep earthquake foci contoured at 1-km intervals. Compare with Fig. 17-11. (*After D. L. Turcotte and E. R. Oxburgh, 1970.*)

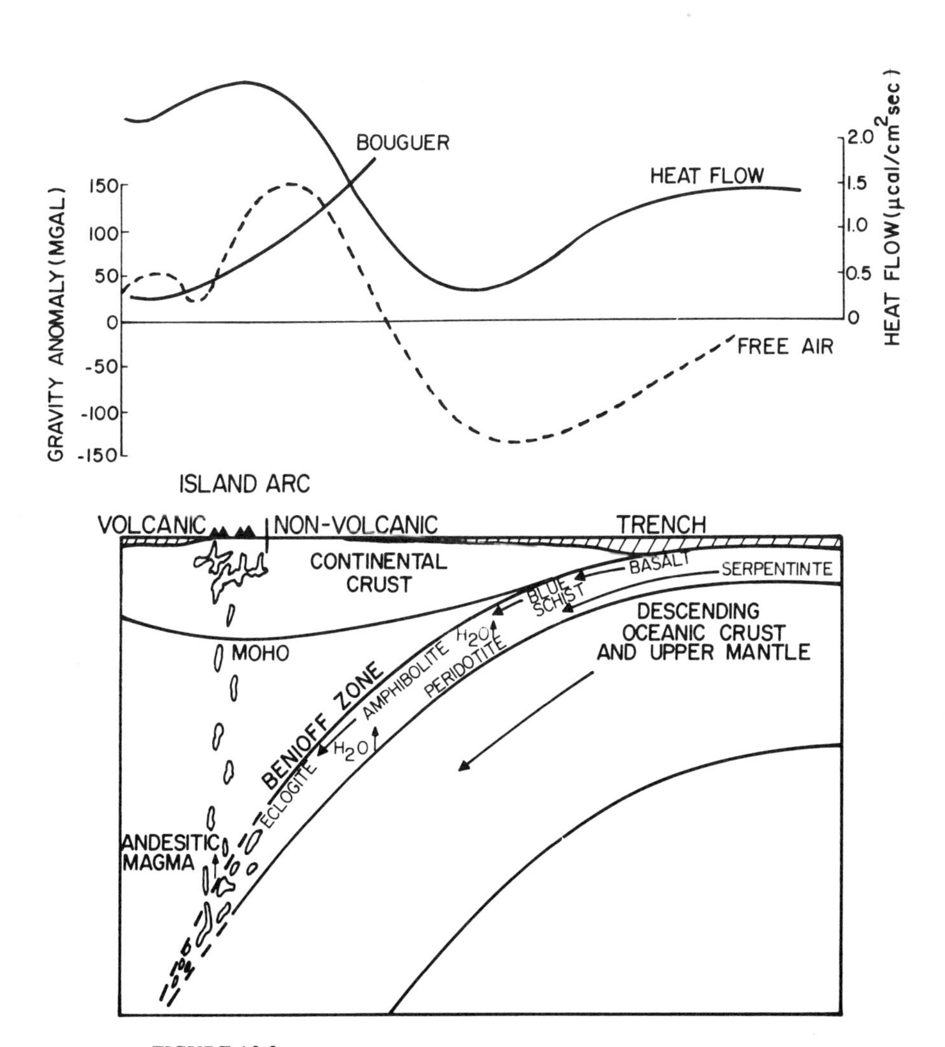

FIGURE 12-2
Cross section of an island arc showing generalized geological structure, heat flow, and gravity anomalies. Hypothetical metamorphic reactions are shown which release water down to the level at which andesitic magma is generated. (*After W. H. K. Lee et al., 1966, and C. B. Raleigh and W. H. K. Lee, 1969.*)

characteristics which are not diagnostic because they are also common to secondary ranges. These features are: (1) most of the elements have the shape of roughly circular arcs; (2) those in the two active belts have all been subject to Cenozoic and Recent folding and uplift.

Arcs may be classified into four types (J. H. F. Umbgrove, 1947, 1949). Those which comprise a curved ocean trench and a parallel line of active volcanic islands are called *single island arcs.*

In some arcs, like the Kuril Islands and Scotia arc, the trench is as long as the arc of volcanic islands, but in others, like the Aleutian Islands, the West Indies, and Indonesia, part of the trench has been replaced by a second line of islands consisting of folded sediments of the eugeosynclinal or greywacke facies. Examples are Kodiak Island off the Aleutians; Trinidad, Tobago, and Barbados islands off the West Indies; and the island of Timor in Indonesia. In every case these islands lie close to a continent, which suggests that sediment from the land has helped to fill the trench (R. L. Chase and E. T. Bunce, 1969). In every case only part of the trench has been replaced by sedimentary islands. These sections are called *double island arcs.*

Where the double arc of the Aleutians and Kodiak Island reaches the coast, the Alaska and Chugach mountains form a continuation having similar volcanic and sedimentary features, so that they have been called a *double mountain arc*. Again the Coast Range of British Columbia and southern Alaska and the offshore islands may be regarded as another double mountain arc, and so may the volcanoes and batholiths of the Cascades and the Sierra Nevada Mountains and the Franciscan sediments of the coastal ranges from California to Washington.

Again the Andes Mountains which have many volcanoes are paired with a

Table 12-1 FEATURES OF REGULAR, ACTIVE PRIMARY ARCS (FROM CONVEX TO CONCAVE SIDE)

1. Outer Sedimentary Part, Primary Arc

A deep oceanic trench or a chain of islands or mountains of sediments of greywacke facies. Sometimes both together, in which case the trench is small and is displaced to lie outside the islands. Shallow earthquakes and a belt of large negative gravity anomalies occur under the islands or mountains or on the inner side of the trench.

2. Intermediate Part, Primary Arc

A valley or shallow trough or slope between the outer and inner parts of primary arcs. Earthquakes at a depth of about 60 km and a belt of positive gravity anomalies occur beneath it.

3. Inner Igneous Part, Primary Arc

A volcanic island arc or mountain arc of andesitic to acid volcanics and granodiorite batholiths, with earthquakes beneath it at about 100 km depth. It parallels the outer part at a distance of about 180 km.

4. Conical Zone of Intermediate to Deep Earthquakes

Intermediate and deep-focus earthquakes, with foci on a conical zone dipping toward the centers of the arcs and extending in some cases to 700 km in depth.

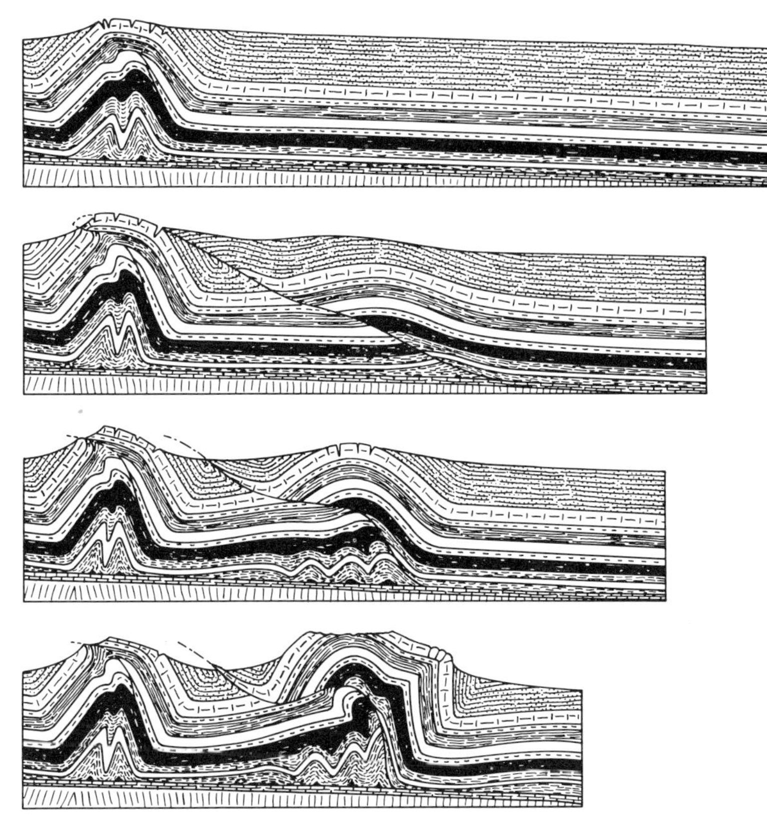

FIGURE 12-3
Stages in the development of the structure revealed in the Grechenberg tunnel, Swiss Jura Mountains. (*After A. Buxtorf, 1916.*)

deep trench offshore to form what has been called a *single mountain arc* (J. T. Wilson, 1954). Here the deserts of Peru and northern Chile produce no large rivers and not enough sediments to fill the trench. Mountain arcs are less regular than island arcs.

In addition to these primary features with deep roots, there are secondary arcuate features with no deep earthquakes, no large gravity anomalies, and no volcanoes. They usually lie on the continental side of primary arcs and are so arrayed that the center of a secondary arc lies opposite to the junction of the two primary arcs (Fig. 12-1). Examples of secondary arcs are the mountains of Taiwan and Kamchatka, the Jura Mountains and the Alps, and the Valley and Ridge province in the Appalachians.

Most of these seem to have slid over the basement (Fig. 12-3), a separation known as *decollement* (A. W. Bally et al., 1966).

Arcs are too numerous and too regular to be due to chance, but their origin remained a puzzle until continental drift became acceptable when F. C. Frank (1968) pointed out a possible explanation. On a sphere "a flexible but inextensible thin spherical shell may be bent inward through an angle θ on and only on a circle whose radius of curvature (expressed in angular measure on the sphere) is $\frac{1}{2}\theta$." He noted that the radius of several trenches is about 22° and that the dip of their Benioff zones is 45°, and he proposed that the latter represent sections of the lithosphere which have been bent inward and pushed down into the asthenosphere, thus explaining the shape, curvature, dip, and seismicity of some island arcs. An indented Ping-Pong ball illustrates the same principle. The reason why arcs lie off some coasts but not off others is discussed in Sec. 17-6.

12-9 THE DEVELOPMENT OF THEORIES OF LATERAL DISPLACEMENT OR CONTINENTAL DRIFT

The theories presented in Secs. 12-1 to 12-7 agree in the importance of periodic uplift; they reject large horizontal motions; and they were conceived before much was known about the sea floor and the earth's interior.

There is another group of hypotheses which agrees with the above hypotheses in recognizing the importance of vertical motion, but differs in that they also emphasize the importance of large horizontal motions and take into account new knowledge of ocean floors and the earth's interior. Collectively, they are called *continental drift*.

If, in the earth, a rigid lithosphere overlies a deformable asthenosphere, the lithosphere can move in any one of three ways. First, it can slide over the asthenosphere by rotating as a whole. This is called *polar wandering*. Second, the lithosphere can break into pieces which move relative to one another. A. Wegener (1966) named this *continental drift*. Third, both types of motion may proceed together.

The origin of the concept of continental drift is old. A. L. du Toit (1937) found evidence for similar ideas in the work of Francis Bacon in 1620 and in that of Buffon and of von Humboldt, but A. V. Carozzi (1970) emphasizes that A. Snider in 1858 was the first to "unequivocally postulate and illustrate a juxtaposition and drifting of the continents."

In any event these uncertain ideas attracted little attention until 1910, when F. B. Taylor, and soon afterward H. B. Baker and Alfred Wegener (1966), independently developed the concept, and Wegener assembled much supporting evidence. He proposed that until Mesozoic time there had been only one supercontinent, called Pangaea. During the Mesozoic era gravitational and tidal forces disrupted it into the present continents and propelled them like rigid ships through a yielding oceanic floor (Fig. 12-4). Oddly enough, he did not consider radioactive heating as a cause of motion and proposed instead some effects of gravity and rotation of the earth which others soon showed to be negligible.

After World War I he extended his arguments and wrote a book, of which the fourth and last edition published in Germany in 1929 has only recently been translated into English (A. Wegener, 1966). This work led a few geologists, including E. Argand (1924), R. A. Daly (1926), R. Staub (1928), and E. B. Bailey (1929), to accept his ideas. A particularly strong supporter was A. L. du Toit (1927 and 1937), who pointed out the importance of radioactivity as a possible cause of motion and advocated not one but two primitive supercontinents, Gondwanaland in the south and Laurasia in the north. A. Holmes (1931 and 1965) revived O. Fisher's ideas about convection currents and related them to continental drift. As a result of the renewed interest, W. A. J. M. van Waterschoot van der Gracht (1928) organized a symposium, which showed the overwhelming antagonism of North American geologists to continental drift.

At the time of the debate the state of knowledge was such that most of the arguments were general and lacked precision, and as opponents soon pointed out, some were untenable. At that time only a few were converted, including L. C. King (1953, 1967), S. W. Carey (1958), F. Ahmad (1961), and R. Maack (1969). Most geologists and geophysicists alike either ignored the hypothesis of continental drift or spoke strongly against it (V. V. Beloussov, 1967, 1968a, 1970; E. N. Lyustikh, 1967; H. Jeffreys, 1969, 1970; P. S. Wesson, 1970; and A. A. Meyerhoff, 1970a,b). It was not until nearly forty years later that new discoveries caused a great swing in opinion toward accepting drift (Sec. 14-10).

12-10 THE UNIFICATION OF EARTH SCIENCES

In 1859, when C. Darwin published "The Origin of Species," geology was perhaps the most highly regarded of the sciences. Geological surveys had begun mapping the earth's surface, and geological observations had provided Darwin with much of the basis for the theory of evolution, and J. L. Agassiz with his concept of ice ages, which displaced notions of a universal flood. Geologists such as B. Silliman, H. O. Rogers, J. W. Powell, and W. Logan were among the leading scientists.

For the next century geology advanced in an absolute way at a slow pace, but relative to other sciences it declined in esteem. It produced no discoveries to compare with those in nuclear physics, organic chemistry, or genetic biology. It failed to become a precise science, and earth science remained divided, as it had been since the time of the early miners, travelers, and surveyors.

Geology, the mapping and interpretation of the earth's surface, has been a great task, but has suffered from two limitations: geological maps and reports present information better than they interpret it, and the land surface is but a small part of the earth as a whole.

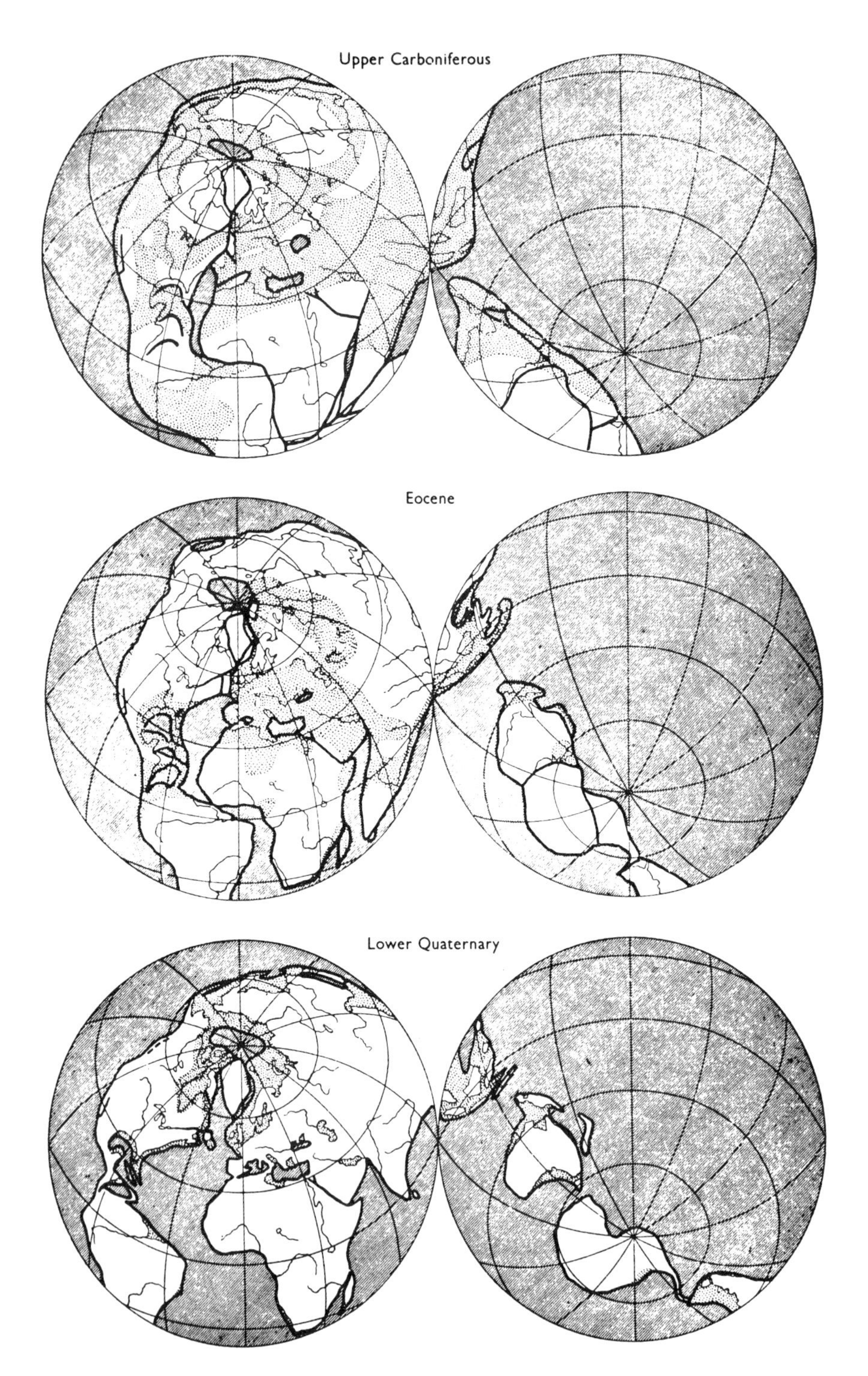
Upper Carboniferous
Eocene
Lower Quaternary

FIGURE 12-4
◀ Reconstruction of the map of the world according to the theory of continental drift for three past periods. Dotted areas represent shallow seas, but the present outlines have been added to aid identification. (*From "The Origin of Continents and Oceans," by Alfred Wegener, Dover Publications, Inc., New York, 1966. Reprinted by permission of the publisher.*)

Geologists developed precise and accurate techniques for identifying minerals, analyzing rocks, and classifying fossils, which they used to map small areas, but they were unable to match these skills in solving detailed problems with any corresponding precision in interpreting the major problems of the earth as a whole. Explanations of such important matters as to whether ocean basins are permanent, how mountains are built, or where ores occur remained vague. Perhaps this is not surprising because geological methods are applicable only to the exposed surface of the earth; one cannot discover all about an egg by even the most careful examination of one-quarter of an eggshell.

Geophysics, the study of the earth by the methods of physics, developed from the early studies by travelers of the weather, oceans, and readings of the magnetic compass. One of the first scientific books was "De magnete," which William Gilbert published in 1600. He showed that the earth behaves as a magnet with a field similar to that which he found by experiment for a small sphere fashioned out of lodestone. Soon afterward Isaac Newton (1642-1727) discovered the inverse-square law of gravitation. Much later other physicists developed theories of seismology, heat flow, and radioactive decay, but none of these elegant theories could be applied to the study of details of the earth until good instruments were developed in the last few decades.

Geology and geophysics remained separate as long as geologists, however excellent their techniques for studying details, lacked precise general theories and as long as geophysicists, however excellent their general theories, lacked methods for making precise measurements of details rapidly. Only very recently have new instruments and the new theory of plate tectonics and continental motion filled these lacunae and made a unified study of the earth possible.

T. S. Kuhn (1970) has pointed out how commonly branches of science have progressed from an early stage concerned with the ordering of commonplace observations by practical man through a scientific revolution to a second stage of far greater precision and power. The observations of the early Ptolemaic astrologers were not wrong, and many of their deductions led to practical uses. It was they who devised our calendar, predicted eclipses, and guided Columbus and Magellan, but their way of thinking about the solar system was incorrect, and it was only after Copernicus had changed the current point of view that the earth was the center of the universe that the new astronomers could begin to comprehend the true nature of the sun and stars. The change from alchemy to chemistry, the discovery of evolution, and the realization of the true nature of the atom were equally significant milestones.

It seems plausible to suggest that the low estate of geology and geophysics is a result of their having remained until now as mere codifications of the observations and ideas of practical men. Now, since 1967, the sudden acceptance of sea-floor spreading and crustal motion offers another scientific revolution and an escape to a much more powerful future earth science of which geology and geophysics will both be essential parts. As in several cases in the past, it seems appropriate to mark the revolution by a change in name to *geonomy*, the science of the study of the solid earth, a name advocated among others by V. V. Beloussov (1964) and R. W. Van Bemmelen (1969). The term *earth science* can be reserved for the broader physical study of the whole earth, including the atmosphere and hydrosphere as well.

13

THE OLDER ARGUMENTS FOR AND AGAINST CONTINENTAL DRIFT

13-1 INTRODUCTION

The fundamental question in tectonics is whether the ocean basins are permanent and the continents have evolved in approximately their present positions or whether ocean basins have opened and closed as continents have moved about. A related problem is whether there have been several cycles of movement so that each continent is a mosaic of fragments repeatedly fractured and rewelded.

Clearly, no conclusions can be reached about the sources of sediments and ore deposits, past climates, the causes of mountain building, or even the full history of evolution of life, until these matters are settled. This chapter presents the older arguments for and against the hypothesis of continental drift which at the time they were advanced were inconclusive.

13-2 THE FIT OF COASTLINES

One strong argument for continental drift which has long appealed to the public is the similar shape of the two opposite coasts of the South Atlantic Ocean.

A. Wegener (1966, p. 17) wrote that

South America must have lain alongside Africa and formed a unified block which was split in two in the Cretaceous. . . . The edges of these two blocks are even today strikingly congruent. Not only does the large rectangular bend formed by the Brazilian coast at Cape São Roque mate exactly with the bend in the African coast at the Cameroons, but also south of these two corresponding points every projection on the Brazilian side matches a congruent bay on the African, and conversely.

Critics pointed out that, because changes in sea level would alter the shape of coastlines, the fit might be a coincidence and that few other fits were apparent (H. Jeffreys, 1970), but E. C. Bullard et al. (1965) showed with the aid of a computer that the steep parts of the continental slopes match well, and other good fits have now been found (R. S. Dietz and J. C. Holden, 1970a,b).

13-3 MATCHING GEOLOGY BETWEEN OPPOSITE CONTINENTS

The need to supplement bathymetric fit with geological fit occurred to A. Wegener (1966, pp. 76-77), who wrote of the need to bring

the continuation of each formation on the farther side into perfect contact with the formation on the near side. It is just as if we were to refit the torn pieces of a newspaper by matching their edges and then check whether the lines of print run smoothly across. If they do, there is nothing left but to conclude that the pieces were in fact joined in this way.

In practice this is difficult for several reasons. Along many coasts, like those of the eastern United States, Tertiary and Cretaceous coastal plains have covered the edge of the continent and spread far inland since drift began. Along others, like the Atlantic coasts of Norway and Greenland, fold mountains or metamorphic belts strike along the coasts with little change in features by which the two sides can be matched without a great deal of detailed work. A. L. du Toit (1927) was among the first to make a careful field study of two opposite coasts. Between South America and Africa he found 15 points of conformity among major geological features and showed that not only did these match across the ocean, but so also did variations within formations (Fig. 13-1). Others have supported these results (H. Martin, 1961; P. M. Hurley, 1968, 1970; R. Maack, 1969). G. O. Allard and V. J. Hurst (1969), having mapped the details of a complex succession of rocks in a large syncline with strikes out to sea in Brazil, crossed to Gabon and found an identical continuation striking into Africa in the place predicted on the basis of topographical fit.

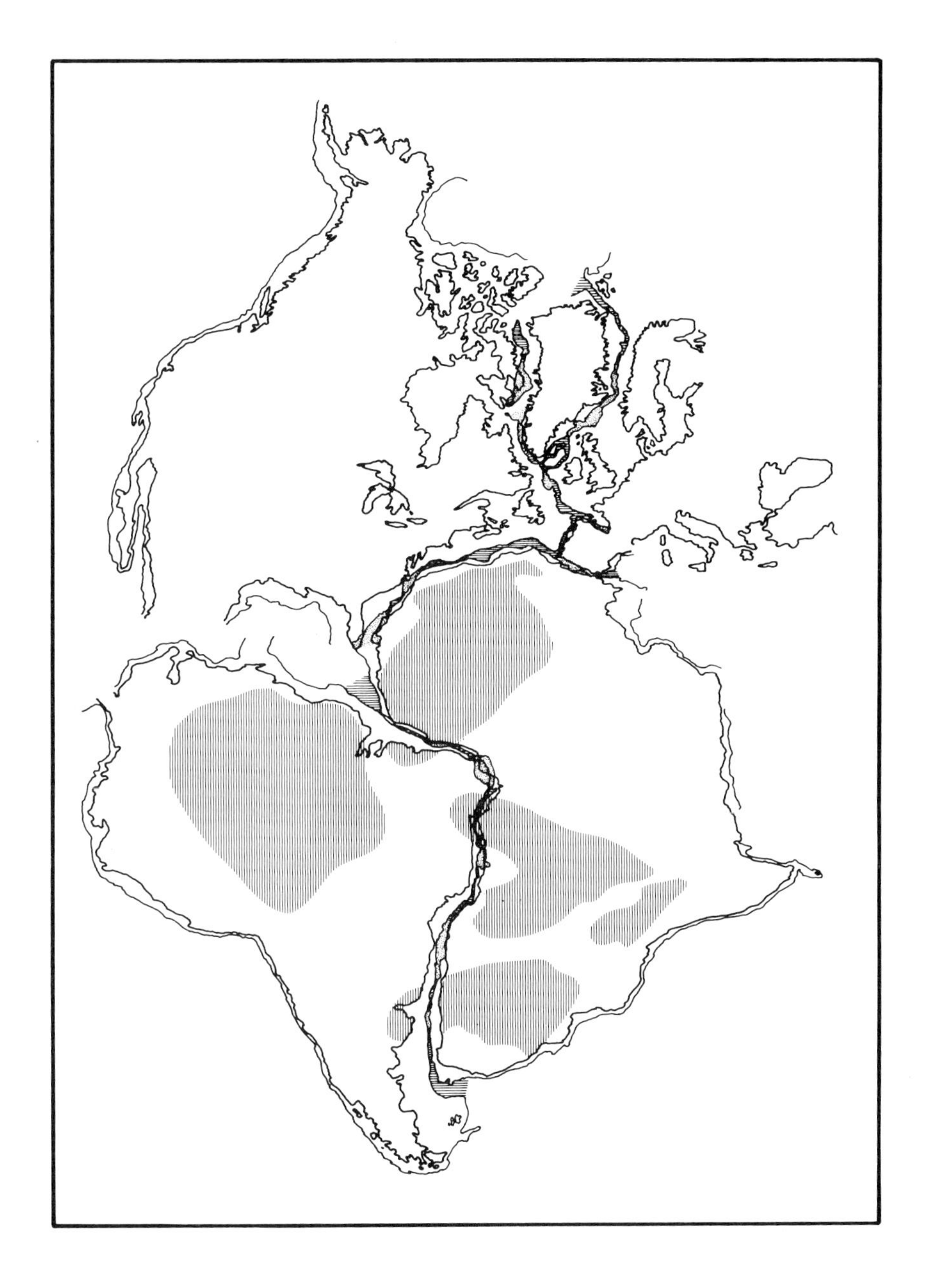

FIGURE 13-1
Best fit on the 500-fm contour of the continents bordering the Atlantic Ocean. Note the match of the geological provinces in Africa and South America. (*Redrawn after E. C. Bullard, J. E. Everett, and A. G. Smith, 1965, and P. M. Hurley, 1968.*)

13-4 EVIDENCE FROM FAULTS AND FOLDS OF GREAT HORIZONTAL DISPLACEMENTS

If continents have moved about, the process should have produced large faults and folds as evidence of horizontal displacements. Large transform faults have been discussed (Sec. 11-3).

Escher von der Linth first recognized great thrust sheets in the Swiss Alps and with A. Heim (1878) developed the concept of nappes, large overturned folds in which nearly horizontal thrusts have moved the upper parts of the folds many kilometers from their original roots. P. Termier (1903), R. Staub (1928), and others extended this concept, while L. W. Collet (1927) and E. B. Bailey (1935) translated the ideas into English. E. Argand (1924), in particular, related the idea of nappes to continental drift and interpreted other mountains of Eurasia in a similar way. R. Trümpy (1960), J. Goguel (1964), M. G. Rutten (1969), and J. Debelmas and M. Lemoine (1970) have given modern summaries.

Although these views are now widely accepted, the early announcements that these two types of features existed and were due to large horizontal movements was greeted with skepticism, and some authorities still question the evidence and oppose the underlying concept. Thus V. V. Beloussov (1962, pp. 43 and 707) admits of 15 to 20 km of displacement in the western Alps but holds that L. Kober's (1928) claim of 200 km shortening is a tenfold exaggeration resulting from "elaborate and complicated nappe structures, nine-tenths of which were the fruit of purely geometrical speculation not based on facts and completely inexplicable from the standpoint of mechanism and history of the structure. The damage done by the theory of the nappe structure of the Alps was exceedingly great."

13-5 PALEOCLIMATIC CHANGES

Today, temperature and climate vary markedly with latitude. In the past the whole earth may have sometimes been warmer or cooler. Ocean and atmospheric currents may have transported heat to modify or broaden climatic zones; barriers of land and mountains may have distorted the shape of zones; but the configuration of the solar system must always have produced a temperature gradient from the equator to the poles. Today the difference in mean annual temperature at sea level, between these extremes, is about 60°C (P. M. S. Blackett, 1961).

Fossil climates are not preserved, but many paleoclimatic indicators are (A. E. M. Nairn, 1961). These include the nature and distributions of past floras and faunas and of their living descendants. Many sedimentary rocks, including evaporites, desert sandstones, coal, bauxite, and carbonate reef deposits can form only in arid or hot conditions associated with low latitudes. Striated pavements and tillites are evidence of continental glaciation and, hence, of high latitudes. A good example of this is the evidence from rocks of Paleozoic age. Evidence suggests that during the Ordovician and Silurian periods a continental ice sheet covered much of northwestern Africa from Morocco to Lake Chad. Another related example in the lower and middle Paleozoic are reefs and evaporite deposits extending from the central United States across the Arctic to Siberia, suggesting that the equator then crossed North America from north to south (M. Schwarzbach, 1963).

Another example which has attracted much attention is the distribution of Carboniferous coal and glacial deposits. During that period coal, coral reefs, and evaporites formed in North America and Europe as far north as Spitsbergen, while at the same time continental ice sheets spread from Antarctica to the present tropics across Africa, India, Australia, and South America. From the large amount of geological data, most authorities consider that the poles have moved (T.-Y. H. Ma, 1960; N. D. Opdyke, 1962; M. Schwarzbach, 1963; R. W. Fairbridge, 1969), but some still think that they have not (F. G. Stehli et al., 1969; A. A. Meyerhoff, 1970a,b).

13-6 THE DISTRIBUTION OF FOSSIL AND LIVING FORMS OF LIFE

Some forms of life are widely spread about the earth and travel readily by swimming or walking or because they are carried by the wind, by currents, or as parasites. Others, like corals, are sensitive to the temperature of the water or the air and have a relatively limited range of distribution. Many, including earthworms, land snails, fresh-water fishes, insects, and some plants, cannot easily cross sea barriers, and in view of this the present distribution of many taxa is peculiar and suggests past movements. For example, the trees of eastern North America resemble those of Europe more than they do those of western North America, while the latter have affinities across the Pacific (R. W. Chaney, 1950). As A. W. Grabau (1936) pointed out, this presents two problems: (1) that of separation by ocean barriers of similar species and (2) that of juxtaposition on the same continent of dissimilar species perhaps separated by a mountain range.

In spite of these anomalies many leading paleontologists have rejected continental drift (R. C. Moore, 1958, 1970; B. Kummel, 1970). They attribute the separation of similar forms to migration and the juxtaposition of different faunas to changes in facies or environments of the past, or to the effectiveness of barriers such as mountains (R. Florin, 1963; D. J. Axelrod, 1964), but it is difficult to explain distributions in the southern continents without drift (P. J. Darlington, Jr., 1965). The argument that animals could reach all continents by walking, swimming, or being rafted has been weakened by the recent discovery of large fossil reptiles and amphibians in Antarctica (A. S. Romer, 1968; D. H. Elliott et al., 1970).

Unlike paleontologists, most biologists who deal with living taxa have long accepted continental drift. For example, L. Brundin (1965 and 1966), from an

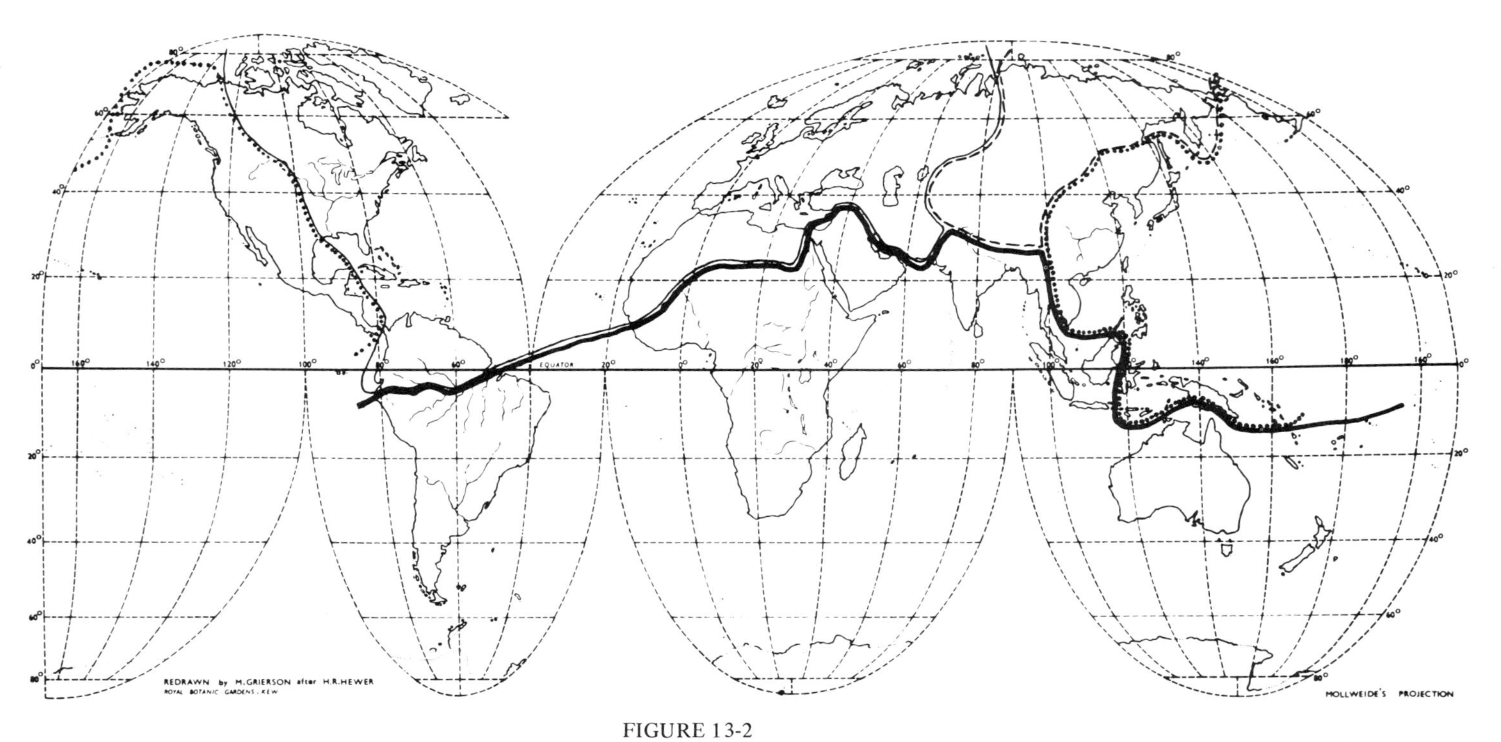

FIGURE 13-2
Paleozoic floral regions. Heavy line, the glossopterid line marking the northern boundary of the fragments of Gondwanaland; thin line, boundary of the Carboniferous coal flora; dotted line, boundary of the *Gigantopteris* flora; broken line, boundary of the *Angara* flora. (*After R. Melville, 1966.*)

exhaustive study of the relationships and probable ancestry of the insects called chironomid midges of all the southern continents, finds the conclusion inescapable that

> the transantarctic relationships developed during the period when the southern lands were directly connected. The phyletic structure and the distribution of the midge groups show that the disruption of the connections started with the separation, relative to Antarctica, of Southern Africa and, probably somewhat later, New Zealand. This separation occurred in the Upper Jurassic-Lower Cretaceous. The separation of Australia followed later. The connections between Patagonia and Antarctica were broken fairly late. New Zealand was never directly connected with Australia. (L. Brundin, 1965, p. 505.)

Many paleobotanists agree that during the later Paleozoic and early Mesozoic four great floral realms can be readily distinguished (N. W. Radforth, 1966). Authorities agree that each of these floras flourished in a different climate. Thus the *Glossopteris* flora in the southern continents is generally indicative of a cold climate, and the Euramerican flora, of a warm to tropical one. The boundaries chosen by different authorities differ only in detail and for the most part follow mountains (Fig. 13-2). What is odd about these realms is that three of them at the present time are subdivided by oceans, which supposedly should form good barriers. Some experts have attributed similarities across oceans to vanished land bridges or sunken islands (C. G. G. J. van Steenis, 1962). Others, such as E. P. Plumstead (1964), R. Melville (1966), and J. M. Schopf (1970), consider that drift provides a better explanation. It is a remarkable fact that the clear separation of many fossil plants into four floral realms is closely followed by the pattern of distribution of some present-day invertebrates, for example, earthworms (P. Omodeo, 1963).

13-7 INSTRUMENTAL MEASUREMENT OF DRIFT

Wegener noted that repeated measurements of latitude and longitude made by theodolite observations on stars from Europe and Greenland suggested that the continents were moving apart at a rate of several meters a year, but these measurements were wrong, and modern indications suggest movements of only a few centimeters a year.

Measurements from one continent to another are difficult to make with the required accuracy, but several methods now under investigation may soon be practical, including triangulation to the moon with lasers and measurements of the rate of change of interference fringes produced when pairs of radiotelescopes observe quasars (W. Markowitz and B. Guinot, 1968; M. H. Cohen et al., 1968).

Within single continents the problem is simpler. Earthquakes often produce visible displacement along faults. (For examples, see H. W. Wellman, 1953; J. Oliver

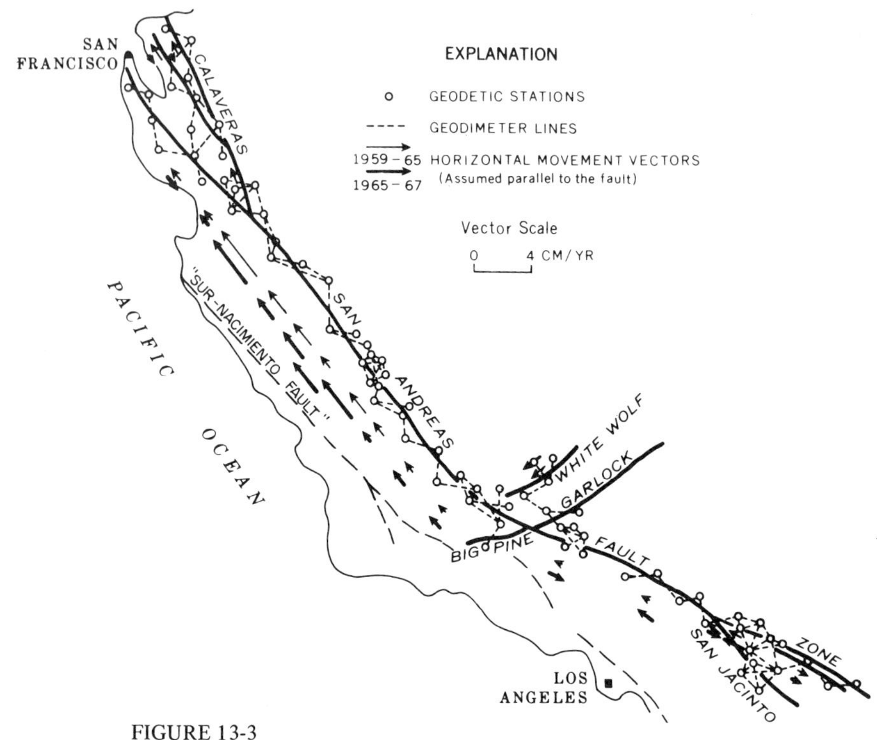

FIGURE 13-3
Average annual fault movement 1959–1965 and 1965–1967 on active faults in California, shown in solid lines. Some older faults now inactive are shown in dashed lines. (*After R. B. Hoffmann, 1968a, and B. M. Page, 1970, reprinted with additions by permission of the California Department of Water Resources.*)

et al., 1970; and N. N. Ambraseys, 1970.) Precise surveys across faults have indicated continuous movements. For example, along the San Andreas fault (Fig. 13-3) the western part of California is moving northward at a rate of as much as 4 cm/year relative to the rest of the continent (W. R. Dickinson and A. Grantz, 1968).

13-8 GEOPHYSICAL ARGUMENTS AGAINST DRIFT

Although recent discoveries have converted many to a belief in drift, it would be quite wrong to dismiss all arguments against it as inconsequential. Many were valid in the context in which they were stated, and it is important to consider them and see whether the objections they raise can be answered. H. Jeffreys (1964, 1969, 1970) has summarized most of the criticisms of a geophysical nature.

1 Jeffreys presents excellent arguments against A. Wegener's (1966) ideas, pointing out, for example, that continents cannot have plowed through the ocean floor like ships through the sea. Supporters of plate tectonics can agree with him, but unfortunately he ignores this new hypothesis, and also that of sea-floor spreading, and he makes no attempt to counter these new ideas.

2 In the same way Jeffreys shows that the Eötvös gravitational force is much too weak to move continents, but again this is beside the point. Supporters of plate tectonics would agree, for they consider that radioactive heating is a much more probable cause of motions.

3 Jeffreys points out that many arguments do not provide valid support but are merely inconclusive. Thus measurements made today which show that two continents are in relative motion cannot establish the truth of drift because the motions may be transient or oscillatory and drift requires that they continue in the same direction for a very long time. To demonstrate the likelihood of this requires other kinds of evidence.

4 It is widely held that the fit of continents on either side of the South Atlantic proves drift. Jeffreys claims that the fit is not very good and in any case may be due to other causes, but E. C. Bullard et al. (1965) have shown that Jeffreys has neglected the broad coastal plains off Patagonia (Fig. 13-1) and that the fit of the continental slopes is good.

5 An argument raised against using paleomagnetic data to support drift is that the earth's magnetic field may not always have been essentially dipolar (J. W. Northrop III and A. A. Meyerhoff, 1963), but N. D. Opdyke and K. W. Henry (1969) concluded from a worldwide study that during the Pleistocene the average field has been dipolar, while A. N. Khramov et al. (1966) reached the same conclusion from studies of Paleozoic rocks in Siberia. The argument that hammering and shocks have invalidated magnetic results does not seem valid.

6 Jeffreys points out that no satisfactory mechanism has been proposed and stated that "a law of imperfection of elasticity at small stress forbids convection and continental drift" (H. Jeffreys, 1964, 1969), but many other physicists do not agree with him (J. Weertman, 1962; W. M. Elsasser, 1966; D. C. Tozer, 1967; L. Knopoff, 1969).

Although Jeffreys has been a strong opponent of drift, it is only fair to point out that he has considered possible causes and given a good description of how subcrustal currents might move the surface (H. Jeffreys, 1970).

13-9 GEOLOGISTS' ARGUMENTS AGAINST DRIFT

It was inevitable that when so fundamental an idea as continental drift and one so contrary to what geologists had long held was proposed, it would encounter strong opposition. This was voiced by many at a symposium organized by W. A. J. M. van Waterschoot van der Gracht (1928), and others have continued to point out difficulties.

1. F. G. Stehli (1968) has examined several classes of fossils, particularly from the Permian period, and finds that their distribution fits the present climatic pattern suggesting no continental movements; but these fossils are found only in comparatively few places, and other geologists believe that these incomplete data can be fitted by other pole positions (J. B. Waterhouse and S. Piyasin, 1970).
2. V. V. Beloussov (1967, 1968) asks how any process of sea-floor spreading can create oceanic crust of uniform thickness; but it is at least as difficult to understand why the process which he advocates of "oceanization" by heating and intrusion from below should do this, and he ignores H. H. Hess's (1962) proposals and subsequent developments.
3. Beloussov asks why enclosed seas like the Sea of Okhotsk and the Bering Sea have oceanic crust when they contain no conspicuous rifts. This was a good point until some rifts were found (Sec. 17-6).
4. Beloussov asks how continents can move and yet leave some continental margins undisturbed, but this only applies to Wegener's ideas and not to plate tectonics (Sec. 14-10).
5. Beloussov objects to the concept of having two or more rifts side by side without an intervening subduction zone (e.g., the South Atlantic, East African, and Indian Ocean rifts), but this is possible if the rifts are moving apart at the same rate as they spread. Merely to state that such movement is contrary to an "elementary concept of hydrodynamics" is not to disprove it.
6. Beloussov also assumes that tectonic cycles are regular and demands that drift explain this and states that drift only began in the Mesozoic era; but these concepts are by no means proven or agreed upon.
7. Beloussov points out that it is puzzling that the heat flow is the same through ocean floors as through continents although the latter are much richer at the surface in radioactive elements. This observation is surprising, but R. D. Schuiling (1966) has pointed out that motion of continents may be able to explain this more easily than static conditions. A simple analogy is that, if a heated pot has been placed for a time on a cold table and then moved, it will have imparted heat to the table top. Schuiling argues that it may have been the movement of the continents which has heated the ocean floors over which they have moved. Other answers are possible (J. G. Sclater and J. Francheteau, 1970).

8 In a later paper Beloussov (1970) raises more objections and, to a greater extent than other detractors, offers some alternative explanations; but to deny that the pattern of magnetic anomalies is symmetrical is not justified (Fig. 16-2), and his explanation in terms of a succession of lava flows of great regularity is contrary to all observations on the behavior of eruptions.

9 In several cases Beloussov uses old or erroneous data which have been corrected since he wrote. Thus the age of the Cobb seamount which lies near the crest of a mid-ocean ridge has been reduced from 29 to less than 3.5 m yr (J. R. Dymond et al., 1968). J. Ewing and M. Ewing, (1967) have changed their views about plate movements, and the fragmentary observations of A. E. Godby et al. (1968) have been replaced by a complete survey that supports drift (P. R. Vogt et al., 1969a, fig. 14). Beloussov quotes the views of B. C. Heezen et al. (1959) about the shape and slope of mid-ocean ridges, but H. W. Menard (1969a) has different ideas. He holds that a strong magnetic anomaly at the crest and greenschists on the floor make drift impossible, which ignores the explanations given by F. J. Vine (1966), J. R. Cann (1970), and E. Irving et al. (1970). He quotes early objections to the proposed mid-ocean ridges off British Columbia, and the triple-point junction in the Gulf of Alaska, but does not note that F. J. Vine and H. H. Hess (1970) have answered them. He also states that drift cannot explain the nature of sediments in trenches, but does not appreciate that answers have been given by changed views about plate tectonics (Sec. 17-6). He quotes isolated early reports of conflict between the ages of some abyssal sediments and sea-floor spreading, which may have been due to misunderstandings, because far more abundant later work shows no such conflicts (e.g., A. E. Maxwell et al., 1970a,b).

10 Beloussov's reference to volcanic islands in the central part of the Pacific makes sense only if the Eocene age found for Midway Island is also attached to Hawaii, for which there is no evidence (Sec. 16-4).

11 The opponents of drift have shown that many of the arguments about drift are two-sided and that the causes and some of the effects of motion are not well understood. Where they fail is that they have produced no useful alternative. On the other hand, the supporters of plate tectonics have advanced a powerful and integrated hypothesis which is constantly being strengthened as it is used to make predictions and as these are verified.

14

NEW EVIDENCE FOR CONTINENTAL DRIFT FROM THE HYPOTHESES OF SEA-FLOOR SPREADING AND GLOBAL PLATE TECTONICS

14-1 INTRODUCTION

For thirty-five years after the great debate in 1928 and the death of Alfred Wegener in 1930, orthodox geologists and geophysicists continued to regard continents as fixed and ocean basins as permanent. Only a few geologists, most of them working in South Africa, South America, and the Alps, upheld the reality of continental drift. It was not until new techniques, introduced after World War II, had produced new kinds of evidence that a reconsideration began.

Alfred Wegener (1966) believed that the continents moved through a yielding ocean floor like ships plowing through the sea. This chapter describes discoveries which led to abandoning that idea in favor of two new concepts. The first is the hypothesis of *sea-floor spreading*, which is that ocean basins grow by the generation of new oceanic crust and lithosphere along the crest of mid-ocean ridges. The second is the hypothesis of *global plate tectonics*, which extends the first to include the idea that the lithosphere is divided into plates which are rigid and not appreciably deformed except along their common boundaries and which, as well as spreading apart, also slide past one another and are reabsorbed back into the mantle. Section 17-5 considers a possible mechanism which might cause these motions.

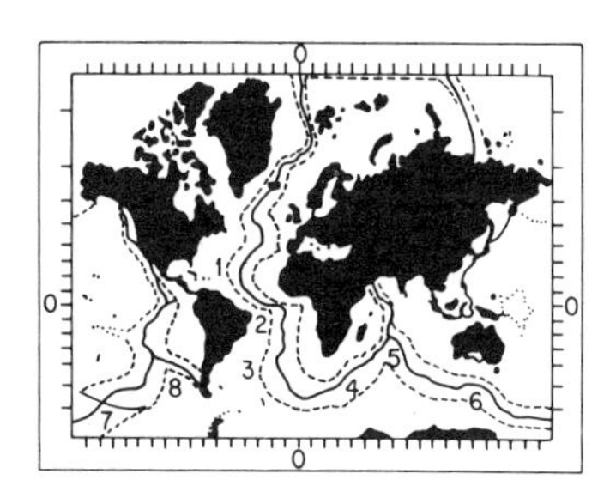

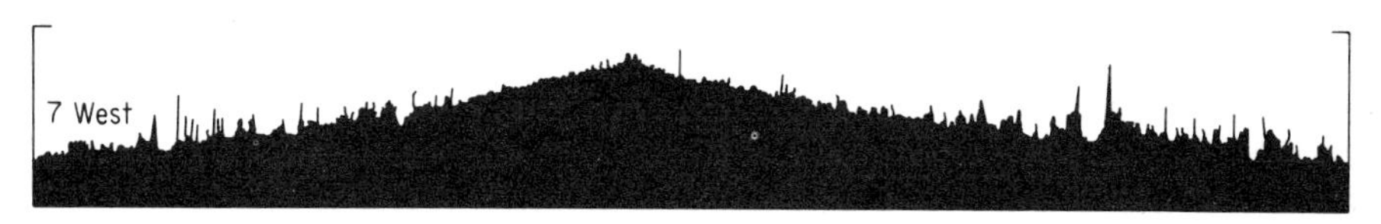

FIGURE 14-1
Profiles across the mid-ocean ridge system showing the rift and rougher topography characteristic of the Mid-Atlantic Ridge and the lack of these features on the East Pacific rise. (*After B. C. Heezen, 1962.*)

14-2 THE INVESTIGATION OF THE OCEAN FLOOR AND BASINS

Foremost among the new kinds of evidence are those arising from the exploration of the ocean floors, which culminated in the demonstration by M. Ewing and B. C. Heezen (1956, 1960) that the world's greatest system of mountains lies hidden beneath the sea (Fig. 14-1). It is remarkable that these investigations began so late and that both geologists and geophysicists had for so long largely ignored the sea floor.

Soundings in coastal regions were made for practical reasons, but until 1840 when Captain John Ross made the first deep sounding, no one knew the depth of the oceans (R. S. Dietz and H. J. Knebel, 1968). By 1855 Commodore M. F. Maury had discovered a broad rise in the Central Atlantic. Occasional deep-sea scientific expeditions beginning with that of the *Challenger* from 1872 to 1875 slowly added more information (G. Wüst, 1964). About 1920 the echo sounder was invented, but its use was laborious and ocean charts remained poor until the introduction of continuous recording during World War II, the development of the precision depth recorder (B. Luskin et al., 1954), and the introduction of satellite navigation (G. L. Johnson and M. K. Jugel, 1968). M. N. Hill (1963), M. Talwani (1964), T. Ichiye (1968), M. J. Keen (1968), A. G. Mourad (1970), and A. E. Maxwell et al. (1970c) have reviewed the development of these and other oceanographic techniques.

The *Challenger* expedition developed dredges to scrape rock and mud from the sea floor, and in 1947 B. Kullenberg invented a piston corer which could collect relatively undisturbed cores up to 25 m long of bottom oozes, but seismic investigations had shown that the oceanic crust was several kilometers thick.

A proposal to drill through it and obtain cores of the crust and underlying mantle was made (the Mohole Project), and in 1961 a preliminary trial was carried out off Guadalupe Island, south of Los Angeles. This demonstrated the feasibility of penetrating the sedimentary cover without changing drilling bits, but when estimates of the cost of penetrating the whole crust proved too expensive, the project was abandoned (P. H. Abelson, 1966) and replaced by the Deep Sea Drilling Project, which continued the type of shallower drilling that had proved successful in the Mohole trial. This is being executed by JOIDES (Joint Oceanographic Institutions for Deep Earth Sampling), and this project has recovered several dozen cores from the deep ocean floor up to 100 m long (M. Ewing et al., 1969a; M. N. A. Peterson et al., 1970; A. E. Maxwell et al., 1970b; R. G. Bader et al., 1970; D. A. McManus et al., 1970). This project has been highly successful and is being developed to drill deeper holes. Sections 3-6.4 and 15-10 mention some results of coring. A drill to recover rock cores has been built (C. Schafer and J. Brooke, 1970), and in 1940 M. Ewing and his colleagues developed the first underwater cameras, now extensively used (J. B. Hersey, 1967).

The first geophysical observations were charts of magnetic declination. E. Halley (1655-1742) was one of the early pioneers in this field, and such charts have long been prepared to assist navigation. During World War II continuously recording magnetometers were developed to tow behind ships or aircraft. They depend on proton magnetic resonance and are sensitive and self-orienting (E. T. Miller and M. Ewing, 1956). Their use greatly improved the charts. P. H. Serson et al. (1957) developed a three-component flux-gate magnetometer for aircraft.

In 1923 F. A. Vening Meinesz (1929) began precise measurements of gravity at sea (Sec. 4-3) and discovered large negative anomalies over deep trenches, although the ocean basins as a whole are in general isostatic balance with the continents.

Seismic waves penetrate the earth and reveal subsurface structure. M. Ewing adapted the methods developed to prospect for petroleum-bearing structures on the continents to explore the sea floor (M. Ewing et al., 1937; J. Ewing and M. Ewing, 1959). Waves generated by small explosions and refracted in the sea floor revealed three layers above the mantle, one of which G. L. Maynard (1970) has recently shown to be double. Later J. B. Hersey and M. Ewing (1949) began to use reflected waves and J. I. Ewing and G. B. Tirey (1961) and J. Ewing and M. Ewing (1967)

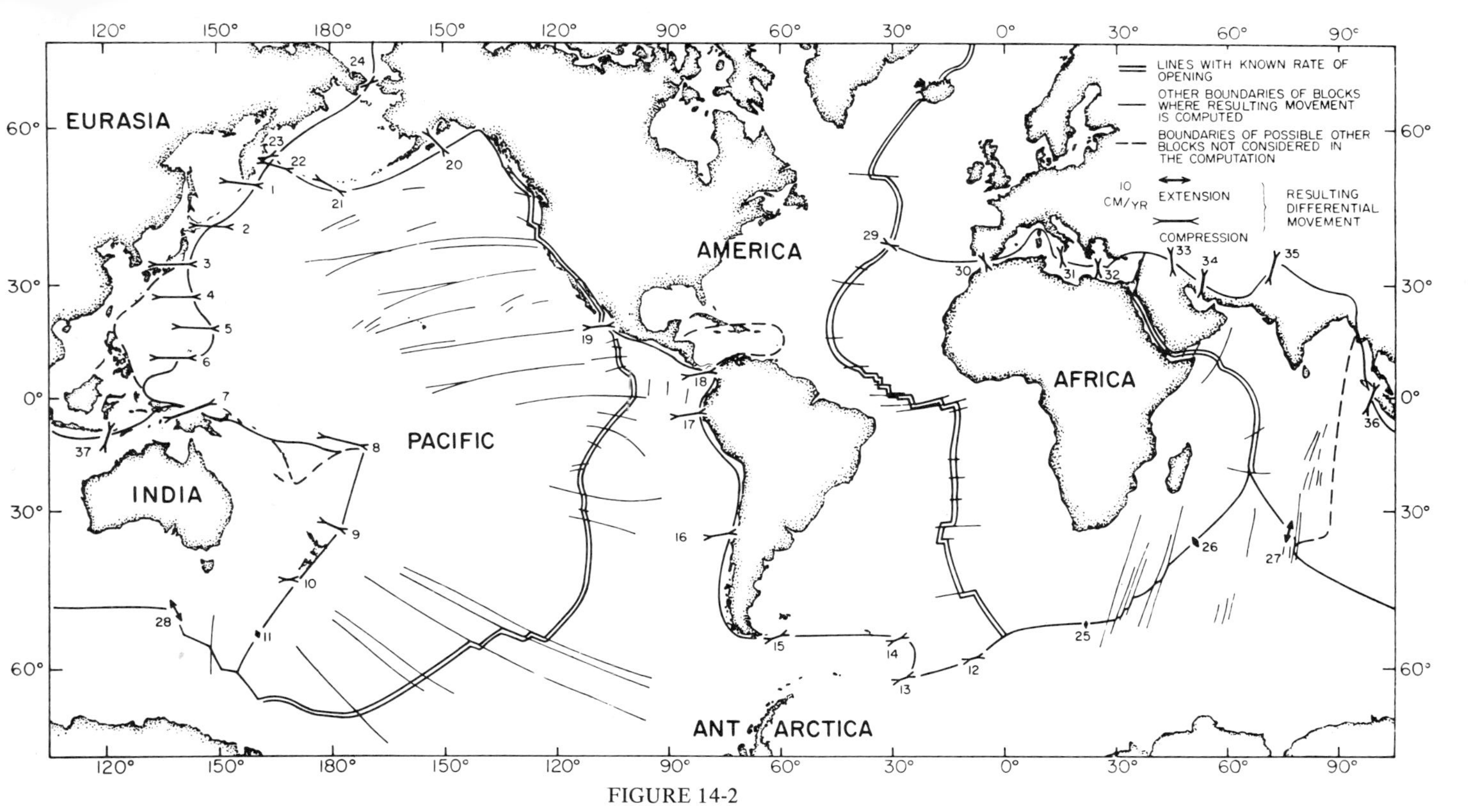

FIGURE 14-2
The six principal lithospheric plates (named) showing junctions along which spreading rates have been measured (double lines) and others along which rates have been computed (single lines with double-headed arrows, of which those in the legend represent rates of 10 cm/year, with others scaled proportionately). (*After X. Le Pichon, 1968.*)

invented today's continuous seismic profiling techniques which have revealed details about the layering of sediments on the bottom (Sec. 3-6.6).

Love and Rayleigh showed that earthquakes can generate the two types of surface waves which bear their names. In 1939 K. E. Bullen developed a theory of surface waves which J. Tinsley Wilson (1940) applied to show that the oceanic crust is everywhere much the same and thinner than continental crust. F. Press and M. Ewing (1955) and M. Ewing et al. (1957) have described further developments.

E. C. Bullard (1939) and A. E. Benfield (1939) made the first measurements of heat flow on land, and E. C. Bullard (1954a) with R. Revelle and A. E. Maxwell (1952) also developed the first thermal probe for use at sea. This was improved by R. Gerard et al. (1962). The results have been given in Sec. 7-2.3.

14-3 FRACTURE ZONES ON THE SEA FLOOR

Naval operations during World War II greatly increased our knowledge of the Pacific basin (H. H. Hess, 1946, 1948; R. S. Dietz, 1954). This interest has continued, and since 1950 H. W. Menard and his colleagues have discovered a series of conspicuous linear features which extend in parallel lines for thousands of kilometers across the floor of the Northeast Pacific Ocean (Fig. 14-2; and H. W. Menard and R. S. Dietz, 1951; H. W. Menard, 1967). Each is expressed by scarps, troughs, and ridges as much as 3 km high extending over a zone up to 200 km wide. They frequently separate provinces in the ocean floor which differ in topography and depth, the latter suggesting vertical displacements. Subsequent investigations have discovered fracture zones in every ocean and shown that they form groups. All the members of any group are circular arcs having a large radius about a distant common center (W. J. Morgan, 1968), suggesting horizontal displacements. That fracture zones offset patterns of magnetic anomalies was discovered by V. Vacquier (1959, 1962), who found an offset of 1150 km on the Mendocino fracture zone (Fig. 14-3 and Sec. 16-2). Many earthquakes occur on fracture zones, but only along the short sections where they offset active mid-ocean ridges (Fig. 14-4). It appears that fracture zones are all large transform faults, chiefly of the ridge-ridge type (Sec. 11-3). Those in the eastern Pacific cannot be traced with certainty beyond the flanks of the East Pacific rise, where they terminate either by dying out or by reaching a boundary between parts of the Pacific floor of different ages (Fig. 14-8; H. W. Menard, 1964, 1967; J. T. Wilson, 1965a). To the east they meet the continent. There the magnetic anomalies lose their character, and whether the fracture zones penetrate the continent or not has been a matter of controversy (W. N. Gilliland, 1962).

In the equatorial Atlantic B. C. Heezen et al. (1964) mapped another set of parallel fracture zones which cross the Mid-Atlantic Ridge at right angles and extend toward the coasts of Africa and South America (Fig. 14-4). Again the fracture zones are confined to the ocean basins, although they displace the ridge by

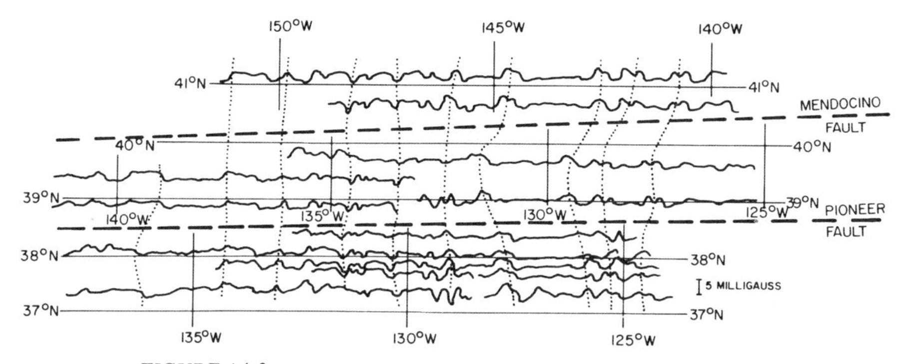

FIGURE 14-3
Showing that the total magnetic intensity profiles measured off California can be matched by a left-lateral displacement of 265 km along the Pioneer fault and 1150 km along the Mendocino fault, as also shown in Fig. 14-14. (*After V. Vacquier, 1962.*)

hundreds of kilometers, and earthquakes along them are confined to the offsets (D. C. Krause, 1964; K. Burke, 1969; J. P. Fail et al., 1970; Sec. 15-11).

The discovery of these large submarine faults with great offsets was surprising, and suggested large horizontal displacements in the earth's crust.

Observation shows that fracture zones are consistently at right angles to the directions of the mid-ocean ridges and magnetic anomalies which they cross. This perpendicular configuration is not required by the geometry of moving blocks, but does appear to be quickly established, as seems to be happening in the Gulf of Aden (A. S. Laughton, 1966).

It follows that a change in direction of spreading rapidly produces a change in the direction of both fracture zones and mid-ocean ridges, which is recorded in the topography and pattern of magnetic anomalies (H. W. Menard and T. M. Atwater, 1968, 1969; H. W. Menard, 1969b; P. R. Vogt et al., 1970a; Fig. 14-5).

14-4 THE MID–OCEAN RIDGE SYSTEM

After M. F. Maury (1855) and the *Challenger* expedition had vaguely identified a mid-ocean ridge in the North Atlantic, no others were found until 1930, when the German *Meteor* expedition charted another rise in the South Atlantic, and Danish and British expeditions found the Carlsberg ridge in the Indian Ocean. World War II stimulated investigation of the ocean basins, which led B. C. Heezen, M. Tharp, and M. Ewing (1959) to map the Mid-Atlantic Ridge in more detail, while others recognized the existence of a broad rise in the East Pacific (H. W. Menard, 1960).

Meanwhile, B. Gutenberg and C. F. Richter (1954) noted that earthquakes in

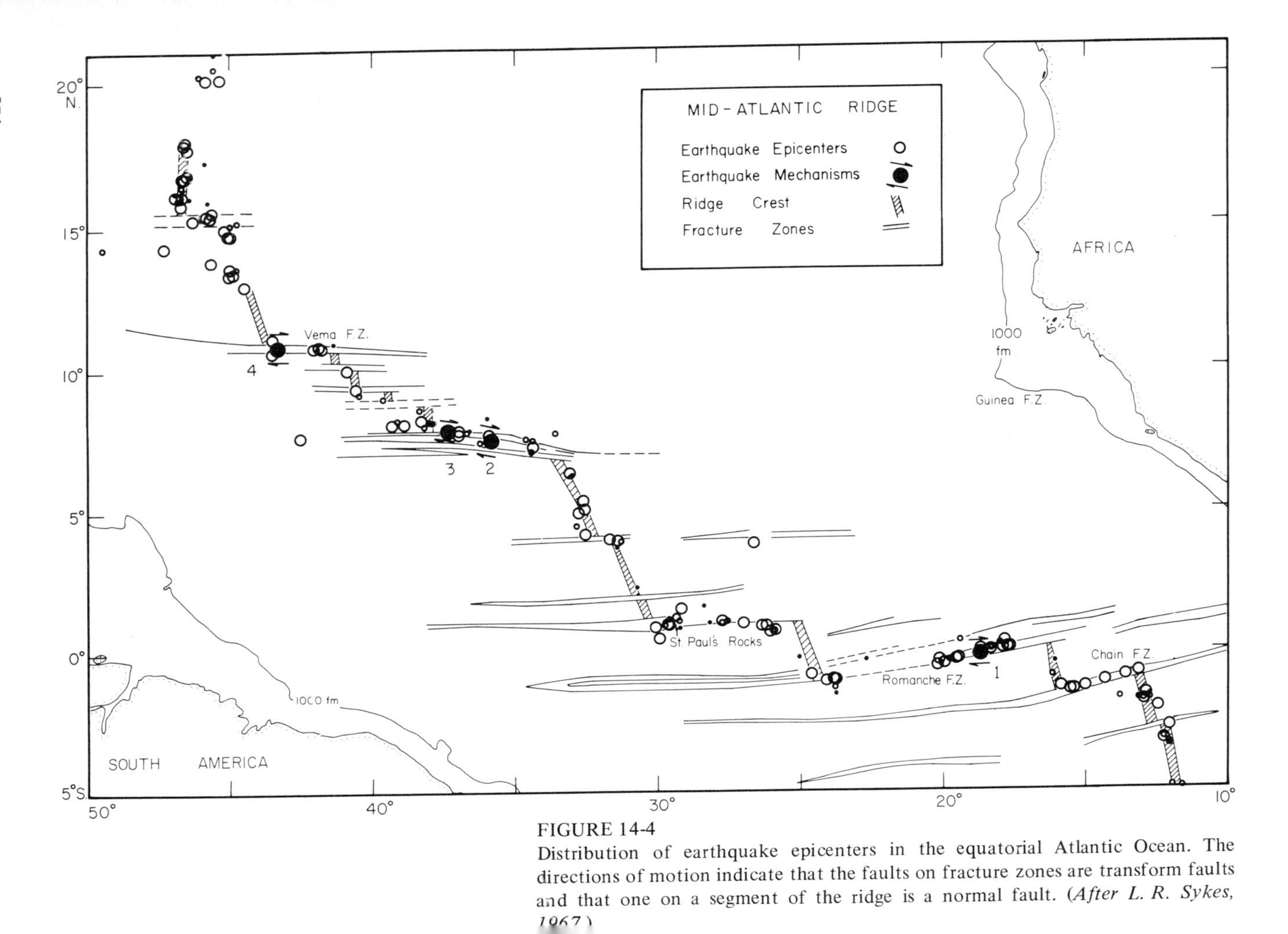

FIGURE 14-4
Distribution of earthquake epicenters in the equatorial Atlantic Ocean. The directions of motion indicate that the faults on fracture zones are transform faults and that one on a segment of the ridge is a normal fault. (*After L. R. Sykes, 1967.*)

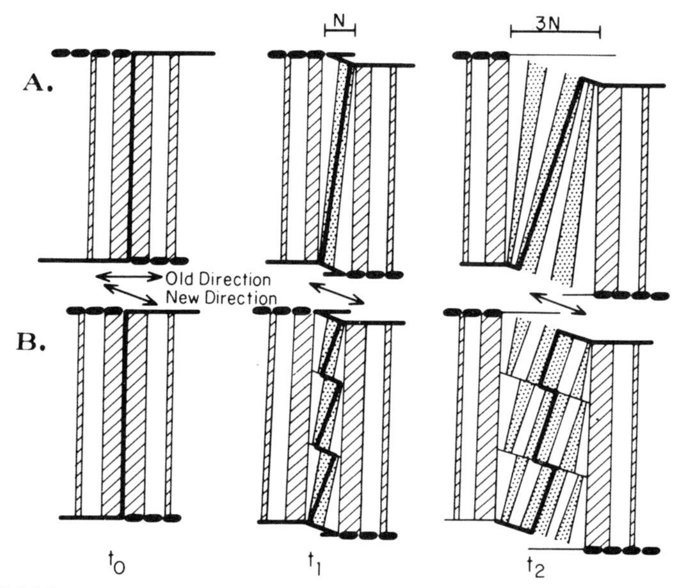

FIGURE 14-5
Two possible modes for adjustment of a ridge to a change in spreading direction. (*A*) The ridge readjusts as a unit between existing fracture zones, taking on a new direction by time t_2, after spreading through a distance indicated by $3N$ has occurred. (*B*) The ridge breaks into segments by the formation of new fracture zones, each piece becoming realigned by time t_1, after the ridge has spread through a distance of only N. (*After H. W. Menard and T. M. Atwater, 1968.*)

the ocean basins are concentrated along narrow belts which are continuous about the earth and follow the mid-ocean ridges. In 1956 M. Ewing and B. C. Heezen proposed that the ridges were continuous and form the largest mountain system on earth. The failure to have recognized this sooner is remarkable and indicates the ineffectiveness of early investigations and the barrenness of older tectonic theories. Fortunately, the hypothesis of sea-floor spreading from an active mid-ocean ridge system is providing a satisfactory explanation for many observations and has been widely accepted. Figure 14-1 illustrates two characteristic profiles of the ridge system. All subsequent work has shown that earthquakes in the centers of the ocean basins are shallow and are concentrated very precisely beneath the median rifts and offset portions of the fracture zones which cut across the mid-ocean ridges (M. Barazangi and J. Dorman, 1969; Fig. 14-4).

The highest and most rugged topography is found along the crests which rise to an average depth of about 3000 m. The East Pacific rise is smooth, but the Mid-Atlantic Ridge is more rugged, with a median rift. This rift is about 30 km wide and lies between 1000 to 3500 m below the rift mountains which bound it. Particularly along the Mid-Atlantic Ridge, faults parallel with the crest have broken

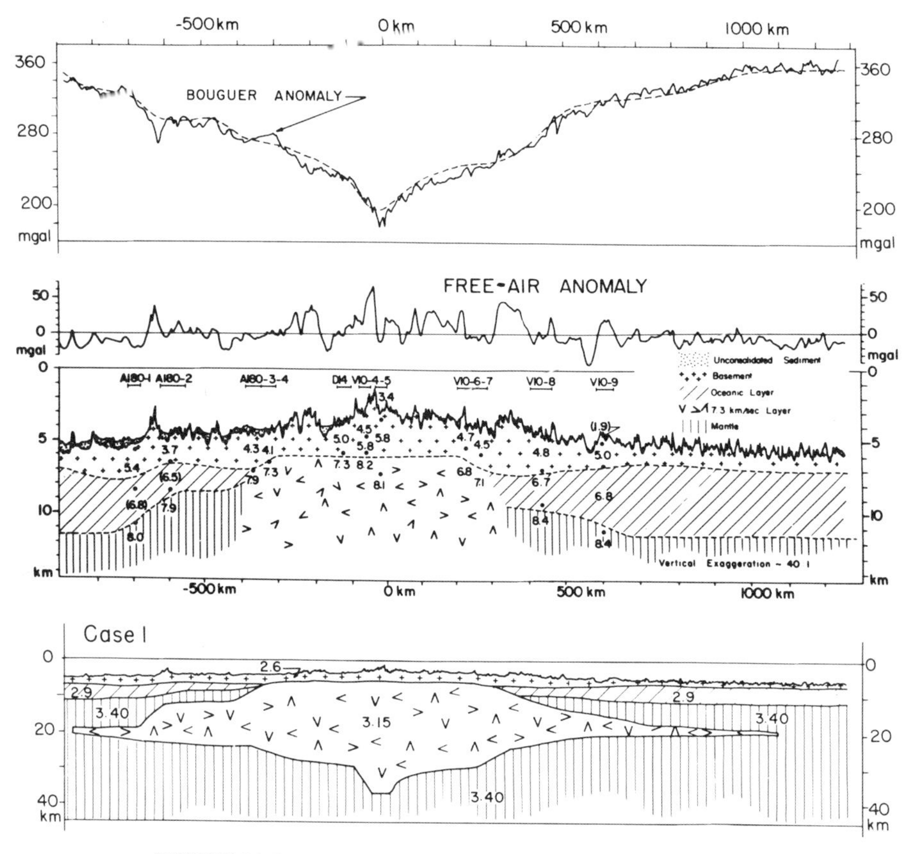

FIGURE 14-6
Gravity anomalies (observed in full line and calculated in dashed line) and seismically determined structure across the Mid-Atlantic Ridge and one possible model of shallow depth which is in accord with the data. (*After M. Talwani et al., 1965.*)

the oceanic crust into blocks which are tilted so that the scarps face toward the crest and the original surface of the ocean floor dips outward.

H. W. Menard (1969a) suggests that a moving oceanic plate is elevated where it is created over a mid-ocean ridge and subsides as it moves away from the crest and becomes older. He finds rates of subsidence in the Pacific basin of 9 cm per 10^3 years for the first 10 m yr, 3.3 cm per 10^3 years for the next 30 m yr, and about 2 cm per 10^3 years for the next 30 m yr, so that after 70 m yr the old surface of the ocean floor is mantled with sediment and lies at a depth of 2500 m.

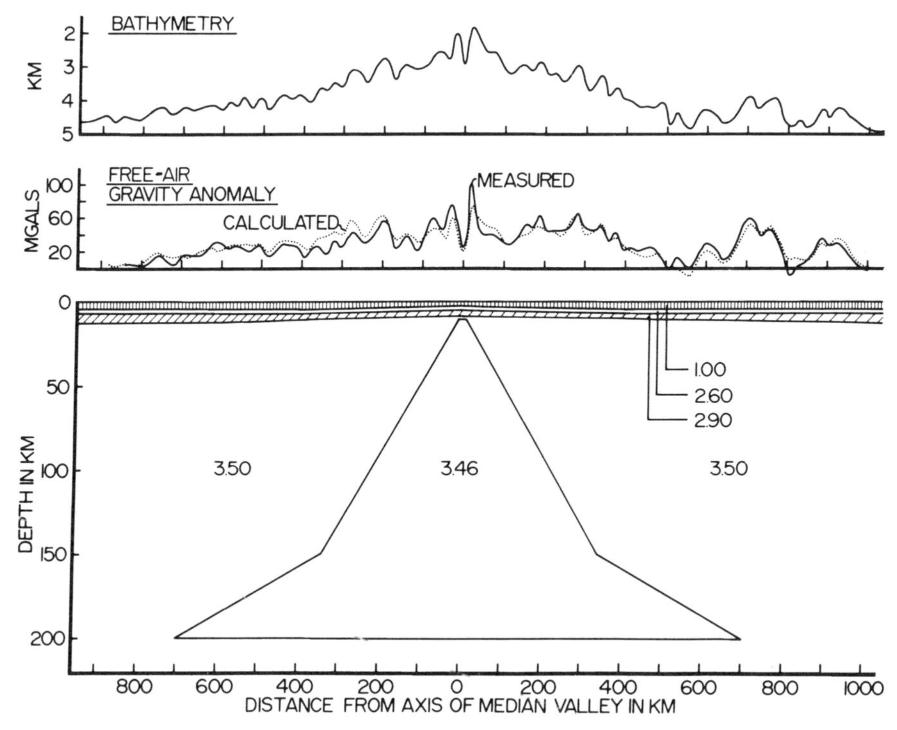

FIGURE 14-7
Bathymetry and computed and observed gravity anomalies across the Mid-Atlantic Ridge with one possible model which is in accord with these data and seismic observations. The anomalous mantle is assumed to extend to a depth of 200 km into the asthenosphere. The numbers are assumed densities. (*After C. E. Keen and C. Tramontini, 1970.*)

In contrast to the variability of the continental crust, the oceanic crust is surprisingly uniform, as M. Ewing and M. N. Hill found for much of the Atlantic basin in the early 1950's and R. W. Raitt found in the Pacific. Also, the crust of the East Pacific rise has much the same structure as the rest of the Pacific. For some time the crestal region of the Mid-Atlantic Ridge was considered to form a puzzling exception, with a different, if uncertain, layering of the lower crust and upper mantle (M. Talwani et al., 1965; Fig. 14-6). However, this interpretation was based upon sparse and scattered observations, and a recent intensive survey of a smaller area suggests that the shallow structure in the Mid-Atlantic is the same as everywhere else and that spreading from the ridge could generate the ocean floor (C. E. Keen and C. Tramontini, 1970; Fig. 14-7).

14-5 THE AGES OF ISLANDS AND OF CORES

Besides hundreds of small islands, the ocean basins contain thousands of submerged peaks, which by definition are called *seamounts* if they rise more than 500 fathoms (about 1000 m) from the sea floor. Both islands and seamounts are most numerous in the Pacific and scarcest in the Atlantic. It is convenient to distinguish between those islands which rise from continental shelves, those which form *island arcs* around the borders of the oceans, and those *oceanic islands* which rise in isolation from the deep-sea floor.

With few exceptions all oceanic islands in the deep basins appear to have originated as basaltic volcanoes. In the tropics young and active volcanic islands are customarily ringed with *fringing reefs* of coralline and algal limestone. In 1842 Darwin (1962) proposed that volcanic islands sink and that, as they do, the reef continues to grow upward. At the same time erosion and the conical shape of the central island combine to reduce its size so that the distance to the reef increases and it becomes a *barrier reef*, and finally, when the volcano is completely submerged, an atoll (Fig. 3-14). An *atoll* is thus a ring of coralline islands and shallow reefs enclosing a *lagoon* of calm and shallow water.

H. H. Hess (1946) discovered a few flat-topped seamounts, or *tablemounts*, which he called *guyots* after Professor Arnold Guyot (1807-1884), scattered in the Western Pacific. He suggested that the seamounts were islands which had been truncated when the sea level was lower during the Precambrian.

The dredging of rounded cobbles and shallow-water fossils of middle Cretaceous age from the top of guyots confirmed the drowned-island hypothesis, but showed the truncation to be young (E. L. Hamilton, 1956).

H. H. Hess (1962) and H. W. Menard (1964) showed that the depths of guyots can be contoured. They claimed that this pattern reveals a former mid-ocean ridge which has sunk, and they named it the Darwin rise (Fig. 14-8).

Most of the active volcanoes in the ocean basins are either close to the crest of the mid-ocean ridges, as on Iceland and the Azores, or close to the borders with continents, as on the Canary and Cape Verde Islands and many island arcs. It is also true that although young rocks are found on all oceanic islands, no old rocks are found on islands near the crest of mid-ocean ridges, and that the ages of the oldest rocks found tend to increase with distance from the crest (J. T. Wilson, 1963a, 1965a).

The age of the oldest rock yet found on any volcanic oceanic island is basal Cretaceous or perhaps Upper Jurassic on Maio Island in the Cape Verde Islands (J. M. Pires Soares, 1949; A. Serralheiro, 1968). The same pattern is better seen in the results from coring and drilling on the sea floor (T. Saito et al., 1966a), B. M. Funnell and A. G. Smith (1968), A. E. Maxwell et al. (1970b), and B. M. Funnell and W. R. Riedel (1971).

Arguments still arise about whether the ages of islands and shallow cores are representative, and whether greater ages are present and buried (M. L. Keith, 1970),

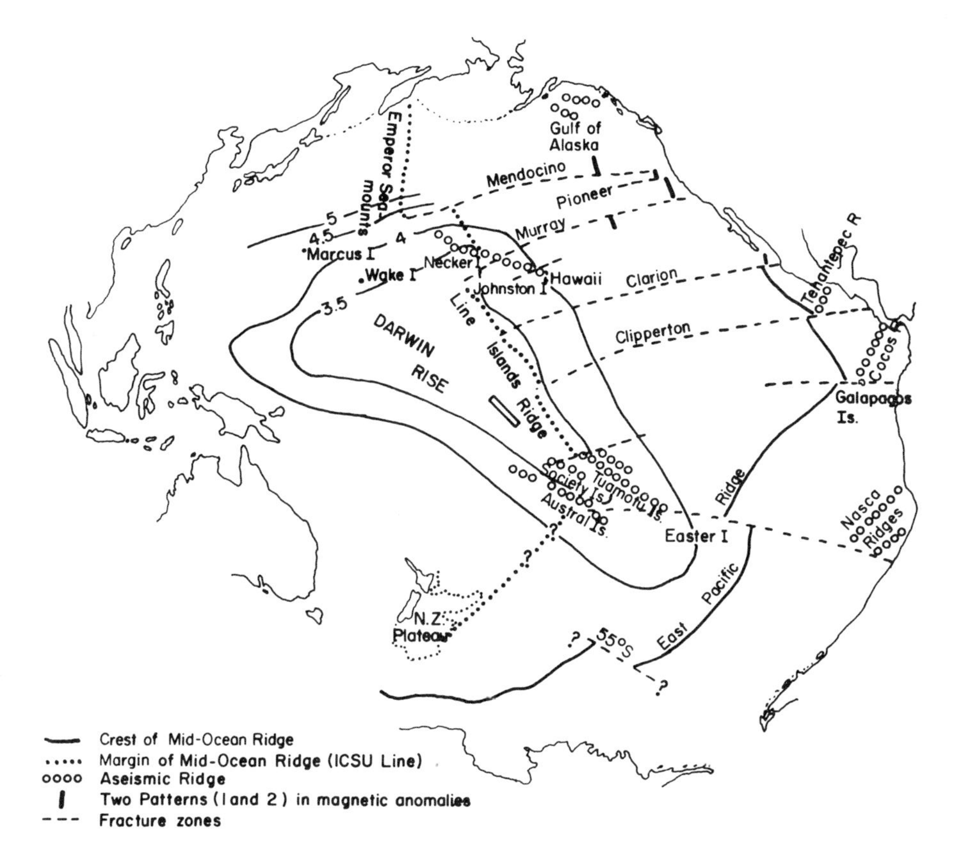

FIGURE 14-8
Sketch map of the Pacific Ocean showing the crest of the East Pacific rise, the Darwin rise with paleobathymetry contours, and by a dotted line the ICSU ridge, which may form the western margin of the East Pacific rise. Note the symmetrical arrangement about the East Pacific rise of Nasca and Tuamotu ridges and their adjacent ridges. (*After H. W. Menard, 1964, and J. T. Wilson, 1965a.*)

but this has been largely answered by the JOIDES program of drilling in the oceans. Many of these deeper drill holes have penetrated through sediments to basalt, and it has been shown that the age of the basal sediments resting on the basalts increases with distance from the mid-ocean ridges (M. Ewing et al., 1969a; M. N. A. Peterson et al., 1970). The chief exception is that some marginal seas between island arcs and continents, for example the Philippine Sea, appear to be young (Sec. 17-10; A. G. Fischer et al., 1969). Nevertheless, the general increase in maximum ages of islands and cores with distance from the crest of mid-ocean ridges is an argument in favor of sea-floor spreading.

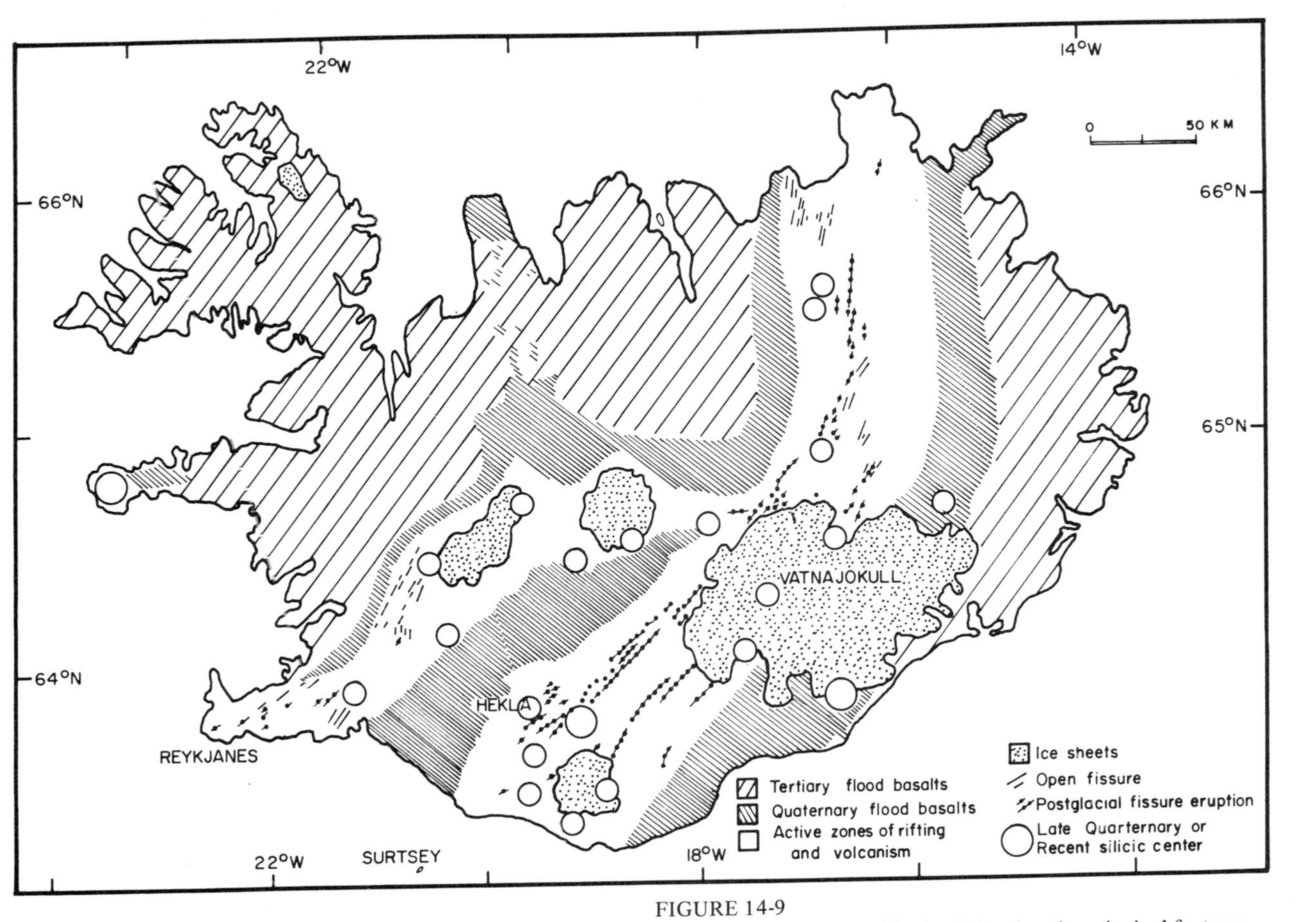

FIGURE 14-9
Geological sketch map of Iceland showing the principal features.

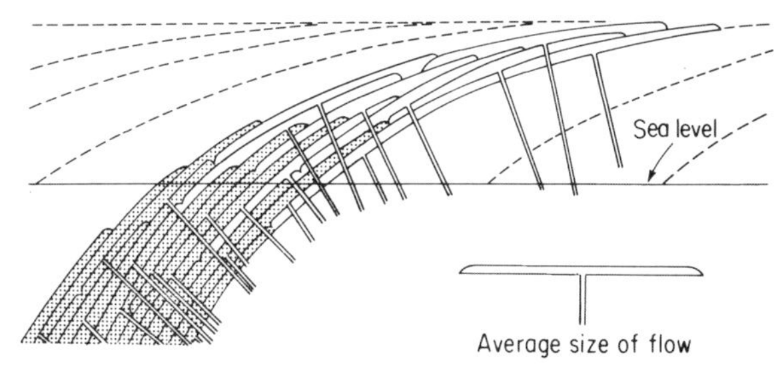

FIGURE 14-10
Schematic section showing thinning up dip of a group of basalt lava flows as found in eastern Iceland. Flows fed from dikes concealed below sea level are stippled. (*After G. P. L. Walker, 1965.*)

14-6 THE SPREADING OF ICELAND

Iceland is the largest oceanic island and the only one lying athwart a mid-ocean ridge (S. Bjornsson, 1967; S. Thorarinsson, 1966). In 1927 K. Sapper estimated that one-quarter of all lava and ash which had poured out on land during historical time was in Iceland (S. Thorarinsson, 1967). In 1783-1784 flood basalts poured from the Laki fissure to cover 565 km^2.

The predominant rocks are an immense number of gently dipping basalt flows fed by dikes. Protruding at intervals through the flows are a few score central volcanoes which have erupted lavas of more siliceous and alkalic composition (Fig. 14-9).

The structure of Iceland is synclinal. Older rocks in the east and west dip under a median zone in which lie younger rocks and most of the active volcanoes.

Two views have been held concerning the origin of the island. The classical view is that Iceland forms part of the Brito-Arctic, or Thulean, basalt province of Eocene age which extends from Scotland and Ireland through Iceland to Greenland and even Baffin Island (S. Thorarinsson, 1960) and that the dip of the lavas is due to faulting and tilting. A strong argument against this is that the oldest well-dated fossils found in Iceland are Pliocene, and the oldest age determinations are not Eocene but Miocene (S. Moorbath and G. P. L. Walker, 1965).

As a result of ten years of detailed field mapping in eastern Iceland, G. P. L. Walker (1960, 1965) has produced the alternative view that the dip of the flows toward the central zone is original. He showed that up dip the number of flows falls off rapidly, so that mappable horizons trend toward an asymptotic level only a short distance above the present summits (Fig. 14-10).

He then pointed out that the flows were fed by dikes and that their intrusion must have caused a dilation of the island and its spreading away from the median zone. In one section of coastline 55 km in length, he counted 1000 dikes at sea level, with an aggregate thickness of 3 km, which fed a prism of lava flows 1500 m

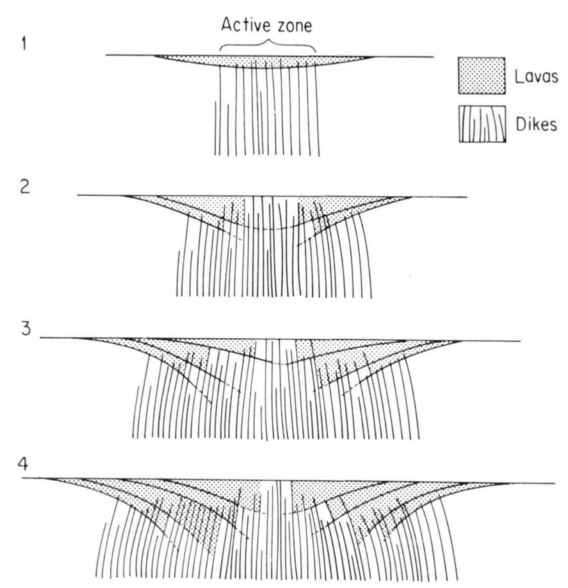

FIGURE 14-11
Crustal drift by dike injection: four stages in the evolution of Iceland. (*After G. Bodvarsson and G. P. L. Walker, 1964.*)

high. Although every dike did not produce a flow, the number of dikes at any level is related to the number of flows above that level. With increasing elevation the number of dikes falls off, and below sea level, toward the bottom of the pile, the number of dikes increases. Therefore the thicker the pile, the more dikes it should contain and the greater the extent of dilation. G. Bodvarsson and G. P. L. Walker (1964) used seismic-refraction surveys to estimate that the thickness of the lavas in the crust was about 10 km, and from this and a knowledge of the frequency of dikes in the outcrops, concluded that intrusion was causing Iceland to spread at a rate of the order of 0.5 cm/year and that the whole island had developed by spreading from the central axis (Fig. 14-11). The idea that the island is spreading is supported by the existence on it of many small rifts which are parallel to the Mid-Atlantic Ridge. These open cracks in the rocks are like crevasses in glaciers and are uncommon elsewhere (B. C. Heezen, 1960; Fig. 14-10).

Since volcanism and seismicity are chiefly confined to the median zone, and since it connects at either end with the Mid-Atlantic Ridge (P. L. Ward et al., 1969), the idea that Iceland is spreading supports and confirms the concept of sea-floor spreading from mid-ocean ridges.

14-7 PALEOMAGNETISM

Early this century B. Brunhes and R. Chevallier discovered that some rocks are reversely magnetized (Secs. 8-9 and 8-10). E. Irving (1964) and D. Strangway (1970) have reviewed the development of paleomagnetism.

The first person to realize that this discovery might be used as an aid in geochronology and to study the continental drift and polar wandering was P. L. Mercanton (1926). E. A. Johnson et al. (1948) and J. W. Graham (1949, 1955) pursued his ideas, but just after the latter had begun to discover evidence of polar shifting, he turned his attention to other matters. It seems likely that the strong opposition prevalent in North America to any idea of continental drift discouraged him.

In consequence it was J. Hospers (1951), P. M. S. Blackett (1956), S. K. Runcorn (1956), and their colleagues who first implemented these ideas. Their success owed much to a sensitive astatic magnetometer devised for another purpose by P. M. S. Blackett (1952) and to the development by R. A. Fisher (1953) of a statistical procedure for dealing with observations on a sphere.

It was members of this group who first realized that, although a single paleomagnetic determination fixes only latitude and azimuth, a succession of pole positions for rocks spanning a range of ages in a single district may be joined to give a *polar-wandering curve* (Fig. 14-12). Two continents which have moved as a unit without relative motion between them will have similar polar-wandering curves during the time they were together. If the curves diverge, the time they begin to do so indicates when the continents began to move independently of one another. In addition, if similar curves for two continents are superimposed by rotation, the continents may be brought back into their former juxtaposition (Fig. 14-13). K. M. Creer et al. (1954) first established a polar-wandering curve for Europe during part of Paleozoic time. In 1956 several authors showed, by comparing the curves they had found for different continents, that continental drift had probably taken place (S. K. Runcorn, 1962b; E. Irving, 1964).

Since then K. M. Creer (1970), M. W. McElhinny et al. (1968), and K. M. Creer et al. (1969, 1970) have shown that Africa and South America were probably in contact from Cambrian to Triassic time, but had begun to separate by the time the Serra Geral lavas were poured out during the Cretaceous period (110 to 130 m yr ago). Unfortunately, data are lacking for the Jurassic period. K. M. Creer (1970) and J. D. Phillips and D. Forsyth (1972) have found that throughout Devonian to Permian time Europe and North America were probably in contact and the North Atlantic closed. Separation may have started during the Triassic period but had not proceeded far. Data are again lacking for the Jurassic, but by Cretaceous time the continents were well separated. Thus the latest paleomagnetic data for the continents surrounding the Atlantic support other evidence for reconstructions, such as that based upon topographical fit, and agree with the dates of breakup found by geologists. The problem of using paleomagnetic data to reassemble the

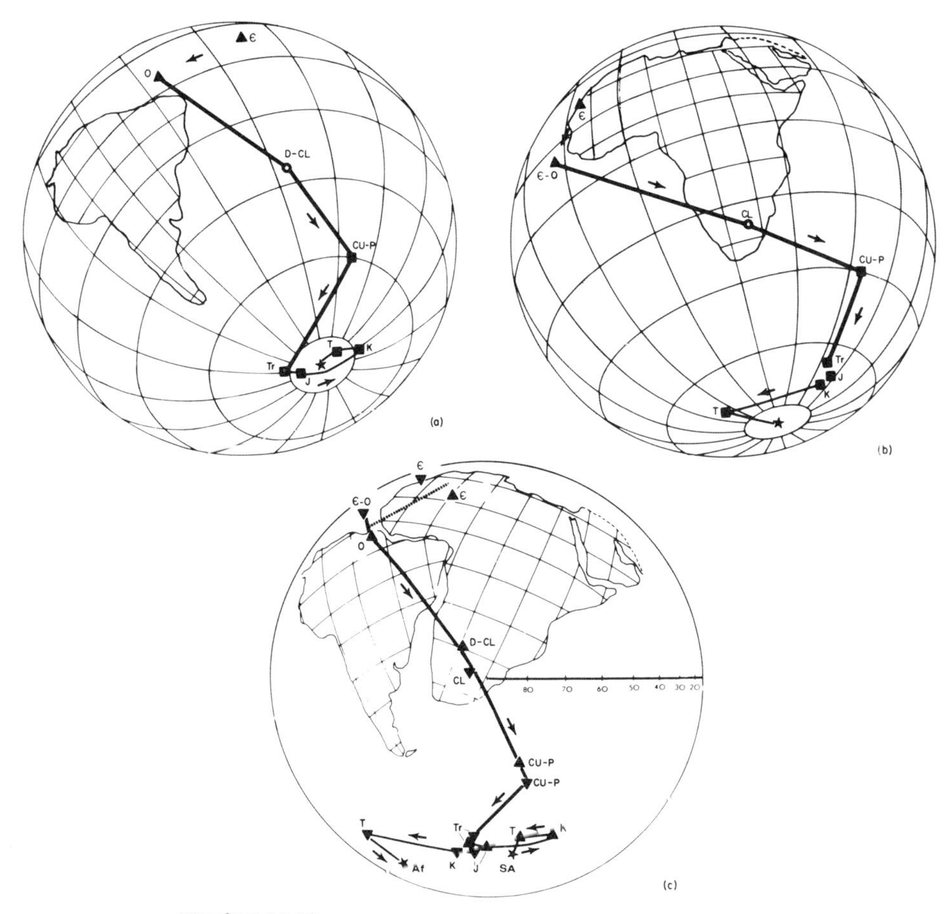

FIGURE 14-12
Polar-wandering curves for paleomagnetic South Poles of (*a*) South America and (*b*) Africa and (*c*) both curves superimposed. The following abbreviations are used: Cambrian (ϵ), Ordovician (O), Devonian (D), lower Carboniferous (CL), upper Carboniferous (CU), Permian (P), Triassic (Tr), Jurassic (J), Cretaceous (K), and Tertiary (T). In figure *c* the erect triangles are for South American poles, the triangles with apices down are for African. (*After K. M. Creer, 1970.*)

other southern continents is more elusive. Figure 14-13 shows a revision of J. C. Briden's 1967 reconstruction of the Gondwanaland continents throughout the Paleozoic era with the path of the South Pole across them during that time. As the diagram shows, he considers that while these continents were together, they were static, except for two periods of rapid and extensive polar wandering during Devonian and upper Carboniferous time. M. W. McElhinny et al. (1968) and M. W. McElhinny (1970) agree with this reconstruction and consider that the main

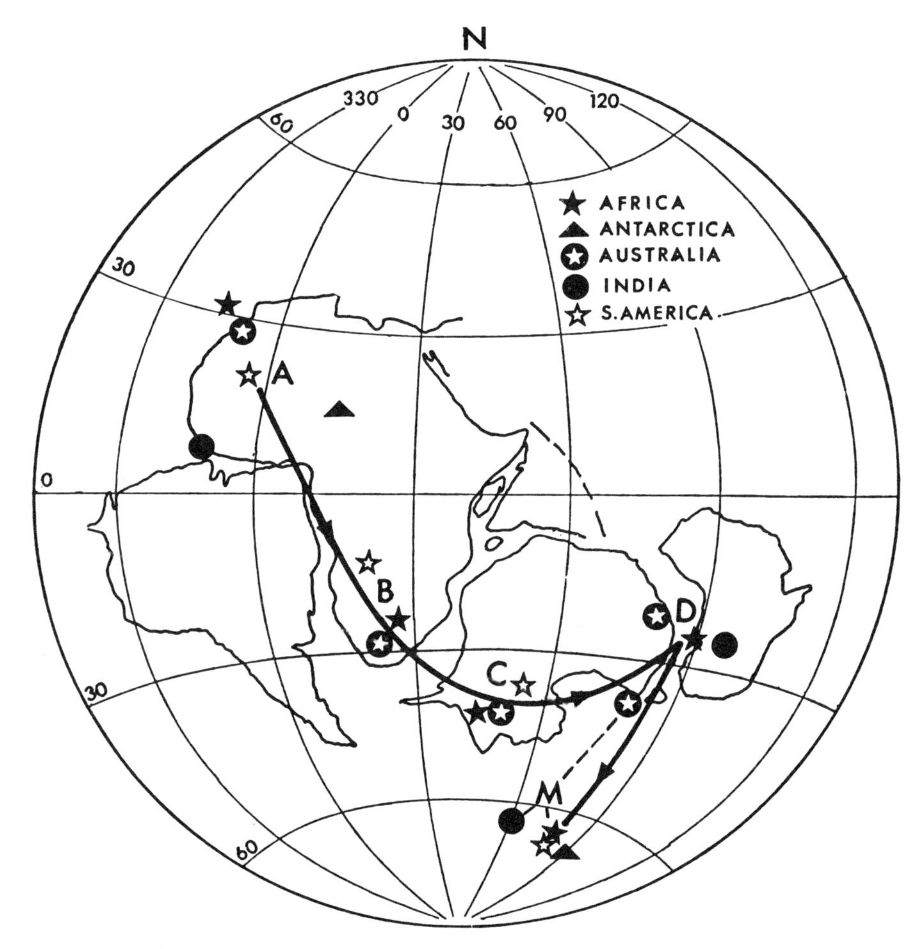

FIGURE 14-13
Reconstruction of Gondwanaland during the Paleozoic era. (*After M. W. McElhinny, 1970.*)

breakup of Gondwanaland occurred during the Cenozoic era, although some disruption had started earlier. In general, these paleomagnetic results agree with the ideas of A. L. du Toit (1937), L. C. King (1967), and A. G. Smith and A. A. Hallam (1970), based upon geology. McElhinny et al. (1968, 1970) also give paleomagnetic data for the Precambrian of Africa during the periods 2700 to 2100 m yr ago, 1950 to 1300 m yr ago, and since 600 m yr ago. They find evidence of extensive polar wandering, but comparisons with Canadian data show that the Canadian and South Africa shields did not move relative to one another by a detectable distance during the time 1950 to 1300 m yr ago.

Besides the major problems of continental reconstructions and polar wandering, several local tectonic problems have been studied and reviewed by J. Hospers and S. I. Van Andel (1969). They conclude that the following ideas are among those which should be accepted on the basis of the available paleomagnetic evidence: the anticlockwise rotation of the Iberian Peninsula to open the Bay of Biscay, the rotation of Corsica and Sardinia, the bending of the Japanese islands and of the Columbia arc in the northwestern United States, and a lack of any bending of the Appalachian Mountains in Pennsylvania and of the Alberta Rocky Mountains. Still undecided are the questions of the rotations of Newfoundland (G. W. Pearce and G. N. Freda, 1969) and of Arabia and the claim for a great shear through the Mediterranean area (Sec. 16-11).

14-8 THREE IDENTICAL GEOMAGNETIC RATIOS

Section 8-10 has already described the discovery that of all rocks examined, approximately one-half are magnetized in one direction and one-half in the opposite direction. The explanation offered is that, although a few reversals may be due to other factors, most are a consequence of reversals in polarity of the earth's main magnetic field. This provides a unifying explanation for the discovery during recent years that measurements of three quite different kinds of magnetic properties of rocks all reveal scales of times or distances that are in identical ratios.

The discovery of the first of these scales was due to the interest of J. Verhoogen, who, in 1953, linked the paleomagnetic work in England with the development in California of the K-Ar method of age determination by P. H. Reynolds. This enabled A. Cox, R. R. Doell, and J. B. Dalrymple (1964) and A. Cox (1969a,b) to establish a time scale of geomagnetic reversals for the past several million years, which was also found by I. McDougall and D. H. Tarling (I. McDougall and F. M. Chamalaun, 1966). Measurements of the polarity and age of young lava flows show that the reversals occurred simultaneously everywhere (Fig. 8-15). This time scale of reversals measured in lavas on land was the first of the three identical geomagnetic ratios to be established.

The second discovery of this same ratio was made in measurements of distance over the sea floor. In 1953, B. C. Heezen et al. noted that a prominent anomaly follows the crest of the Mid-Atlantic Ridge, and in 1959, V. Vacquier and A. D. Raff and R. G. Mason (1961) found a striking pattern of long parallel anomalies over the Pacific floor off western North America (Fig. 14-14).

F. J. Vine and D. H. Matthews (1963) carried out another magnetic survey in the Indian Ocean and found linear anomalies parallel to the crest of the Carlsberg Ridge. They suggested that this pattern might be one consequence of the sea floor spreading away from the ridge during periodic reversals of the earth's magnetic field, and that during each period of normal or reversed magnetization, new lava rising along the crest cooled through its Curie point and became magnetized in the

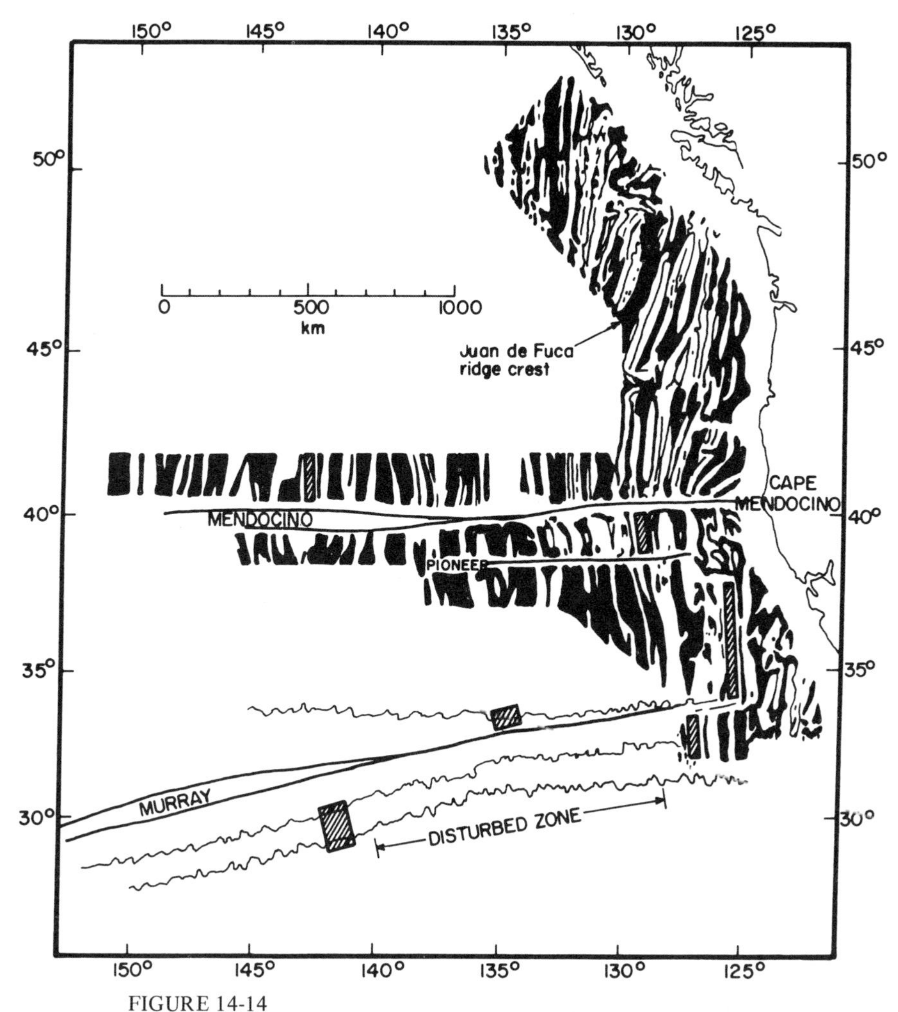

FIGURE 14-14
Pattern of linear magnetic anomalies off the west coast of North America, with positive anomalies in black and key anomalies shaded to indicate offsets on the Mendocino, Pioneer, and Murray fracture zones. (*After V. Vacquier, 1962, and A. D. Raff and R. S. Mason, 1961.*)

appropriate direction. Depending upon which side of the crest the lava lay, it would have formed part of one of two blocks, both similarly magnetized but carried away in opposite directions. As the field reversed, successive blocks formed strips magnetized in alternate directions. L. W. Morley and A. Larochelle (1964) and S. Uyeda arrived at the same conclusions independently.

In the area off the west coast of North America surveyed by Raff and Mason

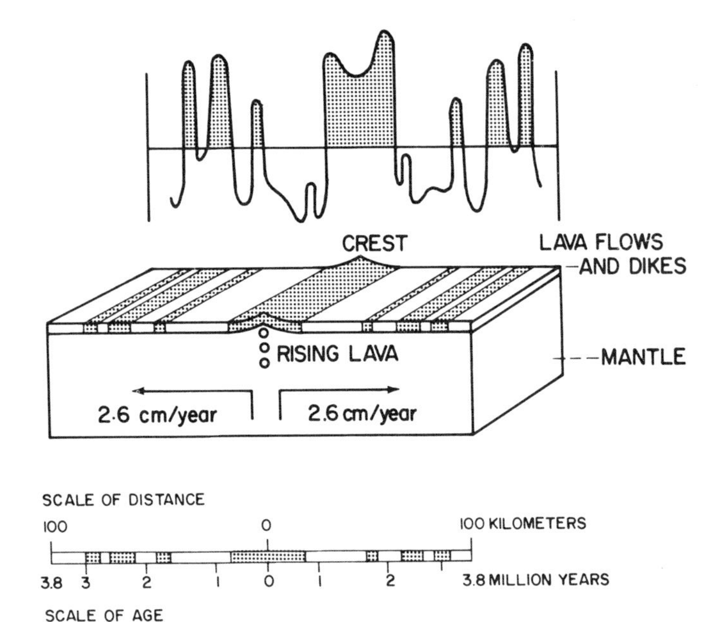

FIGURE 14-15
A diagrammatic sketch indicating, at the top, the type of magnetic profile observed when crossing the crest of a mid-ocean ridge; in the center, the upwelling which, in a naturally reversing geomagnetic field, produced strips of lava on the ocean floor that are alternately magnetized in normal and reverse directions; at the bottom, the time scale of magnetic reversals.

(1961), the distribution of earthquakes is peculiar. Many earthquakes occur along the San Andreas fault on land and for some distance along its submarine extension northwesterly. They then stop. Another line of foci lies parallel with the coast off British Columbia, but there is a gap between the two (H. Benioff, 1962). H. W. Menard (1964) suggested that there might be a mid-ocean ridge in the vicinity. F. J. Vine and J. T. Wilson (1965) showed that, if so, the widths of the anomalies from the crest were in the same ratio as the time scale of reversals. Thus the Vine and Matthews hypothesis could explain the anomalies if the spreading rate had been uniform and at a rate of 2.9 cm/year on each limb (Fig. 14-15). The San Andreas fault is thus a large ridge-ridge transform fault connecting the East Pacific rise to the Gorda ridge. Other surveys soon showed that magnetic imprinting of layer 2 on the sea floor could explain the anomaly patterns found in many parts of the world (J. R. Heirtzler et al., 1966, 1968; W. C. Pitman et al., 1966).

The third ratio was discovered by N. D. Opdyke et al. (1966) and D. Ninkovich et al. (1966), who measured the directions of magnetization of samples taken at frequent intervals along undisturbed deep-sea cores. They also found reversals, and that the ratio of distances between reversals measured downward from the surface was the same as the other ratios (Fig. 8-16).

Thus three quite different measurements all reveal the same ratio regardless of where in the world they are made. The only explanation yet offered of why the ratio of lengths of time between reversals measured in lava flows on land, the ratio of widths of anomalies on the sea floor, and the ratio of distances between reversals in cores should all be equal is that the earth's field reverses and that the sea floor is spreading. Only the latest reversals have been dated precisely, but J. R. Heirtzler et al. (1968) and P. R. Vogt et al. (1969a) have assigned ages and arbitrary numbers to well-marked anomalies which the results from drilling suggest to be correct (A. E. Maxwell et al., 1970a,b).

F. J. Vine and H. H. Hess (1970), C. H. Helsey and M. B Steiner (1969), J. E. Nafe and C. L. Drake (1969), and B. E. McMahon and D. W. Strangway (1967) have discussed the subject and presented minor variations in the time scale and other details.

14-9 SEA–FLOOR SPREADING

The year 1956 marked a turning point in the story of continental drift. Not only did M. Ewing and B. C. Heezen publish their suggestion that the mid-ocean ridges constitute a continuous system, and the students of paleomagnetism produce evidence that each continent had a different polar-wandering curve, but S. W. Carey (1958) organized the first major symposium for thirty years to consider continental drift.

He maintained that the earth has expanded greatly, opening all the oceans, including the Pacific, during the latter part of geological time. Many now doubt this theory (Sec. 12-6), but he clearly embodied in it the idea of sea-floor spreading. In another respect his interpretation was ambivalent. While he used the excellence of fit of some continents as an argument in favor of drift (S. W. Carey, 1955), on the other hand he maintained that other continents had been greatly distorted, because he held that many mountain ranges had been bent since they had been formed. Examples are the Alaska range and the Bolivian Andes. He called these *oroclines.* This form of sea-floor spreading did not involve rigid lithospheric plates. Nevertheless, Carey made an important contribution in reopening the question of drift and in insisting that it be examined with a geometric approach on a worldwide scale.

In 1960 H. H. Hess related the hypothesis of sea-floor spreading to the newly discovered mid-ocean ridges, refining the ideas of O. Fisher (1889) and A. Holmes (1931) and proposing a hypothesis for the continuous formation of oceanic crust which seismic investigation had shown to be thin, uniform, and layered:

> The Mid-Atlantic Ridge is median because the continental areas on each side of it have moved away from it at the same rate–a centimeter a year. This is not exactly

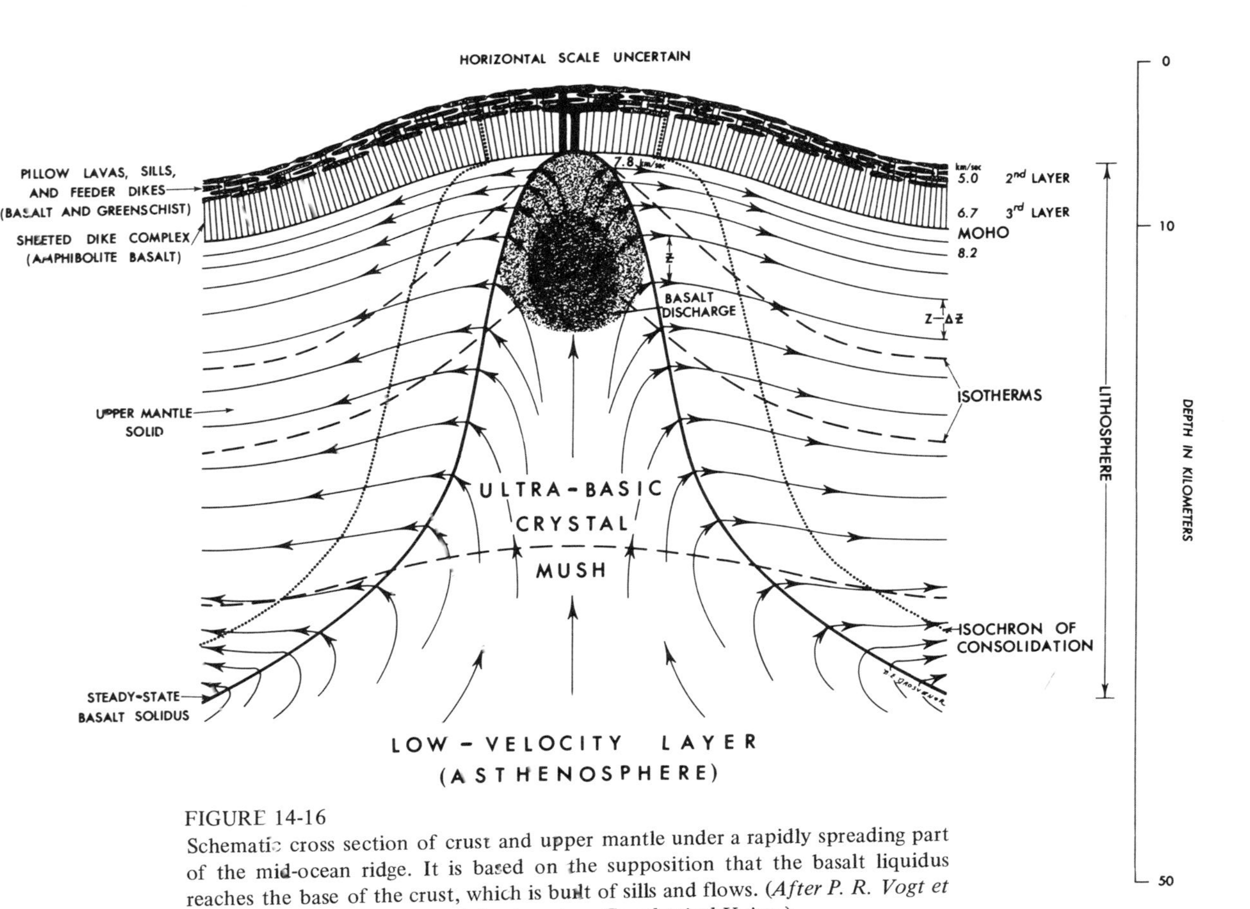

FIGURE 14-16
Schematic cross section of crust and upper mantle under a rapidly spreading part of the mid-ocean ridge. It is based on the supposition that the basalt liquidus reaches the base of the crust, which is built of sills and flows. (*After P. R. Vogt et al., 1969b, with permission of the American Geophysical Union.*)

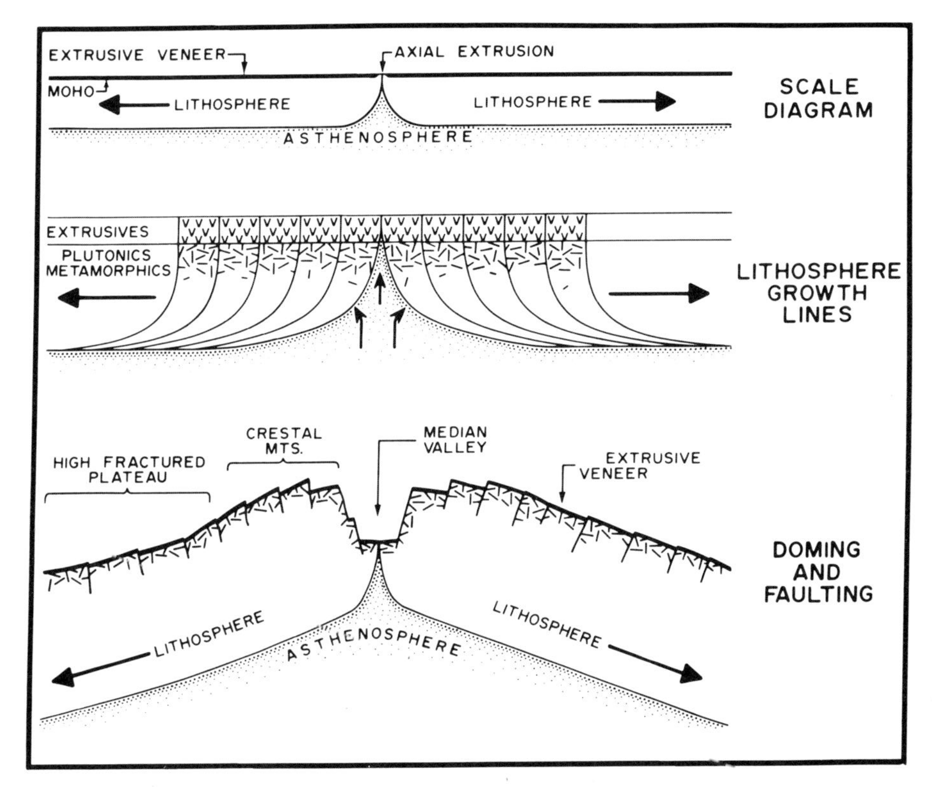

FIGURE 14-17
Proposed model for the Mid-Atlantic Ridge at 45°N. The two top diagrams illustrate the growth of the lithosphere, and the third diagram the effect of doming and faulting. The thin volcanic veneer responsible for the marine magnetic anomalies is about 200 m thick, and relatively nonmagnetic plutonic and metamorphic rocks are exposed along the infacing escarpments. (*After E. Irving et al., 1970.*)

the same as continental drift. The continents do not plow through the oceanic crust impelled by unknown forces, rather they ride passively on mantle material as it comes to the surface at the crest of the ridge and then move laterally away from it (H. H. Hess, 1962; R. S. Dietz, 1961).

Figures 14-16 and 14-17 illustrate this concept of upwelling and spreading, which is further discussed in Sec. 3-6.6 and in Chap. 17.

14-10 THE HYPOTHESIS OF GLOBAL PLATE TECTONICS

The extension of the concept of sea-floor spreading to that of plate tectonics required two further steps. The first was the demonstration that most of the earth's surface is at any time underlain by lithospheric plates which are not being appreciably deformed and that movement and distortion are essentially confined to the boundaries between them. The second was an explanation of how the expansion due to sea-floor spreading was compensated.

The idea that the greater part of the earth's surface is not being deformed arose from the observation that important patterns have been preserved for long periods of time. These include the undisturbed Paleozoic and Mesozoic cover on continents, the excellent fit of some opposite coastlines, the patterns of magnetic anomalies over the ocean basins, and the great fracture zones and their interpretation as transform faults. At the same time the preservation of formerly mobile, ancient mountains and mid-ocean ridges suggests that the mobile belts have changed positions from time to time.

The complementary idea that deformation has always been concentrated in narrow belts arose from observations in mountains and was extended by studies of seismicity.

Using new data from the World-Wide Standardized Seismograph Network (L. M. Murphy, 1966) and other modern arrays, M. Barazangi and J. Dorman (1969) have confirmed that beneath the ocean basins these belts are very narrow and continuous. They lie exactly beneath the crest of mid-ocean ridges, and in a more complex fashion follow faults in island arcs and mountains (B. Isacks et al., 1968; J. P. Eaton et al., 1970; N. N. Ambraseys, 1970). These active belts divide the earth's surface into six large and several small lithospheric plates (Figs. 11-4 and 14-2).

In 1967 D. P. McKenzie and R. L. Parker used seismological evidence to show that the whole northwestern part of the Pacific basin appears to be moving as one plate bounded by narrow belts of faults, and in 1968 W. J. Morgan stated the basis for the present hypothesis of global plate tectonics.

He pointed out that Euler (1707-1783) had proved that any relative motion between two rigid plates on the surface of a sphere can be represented by a rotation about an axis. The motion can be completely described by three parameters, two to specify the location of the pole and one the magnitude of the angular velocity. He identified fracture zones and transform faults as small circles about poles of rotation (Fig. 14-18). He recognized that mid-ocean ridges are zones of spreading and that they are generally oriented radially to a pole. He showed that a pole can be located by drawing perpendiculars to transform faults and finding their point of intersection. F. J. Vine (1966), by interpreting magnetic anomalies, provided a precise method of measuring the rate of rotation to which J. B. Heirtzler et al. (1968) added a time scale. It should be noted that in a rotation the velocity between two plates varies from zero at their common poles to a maximum halfway between, so that a statement that "the Atlantic Ocean is opening at a rate of 1 cm per year" has little meaning unless the location of the observation is stated. The modern trend is to use radial velocities conveniently measured in 10^{-7} deg/year (J. D. Phillips and B. P. Luyendyk, 1970).

X. Le Pichon (1968, 1970) applied these methods to the six major

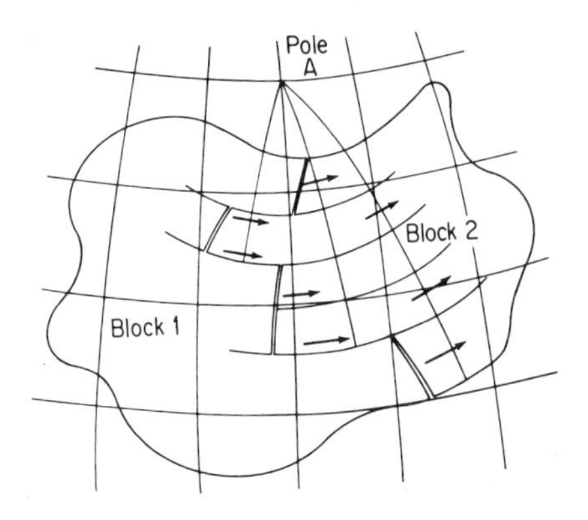

FIGURE 14-18
On the surface of a sphere the motion of block 2 to block 1 must be a rotation about some pole. It is observed that fractures commonly form as pure shears in the shape of small circles (fracture zones) or as pure tension features radial to the pole (sections of mid-ocean ridges). (*After W. J. Morgan, 1968.*)

lithospheric plates. By studying the boundaries along which pairs of plates separated, he found the poles and established the relative motions between them all. By using the motions which he had found along boundaries where plates diverge, he was able to calculate the directions and rates at which each pair of plates converged (Fig. 14-2).

Then, ignoring the effects of minor plates and of probable changes in pole positions and in rates of rotation, he calculated to a first approximation the positions of the plates 80 m yr ago (Fig. 14-19). Although admittedly inexact, the method illustrates a technique which can be applied with increasing accuracy as more details become available. Minor seismic zones cross most continents (J. Oliver et al., 1970; J. R. Cleary and D. W. Simpson, 1971). Active continental areas such as Southeast Asia offer particular difficulties (P. Molnar and L. R. Sykes, 1969; P. R. Vogt et al., 1969b; D. P. McKenzie, 1970).

B. Isacks et al. (1968) reviewed these discoveries and showed that they provide precise explanations for many seismological observations of which the cause had previously been obscure.

The general view of global plate tectonics which has emerged is that of six great major plates rotating slowly relative to one another. From time to time the patterns of plates and their motions must change, but it seems that for periods of millions to tens of millions of years motions may be steady.

Along the common boundary between any pair of rotating plates five types of motion are possible: The plates may approach normally or with a component of shear, they may separate normally or with shear, or they may move past one another in pure shear. Fortunately, observation suggests that only three motions are common: (1) pure separation normal to mid-ocean ridges, (2) pure shear along transform faults, and (3) approach, compression, and overlap, with a component of shearing.

The first common form of motion, the separation of plates, is recorded in great detail by magnetic imprinting. As W. J. Morgan (1968) pointed out, mid-ocean ridges are radial to the poles of rotation, and the separation involves no shearing along the ridges.

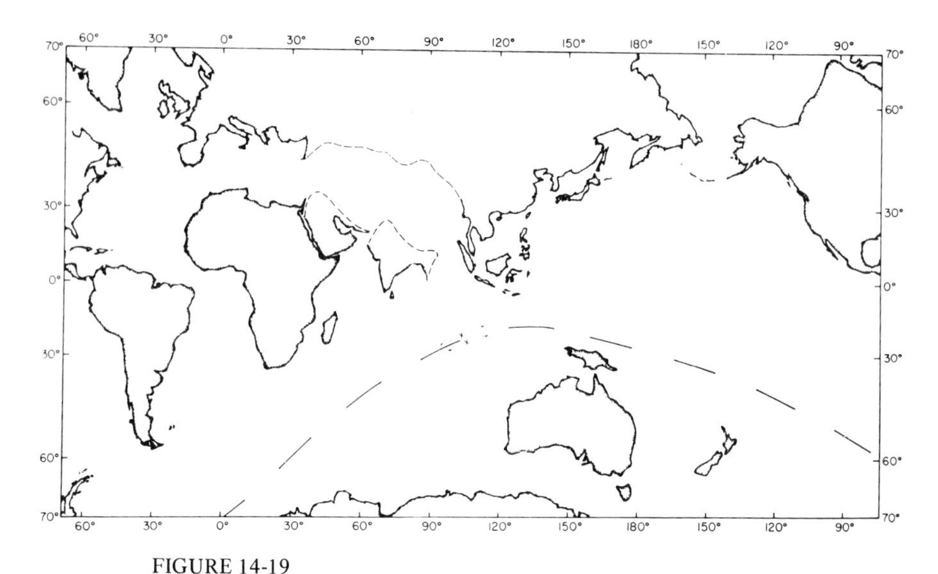

FIGURE 14-19
The approximate positions of the continents at the time of anomaly 31 (71 m yr ago, near the end of the Cretaceous period). The relative positions have not been traced across the dashed line. Africa and Antarctica are assumed to have been in their present positions. (*After X. Le Pichon, 1968.*)

H. W. Menard and T. M. Atwater (1969) and P. R. Vogt et al. (1969a) have shown that a change in pole position immediately produces an adjustment of ridge crests until they become radial to the new pole (Fig. 14-5). This motion generates oceanic crust in a symmetrical fashion and beautifully displays the record of past movements (H. W. Menard, 1969b). Earthquakes are few and shallow along crests.

The second common direction of relative motion is that between plates which are sliding past one another in pure shear along transform faults. These faults are small circles about poles of rotation. They leave a historical record of their motions recorded in the different distances by which rocks of different ages have been offset. Along these boundaries crust is not generated or destroyed. Earthquakes are important, and all are shallow.

The third basic direction of relative motion is that between plates which are coming together by overriding (Fig. 12-2). Mountains, island arcs, and ocean trenches have long been regarded as *subduction zones* of compression and downward motion (e.g., F. A. Vening Meinesz, 1929; A. Amstutz, 1951; D. A. White et al., 1970.) Beneath them occur all the world's deep earthquakes, and B. Isacks et al. (1968) proposed that these were associated with plates of cool and still brittle lithosphere which had been thrust down. They pointed out that this bending of plates might explain the normal faults observed along the convex edge of trenches and in earthquakes (W. J. Ludwig et al., 1966), while the possibility that the crust immediately beneath ocean trenches is in tension may explain why sediments in ocean trenches are not contorted (D. W. Scholl and R. von Huene, 1970; C. K. Seyfert, 1969).

B. Isacks et al. (1969) interpreted other shallow earthquakes as evidence of shearing as one plate overrides another. Deeper earthquakes to 700 km provide evidence of the location and behavior of plates being thrust downward. Anomalous seismic velocities, the attenuation of seismic waves, and large gravity anomalies provide other clues (P. Molnar and J. Oliver, 1969).

Where two plates come together and overlap, the area of the lower plate is diminished while that of the upper plate is little changed. In this process large parts of the sedimentary rocks which held the record of the geological history are likely to be carried down or to be contorted, metamorphosed, and made difficult to decipher. This destruction is the reason why the geological history of mountain belts is so much more difficult to decipher than the magnetic pattern imprinted during the generation of ocean floors. For example, A. Gansser (1966) suggests that the Indus Line in the Himalayas marks the trace of a subduction zone along which the deep-water marine sediments of a former basin completely disappeared during the Himalayan orogeny. This upsets older views which never envisaged that whole geosynclines might be carried down and vanish. The loss of deep-water sediments also invalidates the claim of some geologists that since only shallow-water sediments are now found, the Tethys Sea was never more than a shallow incursion over the continent.

We have noted that plates grow perpendicularly away from ridges and slide past one another in pure shear. It follows from geometry that plates of irregular and constantly changing shapes are unlikely to come together in pure compression without an element of shear motion. This may explain why longitudinal strike-slip faults are common in mountain systems (see examples in the Cordillera and Appalachians in P. B. King, 1969a,b).

J. F. Dewey and J. M. Bird (1970a,b) have pointed out that there are four possible causes of overlap between continents and ocean floors, which may be described thus:

1 Oceanic crust overriding oceanic crust along an island arc
2 Oceanic crust overriding continental crust along an island arc
3 Continental crust overriding oceanic crust along a continental margin
4 Continental crust overriding continental crust along a continental margin

They have cited present-day examples and given possible evolutionary histories for each of these cases (Fig. 14-20).

Because oceanic crust can be subducted and consumed readily, the first and third cases are the common ones. To carry down continental crust against the force of gravity requires much work. D. P. McKenzie (1970) suggests that when this occurs, the onset of extra work causes the pattern of motion of plates to change, and he cites the recent behavior of plates in the eastern Mediterranean as an

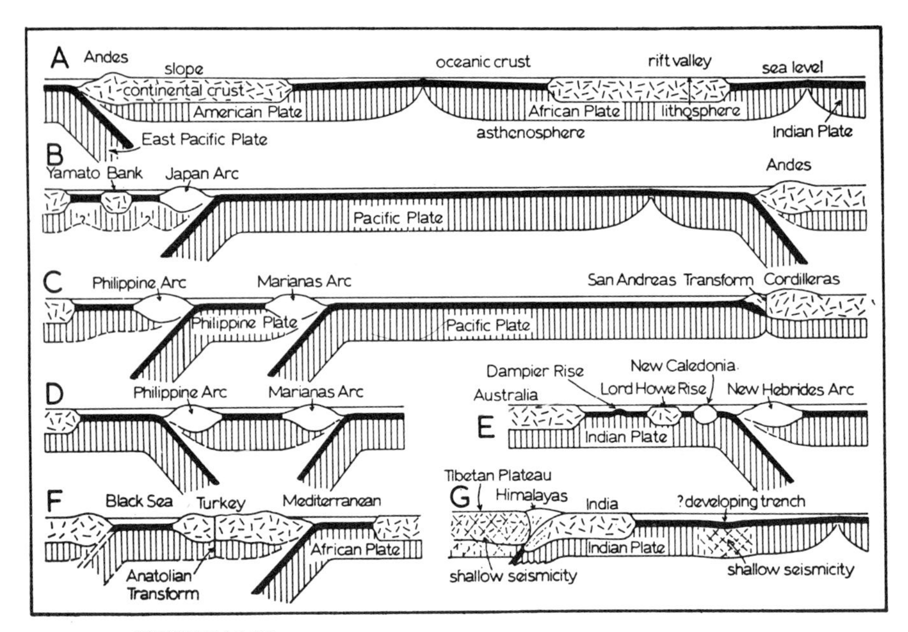

FIGURE 14-20
Schematic sections showing some assumed relationships between oceans and continents across subduction zones. (*The developing trench (?) in Figure G is from L. R. Sykes, 1970. After J. F. Dewey and J. M. Bird, 1970a.*)

example. Furthermore, because continents have irregular shapes, some parts will come into contact with island arcs or other continents before other parts. This will affect the timing and degree of deformation in different parts of the ensuing mountain belts (J. T. Wilson, 1966).

The problems of why island arcs are circular arcs, and possible mechanisms, are discussed in Chap. 17.

15
THE LIFE CYCLE OF OCEAN BASINS: STAGES OF GROWTH

15-1 STAGES IN THE GROWTH AND DECLINE OF OCEAN BASINS

If the hypothesis of plate tectonics is valid, one can visualize the life cycle of a typical ocean basin which grows from small beginnings to a maximum and then shrinks, perhaps ultimately closing completely.

Evidence from geology and from magnetic imprinting suggests that some parts of oceans are younger than others and that narrow branches like the Red Sea have opened more recently than the main basins. A. L. du Toit (1937, p. 252) recognized this relationship when he wrote of "disjunctive ocean basins, which are merely rift valleys of unusual width." During the Permian and Triassic periods the maximum sizes of ocean basins and of continents were reached when the single ocean Panthalassa and the single supercontinent Pangaea covered all the earth's surface. Since then the Atlantic and Indian Oceans have grown and the pacific is the shrunken remnant of Panthalassa (Fig. 12-4). E. Suess (1904-1924) indicated still later stages when he described the region where the Mediterranean Sea and the Alpine and Himalayan Mountains now lie as a former seaway, the Tethys Ocean. The existing oceans and related features can thus be arranged in six stages of a generalized life cycle of ocean basins (Table 15-1; Fig. 3-6; J. T. Wilson, 1968; E. D. Schneider, 1969; P. R. Vogt, 1970; J. F. Dewey and J. M. Bird, 1970a,b).

Because ocean basins are the largest features of the earth's surface, their life cycle may control other phenomena, and this may explain why each stage has characteristic directions of vertical and horizontal movement, associations of igneous and sedimentary rocks, degrees of metamorphism, structural features, economic deposits, and landforms. This chapter describes the stages of growth in terms of three existing types: the East African rift valleys, the Gulf of Aden, and the Atlantic Ocean.

15-2 THE TYPE EXAMPLE OF STAGE 1: THE EAST AFRICAN RIFT–VALLEY SYSTEM

The East African rift-valley system has so many features in common with mid-ocean ridges that it is considered to be an embryonic ocean and it has been chosen as the type example of Stage 1. E. Suess (1904-1924) first described this great fracture system which extends for more than 4000 km from Asmara on the Red Sea to old rifts south of the Zambesi River (Fig. 15-1). J. W. Gregory (1920) coined the term *rift valley* for its characteristic elements, each of which is an elongated, downfaulted area over 100 km long and from 30 to 60 km wide. E. J. Wayland (1930) observed "the rise to the rift," noting that the rifts are fracture zones along the crest of broad upwarps. Although large and deep lakes fill some rift valleys, the floors of most are above the level of the surrounding continent, and B. Willis (1936) and H. Cloos (1939) emphasized the primary role of doming, which is accompanied by rifting and the upwelling of lavas. B. H. Baker (1965) showed that many rift valleys are asymmetrical (Fig. 15-2).

All the rifts lie along the crests of a connected series of broad swells. The most northern dome in Ethiopia and southwestern Arabia is the highest, reaching 4620 m. To the south the altitude of the crest drops to less than 1000 m near Lake Rudolf, and then rises again in domes on either side of Lake Victoria. High ground continues in an irregular way to the south, eventually merging with plateaus in Rhodesia and South Africa.

The rifts are seismically active, but the earthquakes seem to be more widely spread than along mid-ocean ridges (D. G. Tobin et al., 1969; J. Wohlenberg, 1969). J. R. Vail (1965, 1967) and J. D. Fairhead and R. W. Girdler (1969) have pointed out that small earthquakes, old faults, and Mesozoic volcanism may mark an extension of the rift valleys for at least 500 km south of the Zambesi valley to the northern border of South Africa. J. B. Wright and P. Rix (1967) have found a rift to the east.

Recently, W. T. C. Sowerbutts (1969) showed that, while the East African Plateau is in approximately isostatic equilibrium, a 60- to 70-mgal negative Bouguer anomaly requires that the uplift of the region be compensated at depth by a

Table 15-1 STAGES IN THE LIFE CYCLE OF OCEAN BASINS AND THEIR PROPERTIES

Stage	Examples	Dominant motions	Characteristic features	Typical igneous rocks	Typical sediments	Metamorphism
1. Embryonic	East African Rift Valleys	Uplifts	Rift valleys	Tholeiitic flood basalts, alkalic basalt centers	Sedimentation minor	Negligible
2. Young	Red Sea, Gulf of Aden	Spreading	Narrow seas with parallel coasts and central depression	Tholeiitic flood basalts, alkalic basalt centers	Shelf and basin deposition; evaporites possible	Negligible
3. Mature	Atlantic Ocean	Spreading	Ocean basin with active mid-ocean ridges	Tholeiitic flood basalts, alkalic basalt centers but activity concentrated at center	Abundant shelf deposits (miogeosynclinal)	Minor
4. Declining	Pacific Ocean	Shrinking	Island arcs and adjacent trenches around margins	Andesites, granodiorites at margins	Abundant deposits derived from island arcs (eugeosynclines)	Locally extensive
5. Terminal	Mediterranean Sea	Shrinking and uplifts	Young mountains	Volcanics, granodiorites at margins	Abundant deposits derived from island arcs (eugeosynclines) but evaporites possible	Locally extensive
6. Relic scar (geosuture)	Indus Line in the Himalayas	Shrinking and uplifts	Young mountains	Minor	Red beds	Extensive

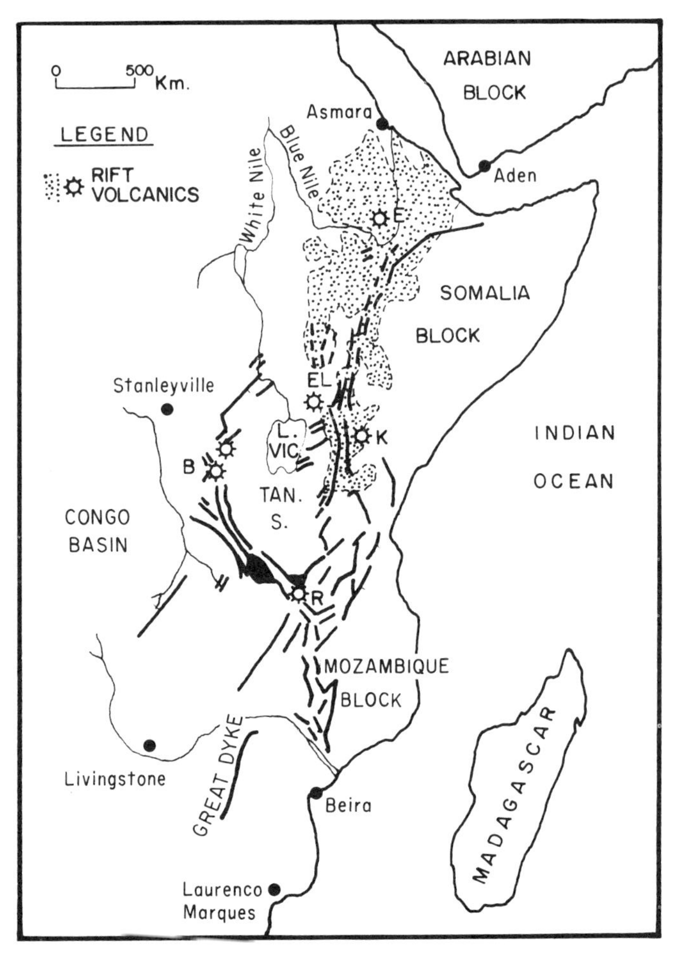

FIGURE 15-1
Structural sketch map of East African rift system showing major rift scarps and main areas of volcanic rocks. The following are the principal volcanic centers: E – Ethiopia, El – Elgon, K – Kilimanjaro, B – Birunga, and R – Rungwe. (*After R. B. McConnell, 1967, and J. R. Vail, 1967.*)

relative mass deficiency. He concluded that

The quantity of low-density material beneath East Africa required to account for the gravity anomalies can only have formed from higher density material by heating and expansion or by change in bulk chemical composition. . . . The formation of many features of the Rift System in East Africa, but not all, can be explained if a single convective upwell is postulated beneath the East African Plateau as a heat transfer mechanism.

Late Miocene - Early Pliocene

1. Phonolites erupted on crest on uplift
2. Faulting on west side of rift, monoclinal flexuring on east side

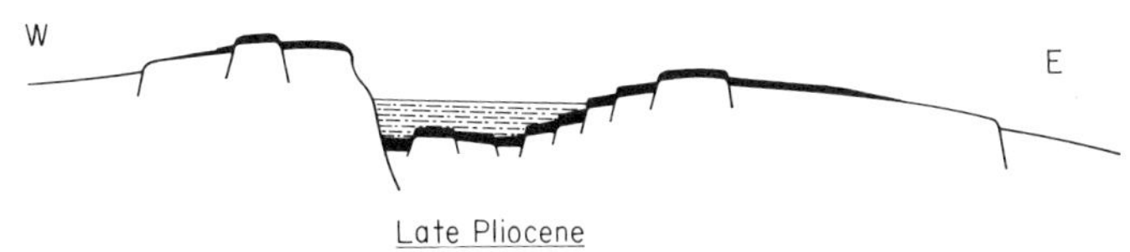

Late Pliocene

3. Faulting of floor of rift; renewal of movement on main fractures; new fractures on rift shoulders
4. Trachytic-basaltic volcanicity in rift floor

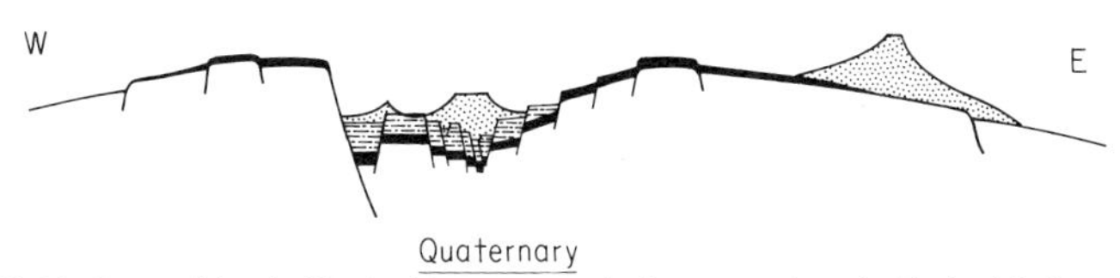

Quaternary

5. Further uplift of rift shoulders; renewal of movement on faults in rift floor; new closely spaced fractures develop in median zone
6. Small plugs and larger calderas built in rift floor; Some central volcanoes on the rift shoulders

50 miles

FIGURE 15-2
Diagrammatic sections showing development of the central rift in Kenya. (*After B. H. Baker, 1965.*)

R. W. Girdler et al. (1969) found that, as under mid-ocean ridges, this lighter material extends to a considerable depth and has small density contrast, so that it could be a diapir rising from the asthenosphere to fragment the lithosphere (Fig. 14-16). This interpretation has been supported by further work, including seismic investigations, and is compatible with plate tectonics (R. C. Searle, 1970; D. H. Griffiths et al., 1971; M. A. Khan and J. Mansfield, 1971).

Beneath the eastern rift at the equator a much smaller superimposed positive anomaly may mark a shallow body of basalt magma feeding Kilimanjaro and the neighboring volcanoes. The view that the rise of hot material is the cause of doming and volcanism is supported by the concentration of fumaroles, flood basalts, and alkali volcanic centers along the rift valleys (B. C. King and D. S. Sutherland, 1960). D. K. Bailey (1964) attributes the magmas to the beginning of partial melting and rifting, not to the concluding stages of fractional crystallization. C. G. Murray

(1970) and P. A. Mohr (1971) note that oceanic lavas are different, and P. G. Harris (1969) finds that volcanism decreases and becomes more alkalic away from rifts.

The boundaries of basement provinces, trends in their foliation, and alignment with large Archean intrusives to the south seem to have influenced the location of rifts and uplifts, but this relationship does not necessarily demand that the rifts be old (R. B. McConnell, 1967; R. B. Hargraves, 1970).

The rift-valley system consists of several principal domes in Ethiopia, Kenya, and Transvaal, in each of which similar events occurred, but at different times. Each dome began with uplift accompanied and followed by extensive basalt flows, central volcanism of varied types, and rift faulting. Sediments collected in grabens, several of which follow metamorphic belts subdividing the ancient Precambrian rocks.

The earliest episode began with the deposition of Karroo (Carboniferous to Triassic) sediments, followed by late Jurassic uplift, opening of the Indian Ocean, and the extrusion of basic lavas and igneous complexes from 190 to 155 m yr old (L. C. King, 1967; I. McDougall, 1963).

During Eocene and Oligocene time a second episode of uplift, fracturing, and fissure volcanism created the Ethiopian dome and the valleys in which lie the East African rift and the Gulf of Aden (Secs. 15-4 and 15-5).

The youngest areas of uplift are in equatorial Africa. In Kenya the main rift bisects an oval region which was raised in late Cenozoic time to an elevation of 3000 m. A subsidiary uplift formed on the west side of Lake Victoria (B. H. Baker and J. Wohlenberg, 1971; B. H. Baker et al., 1971).

All had similar histories and are held to be examples of the first stage of the life cycle of ocean basins. Although it remains a problem to understand how a series of domes separated in space and time could be joined and extended to form a mid-ocean ridge, it does seem that the Mesozoic uplift of the coastal regions was connected with the breakup of Gondwanaland, that the Ethiopian dome was related to the opening of the Red Sea and of the Gulf of Aden, and that the latest uplifts in East Africa may be part of a process which will join these domes and initiate the opening of a new ocean. I. G. Gass believes that the isolated volcanic uplifts of the Hoggar and Tibesti in the Sahara may be examples of still independent young domes (R. Furon, 1963; T. N. Clifford and I. G. Gass, 1970; N. L. Falcon et al., 1970).

15-3 OTHER POSSIBLE EXAMPLES OF STAGE 1

Many other rift valleys are known, but none of them are so clearly related to the mid-ocean-ridge system. Consideration of which ones, if any, represent an embryonic ocean is therefore a highly speculative matter.

A case has been made for the Basin and Range province in western North America, but it differs from the East African rift valleys in that it seems to be due to the continent having overridden an established mid-ocean ridge, the East Pacific rise, rather than to the start of a new ridge (Sec. 16-6). Several large rift valleys in the Adelaide, Amadeus, and Fitzroy basins of Australia may be connected to cross that continent (J. T. Wilson, 1968; J. R. Cleary and D. W. Simpson, 1971), but the rift-valley system which seems most analogous to that in East Africa crosses eastern Siberia and contains Lake Baikal and other valleys surrounded by uplifts of over 3000 m. Another may cross China (J. T. Wilson, in press).

V. V. Lamakin (1968) and M. E. Artemyev and E. V. Artyushkov (1971) consider that, like the African rifts, the Baikal rift zone has high seismicity, high heat flux, negative gravity anomalies, and comparable geology. D. V. Nalivkin (1960) traces the rift valleys to the Sea of Okhotsk. Earthquakes scattered across Mongolia suggest connections with the Pamir and Himalayan Mountains. J. F. Dewey and J. M. Bird (1970a,b) show the system as the northern boundary of the Southeastern Asia lithospheric plate. On the other hand, A. P. Bulmasov (1967) holds that this rift has no connection with others and that it is a foredeep due to overriding of the Transbaikalian Mountains on to the east Siberian shield. N. A. Florensov (1969) agrees that this rift is independent, but holds that the motion of earthquakes shows that the region is in a state of tension.

The valley of the lower Rhine is a clear and well-studied example of a rift valley (C. Sittler, 1969; J. H. Illies and St. Mueller, 1970). J. H. Illies (1969) has given a recent description and tried to connect it to the East African rifts, but it is much smaller, the gaps are great, and it seems more closely related to the adjacent Alps. Other small rift valleys which can be regarded as branching from mountain systems are the Chiquitos graben in Bolivia (R. P. Morrison, 1973) and several grabens along the western side of the Appalachians (J. T. Wilson, 1954; P. S. Kumarapeli and V. Saull, 1966; R. Doig, 1970). Other small rift valleys branch from the shores of ocean basins. These include the Bahia and other grabens along the Atlantic coast of Brazil (Sec. 15-11), the graben through which the Benue and lower Niger Rivers flow in West Africa (J. B. Wright, 1968), the Carnarvon-Perth basin in Western Australia (R. F. Thyer, 1963), and the Cambay depression extending north from Bombay, India (N. K. Kalinin, 1964).

15-4 THE TYPE EXAMPLE OF STAGE 2: THE GULF OF ADEN

Unlike the East African and other rift valleys whose connection with the ocean-building cycle is based on indirect arguments, the Gulf of Aden and the Red Sea both appear to be young and narrow ocean basins showing many features common to larger oceans. Their geology and geophysics have been well described in two collections of papers (N. L. Falcon et al., 1970; T. N. Clifford and I. G. Gass, 1970).

A deep axial trough is marked by a belt of earthquake epicenters, high heat flows, and strong linear magnetic anomalies. The Gulf of Aden is crossed and offset by several parallel faults, which L. R. Sykes (1968) has shown to be transform. The whole floor is very rough and broken, and some of the transform faults are not at right angles to sections of the mid-ocean ridge and the associated linear magnetic anomalies (D. P. McKenzie et al., 1970). This unusual example of a ridge spreading obliquely to the direction of motion of the bounding plates may be a result of its youth. A. S. Laughton and C. Tramontini (1969) have shown by seismic-refraction studies that the whole gulf has a floor of oceanic crust underlain by anomalous mantle of low density which can be interpreted as an upwelling (Fig. 15-3). The two coasts can be matched closely along the 1000-m submarine contour, and pre-Miocene geological features are continuous across the reassembly. Z. R. Beydoun (1969) has described the geology of the region, pointing out that the uplift of some of the highlands near the Indian Ocean began during the Jurassic period, but that Ethiopia was not raised until the Eocene, and that the Gulf of Aden did not start to open until the Miocene. Socotra Island at the entrance to the gulf is continental (Z. R. Beydoun and H. R. Bichan, 1970).

Connections important to our thesis that ocean basins have grown were established when D. H. Matthews et al. (1967) linked the central ridge of the Gulf of Aden to the Carlsberg ridge in the Indian Ocean, and P. A. Mohr (1967) and G. Peter and O. E. DeWald (1969) showed that the Red Sea, Gulf of Aden, and African rift valleys are all connected. H. Tazieff (1969, 1970) and F. Bonatti et al. (1971) have described the peculiar Afar desert at the junction as a depression floored only by salt deposits and basalts and in places lying below sea level.

Most authorities now agree that the opening of the Gulf of Aden, Red Sea, and East African rift valleys involved the main African (Nubian) plate, the smaller Arabian and Somalian plates, and two fragments, the Danakil and Aisha Horsts, but they have not yet agreed upon the precise timing or upon the poles of rotation (D. G. Roberts, 1969; D. P. McKenzie et al., 1970; P. A. Mohr, 1970).

15-5 THE RED SEA, AN EXAMPLE OF STAGE 2

The Red Sea is considered another good example of Stage 2. Topographically this depression is over 2200 km long and up to 450 km wide, so that it is 10 times the size of a typical East African rift valley. The relief is great, ranging from 2091 m above sea level north of Port Sudan to a depth of 2341 m.

Many papers in N. L. Falcon et al. (1970) and T. N. Clifford and I. G. Gass (1970) have reviewed the geology and geophysics of the Red Sea region. The oldest rocks overlying the Precambrian are Carboniferous to Triassic strata which occur only in the Gulf of Suez area, but Jurassic, Cretaceous, and early Tertiary seas spread over both the northern and southern ends of the depression. In late Eocene time a great dome uplifted Ethiopia and raised Eocene marine sediments to over 1800 m in Somalia. Basalt eruptions which had begun during the Cretaceous period in Yemen continued to pour out over a wide area, reaching their climax in

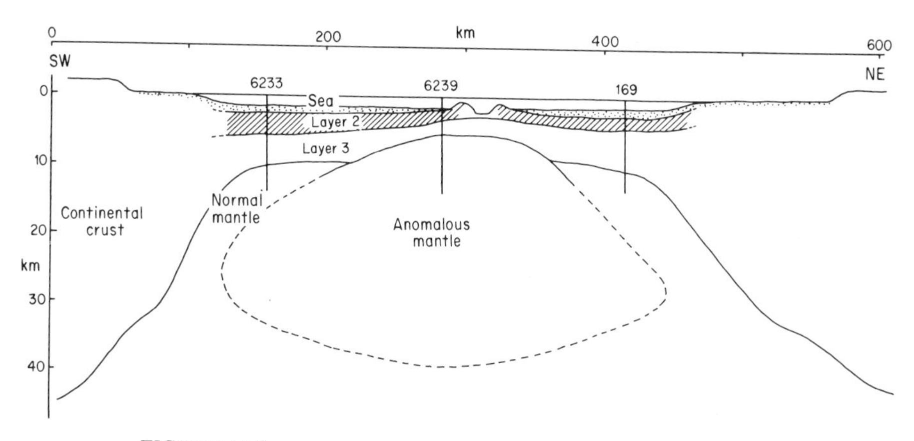

FIGURE 15-3
Section across Gulf of Aden indicating postulated low-density mass of anomalous mantle. (*After A. S. Laughton and C. Tramontini, 1969.*)

Oligocene time. Concurrent faulting produced the Red Sea depression, and during Miocene time alternating connection and isolation from the main oceans led to lagoonal conditions and the formation of evaporites.

There has been much debate about whether the Red Sea depression is due to (1) compression; (2) downfaulting of a continental block; (3) spreading and separation only along a narrow axial trough, with downward flexuring along the coast; or (4) large-scale separation to form the whole sea. The first two views are now generally abandoned. The third was advanced by C. L. Drake and R. W. Girdler (1964). They used magnetic, gravity, seismic, and bathymetric data to suggest that the axial trough alone is a narrow fissure up to 70 km wide in the continental crust, filled with basic igneous intrusives, and that the marginal zones are underlain by downfaulted blocks of the bordering Precambrian shields, but it was later realized that this view is difficult to reconcile with the geology of the Levant, so that the fourth explanation involving wide separation is now favored. This also fits the Tertiary tectonic history of Arabia (P. J. Burek, 1970).

At the north end the Red Sea joins a fault zone through the Dead Sea and the Jordan Valley, and the geological record is more complete than farther south. A. M. Quennell (1958) and L. Dubertret (1959) interpreted this fault zone as a left-handed shear with a displacement of about 105 km, and most Israeli geologists have accepted this interpretation (R. Freund et al., 1968), although L. Picard (1970) still attributes the rift to predominantly vertical movements.

In 1966 R. W. Girdler pointed out that this displacement along the Dead Sea rift should have produced about 215 km (instead of 70 km) of separation in the Red Sea at 16°N latitude. In 1969 C. Tramontini and D. Davies made detailed refraction surveys and found that a great thickness of sedimentary and coralline cover underlies all but the axial trough of the Red Sea. They claim that beneath this cover the underlying rocks are not Precambrian shield but basic ocean floor, formed during wide separation.

Three deeps in the Red Sea are unique in that they contain hot, concentrated brines with salinities up to 25.6 percent (the normal Red Sea water has 4.0 percent) and temperatures as high as 56°C (E. T. Degens and D. A. Ross, 1969). Both the brines and the sediments beneath them are rich in metals. Many believe that the heat is due to volcanism, that the excess salt is from the evaporites, and that the hot brines have leached the metals from surrounding continental rocks.

15-6 THE OPENING OF THE ATLANTIC OCEAN

Many advocates of drift, noting the lack of marine sedimentary rocks of Permian and Triassic age in the Atlantic region, have concluded that the ocean was not then in existence and that the resulting supercontinent was a desert. They consider that uplift, rifting, and volcanism began during the Triassic period, but that the ocean did not open appreciably until the Jurassic (J. D. Phillips and D. Forsyth, 1972).

If any pair of continents separate, this must be by rotation, but the poles and rates of rotation may change. Fortunately, the evidence suggests that motions tend to remain steady in direction and rate for periods of tens of millions of years, and E. C. Bullard et al. (1965) found a single pole for the separation of Europe and North America and another for Africa and South America, but did not consider rates. W. J. Morgan (1968), Le Pichon (1968), and M. M. Ball and C. G. A. Harrison (1970) found the same general pattern, but about slightly different poles, although at steady rates. Others have regarded the situation as more complex (B. M. Funnell and A. G. Smith, 1968; P. J. Fox et al., 1969; P. R. Vogt et al., 1969a), but J. D. Phillips and B. P. Luyendyk (1970) have reviewed the matter and support steady but independent opening of the North and of the South Atlantic for the past 40 m yr.

The matter is made more complex by the existence and independent motions of small plates, including Greenland, the Canadian Arctic islands, the Canary Islands and Rockall Bank. As more detailed investigations are undertaken, it is becoming apparent that the opening of oceans produces continental fragments ranging in size from Eurasia to the Seychelles, and in degree of separation from the width of an ocean to a minor rift. Account must be taken not only of the exposed fragments, but also of submarine banks, some of which have been shown by geophysical studies or drilling to be continental. Examples are Rockall, Orphan Knoll near Newfoundland, and Agulhas Bank. In many other places subsided blocks are likely to be buried in continental shelves, but some islands and banks like Iceland, the Faroes, and perhaps Flemish Cap are entirely oceanic (M. Bacon and F. Gray, 1971).

The main Atlantic Ocean has been chosen as the type example of Stage 3, but the Greenland and Norwegian Seas and Baffin Bay seem to be younger and still in Stage 2.

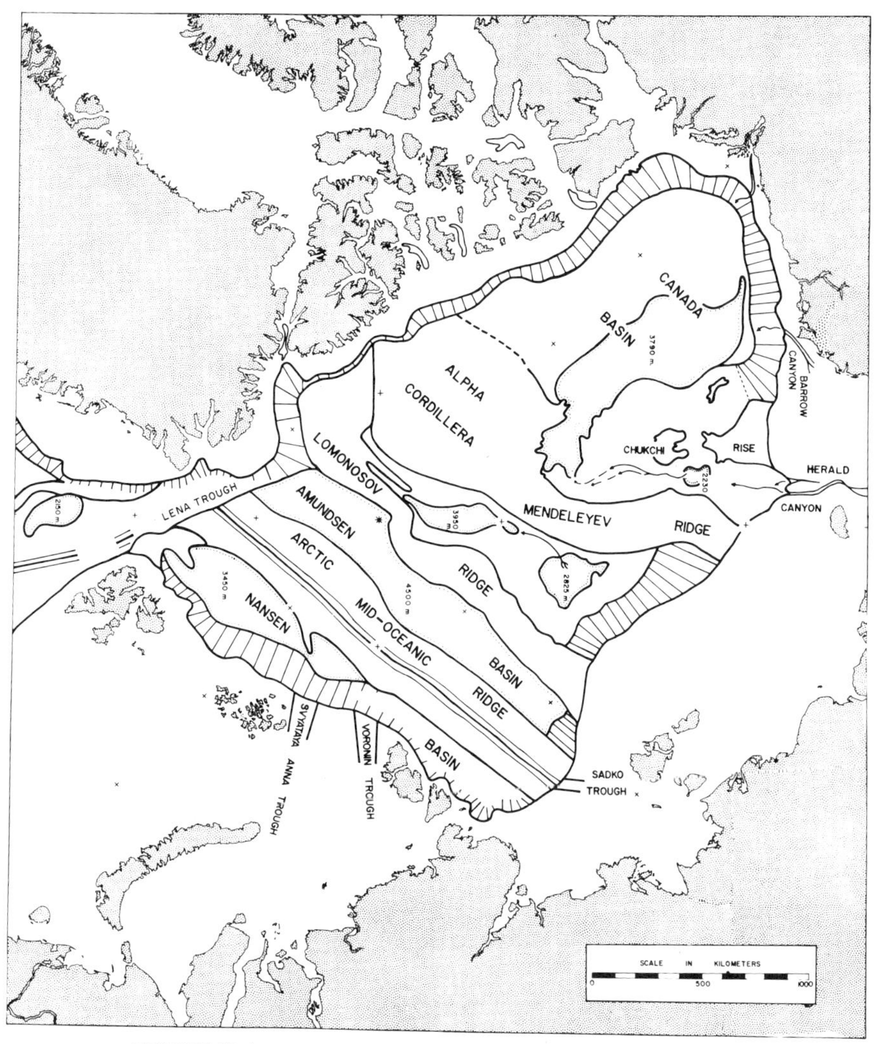

FIGURE 15-4
Arctic Ocean physiography. (*After R. M. Demenitskaya and K. L. Hunkins, 1970.*)

15-7 THE ARCTIC BASIN, WITHIN WHICH THE EURASIAN BASIN APPEARS TO BE AN ACTIVE EXAMPLE OF STAGE 2

The Arctic Sea forms an extension of the Atlantic Ocean, offset to the northwest along the deep Lena trough between Greenland and Spitsbergen (Fig. 15 4). The straight shape, the great depth of 4848 m, and the distribution of earthquakes along the offset part of this trough mark it as a transform fault zone (C. E. Wegmann, 1948; W. T. Horsfield and P. I. Matou, 1970). A similar but older fault may have formed Nares Strait between Greenland and Ellesmere Island (Sec. 15-8).

The other openings to the Arctic Sea are shallow channels across the wide continental shelves which border the Barents Sea, the Canadian arctic islands, and Bering Strait. This strait is of minor geological importance because it did not open until late Tertiary time, and the main change between Asian and North American structures occurs in the Verkhoyansk range, which was a seaway until the Pliocene (A. P. Markovsky, 1961; D. M. Hopkins, 1967). The Arctic Sea contains no oceanic islands, but is surrounded by large archipelagoes, all continental in character. A. J. Eardley (1962) and A. P. Puminov (1967), noting these islands and also that Devonian clastic sediments appear to have been derived from a source offshore, maintained that the basin is underlain by a sunken block of continental crust, but seismologists have shown that the deep floor is oceanic (J. Oliver et al., 1955; T. A. Santo, 1962; J. N. Brune, 1969).

The Lomonosov ridge crosses the Arctic basin for 1700 km from Ellesmere Island to the New Siberian Islands to divide it into two parts, the Eurasian and Amerasian basins. This ridge rises 3000 m above the surrounding floors to a flat top about 30 km broad with a shallowest sounding of 954 m. It is asymmetric, with steep sides, more precipitous on the Amerasian side. It may be a slice of the Eurasian continental margin separated by the opening of the Eurasian basin (J. T. Wilson, 1965c; R. M. Deminitskaya and K. L. Hunkins, 1970).

To the east of the Lomonosov ridge the Eurasian basin in turn is bisected by the Gakkel ridge into the Nansen basin beside Siberia and the Amundsen basin, which includes the North Pole. Both contain abyssal plains over 4000 m deep. As B. C. Heezen and M. Ewing (1961) noted, the Gakkel ridge has all the properties of an extension of the Mid-Atlantic Ridge, including a median position in the Eurasian basin, a rift along the crest, thin to absent sedimentary cover, numerous shallow earthquakes, and a pattern of linear magnetic anomalies symmetrically arranged about it (L. R. Sykes, 1965; R. M. Deminitskaya and A. M. Karasik, 1969; G. L. Johnson, 1969; M. Barazangi and J. Dorman, 1970).

On the western side of the Lomonosov ridge lies the shallower Amerasian basin overlain by thick sediments and again subdivided by the little known Alpha-Mendeleyev ridge into the large Canada basin against the North American coast and the smaller Makarov basin (M. M. de Leeuw, 1967).

P. R. Vogt et al. (1970b) suggest that during early Tertiary time Baffin Bay and the Amerasian basin were linked by a fault through Nares Strait and were both spreading, the latter about the Alpha-Mendeleyev ridge. Since this ridge is not symmetrically situated, another boundary ridge like the Lomonosov ridge may cross the Arctic basin, probably through Chukchi Cap, which is continental (K. Hunkins, 1962) Together these two boundary ridges would divide the Arctic basin into three parts of different ages, each with its own section of active or former mid-ocean ridge. The three parts are the Eurasian basin against Siberia, which is actively spreading today, a central basin which spread in early Tertiary time, and a much older basin against Alaska and Canada. These latter coasts are bordered by a shelf with Triassic strata cut by diapirs of Pennsylvanian evaporites. These suggest to some that the sea began to open at that time (E. T. Tozer and R. Thorsteinsson, 1964; J. T. Wilson, 1963b and c, 1968).

Some Alaska geologists follow S. W. Carey's (1958) view that northern Alaska swung away from the Canadian Archipelago, bending and folding Alaska in an orocline. If this happened during the Tertiary period, it leaves the magnetic patterns in the oceans unexplained and introduces a special origin for the Aleutian Islands, which otherwise seem to form part of the general Pacific pattern. If it happened before that period, it cannot be the explanation of the Tertiary faulting, folding, and volcanism in Alaska and the Aleutian Islands. M. Churkin, Jr. (1969) has reviewed the problem and rejected oroclinal folding. The matter is of great economic interest to petroleum geologists.

15-8 THE GREENLAND AND NORWEGIAN SEAS AND BAFFIN BAY: EXAMPLES OF STAGE 2

The North Atlantic Ocean between Spitsbergen and Iceland has been well studied (W. B. Harland, 1969; G. L. Johnson and B. C. Heezen, 1967; P. R. Vogt et al., 1970a). It readily divides into three equal sections (Fig. 15-5).

In the northern section a mid-ocean ridge extends from the Lena trough off the west coast of Spitsbergen to latitude 73°N, where it joins the aseismic and lateral Greenland ridge and turns through nearly a right angle to become the Mohns ridge.

In the central section, from latitude 73°N to Jan Mayen Island, the Mohns ridge has all the usual features of an active mid-ocean ridge.

The southern section extends from a major fracture zone beside Jan Mayen Island to Iceland. It contains three ridges striking north and south, of which the central Jan Mayen ridge is aseismic and may, like the Lomonosov ridge, be a sliver of continental shelf (O. E. Avery and G. D. Burton, 1968). If so, it was probably separated from Greenland by the growth of the active ridge which now lies between them. The third ridge is a line of seamounts between Norway and the Jan Mayen ridge, which P. R. Vogt and his colleagues hold to be a formerly active but now extinct mid-ocean ridge, as shown in Fig. 15-6.

Jan Mayen Island is an active volcano built during the Pleistocene and displaying an interesting succession of lava flows and glacial deposits (F. J. Fitch et al., 1965; T. Gjelsvik, 1970).

Iceland lies at the intersection of the mid-ocean ridge, with broad lateral ridges extending past the Faroe Islands to volcanic rocks on the coasts of Greenland and Scotland (Sec. 14-6). The Faroe Islands have been carved from a pile 2900 m thick of iron-rich tholeiite lava flows which were extruded in Eocene time along

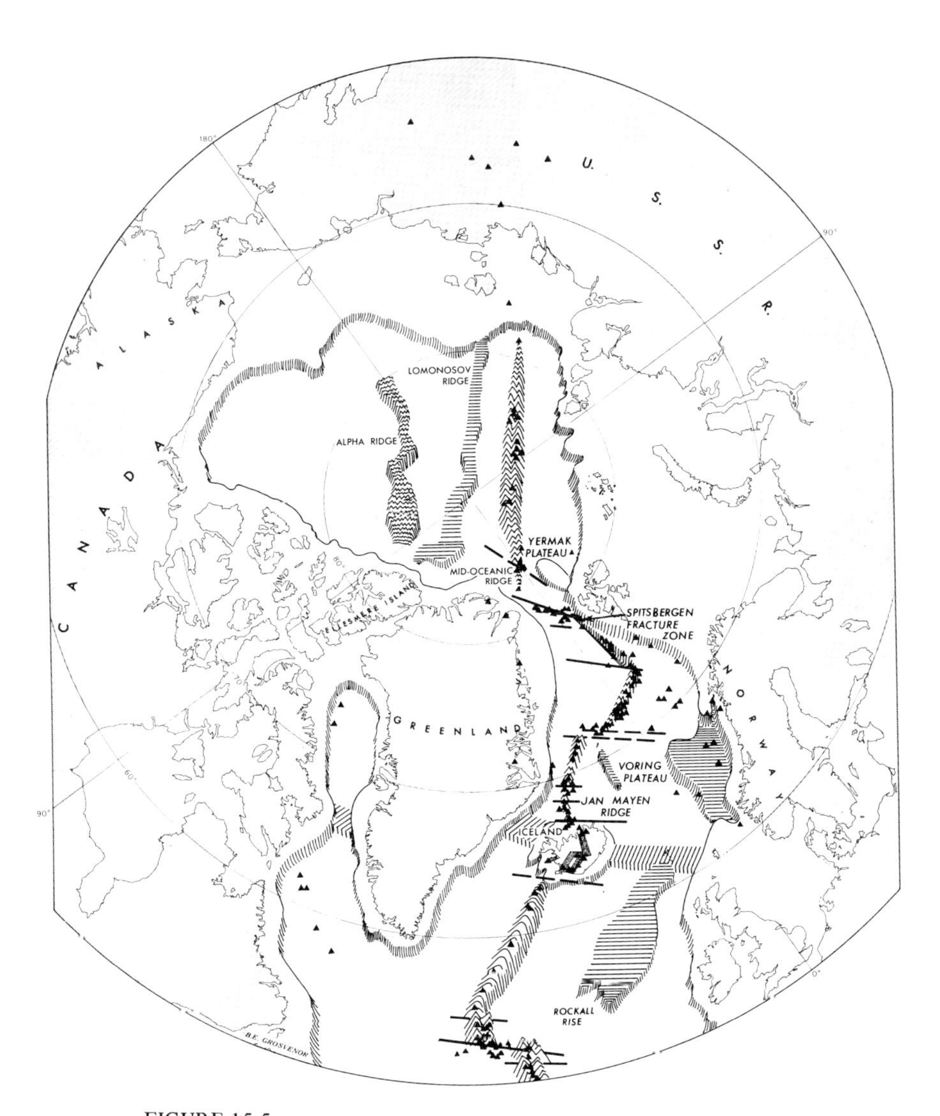

FIGURE 15-5
Schematic chart of physiographic features in deep-water areas of the Arctic. Triangles denote earthquake epicenters; horizontal ruling indicates micro-continents. Close-spaced and open-peaked patterns show active and former (Alpha ridge) mid-ocean ridges. (*After G. L. Johnson, 1969, and P. R. Vogt, N. A. Ostenso, and G. L. Johnson, 1970a.*)

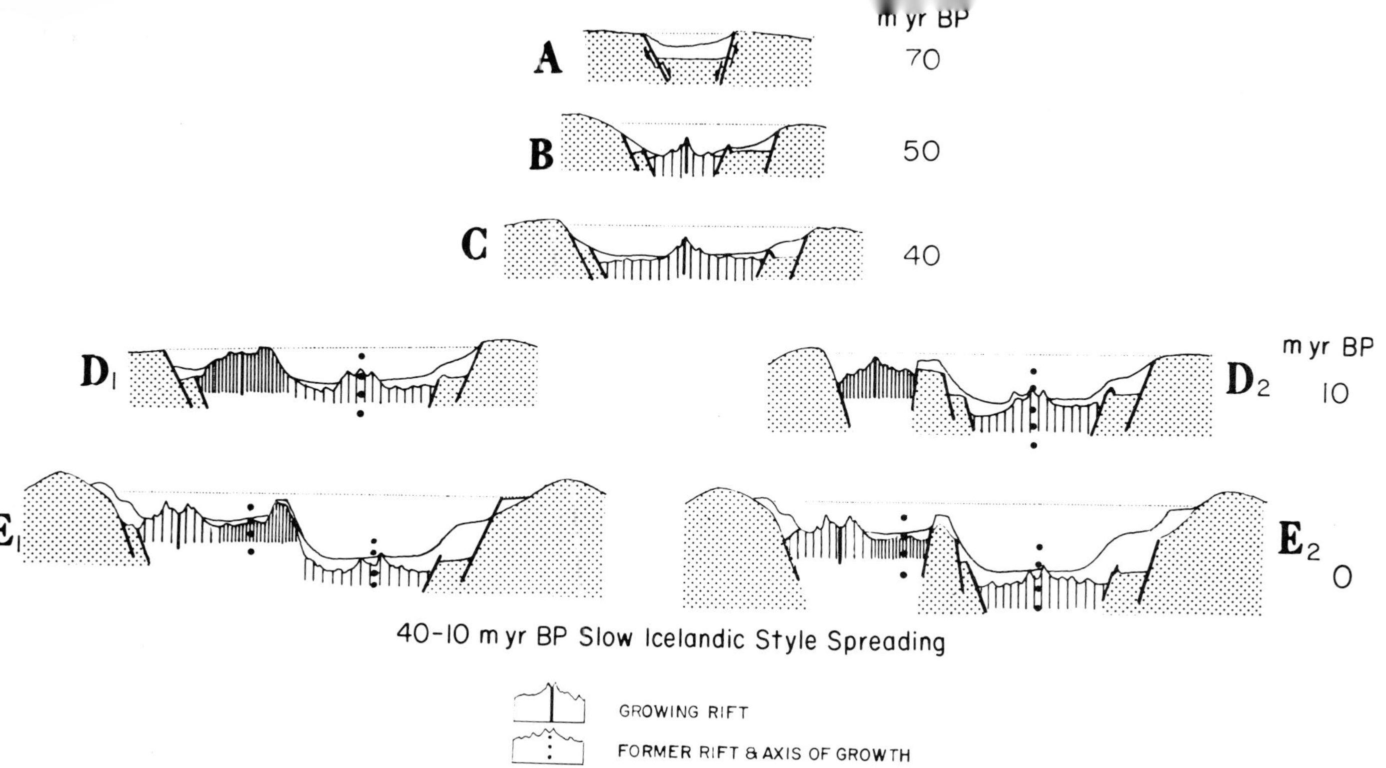

FIGURE 15-6
Evolution of the Norwegian basin north of Iceland. From 70 to 10 m yr ago the basin opened by spreading about a ridge between Norway and the Voring Plateau (Fig. 15-5). Two possible interpretations of the growth from 10 m yr ago to the present are given. Both involve spreading about two crests between the Voring Plateau and Greenland. (*After P. R. Vogt et al., 1970a.*)

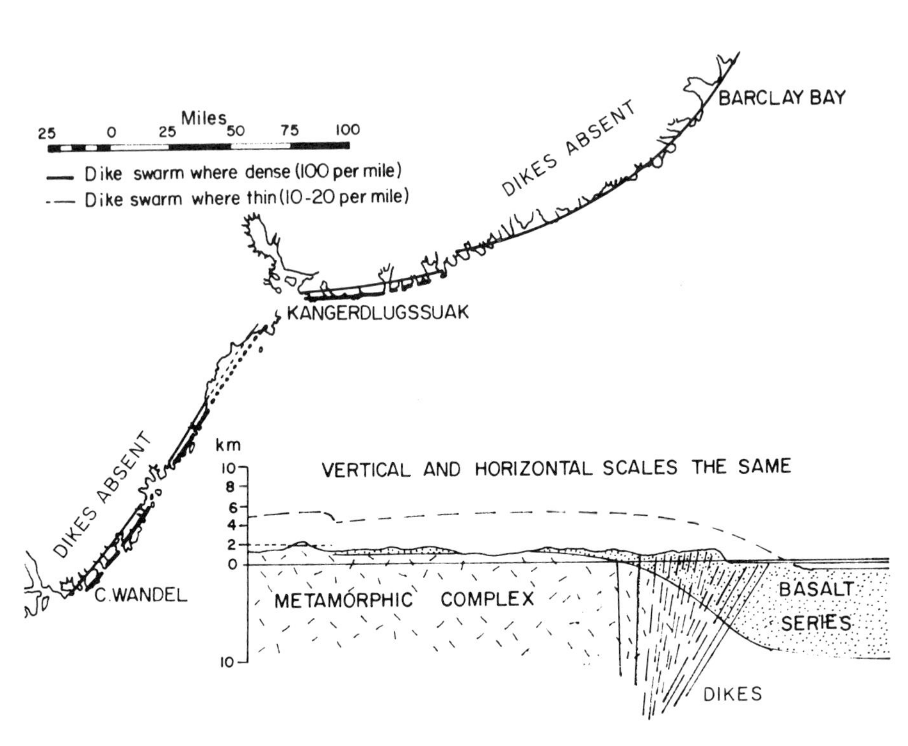

FIGURE 15-7
Sketch map and cross section showing the concentration of dikes close to the coast of east Greenland and their change in dip. Their curved strike is unlike the shape of mid-ocean ridge, suggesting that the orthogonal pattern of fracture zones and sections of ridge develops after the initial rifting of a continent. (*After L. R. Wager and W. A. Deer, 1938.*)

fissures and later warped into a gentle fold (A. Noe-Nygaard, 1968; J. Rasmussen and A. Noe-Nygaard, 1970). Measurements of gravity and of crustal thickness suggest that they and the ridges are oceanic. In contrast Rockall Bank and Islet, which extend to the southwest, are probably a microcontinental fragment (P. A. Sabine, 1965; S. Moorbath and H. Welke, 1968; A. S. Laughton and W. A. Berggren, 1970).

Two recent symposiums discuss the geological evidence for the opening of the Northern Atlantic (P. M. S. Blackett et al., 1965; S. Bjornsson, 1967). In the Triassic period shallow seas encroached upon the area, and in Upper Jurassic time faulting began along the coasts. Some separation and volcanism occurred during the Cretaceous period, and continental shelves began to form (V. D. Dibner et al., 1963; O. Holtedahl, 1970), but in early Tertiary time, between 54 and 60 m yr ago, extensive dikes and flows were emplaced along the present coasts with accompanying major uplift (A. Harker, 1904; W. S. Watt, 1969; I. Madirazza and S. Fregerslev, 1969; P. R. Vogt, 1970; D. B. Clarke and B. G. J. Upton, 1971). The curved shape of faults and dikes along the Greenland coast, if they were intruded along rifts as separation began, suggest that the orthogonal patterns of mid-ocean ridges develop after the original break (L. R. Wager and W. A. Deer, 1939; Fig 15-7).

P. R. Vogt et al., (1969a), by extensive magnetic surveys, have traced the subsequent history of sea-floor spreading, pointing out that from 60 to 20 m yr ago the Norwegian Sea, Baffin Bay, and the main North Atlantic were all spreading, although the rates and directions changed 42 m yr ago (Fig. 15-8). For the last 20 m yr Baffin Bay has been inactive. This interpretation is compatible with the bathymetry (B. R. Pelletier, 1966), with geophysical work (C. L. Drake et al., 1963; D. L. Barrett et al., 1971), with the later geological interpretation of J. W. Kerr (1967, 1970), with the paucity of earthquakes there (W. G. Milne et al., 1970), and with the formation during Pleistocene time of a great undisturbed submarine canyon along the axis of the bay (B. C. Heezen et al., 1970a), but debate continues about the amount of offset in Nares Strait due to faulting.

During the Tertiary period the northern coasts were uplifted to form the existing mountains of Norway, east and west Greenland, Baffin Bay, and Labrador. According to the hypothesis of a cycle of ocean building, these uplifts, faults, and intrusions in the North Atlantic region formed during the preliminary doming, rifting, and volcanism prior to opening of the ocean, and they have the same origin as the rise to the rift and the flood basalts of the East African and Red Sea regions. This suggests that Iceland has long been a source of excess lavas which formed the lateral ridges joining it to the uplifted coasts. Farther south the ocean basin opened earlier, and the coastal uplifts have subsided as the shores spread away from the broadening Mid-Atlantic Ridge (Table 15-1 and Fig. 3-6).

15-9 STAGE 3: THE ATLANTIC OCEAN BETWEEN EUROPE AND NORTH AMERICA

B. C. Heezen et al. (1959 and 1968a) have published charts of this region, and a recent symposium on the opposite coasts of the North Atlantic contains many important papers. The editor found that "the similarities in the stratigraphic-structural belts of Newfoundland and the British Isles are so great that the probability of their being accidental is trivial" (M. Kay, 1969, p. 967). Either the belts continue across the ocean floor, for which there is not the slightest evidence, or beginning in early Mesozoic time the continents have spread apart, as indicated by paleomagnetism and sea-floor spreading (J. Hospers and S. I. Van Andel, 1968; F. J. Vine, 1966; J. R. Heirtzler et al., 1968; P. R. Vogt et al., 1969a).

The continental shelves have been thoroughly studied. On the European side A. H. Stride et al. (1969) believe that this section of the Atlantic had begun to open by Cretaceous time and that it may have done so in two stages, with an intervening

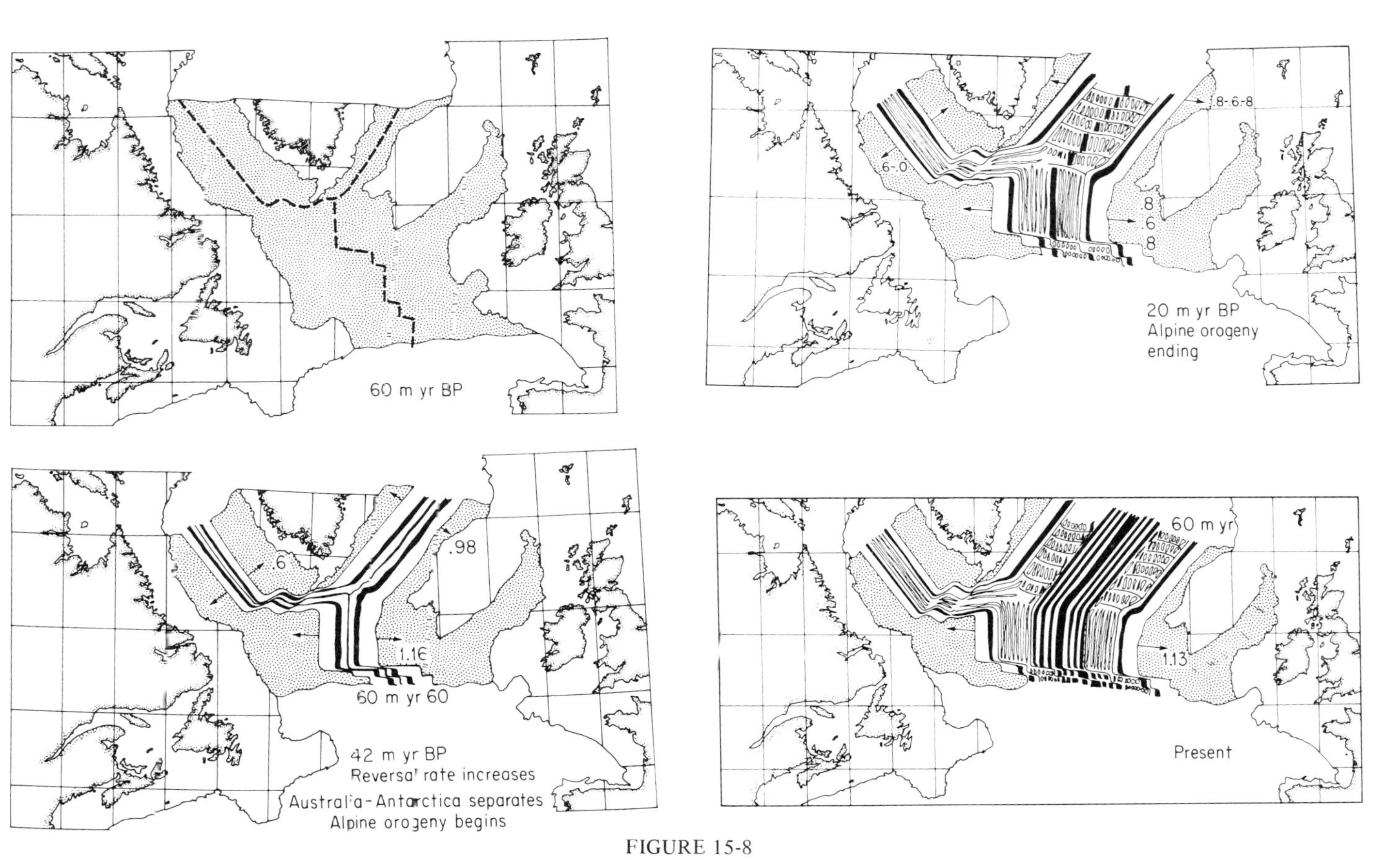

FIGURE 15-8
Stages in the opening of the North Atlantic Ocean showing mapped magnetic anomalies in diagrammatic fashion. (*After P. R. Vogt et al., 1969a.*)

episode of erosion of the continental margin in Late Cretaceous to early Tertiary time. Intrusives on the coast of central Portugal are Upper Jurassic in age, and near Lisbon Upper Cretaceous (85 ± 8 m yr), and thus older than those in the Brito-Arctic province (G. Zbyszewski, 1961; J. B. Wright, 1969).

Off eastern Canada A. M. Dainty et al. (1966), M. J. Keen (1968), and M. A. Mayhew et al. (1970) have shown that both the Appalachian mobile belt and the continental crust end near the edge of the shelf. They are overlain by Cretaceous and Tertiary strata which have been drilled (A. C. Grant, 1970; G. A. Bartlett and L. Smith, 1971; Fig. 15-9).

P. Allen (1969) used the distribution and source of sediments and the age of intrusives to argue that the separation of the Canadian Appalachians from Spain and the British Isles took place in Lower Cretaceous time between 135 and 110 m yr ago. On the other hand, oil geologists have dredged Upper Jurassic sedimentary rocks off the coast of Labrador near 55°N, but these may be shallow-water deposits.

Separation may be related to faulting and possible intrusions along the south side of the Grand Banks, to rifting in Nova Scotia in the St. Lawrence Valley and perhaps in the Strait of Belle Isle, and to the intrusion of many small bodies about 120 m yr old in the Monteregian Hills and in New England (P. A. Christopher, 1969; R. Doig, 1970; L. H. King and B. MacLean, 1970a,b).

To reconcile the fit of Europe and North America, it is necessary for Spain to have rotated and the Bay of Biscay to have opened, and much geophysical work supports this conclusion (D. H. Matthews and C. A. Williams, 1968; E. W. J. Jones and B. M. Funnell, 1968; M. Bauer et al., 1969; R. van de Voo, 1969), but some geologists working in the Pyrenees object (M. Mattauer, 1969).

J. R. Cann and B. M. Funnell (1967) and A. T. S. Ramsey (1970) have related this rotation to the uplift of Eocene oceanic basement containing much serpentine to form the submarine Palmer ridge off Spain. Other problems concern the possible rotation of Newfoundland and the fitting of Flemish Cap and Galicia Bank if they are not oceanic (P. J. Hood, 1966). The intervening section of the Mid-Atlantic Ridge has been well studied near 45°N by the Bedford Institute of Oceanography and near the Azores islands (E. Irving, 1970). These islands and the seismically active east Azores fracture zone extending to Gibraltar are unusual and complex features. The oldest rocks are Miocene marine strata on Santa Maria, the island farthest east of the mid-ocean ridge (G. Zbyszewski et al., 1961, 1962). F. Machado et al. (1962) have described a recent eruption on Fayal, which is near the ridge. D. C. Krause and N. D. Watkins (1970) have interpreted bathymetric and geophysical data to produce a model of the history of the development of this region. They suggest that the present system began 45 m yr ago as a triple junction, but K. Burke et al. (in press) have proposed a younger alternative explanation.

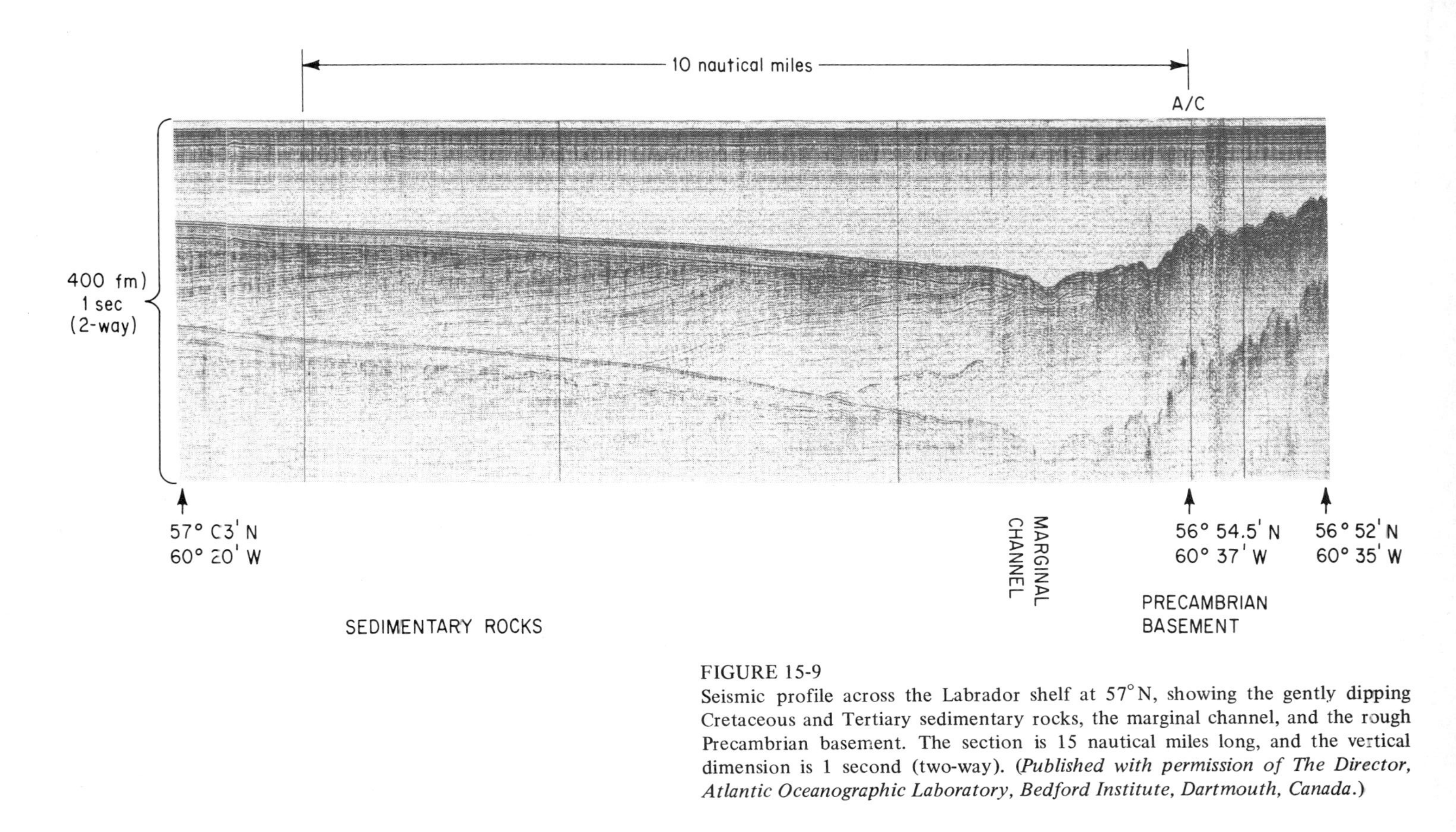

FIGURE 15-9
Seismic profile across the Labrador shelf at 57°N, showing the gently dipping Cretaceous and Tertiary sedimentary rocks, the marginal channel, and the rough Precambrian basement. The section is 15 nautical miles long, and the vertical dimension is 1 second (two-way). (*Published with permission of The Director, Atlantic Oceanographic Laboratory, Bedford Institute, Dartmouth, Canada.*)

15-10 STAGE 3: THE ATLANTIC OCEAN BETWEEN THE UNITED STATES AND AFRICA

The Atlantic basin between 40 and 15°N lies between the United States and Africa. If the Atlantic first opened from the Tethys Sea near Gibraltar to the Gulf of Mexico to divide the older supercontinent Pangaea of Wegener into Laurasia and Gondwanaland, this may be the oldest part (R. S. Dietz and J. C. Holden, 1970a,b). Permian and Triassic rocks around the Atlantic are nonmarine sediments with evaporites, suggesting continental desert conditions. The first indication of opening is the Permian uplift identified in New England, which may extend from Nova Scotia to the southern Appalachians (R. E. Zartman et al., 1970). During the Upper Triassic, faulting formed grabens in which clastics and basalts collected and a Pacific sea crossed into the Gulf of Mexico to spread up the Atlantic coast (G. de V. Klein, 1969; B. M. Funnell and A. G. Smith, 1968). Offshore the east-coast magnetic anomaly closely follows the edge of the continental shelf as far south as Georgia, where it turns inland, and diabase intrusives at least 140 m yr old have been drilled (R. L. Grasty and J. T. Wilson, 1967). P. T. Taylor et al. (1968) consider the anomaly to be due to a large intrusive connected with the opening. D. R. Bracey (1968) found a similar anomaly off the northern margin of the Bahamas and Antilles islands. Many dikes and sills of about the same age crop out along the coast from Nova Scotia to Virginia, including the Palisades sill in New Jersey (K. R. Walker, 1969), but the radiometric dates are suspected of being too young, and all the intrusives may be Triassic or even older (R. L. Armstrong and J. Besancon, 1970). This would agree better with the interpretations of magnetic anomalies on the sea floor which suggest an earlier opening. J. R. Heirtzler and D. E. Hayes (1967) consider that the weak anomalies near the coast were formed by spreading during the long Kaiman interval without reversals in upper Carboniferous and Permian time, but P. R. Vogt et al. (1969a, 1970a) consider them to be Upper Permian. R. L. Larson and W. C. Pitman III (1972) have dated the Keathley lineations between Bermuda and the Bahama islands as 150 to 110 m yr old.

Few coasts have been so thoroughly examined as the east coast of the United States (B. C. Heezen, 1968; J. L. Worzel, 1968; J. E. Nafe and C. L. Drake, 1969; H. D. Hedburg, 1970; K. O. Emery et al., 1970). The thick wedges of sediment which lie along the coast have grown outward and upward as the continental margin sank. They are shallow, well-sorted shelf deposits of a miogeosynclinal nature (Fig. 15-9). C. L. Drake et al. (1959) found a double trough off New York, but this appears to be a local irregularity. The intervening ridge could be a slumped and buried block. The deep Blake Plateau off Florida is partly due to lack of deposition and partly due to settling and has an unusual outer ridge beyond it on the sea floor (R. G. Markl et al., 1970).

The opposite coast of Northwestern Africa is less known, but P. A. Rona (1969, 1970) has shown that its pre-Mesozoic geology resembles that of the United States, and that the Tertiary geology differs. Salt domes southwest of the Canary Islands resemble those off Newfoundland and in the Gulf of Mexico. R. E. Sheridan et al. (1969) found that the ocean floor has been faulted down along the African coast (Fig. 15-10).

The Bermuda, Bahama, Madeira, and Canary and Cape Verde islands are all oceanic, and the last two contain active volcanoes (A. R McBirney and I. G. Gass, 1967). Bermuda is a basalt volcano on a large rise capped with coralline limestone and dune deposits (G. Blackburn and R. M. Taylor, 1969; R. A. Gees, 1970). The Bermuda rise has been uplifted since Cretaceous time (H. W. Menard, 1969a; G. B. Engelen, 1964). Many volcanic seamounts extend north from it to the New England seamount chain (J. Northrop et al., 1962; J. E. Walczak and T. Carter, 1964).

Beneath the Bahamas drilling has penetrated 5 km of shallow-water coralline limestone which rests upon more sediments and basalt (E. T. Miller and M. Ewing, 1956; J. E. Andrews et al., 1970; R. S. Dietz et al., 1970a, P. R. Vogt et al., 1970).

Differences in the geology of the eastern and western Canary Islands suggest that the two Canary islands closest to Africa are underlain by a detached continental fragment (P. Rothé and H.-U. Schmincke, 1968; D. T. McFarlane and W. E. Ridley, 1969; R. S. Dietz and W. P. Sproll, 1970a). Maio in the Cape Verde Islands has Lower Cretaceous and perhaps Upper Jurassic marine fossils, the oldest found on any oceanic island. Carbonatitic dikes have also been reported (A. Serralheiro, 1970).

Petrological studies show that the sea floor is composed predominantly of tholeiite basalt similar to continental flood basalts but lower in potassium (A. E. J. Engel and C. G. Engel, 1964; W. G. Melson et al., 1968; A. Miyashiro et al., 1969; R. Kay et al., 1970). Sediments on the ocean floor increase in thickness from zero at the crest of the Mid-Atlantic Ridge toward the coasts, where there are many large abyssal plains (J. E. Nafe and C. L. Drake, 1969; M. Ewing et al., 1969a; B. J. Collette et al., 1969). Three widespread reflecting horizons near Bermuda furnish some stratigraphic control. M. Ewing et al. (1966a) and T. Saito et al. (1966b) first identified horizon *A*, which JOIDES drilling found to be the top of a chert layer of Eocene age, probably from volcanic sources (M. N. A. Peterson et al., 1970, p. 32; T. G. Gibson and K. M. Towe, 1971). Horizon β appears to mark an unconformity near the base of the Upper Cretaceous, and horizon *B* is considered to lie within the Jurassic, the oldest series found (C. C. Windisch et al., 1968, p. 237).

Along the north shore of the Gulf of Mexico drilling has penetrated diapirs of the Louann evaporites believed to be of Upper Triassic age, and geophysical work in the Gulf of Mexico by M. Ewing and his colleagues discovered the Sigsbee and other knolls in water about 3000 m deep. JOIDES drilling in one has penetrated Middle to Upper Jurassic cap rock, confirming that the knolls are salt domes and that much of the Gulf of Mexico was an evaporite basin at that time, suggesting that the Atlantic had opened as a rift (M. T. Halbouty, 1967; W. R. Bryant et al., 1968, M. Ewing et al., 1969a). Since the East Pacific rise had not then opened, it is interesting to speculate whether the mid-ocean ridge which opened the Atlantic south of Gibraltar passed through the Gulf of Mexico to join the Darwin rise (Sec. 16-8).

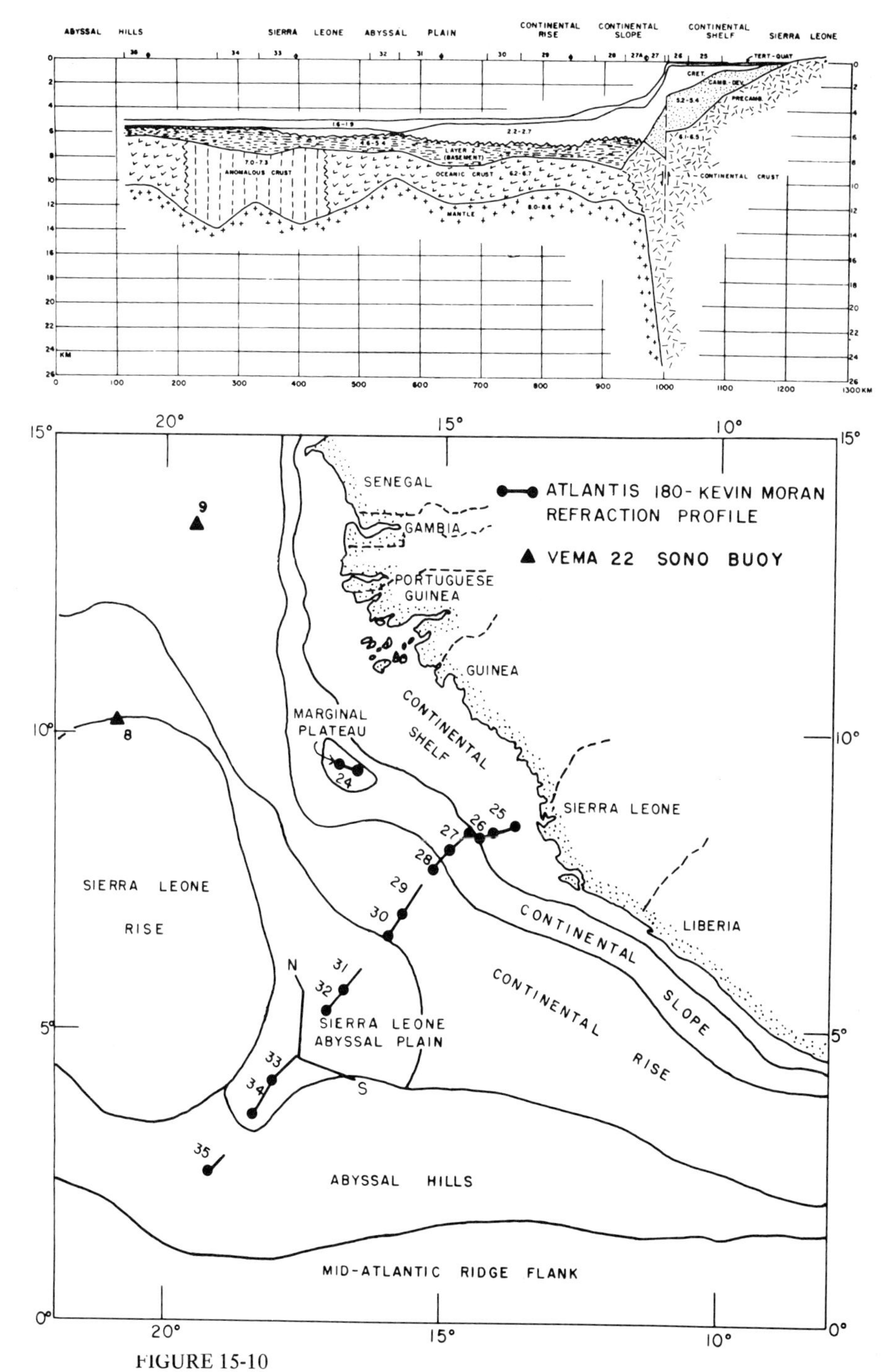

FIGURE 15-10
Cross section from seismic-refraction data of the continental margin off Sierra Leone. (*After R. E. Sheridan et al., 1969.*)

15-11 THE TYPE EXAMPLE OF STAGE 3: THE ATLANTIC OCEAN BETWEEN AFRICA AND SOUTH AMERICA

In this region opposite coasts were first noticed to have similar shapes, and improved bathymetry has shown that the deep contours of continental slopes match well (E. C. Bullard et al., 1965; B. C. Heezen and M. Tharp, 1968). The geology along the two shores also corresponds, and A. L. du Toit (1927) listed 15 matching features which more recent work has confirmed. Thus P. M. Hurley and J. R. Rand (1969) have shown that the boundaries of the basement provinces fit together, and G. O. Allard and V. J. Hurst (1969), after mapping a succession of Precambrian strata in Brazil, found their apparent continuation with the same strike in Gabon (Fig. 13-1).

Drilling for petroleum in basins along both coasts has shown that terrestrial and lacustrine Jurassic and Lower Cretaceous strata correspond closely. The Serra Geral intrusives of younger age, 125 m yr, in the Parana basin are matched on the opposite coast by the Kaoko volcanics (I. McDougall and N. R. Ruegg, 1966; G. Amaral et al., 1966; G. Siedner and J. A. Miller, 1968). In both formations, dikes strike parallel to the coast.

On the other hand, younger beds are predominantly marine and show increasing divergences. Upper Aptian-Albian (mid-Cretaceous) evaporite deposits on both coasts suggest that separation into a narrow sea occurred at that time (W. Q. Kennedy et al., 1965; R. A. Reyment, 1966; F. F. M. de Almeida, 1969; J. J. Zambrano and C. M. Urien, 1970; L. W. Butler, 1970). Some evidence suggests that the opening spread from south to north (J. B. Wright, 1968; R. B. McConnell, 1969).

In the axial part of the South Atlantic basin the magnetic evidence for sea-floor spreading is remarkably clear. Along the west side G. O. Dickson et al. (1968) have traced 31 key anomalies and A. E. Maxwell et al. (1970a,b) have confirmed that anomaly 31 is 75 m yr old. Near both coasts the anomalies are small. Extrapolation at a constant rate of spreading would put the continents together about 110 m yr ago (R. L. Larson and W. C. Pitman III, 1972).

In the South Atlantic basin aseismic lateral ridges form a herringbone pattern. Most conspicuous are the Walvis and Rio Grande ridges. They meet the coasts at points which would touch one another if the Atlantic were closed and near which the Serra Geral and Kaoko volcanics are most abundant. Both ridges strike toward the active volcano Tristan da Cunha close to the Mid-Atlantic Ridge. Several other parallel ridges lie off Sierra Leone, the Gulf of Guinea, Cape Town, and off South America, connecting flood basalts on shore with islands on the ridges (F. F. M. de Almeida, 1960, 1969; L. A. Barros, 1960; Fig. 17-7; Sec. 17-5).

M. Ewing et al. (1966a) proposed that the Walvis and Rio Grande ridges are microcontinents or sections of ocean floor which have been upfaulted, but they could also be trails left by volcanic hot spots due either to enduring points of weakness or long-lived hot upwellings on the mid-ocean ridge which have continually generated an excess of lava (J. T. Wilson, 1965a; R. S. Dietz and J. C. Holden, 1970a,b).

In the South Atlantic all the islands with two exceptions consist of tholeiites and alkali basalts of Cenozoic age with minor trachyte intrusives. One exception is Rocas, an atoll of algal limestone on the coast of Brazil (F. Ottman, 1962), and another is St. Paul Rocks, which are ultrabasic rocks faulted up from the mantle (W. G. Melson et al., 1967a,b, 1971).

The basaltic islands are Fernando de Noronha, Trinidade, Martim Vas, and Albrohos off Brazil (F. F. M. de Almeida, 1955, 1961; A. Richardson and N. D. Watkins, 1967); four Spanish and Portuguese islands aligned southwesterly through the Gulf of Guinea; and Ascension, St. Helena, Tristan da Cunha, Gough, and Bouvet islands, which all lie near the crest of the Mid-Atlantic Ridge (F. B. Atkins et al., 1964; I. Baker, 1969; I. G. Gass, 1967; R. W. Le Maitre, 1962; P. E. Baker, 1967a). Off Cape Town the Vema seamount rises almost to the surface (J. R. Heirtzler and M. L. Hadley, 1966; J. A. Cooper et al., 1966). Bouvet, Tristan, and Ascension are active volcanoes.

Near the equator the Atlantic basin swings through an S-shaped bend, to which the mid-ocean ridge accommodates by a series of orthogonal offsets (B. C. Heezen et al., 1964). If these were left-handed transcurrent faults, one would expect that (1) they would cross the ocean and offset the geology of both continents; (2) earthquakes would occur all along them; and (3) recent earthquakes would show left-handed directions of motion. Except for slight seismic activity where the fracture zones reach the African coast, none of these criteria are true, and they are now generally regarded as right-handed transform faults (J. T. Wilson, 1965b; L. R. Sykes, 1967; Sec. 14-3).

To the south the Atlantic Ocean passes on to the Antarctic plate, which meets the American and African plates in a triple point near Bouvet Island (Fig. 14-2). Adjacent to the boundary is the young Scotia arc, consisting of seven very active volcanoes with a deep trench and many earthquakes (D. H. Matthews, 1959; H. H. Lamb, 1970). B. C. Heezen and G. L. Johnson (1965) consider it to be similar to the West Indies, and like that arc to be joined at either end to the adjacent continents by large systems of strike-slip faults. P. F. Barker (1970), noting how very much more active the arc appears to be than the connecting fault zones, used geophysical data (D. H. Griffiths et al., 1964, 1967) to propose that the arc is part of a separate small lithospheric plate which is being propelled into the Scotia trench by growth of a short section of mid-ocean ridge close to the arc and nearly parallel to it. He also shows a fault across Drake Strait, whereas Griffiths suggested that the relic of an inactive ridge passed through the strait.

15-12 THE INDIAN AND SOUTHERN OCEANS

The Indian Ocean is the least explored ocean, but B. C. Heezen and M. Tharp (1965, 1966, 1967) and G. B. Udintsev (1962) have compiled bathymetric charts, and general accounts have been published of the distribution of sediments (M. Ewing et al., 1969b), seismicity (L. R. Sykes, 1970), gravity anomalies (X. Le

Pichon and M. Talwani, 1969), heat flow (R. P. von Herzen and W. H. K. Lee, 1969), and magnetic anomalies (X. Le Pichon and J. R. Heirtzler, 1968; Fig. 15-11).

The history of this ocean is complex and involves determining the correct reassembly of Gondwanaland and a description of the times and manner in which the many land masses involved separated. Most authors agree that the ridge which crosses the Indian Ocean diagonally from the Gulf of Aden to the East Pacific rise has been spreading recently (F. J. Vine, 1968), and most favor A. L. du Toit's (1937) original reassembly (Fig. 14-13; A. G. Smith and A. Hallam, 1970; M. W. McElhinny, 1970; R. S. Dietz and J. C. Holden, 1971), but other views are still tenable (D. H. Tarling, 1971; J. J. Veevers et al., 1971).

The ocean can conveniently be considered as a stubby diagonal cross. We will describe the four arms in turn, separating them by imaginary boundaries running south from Ceylon and east from the northern tip of Madagascar.

The southeastern quadrant is the simplest and youngest, for it has few islands, and a single active section of mid-ocean ridge bisects it. Paleomagnetic data suggest that this ridge may have remained fixed in latitude while Australia, Antarctica, and oceanic islands moved away from it (P. Wellman et al., 1969; E. Irving and W. A. Robertson, 1969). X. Le Pichon and J. R. Heirtzler (1968) found that magnetic anomaly 18 lies against the Australian continental shelf, suggesting that these continents began to separate 43 m yr ago by rotation about a pole in Libya. Topography supports this fit and also suggests that the aseismic Kerguelen-Gaussberg and Broken ridges were once together (W. P. Sproll and R. S. Dietz, 1969). Both may be volcanic boundary ridges formed where new spreading started to divide an older plate (B. C. Heezen and M. Tharp, 1965; G. B. Udintsev, 1966). The Wharton basin north of Broken ridge appears to be old, for the sediments on its floor are thick and in part Cretaceous, and the Sahul shelf north of Australia is one of the widest in the world (R. S. Dietz and J. C. Holden, 1971; T. H. van Andel and J. J. Veevers, 1965; N. D. Opdyke and B. P. Glass, 1969).

The ridges linking the western ends of Kerguelen and Broken ridges to Amsterdam and St. Paul Islands may be a pair of lateral ridges meeting in a common hot spot (J. T. Wilson, 1963c). The latter is a volcanic cone active in 1821, not to be confused with St. Paul Rocks off Brazil (R. L. Fisher, 1966). Another pair of lateral ridges link Tasmania to Cape Adare, Antarctica.

An Eocene marine invasion along the south coast of Australia supports that date of separation (R. C. Sprigg, 1952; D. A. Brown et al., 1968). On both coasts igneous rocks are found only near the ends of the lateral ridges in Tasmania and at Cape Naturaliste, Cape Adare, and Gaussberg (H. J. Harrington, 1965; H. J. Harrington et al., 1967; D. A. Brown et al., 1968). The Jurassic volcanics of Tasmania and the Ross Sea are much older and were probably connected with tectonic events in the Pacific. There is Tertiary and active volcanism in the Balleny Islands and in McMurdo Sound (T. Hatherton et al., 1965).

On the Kerguelen-Gaussberg ridge, Heard Island has marine Eocene limestones interbedded with its lower volcanics (P. J. Stephenson, 1963; E. Irving et al.,

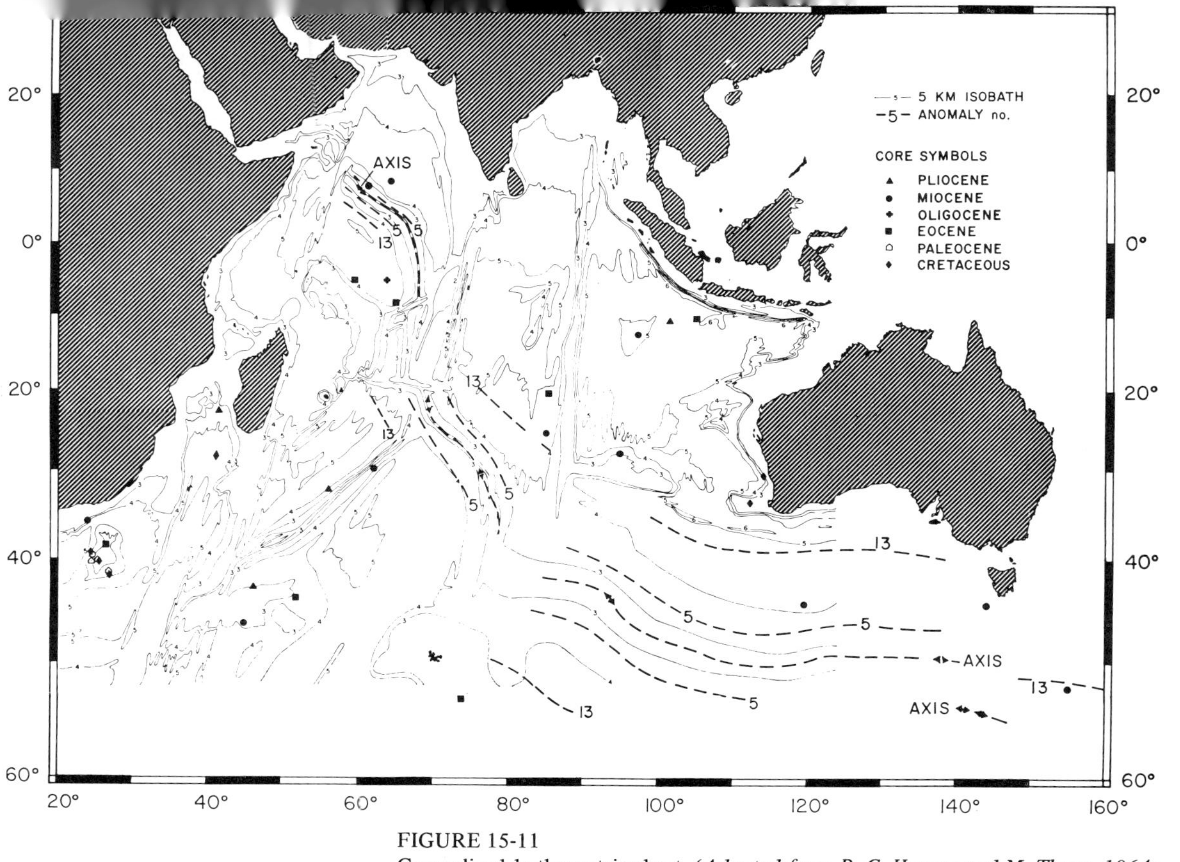

FIGURE 15-11

Generalized bathymetric chart. (*Adapted from B. C. Heezen and M. Tharp, 1964; pre-Pleistocene cores are from T. Saito, 1967. Dashed lines are magnetic anomalies identified according to the nomenclature of W. C. Pitman et al., 1968; arrowheads indicate magnetic axis of ridge where identified. After X. Le Pichon and J. R. Heirtzler, 1968.*)

1965), and the Kerguelen Islands are a 6500-km^2 archipelago in the area, built of Eocene to Recent flood basalts (A. B. Edwards, 1938; J. Nougier, 1969).

The southwestern quadrant of the Indian Ocean between Africa and Antarctica is older, and much of it is thickly covered with sediments partly of Cretaceous age (Y. Herman, 1963; M. Ewing et al., 1969b). Paleomagnetic evidence also suggests that it has not spread since Eocene time (M. W. McElhinny and P. Wellman, 1969). R. S. Dietz and W. P. Sproll (1970b) have fitted Africa against Antarctica. The mid-ocean ridge which divided them is offset by many large fracture zones but is scarcely active at present (F. J. Vine, 1968; A. R. Banghar and L. R. Sykes, 1969; D. P. McKenzie and J. G. Sclater, 1971).

This quadrant contains Isles Crozet, Marion, and Prince Edward and Mascarene groups of volcanic islands, some rising from poorly defined aseismic ridges (P. Bellair, 1963; J. F. Truswell, 1965; W. J. Verwoerd and O. Langenegger, 1967).

Of the Mascarene Islands, Réunion is a double volcano, one active and the other 2 m yr old, with geology resembling that of the Hawaiian Islands (F. H. Chamalaun, 1968; B. G. Upton and W. J. Wadsworth, 1970). Mauritius is a shield volcano built subaerially in two stages in Pliocene and Pleistocene time (E. S. W. Simpson, 1951; F. Walker and L. O. Nicolaysen, 1954; I. McDougall and F. H. Chamalaun, 1969). Rodriguez is a Pleistocene basaltic island (B. G. J. Upton, W. J. Wadsworth, and T. C. Newman, 1967). All rise from the large and shallow Mascarene Plateau, most of which is a volcanic ridge (R. L. Fisher et al., 1967), but the Seychelles Islands are unmistakably continental, being granite hills 600 m yr old, cut by Eocene diabase dikes (B. H. Baker, 1963; B. H. Baker and J. A. Miller, 1963; D. Davies, 1968). So also is the Agulhas Bank off South Africa (K. W. T. Graham and A. L. Hales, 1965; R. W. E. Green and A. L. Hales, 1966) and Socotra Island (Sec. 15-4).

Madagascar is a Precambrian plateau with a monocline of Karroo to Tertiary sediments along the Mozambique Channel. There are Miocene volcanics in the northern part, and the east coast has been uplifted in a half dome and faulted (R. Furon, 1963). Three main views have been advanced about the movements of Madagascar relative to Africa: (1) that it has been fixed and that Mozambique Channel is a sedimentary basin which has been the source of sedimentary inclusions found in the volcanic rocks of the Comores Archipelago in which Grande Comore is active (F. Dixey, 1960; J. Esson, 1968; D. F. Strong and M. J. F. Flower, 1969; D. F. Strong and C. Jacquot, 1970); (2) that it has moved south from a position against the coast of Tanzania; and (3) that it has moved northeast from a position against Mozambique.

The northwestern quadrant of the Indian Ocean is best known. The young and active Carlsberg mid-ocean ridge crosses it to end in the north along the Owen fracture zone, an active ridge–concave arc transform fault extending from the mid-ocean ridge to the Hindu Kush (J. T. Wilson, 1965b; D. H. Matthews, 1966; R. L. Fisher et al., 1971). The sea-floor basalts and spilites are normal (R. Hekinian, 1968; J. R. Cann, 1969; C. G. Engel and R. L. Fisher, 1969). The sediments thicken rapidly away from the crest into a submarine fan formed by the Indus River since the mid-Miocene uplift of the Hindu Kush and Himalayas and a thick shelf off Kenya (B. H. Bungenstock et al., 1966; T. J. G. Francis et al., 1966; M. Ewing et al., 1969b).

Geophysical surveys have shown the Laccadive-Maldive and Mascarene ridges to be composed of thick coral deposits resting on basalt, and they have been held to be either boundary or, more likely, lateral ridges connected with the Paleocene separation of India, Seychelles, and Madagascar (E. A. Glennie, 1936; J. T. Wilson, 1963c; D. Davies and T. J. G. Francis, 1964; R. L. Fisher et al., 1967; D. Davies, 1968). In deciphering the structure, account must be taken of the place of the Vema trench, an isolated arcuate deep resembling other ocean trenches but without an accompanying island arc (B. C. Heezen and J. E. Nafe, 1964).

Banks in this quadrant bear many reefs, atolls, and islands, among which Aldabra is the best known (W. M. Davis, 1928; J. S. Gardiner, 1926–1936; T. S. Westoll and D. R. Stoddard, 1971). All outcrops are recent except for Oligocene or Eocene limestone dredged off Providence Island north of Madagascar (J. D. H. Wisemen, 1937).

The northeast quadrant has no active mid-ocean ridge, and its most striking feature is the long straight aseismic Ninety East ridge, discovered during the International Indian Ocean Expedition, 1959–1965. It divides this quadrant into the Wharton basin already described and the mid-Indian basin.

The mid-Indian basin has thin sedimentary cover except in the extreme north where rapid erosion of the rising Himalayas has built the Ganges delta and a great submarine fan, marked by large submarine channels (J. R. Curry and D. G. Moore, 1971). This and other geological evidence from the mountains support paleomagnetic data for rapid growth of the mid-ocean ridge to the south and rapid northward movement of India from 30°S to 20°N since Miocene time (R. S. Dietz, 1953; A. Gansser, 1966; M. W. McElhinny, 1968).

In the Wharton basin lie the Cocos Islands, where C. Darwin (1962) obtained his classic ideas on the formation of coral islands and atolls, and Christmas Island, a former atoll uplifted 360 m to expose Eocene basalts intruding coralline strata (N. A. Trueman, 1965). The uplift has been ascribed to bending of a lithospheric plate in front of the Indonesian trench (J. T. Wilson, 1963a).

The Indonesian islands in this region form a characteristic island arc (F. A. Vening Meinesz, 1929; J. H. F. Umbgrove, 1949; R. W. van Bemmelen, 1949; A. Holmes, 1965). Although the most southerly islands are Tertiary in age, successively older ranges succeed one another toward Borneo, where Devonian rocks crop out, as though the region had been growing outward. L. R. Sykes (1970) has proposed that earthquakes in the ocean may mark the start of yet another arc. Great strike-slip faults through Sumatra, Malaysia, and the Philippines indicate relative movement of Indonesia toward the southeast. This motion appears to be related to spreading of the crust and the generation of this and young ocean floor in the Andaman, Celebes, and Sulu basins and the South China Sea (Sec. 16-9; K. S. Rodolfo, 1969; J. A. Katili, 1970; M. F. Ridd, 1971; D. E. Karig, 1971).

16

THE LIFE CYCLE OF OCEAN BASINS: STAGES OF DECLINE

16-1 THE TYPE EXAMPLE OF STAGE 4: THE PACIFIC OCEAN BASIN

This chapter is devoted to the three stages of shrinking ocean basins (Table 15-1 and Fig. 3-6). The type example for Stage 4 is the Pacific, which is regarded as a remnant of the former ocean Panthalassa, which in Permian and Triassic time covered all the earth's surface except for the former supercontinent of Pangaea (Fig. 12-4).

It may seem a paradox to suggest that this great basin, which contains the world's largest ocean and its most rapidly spreading mid-ocean ridge, is diminishing, but a moment's consideration will convince one that this is possible if the combined rate of shrinking of all the margins is greater than the rate of growth of the East Pacific rise. The deep trenches, the young folded mountains, and the great faults which surround the Pacific floor are evidence that the surrounding continents are overriding it.

G. B. Udintsev et al. (1964), B. C. Heezen and M. Tharp (1969, 1971), T. E. Chase and H. W. Menard (1969a), H. W. Menard (1964), and J. Z. Fraser et al. (1972) have published general charts and accounts. Many pertinent papers are included in symposiums and collections of papers edited by T. Matsumoto (1967), L. Knopoff (1968), P. J. Hart (1969), L. Knopoff et al. (1969), and A. E. Maxwell et al., (1970c).

H. W. Menard (1960) and H. H. Hess (1962) early divided the basin into the active East Pacific rise, which forms the floor of the eastern and southern parts, and an older northwestern part (Fig. 14-8).

16-2 THE EAST PACIFIC RISE

The East Pacific rise is a vast swell on the sea floor comparable in area with North and South America combined (Fig. 16-1). It extends for 13,000 km from Mexico to south of New Zealand, and it is between 2000 and 4000 km wide. The surface slopes gently up to the crest, which is between 2 and 3 km above the surrounding abyssal plains. The volume of the ridge above the level of these plains is about 5×10^7 km^3, and if the ridge did not exist, sea level would be about 130 m lower.

M. Ewing and B. C. Heezen (1956) first suggested that it might form part of a world-encircling system of mid-ocean ridges. Unlike mid-ocean ridges elsewhere, it has no central rift and the surface is relatively smooth. This may be a consequence of its spreading at a rapid rate.

At intervals of a few hundred kilometers, major fracture zones cross it. Characteristically, each is several thousand kilometers long, up to a hundred kilometers wide, with a relief of a few kilometers expressed in cliffs, grabens, or lines of inactive submarine volcanoes. They have been identified as ridge-ridge transform faults. Many of them offset the patterns of magnetic anomalies lying parallel to the crest by several hundred kilometers (R. G. Mason, 1958; V. Vacquier, 1962). Bathymetry shows vertical displacements of up to 3 km between the two sides and different concentrations of volcanic seamounts on the two sides (H. W. Menard, 1964; H. W. Menard and J. Mammerickx, 1967; T. E. Chase and H. W. Menard, 1969a). D. E. Hayes and W. C. Pitman III (1970) have reviewed the southwestern part of the rise.

The pattern of magnetic anomalies is particularly well developed over the East Pacific rise and shows the same sequence of anomalies on both sides of the ridge as in the other oceans (Figs. 16-1 and 16-2). The interpretation of the anomalies suggests rotation about an axis with poles in Baffin Bay and in Antarctica, with a maximum spreading rate off Peru of 6 cm/year for each limb (W. C. Pitman III and J. R. Heirtzler, 1966; F. J. Vine, 1966; X. Le Pichon, 1968; F. J. Vine and H. H. Hess, 1970; Fig. 16-2).

Earthquakes are clustered on the intercepts of fracture zones or transform faults which offset the East Pacific rise (L. R. Sykes, 1967; M. Barazangi and J. Dorman, 1969).

The oceanic crust is uniform, except over the crest, where it appears to thin slightly and where the underlying mantle has an anomalously low velocity of

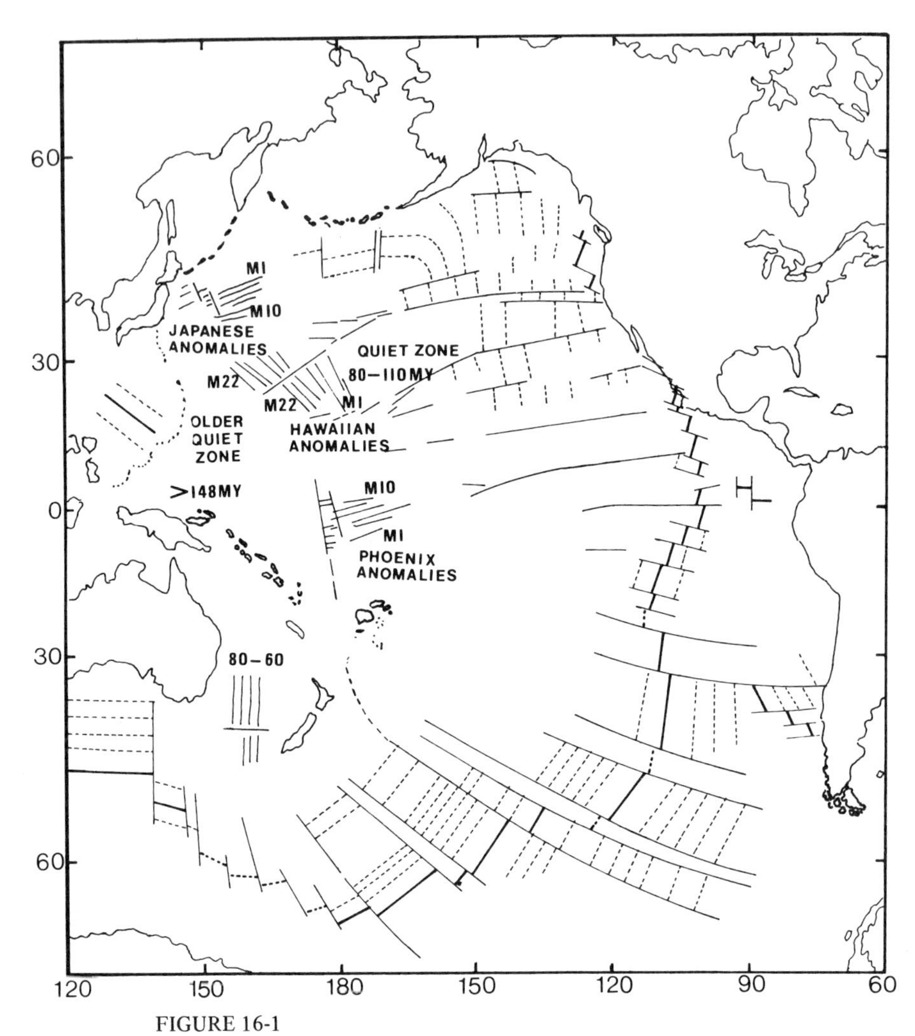

FIGURE 16-1
The crest of the East Pacific rise is marked by heavy lines and fracture zones by light lines. Magnetic anomalies associated with the rise are dashed. Those in the Western Pacific are light lines and are from R. L. Larson and W. C. Pitman III. *(After W. C. Pitman III et al., 1968, H. W. Menard and T. M. Atwater, 1968, J. Francheteau et al., 1970, R. L. Larson and W. C. Pitman III, 1972.)*

7.5 km/sec (H. W. Menard, 1960). G. B. Morris et al. (1969) and G. G. Shor and R. W. Raitt (1969) found the usual three crustal layers, but G. L. Maynard (1970) has since reported finding that the deepest layer is really double. Layer 1 consists of unconsolidated sediments which are lacking on the crest and thicken along the equatorial belt of high organic productivity (G. R. Heath, 1969) and near some

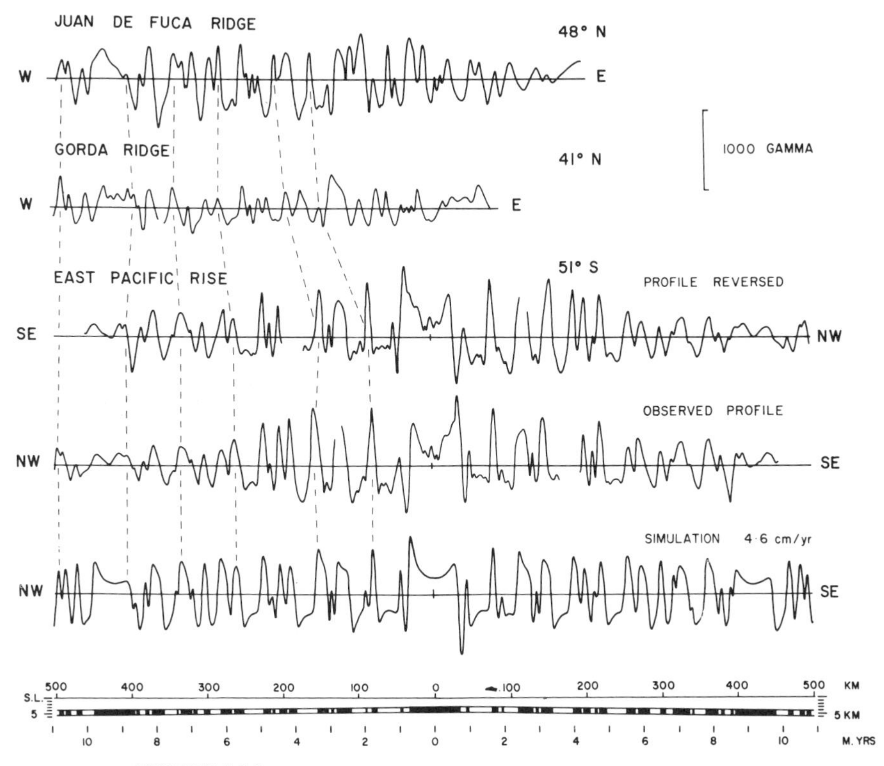

FIGURE 16-2
Comparison of anomaly profiles in the North and South Pacific with a simulation. *(After F. J. Vine, 1968.)*

margins where turbidites are extensive in the Gulf of Alaska and in the Bellinghausen basin off Antarctica (M. Ewing et al., 1966b,c, 1969c; J. I. Ewing et al., 1968).

J. I. Ewing and M. Ewing (1967) found a sudden thickening near anomaly 5. This was at first attributed to a change in spreading rate about 10 m yr ago, but later explained by a change in the rate of deposition brought about by an alteration in ocean currents (P. R. Vogt et al., 1969b; A. E. Maxwell, 1970c). Dredging and deep-sea drilling have shown that the maximum age of fossils increases away from the crest (L. H. Burckle et al., 1967; J. D. Hays, 1970). Because layer 1 is thin with an average thickness of only 300 m, and because paleomagnetic measurements indicate a rate of accumulation of 2 m/m yr or more, the Pacific Ocean seems to be young, and none of it is probably older than Mesozoic in age (N. D. Opdyke et al., 1966; D. Ninkovich et al., 1966; J. D. Hays et al., 1969).

Heat flow is high over the rise (Sec. 7-2.3), and a belt of negative gravity anomalies follows it (Secs. 4-7 and 14-11). Nevertheless, the elevation of the rise must be partly compensated by lighter material beneath it, and since light crust does not thicken, this must be due to a decrease in density in the upper mantle. All this supports the view that the sea floor is spreading away from the rise because of an upwelling beneath it of hot and light mantle material. No doubt the lowered velocities observed there and a marked attenuation in seismic curves reflect this.

H. H. Hess (1964) suggested that the upper mantle under the rise is anisotropic, and G. B. Morris et al. (1969) and R. W. Raitt et al. (1969) have confirmed that the anisotropy is considerable. Near Hawaii the velocity of *P* waves is 8.45 km/sec in an east-west direction and 7.85 km/sec in a north-south direction. Weaker anisotropy has been detected along the Mid-Atlantic Ridge (C. E. Keen and C. Tramontini, 1970). Hess suggested that deformation of serpentine may have caused this, but the problem has not yet been fully resolved.

16-3 THE WESTERN BOUNDARY OF THE EAST PACIFIC RISE

In the Southwestern Pacific the crest of the East Pacific rise lies midway between Antarctica and the New Zealand Plateau. New Zealand is thus likely to have once been attached to Antarctica, and the edge of the plateau forms part of the northwestern border of the rise, but the continuation of the boundary has not been agreed upon. H. W. Menard (1964) used bathymetry to define it as the line along which the rough and gently sloping surface of the rise merges with the smooth and flat sea floor approximately along longitude 150°W. Farther west he and H. H. Hess (1946, 1962) noted many submarine flat-topped tablemounts, or guyots, which they considered to mark the location of a former mid-ocean ridge that had become quiescent and sunk. This they named the Darwin rise (Fig. 14-8). Deep-sea drilling has penetrated Upper Jurassic strata on the Shatsky Plateau off Japan (A. G. Fischer and B. C. Heezen, 1969; A. G. Fischer et al., 1970), confirming their view that the Western Pacific is older than the East Pacific rise. Fischer and his colleagues also discovered that the seismic reflectors in the ocean floor are layers of chert, not clastic turbidites derived from the former Darwin rise as Menard has proposed (Fig. 16-3). This led them to abandon the concept of a twofold division of the Pacific basin and to place the border of the East Pacific rise at the trenches off the coast of East Asia.

This proposal leaves several questions unanswered. It fails to explain why south of New Zealand the East Pacific rise is of Tertiary age with a definite northwestern border, whereas in the equatorial Pacific, part of the rise would be of Upper Jurassic age with no border except subduction zones. It does nothing to explain the presence of so many guyots in the Western Pacific or the origin of many large ridges which elsewhere mark boundaries (Secs. 15-7 and 15-12). One such line of ridges which might mark the border lies close to Menard's boundary and has been

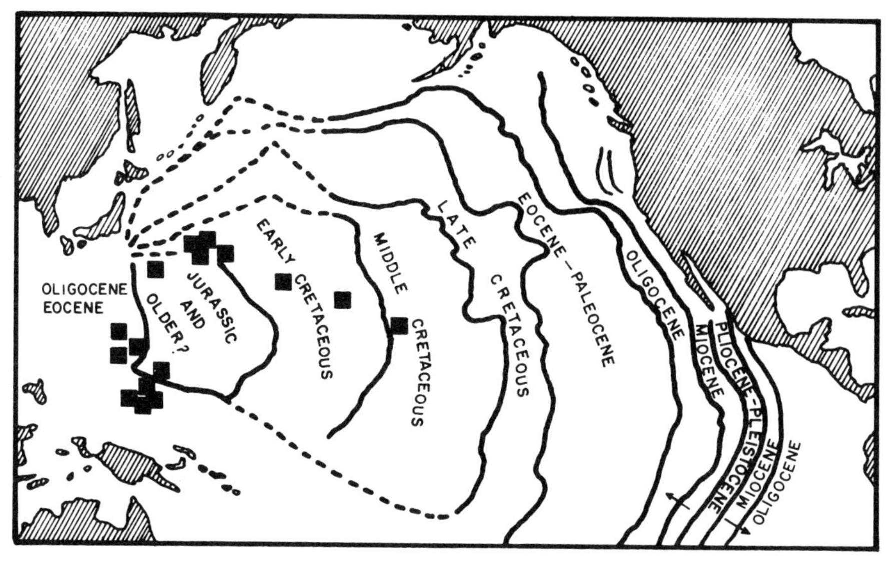

FIGURE 16-3
Basement ages in the North Pacific and drill site of leg 6 of the cruises of the *Glomar Challenger*. (*After A. G. Fischer et al., 1970.*)

called the ICSU ridge.[1] Its chief components are the Emperor seamounts, part of the Mid-Pacific Mountains, the Line Islands, and the eastern edge of the New Zealand Plateau (J. T. Wilson, 1965a; R. S. Dietz, 1954; E. L. Hamilton, 1956; H. Kuno et al., 1956; K. O. Emery, 1956; D. J. Cullen, 1969; C. P. Summerhayes, 1969; Fig. 16-1). On the other hand, B. H. Erickson et al. (1970) suggest that a fracture zone east of the Emperor seamounts is the boundary.

If the ICSU ridge is the boundary, the theory of transform faulting would require it to mark the end of fracture zones and to have the same shape and corresponding offsets as the East Pacific rise. This seems to be the case, and Sec. 16-4 gives other arguments.

As F. J. Vine and H. H. Hess (1970) have pointed out, magnetic mapping, which is still incomplete, may suggest a solution to this problem. Some indications of a confused pattern north of Hawaii may indicate a change there in Cretaceous time (S. Uyeda et al., 1968; A. Malahoff and G. P. Woollard, 1968; P. J. Grim and B. H. Erickson, 1969; D. K. Rea, 1970; R. L. Larson and W. C. Pitman III, 1972).

[1] Named after the International Council of Scientific Unions, which sponsored the International Geophysical Year and much oceanographic research.

16-4 THE ISLANDS OF THE EAST PACIFIC RISE AND SOME SYMMETRIES ASSOCIATED WITH THEM

A. R. McBirney and I. G. Gass (1967) and A. R. McBirney and H. Williams (1969) have described the geology of most islands lying on the East Pacific rise. Some which they did not mention are Peter the First Island (G. L. Johnson, 1966), the Balleny Islands (T. Hatherton et al., 1965), and Scott Island, all close to Antarctica; Clipperton Island off Mexico, which has only one altered outcrop of igneous rock (M.-H. Sachet, 1962; A. C. Obermüller, 1959); and Macquarie Island, which may be upthrust ocean floor related to the Tonga–New Zealand chain (R. Varne et al., 1969). Like oceanic islands elsewhere, all these islands are basaltic, some mantled according to their latitude with ice or coralline limestone.

There is an immense literature on the geology of the Hawaiian Islands which H. T. Stearns (1966) and G. A. Macdonald (1968) and G. A. Macdonald and A. T. Abbott (1970) have summarized, while J. P. Eaton and K. Murata (1960) and D. A. Swanson et al. (1971) have described the rise of lava in an eruption. H. S. Ladd et al. (1970) drilled pre-Miocene basalt on Midway Island, and B. C. Schreiber (1969) recovered Eocene sediments from a position which implied that the Hawaiian Islands had not then developed.

The symmetry of the pattern of magnetic anomalies over the East Pacific rise (Fig. 16-1) is matched by a certain symmetry in the arrangement of its islands and seamounts. Toward the center of the ocean the Hawaiian, Marquesas, Tuamotu, Society (A. R. McBirney and K.-I. Aoki, 1968), and Austral (R. H. Johnson and A. Malahoff, 1971) islands have long been recognized to be arranged in straight parallel chains which grow older from southeast to northwest. H. W. Menard and R. S. Dietz (1951) have described two more parallel chains of seamounts in the Gulf of Alaska. The four chains in the central Pacific and the two Nasca ridges of seamounts off the coast of South America are in mirror-image relationship to one another on either side of Easter Island on the East Pacific rise (P. E. Baker, 1967b; J. Booker et al., 1967; Fig. 14-8).

F. Betz, Jr., and H. H. Hess (1942) proposed that the Hawaiian Islands are straight because they lie along a strike-slip fault, but the faults on the islands are not so aligned. L. J. Chubb (1957) suggested that the chains formed along the crests of parallel anticlines. All might have been formed by the spreading of plates away from a hot source in the mantle, as has been suggested for Iceland and Tristan da Cunha (Sec. 15-10; J. T. Wilson, 1963e, 1965a). R. I. Walcott (1970) has combined this view with that of Betz and Hess. Another view is that the Marshall, Gilbert, and Ellice Islands, the Line Islands, and the Emperor seamounts, respectively, are extensions of the Austral, Tuamotu, and Hawaiian Islands and are three old chains formed over three hot spots by plate movements. Still another view is that some of these chains represent plate boundaries (J. Francheteau et al., 1970).

16-5 THE RELATIONSHIP OF THE EAST PACIFIC RISE TO SOUTH AND MIDDLE AMERICA

The crest of the East Pacific rise joins North America in the Gulf of California, which may be regarded as being in Stage 2 of development. To the south the rise diverges from the Americas, and three elements constitute the boundary. These are successively the Middle America trench, extending from Mexico to Costa Rica, a central region with no trenches, and the Peru-Chile trench (Fig. 16-4).

Many now believe that the Americas have overridden the eastern part of the rise, but R. L. Fisher (1961), G. G. Shor, Jr., and R. L. Fisher (1961), and D. E. Hayes (1966) wrote their important descriptions of the trenches before sea-floor spreading was considered.

Hayes concluded from his seismic and gravity data that the crustal structure varies little along the entire west coast of South America. C. Lomnitz (1969) has given a recent interpretation (Fig. 16-5).

The Peru-Chile trench has three main parts. From 8 to 32°S, a central province has a deep trench without sedimentary fill. To the south the trench is increasingly filled with sediments, but a strip of negative gravity anomalies continues as far south as Drake Passage. In the north the trench becomes irregular and partly filled as it dies out. Hayes and also C. Galli-Oliver (1969) believe that the arid climate and lack of erosion on the adjacent coast explain why sediments have not filled the main trench. This is certainly a major factor, but changes in the structure of the Andes suggest that the full explanation may be more complex. These occur opposite to the two ends of the trench near Guayaquil and Santiago. Where the trench is open, the coast ranges are poorly developed, but where the trench is filled, the coast ranges are conspicuous. This suggests that the ranges may be former trench fillings piled up onshore and that the Bolivar geosyncline in Colombia may be an example of this of Miocene age (D. E. Hayes, 1966; H. Bürgl, 1967).

A problem which has given rise to considerable debate concerns the lack of crumpling of the sediments in the Peru-Chile and other deep trenches (D. W. Scholl et al., 1968, 1970; C. K. Seyfert, 1969). It has often been supposed that the trenches over subduction zones should be subject to compression, but B. Isacks et al. (1968, 1969) and W. J. Ludwig et al. (1966) have found evidence of tension at the surface of trenches, so that the compression may only begin at depth or the overridden lithosphere may be so dense that it is sinking and pulling the floor of the trench down.

The structural history of the ocean floor to the north, west, and south of South America is complicated by the presence of several small plates and ridges and by controversy about some details. The Scotia arc to the south has already been discussed (Sec. 15-11).

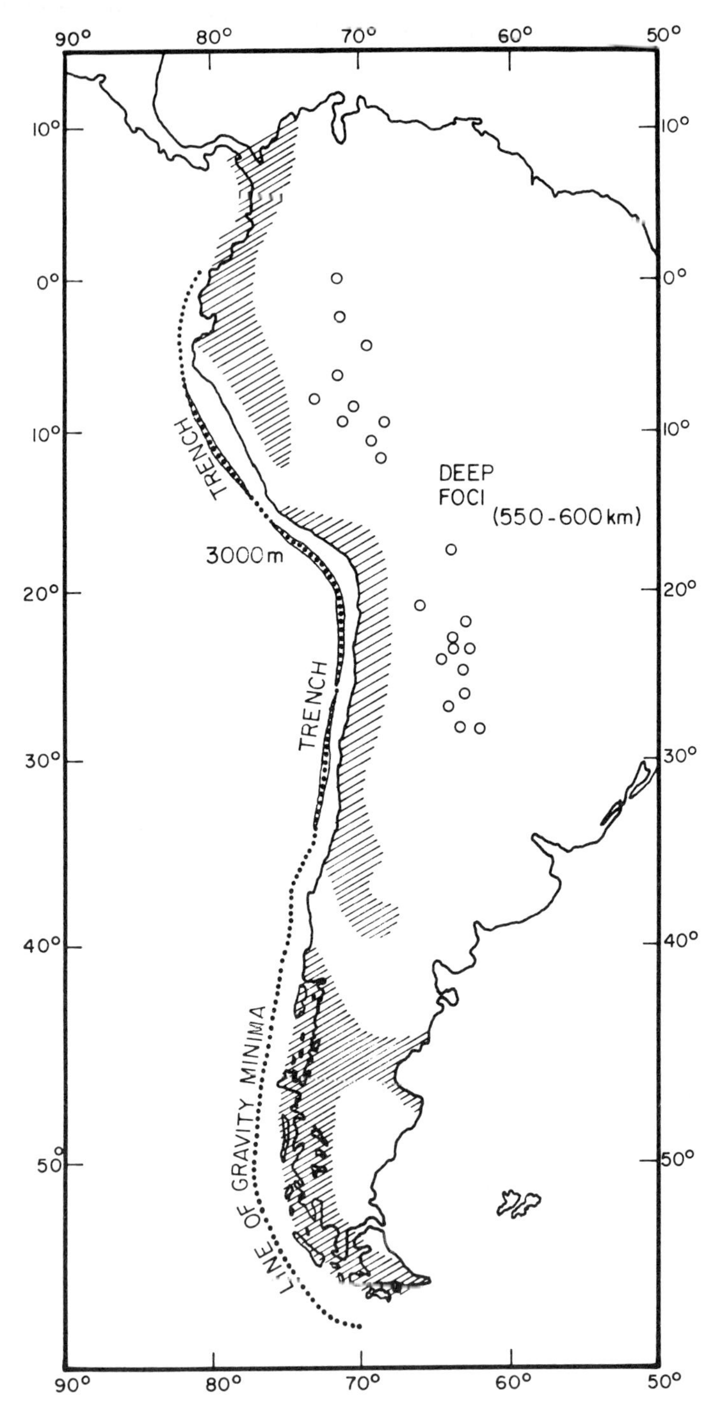

FIGURE 16-4
Sketch map of some tectonic features of South America. Depths of 3000 m in the Peru-Chile trench are contoured. The dotted line shows location of accompanying gravity minima. Hatching indicates extent of marine basins of Lower Jurassic time. Dots indicate epicenters of earthquakes with foci at depths 550 to 660 km. (*After D. E. Hayes, 1966, and H. J. Harrington, 1962.*)

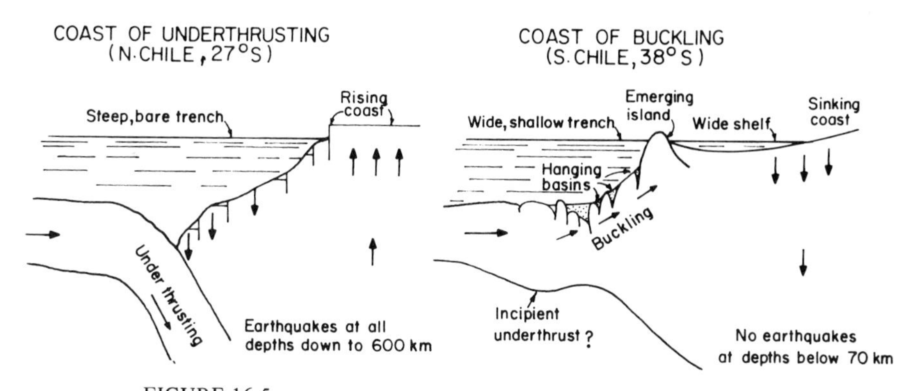

FIGURE 16-5
Two types of ocean-continent interaction. Vertical exaggeration about 11:1. (*After D. W. Scholl et al., 1968, and C. Lomnitz, 1969.*)

North and west of South America the Caribbean, Cocos, and Nasca plates lie between the Pacific, American, and Antarctic plates separated by the Panama, Carnegie, and Chile rises (P. Molnar and L. R. Sykes, 1969; D. P. McKenzie and W. J. Morgan, 1969; Fig. 16-6).

Next to the Antarctic plate the Nasca plate is being pushed east into the Chile trench while the Americas plate moves west. The descent into the subduction zone is marked by many and deep earthquakes which stop almost completely at its southern boundary with the Antarctic plate. This boundary is the Chile rise, for which E. M. Herron and D. E. Hayes (1969) and W. J. Morgan et al. (1969) have provided two different interpretations, both based on scanty data.

H. J. Harrington (1962, 1963) has discussed the relation of the deep earthquakes to continental drift and to the uplift of the Andes during the Cenozoic era. He separates the rocks and history of the coast of Chile from those of the rest of South America and shows that they have affinities with those of New Zealand. A. Windhausen (1921) had already noted that the Paleozoic rocks of Patagonia and of the Falkland Islands are different from those of the rest of South America. He considered that Patagonia had been part of Antarctica until joined to South America during the Jurassic period. E. Kausel and C. Lomnitz (1969) and R. P. Morrison (1973) agree with this viewpoint.

The Antarctic plate is the only major plate not bordered for a considerable length by a subduction zone. This may be why the seismicity of Antarctica is so low. When Gondwanaland began to break up, the crest of the ring of mid-ocean ridges which surrounds Antarctica must have lain along the coast, and hence since then it must have been moving north.

Thus the Nasca plate has a northward component of motion as the direction of strike-slip faults and earthquakes suggest, as well as having an eastward one (P. St. Amand, 1961; G. P. Plafker and J. C. Savage, 1970). The Chile rise is also

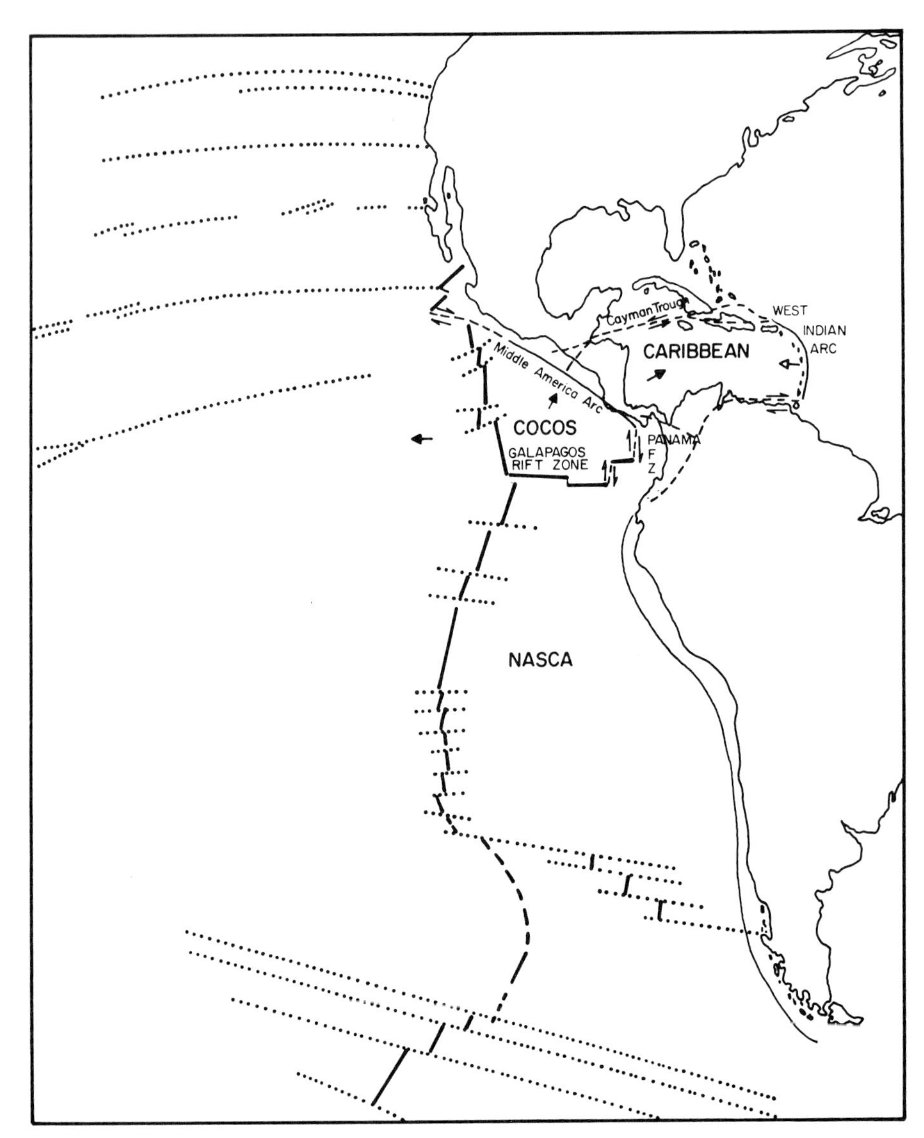

FIGURE 16-6
Sketch map showing the relationship of plates in the Western Pacific and Caribbean regions.

moving northward, and perhaps it formerly passed through Drake Strait and assisted in separating Patagonia from Antarctica.

This plate takes its name from the submarine Nasca ridges. They lie on its eastern margin, but Hayes, unlike W. Rüegg (1960), failed to find any continuation of them in the Andes (Fig. 16-1). This supports the view that they are in process of being overridden, but their symmetrical position relative to the Society Islands and other chains on the western edge of the East Pacific rise suggests that South America has not yet overridden much of the East Pacific rise (Sec. 16-3).

The Nasca plate is separated from the smaller Cocos plate by the Carnegie ridge, which extends from Ecuador past the Galapagos Islands to a triple-point junction with the East Pacific rise. E. M. Herron and J. R. Heirtzler (1967) and A. D. Raff (1968) agree that the Carnegie ridge is a spreading mid-ocean ridge. The Galapagos Islands are a large archipelago with at least five active volcanoes, of which one had a caldera collapse in 1968 (T. Simkin and K. A. Howard, 1970; A. R. McBirney and H. Williams, 1969).

The simultaneous growth of the East Pacific rise and the Carnegie ridge are driving the Cocos plate northeastward into the Middle America trench. G. G. Shor, Jr., and R. L. Fisher (1961) investigated its topography and geophysics. It is a very active trench, and D. A. Ross (1971) has calculated that the present rate of sedimentation could fill it in 10 m yr if its motions do not keep it open. M. Maldonado-Koerdell (1966) and A. R. McBirney and H. Williams (1965) have described the volcanoes which border the southern half of the trench, and F. Mooser (1969) has shown how the active belt turns east to cross Mexico. G. Dengo (1967), Z. de Cserna (1969), and G. Dengo et al. (1970) have described major faulting along this coast.

On the other side of Central America the Caribbean plate appears to be moving east relative to and overriding the Atlantic floor along the West Indies arc (R. C. Chase and E. T. Bunce, 1969; P. Molnar and L. R. Sykes, 1969; Fig. 16-7). Transform faults lie along the Cayman trough and through northern Colombia and Venezuela (E. Rod, 1956; C. O. Bowin, 1968).

The geology of the Caribbean region is complex, and most published accounts of it do not incorporate the new ideas (P. H. Mattson et al., 1966; H. H. Hess, 1966; J. R. Saunders, 1968; V. Skvor, 1969; C. W. Hatten and A. Meyerhoff, 1970). The Greater Antilles islands of Cuba, Hispaniola, Jamaica, and Puerto Rico, and also Trinidad, have Cretaceous and some Jurassic strata, often resting upon serpentinite massifs, but the Lesser Antilles are apparently younger. W. H. Monroe (1968) believes that the Puerto Rico trench formed in late Eocene or Oligocene time. J. Ewing et al. (1967) cored Eocene ooze from a fault scarp on the Beata ridge south of Hispaniola, and JOIDES drilling has penetrated Eocene beds in the Caribbean Sea and in the adjacent Atlantic basin. The Puerto Rico trench has been much studied (C. A. Burk, 1964). Although sometimes referred to as being associated with an island arc, it seems more correctly to be regarded as a shear zone.

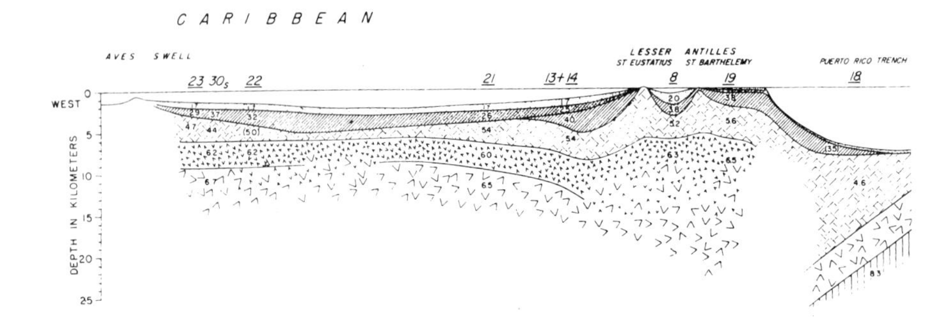

16-6 THE RELATIONSHIP OF THE EAST PACIFIC RISE TO WESTERN NORTH AMERICA

The fragmentation of Gondwanaland formed the Atlantic, Indian, and Southern Oceans by the spreading of the mid-ocean ridges which lie along their axes, but the East Pacific rise is not centrally situated. One reason is that the Americas appear to have overridden part of the rise, and another possible reason is that the western part of the Pacific may be older than the rise.

The geology of the ocean floor is quite different from that of North America, and opinions about the degree of interaction between the two vary widely. Thus R. von Huene (1969) emphasizes the lack of connection between fracture zones on the sea floor and faulting on land, and W. H. Taubeneck (1966) the greater age of most continental disturbances, but D. U. Wise (1963), R. W. Pease (1969), and R. G. Yates (1968) have suggested that effects from the ocean spread far inland, causing great dextral shears striking northwesterly and possibly rotating part of the northwestern United States (N. D. Watkins, 1967; F. A. Wright and B. W. Troxel, 1970; J. H. Stewart et al., 1970).

H. W. Menard (1960) considered that the overridden part of the rise still persists beneath the Mexican and Colorado Plateaus and the Basin Ranges, and G. A. Thompson (1966) and K. L. Cook (1969) claimed it to be the cause of earthquakes, although J. Gilluly (1963) and E. R. Kanasewich (1965) related the seismicity to movements between old continental blocks. H. Palmer (1968) related mountain building to the rise, in spite of a Devonian start to tectonic activity in the southwestern United States. D. P. McKenzie and W. J. Morgan (1969) and T. Atwater (1970) discussed what happened to the ocean floor, including the small plates off Oregon and British Columbia, which F. J. Vine and J. T. Wilson (1965)

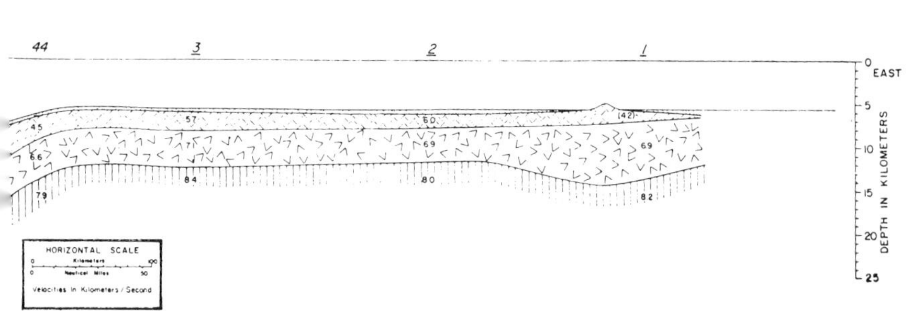

FIGURE 16-7
West-east structure section from the Aves swell in the Caribbean basin through the Lesser Antilles and their trench to the outer ridge. (*After C. B. Officer et al., 1959.*)

interpreted in terms of sea-floor spreading, although N. Pavoni (1966) and G. Peter and R. Lattimore (1969) have proposed alternative solutions for the magnetic maps produced by A. D. Raff and R. G. Mason (1961). A. S. Brown (1968), P. Dehlinger (1969b), C. R. B. Lister (1970b), and K. S. Deffeyes (1970) have discussed work done on the Juan de Fuca and Gorda rises and vicinity, while G. A. Davis (1969), E. M. Moores (1970b), J. F. Dewey and B. Horsfield (1970), W. G. Ernst (1970), E. H. Bailey et al. (1970), J. G. Souther (1970), and E. A. Silver (1971) have reviewed the history of the coastal Cordillera. J. Gilluly (1969) noted evidence that overriding had probably consumed sedimentary rocks.

Central to this discussion is the problem of what happens when a continent bordered by a subduction zone rides over an ocean basin containing an upwelling rise (Figs. 16-8 and 16-9). At the continental margin the continental and oceanic plates approach each other with opposite directions of motion, and one or both must be sliding along the asthenosphere over the deeper mantle. To accommodate these motions the leading edge of the oceanic plate moves down, forming a subduction zone, which is reabsorbed.

After the continental margin crosses the crest of the rise, it will ride onto another plate of oceanic crust. Since this oceanic plate is moving in the same direction as the continental plate, the subduction zone is no longer needed and disappears. On the other hand, the upwelling plume that produced the rise may be a cause, not a result, of plate motion, and has no need to stop, but the rest of the ridge system is shallow and may die out. A shear motion of the continental plate over the deep mantle must be accepted.

Suppose that a plume of the East Pacific rise is still active beneath North America; then a precise method of locating its present position is to assume that overriding of the rise by the continent did not greatly disturb the regularity of its

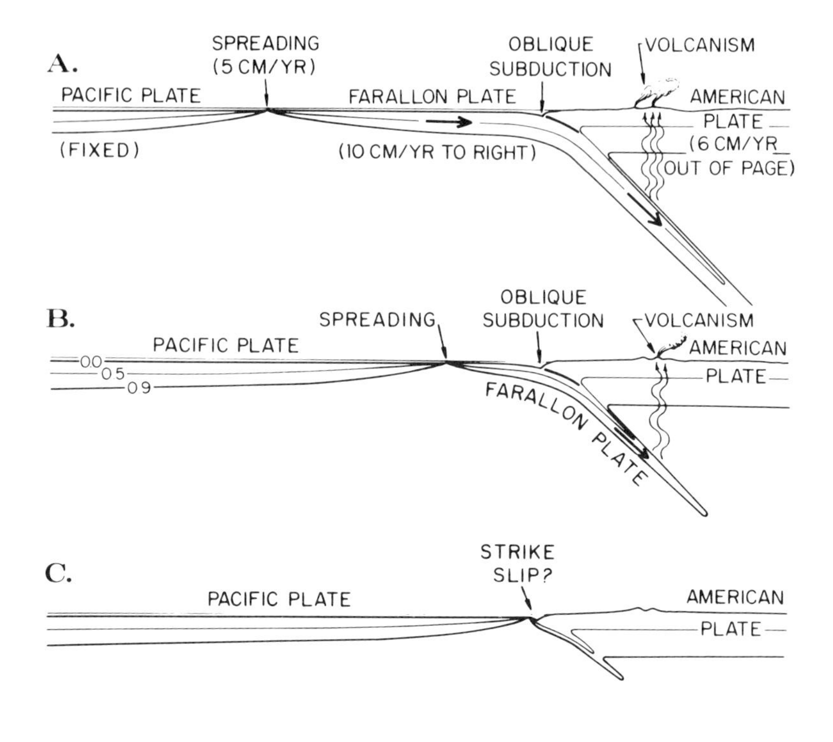

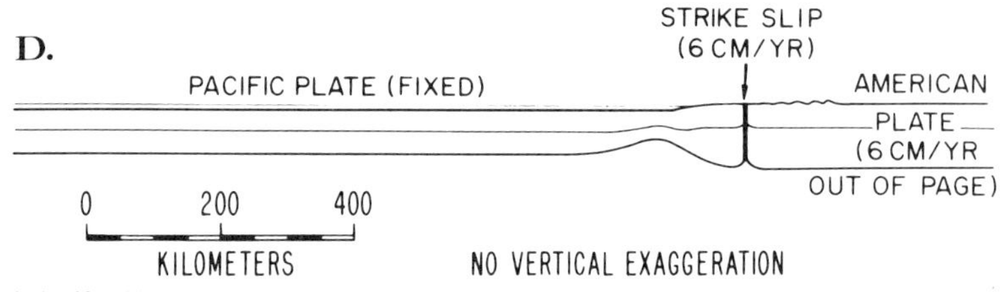

FIGURE 16-8
Diagrammatic cross sections of collision of plates off the west coast of North America. The Pacific plate is regarded as fixed. The ridge has a half-spreading rate of 5 cm/year. If the Pacific plate is regarded as fixed, the ridge moves to the right at 5 cm/year and the Farallon plate at 10 cm/year. Both plates move into the page relative to the American plate at 6 cm/year, so that the trench accommodates both shearing and subduction. *A* represents the situation in early Tertiary time, *C* at the time of collision of ridge and trench, and *D* at the present time. (*After T. Atwater, 1970.*)

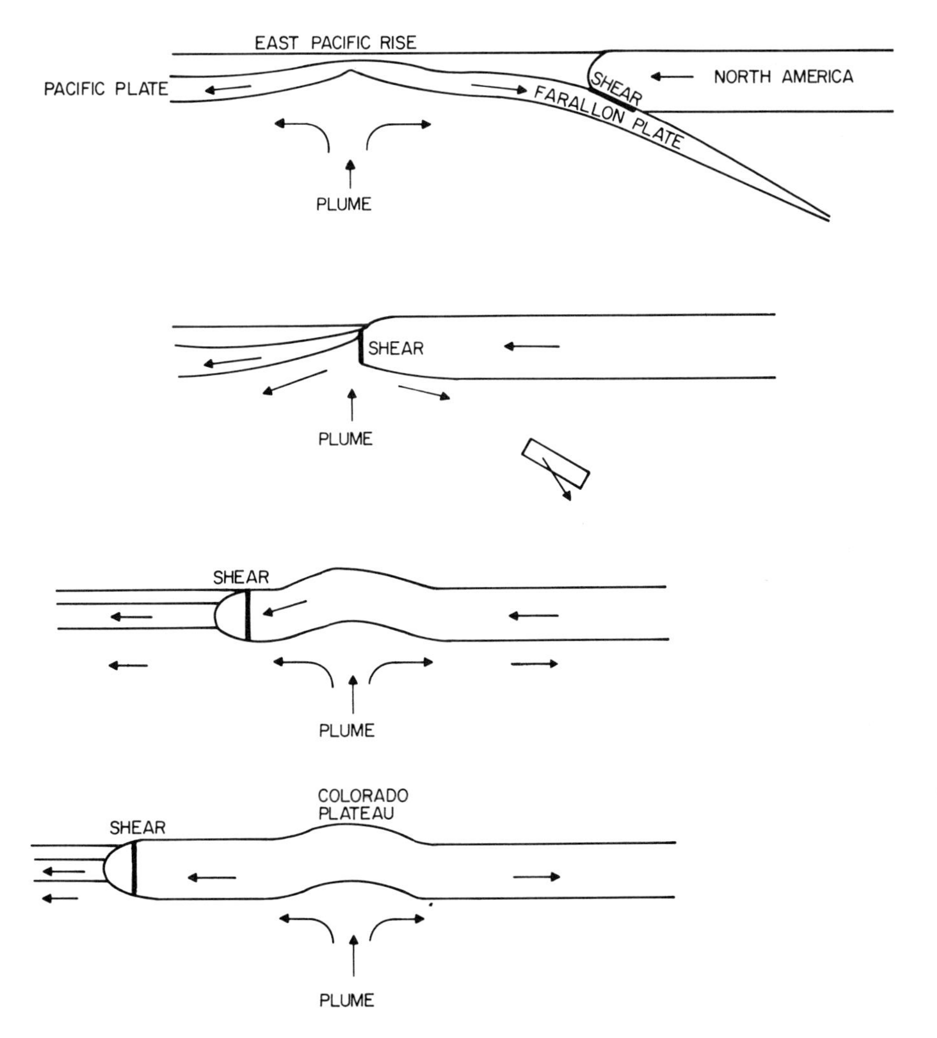

FIGURE 16-9
Four diagrammatic sketches showing one interpretation of stages during the Tertiary overriding of the East Pacific rise by North America. Mesozoic mountains have been omitted. As the overriding occurred, the changes in direction of motion eliminated the trench and subduction zone, but not the deep-seated plume. The disappearance of the trench which had accommodated the shear motion of the Pacific plate necessitated the formation of a fault, now the San Andreas fault, through the margin of the continent.

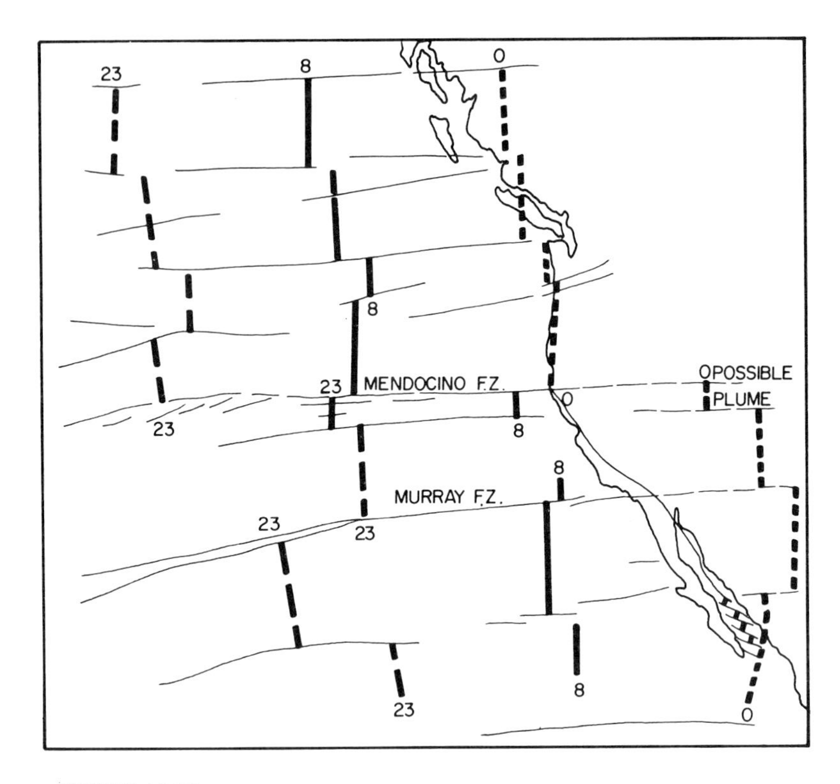

FIGURE 16-10
The location of magnetic anomalies 8 and 23 is shown in the Eastern Pacific Ocean. They are, respectively, 29 and 58 m yr old. At the bottom of the map their distance apart is equal to the distance between anomaly 8 and the crest of the East Pacific rise (anomaly 0). Assuming this to be a general relationship, the possible present position of the hidden crest has been plotted. The general location of a possible deep upwelling is indicated at the postulated junction of the Mendocino fracture zone and the rise. (*Anomalies after H. W. Menard, 1969.*)

upwelling. This is admittedly speculation, but may serve as a first approximation. If overriding has caused little distortion, extrapolation from magnetic anomaly maps of the Pacific Ocean can indicate the position of the rise beneath North America. In Fig. 16-10 anomalies 23, 8, and 0 (the crest) have been copied from T. Atwater and H. W. Menard (1970) and H. W. Menard (1969b) and the major fracture zones projected under the continent. These anomalies were chosen because anomalies 23 and 8 are everywhere exposed and the distance from anomaly 23 to anomaly 8 is equal to that from anomaly 8 to the crest, and hence the position of the hidden parts of the crest can be estimated. These lie close to the Pacific coast, except between the extensions of the Mendocino and Murray fracture zones, where the crest has been offset far to the east to lie under the Colorado Plateau. The junction of the rise with the Mendocino fracture zone is marked by an offset of over 1000 km, and so seems the most likely place for a plume. We shall assume that it is there, even if elsewhere beneath the continent the rise has vanished. Close to the coast it is probably in the process of dying away. W. B. Hamilton and W. B. Myers (1968) and W. B. Hamilton (1969) reached a similar conclusion for the whole rise for other reasons.

Consider the probable history of the region. During the Mesozoic era neither the Atlantic Ocean nor the East Pacific rise were as wide as they are today, and both were expanding slowly. The entire crest of the East Pacific rise then lay in the Pacific Ocean basin off the west coast of North America, and expansion of the Atlantic was slowly driving North America westward over the rise. Contemporary expansion of the mid-ocean ridge around Antarctica was also pushing the Pacific floor northward (Sec. 16-5). Thus the combined movements caused the rise to approach North America like a wave of uplift moving northeastward.

The crest was not straight, and the offset part centered on the supposed plume would have struck the California coast first and passed inland. According to D. P. McKenzie and W. J. Morgan (1969), this happened about 32 m yr ago during Oligocene time, although uplift of the continent would have started sooner. T. Atwater (1970) suggests that the rise died away, and this may be true for all parts except the plume. Its continued upwelling could account for the elevation of the coast and of the Basin and Range province during the Eocene, when the rest of the western states was being reduced to a low peneplain (C. O. Dunbar and K. M. Waage, 1969; J. J. Anderson, 1971; Fig. 16-11). By Miocene time this part of the crest would have moved farther east, allowing the Basin and Range province to sink while raising the Colorado Plateau (Fig. 16-12). J. H. Stewart (1971) has interpreted the Basin and Range structure as horsts and grabens in which "extension has taken place in the last 17 m yr or perhaps even in the last 7-11 m yr indicating a rate of extension in the range of 0.3 to 1.5 cm/yr."

About 5 m yr ago other parts of the crest would have reached the coast, where they could have produced the Pliocene uplift called the Cascadian Revolution in the north. Farther south in Mexico, R. L. Larson et al. (1968) and D. G. Moore and E. C. Buffington (1968) have interpreted the magnetic anomaly pattern to suggest that the Gulf of California started to open 4 m yr ago. Complications and minor distortions occurred, which C. G. Chase et al. (1970) and T. Atwater (1970) have recently discussed.

If a plume of the East Pacific rise does indeed underlie the Colorado Plateau, its presence can explain many features, for example the notable differences in the topography of the western parts of North and South America. The whole western half of the United States and the eastern side of the North Pacific slope upward from either side toward the Cordillera. No such rise is present in South America, where the plains are low to the foot of the Andes, and the Pacific floor does not rise but rather sinks into a trench.

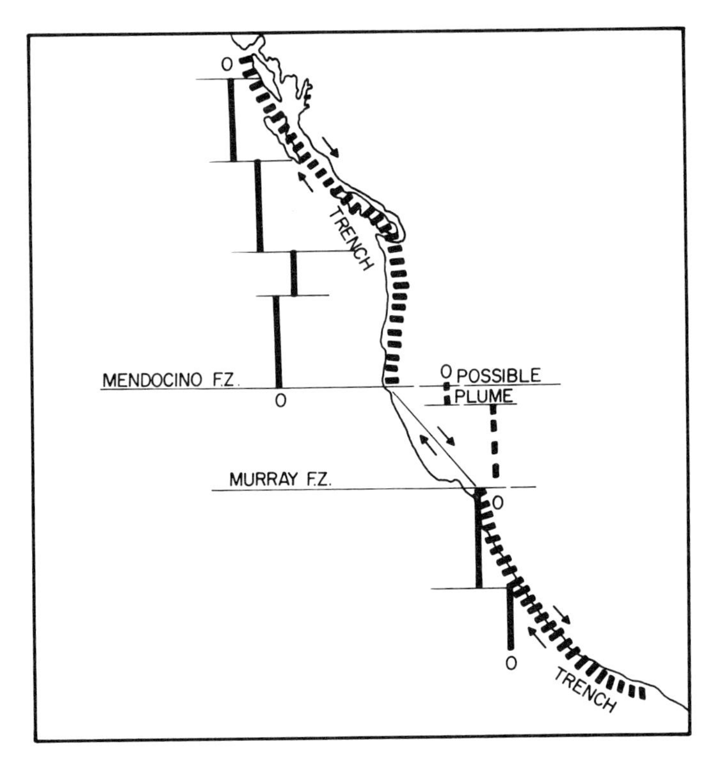

FIGURE 16-11
The assumed situation about 20 m yr ago when North America had overridden only a small part of the East Pacific rise. Trenches existed along most of the coast.

The influence of the hidden rise may be detected in the evidence for an extension inland of the Mendocino fracture zone (W. N. Gilliland, 1962, 1964). It could account for the uplift of the Oligocene peneplain, of which the eroded remnants form the accordant, flat-topped summits of the middle and southern Rocky Mountains, and for the absence of flat summits in the north in the Canadian Rockies, where both the mountains and the interior plains are lower.

Observations of high values of heat flow in the southwestern United States fit this interpretation. G. Simmons and R. F. Roy (1969) have suggested that these are due to injections 10 m yr ago of hot material beneath the Basin and Range province and to other injections 2 m yr ago beneath the Colorado Plateau. In the northwestern United States a narrower zone of high values lies along the coast, as does the crest of the rise (D. D. Blackwell, 1969).

The map of seismicity by M. Barazangi and J. Dorman (1969) shows that besides shallow activity along the San Andreas fault, which forms the boundary between the plates at the surface, there is also a seismic zone extending northward

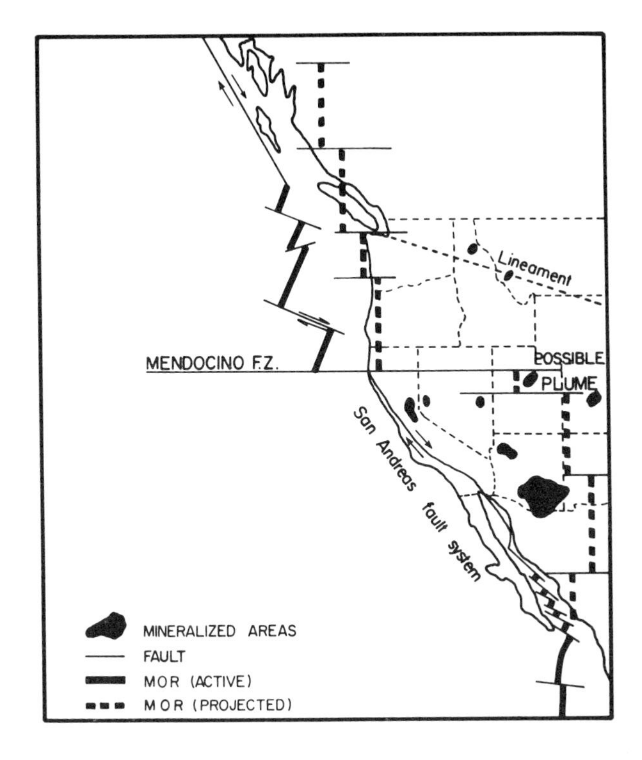

FIGURE 16-12
The present calculated position of the crest of the East Pacific rise in dashed lines and the location of existing spreading crests in full heavy lines. The dark patches mark the location of major mining districts. (*After J. A. Noble, 1970.*)

from Mexico to the vicinity of the Colorado Plateau, where it stops abruptly. This could be following the hidden crest. L. C. Pakiser and J. S. Steinhart (1964) found crustal thickness and mean crustal velocities in the western states to be low, and E. Herrin (1969) found low P_n velocities, which he associated with high temperatures in the mantle and high heat flow (Fig. 16-13). P. Molnar and J. Oliver (1969) noted that the upper mantle beneath the western states transmits S_n waves anomalously. Two profiles published by G. P. Woollard (1968) show that in the southwest, where the crust is thin and the mantle near the surface, there is a regional negative gravity anomaly, which suggests that the mantle is less dense near the proposed extension of the rise.

Variations in magnetic force also suggest a thermal structure under the southern Rockies (D. I. Gough and H. Porath, 1970; S. Uyeda and T. Rikitake, 1970), with the largest anomaly in the Uinta basin in northeast Utah, which in turn lies over the postulated plume. Whether the cause is due to sediments in the basin

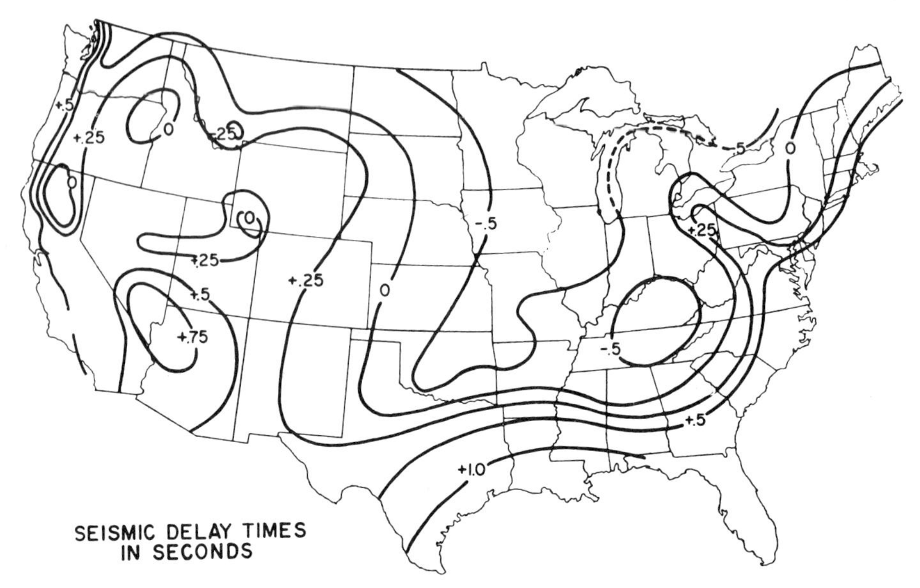

FIGURE 16-13
Relative seismic delay time in the United States, showing the disturbed pattern in the southwestern region which may have overridden an upwelling in the mantle. (*After E. Herrin and J. Taggart, with permission of the American Geophysical Union.*)

or heat in the mantle is debated (H. Porath and D. I. Gough, 1971; H. Porath and A. Dzeiwonski, 1971). Other distinctive features are the Tertiary volcanism and ore deposits of the western United States, which are lacking in the Canadian Rocky Mountains (J. Gilluly, 1963, 1969; J. Gilluly et al., 1970; R. J. W. Douglas, 1970). J. M. Guilbert and J. S. Sumner (1968) and J. A. Noble (1970) have discussed the distribution of metal provinces in terms of sea-floor spreading and properties of the mantle. Both papers conclude that a hot upwelling rise could explain many observations and be the cause of some concentrations of ore, although many are certainly Mesozoic and much older than this proposed event. The location of the principal metallogenic regions of the southwest lies over the region under which the rise is postulated to have traveled, but the ore bodies have been shown to be older, which seems to spoil the relationship (D. E. Livingston et al., 1968; Fig. 16-12).

This overriding could also explain the great changes observed in the tectonic regime of California. During the Mesozoic era a crest of the rise lay off the coast, which was bounded by a trench. The trench, like that off Chile today, accommodated strike-slip faulting. When part of the crest passed under the continent, the direction of flow at the coast in that section would have reversed, and the trench would have closed, necessitating the initiation of the forerunners of the San Andreas fault system to accommodate the faulting. From the theory of global tectonics and triple junctions, D. P. McKenzie and W. J. Morgan (1969) pointed out that H. W. Menard's maps of magnetic anomalies for the northeastern Pacific basin show that the motion of the main Pacific and American plates has remained constant at about 6 cm/year parallel to the San Andreas fault for the past 32 m yr. This direction and rate agree with present-day seismic and geodetic measurements (B. A. Bolt et al., 1968; D. G. Tobin and L. R. Sykes, 1968; W. R. Dickinson and A. Grantz, 1968). During this time the Pacific has been displaced through 1920 km relative to North America (Fig. 13-3).

It is a paradox that whereas, when M. L. Hill and T. W. Dibblee, Jr. (1953) proposed an offset of 550 km on the San Andreas fault since Jurassic time, few believed that the displacement could be so large, but now that amount seems to be too small and to be spread over too long a period of time. B. M. Page (1970) has suggested that during the Tertiary, movement on other faults and folds has been important, which could account for this large discrepancy. L. C. Pakiser et al. (1969) have described the recent seismicity.

16-7 THE ALEUTIAN ISLAND ARC, THE GULF OF ALASKA, AND THE BERING SEA

The Aleutian and Komandorskii Islands rise from a great submarine ridge which is very close to being circular. W. M. Gibson and H. Nichols (1953) and G. O. Gates and W. M. Gibson (1956) have published good bathymetric charts of it. R. R. Coats (1962), B. S. Steinberg and L. A. Rivosh (1965), D. V. Nalivkin (1960), and C. A. Burk (1965, 1966) have described the geology of these Tertiary and Recent andesitic volcanoes, which are active only in the eastern part (P. L. Ward and T. Matsumoto, 1967).

The Aleutian trench lies about 200 km south of the axis of the ridge. The narrow abyssal plain which forms its floor is 7600 m deep off Tanaga Island and rises in either direction. To the west it joins the Kamchatka-Kuril trench in a right-angle bend. To the east it becomes filled with sediments which have been squeezed and forced up to form Kodiak Island and the Kenai Peninsula and a section of double island arcs (Sec. 12-8). A large negative gravity anomaly follows the north wall of the trench (G. Peter, 1966), and G. G. Shor, Jr. (1964) showed that a root lies below the ridge.

Most geophysicists now believe that this arc and trench are due to subduction of the northward-moving Pacific plate, and they support this with a variety of topographical, magnetic, seismological, and geological arguments (J. D. Hays, 1971).

The topography of the Emperor seamounts and of those in the Gulf of Alaska provides one argument. The most northern seamount of each group lies within the trench. The tops of both are unusually deep and give the appearance of having been tipped over in the process of sliding down into the trench (R. S. Dietz, 1954; H. W. Menard and R. S. Dietz, 1951).

Recently, P. J. Grim and F. P. Naugler (1969) and F. P. Naugler (1970) found two fossil deep-sea channels on the Aleutian abyssal plain created by the flow of turbidity currents southwestward from Alaska. One channel now climbs up the wall of the Aleutian trench before flowing down the sloping Pacific floor. Since turbidity currents could not have flowed uphill out of so deep a trench, the authors conclude that the trench formed after the channel.

The character of the magnetic profiles in the Gulf of Alaska and the paleomagnetism of Pacific seamounts both suggest that in the early Tertiary the Pacific floor was 25° south of its present position. Paleomagnetic measurements suggest that during the Cretaceous, North America lay 20° farther north. The sum indicates great relative motion between the Pacific and American plates (F. J. Vine and H. H. Hess, 1970; J. Francheteau et al., 1970; A. Larochelle, 1968).

In the northeastern Pacific the pattern of magnetic anomalies is strikingly regular, but the Aleutian trench truncates it abruptly. The pattern includes a remarkable change in strike called the Great Magnetic Bight. Sea-floor spreading with subduction of the Pacific floor beneath the Aleutian trench can explain both features, the latter as the remnant of a former triple point, or junction of three plates (D. J. Elvers et al., 1967; W. C. Pitman III and D. E. Hayes, 1968; B. H. Erickson and P. J. Grim, 1969; J. A. Grow and T. Atwater, 1970; F. J. Vine and H. H. Hess, 1970; Fig. 16-1). In fairness to T. Atwater (1970 and last section) it must be pointed out that the supposed overriding of the Great Magnetic Bight has not caused any identifiable uplift of the Alaska region as occurred in the southwestern United States (E. D. McKee and E. H. McKee, 1972).

W. Stauder (1968) found that the behavior of earthquakes in the Aleutian region suggests that the Pacific floor is being overridden. M. Ewing et al. (1965) and D. R. Horn et al. (1969) showed that there are no sediments over the Aleutian ridge, that the thickness of sediments on the south wall of the trench is the same as that on the floor of the Pacific, and that the sediments on the floor of the trench lie horizontally. R. von Huene and G. G. Shor, Jr. (1969) concluded from this lack of deformation of the sediments in the Aleutian trench that there is no large thrust fault zone at the base of the continental slope nor any evidence for compression and subduction. It may be that the trench is not due to compression but to tension caused by the sinking of the lower lithospheric plate because of its high density.

The eastern margin of the Pacific plate lies along the British Columbian and southern Alaskan coasts, where northward motion has formed great strike-slip faults, changing to curved thrusts around the Gulf of Alaska (R. Stoneley, 1967; R. Page, 1969).

The Bering Sea is divided into two parts separated by a steep slope from which Cretaceous sedimentary rocks have been dredged (D. B. Stone, 1968; D. M. Hopkins et al., 1969). The deeper western part appears to have normal ocean floor and connects with the main Pacific basin through a channel along the Kamchatka coast. The eastern part is one of the flattest and largest shelf areas in the world. R. S. Dietz et al. (1964) suggests a slope of only 50 m in 560 km. D. W. Scholl et al. (1968) find the deep structure to be also flat and suggest that the shelf has long been stable. This is in keeping with the view that during the Tertiary, North America has been firmly connected with the eastern tip of Siberia, and that the structural break between Asia and North America occurs in the Verkhoyansk Mountains (Sec. 15-7). A. Cox et al. (1966) have studied the age and paleomagnetism of the Pliocene to Recent lava flows on the Pribilof Islands.

In an alternative interpretation, S. W. Carey (1958) has suggested that the originally straight Cordillera was bent to produce the Alaska range and the curved faults parallel with it and that the Aleutian Islands mark the locus of motion (Sec. 15-7).

16-8 THE FLOOR AND OCEANIC ISLANDS OF THE WESTERN PACIFIC OCEAN

Samples obtained by dredging and by JOIDES drilling suggest that the western part of the Pacific Ocean basin is younger than Paleozoic but older than the East Pacific rise. Three principal hypotheses have been proposed to explain this (Fig. 16-1).

A. G. Fischer et al. (1969) regarded the Western Pacific simply as the western part of the East Pacific rise (Sec. 16-4).

Some Soviet and other scientists hold a second view. They believe in fixed continents and permanent ocean basins and favor vertical motions rather than horizontal ones. They consider that upward migration of heat and transfer of material from the mantle has led to "oceanization," or "basification," of what had been continental crust to oceanic crust. In particular, they believe this to be the origin of the Sea of Okhotsk. Recent statements of this point of view in English include those of V. V. Beloussov and E. M. Ruditch (1961), L. I. Krasny (1967), M. M. Lebedev et al. (1967), V. V. Beloussov and I. P. Kosminskaya (1968), and many papers in L. Knopoff et al. (1968). Dutch scientists have also proposed the oceanization of the western Mediterranean (W. P. de Roever, 1969; Sec. 16-11).

H. H. Hess (1962, 1965) and H. W. Menard (1964) proposed a third alternative, that the Western Pacific is the site of an older, former mid-ocean ridge, the Darwin rise, which became inactive and subsided after the East Pacific rise had formed. If this is so, it is of interest to speculate whether, before the East Pacific rise spread, the Darwin rise was the only mid-ocean ridge (H. W. Menard, 1969a,c) or whether during the Jurassic period it connected with others. One of these could have lain between Australia and India. Another might have joined the Gulf of Mexico to the Tethys, and by opening the oldest part of the Atlantic between

Africa and the United States it could have separated Pangaea into Gondwanaland and Laurasia.

Whatever its origin, the floor of the main basin of the western Pacific Ocean is largely of Mesozoic age, and it contains more islands and seamounts than any other part of the oceans. Because these islands are remote and have few striking geological features, few have been well investigated. Nearly all are coralline atolls, of which Bikini and Eniwetok were studied as sites for testing nuclear devices (K. O. Emery et al., 1954). On Eniwetok, drilling penetrated Eocene reef limestone to basalt at 1400 m (H. S. Ladd and S. O. Schlanger, 1960). Others were investigated during World War II, but only a few of the results are readily accessible (L. E. Nugent, Jr., 1946; T. F. Gaskell and J. C. Swallow, 1954; S. O. Schlanger and J. W. Brookhart, 1955; E. D. McKee et al., 1959). Ocean, Nauru, Niue, Howland, Baker, and Jarvis are near the equator and have economic phosphate deposits (F. R. Fosberg, 1957; W. C. White and O. N. Warin, 1964).

The structure of atolls is particularly well exposed on a few which have been uplifted and eroded. These lie near the western boundary of the Pacific plate or near the crest of the East Pacific rise (J. T. Wilson, 1963d; Sec. 15-12).

Some islands have volcanic rocks exposed and are called high islands, to distinguish them from the low atolls. These include Ponape (871 m high), Kusaie (655 m), and Truk (305 m) in the Caroline group (J. T. Stark and R. L. Hay, 1955), and Rarotonga (639 m), Mangaia (169 m), Atiu (82 m), Mauki (31 m), and Aitutaki (137 m) in the Cook group (P. Marshall, 1927, 1930). Near Fiji three groups of basalt islands have active or recent volcanoes. Samoa is a chain of islands of predominantly Pliocene basalts with recent trachyte intrusives and flows (R. A. Daly, 1924; D. Kear and B. L. Wood, 1959; D. H. Tarling, 1966). The Wallis Islands form a continuation of the same chain (H. T. Stearns, 1945; G. A. Macdonald, 1945). Nuiafou is a small active volcano between Fiji and Samoa (R. W. Fairbridge and H. D. Stewart, Jr., 1960). Fiji itself is described in Sec. 16-10.

Scattered among the islands are many submarine seamounts, including most of the flat-topped guyots, or tablemounts, in the world. H. H. Hess (1946) and H. W. Menard (1969a) proposed that they were volcanic islands which were eroded to sea level on a former mid-ocean ridge and were carried down when it subsided (Fig. 14-8). There are other somewhat larger rises, including the Shatsky rise east of Japan (M. Ewing et al., 1966a) and the Manhiki rise northeast of Samoa (B. C. Heezen et al., 1966b), which H. W. Menard (1969a) has suggested may be due to local heating. In the Atlantic Ocean the Bermuda rise (G. B. Engelen, 1964) and that beside the Scotia arc (P. F. Barker, 1970) appear to be similar features and could be oceanic domes corresponding to the continental domes of East Africa (Sec. 15-2).

The paleomagnetism of about 50 seamounts has now been determined by a method due to V. Vacquier (1962). These range in age from Upper Cretaceous to Recent (J. Francheteau et al., 1970; Fig. 16-1). The results indicate that the northeastern Pacific has moved 30° northward during that time.

16-9 THE EAST ASIAN ISLAND ARCS AND MARGINAL SEAS

The chains of islands bordering East Asia and Alaska are the finest examples of island arcs, and appear to form two belts. The outer and younger of these consists of the Aleutian, Kuril, northern Japanese, Bonin, Mariana, Yap, and Palau Islands, which form the active margin of the Pacific plate. The subduction zone is well marked by many shallow and deep earthquakes, numerous volcanoes, and the deepest ocean trenches in the world. All except northern Japan are small volcanic islands, predominantly of andesite, overlain in the south by Eocene and younger coralline limestones (P. E. Cloud et al., 1956; Sec. 16-7).

An inner and less active belt of older and larger islands extends from northeastern Siberia through Sakhalin Island, northern and southern Japan, the Ryu-Kyu Islands, and Taiwan to the Philippines, touching the younger belt only in northern Japan. One can speculate whether the older belt was formerly a subduction zone, perhaps associated with the Darwin rise, and whether the two belts were forced together (Fig. 12-1).

There is little literature in English about the geology of the Kuril Islands except some papers given at the beginning of the last section, and G. S. Gorshkov (1970) has published a long recent review. They are all young islands arranged in a simple island arc with a deep parallel trench. The Japanese scientists who have studied the Sea of Okhotsk (Fig. 7-10 and Sec. 7-2.3) found a normal oceanic crust with linear magnetic anomalies parallel to those in the Pacific basin and with high heat flows (M. Yasui et al., 1968a,b,c,d).

The Japanese Islands have a complex geology and regional geophysics concerning which there is a vast literature. F. Takai, T. Matsumoto, and R. Toriyama (1963), M. Minato, M. Gorai, and M. Hunahashi (1965), and T. Kimura (1967) have described the geology, which dates back to the Silurian period. M. Gorai (1968) had discussed the structural divisions. T. Rikitake et al. (1968) have reviewed the geophysics. The Fossa Magna, which crosses central Honshu, marks a striking change. North of it the Japanese Islands, although older than the active belt, seem to be part of it and to share its features, which include active volcanoes, deep earthquakes, and a strip of negative gravity anomalies. This section of the Japanese Islands cuts across the trend of the magnetic anomalies on the sea floor. To the south of the Fossa Magna the islands are parallel to the magnetic anomalies (Fig. 16-14), and there is no trench or deep earthquakes and only a few volcanoes. S. Tokuda (1927; see also W. H. Bucher, 1933; J. T. Wilson, 1954) showed that the volcanoes throughout the Japanese Archipelago are arranged in short *en echelon* lines (Fig. 16-15). These are subparallel to the trend of the magnetic anomalies. Paleomagnetic results suggest that the Japanese Islands have been bent (S. Sasajima et al., 1968; J. Hospers and S. I. van Andel, 1969).

H. Kuno (1966a,b) noted a change in petrology from abundant tholeiites on the Pacific side of the Japanese Islands through high-alumina basalts to alkali basalts in the Sea of Japan, which he considered to be due to generation of the lavas at

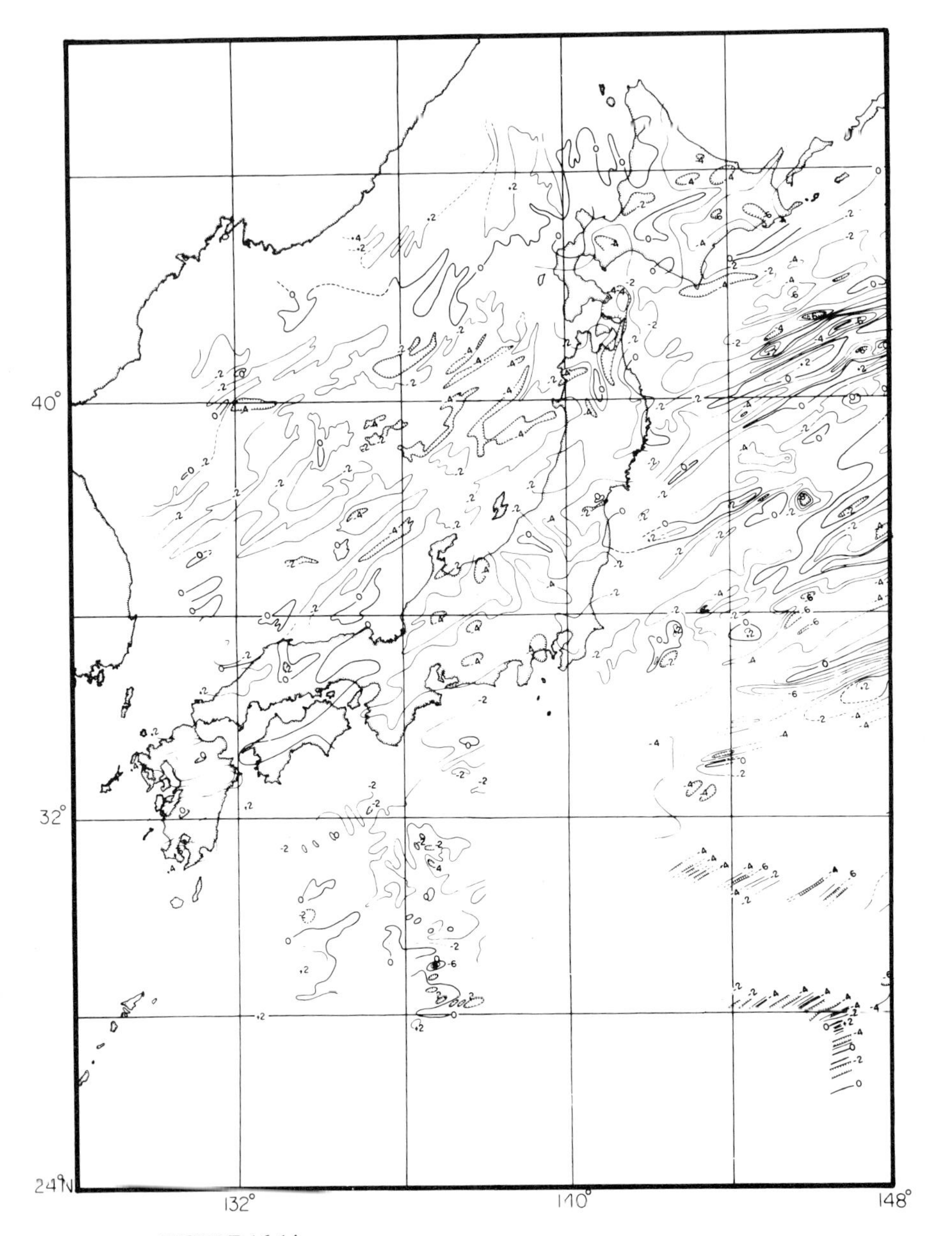

FIGURE 16-14
Chart of magnetic anomalies in the vicinity of Japan in units of 100γ. (*After S. Uyeda and V. Vacquier, 1968.*)

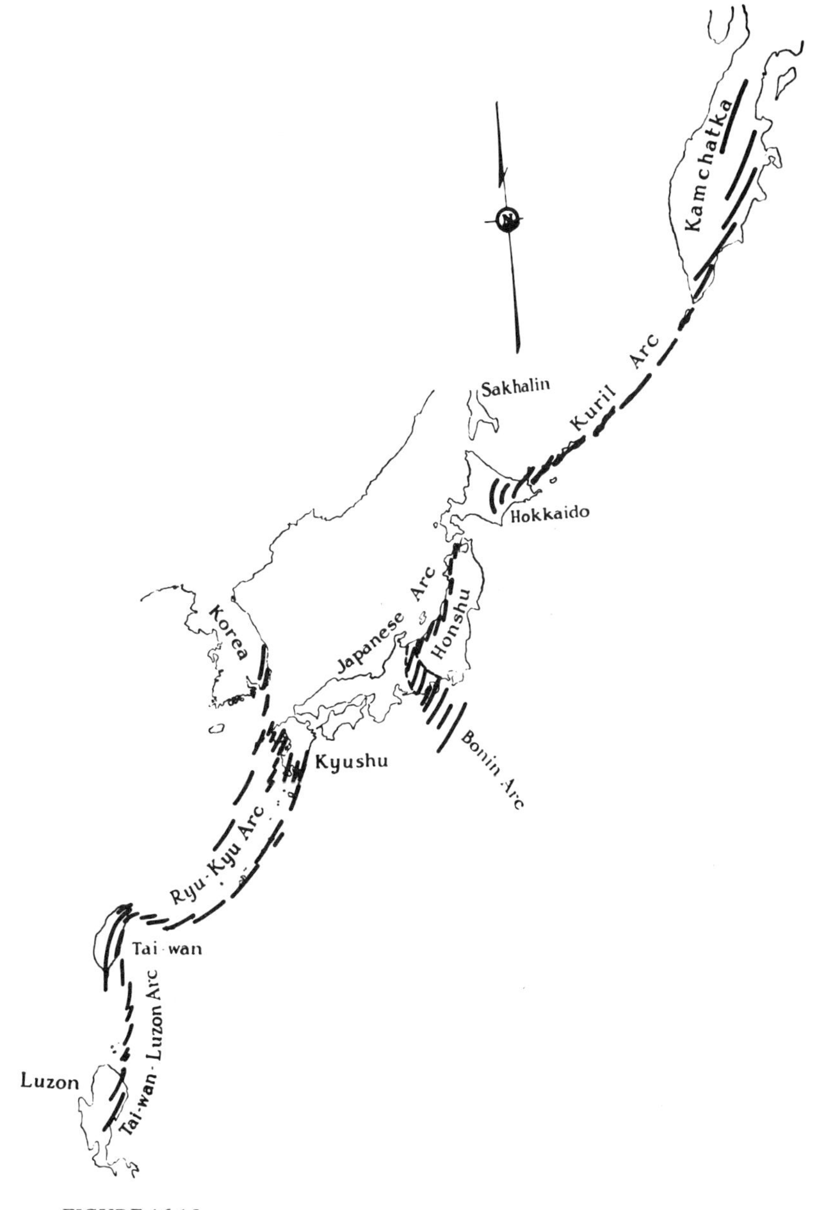

FIGURE 16-15
Structural trends joining volcanic vents in Japan which reveal an *en echelon* pattern. (*After S. Tokuda, 1927.*)

successively greater depths compatible with the underthrusting of Japan by a rigid oceanic plate along Benioff zones of deep earthquakes. M. Tatsumoto (1969) gave the same explanation for a change in the isotopic ratios of lead in primary basalts.

A. Miyashiro (1961) noted the existence of paired metamorphic belts in Japan, the outer of which appears to have been formed under conditions of low temperature and high pressure, while the inner required high temperature and low pressure. W. J. Ludwig et al. (1966) found evidence of tension in the Japanese trench which could be due to bending or active sinking of an oceanic plate (Sec. 17-6 discusses possible explanations for these observations). The Japan Sea resembles the Sea of Okhotsk in having normal oceanic crust, linear magnetic anomalies parallel to those in the main Pacific basin, and high heat flow. The central Yamato Bank is parallel with the anomalies and could have been a center for spreading (M. Yasui and Y. Hashimoto, 1967; S. Uyeda and V. Vacquier, 1968; M. Yasui et al., 1968b).

The Bonin, Mariana, Yap, and Palau Islands are all chains of small and young islands, with the greatest deep in the world off the Marianas (P. E. Cloud, Jr., et al., 1956; D. E. Karig, 1971b).

In the region south from Japan to Taiwan, C. S. Ho (1967), S. Murauchi et al. (1968), T. P. Yen (1968), M. Katsumata and L. R. Sykes (1969), and C. C. Biq (1969) have reviewed the information on crustal structure and the direction and distribution of earthquakes. There are rocks as old as the Permian on Okinawa Island.

The geology of the Philippines has been reviewed by F. C. Gervasio (1967), L. Bryner (1969), and S. Omote et al. (1969). The last authors point out that besides the well-known left-handed strike-slip fault parallel to the Philippine trench along the east side of the islands (C. R. Allen, 1965; R. W. R. Rutland, 1968), there is another left-handed strike-slip fault crossing southern Luzon Island at right angles to the first and parallel with the west coast of the archipelago (J. S. Teves, 1955). This second fault, the Taal Line, divides northern Luzon from the southern Philippines. Luzon forms an arc convex to the west which appears to be overriding the South China Sea. W. J. Ludwig et al. (1967) and D. E. Hayes and W. J. Ludwig (1967) found a trench off the west coast marked by a strip of negative gravity anomalies (Fig. 16-16). The southern Philippines are divided into a stable western block and an eastern arc convex toward the east and overriding the Pacific.

16-10 THE SOUTHWESTERN BORDER OF THE PACIFIC OCEAN

From the Philippines and Indonesia to New Zealand, the Pacific is bordered by islands many of which are large and most of which are arranged in linear chains. They are the site of many large earthquakes and volcanic eruptions. The general relationship between age and size of islands suggests that islands grow; according to plate tectonic theory, linearity is due to faulting.

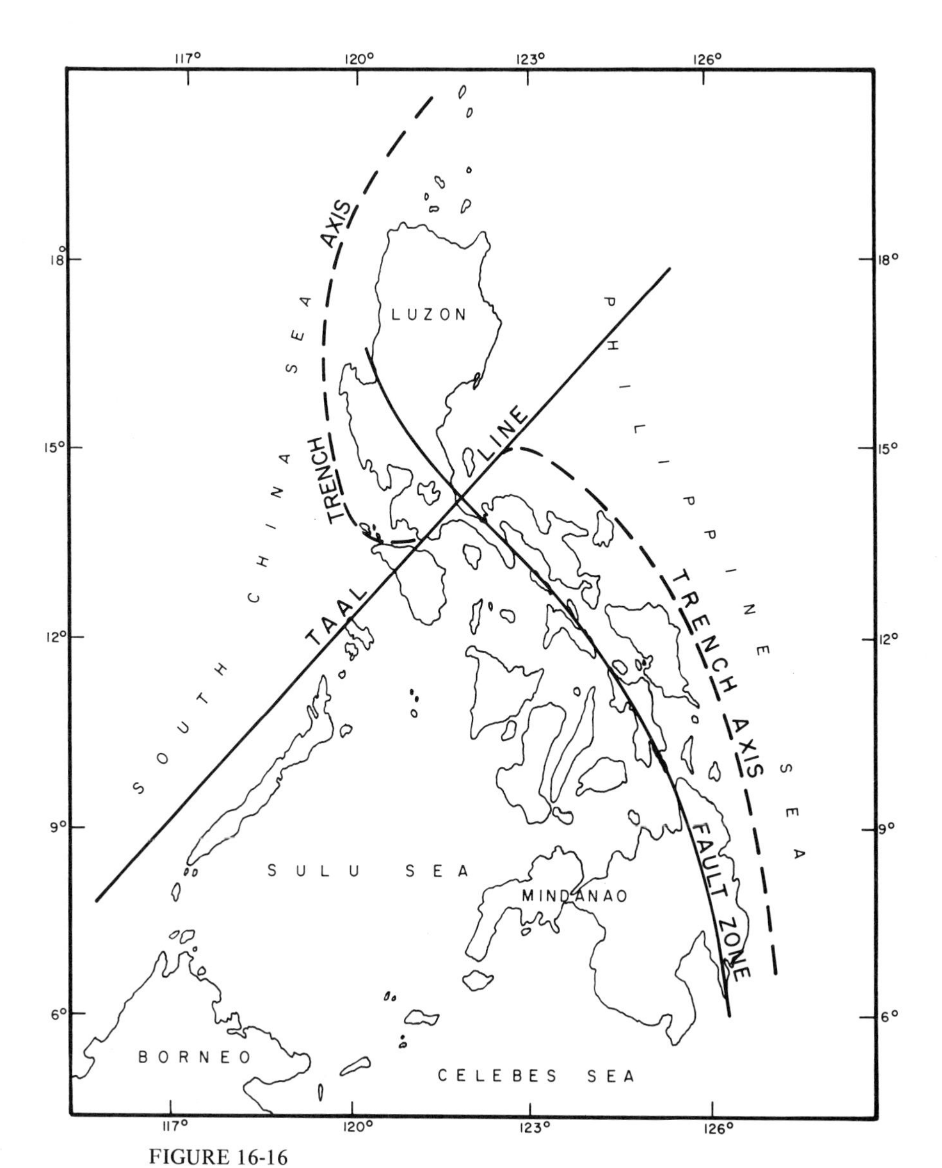

FIGURE 16-16
Sketch map of the Philippines showing the change in the location of trenches and the reversal in the direction of apparent motion across the Taal Line, although the active fault zone crosses that line without apparent offset. (*After J. S. Teves, 1955, and L. Bryner, 1969.*)

W. A. Visser and J. J. Hermes (1962) and J. J. Hermes (1968) have described the geology of western New Guinea in great detail. They concluded that the Pacific floor is sliding west past the island, producing great strike-slip faults. J.E. Thompson and N.H. Fisher (1965) have given an account of eastern New Guinea, of which a striking feature is the Papuan ultramafic belt 400 km long. It is layered, passing upward from peridotite, through gabbro and minor diorite, to basaltic lavas (Fig. 16-17). H. L. Davies (1968) has interpreted this belt as a slice of Cretaceous ocean floor thrust westward on to New Guinea.

A. Malahoff (1970), P. J. Coleman (1970), and several papers in L. Knopoff (1968) have reviewed the geology of the Solomon and New Hebrides Islands, which are Tertiary blocks of mixed basalts and andesites lying along great strike-slip faults. J.R. Richards et al. (1966) found an age of 50 m yr for the basal schists. G.G. Lowder and I.S.E. Carmichael (1970) have discussed the petrology of some New Britain volcanoes. D. Denham (1969) has summarized the main structural units of the region and related the seismicity to them in terms of plate tectonics. Deep earthquakes lie almost vertically below the north coast of New Guinea and beneath New Britain and the Solomon Islands. A belt of shallow earthquakes follows the Admiralty Islands to enclose the Bismarck Sea.

The New Hebrides condominium consists of a Y-shaped group of islands, predominantly andesites. The oldest rocks are pre-Miocene, probably Oligocene, and they have been cut by upper Miocene and Pliocene intrusives and active volcanoes (D.H. Tarling, 1967a; D.I.J. Mallick, 1970). L.R. Sykes (1964) has given an account of the many deep earthquakes whose foci lie vertically below the islands.

D. C. Krause (1965, 1966, 1967), M. Ewing et al. (1970), J.V. Gardner (1970), and J.I. Ewing et al. (1970c) have described the bathymetry, crustal structure, and sediments of the Solomon and Coral Seas. The floor of the Coral Sea is normal oceanic crust covered with 2.5 km of sediment, while the Queensland Plateau appears to be a submerged extension of the Australian continent.

New Caledonia is a French island remarkable for the largest body of ultrabasic rocks on land, which forms its northeastern side (A.R. Lillie and R.N. Brothers, 1970). These rocks appear to be of Oligocene age since they overlie Eocene basalts and cut Permian tuffs and Triassic and Jurassic greywackes. Contrary to the view that the intrusion of ultrabasics marks the start of a geosyncline (H.H. Hess, 1955), New Caledonia was already old when the ultrabasics were emplaced, and many now believe that they represent a sheet of ocean floor thrust up on to the island. J. B. Waterhouse (1967) has pointed to great similarities with rocks of New Zealand on the same Norfolk ridge. He and B. Tissot and A. Noesmann (1958) and C. Kieft (1962) have reviewed earlier work by Piroutet, who believed that the periodites were intruded along steep faults; by J. Avias, who considered them to be formed by extreme metamorphism; and by P. Routhier, who visualized them as nappes.

To the east lies Fiji, which consists of a complex of primarily andesitic

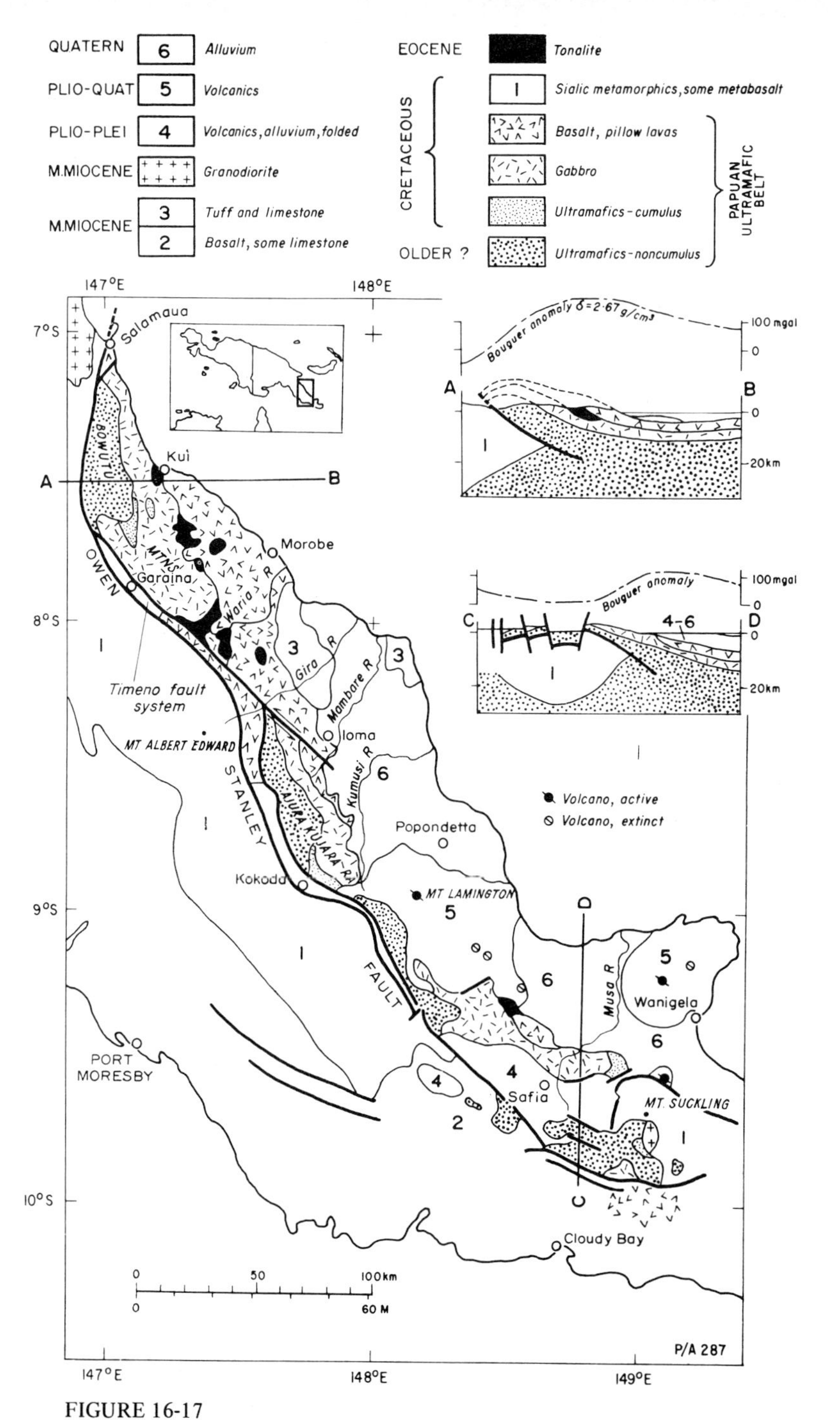

FIGURE 16-17
Geological sketch map of the Papuan ultrabasic belt with two cross sections, suggesting that oceanic crust may have been thrust south over eastern New Guinea. (*After H. L. Davies, 1970, with permission of the Director, Bureau of Mineral Resources, Australia.*)

volcanic rocks and associated clastics of early Tertiary age, cut by numerous basic to granitic intrusives and surrounded by coral reefs (W. R. Dickinson, 1967; R. F. Duberel and P. Rodda, 1968). Gold has long been mined in Viti Levu Island (H. H. Dunkin, 1968). Paleomagnetic work has proved disappointing (D. H. Tarling, 1967b), but heat flow has been shown to be high over the submarine Fiji Plateau, a characteristic shared with other seas just inside the Andesite Line (J. G. Sclater and H. W. Menard, 1967; D. P. McKenzie and J. G. Sclater, 1968).

The active andesitic volcanoes and islands of the Tonga and Kermadec groups lie along a straight line from Samoa to New Zealand, with a deep trench and a strip of large negative gravity anomalies to the east (M. Talwani et al., 1961). About 70 percent of the world's deep-focus earthquakes occur in the seismic zone, which dips at about 45° westward beneath the islands. Extensive studies by B. I. Isacks et al. (1968, 1969), W. Mitronovas et al. (1969), and L. R. Sykes et al. (1969) have shown that this zone is not more than 25 km thick and that shallow earthquakes can be explained by tension where one lithospheric plate bends down, slightly deeper earthquakes by shearing between the two plates, and deep-foci earthquakes by compression of that plate which is being forced down. Transform and hinge faults occur at the end of the seismic zone (Figs. 16-18 and 16-19).

New Zealand, like Japan, is a large and complex island with formations of all periods since the Cambrian. Most strata are folded into a great syncline across which the Alpine fault cuts (H. W. Wellman, 1952, 1956; R. P. Suggate, 1963). D. P. McKenzie and W. J. Morgan (1969) have interpreted this as a transform fault joining the Kermadec trench to another trench which lies to the south and extends to Macquarie Island. Geological and gravity mapping is complete (D. A. Brown et al., 1968; W. I. Reilly, 1965a,b), and the seismicity and tectonics are becoming understood (R. M. Hamilton and A. W. Gale, 1969). J. B. Waterhouse and P. Vella (1965) and J. B. Waterhouse and S. Piyasin (1970) have recently found fossils of late Paleozoic age belonging to two faunal realms. They suggest that a small part of New Zealand having Gondwanaland fossils was once connected to Australia and the remainder to Chile and South America. D. S. Coombs (1961) has discussed the regional metamorphism.

Several large and shallow ridges form an unusual feature of this whole region (Fig. 16-20). Largest and best known are the Campbell Plateau and Chatham rise, which extend eastward from New Zealand as shoal areas. They seem to be submerged fragments of a continent, and six groups of continental and volcanic islands rise from them (C. P. Summerhayes, 1969). On the other side of New Zealand the Lord Howe and Norfolk ridges extend northwest, each capped by a single small basaltic island. W. J. M. van der Linden (1969) interpreted these as slabs of stretched continent, and the small Dampier ridge toward Australia as a mid-ocean ridge. It is bordered by two rows of seamounts (J. R. Conolly, 1969). D. J. Cullen (1970a) considers on the basis of additional geophysical data that all three ridges, and also probably the Three Kings–Loyalty Island ridge, are continental fragments stretched apart with the creation of fresh ocean floor in the

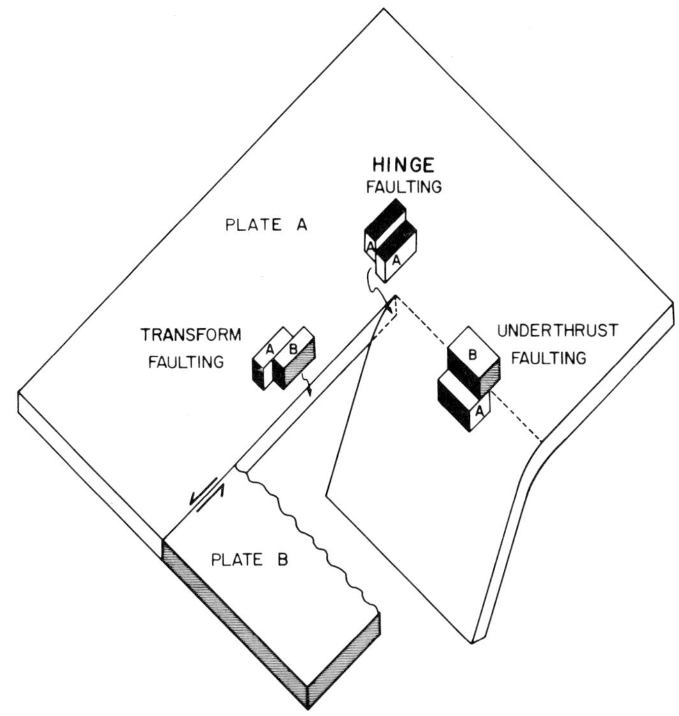

FIGURE 16-18
Types of faulting in the lithospheric plate model of the termination of an island arc by a transform fault. The model is based upon the supposed relationships at the northern end of the Tonga-Kermadec arc. (*After B. Isacks et al., 1969.*)

intervening basins. P. J. Grim (1969) has shown that, unlike other basins lying immediately inside the Andesite Line, the Tasman Sea has low values of heat flow. The suggestion that these ridges are fractured strips of continent which convection currents have opened like spreading fingers is an interesting speculation, but implies a closer connection between New Zealand and Australia than the geology warrants.

This concludes a brief review of the present state of knowledge of the vast Pacific basin. In the past twenty years much has been discovered, but much more remains to be investigated, especially in the south. Although there is evidence of expansion in some regions, notably along the East Pacific rise, most of the evidence appears to be compatible with shrinking of the basin as a whole as it is overridden by the surrounding lithospheric plates (Sec. 17-6).

An interesting feature of the borders of the Pacific are the close biological connections between opposite sides. The similarities between some living insects and some Permian fossils of Chile, New Zealand, and Antarctica have been mentioned (J. B. Waterhouse and P. Vella, 1965; L. Brundin, 1965), and so have

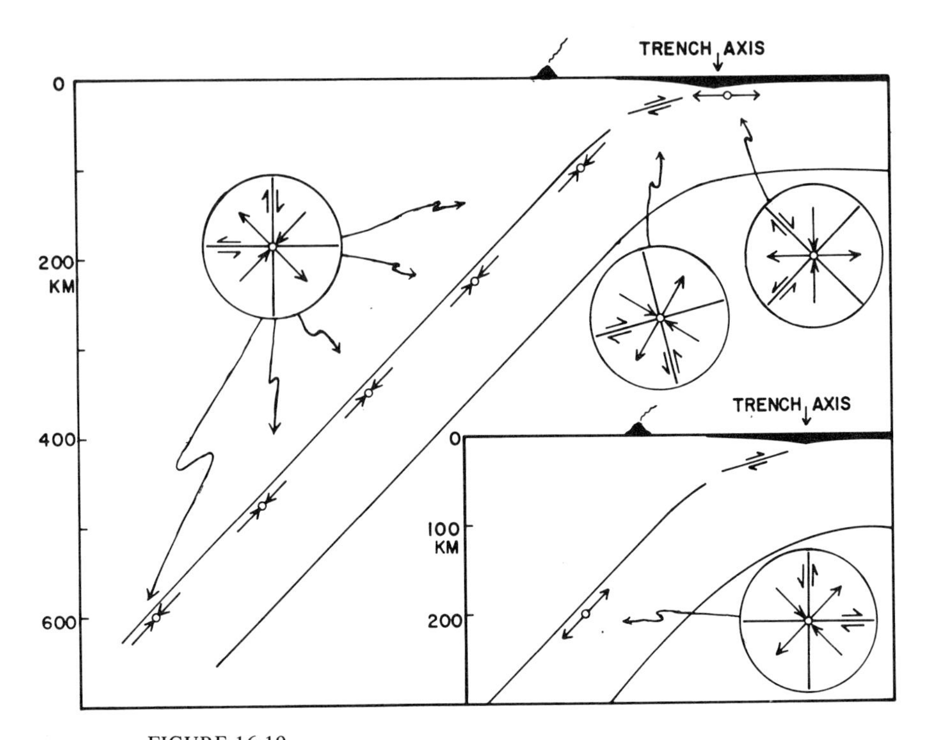

FIGURE 16-19
Vertical sections perpendicular to the strike of an island arc showing schematically possible orientation of double-couple focal mechanisms. The vertical and horizontal scales are equal, and null axes are perpendicular to the section. In the circular enlargements the sense of motion is shown for both possible slip planes. The insert shows the orientation of a focal mechanism, which could indicate extension instead of compression parallel to the dip of the zone. (*After B. Isacks et al., 1969.*)

the relationships between some East Asian and North American plants and invertebrates (Sec. 13-6; J. W. H. Monger and C. A. Ross, 1971). It is interesting to speculate whether a small continent once existed in the Pacific which broke up during the Mesozoic era due to the growth of either the supposed Darwin rise or the Great Magnetic Bight, so that fragments, once together, were driven to opposite margins of the Pacific.

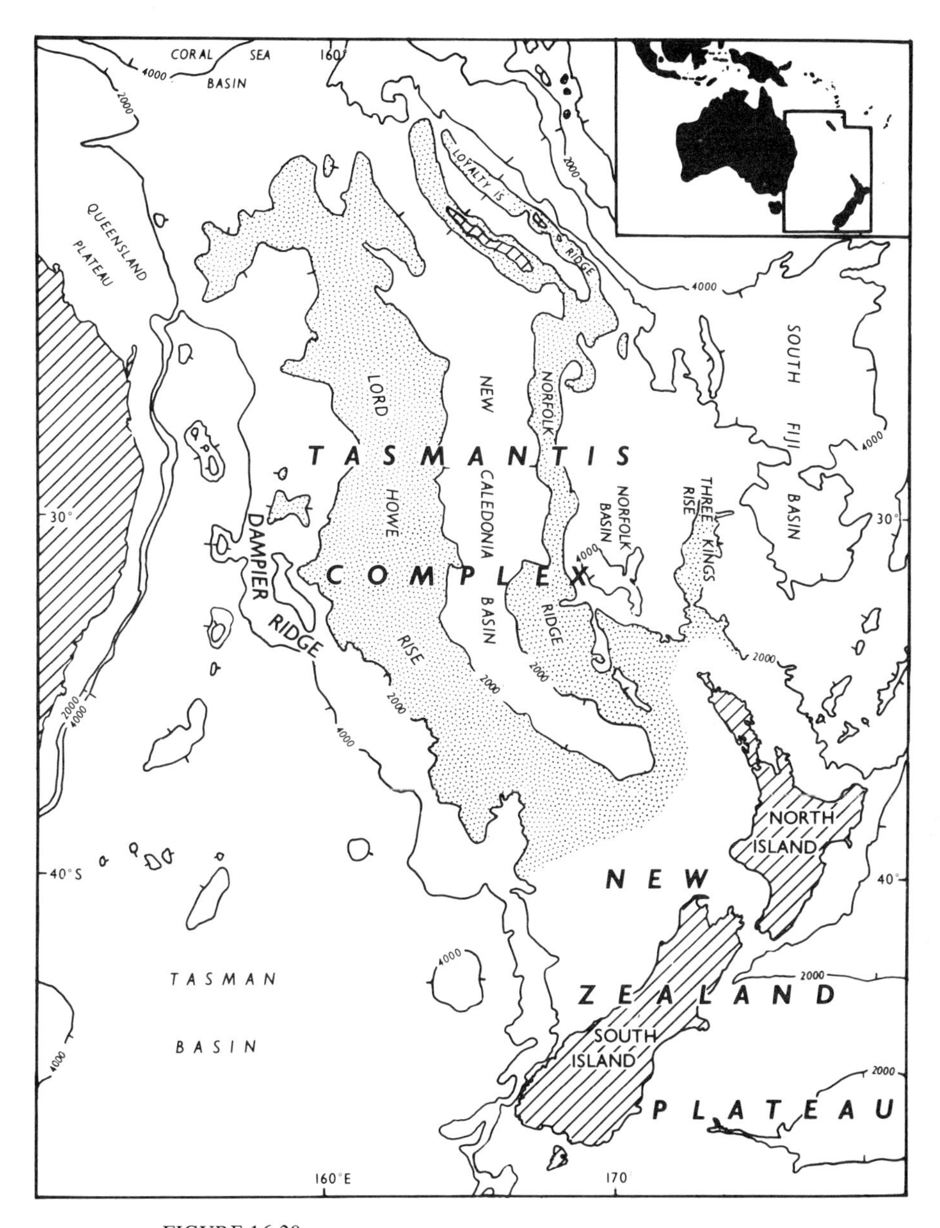

FIGURE 16-20
Sketch map of the submarine morphology of the Tasman Sea, showing the main ridges (stippled) and basins of the Tasmantis complex. Land masses shaded. Only the 2000- and 4000-m isobaths are shown. (*After D. J. Cullen, 1970a.*)

16-11 THE TYPE EXAMPLE OF STAGE 5: THE MEDITERRANEAN SEA

The Mediterranean Sea extends for 3700 km from Gibraltar to the Levant. Some of the surrounding lands have been well studied for over a century, but the basin itself and other parts of the coasts are only now being thoroughly explored (B. C. Heezen and I. P. Kosminskaya, 1970b; P. R. Vogt et al., 1971). A classic paper by H. D. Klemme (1958) has been superseded in places by later work, much of which has been reviewed in P. Hepple (1969), M. G. Rutten (1969), and guide books of the Petroleum Exploration Society of Libya (E. E. Hotz, 1965; J. J. Williams, 1966; L. Martin, 1967; F. T. Barr, 1968). J. A. Watson and G. L. Johnson (1969) have published a physiographic chart (Fig. 16-21).

The structure of the Mediterranean basin is complex, and there is no general agreement upon its tectonics or history. One way of regarding it is to consider that the Apennine and Atlas Mountains, together with Sicily and the broad, shallow extensions of the continental shelves which link them, originated as an island arc, while the Dinaric coast of Yugoslavia, some volcanic Greek islands, and part of Turkey were another (J. T. Wilson, 1954). The active and Pleistocene volcanoes of the Mediterranean and many eugeosynclinal rocks lie along these supposed arcs. The Alps, Carpathians, and Crimean Mountains are largely composed of miogeosynclinal rocks and are considered to be secondary arcs.

This oversimplification serves to emphasize the great difference between the Western and Eastern Mediterranean basins, for it suggests that the western basin and the small and shallow Aegean Sea are marginal basins and that by analogy with the Pacific they might be expected to be opening to form a young ocean floor, while the Eastern Mediterranean should be older and in the process of being overridden.

In the Western Mediterranean the continental margins and the three ridges on which stand Corsica and Sardinia, Alboran, and the Balearic Islands descend to deep basins floored with abyssal plains underlain by sedimentary deposits, which seismic investigations show exceed 1 km in several places and to have been largely transported by turbidity currents from the adjacent coasts (D. J. Stanley et al., 1970). One such current induced by the Orleansville, Algeria, earthquake of 1954 broke neighboring submarine cables (B. C. Heezen and M. Ewing, 1955). The large Rhone fan is marked by canyons (H. W. Menard et al., 1965) and buried diapiric structures, believed to be salt domes (J. B. Hersey, 1965; L. Glangeaud et al., 1967). Similar domes are reported over most of the Western Mediterranean, suggesting that the whole may be an evaporite basin (J. A. Watson and G. L. Johnson, 1968). The base of the crust beneath the basin was found to be at 30 km by M. J. Berry and L. Knopoff (1967), and they term the crust transitional rather than oceanic in nature. The basin is also underlain below a depth of 80 km by an ultralow velocity channel reminiscent of those under part of the Alps.

The solitary islet of Alboran rises in the basin east of Gibraltar. It is only 600 m long, the remnant of a Pliocene basaltic volcano. It is probably connected

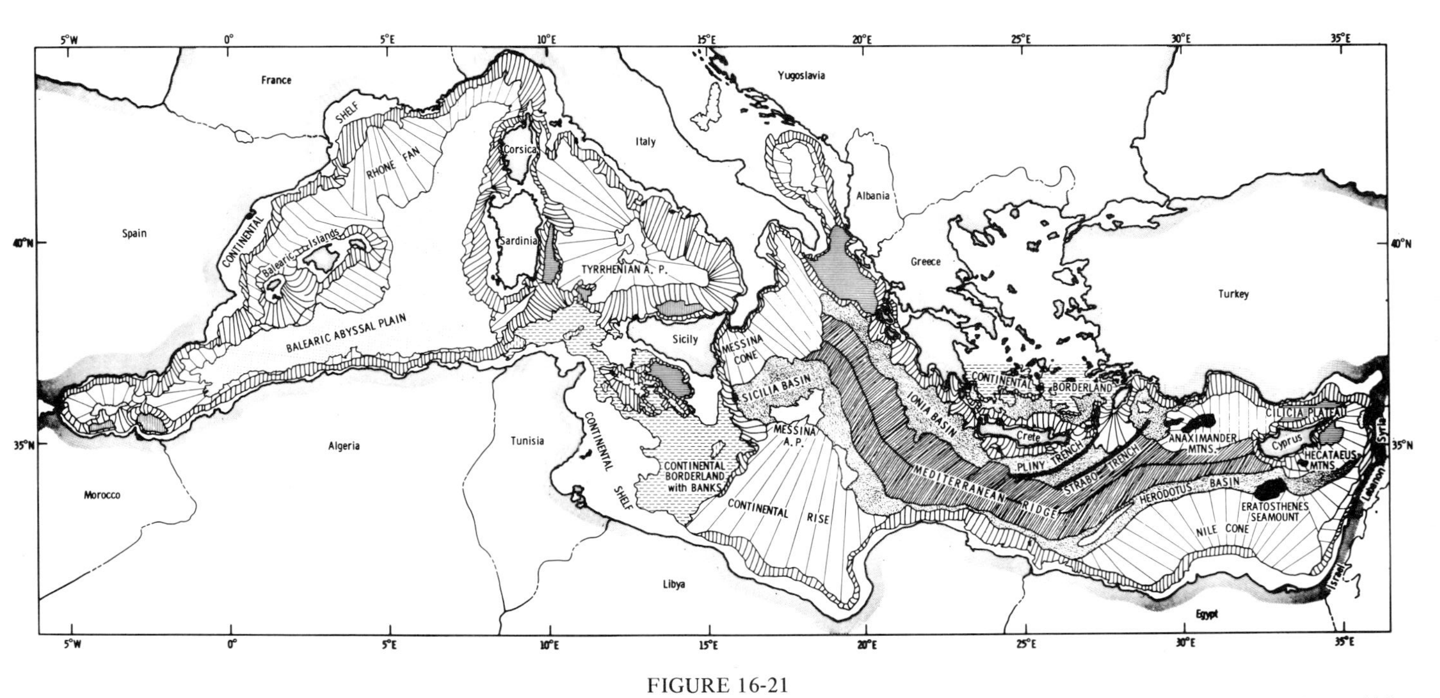

FIGURE 16-21
Outline chart of the Mediterranean basin showing physiographic provinces. (*After J. A. Watson and G. L. Johnson, 1969.*)

with other similar volcanic centers in Algeria, but could be oceanic (S. Calderon, 1882; K. G. Eriksson, 1961).

The Western Mediterranean is bordered on the south by the Atlas Mountains, a part of the Alpine system which was chiefly deformed and thrown into thrusts and nappes during the Miocene epoch (L. Martin, 1967; P. Hepple, 1969; M. G. Rutten, 1969). To the south a great fundamental fault separates the Atlas Mountains, which are European and Alpine in character, from the rest of Africa. It extends from the Gulf of Gabes in Tunisia to Agadir on the Atlantic coast. This fault is active and characterized by facies changes, enormous right-handed displacements, great unconformities, severe earthquakes, and some Tertiary volcanics (L. V. de Sitter, 1964, pp. 165 and 466). Beneath the Atlas Mountains and other Mediterranean basins, evaporites in the Triassic, near the base of the sedimentary succession, form a surface of decollement above which movements have greatly increased the tectonic complexity of the overlying rocks (W. D. Gill, 1965).

Besides possibly opening during the Miocene, the Western Mediterranean may have been sheared either through several hundred kilometers in a right-handed direction (E. Irving, 1967) or possibly through a shorter distance by a left-handed fault (J. D. A. Zijderveld et al., 1970). J. Hospers and S. I. van Andel (1969) do not consider this as proved, but accept rotation of Corsica and Sardinia. K. A. de Jong and R. van der Voo (1970) reported that the latter has rotated 50° counter-clockwise relative to most of Europe since Permian time, as J. D. A. Zijderveld et al. (1970) find is the case for the southern Alps.

R. W. van Bemmelen and other Dutch geologists have given another interpretation, suggesting that oceanization and vertical sinking are the cause of the young ocean floor and rotation in the Western Mediterranean basin (W. P. de Roever, 1969). Foundered continental blocks have indeed been found in the deep floor, so that both processes may have operated (B. C. Heezen et al., 1971).

Pleistocene volcanoes along the Dinaric coast of Yugoslavia, two active volcanoes in the Greek islands north of Crete, and more Pleistocene volcanoes in Turkey may mark a former island arc. The most notable Greek volcano is Santorini, or Thera, which erupted with the greatest violence in about 1400 B.C., probably destroying the Minoan civilization, and certainly depositing the uppermost of several tephra layers over the floor of the surrounding Mediterranean (D. Ninkovich and B. C. Heezen, 1965; A. G. Galanopoulos and E. Bacon, 1969). South of these volcanic islands the rugged Ionia basin is the deepest part of the Mediterranean (6087 m). With its continuations, the Pliny and Strabo trenches, it forms the southern boundary of Greece and the shallow Aegean Sea, and according to W. B. F. Ryan and B. C. Heezen (1965), fresh scarps on the Messina cone south of Sicily suggest that it is an actively subsiding trench. D. P. McKenzie (1970) suggests that this trench and the coasts of Albania and Yugoslavia form the active boundary of a small tectonic plate under the Aegean Sea which is overriding the Ionia trench and pushing it down (Fig. 16-22). He considers that another small plate lies under

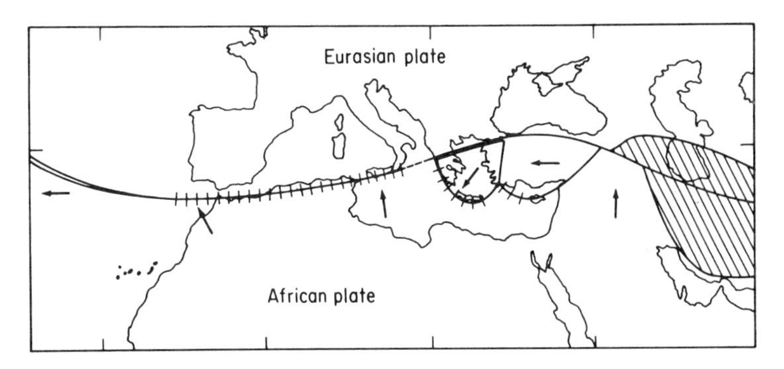

FIGURE 16-22
Approximate positions of plate boundaries at present active, with arrows marking the directions of motion relative to the Eurasian plate. Boundaries creating lithosphere are shown with double line, boundaries consuming plates with short lines at right angles to them. The crosshatched region in eastern Turkey and Iran is seismically active throughout. Fault-plane solutions here are all overthrusts and show that the crust is being thickened all over this region. Most major shocks within the crosshatched area occur on major active faults mapped by Wellman from aerial photographs and shown as solid lines. (*After D. P. McKenzie, 1970.*)

Turkey and that the north Anatolian fault zone forms its northern boundary. This region is notorious for frequent and disastrous shallow earthquakes due to dextral motion of about 11 cm/year (J. N. Brune, 1968; N. N. Ambrasseys, 1970). The movement of these plates has contributed to the complex geology of Greece and Turkey and has thrust ultrabasic bodies up from the mantle (J. Aubouin, 1964; E. E. Hotz, 1965; E. M. Moores, 1969). McKenzie believes that these plates are moving west and south over the Mediterranean because it requires less work for them to move in that direction and override oceanic crust than to move directly south to override the continental plate of Arabia.

Immediately to the south of the Ionia basin and the Pliny and Strabo trenches lies the Mediterranean Ridge. It crosses Cyprus and forms a dominant feature across the Eastern Mediterranean basin. The submarine part is covered with sediments, partly indurated, and it has been variously likened to a continuation of the Apennines, to a mid-ocean ridge, and to an upthrust section of ancient ocean floor. J. C. Harrison (1955) found negative isostatic anomalies over it, and P. R. Vogt and R. H. Higgs (1969) have shown that the magnetic pattern is an undisturbed one. They conclude that the Mediterranean Ridge may be ancient oceanic crust, part of which has been thrust to the surface and exposed across Cyprus in the Permo-Triassic Troodos complex (I. G. Gass and D. Masson-Smith, 1963; Fig. 3-13). They believe that the lack of conspicuous magnetic anomalies may be due to (1) deep burial and metamorphism, (2) formation in low magnetic latitudes, or (3) formation during a long period without geomagnetic reversals, presumably the Permian Kaiman interval (Sec. 8-10). J. Woodside and C. Bowin (1970) and I. G. Gass (1968) support this view, pointing out that the sheeted dike swarm in the Troodos massif strikes north-south and hence cannot form part of an east-west mid-ocean ridge. P. D. Rabinowitz and W. B. F. Ryan (1970) largely agree, but find evidence of crustal thickening and of nappes on the sea floor, and hold that a possible middle Cretaceous age of the sea floor in the Eastern Mediterranean is incompatible with its having been part of the older Tethys floor.

G. Steinmann (1905) first drew attention to the association in many parts of the world of serpentinized ultramafic rocks, pillow lavas, and radiolarian cherts. E. B. Bailey and W. J. McCallien (1953) called these the "Steinmann trinity." Many examples have now been found, especially in the Tethyan belt. In spite of the opinion of R. Trümpy (1960), J. Aubouin (1964), and others that the Tethys was a shallow epicontinental sea, the opinion is gaining ground that it marks the closure of a great ocean which swallowed its deep-sea strata (A. Gansser, 1966). If so, the ophiolitic rocks in these complexes may well be upthrust section of old sea floor, a view supported by isotopic studies (S. Graeser, 1969; A. Bezzi and G. B. Picardo, 1971).

To the south of the Mediterranean Ridge the structure is less disturbed. Great oil basins rest upon the northern part of the African shield in Libya, Tunisia, and Algeria (J. J. Williams, 1966; F. T. Barr, 1968; P. F. Burollet, in P. Hepple, 1969), and the Nile fan covers much of the sea floor (H.-K. Wong and F. K. Zarudzki, 1969).

16-12 OTHER EXAMPLES OF STAGE 5: THE BLACK AND CASPIAN SEAS

Both these seas lie in large deep basins. The area of each is many times greater than that of any fresh-water lake, and their depths approach those of the oceans, reaching a maximum of 2685 m in the Black Sea and 1097 m in the South Caspian basin.

Geophysical data show that the crust beneath these basins is thin and intermediate in character between continental and oceanic (Y. P. Neprochnov et al., 1970; I. A. Rezanov and S. S. Chamo, 1969). As Fig. 16-23 shows, below the floors there are layers 10 to 15 km thick in the Black Sea and 25 to 28 km thick in the Caspian in which the velocity of longitudinal seismic waves is 3 to 5 km/sec. These velocities are characteristic of sedimentary formations which are folded in the Caspian basin but not in the Black Sea. In both cases the second layer is 12 to 18 km thick, with velocities of 6.4 to 6.8 km/sec, which is the same as that in the basaltic layer of ocean basins. Indeed, this type of crust is also present in the Bering Sea and in the Sea of Okhotsk. The continental granitic layer is either absent or very thin.

There are two principal theories about the origin of these basins. One view,

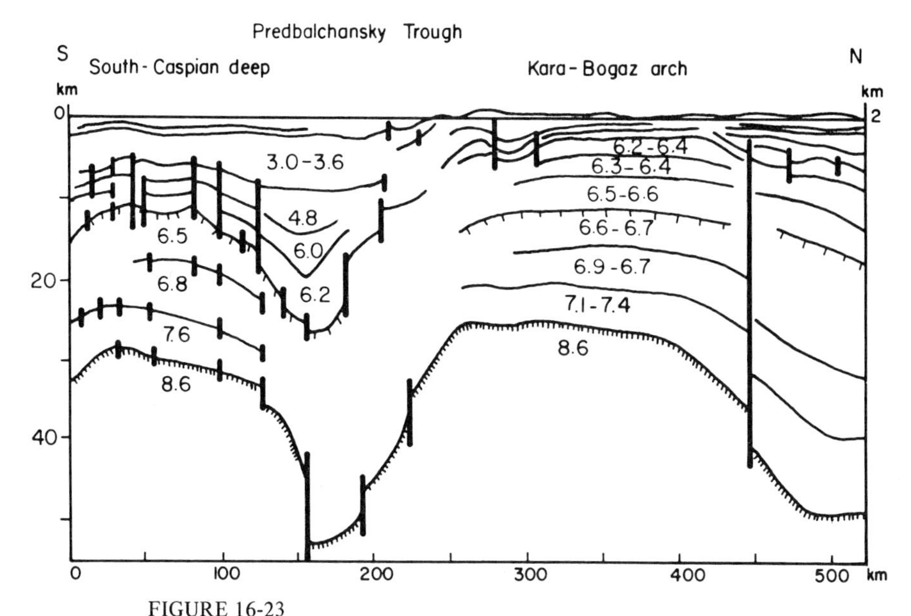

FIGURE 16-23
Deep seismic sounding profile along the southern half of the Caspian Sea near its eastern coast. Figures indicate velocity of longitudinal waves in kilometers per second. Vertical lines are faults. (*After I. A. Rezanov and S. S. Chamo, 1969.*)

popular but not universally held in the Soviet Union, is derived partly from studies of the geology in the neighboring Caucasus Mountains. This is that motions of the crust are predominantly vertical and that occasionally inversion takes place whereby regions which have for a long period been sinking are vertically raised into mountains, and other regions long elevated are depressed. In the case of these basins such authorities as V. V. Beloussov (1962) hold that throughout most of Paleozoic and Mesozoic time these basins were uplifted and subject to such deep erosion that most of the continental crust was worn away. Inversion took place, and during the late Tertiary, the basins were depressed and partly filled with poorly consolidated sediments, while the remaining basement was highly metamorphosed to give velocities in the same range as basalts. The causes of these processes have never been made clear (A. R. Ritsema, 1970). Some Soviet authorities (P. Peyve, 1969) reject this hypothesis.

The other theory is that these basins for some reason were never underlain by a granitic crust, and that the sediments were deposited upon an oceanic crust. Many accept this second view and also believe that Africa, Turkey, Arabia, Iran, and India are all rapidly closing upon the rest of Eurasia. Therefore these basins are, like the Mediterranean, remnants of the former Tethys Ocean. Africa appears to be rotating relative to Eurasia about a pole off the coast of French Guiana. This is producing compression in the area of the Caspian Sea, but farther west along the Azores to the Gibraltar ridge the same motion takes the form of dextral shearing.

D. P. McKenzie (1970) has shown that the Turkish plate forms the south side of the Black Sea from the Dardenelles to the Caucasus. Earthquakes are scattered widely over Iran and are not easy to interpret, although H. W. Wellman (1966) has shown that many lie on the traces of active faults which he has plotted, and N. Canitez (1969) has divided them into three groups. Their distribution makes it seem plausible that another small plate of continental crust underlies the stable Lut block of central Iran (J. Stöcklin, 1968). If so, its northern boundary forms the south side of the Caspian Sea, abutting against the Eurasian plate from the Caucasus to the Elburz Mountains. A Gansser (1969) has shown that two lines of ophiolitic rocks extend around both sides of the Lut block.

According to this view, the layers on the floor of the basin are old oceanic crust overlain by thick sediments. Most of the latter has collected since the southern lands collided with the main mass of Eurasia and raised the surrounding mountains, providing an abundant source.

16-13 THE TYPE EXAMPLE OF STAGE 6: THE HIMALAYA MOUNTAINS

The preceding discussions lead inevitably to the view that folded mountains are the scars left where oceans have closed, but their history is harder to interpret than that of oceans. The reason for this is a fundamental one and is due to the very great difference between an upwelling zone and a subduction zone. Ocean basins, being the site of upwellings, grow by spreading and preserve on their expanding floors the history of their growth. Folded mountains, on the other hand, being over subduction zones, preserve, and then in much disturbed form, only those rocks which the subduction zones have not consumed. For these reasons the theory of plate tectonics when applied to neighboring oceans usually provides a more precise basis for calculating the history of the rates and directions of closure of Tertiary mountain systems than does geological mapping of the mountains themselves (A. Gansser, 1966; X. Le Pichon, 1968).

Consider what happens when an ocean basin closes. Two coasts will normally be involved, both of which will have shelf deposits and one, if not both, of which will have a trench over a subduction zone. Associated with them will be island arcs or primary mountains. The extent of these deposits will vary with the age, climate, and surroundings of the coasts.

As long as the coasts are far apart and belong to different continents, the plants, land animals, and many shallow-water invertebrates will be distinct, but when the continents join, species from each side will invade the other continent, struggle for survival, and produce a common, or at least partly common, fauna and flora.

As they join, one continental block will ride up over the other, producing a sudden and great uplift from which alluvial fans of clastic rocks and red beds will pour down. Depending on the extent of overriding, part or all of the former island arc and shelf deposits may be carried down into the subduction zone and lost. Because the shape of the two coasts will not be the same, some places will meet sooner than others, and some will be farther overridden. Thus episodes of mountain building along a single system will neither be synchronous nor of equal intensity in different parts (J. T. Wilson, 1966). The two continental slabs are likely to be of different ages.

Because there is no reason why the direction of motion of a plate should be normal to its leading edge, closure will frequently involve a component of shearing, as well as compression, which produces the longitudinal strike-slip faults seen in many folded mountains.

The surfaces of decollement upon which overriding takes place will be the bottom of the lithosphere, not the base of the crust, and thus one would expect to find some slices of mantle material sandwiched between the slabs of crust of the two continents (B. M. Reinhardt, 1969).

All these features can be readily recognized in folded mountains. Miogeosynclines are old shelf deposits; eugeosynclines and flysch are old island arcs; molasse deposits are old clastic fans; ophiolites are patches of the mantle; longitudinal faults are common (J. F. Dewey and J. M. Bird, 1970a,b).

As a type example of this stage we may take the Himalaya Mountains. E. Suess (1904-1924) recognized that they had in large measure formed from the sedimentary rocks deposited in a great trough, which he named Tethys. This former ocean not only preceded the Himalayas, but also formed the boundary between the late Paleozoic–early Mesozoic floras and faunas of Gondwanaland and Laurasia, and its trace still marks an important floral boundary (N. W. Radforth, 1966; Fig. 13-2).

There has been much debate about its nature. Some have held it to be the remnant of a great and deep ocean which closed when India, Arabia, and Africa moved northward into collision with the rest of Eurasia. On the other hand, B. Kummel (1970), noting the absence of deep-water sediments, has maintained that it was merely a shallow sea formed by invasion of the continent.

A. Gansser (1966, pp. 842-843, and 1964) has offered an explanation for the absence of deep-water sediments which reconciles both views.

All these facts seem to indicate that a large amount of the crustal layer must have disappeared along the present Indus Line. The crustal shortening through down-buckling must account for the sharp facies differences in the sediments and the presence of Upper Cretaceous ophiolites which are otherwise unknown in the Himalayas. . . . The greater part of the Himalayan range consists of shield material which has been thrust over shield material, or rather underthrust by the northwards drifting mass of the main Indian Shield. The greater part of the largest mountains on our globe thus do not form a geosynclinal range, and have not evolved through the "classical geosynclinal theory." Only along the Indus Line were deeper water marine sediments deposited in a "geosyncline basin" which has now completely disappeared during the orogeny.

A quite different view, interpreting the region north of the Himalayas in terms of vertical movements, has been presented by V. N. Krestnikov and I. L. Nersesov (1964) and M. N. Qureshy (1969). The Soviet point of view is better understood if one realizes that the underthrusting occurs on the Indian, not on the Soviet, side of the mountains and that what the Soviet geologists see is an uplifted range off which thrust sheets and severe landslides are sliding under gravity (I. E. Gubin, 1967).

Most of those who have studied the paleomagnetic data support the view that, since the Tertiary, India has been moving rapidly north and converging upon the rest of Eurasia (M. W. McElhinney, 1970). This is also the conclusion of most students of the Indian Ocean, who find no evidence that it is old and much that it has expanded rapidly (Sec. 15-12). For example, X. Le Pichon (1968) finds from studies of magnetic anomalies and plate tectonics that near the junction of the Hindu Kush and the Himalayas the Indian and Eurasian blocks are closing at a rate of 5.6 cm/year in a direction N13°E. A. R. Ritsema's (1966) and T. J. Fitch's (1970) analyses of earthquake mechanisms in the Hindu Kush and Himalayan and Burmese regions support horizontal compression and the thrusting of India under the rest of Eurasia, and their results agree with Le Pichon's conclusion. L. D. McGinnis (1971) has interpreted his gravity studies to agree with this. The eastern extension of the Himalayan Mountains bends sharply and enters the Indian Ocean to connect with the Sunda island arc of the Andamans and northwestern Indonesia.

17

THE HISTORY OF THE EARTH AND A POSSIBLE MECHANISM FOR ITS BEHAVIOR

17-1 INTRODUCTION

The two preceding chapters have described the world's oceans and related features in terms of a supposed life cycle of ocean basins based upon the hypothesis of plate tectonics. Theory and the fact that the present ocean basins are all Cenozoic or Mesozoic in age suggest that ocean basins are relatively ephemeral, lasting no more than a few hundred million years. We now discuss, in the briefest and most general fashion, the earlier and greater part of the earth's history.

According to our hypothesis, when oceans close they leave behind folded mountains as their enduring scars, so that it is to mountains and continental shields that we should look for much of the history of the past. This immediately introduces the concept of geosynclines, because geologists, although they differ widely about the causes of mountain building, have pointed out that many folded mountains formed in troughs or basins of thick sediments, called *geosynclines*. The numerous mountains in which no such trough is present are less often mentioned, although these include such varied types as the Himalayas, the Rocky Mountains of Wyoming and Colorado, the coastal uplifts of Labrador and Baffin Island, and the inner belt of Japan (T. Matsuda and S. Uyeda, 1971).

We will examine the concept of geosynclines in terms of the hypothesis of plate tectonics and attempt to trace the history of mountains back in time, observing that marked variations in the style of mountain building occurred about 600, 2500, and 3500 m yr ago. These are approximately the dates which divide Phanerozoic, Proterozoic, Archean, and pre-Archean times. Finally, we will present an opinion upon what at the present time appears to be the probable cause of the behavior of the earth's upper mantle and crust.

17-2 GEOSYNCLINES

The hypothesis of plate tectonics can immediately explain G. K. Gilbert's (1890) distinction between *epeirogenesis*, which is uplift or subsidence of land or ocean floor without important deformation, and *orogenesis*, which is the formation of mountains by a combination of uplift and profound deformation. Epeirogenesis is uplift due in most cases to the upwelling and spreading action beneath mid-ocean ridges, and some other places which are believed to be tectonically related, including the East African rift valleys, the coasts of narrow seas, and the Colorado Plateau, while orogenesis is associated with the junction of plates in compression over subduction zones. How does this second concept fit the traditional view of *geosynclines*?

In 1857 J. Hall (1859, 1883) observed that the Appalachian Mountains had been formed by the folding and faulting of shallow-water sediments which had collected in an elongated trough to a much greater thickness than elsewhere. In 1873 J. D. Dana accepted this proposal and coined the name "geosynclinal" for it, but pointed out that Hall's view that the sediments sank under their own weight failed to explain either the sinking or the subsequent deformation and uplift. As he wittily remarked, Hall had promoted "a theory for the origin of mountains with the origin of mountains left out." Both men understood that geosynclines formed on continental margins, but C. Schuchert (1923), in extending the study to the whole of North America and beyond, included basins lying on continents, and began a classification which H. Stille (1940) and M. Kay (1947, 1967) elaborated. In particular, they distinguished between *eugeosynclines*, deposits apparently formed around former island arcs and characterized by the prevalence of shale, chert, volcanic rocks, and ophiolites, and *miogeosynclines*, old shelf deposits which lie between cratons and eugeosynclines and are characterized by shallow-water sediments and few volcanic rocks. M. F. Glaessner and C. Teichert (1947) and J. G. Dennis (1967) have reviewed the subject.

On the basic idea that folded mountains have often been associated with thick and gently sinking elongated troughs there is no doubt, but the original idea did not envisage drift. Now plate tectonics provides the mechanism the absence of which had been noted by Dana. J. F. Dewey and J. M. Bird (1970a,b) and W. R. Dickinson (1971) have tried to adapt Kay's terminology to the phases of collisions between moving lithospheric plates, but their results are not satisfactory.

When they were named, geosynclines were regarded as basins subject only to vertical motions. It is now realized that a single geosyncline may contain many elements brought together by great horizontal displacements as well as vertical movements. These may include shelves formed on two continents which were once far apart, one or more island arcs, and other fragments as well. The complexity of the Mediterranean, a closing geosyncline, shows the impossibility of defining the subject closely or expecting examples to be repeated exactly. A. Gansser (1966) has pointed out that in the Himalayas most of the geosyncline has vanished into a subduction zone, a complication not foreseen in the earlier terminology. For the present it seems better to use descriptive names. Is it not simpler and shorter to call a miogeosyncline a shelf deposit, and a eugeosyncline an island-arc deposit?

17-3 INACTIVE FOLDED MOUNTAINS, THE SCARS OF VANISHED OCEANS

If the Himalayas represent the trace of the vanished Tethys Ocean, do all other folded mountains and even the provinces of Precambrian shields represent the scars left by former oceans? The answer to this question in terms of plate tectonics has scarcely been tackled, for it involves re-examining the whole of world geology and much geophysics. This places the matter beyond the scope of this book, and it will be touched upon only in the briefest manner as an indication of future problems rather than in an attempt to provide solutions.

The active folded mountains of the world have already been described in very summary fashion in connection with the oceans. Their study leads directly to older and inactive mountains, because geologists have never hesitated to use the same terminology and attribute the same origin to both, for example to the active Cordillera and the inactive Appalachians, or to the active Alps and the inactive Caledonides.

For over a century it has been recognized that the Cambrian strata of the Appalachian and Caledonian mountain systems are each divided into two realms marked by different faunal assemblages which are "amazingly uniform throughout each realm," but so different as to "make correlation between them very difficult" (R. D. Hutchinson, 1952; Fig. 17-1). This has been attributed to the effect of different environments, but this seems unlikely to be the reason, because in three places in Scotland, representatives of the same faunal assemblages occur in markedly different types of rocks (G. Y. Craig, 1965).

Strata of both realms lie side by side on each side of the Atlantic Ocean. In 1936 A. W. Grabau pointed out that continental drift could explain why similar faunas had become separated, and in 1966 J. T. Wilson proposed that the closing of an earlier proto-Atlantic Ocean could explain why contemporaneous but different faunas were brought together (Figs. 17-2 to 17-4). R. J. Ross, Jr., and J. K. Ingham (1970) and J. M. Bird and J. F. Dewey (1970) have supported the notion that a proto-Atlantic Ocean existed at the beginning of Paleozoic time, closed to form the

FIGURE 17-1
The North Atlantic region showing the present distributions of the "Atlantic" faunal realm (horizontal shading) and the "Pacific" faunal realm (vertical shading). (*After J. W. Cowie, 1960, A. W. Grabau, 1936, and R. D. Hutchinson, 1952.*)

Appalachian-Caledonides, and reopened in much the same place to form the present Atlantic. Recent descriptions of these mountains with references have been given by M. G. Rutten (1969), M. Kay (1969), J. Rodgers (1970), G. Y. Craig (1965), L. Størmer (1967), M. Brooks (1970), and H. B. Whittington and C. P. Hughes (1972).

If the Appalachian Mountains were so formed, it seems probable that their apparent continuations, or at least counterparts in the Ouachita and Marathon Mountains, should have been formed by a collision between North and South America before the opening of the Gulf of Mexico and Caribbean Sea (P. B. King, 1959; W. J. Antoine and T. E. Pyle, 1970; J. I. Ewing et al., 1970d). The same idea has been suggested for building the Innuitian Mountains in Arctic Canada and the Cordillera by impact of parts of Asia (J. T. Wilson, 1968). The Cordillera may be a double system of mountains, a view which J. W. H. Monger and C. A. Ross (1971) support in British Columbia by stating that diversity in upper Paleozoic faunas may have been brought about "by major crustal movements juxtaposing originally isolated biogeographic provinces" (Figs. 17-5 and 17-6). It is uncertain whether the additions along the outer coast were island festoons or truly continental fragments.

Some Soviet earth scientists have long believed that, like the Himalayas, the Ural Mountains lie at the junction of formerly separate continents. P. N. Kropotkin (1971) has stated this, basing his arguments largely upon paleomagnetic results obtained by A. N. Khramov et al. (1966) which show large differences in position

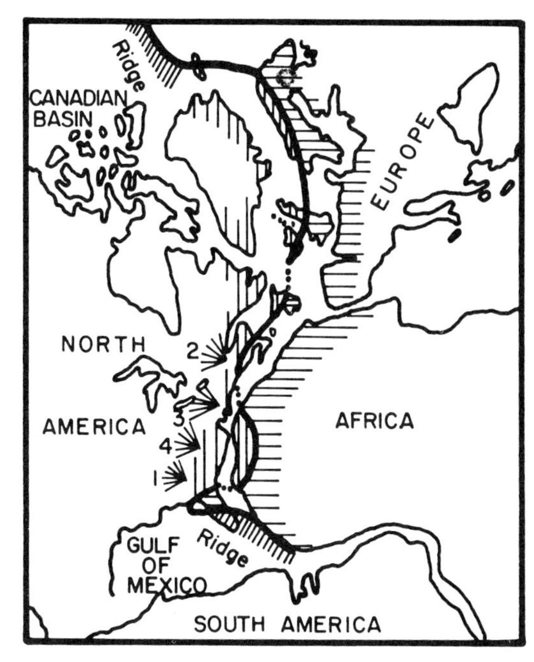

FIGURE 17-2
The North Atlantic region in Upper Paleozoic and Lower Mesozoic time showing that, of the present Atlantic Ocean, only the Canadian basin and perhaps the Gulf of Mexico then existed. Four fans are shown which were formed (1) during Middle Ordovician, (2) during Upper Ordovician, (3) during Upper Devonian, (4) during Pennsylvanian. The heavy line separates "Pacific" and "Atlantic" faunal realms. The two marked ridges are considered to have formed when the modern Atlantic started to open. (*After J. T. Wilson, 1966.*)

between Europe and Siberia in early Paleozoic time (Fig. 17-7). He also believes that other inactive mountains in Siberia may have had similar origins. These include the Verkhoyansk Mountains in northeastern Siberia, of which the rocks were formed in a sea that closed during Tertiary time as the North Atlantic Ocean opened (A. P. Markovsky, 1961; J. T. Wilson, 1963c; P. R. Vogt and N. A. Ostenso, 1970b). This range provides a clear tectonic break across Siberia from the Arctic Sea to the Pacific and forms a more significant break between the geology of most of Eurasia and that of North America than does the Bering Strait across which geological structures are continuous from Alaska into the eastern tip of Siberia.

There seems to be no reason why similar ideas cannot be applied to other folded mountains of Phanerozoic age in Australia (D. A. Brown et al., 1968), Southern Africa (S. H. Haughton, 1969; L. R. M. Cocks et al., 1969), near Buenos Aires (R. P. Morrison, 1973), and across China and southern Siberia (Ch'ang Ta, 1963; D. V. Nalivkin, 1960). But this approach has received little consideration.

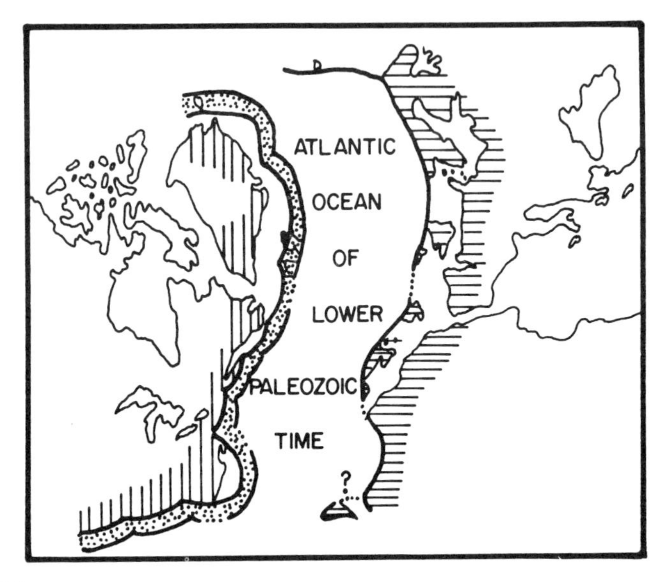

FIGURE 17-3
The North Atlantic region in Lower Paleozoic time. The proto-Atlantic Ocean would have formed a complete barrier between two faunal realms (shaded). Island arcs (dotted) probably lay along the North American coast. The floor of this ocean could have been absorbed in the trenches associated with these arcs as the ocean closed. (*After J. T. Wilson, 1966.*)

17-4 PROBLEMS OF THE PRECAMBRIAN

The earth is now believed to be about 4550 m yr old, but the oldest rocks preserved, unlike those of the moon, are 1000 m yr younger. Phanerozoic time is barely 600 m yr long, so that Precambrian rocks represent about 3000 m yr, or two-thirds of the earth's history. The problems of subdividing and classifying them are severe and are only beginning to be unraveled, although with nothing yet approaching the precision achieved in the Phanerozoic. On the other hand, radiometric dating has the same precision for rocks of all ages, and its application has enabled some progress to be made.

17-4.1 Difficulties of Subdivision and Correlation

By convention the oldest strata with shelly fossils mark the base of the Cambrian, so that in the Precambrian there are no index fossils. Their presence in younger rocks has enabled the last 600 m yr to be divided into over 200 stages, with an average length of less than 3 m yr each (R. C. Moore, 1958). No scheme of relative dating and correlation approaching this in accuracy has been found for the Precambrian. Few creatures produced hard parts, and fossil remains or evidence of

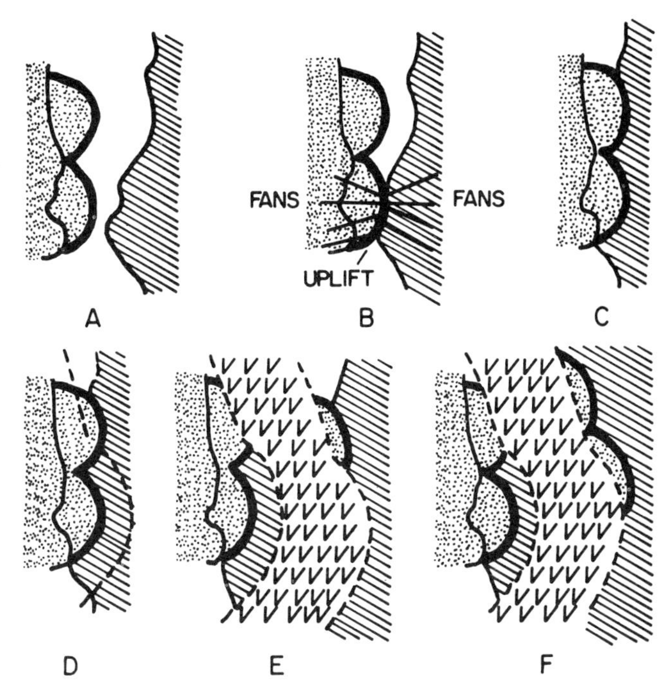

FIGURE 17-4
(*A*) A closing ocean, with island arcs on one coast, separating two different faunal realms. (*B*) First contact between two opposite sides of a closing ocean. (*C*) The ocean closed by overlap of the opposite coasts. (*D*) A possible line (dashed) along which a younger ocean could reopen. (*E*) A new ocean (checked) opening in an old continent. (*F*) A geometrically impossible way for a younger ocean to open because the arcs do not fit together. (*After J. T. Wilson, 1966.*)

organisms are rare. None provide the basis for a precise time scale (P. E. Cloud, Jr., and M. A. Seminkhatov, 1969; M. F. Glaessner, 1971).

Other difficulties are numerous. With increase in age the proportion of unaltered sedimentary strata preserved decreases and the proportion of metamorphosed rocks increases, so that interpretation is made more difficult (D. V. Higgs, 1949; R. M. Garrells and F. T. Mackenzie, 1969). The occurrence of sedimentary strata in many separate basins between which correlation is uncertain or impossible accentuates the problem. Other Precambrian rocks crop out in the isolated cores of mountains. Although radiometric dating has shown that many metamorphic areas formerly considered Precambrian are younger, great thicknesses of unmetamorphosed Precambrian strata are still recognized in the Cordilleran, Appalachian, and Innuitian orogens and on other continents.

Attempts to compare Precambrian sedimentary basins with later geosynclines have usually failed to take into account the possibility recognized by A. Gansser

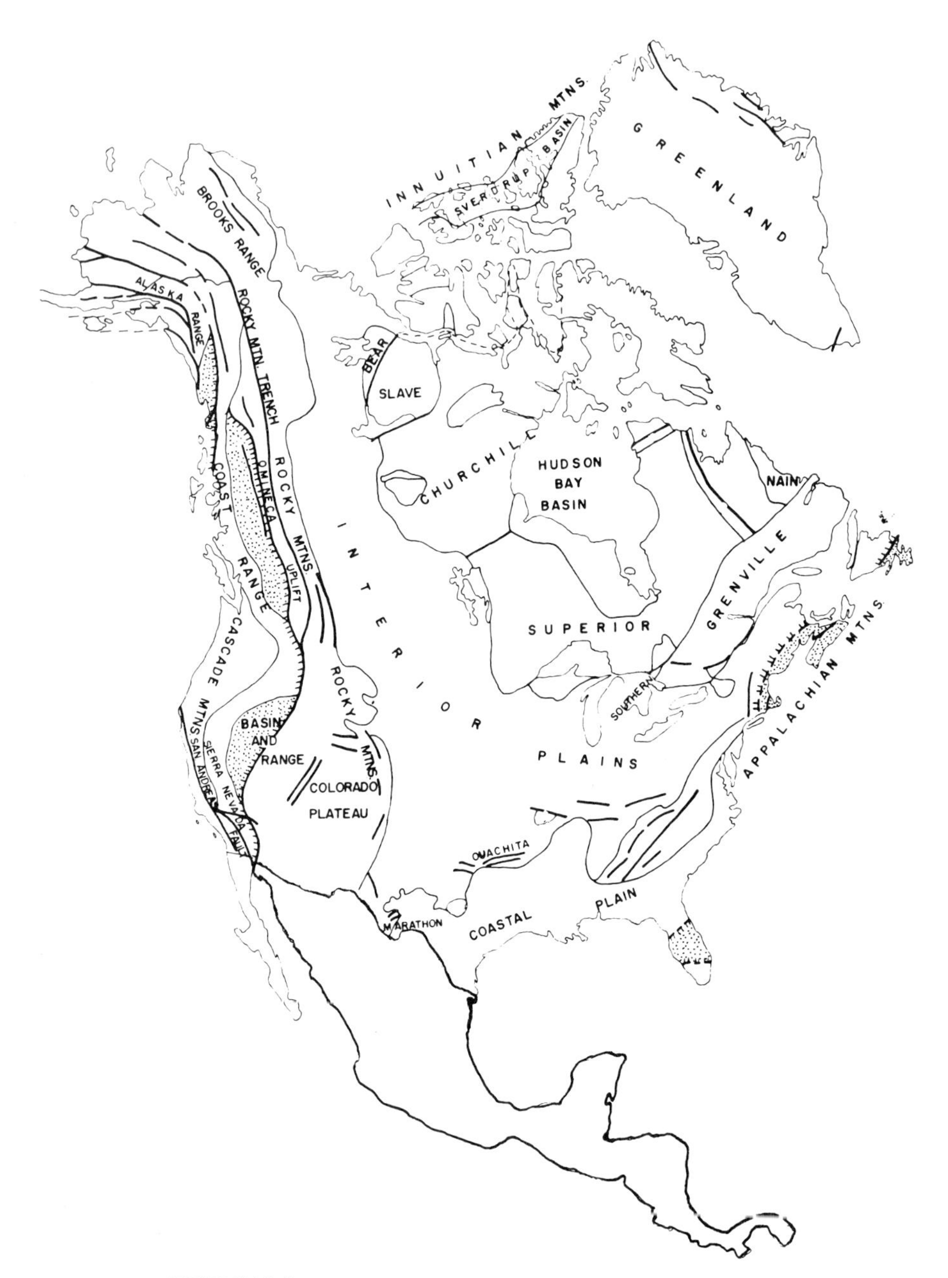

FIGURE 17-5
Sketch map of North America showing junction of formerly separate continents. Fragments in Florida, between Newfoundland and New England and from Nevada to Central British Columbia, are considered to have perhaps once been parts of Africa, Europe, and Asia, respectively. (*After J. T. Wilson, 1968.*)

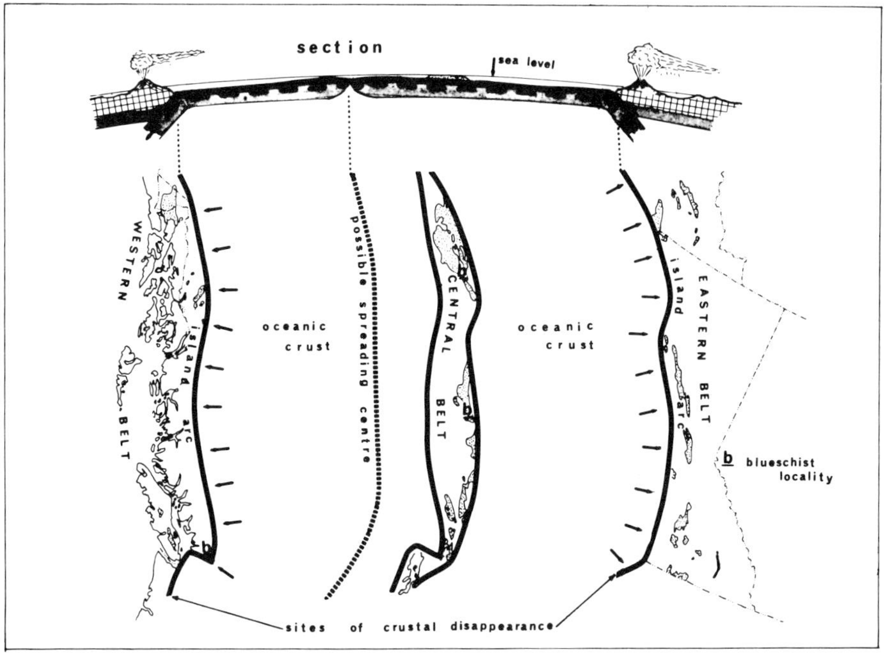

FIGURE 17-6
Schematic diagram showing the possible distribution of crustal segments in a simple plate tectonic model of the Canadian Cordillera in late Paleozoic time. (*After J. W. H. Monger and C. A. Ross, 1971.*)

(1966) and in the theory of plate tectonics that parts or the whole of geosynclines can be swallowed up and disappear in subduction zones. Whether or not this process affected Precambrian basins is unknown. Much has been written about recurrent cycles, but the validity, universality, and significance of these are debated. Radioactive methods offer the only good way to date the Precambrian. In the case of metamorphic rocks these frequently give two dates. The one which is more easily obtained gives the date of metamorphism, and the other gives a date of accumulation, which may be much older. The two were at first often confused (P. E. Kent et al., 1969). The problems of how to cope with all these difficulties have been well reviewed by A. F. Trendall (1966), F. J. Pettijohn (1970), and P. E. Cloud, Jr. (1971).

Most of the ideas about the division of Precambrian time which have been current until recently owe their origin to the reports of two committees established early this century when it was still thought that the earth had cooled quickly from a hot beginning and there was no idea that the Precambrian time was immensely long (J. T. Wilson et al., 1956). The first of these joint American-Canadian committees examined the region around Lake Superior and divided the rocks into

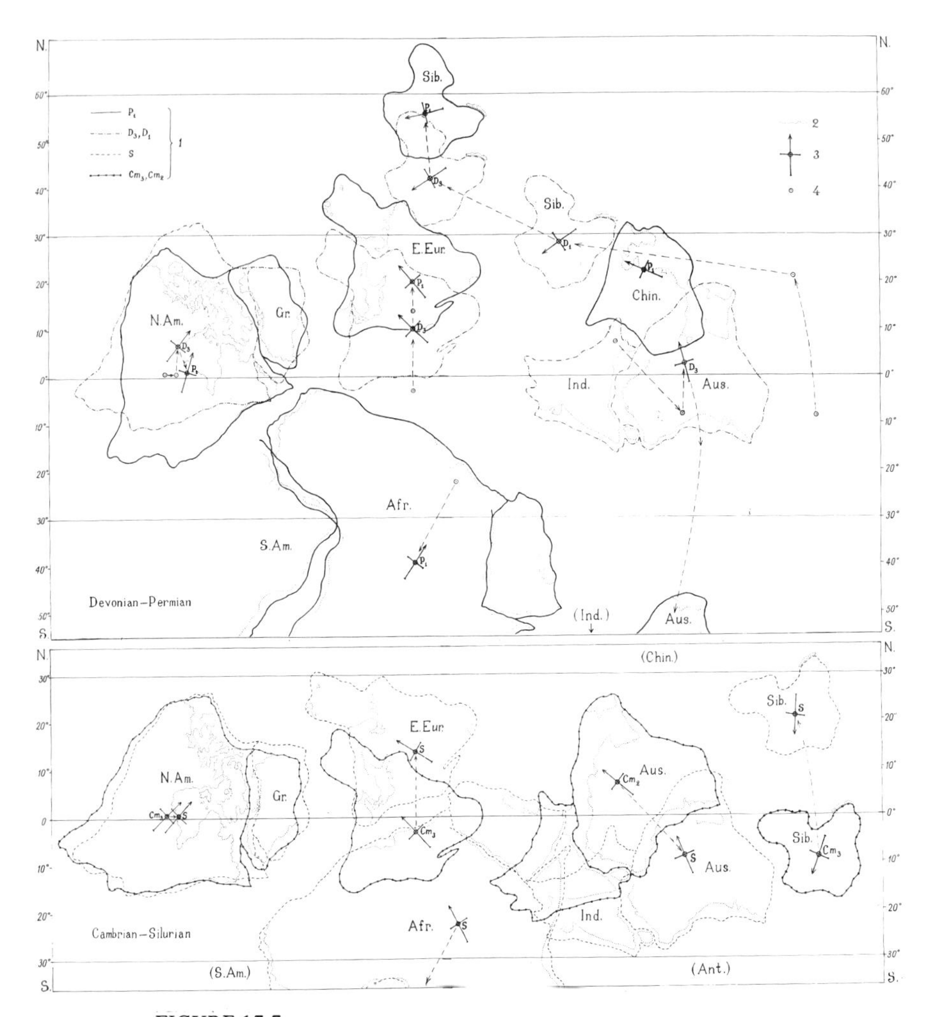

FIGURE 17-7
Relative position of Precambrian platforms during different periods of the Paleozoic deduced from paleomagnetic data. (1) Tectonic boundaries and probable positions of platforms during Upper Cambrian (Cm_2, Cm_3), Silurian (S), Devonian (D_1, D_3) and Lower Permian (P_1). For the Cm_3 of the Russian platform and D_3 of the North American platform, pole positions interpolated from earlier and later geological epochs are used. (2) Recent shorelines on platforms and outline of Lake Baikal. (3) Central points of platforms with present orientation of meridians and parallels. (4) Central points of platforms in lower Paleozoic times. *(After P. N. Kropotkin, 1971.)*

an older Archean era represented largely by plutonic rocks and lavas and a younger Proterozoic era of little metamorphosed sedimentary strata. These results were widely publicized and accepted, although such a classification based upon degree of metamorphism was known not to apply to younger rocks, and although the second committee meeting east of the Great Lakes found that it did not apply to their region. The Adirondack and Laurentian regions are underlain by metamorphic Grenville rocks which do not resemble the Archean rocks around Lake Superior and are younger than some of the unaltered strata. Therein lies a major cause of confusion in the classification of Precambrian rocks. There are three major groups of Precambrian rocks, but only two dominant names and different workers have applied the names in different ways.

17-4.2 The Three Major Classes of Precambrian Rocks

Much work has shown that all shields are built of three main classes of rocks. Adopting and extending the terminology of C. R. Anhaeusser et al. (1969), these are (1) small stable *cratons* crossed by narrow synclinal greenstone belts which are distorted but not highly metamorphosed, (2) *mobile belts* lying between and around cratons and composed of highly metamorphosed gneisses, and (3) *folded belts* of strata which rest upon the other rocks and are neither highly deformed nor metamorphosed.

Pettijohn accepts the first two classes, but subdivides the folded belts into those which resemble later geosynclines and those which do not. All are cut by minor intrusions, of which basic dike swarms are proving important in correlation (W. F. Fahrig et al., 1971).

The word craton is used in a variety of ways, but it seems best to apply it only to those provinces of shields which are both the oldest and which are marked by many small sinuous belts of greenstones. These constitute the true Archean. There is also general agreement that the unmetamorphosed folded belts are Proterozoic and younger than the cratons. It is the mobile belts such as the Grenville which have caused the trouble, for their nature was long misunderstood. At first they were thought to be entirely older than the folded belts, so that the mobile belts were regarded as Archean. It is now recognized that they represent the same wide range in ages as do the folded belts, and that the mobile belts were either deposited or have at least been entirely reworked since the Archean, and that both folded and mobile belts are Proterozoic.

The start of modern ideas began after 1930 with the advent of air photographs and the development of methods of age determination and isotopic study (H. C. Cooke, 1948). These methods enabled the shields to be divided into provinces with different structural trends which could be dated with at least some semblance of modern ideas by such geologists as F. J. Pettijohn (1943), H. E. McKinstrey (1945), A. Holmes (1948), J. E. Gill (1948, 1949), J. T. Wilson (1949), A. M. Macgregor (1951), J. T. Wilson et al. (1956), and A. W. Joliffe (1952). (See Fig. 17-8.)

FIGURE 17-8
Provinces and structural trends in the Canadian shield. The two principal cratons are the Superior province, surrounding Hudson Bay north of Lake Superior, and the smaller Slave province, north of Great Slave Lake. Note how structural trends change at boundaries of provinces. Some folded belts, notably the Labrador trough, are not distinguished from metamorphosed provinces. (*After C. H. Stockwell, in R. J. W. Douglas, 1970.*)

Their work showed that the histories of all continents have some similarities, which can be summarized in this way: The earth probably formed by cold accretion just prior to 4550 m yr ago, after which it rapidly became heated, so that the initial crust and the craters which had presumably been formed upon it, as on the moon, were destroyed.

The cratons Before 3300 m yr ago a new crust began to form. The age of the oldest preserved rocks is somewhat uncertain and may be as much as 500 to 300 m yr older than this. Most authors regard the early crust as thin, but if so, the observation in Africa that it was soon strong enough to bear the enormous thicknesses of Witwatersrand and Transvaal strata suggests that it thickened quickly (S. R. Hart et al., 1970; P. E. Cloud, Jr., 1971). Upon it the great greenstone accumulations of the Archean poured out on all continents. The nature of this crust is uncertain, because the base of the greenstones has generally been invaded by younger or perhaps reactivated granites (P. Eskola, 1948; J. B. Thompson, Jr., et al., 1968), but in some places it seems to be a soda granite, elsewhere an oceanic crust. Both views may be correct, because the primitive continental crust may have been formed as, or been broken into, small blocks, with the main accumulation between but sometimes overlapping the fragments (Figs. 17-9 and 17-10). These greenstone belts and granitic batholiths are readily identified, and it seems best to confine the term craton to such provinces. Summary accounts of cratons have been given by H. D. B. Wilson (1967, 1970), A. M. Goodwin (1968), C. S. Pichamuthu (1968), M. J. Viljoen and R. P. Viljoen (1970), A. Y. Glikson (1970), A. J. Baer (1970), J. T. Wilson (1972), and S. Moorbath et al. (1972).

The oldest rocks of the greenstone succession occur in Greenland and in Africa and are ultramafic and basic lavas with minor sedimentary rocks of chemical origin. Anhaeusser and colleagues (1969) have stated that the "Greenstone Group consisting of basalt, andesites, dacites and rhyolites, with interbedded clastic and chemical sedimentary rocks, usually dominate the stratigraphy of any greenstone belt and this group represents a progressive sequence from mafic to salic volcanic rocks." The basalts are most abundant and are frequently pillowed from extrusion under water (Fig. 17-11). Unconformably overlying the volcanic phases, related sedimentary rocks are found in most belts, consisting chiefly of graded beds, greywackes, and conglomerates, often with iron formation and only a little limestone. Sometimes there are two such successions, but they have always been tightly folded, faulted, and bent into narrow sinuous synclines lying between upwelling granite batholiths some tens of kilometers across. Occasionally, shelf deposits of shales occur without volcanics. The movements seem to have been essentially vertical and to be marked by different thicknesses of crust below the volcanic and the granite-gneiss belts. The metamorphism is low-grade, often in the greenschist facies. Quartz veins carry most of the world's primary gold deposits. These assemblages are similar to those found in island arcs today and have led many to regard the Archean as built of parallel arclike belts at intervals of scores of miles. In

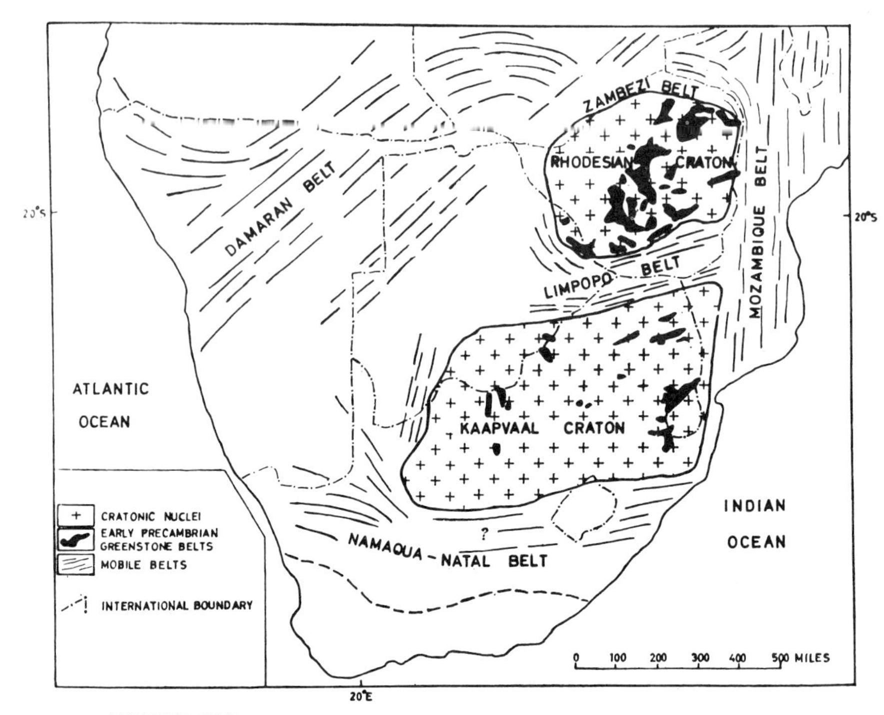

FIGURE 17-9
The Southern African crystalline shield stripped of its younger cover of folded belts and Phanerozoic strata to show two ancient greenstone-granite cratons and the encircling younger mobile belts. (*After C. R. Anhaeusser, R. Mason, M. J. Viljoen, and R. P. Viljoen, 1969.*)

each craton the belts show a generally uniform strike, east-west for example in the Superior province of Ontario, but north-south in the Kalgoorlie craton of Western Australia.

Cratons having similar distinctive properties occur on every continent, except perhaps Antarctica, and are readily distinguishable. Although they range in age from about 3500 m yr to about 2500 m yr, no one has yet submitted any worldwide method of subdividing them. No other rocks except those of cratons have been found of that age.

The question of when cratons ceased to form is debated. Many geologists like to think that about 2500 m yr ago a major event in the earth's history marked the end of craton building. If so, then it is convenient to equate cratons with the Archean and to end both at that date. On the other hand, although the change in geology is clear, it is not certain whether it occurred at the same time in all places. Most assigned dates are between 2700 and 2400 m yr, but in Arizona, P. E. Cloud, Jr. (1971) has suggested that the change did not occur until about 1700 m yr ago.

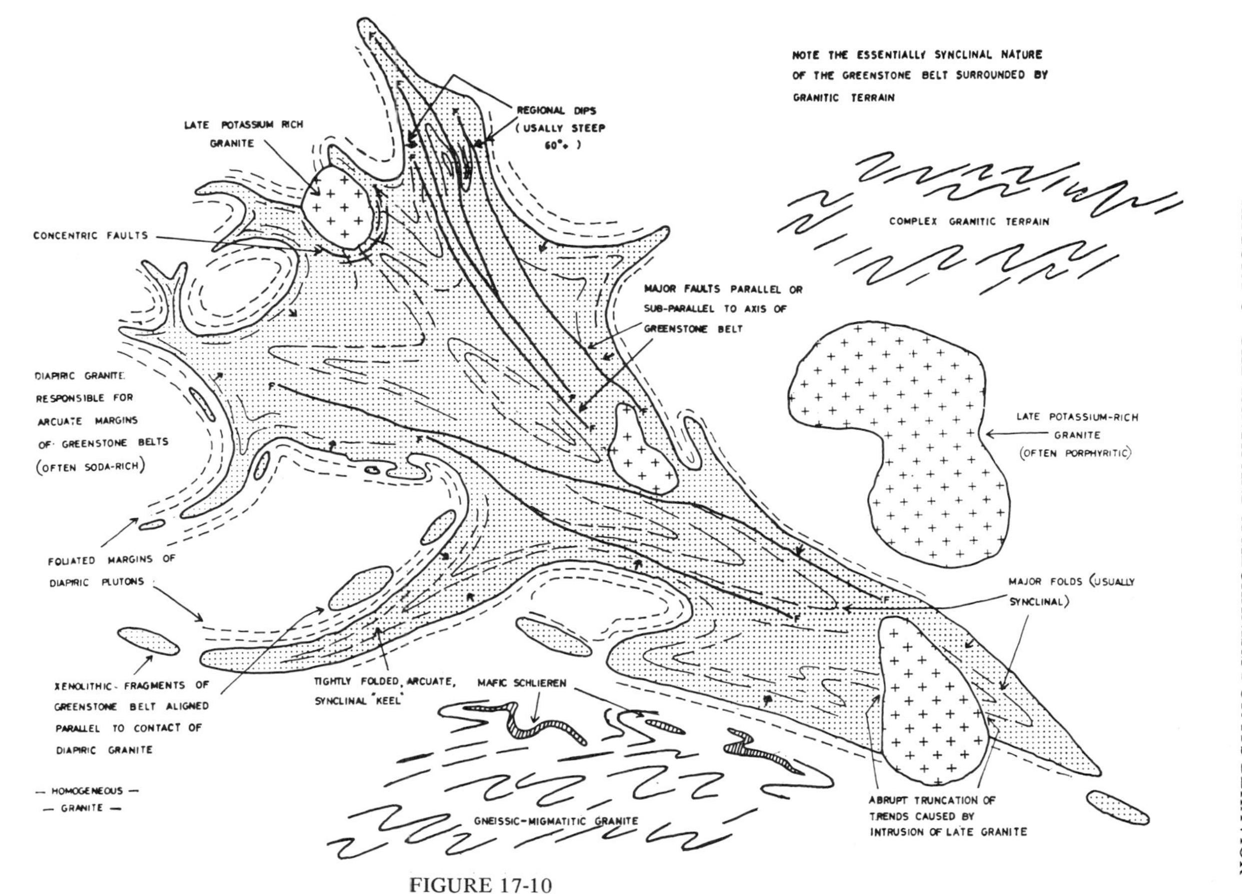

FIGURE 17-10
Diagrammatic sketch map illustrating the distinctive and characteristic features of the early Precambrian greenstone belts and associated granites of cratons in Southern Africa. (*After C. R. Anhaeusser, R. Mason, M. J. Viljoen, and R. P. Viljoen, 1969.*)

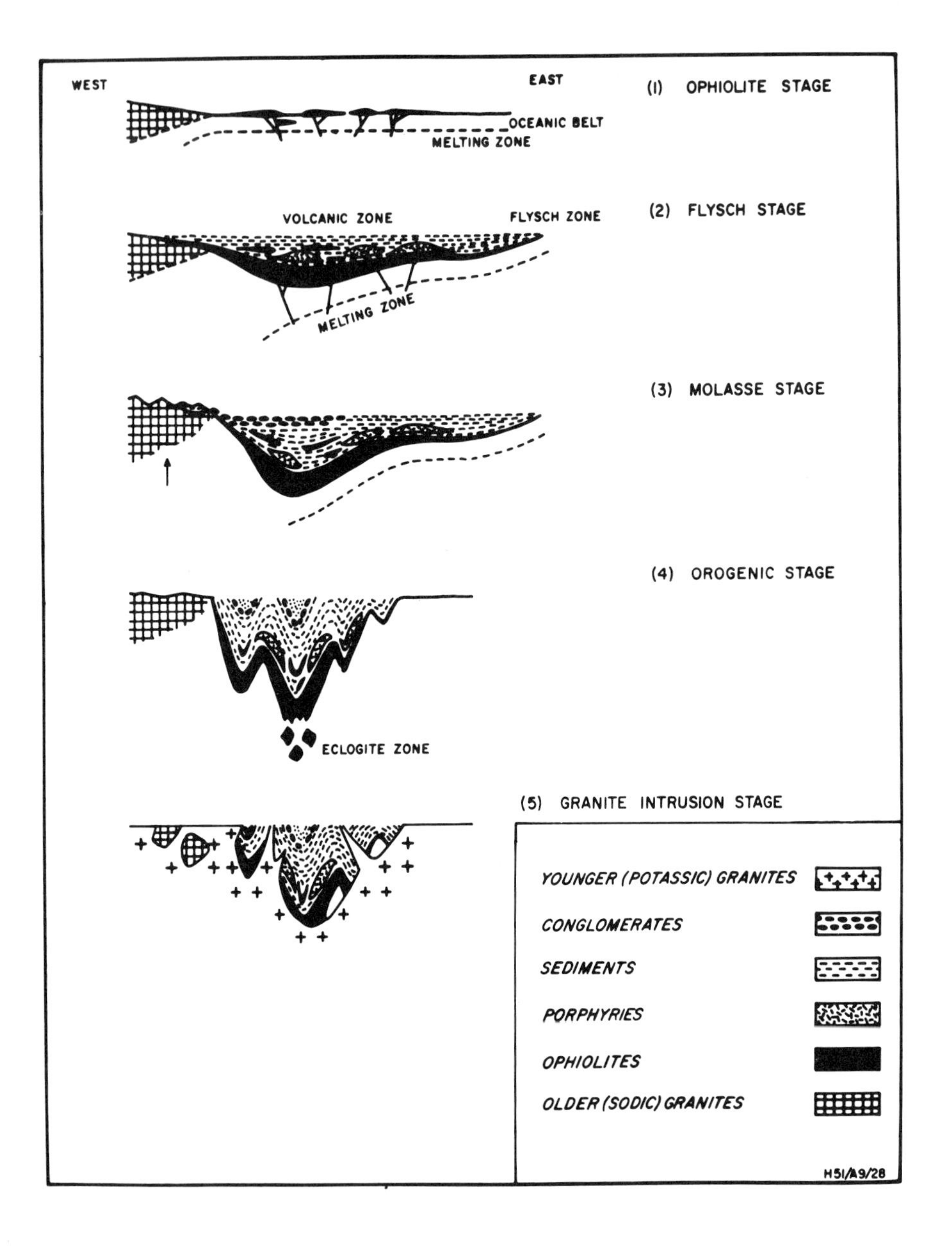
WEST
EAST
(1) OPHIOLITE STAGE
OCEANIC BELT
MELTING ZONE
VOLCANIC ZONE
FLYSCH ZONE
(2) FLYSCH STAGE
MELTING ZONE
(3) MOLASSE STAGE
(4) OROGENIC STAGE
ECLOGITE ZONE
(5) GRANITE INTRUSION STAGE
YOUNGER (POTASSIC) GRANITES
CONGLOMERATES
SEDIMENTS
PORPHYRIES
OPHIOLITES
OLDER (SODIC) GRANITES
H51/A9/28

This presents one of the most fundamental dilemmas in Precambrian geochronology. It is relatively easy to distinguish Archean cratons from younger rocks, but unless Cloud is mistaken, it is open to question whether the distinction has more than local value as a precise indicator of age.

Mobile belts If we accept the universal nature of this change in the construction of continents, Proterozoic rocks are divided between mobile belts and folded belts, chiefly the former. These mobile belts are now complexes of highly metamorphosed plutonic rocks, most of which, it is now agreed, are altered sedimentary and volcanic successions (H. J. Zwart, 1967; H. Wynne-Edwards, 1969; T. N. Clifford and I. G. Gass, 1970; A. J. Baer, 1970). Some small mobile belts, like that along the Limpopo River in Africa, seem to be entirely made of Archean cratonic rocks which have been metamorphosed and downfaulted into a graben. Larger belts, like the Grenville and Mozambique belts, contain some reworked Archean cratonic rocks on their inner margins, but probably also contain additional material which has enabled the continents to grow. Even those who maintain that the Archean continents covered as large an area as today's allow that they may have been thinner. It is difficult to account for rates of erosion and the ratio of isotopes in some elements without continental growth (J. T. Wilson, 1967). It is of course reasonable to expect nearly half the growth of continents to have occurred in Archean time, since that represents a major, and probably a more active, part of geological time. The mode of origin of these mobile belts is obscure and has been the subject of much debate.

The first student of the Canadian shield, W. Logan (1863), recognized the sedimentary origin of the Grenville gneisses, but many later geologists considered that the gneisses were igneous rocks which had been molten, and the question of

FIGURE 17-11
Hypothetical sketch of the evolution of the Kalgoorlie system. (*After A. Y. Glikson, 1970.*)

1 Ophiolite stage. Submarine eruption of ophiolites took place near a sialic nucleus composed mainly of sodic granites and porphyries.

2 Flysch stage. The subsidence of the ocean crust was associated with eruption of ophiolites and porphyries, issuing from increasingly deeper levels of the crust. The concentration of igneous activity along the axial zone was associated with the development of an external sedimentary zone. The sediments are believed to have been derived mainly from the porphyries.

3 Molasse stage. The isostatic rise of the older sialic nucleus resulted in the deposition of conglomerates [which near Kalgoorlie include sodic granite pebbles dated at 3000 to 3100 m yr (W. Compston and P. A. Arriens, 1968)].

4 Orogenic stage. Folding, metamorphism, and the formation of eclogite at the roots of the geosynclinal pile.

5 Generation and intrusion of granitic magmas, which possibly originated through the partial melting of eclogites (A. E. Ringwood and D. H. Green, 1966.)

the origin of granite became the subject of a long debate (J. Gilluly, 1948; H. H. Read, 1957).

The excellent crystallization of minerals seen under the microscope contributed to the view that the rocks had melted and formed a "sea of granite" (F. D. Adams and A. E. Barlow, 1912; F. F. Grout, 1948), an opinion seemingly confirmed by early work in experimental petrology (O. F. Tuttle and N. L. Bowen, 1958). Only after a great effort were these views reversed, so that the rocks of mobile belts are again accepted as being largely composed of highly altered sediments and volcanics (Sec. 3-5.1).

The confusion probably arose because these rocks have indeed been recrystallized, but without becoming molten, and they have indeed been extensively deformed, but only at very slow rates. Many have regarded these gneisses as having become altered while they were pushed down in the roots of mountains. Thus M. B. Katz (1969) suggests that the metamorphism of part of the Grenville province was due to burial of sediments to depths of 19 to 42 km and temperatures of 600 to 800°C, but Anhaeusser and his colleagues disagree and consider that the Proterozoic mobile belts are not like Alpine belts. Perhaps they formed by heating *in situ* before continental drift, but there is now evidence for an early start for continental motions (M. W. McElhinny et al., 1968; H. Spall, 1971). Whatever the cause of metamorphism, the date at which it affected different belts ranges from over 2300 m yr to less than 600 m yr, i.e., the full range of the Proterozoic.

Folded belts The other main assemblages of the Proterozoic rocks are the folded belts of comparatively unaltered sedimentary strata and volcanics. These always rest on a metamorphosed basement which may be either a craton or an older mobile belt, but different mobile belts range in age throughout the whole span of the Proterozoic, so that the stratified rocks constituting the oldest folded belts were deposited over 1500 m yr before the youngest mobile belts were metamorphosed. Much has been learned of the structure of folded belts in recent years (W. N. MacLeod et al., 1963; S. M. Roscoe, 1969; E. Dimroth, 1970; A. J. Baer, 1970; H. C. M. Whiteside, 1970).

F. J. Pettijohn (1970) distinguishes between folded belts which have the structure and types of rocks normal for Alpine geosynclines and those which resemble basins of platform rocks. The former often appear to be paired with a parallel mobile belt on the seaward side. For example, a succession of folded belts, some of which contain Huronian rocks, lies along the inner side of the Grenville mobile belt. The northern edge of the Grenville contains some reworked rocks apparently once part of the Huronian. The boundary is marked by a rapid increase in metamorphic grade and by faulting. These folded belts contain much less lava than do the Archean belts, but the method of deposition of banded iron formations and of uranium has suggested to some that the atmosphere was anoxygenic during the early Proterozoic.

17-4.3 Precambrian Time Scales and Method of Correlation

Many stratigraphers accustomed to the ease of correlating Phanerozoic strata by their fossils have tried to do the same for the Precambrian, but as A. F. Trendall (1966) points out, this is impossible. In attempting to subdivide the Precambrian rocks, many methods have been tried. Most of them depend upon using radiometric methods of age determination to find global events, but the simplest method is to use the radiometric age determinations to refer to events in absolute terms. Thus Precambrian rocks span the periods 600 to 3500 m yr ago, or the seventh to thirty-fifth megacenturies. This has universal meaning, but the dates are subject to revision as techniques improve.

A second method is to date and use a few of the most important events. Of these the most widely (but not universally accepted) system of subdividing the Precambrian is illustrated in Table 17-1. The date of origin of the earth is generally regarded as being fairly accurate. The age of the oldest rocks is less well established, and there is even less agreement about whether there was a worldwide change in the manner of formation of rocks 2500 m yr ago. The date of the first shelly fossils is an event of a different nature, but generally accepted.

A third method of subdivision, favored in Australia and South Africa, is based on the great areas and thicknesses of sedimentary strata which are well preserved on those continents. In most folded belts the nature of the strata and their structure is such that the stratigraphy can be worked out in detail. This has been done and has produced a vast number of local stratigraphic names, each with a precise but only local significance. These classifications have proved of great economic value in the Rand goldfields in Africa. What is frustrating is that every belt is limited, and it is usually impossible to correlate from one belt to another except by radiometric ages, which do not begin to approach the accuracy with which each basin can be subdivided.

In Australia an attempt has been made to divide the Proterozoic into Nullaginian, Carpentarian, and Adelaidian time, based on the thick stratigraphic sequences of sedimentary rocks found in three separate basins in Northwestern, North Central, and South Central Australia, respectively.

The Nullaginian (approximately Lower Proterozoic) strata rest upon the Pilbara Archean rocks and include rocks dated at 2100 and 1720 m yr old (P. J. Leggo et al., 1965). The Carpentarian (Middle Proterozoic) sequence of 13,000 m of slightly deformed sedimentary and volcanic rocks spans a period longer than 1800 to 780 m yr (I. McDougall et al., 1965). The Adelaidian (mainly Upper Proterozoic) extends upward to the Lower Cambrian from a basement that is 1500 m yr old (B. P. Thomson, 1964). Thus the three basins overlap in time, and it is not clear where the boundaries between Lower, Middle, and Upper Proterozoic should be placed.

These dates and divisions are useful in the three type-basins, but the sparsity and inaccuracy of dates is in striking contrast to the elegant stratigraphic successions worked out within each basin. Without index fossils the attempt to follow stratigraphic principles used in Phanerozoic rocks achieves very little.

A fourth method, adopted in the Canadian shield, which has a smaller proportion of sedimentary folded basins preserved, is to use the age of metamorphism of mobile belts as subdivisions. Thus C. H. Stockwell (1964, and in

Table 17-1 TIME SCALES FOR THE SUBDIVISION OF GEOLOGICAL TIME

Time before present, m yr	Event	Era and duration (traditional units)	Name of eon (units more nearly equal in age)	Characteristics
0	*Present*			
		Cenzoic (70 m yr)		
70	Sudden cooling; change in fossil life	Mesozoic (130 m yr)	Phanerozoic (570 m yr)	Precise relative correlation by shelly index fossils
200	Union of continents	Paleozoic (370 m yr)		
570	*First shelly fossils* (union of continents?)		Upper Proterozoic (430 m yr)	
c. 1000	Widespread metamorphic event	Proterozoic (1930 m yr)	Middle Proterozoic (800 m yr)	Formation of folded belts and mobile belts of shields
c. 1800	Widespread metamorphic event		Lower Proterozoic (700 m yr)	
c. 2500	*Changes in rock types*	Archean (2050 m yr)	Archean (1000 m yr)	Formation of cratons
c. 3500	*Oldest preserved terrestrial rocks*		Pre-Archean (1050 m yr)	No rocks preserved
4550	*Origin of the earth*			

NOTES: The more important events are italicized. See T. R. Worsley, 1971, for an account of Mesozoic terminal event.

R. J. W. Douglas, 1970) plotted histograms of K-Ar and Rb-Sr ages for eight mobile belts. The results for the three K-Ar plots are clear, and he chose four type-orogenies, to which he assigned dates and names. In an endeavor to follow "the principles recommended by the American Commission on Stratigraphic Nomenclature, the orogenies are defined by concrete rock in type regions rather than by abstract time." Thus he defines the Kenoran orogeny in terms of the "Superior Province, which is chosen as the type region."

It is very much open to question whether rocks really do provide a better unit of time than do years, as his statement implies. Even if they do, the difference between the type section which a stratigrapher would select on a particular hillside and the province of half a million square miles of unfossiliferous rock chosen by Stockwell is a good measure of how vague is Precambrian geochronology. Figure 17-8 shows the provinces of the Canadian shield.

Stockwell uses mean dates of metamorphic events in different places to subdivide Precambrian time into intervening periods which correspond to Archean and Lower, Middle, and Upper Proterozoic time. For the three later intervals he invented the names Aphebian, Helikian, and Hadrynian, respectively, but it is open to question whether this is an advantage. It is open to doubt whether geologists in other continents will use these names, and even in Canada they are still unfamiliar. For example, E. Dimroth (1970, p. 2718), on introducing the term Aphebian, immediately explains that it means "Early Proterozoic, defined as younger than 2500 m.y. and older than 1600 m.y." He should go further and add, "until better age determinations change the dates of metamorphism of the type Superior and Churchill provinces."

Some folded belts contain well-marked glacial beds, notably the Huronian of Canada and the Adelaidian of Australia. Attempts have been made to use these for correlation. They may be useful within single continents, but the notion that Precambrian glacial periods were synchronous all over the world has been questioned (A. R. Crawford and B. Daily, 1971).

A fifth method of subdivision is based on the concept that geological history may have been divided into regular cycles. In Phanerozoic time the only divisions which seem to be worldwide are the great regressions of the sea near the close of Precambrian time and the end of the Paleozoic era (M. Kay, 1951; R. L. Grasty, 1967). They were probably due to widespread orogenies, and the later event coincided with the final consolidation of the supercontinent of Pangaea. J. W. Valentine (1970) and E. M. Moores (1970a) have pointed out the connection between orogeny, continental drift, and the diversity and abundance of life. When continents have recently been welded together, coastlines are short, latitudinal spread of land is reduced, sea level is low, and habitats are restricted, so that there is less diversity and fewer taxa than when continents are dispersed. Hence the close of Precambrian time may have been significant both as an event in mountain building, which was widespread especially in the southern continents, and as an important boundary in evolution.

Extrapolating from these two dates of about 200 and 600 m yr ago, several geologists claim to have found recurrent cycles at intervals of about 200 m yr throughout geological time (J. H. F. Umbgrove, 1947; L. L. Sloss, 1963; A. Holmes, 1965), but most others have doubted this (Fig. 3-7).

The details of plate tectonics have shown that different oceans and seas started to open at different times and that others closed over long periods of 300 or 400 m yr. If the cycles are not clearly marked in Phanerozoic time, they are even more vague in the Precambrian.

A better case has been made for a few longer cycles. G. Gastil (1960), A. P. Vinogradov and A. E. Tugarinov (1961), J. Sutton (1963), and R. Dearnley (1966), in particular, have noted that frequency curves of radiometric ages show peaks close to 200, 600, 1000, 1800, and 2500 m yr. S. K. Runcorn (1965) tried to relate these to changes in the mode of mantle convection, but this now seems unlikely on theoretical grounds. The first two dates seem to mark times of coming together of continents, but too little is known of Precambrian history to understand the significance, if any, of the 1000 and 1800 m yr events, and it is by no means certain that, if there were such events, they were everywhere synchronous (R. A. Burwash, 1969). In any case each major metamorphic event might have lasted throughout an interval of perhaps 100 m yr, which makes it inherently imprecise. Nevertheless, the view that Proterozoic time can be divided into lower, middle, and upper parts by events occurring about 1800 and 1000 m yr ago is so widespread that these dates can be useful, added to the Precambrian time scale, as Table 17-1 shows.

When this is done, the need to name the units of time arises. Traditionally, Cenozoic, Mesozoic, Paleozoic, Proterozoic, and Archean have been called *eras*, although the older are now seen to be much longer than the younger. Stockwell would call each of the three parts of the Proterozoic an era, and gives the name *eon* both to the whole Proterozoic and to the Archean. It seems unlikely that this change will be followed, and it is therefore better to leave the eras as they are and use the new term eon for the more nearly equal divisions shown in Table 17-1. The divisions and names proposed have the advantage that geologists everywhere are familiar with the terms and can use them readily, but the disadvantage that the boundaries between units are not very precisely defined and may not be of the same age on different continents. Although purists may object that eons and the older eras are not precise units of time, the fact is that our present state of knowledge of the Precambrian does not enable us to be precise and definite.

17-5 A POSSIBLE CAUSE AND MECHANISM FOR THE MOTIONS OF THE EARTH'S SURFACE PLATES

Many mechanisms have been proposed in explanation of those internal motions of the earth which affect its surface, and most of them have already been mentioned in previous chapters. To be acceptable, any solution must meet at least four criteria: The forces must be adequate. The mechanism must follow the basic principles of physics, such as the laws of hydrodynamics, thermodynamics, and mechanics. The behavior must be in accord with geophysical observations about the nature of the interior of the earth. The proposed motions must produce surface effects which relate to the present configuration and recorded geological history of the earth's surface. Most proposals fail to do this. Thus, although gravity is indeed a strong force in the earth, Wegener's "*Polfluchtkraft*" has been shown to be totally inadequate to move continents about. Neither contraction nor expansion can explain why large horizontal displacements occur. The contraction theory, framed to provide a mechanism for forming folded mountains, has never been extended to encompass the formation of mid-ocean ridges. The expansion theory, while meeting some of the requirements for forming ocean basins, has encountered apparently insuperable physical objections.

The principal possibilities remaining are subsurface convection currents and vertical uplifts. Both could be driven by either chemical or thermal forces. The two theories grade into each other, because convection currents cannot affect the surface unless they also include vertical motions, and vertical uplifts, unless due to purely local expansions, would require some compensating horizontal flow.

For over a century physicists have advanced various theories of convection within the earth. Most of these theories have involved the whole mantle and have taken the form of regular patterns which could be stated in manageable mathematical form. Recent developments in seismology have suggested that flow is concentrated in the shallow asthenosphere, while the hypothesis of plate tectonics suggests that it has been steady for tens of millions of years and the effects uniform over horizontal distances of thousands of kilometers (H. Kanamori and F. Press, 1970). No such theory has ever been related in detail to the earth's surface features. These requirements for extremely steady, shallow flow, fitted to the complexities of geology, place such limitations upon convection currents as a primary driving mechanism that E. Orowan (1969b) has advocated abandoning the hypothesis in favor of the concept that lithospheric plates may be sliding off ridges formed by local hydration and expansion in the mantle. The primary uplift could equally well be due to thermal, instead of chemical, causes, or to both. A. L. Hales (1969) and W. R. Jacoby (1970) have shown that this hypothesis appears to be physically acceptable. This section will show that the concept may fit geological observations. It is interesting that thermal upwellings would still involve a return flow, and hence convection of a sort, and it would also meet the emphasis of the Soviet geologists on the importance of vertical motions (V. A. Magnitsky and I. V. Kalasknikova, 1970). Thus it may provide the basis for an acceptable theory for the behavior of the earth's crust and mantle.

The idea that heated upwellings probably rise in the mantle occurred to physicists in the last century. It gained credence from the discovery of radioactivity and continues to have strong support, for example, from W. M. Elsasser (1968), who considers the mantle to be thermally unstable; from D. L. Anderson et al.

(1971), who have described the mobility of the asthenosphere or rheosphere; and from W. H.-K. Lee (1970), who has shown the lack of uniformity in the earth's surface heat flow.

An apparent recognition by a geologist of an example of this phenomenon is H. Cloos' (1939) view that the Ethiopian dome was formed by uplift, rifting, and volcanism. In 1963 R. Furon noted the large number of domal uplifts capped by volcanoes in Africa and remarked that such features were confined to Africa; it may not be quite true of other continents and is not at all true of ocean basins.

In 1963 another quite distinct line of investigation led to the proposal that nine linear chains of islands and seamounts in the Pacific Ocean had been formed by the passage of moving lithospheric plates over plumes rising from relatively fixed sources deep in the mantle below the moving asthenosphere. These plumes were regarded not as linear upwellings, but as localized "hot spots." Later the concept was extended to explain the formation of pairs of lateral ridges which spread to the adjacent continents from Iceland, Tristan da Cunha, and Amsterdam Island, all islands on the mid-ocean-ridge system (J. T. Wilson, 1963a,b, c, 1965a).

This view would seem to have the support of K. E. Torrance and D. L. Turcotte (1971) and E. R. Oxburgh and D. L. Turcotte (1968), who, from theoretical two-dimensional studies, proposed the movement of plumes around the stationary cores of convection cells.

W. J. Morgan (1968) and J. C. Maxwell (1968) also proposed that "sinkers" may have formed beneath trenches to pull them down, and "floaters" beneath the crest lines of ridges to raise them up. These ideas have now been combined to suggest that a few localized upwellings arise from deep in the mantle to uplift domes on continents, cause rifting, generate mid-ocean ridges, and provide the driving mechanism for plate tectonics. This concept appears to be fruitful and capable of explaining many phenomena (W. J. Morgan, 1971; J. T. Wilson, in press).

It is suggested that the driving mechanism of plate tectonics may be as follows. In the lower part of the mantle the radioactive heat generated there escapes upward by conduction and radiation because at high temperatures silicates are transparent to heat. At a level of about 400 to 700 km, the mantle becomes opaque, so that heat slowly accumulates until, due to local irregularities, cylindrical plumes start to rise like diapirs in the upper part of the mantle. These plumes reach the surface, which they uplift, while their excess heat gives rise to volcanism. The lavas at these uplifts are partly generated from material rising from depths of several hundred kilometers, which is thus chemically distinct from that generated at shallower levels. The plumes are considered to remain steady in the mantle for millions of years.

The first visible stage of such uplifts takes the form of isolated uplifts capped by volcanoes, of which the Hoggar massif in the central Sahara is an excellent example. The Precambrian surface of the Sahara which had been flat in early Tertiary time was uplifted into a dome 2.3 km high and nearly 100 km in diameter, with volcanoes reaching to 2918 m on top. The rifts which cross the

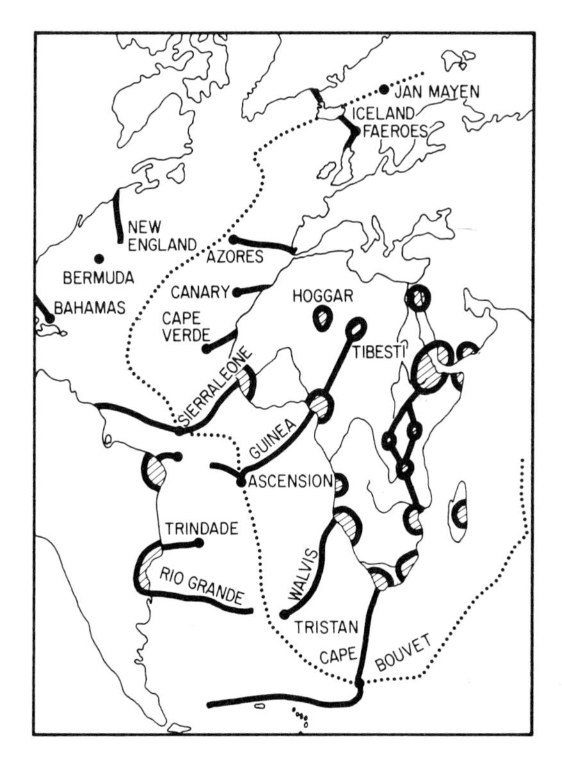

FIGURE 17-12
The Atlantic Ocean and adjacent continents showing the mid-ocean ridge (dotted line), some lateral ridges (full lines in the oceans), domes and half domes (hatchured) on the continents connected by rift valleys (full lines on the continents). Note the isolated Hoggar massif and how many great rivers rise in domes often on the opposite side of the continent. *(After J. T. Wilson, in press.)*

dome from north to south are not known to connect with any others (T. N. Clifford and I. G. Gass, 1970).

A second stage is reached when a series of plumes becomes linked by rifts. Many of the plumes form triple points. Those for the East African rift valleys (B. H. Baker and J. Wohlenberg, 1971) are shown in Fig. 17-12. These domes, which formed at various times during the Tertiary period (with the names of associated active volcanoes in brackets) include Ethiopia (several active volcanoes in Ethiopia and in islands in the Red Sea), the Uganda-Sudan border, Kenya (Mt. Kilimanjaro and others), eastern Congo (Mts. Nyamuragira and Nyiragongo of the Birunga group), the dome at the southern junction of the eastern and western rifts (Mt. Rungwe), and possibly another dome farther south. The Tibesti uplift and the Cameroons volcanoes form a separate pair of linked domes joined by rifts and marked by active volcanism (T. N. Clifford and I. G. Gass, 1970).

The concentration of volcanoes over domes, their isostatic equilibrium, and geophysical observations, all support the contention that hot, light diapirs in the mantle uplift the domes.

In the Red Sea and Gulf of Aden the splitting of domes by widening rifts marks the third stage.

The force that causes the rifts to form is that proposed by A. L. Hales (1969), W. R. Jacoby (1970), and particularly by W. J. Morgan (1971). It is that the lithospheric plates are sliding downhill off the uplifted domes (P. A. Rona, 1971; J. G. Sclater and C. G. A. Harrison, 1971).

It follows that the rifts are generated in quite a different manner from the domes. In the domes it is rising matter that causes the rifting; along the rifts it is fracturing which allows the lava to well out. Not surprisingly, the chemistry in the two cases is quite different, for the domes can draw their lava by partial melting in their deep roots, while the rifts never penetrate below the shallow asthenosphere (R. A. Zielinski and F. A. Frey, 1970). P. A. Mohr (1971) has contrasted the oceanic tholeiites of the rifted floors of the Red Sea and Gulf of Aden with the Tertiary and Quarternary volcanics of Ethiopia and Yemen, which comprise perhaps the world's largest province of alkaline extrusives. The latter are much more undersaturated and enriched in potash and in aluminum relative to iron. They include about 50,000 km^3 of rhyolitic ignimbrites, not associated with any contemporaneous basalts, so that an origin by fractionation is ruled out. Mohr also rules out anatexis of continental crust, and believes with M. H. Battey (1966), P. G. Harris (1969), and P. W. Lipman (1969) that the bulk of the two suites comes from two sources at different depths, the one beneath the rifts being shallower. This is precisely what the proposed mechanism requires.

This mechanism agrees with the conclusion of W. M. Elsasser (1971) that rifts are the result of horizontal tensile stresses and, by substituting rising cylindrical plumes for rising curtains along ridges, escapes from a physical difficulty. Diapirism is a very common geological phenomenon, being known in salt domes, ice pingoes, volcanic pipes, laccoliths, batholiths, and even in shale diapirs, of which R. P. Coats (1964) has described a remarkable example. At least three localities provide good arguments for not accepting Elsasser's proposal that sinking of lithospheric plates along Benioff zones is the chief driving mechanism. If this were so, what would cause the spreading of the Atlantic to drive South America westward to raise up the Andes? Also, the tiny Cocos plate lies beside the very active Middle America trench, and is separated from the other plates by faults, so that it would be expected to be pulled down very quickly, which is not happening. J. A. Mendiguren (1971) has shown from a study of the focal mechanism of an earthquake shock that the Nasca plate appears to be in compression.

The Atlantic Ocean is in the next stage. Continental plates once joined have slid apart, leaving half domes on the coast of Africa and their counterparts in South America still linked by boundary ridges to islands over domes along the Mid-Atlantic Ridge. Thus local highlands marking half domes in Liberia, in the Cameroons, near the Angola–South West Africa border, and at the Cape are linked by the Sierra Leone, Guinea, Walvis, and Cape rises to St. Peter and St. Paul Rocks, Ascension (perhaps St. Helena), Tristan da Cunha, and Bouvet Islands, respectively. The counterparts in the Western Atlantic are less clear, but there is some indication of rises extending from these islands toward possible half domes in French Guiana, Recife, and near Rio. The Scotia arc has disturbed the southern part (Fig. 17-12).

R. S. Dietz and J. C. Holden (1970a,b) have supported the original proposal that Tristan lies over a source which has been fixed in the mantle since mid-Cretaceous time, and that the lateral ridges were formed by a combination of sea-floor spreading in an east-west direction, with northward movement of both plates (J. T. Wilson, 1965a).

Isotopic studies of oceanic islands show evidence of two sources of lead. Thus V. M. Oversby and P. W. Gast (1970) found that "the phonolitic rocks on St. Helena have 207/204 ratios that are similar within the group, but are very different from the 207/204 ratios of basaltic rocks on St. Helena." They endeavored to find a two-stage model from which to derive the leads, with different present-day ratios from a single common source by differentiation in the past, but were forced to conclude that the leads could have come from sources in the mantle which had retained their separate identities for billions of years. They also concluded that the lead in lavas from the 1961 eruption on Tristan da Cunha was a mixture from two different sources.

Z. E. Peterman and C. E. Hedge (1971), from an examination of strontium isotopes and related chemical variations in basalts, concluded that "Potassic lavas of Gough, Tristan da Cunha, Samoa and Tahiti are derived from mantle material that has undergone the least depletion of alkali and from this aspect, such lavas are more primitive than the much more voluminous ocean-ridge tholeiites which are derived from highly depleted mantle." This suits the proposed mechanism, except that they suppose that the ultimate source of the ocean-ridge tholeiites is in a deeper zone than that of the island volcanic rocks. S. R. Hart et al. (1970) found similar chemical distinctions and would derive alkali basalts from deep in the mantle.

In another chemical investigation of rare-earth elements (REE) in xenoliths from Hawaii, J. B. Reid, Jr., and F. A. Frey (1971) conclude that if garnet pyroxenite xenoliths crystallized from some liquid, the liquid was chemically dissimilar to Hawaiian Island lavas, but that these xenoliths might be related to oceanic tholeiites. This again brings out a chemical distinction between island lavas formed over a probably rising plume which are a mixture from deep and shallow sources and oceanic tholeiites formed along ocean ridges between domes which are derived from shallow sources only.

W. G. Melson et al. (1970) noted a similar distinction when they wrote, "Certainly the brown hornblende mylonites of the St. Paul Rocks and the young basalts on its north slope differ from ridge rocks in their critical undersaturation, and in their characteristic alkaline 'basalt' trace element characteristics." (See Sec. 3-6.6.)

It is thus concluded that the Atlantic Ocean has opened because of upwellings fixed in the mantle beneath or near Jan Mayen Island, Iceland, the Azores, St. Paul Rocks, Ascension, Tristan, Bouvet, and perhaps also St. Helena and Gough Islands. Most of them have produced more lava than the rest of the ridge, and this now forms lateral ridges linking them to the continents. Several of them, including the Azores, St. Paul Rocks, and Bouvet, lie at present or former triple points. The rest of the Mid-Atlantic Ridge consists of a series of shallow fractures joining the domes.

A similar story seems to apply to the East Pacific rise.

The trench which formed off the coast of Chile and Peru and others around the Pacific basin mark subduction zones. These have uncoupled the motions of the Pacific plate from other plates and opened the possibility that the East Pacific rise may have also remained fixed relative to the mantle. If so, Macquarie, Easter, Galapagos, and Revilla Gigedo Islands, being the only islands near the ridge, may be presumed to be centers which have remained on or near the crest. Macquarie Island, like St. Paul Rocks, is uplifted sea floor.

This does nothing to explain the Hawaiian Islands, which appear to be due to an isolated dome. Perhaps, as J. T. Wilson (1963e) and R. I. Walcott (1970) have suggested, the chain has been maintained because a diapir beneath the island of Hawaii has created an isolated fault which it is extending across the Pacific plate, creating volcanoes along the line of motion as the plate moves over a diapir which is now beneath the island of Hawaii.

Rather different arguments apply to the Indian Ocean. If the southern Mid-Atlantic Ridge is fixed relative to the mantle, then, according to X. Le Pichon (1968), Africa has been moving northeast over the deep mantle by rotation about an axis with poles near Tasmania and Iceland. Hence the domes in Africa have not remained fixed over the mantle, which may perhaps explain why R. Furon shows such a pronounced northeast-southwest lineation in African rifts and why the African rifts have not been strong enough to develop into an ocean (Fig. 17-12), but an apparent westward shift in volcanic centers poses a problem (B. H. Baker et al., 1971).

Since Africa has no related subduction zones, its northeastward movement can be presumed to be transmitted to the Indian Ocean basin and to the Mid-Indian Ocean Ridge, which extends from the Gulf of Aden to south of Australia. This ridge cannot therefore have remained over its original upwellings. This may be why Réunion and Kerguelen Islands both lie to the west of the Mid-Indian Ocean Ridge. They both appear to be over rising plumes, because the geology of Réunion is like that of Hawaii, while Kerguelen has important and highly siliceous rhyolitic centers and has had active volcanoes since at least mid-Tertiary and perhaps late Cretaceous time (Sec. 15-12).

The postulated rotation about a pole near Tasmania would have carried the Mid-Indian Ocean Ridge away from Réunion and Kerguelen, forming the lateral ridges from Réunion to Rodriguez Islands and from Kerguelen to Amsterdam and St. Paul Islands (R. L. Fisher et al., 1971). The latter have continued to be an active center. If, in the early Tertiary, Australia lay against Antarctica, then the Gaussberg-Kerguelen and the Broken or Diamantina ridges may be boundary ridges marking where the Tertiary spreading split an older part of the sea floor. When this happened, Kerguelen would have been at the triple point between these ridges and the Mid-Indian Ocean Ridge and the Ninety East ridge. This location at or close to a triple point is also true of the Azores, Ascension Island, Bouvet, Ethiopia, Rungwe, and Galapagos centers, which suggests the dominant role of these centers, as W. J. Morgan has emphasized.

These considerations lead to the conclusion that lavas–at least those associated with ocean basins–fall into four kindreds, three of which are basalt kindreds which have been confused together:

1 The andesite kindred long recognized to be characteristic of island arcs.

2 The tholeiites of ocean floors which form along mid-ocean ridges and which are uniform except where large layered intrusives beneath the ocean floor have differentiated (Sec. 3-6.4). They are believed to be due to upwelling from shallow sources in the asthenosphere which have been depleted in some elements.

3 The alkali and silicic lavas formed from undepleted sources deep in the mantle on islands fixed on mid-ocean ridges and lying over plumes rising from deep in the mantle, e.g., Iceland.

4 The lavas of other islands which are intermediate between kindreds 2 and 3, for example, Hawaii and the Canaries (V. M. Oversby et al., 1971).

A. R. McBirney and I. G. Gass (1967) found that near mid-ocean ridges islands tend to have siliceous rocks and that undersaturation and alkalinity tend to increase with distance from the ridge. This they attributed to the islands forming at various distances from the crest, ignoring the evidence that ages of the oldest islands increase away from the crest and that no old islands occur on mid-ocean ridges. They did not explain why recent work on many islands entirely supports the view that many islands remote from mid-ocean ridges are old.

These two somewhat contradictory views can be rationalized if, instead of attempting to arrange the islands in sequences for which the evidence is not clear, they are merely divided into two classes, each rather variable (Fig. 17-13).

With one exception all the islands which have highly siliceous rocks lie on mid-ocean ridges supposedly over plumes, while the other islands do not. Islands like Hawaii, which are not generating a mid-ocean ridge, apparently cannot differentiate to produce silicic lavas.

A very important point which arises is that, in considering the relative motions of two plates, one should realize that three poles of rotation, not just one, are involved. This is because each of the two plates may be moving relative to the mantle as well as to one another. Consider, for example, the opening of the South Atlantic. The African and South American lithospheric plates are moving away

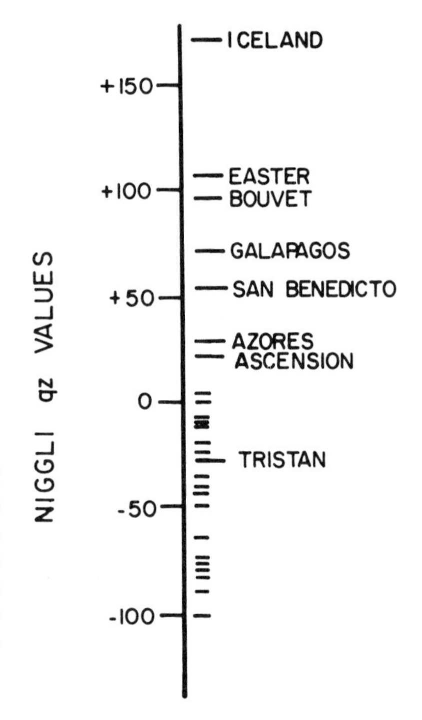

FIGURE 17-13
Niggli qz values reported for each of many islands in the Pacific and Atlantic Oceans. Each dash represents an island, but only the islands on the mid-ocean ridge are named. With one exception these islands are more siliceous and form a separate group from islands elsewhere. (*Adapted from A. R. McBirney and I. G. Gass, 1967.*)

from one another. Such shallow features as fracture zones and sections of the Mid-Atlantic Ridge reveal the relative motions between these plates.

The fracture zones are small circles about the pole of rotation of these lithospheric plates. The segments of ridge are radial to the same pole. The (northern) pole, about which the African and the Americas plates are rotating, is near Iceland and the southern tip of Greenland (X. Le Pichon, 1968).

The motion of Africa relative to the mantle is also a rotation, but not about the same pole; it is about another, which deeper-seated features should reveal. If the Kaoko volcanics and the half domes near the border of Angola and South-West Africa owe their origin to having been formed over a rising plume whose present position is now marked by Tristan or Gough Islands, then the Walvis ridge marks the locus of motion. The same is true of the other ridges in the eastern part of the South Atlantic which connect St. Paul Rocks, Ascension, and Bouvet Islands to Africa. These are approximately small circles about a center near Montreal, which is thus the pole of rotation of Africa over the mantle. This rotation would have moved a point on the coast beside the Canary Islands to the Cape Verde Islands.

In a similar manner the Rio Grande ridge may indicate the motion of the Americas plate over the mantle. The bend in the middle of the ridge may indicate a change in direction near the end of Cretaceous time. The western part of this lateral ridge strikes east and west parallel to several other lines of seamounts off Brazil (F. F. M. de Almeida, 1960), suggesting a pole near Iceland. Because this is also the pole of rotation for the two lithospheric plates, the fracture zones are here parallel to these parts of the lateral ridges.

The rigidity of the Americas plate as it rotates about Iceland demands that all places on that plate or on the Mid-Atlantic Ridge to the north of Iceland rotate clockwise over the mantle. This includes the pole of rotation between the Americas and Eurasian lithospheric plates, which are near the New Siberian Islands. It should be noted that this is different from the pole of rotation for the Americas and African plates. Relative to the mantle, this pole is not fixed, but moving westward. The conclusion that the poles of rotation of pairs of plates can be moving over the mantle is important (Fig. 17-14). In dealing with phenomena with deep roots, one must always be careful to distinguish the motion of two plates relative to one another from the motion of each plate relative to the mantle (J. T. Wilson, in press).

17-6 THE FORMATION OF ISLAND ARCS AND MARGINAL SEAS

It is more than a century since men first noticed that island arcs are too numerous and too regularly shaped to be a result of chance, and many efforts have been made to discover an explanation (A. E. Scheidegger, 1963). These were unsuccessful until F. C. Frank (1968) pointed out that geometry dictates that a thin spherical shell can be folded in upon itself without distortion only if the hinge zone is a circular arc, of which the radius in degrees measured on the sphere is half the dip of the infolded portion. The skin of a tennis or Ping-Pong ball which has been pushed in provides a familiar example.

Many subduction zones are marked by trenches parallel to island arcs, which are indeed circular. Frank suggested that for many arcs the conical Benioff zones upon which the foci of deep earthquakes lie have approximately the correct dip, but two problems remain. For many arcs the dip of the Benioff zones is not precisely what theory predicts, being generally too steep, and along some coasts trenches and subduction zones form without any intervening offshore arcs or marginal seas. Chile is an example.

The first problem is largely answered if the rigid lithosphere is denser than the underlying asthenosphere. Many seismologists believe this to be true, for below subduction zones they find evidence of downfolded, and even of detached, blocks of lithosphere sinking under gravity into the asthenosphere. If this is so, then, wherever a section of the lithosphere has been bent inward, gravity, acting upon the depressed conical fragment, pulls it downward, stretching the cone so that it breaks in tension (as rock can easily do) into separate pendant fingers.

If the lithosphere is denser than the asthenosphere upon which it rests, one may well ask why this inversion does not rapidly correct itself by the lithosphere

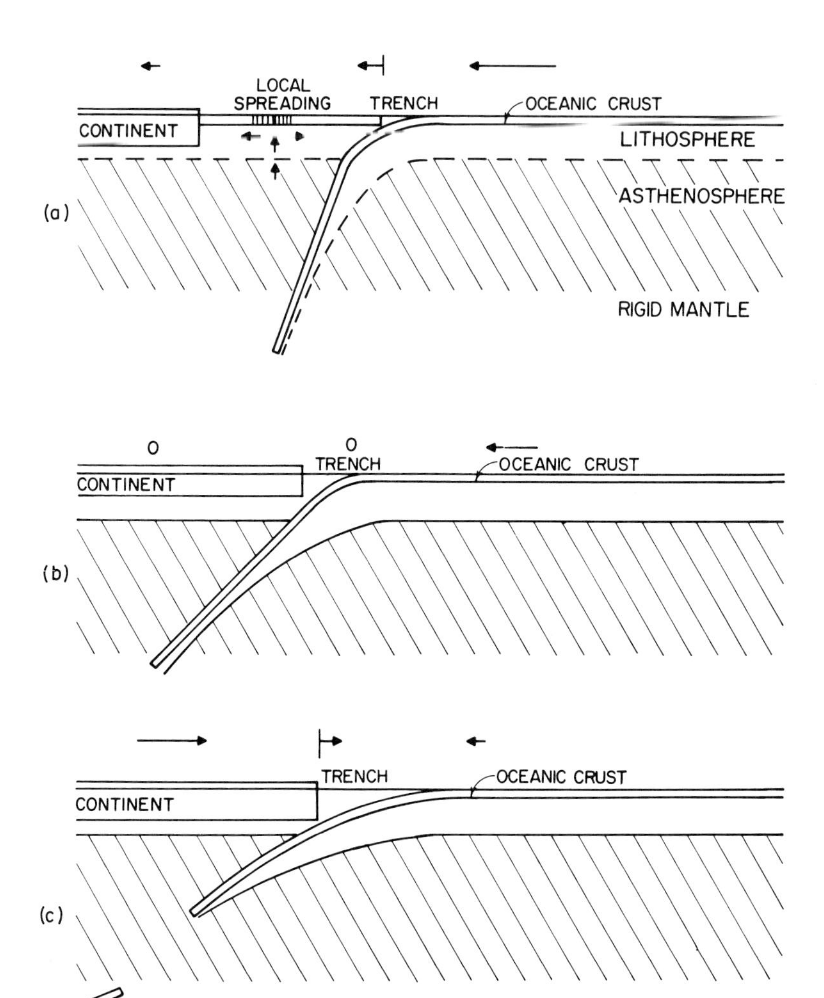

FIGURE 17-14
Three cases are illustrated which are all similar in that at a trench or subduction zone one lithospheric plate with a continent is overriding another of oceanic crust at the same rate. The cases differ in their directions of motion over the lower mantle. Arrows above the continental block, trench, and oceanic crust indicate their respective directions of motion relative to the lower mantle. Postulated effects on the subducted lithosphere and Benioff zone are indicated.

sinking to the bottom of the asthenosphere. The answer appears to be that the lithosphere stays afloat for the same reason that a steel ship does. The ship remains above water because it is full of lighter air, and the ocean floor floats because any cross section of it from the crest of a mid-ocean ridge to a continent resembles the shape of a boat, and the sea water within any basin is lighter than the asthenosphere below it. It is essential that a steel ship be leakproof, but this is also true for the lithosphere, because most cracks forming in the ocean floor promptly fill with lava, which solidifies in a time very short compared with the sluggish flow of the asthenosphere. Any cracks that do not become sealed immediately build volcanoes which soon reach above sea level.

The second problem of why arcs form off some coasts but not off others is believed to be explicable in terms of the mechanism proposed in the last section, although the details have not yet been fully explored. The key to the argument lies in the fact that, whenever two lithospheric plates are in motion over the deeper mantle, relative motions must be considered of the three masses taken in pairs, giving three poles of rotation, one between the two plates and one between each plate and the mantle, respectively (J. T. Wilson, in press).

Consider how this applies in certain specific cases. Under the Mid-Atlantic Ridge, plumes have been postulated to rise in positions fixed in the deep mantle and marked on the surface by islands. These upwellings have caused the two lithospheric plates of the Americas and Africa to separate in a direction indicated by the fracture zones, and they have also caused the Americas plate to move westward over the mantle along a path indicated by the Rio Grande rise, while the African plate has followed a path over the mantle indicated by the Walvis ridge (Sec. 17-5; Fig. 17-12; J. T. Wilson, 1965a).

In a similar manner the Pacific and Nasca plates are moving apart along the East Pacific rise in easterly and westerly directions, while the Nasca plate is moving northeasterly over the deep mantle (assuming that the Nasca ridge gives a proper indication of that direction of motion). As the leading edge of the Nasca plate enters the subduction zone along the Peru-Chile trench, it bends down along a sloping channel to a depth of about 700 km, where the leading edge presumably melts away to become reabsorbed into the mantle. One might expect that as soon as this channel had become established, it would remain fixed in the mantle, but this has not happened, because the Americas plate is advancing over the mantle, pushing the subduction zone ahead of its leading edge, which here is the continental block of South America. Hence the subduction zone is not free to develop, but is possibly constrained to take the shape of the edge of the continental block. Because of this motion the dip of the Benioff zone is less than Frank's theory suggests. The component of shear motion between the plates may be another factor to be taken into account.

Consider, on the other hand, the relative motions of the Eurasian and Pacific plates. The strike of the Hawaiian chain is believed to show the direction of motion of the Pacific. If so, the plate is moving west-northwest across a plume now situated

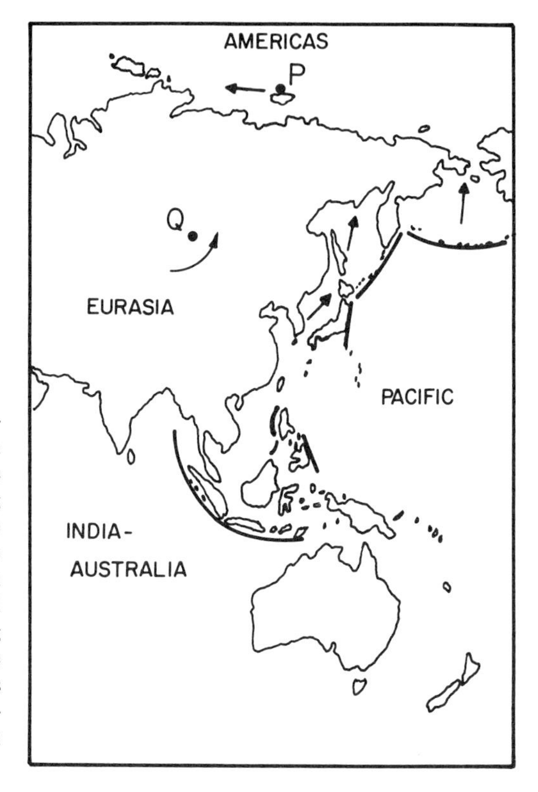

FIGURE 17-15
The Eurasian lithospheric plate is rotating relative to the Americas plate about a pole at *P*, but because both plates are sliding over the mantle, *P* also is moving westward relative to the mantle. Hence the pole of rotation of Eurasia relative to the mantle is not at *P* but at *Q*. Rotation about *Q* (but not about *P*) allows island arcs to form where the plate is moving away from subduction zones in the mantle (case *a* in Fig. 17-14). No arcs form where a continental plate is overriding a subduction zone (e.g., case *c* in Fig. 17-14) and the Mindanao trench.

beneath the island of Hawaii (J. T. Wilson, 1963e). Its leading edge is moving toward and into the subduction zones marked by the trenches of the East Asian arcs.

The motion of Eurasia is not so clearly marked, but one can speculate upon its motion from the following argument. The pole of rotation for motion of the Eurasian and the Americas lithospheric plates is near the New Siberian Islands. At first glance this would appear to be driving Eurasia eastward over the East Asian arcs, but as was noted in the last section, this pole is connected only with the lithosphere, and this pole and both plates are in motion together in a westward direction relative to the mantle. The combination of these movements seems likely to be taking Eurasia westward away from the Bering, Okhotsk, and Japan Seas (Fig. 17-15). It is tempting to speculate about the arcs farther south, but the Philippine Sea and Indian-Australian plates then become involved, which makes the problem even more complex. S. Uyeda and A. M. Jessop (1970), D. E. Karig (1971a), and G. H. Packham and D. A. Falvey (1971) have provided excellent summaries of the origin and development of marginal basins. Karig distinguishes between those parts of basins which are now actively spreading and those which are older (Fig. 17-16). He believes that along the upper surface of downgoing slabs of lithosphere under arcs, friction may heat the mantle, which then rises in a diapir, causing extension and high heat flow, contributing to the formation of the arcuate pattern. Basalts are thought to rise along the axial ridge in marginal basins to form new sea floor. Such magnetic data as exist suggest that extension is parallel to the axial ridge (Fig. 16-14) and to the *en echelon* lines of volcanoes noted by S. Tokuda (1926) and W.H. Bucher (1933). T. Matsuda and S. Uyeda (1971) and E. R. Oxburgh and D. L. Turcotte (1971) also hold that this theory is capable of explaining the paired nature of the Japanese and other orogenic zones around the Pacific. The outer belts, which are 200 km or less wide, lie on the oceanward side, and are formed in eugeosynclinal rocks, with metamorphism due to high pressures and temperatures and ophiolitic suites but no granitic intrusions. The inner belts, which are considered to lie over the heated zones, have been characterized by repeated intrusions of acidic to intermediate magmas throughout their history, but lack evidence of geosynclinal subsidence.

These ideas apply well to the East Asian arcs, but fail to explain why arcs and marginal seas did not develop off Chile and Peru. It is considered that the direction of motion of lithospheric plates over the mantle is an additional factor which must be taken into account (Fig. 17-14; J. T. Wilson, in press).

In support of this argument is the contrast in dips of the Benioff zone beneath the East Asian arcs and Chile. According to Frank's theory, the dip under the East Asian arcs should be about 45°, but as H. Benioff (1949) and M. Katsumada and L. R. Sykes (1969) point out, the dip steepens below 150 km, until it is vertical in many places. On the other hand, the dip under Chile, which should be nearly vertical, is only about 45°, and the deepest part seems to have become detached, yet it is not in the position expected if it had been sinking (E. Kausel and C. Lomnitz, 1969). All this can be readily explained if the dips follow Frank's theory within the lithosphere, and if the dips become altered below the lithosphere. The effect of the advance of the continental crust overriding the oceanic crust and pushing the trench ahead of it is different from the effect of the oceanic crust of the Pacific, which is underriding the Japanese arcs and moving the Japan trench westward in the wake of Eurasia.

Some of the chief uncertainties concern the behavior of the supposed plumes. How long do they endure? Is it true that they stay in the same place, or do they slowly move about in the mantle? How can one locate accurately the pole of rotation of a lithospheric plate over the mantle? Does that pole remain fixed?

The above theory is probably also able to explain the very different habits of different ocean basins. Compare, for example, the South Atlantic and the Arctic Sea. Although both are bounded on the west by the same Americas plate, their behavior has been very different. Whereas the South Atlantic appears to have expanded steadily for about 100 m yr, the Arctic and adjacent seas have changed their loci of expansion three times in the past 60 m yr (Secs. 15-7 and 15-11).

This is understandable if the several plumes indicated in the South Atlantic

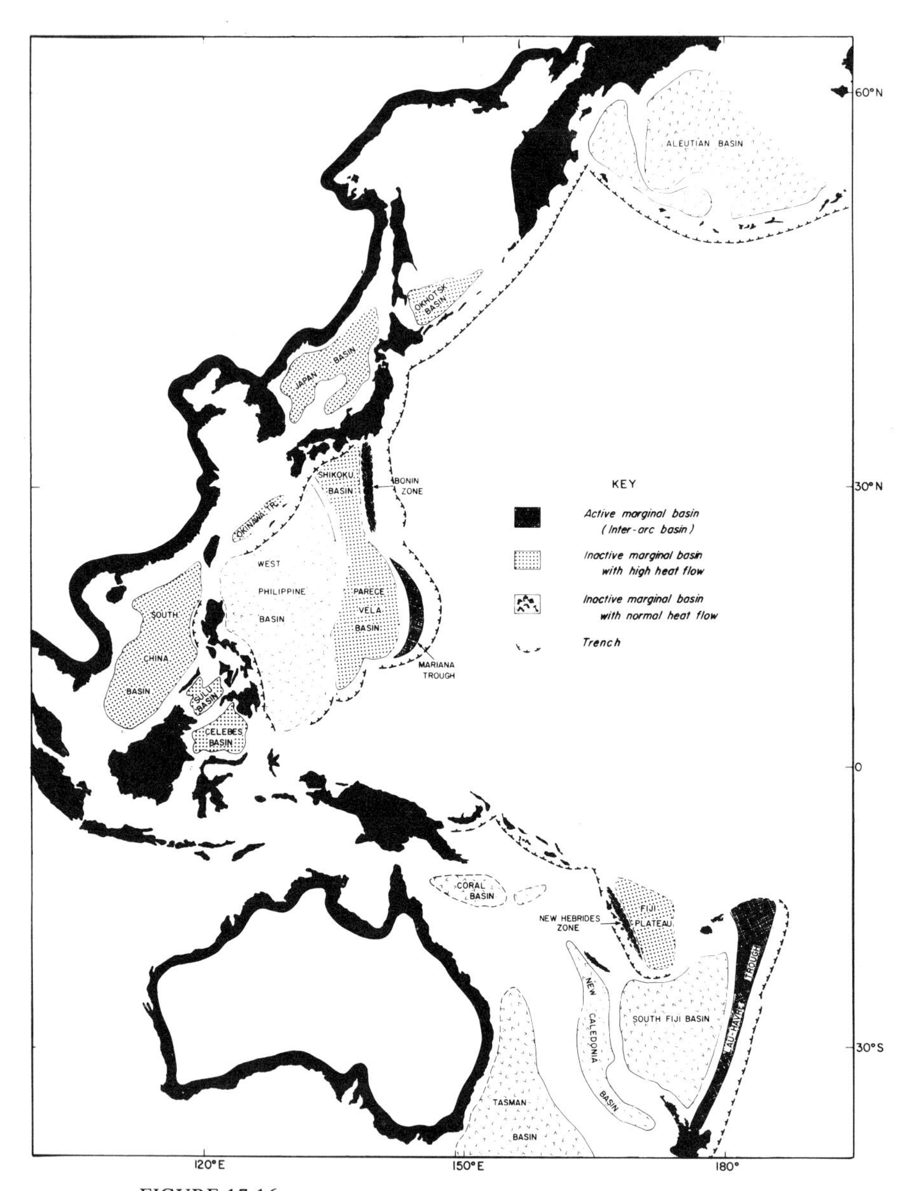

FIGURE 17-16
Marginal basins of the Western Pacific showing, in darker tone, the more active basins and, in intermediate tone, those which are inactive but have higher heat flows than normal. (*After D. E. Karig, 1971.*)

by islands are dominant and have been able to drive South America and Africa steadily apart about a pole near Iceland. The fact that Iceland is near the pole of rotation would tend to keep the Americas plate from touching Iceland. Now consider the motion of Eurasia relative to the Americas. In this case the pole for the lithospheric plates is at the New Siberia Islands, and Iceland is presumably the site of a major plume tending to drive Eurasia and the Americas plates apart. This double role of Iceland involves a contradiction. It is proposed that this is the cause of the erratic motion of the spreading ridges and ocean basins in the Arctic, Baffin Bay, and the Norwegian Sea, which are having to adjust their positions to meet new situations created by the main drive in the south.

The slow-spreading rates in that region also explain why the plume under Iceland has piled up a large island. With faster rates the equivalent volumes of lava would have been spread out along lateral ridges, as has happened in the cases of Hawaii and Tristan da Cunha.

It is considered that similar arguments may be able to explain the contrast between the smooth and steady growth of the East Pacific rise and the complexities of the Tasman Sea with its many ridges (D. J. Cullen, 1970a,b; Fig. 16-20).

It appears that there may be two kinds of ocean basins and two kinds of continents–those fixed over the deep mantle and those moving relative to it. This concept leads to that of two extreme types of mountain building–island arcs in which the continent is fixed and the basin is moving and the Andean type in which the reverse is true.

17-7 CONCLUSION

This tentative outline of a hypothesis for the driving mechanism of the earth's surface motion is as far as it is possible to go at present. Much remains to be done before this or some other theory is established, but it is at least clear that any theory must consider both continents and ocean basins and both geology and geophysics. These have too long been kept apart, and it seems essential that the teaching of both these subjects be united. In few cases do scientists initially trained as classical geologists ever obtain a full grasp of, and familiarity with, geophysics; it is equally true that few physicists ever comprehend the full complexities of geology.

It would appear to be time to break down these barriers and treat both geological and geophysical aspects of the solid earth as one science and give it a new name–geonomy. Earth sciences can then be reserved for the study of the whole earth, including the atmosphere and oceans.

Appendix **A**

DERIVATION OF VELOCITY–DEPTH CURVES FROM TRAVEL–TIME TABLES

Let (r,θ) be the polar coordinates of any point P of a seismic ray which originated at the surface at a point P_0, as indicated in Fig. A-1. Let v be the velocity of the ray at P, v being a function of the distance r only. The time taken to reach P is

$$T = \int \frac{ds}{v} = \int \frac{1}{v} \left[\left(\frac{dr}{d\theta} \right)^2 + r^2 \right]^{1/2} d\theta \qquad \text{(A-1)}$$

where ds is an element of length along the path. The actual path is such that the integral is stationary for small variations of the path. Writing V for the integrand in

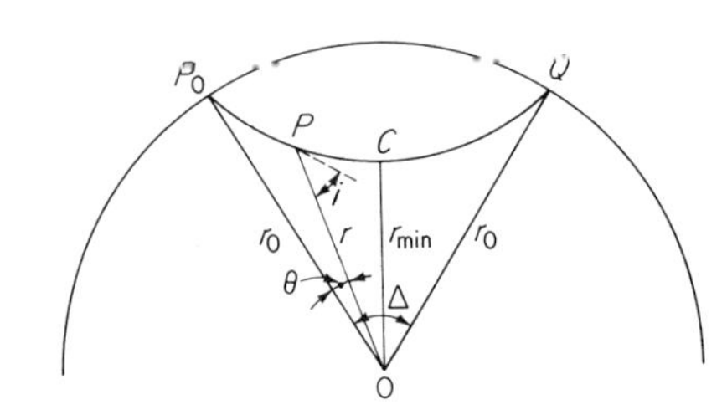

FIGURE A-1

Eq. (A-1) and r' for $dr/d\theta$, it follows from the calculus of variations that r must satisfy the differential equation

$$\frac{d}{d\theta}\frac{\partial V}{\partial r'} - \frac{\partial V}{\partial r} = 0 \qquad \text{(A-2)}$$

a first integral of which is

$$V - r'\frac{\partial V}{\partial r'} = p \qquad \text{(A-3)}$$

where p is a constant for a given ray. Substituting for V, we have

$$\frac{r^2}{v} = p(r^2 + r'^2)^{1/2} \qquad \text{(A-4)}$$

Hence, writing $\eta = r/v$,

$$d\theta = \pm pr^{-1}(\eta^2 - p^2)^{-1/2}\, dr \qquad \text{(A-5)}$$

If θ begins by increasing, r will begin by decreasing, and the negative sign must be taken.

The parameter p has a simple physical interpretation. If i is the angle that the ray makes with the radius OP, then for small displacements along the ray,

$$r\, d\theta = \sin i\, ds = \sin i (r^2 + r'^2)^{1/2} d\theta \qquad \text{(A-6)}$$

so that from Eq. (A-4),

$$p = \frac{r \sin i}{v} \qquad \text{(A-7)}$$

which is a generalized form of Snell's law. When the ray reaches its deepest point C, where $r = r_{\min}$, r' will vanish. The ray then bends upward, remaining symmetrical about OC and reaching the surface again at the point $Q(r_0, \Delta)$. In particular, it follows from Eq. (A-7) that

$$p = \frac{r_{\min}}{v_{r_{\min}}} = \eta_{r_{\min}} \qquad \text{(A-8)}$$

There is another interpretation of the parameter p. Let the ray emerge at Q at time T, let Q' be a neighboring point $(r_0, \Delta + d\Delta)$ in the same plane, and let $T + dT$ be the time required to reach Q'. Draw QR perpendicular to the ray that reaches Q' (Fig. A-2). Then to the first order of small quantities

$$RQ' = v\, dT \quad \text{and} \quad QQ' = r_0\, d\Delta$$

and hence, with the values of i and v at the surface in Q,

$$\sin i = \frac{v\, dT}{r_0\, d\Delta} \qquad \text{(A-9)}$$

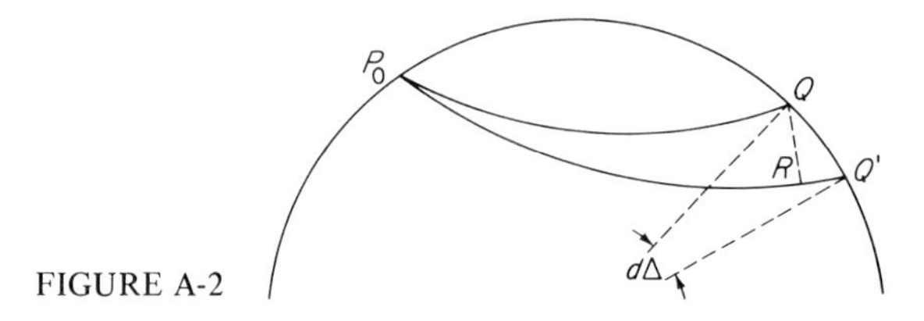

FIGURE A-2

Thus from Eq. (A-7) the ray parameter p may be identified as $dT/d\Delta$. Integrating Eq. (A-5) between P_0 and C,

$$\tfrac{1}{2}\Delta = \int_{r\min}^{r_0} pr^{-1}(\eta^2 - p^2)^{-1/2}\, dr \qquad \text{(A-10)}$$

i.e., using Eq. (A-8),

$$\Delta = \int_{p}^{\eta_0} 2pr^{-1}(\eta^2 - p^2)^{-1/2} \frac{dr}{d\eta}\, d\eta \qquad \text{(A-11)}$$

which may be regarded as an integral equation whose solution determines η, and hence v, as a function of r.

The following solution, due to G. Rasch, was given to H. Jeffreys in a private communication. Let the subscript 1 denote values of the variables at the level r_1 and let Δ_1 be the value of Δ for the ray whose deepest point is at the level r_1. Multiply both sides of Eq. (A-11) by $(p^2 - \eta_1{}^2)^{-1/2}$ and integrate with respect to p over the range η_1 to η_0. Then

$$\int_{\eta_1}^{\eta_0} \Delta\,(p^2 - \eta_1{}^2)^{-1/2}\, dp = \int_{\eta_1}^{\eta_0} dp \int_{p}^{\eta_0} 2pr^{-1}\,[(\eta^2 - p^2)(p^2 - \eta_1{}^2)]^{-1/2} \frac{dr}{d\eta}\, d\eta \qquad \text{(A-12)}$$

Change the order in the double integration. The limits for p become η_1 to η, and those for η run from η_1 to η_0. But if $\eta > \eta_1$,

$$\int_{\eta_1}^{\eta} \frac{p\, dp}{[(\eta^2 - p^2)(p^2 - \eta_1{}^2)]^{1/2}} = \frac{\pi}{2} \qquad \text{(A-13)}$$

so that the right-hand side of Eq. (A-12) becomes

$$\int_{\eta_1}^{\eta_0} \pi r^{-1} \frac{dr}{d\eta}\, d\eta = \pi \log \frac{r_0}{r_1} \qquad \text{(A-14)}$$

Integrating the left-hand side of Eq. (A-12) by parts leads to

$$\left[\Delta \cosh^{-1} \frac{p}{\eta_1}\right]_{\eta_1}^{\eta_0} - \int_{\eta_1}^{\eta_0} \frac{d\Delta}{dp} \cosh^{-1} \frac{p}{\eta_1}\, dp = \int_{0}^{\Delta_1} \cosh^{-1} \frac{p}{\eta_1}\, d\Delta \qquad \text{(A-15)}$$

Thus Eq. (A-12) finally reduces to

$$\int_0^{\Delta_1} \cosh^{-1} \frac{p}{\eta_1} \, d\Delta = \pi \log \frac{r_0}{r_1} \qquad \text{(A-16)}$$

p is a known function of Δ, and η_1 is the known value of $dT/d\Delta$ at Δ_1. Hence r_1 is determined by Eq. (A-16) in terms of Δ_1, and thus in terms of η_1 or r_1/v_1. It is thus possible to determine numerically the velocity as a function of the depth below the surface.

Appendix B
CLAIRAUT'S THEOREM

Let a and b be the major and minor semiaxes of an oblate spheroid of revolution and write $b = a(1 - f)$; that is, $f = (a - b)/a$ is the oblateness, or ellipticity, of the spheroid. Neglecting the square of the ellipticity, the equation of the surface is

$$\frac{x^2 + y^2}{a^2} + \frac{z^2}{a^2(1 - 2f)} = 1$$

that is,

$$x^2 + y^2 + z^2(1 + 2f) = a^2$$

If θ is the colatitude, this may be written

$$r^2(1 + 2f \cos^2 \theta) = a^2$$

or

$$r = a(1 - f \cos^2 \theta) \qquad \text{(B-1)}$$

If V is the potential due to the mass of the earth, the total force $\mathbf{F}$ on a particle of unit mass at $P(x, y, z)$ on the surface is derivable from a potential U, where

$$U = V - \tfrac{1}{2}\omega^2(x^2 + y^2) \qquad \text{(B-2)}$$

and ω is the angular velocity of the earth. The second term is due to the centrifugal action of the rotating earth. The surface of the earth will be an equipotential

$$U = \text{a constant}, C \tag{B-3}$$

If M is the mass of the earth, its potential at a great distance is, to a first approximation, $-GM/r$. This must be corrected because of the term $f \cos^2 \theta$ in Eq. (B-1) by adding a solution of Laplace's equation containing $\cos^2 \theta$, that is, a term GAP_2/r^3, where P_2 is the Legendre polynomial of degree 2 and A is a constant of order f. Thus take

$$V = -\frac{GM}{r} + \frac{GAP_2}{r^3} \tag{B-4}$$

and determine the constant A so that Eq. (B-3) is satisfied by Eq. (B-4) when r is given by Eq. (B-1), that is,

$$-\frac{GM}{a}(1 + f\cos^2\theta) + \frac{GA}{a^3}\frac{1}{2}(3\cos^2\theta - 1) - \tfrac{1}{2}\omega^2 a^2(1 - \cos^2\theta) = C \tag{B-5}$$

a is a sufficient approximation in the second and third terms, since A and $\omega^2 a/g$ are of order f. Since Eq. (B-5) is to be satisfied for all values of θ, equating to zero the coefficient of $\cos^2 \theta$ gives

$$-\frac{GMf}{a} + \frac{3GA}{2a^3} + \tfrac{1}{2}\omega^2 a^2 = 0$$

that is,

$$\frac{GA}{a^3} = \frac{1}{3}\left(\frac{2GMf}{a} - \omega^2 a^2\right) \tag{B-6}$$

and the terms independent of θ determine the constant C. Hence

$$V = -\frac{GM}{r} + \frac{a^3}{r^3}\left(\frac{GfM}{a} - \tfrac{1}{2}\omega^2 a^2\right)(\cos^2\theta - \tfrac{1}{3}) \tag{B-7}$$

The radius from the earth's center to P makes an angle θ with Oz. The force of gravity g at P is the resultant of the attraction and the centrifugal force, and if it makes an angle ν with the radius vector, the radial force is $-g \cos \nu$. Since ν is small, its square may be neglected, and

$$g = \frac{\partial}{\partial r}(V - \tfrac{1}{2}\omega^2 r^2 \sin^2\theta) \qquad \text{at the surface}$$

$$= \frac{GM}{r^2} - \frac{3a^3}{r^4}\left(\frac{GfM}{a} - \tfrac{1}{2}\omega^2 a^2\right)(\cos^2\theta - \tfrac{1}{3}) - \omega^2 r \sin^2\theta$$

where $r = a(1 - f\cos^2\theta)$ and $a\omega^2/g$ is of order f. Hence

$$g = \frac{GM}{a^2}(1 + 2f\cos^2\theta) - \frac{3}{a}\left(\frac{GfM}{a} - \tfrac{1}{2}\omega^2 a^2\right)(\cos^2\theta - \tfrac{1}{3}) - \omega^2 a \sin^2\theta$$

$$g = \frac{GM}{a^2}(1+f) - \tfrac{3}{2}a\omega^2 + \left(\tfrac{5}{2}a\omega^2 - \frac{GfM}{a^2}\right)\cos^2\theta \tag{B-8}$$

If g_e denotes the value of g at the equator ($\theta = \pi/2$) and c is the ratio of the centrifugal force to gravity at the equator, so that $a\omega^2 = cg_e$, then

$$g_e = \frac{GM}{a^2}(1+f) - \tfrac{3}{2}a\omega^2$$

that is,

$$g_e(1 + \tfrac{3}{2}c) = \frac{GM}{a^2}(1+f)$$

Hence $GM = a^2 g_e(1 + \tfrac{3}{2}c - f)$ and Eq. (B-8) becomes

$$g = g_e(1 + \tfrac{3}{2}c - f)(1+f) - \tfrac{3}{2}cg_e + [\tfrac{5}{2}cg_e - fg_e(1 + \tfrac{3}{2}c - f)]\cos^2\theta$$

$$= g_e[1 + (\tfrac{5}{2}c - f)\cos^2\theta] \tag{B-9}$$

which result is known as Clairaut's theorem (1743). It has been extended, taking into account terms of higher order, by G. G. Stokes (1849) and by F. R. Helmert (1884).

Appendix C

ISOTOPIC EQUILIBRIA[1]

H. C. Urey (1947) has used a relatively simple method for the determination of partition function ratios of isotopic molecules. By a combination of these ratios, equilibrium constants for specific exchange reactions are obtained. The method is essentially as follows: A typical exchange reaction may be written

$$aA_1 + bB_2 \rightleftharpoons aA_2 + bB_1 \tag{C-1}$$

where A and B are molecules which have some one element as a common constituent, and subscripts 1 and 2 indicate that the molecule contains only the light or the heavy molecule, respectively. It is well known that the equilibrium constant K for any reaction is given by

$$-RT \ln K = \Delta F^0 \tag{C-2}$$

where F^0 is the standard free energy. For a reaction of the type (C-1), one obtains

$$-RT \ln K = aF^0_{A2} + bF^0_{B1} - aF^0_{A1} - bF^0_{B2} \tag{C-3}$$

[1] Abstracted with permission from A. P. Tudge and H. G. Thode, 1950.

The free energy is related to the partition function by the following equation:

$$F = E_0 + RT \ln N - RT \ln Q \qquad \text{(C-4)}$$

where E_0 = "zero-point energy" of molecule
N = Avogadro number
Q = partition function of molecule

On substituting Eq. (C-4) in Eq. (C-3) and simplifying, one gets the relation

$$K = \frac{(Q_{A2}/Q_{A1})^a}{(Q_{B2}/Q_{B1})^b} \exp\left(\frac{-aE_{0A2} + bE_{0B1} - aE_{0A1} - bE_{0B2}}{RT}\right) \qquad \text{(C-5)}$$

Instead of taking E_0 as the zero-point energy (which is unique for any molecule), one can take E_0 as the bottom of the "potential-energy curve" for the molecule. Since the potential-energy curves are practically identical for isotopic molecules, E_{0A2} equals E_{0A1} and E_{0B2} equals E_{0B1}. Thus the exponential term in Eq. (C-5) becomes unity, and

$$K = \frac{(Q_{A2}/Q_{A1})^a}{(Q_{B2}/Q_{B1})^b} \qquad \text{(C-6)}$$

Therefore, in order to calculate the value of K, it is first necessary to determine the Q_2/Q_1 ration for substances A and B.

The ratio Q_2/Q_1 for a chemical compound (diatomic) is given by the equation

$$\frac{Q_2}{Q_1} = \frac{I_2}{I_1} \frac{\sigma_1}{\sigma_2} \frac{M_2^{3/2}}{M_1^{3/2}} \frac{e^{-U_2/2}(1 - e^{-U_1})}{(1 - e^{-U2})e^{-U_1/2}} \qquad \text{(C-7)}$$

where σ_2 and σ_1 are the symmetry numbers of the two related molecules, M_2 and M_1 are the molecular weights, U_2 and U_1 are related to the fundamental vibrational frequencies by relations of the kind $U_i = hc\omega_i/kT$, and I_2 and I_1 are the moments of inertia.

The following reasoning is then used by Urey to obtain further simplifications. If the right- and left-hand sides of Eq. (C-7) are multiplied by $(m_1/m_2)^{3/2n}$ (where m_1 and m_2 are the atomic weights of the isotopic atoms being exchanged and n is the number of isotopic atoms being exchanged) and if the right-hand side is multiplied and divided by the ratio U_1/U_2, it is possible to simplify the expression. Thus Eq. (C-7) becomes

$$\frac{Q_2}{Q_1} \frac{m_1}{m_2}^{3/2n} = \frac{\sigma_1 U_2 e^{-U_2/2}(1 - e^{-U_1})}{\sigma_2 U_1 (1 - e^{-U_2})e^{-U_1/2}} \qquad \text{(C-8)}$$

since

$$\frac{I_2 M_2^{3/2} \; m_1^{3/2n} U_1}{I_1 M_1^{3/2} \; m_2^{3/2n} U_2} = 1 \qquad \text{(C-9)}$$

according to the Teller and Redlick theorem. New partition functions are then defined where[1]

$$\frac{Q_2'}{Q_1'} = \frac{Q_2}{Q_1}\left(\frac{m_1}{m_2}\right)^{3/2\,n} \tag{C-10}$$

It is obvious that the equilibrium constant K is given by

$$\left(\frac{Q_{A2}'}{Q_{A1}'}\right)^a \left(\frac{Q_{B2}'}{Q_{B1}'}\right)^b = K \tag{C-11}$$

Equation (C-8) can be put in a more convenient form for purposes of calculation. Defining

$$\chi = \frac{U_1 + U_2}{4} \qquad \text{and} \qquad \delta = \frac{U_1 - U_2}{2} \tag{C-12}$$

and expanding in terms of δ, Eq. (C-8) becomes

$$\ln \frac{Q_2'}{Q_1'} = \ln \frac{\sigma_1}{\sigma_2} + \ln \frac{U_2}{U_1} + (\coth \chi)\ \delta \tag{C-13}$$

Although the entire preceding treatment has been given for diatomic molecules, it is possible to go through similar reasoning for polyatomic molecules and obtain equations similar to Eqs. (C-12) and (C-13)

$$\ln \frac{Q_2'}{Q_1'} = \ln \frac{\sigma_1}{\sigma_2} + \sum_i \ln \frac{U_{2i}}{U_{1i}} + \sum_i (\coth \chi_i)\ \delta_i \tag{C-14}$$

$$\chi_i = \frac{U_{2i} + U_{1i}}{4} \qquad \delta_1 = \frac{U_{1i} - U_{2i}}{2} \tag{C-15}$$

U_i are related to the various fundamental vibrational frequencies (in cm^{-1}) of the isotopic molecules as shown above. Therefore, although the calculation for a single partition function would be extremely complicated, the *partition function ratio* for *isotopic molecules* is easily obtained from a *knowledge of vibrational frequencies alone*.

To determine Q_2'/Q_1', it is necessary to obtain the vibrational frequencies of the molecules containing the most abundant isotopes from spectroscopic data and then to calculate the frequencies of the rare isotopic molecules by means of well-known "normal vibration equations." This method must be used because, in general, the rare isotopic molecule is in such low concentration that its vibrational frequencies cannot be experimentally determined.

"Normal vibration equations" have been known for many molecules for some time, and these equations approximately relate the frequencies to "force constants"

[1] Our Qs and Q's are just the reverse of those used by Urey.

and atomic weights. If one assumes these same "force constants" to hold for both the abundant and the rare molecule, it is possible to calculate frequencies for both isotopic molecules (by putting in the appropriate atomic weights) and so find the differences in frequencies. While these calculated frequencies may be slightly in error (since the normal vibration equations are only approximate), the *differences* can be evaluated quite accurately by this method. Using these differences and the experimentally observed fundamental frequencies for the abundant molecule, it is possible to determine the fundamental frequencies for the molecule containing the rare isotope.

As pointed out by Urey, the partition functions for polyatomic molecules may be slightly in error because only the fundamental frequencies are known (i.e., no anharmonic terms). Also, error will be introduced when we apply statistical mechanical formulas relating to ideal gaseous substances to condensed phases or solids in solution.

Appendix D

EQUATIONS OF THE LINES OF FORCE OF A UNIFORMLY MAGNETIZED SPHERE

Consider the field of a uniformly magnetized sphere whose magnetic axis runs north-south, and let P be any external point distant r from the center 0, and θ the colatitude (Fig. D-1). The magnetic potential V at P is given by

$$V = \frac{-M \cos \theta}{r^2} \qquad \text{(D-1)}$$

where M is the magnetic moment.

The radial and cross-radial components of force (Z, H) are given by

$$Z = \frac{\partial V}{\partial r} = \frac{2M \cos \theta}{r^3} \qquad \text{(D-2)}$$

and

$$H = \frac{1}{r}\frac{\partial V}{\partial \theta} = \frac{M \sin \theta}{r^3} \qquad \text{(D-3)}$$

The lines of force by symmetry lie on surfaces of revolution about the axis, and in any axial plane are given by

$$\frac{dr}{Z} = \frac{r d\theta}{H} \qquad \text{(D-4)}$$

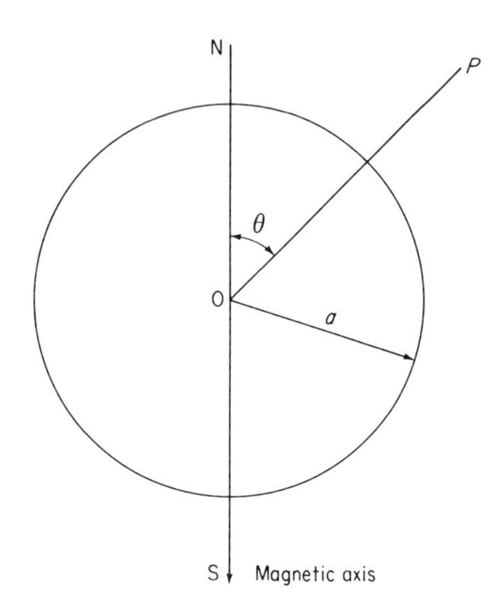

FIGURE D-1

that is, from Eq. (D-2) and (D-3)

$$\frac{dr}{2M \cos \theta / r^3} = \frac{r d\theta}{M \sin \theta / r^3} \tag{D-5}$$

$$\frac{dr}{2 \cos \theta} = \frac{r d\theta}{\sin \theta}$$

or

$$\frac{dr}{r} = \frac{2 \cos \theta}{\sin \theta} d\theta \tag{D-6}$$

which integrates to give

$$\ln r = 2 \ln \sin \theta + \text{constant}$$

or

$$r = C \sin^2 \theta \tag{D-7}$$

where different values of C correspond to different lines of force.

Equation (D-7) may be written in the form

$$\frac{r}{a} = \frac{\sin^2 \theta}{\sin^2 \theta_0} \tag{D-8}$$

where θ_0 is the value of θ at which the line of force meets the sphere $r = a$. The maximum distance of a line of force from the sphere is obtained when $\theta = 90°$, that is, above the equator, when $r = r_{max} = a/\sin^2 \theta_0$. Thus, if $\theta_0 = 30°$, $r_{max} = 4a$, corresponding to a height $3a$ above the equator.

Appendix E

CHEMICAL INHOMOGENEITY IN THE EARTH

The term chemical inhomogeneity is used to include also inhomogeneity arising from phase changes. It follows from Eq. (2-3) that the distribution of $\phi = k/\rho$ throughout the earth can be determined from seismic travel-time curves if V_p and V_s are monotonic functions of the radius r. Assuming hydrostatic pressure [Eq. (2-4)] and an adiabatic-temperature gradient, since

$$k = \rho\phi$$

it follows that

$$\frac{dk}{dp} = \phi \frac{d\rho}{dp} + \rho \frac{d\phi}{dp}$$

$$= \phi \frac{d\rho}{dp} - \frac{1}{g}\frac{d\phi}{dr}$$

For a chemically homogeneous region, $d\rho/dp = \rho/k$ (by definition), and

$$\frac{dk}{dp} = 1 - g^{-1}\frac{d\phi}{dr} \qquad \text{(E-1)}$$

At any point of a region, chemically homogeneous or not, at which $d\rho/dr$ can be assumed to exist, we can write

$$\frac{dk}{dp} = \frac{-\phi}{g\rho}\frac{d\rho}{dr} - \frac{1}{g}\frac{d\phi}{dr}$$

that is,

$$\frac{d\rho}{dr} = \frac{-\eta g\rho}{\phi} \tag{E-2}$$

where

$$\eta = \frac{dk}{dp} + g^{-1}\frac{d\phi}{dr} \tag{E-3}$$

BIBLIOGRAPHY

ABELSON, P. M., et al., The moon issue, *Science*, 167, 447, 1970. [56][1]

ABELSON, P. H., Pennywise, pound foolish, *Science*, 152, 1332, 1966. [370]

ADAMS, F. D., "The Birth and Development of the Geological Sciences," Dover, 1954. [343]

—— and A. E. BARLOW, Geology of the Haliburton and Bancroft areas, *Geol. Surv. Canada, Mem.*, 6, 1912. [95; 488]

ADAMS, L. H., and E. D. WILLIAMSON (see Williamson and Adams), 1923. [39]

ADAMS, R. D., and M. J. RANDALL, The fine structure of the Earth's core, *Bull. Seismol. Soc. Am.*, 54, 1299, 1964. [36]

AGALETZKI, P. N., and K. N. EGOROV, *Izmer. Tekh. Poverochn. Delo*, 6, 29, 1956. [112]

AHMAD, F., Palaeogeography of the Gondwana period in Gondwanaland, with special reference to India and Australia, and its bearing on the theory of continental drift, *Mem. Geol. Surv. India*, 90, 142, 1961. [353]

AHRENS, T. J., D. L. ANDERSON, and A. E. RINGWOOD, Equations of state and crystal structures of high-pressure phases of shocked silicates and oxides, *Rev. Geophys.*, 7, 667, 1969a. [306]

—— and V. G. GREGSON, Shock compression of crustal rocks: Data for quartz, calcite and plagioclase rocks. *J. Geophys. Res.*, 69, 4839, 1964. [306]

——, C. F. PETERSEN, and J. T. ROSENBERG, Shock compression of feldspars, *J. Geophys. Res.*, 74, 2727, 1969b. [306]

[1] These numbers indicate the text page on which this work is cited.

ALDER, B. J., Is the mantle soluble in the core? *J. Geophys. Res.*, 71, 4973, 1966. [318]
ALDRICH, L. T., and G. W. WETHERILL, Geochronology by radioactive decay, *Ann.Rev. Nucl. Sci.*, 8, 257, 1958. [140; 206]
——, ——, G. R. TILTON, and G. L. DAVIS, Half-life of Rb^{87}, *Phys. Rev.*, 103, 1045, 1956. [143]
ALFVEN, H., On the cosmogony of the solar system, *Stockholms. Obs. Ann.*, 14, nos. 2, 5, 9, 1942/1945. [10]
——, "On the Origin of the Solar System," Clarendon, 1954. [10]
ALLARD, G. O., and V. J. HURST, Brazil-Gabon geologic link supports continental drift, *Science* 163, 528, 1969. [358; 420]
ALLEN, C. R., Transcurrent faults in continental areas, *Phil. Trans. Roy. Soc. London, Ser A*, 258, 82, 1965. [454]
ALLEN, J. R., "Physical Processes of Sedimentation," American Elsevier, 1970. [90]
ALLEN, P., Lower Cretaceous sourcelands and the North Atlantic, *Nature*, 222, 657, 1969. [415]
AL'TSHULER, L. V., A. A. BAKANOVA, and R. F. TRUNIN, Shock adiabats and zero isotherms of seven metals at high pressure, English translation, *Soviet Phys. JETP*, 15, 65, 1962. [306]
——, S. B. KORMER, M. I. BRAZHNIK, L. A. VLADIMIROV, M. P. SPERANSKAYA, and A. I. FUNTIKOV, The isentropic compressibility of aluminum, copper, lead and iron at high pressures, English translation, *Soviet Phys. JETP*, 11, 766, 1960. [306]
——, K. K. KRUPNIKOV, and M. I. BRAZHNIK, Dynamic compressibility of metals under pressures from 400,000 to 4,000,000 atmospheres, English translation, *Soviet Phys. JETP*, 34, 614, 1958b. [306]
——, ——, B. N. LEDENEV, V. I. ZHUCHIKHIN, and M. I. BRAZHNIK. Dynamic compressibility and equation of state of iron under high pressure, English translation, *Soviet Phys. JETP*, 34, 606, 1958a. [306]
——, R. F. TRUNIN, and G. V. SIMAKOV, Shock wave compression of periclase and quartz and the composition of the Earth's lower mantle, English translation, *Bull. Phys. Solid Earth*, 10, 657, 1965. [306]
AMARAL, G., U. CORDANI, K. KAWASHITE, and J. H. REYNOLDS, Potassium-argon-dates from basaltic rocks from Southern Brazil, *Geochim. Cosmochim. Acta*, 30, 159, 1966. [420]
AMBRASEYS, N. N., Some characteristic features of the Anatolian fault zone, *Tectonophysics*, 9, 143, 1970. [364; 392; 465]
AMPFERER, O., Über das Bewegungsbild von Faltengebirgen, *Jahrb. Geol. Reichsanatalt*, 56, 529, 1906. [346]
AMSTUTZ, A., Sur l'evolution des structure Alpines, *Arch. Sci.*, 4, 323, 1951. [394]
ANDERS, E., Origin, age and composition of meteorites, *Space Sci. Rev.*, 3, 583, 1964. [5; 21; 23]
——, Fragmentation history of asteroids, *Icarus*, 4, 399, 1965. [5; 21]
ANDERSON, D. L., Recent evidence concerning the structure and composition of the Earth's mantle, in L. H. AHRENS et al. (eds.), "Physics and Chemistry of the Earth," Vol. VI, p. 1, Pergamon, 1966a. [294; 307]
——, Earth's viscosity, *Science*, 151, 321, 1966b. [311; 312]
——, Phase changes in the upper mantle, *Science*, 157, 1165, 1967a. [307]
——, Latest information from seismic observations, in T. F. GASKELL (ed.), "The Earth's Mantle," p. 355, Academic, 1967b. [307]
—— and H. KANAMORI, Shock wave equations of state for rocks and minerals, *J. Geophys. Res.*, 73, 6477, 1968. [306]
——, C. SAMMIS, and T. JORDON, Composition and evolution of the mantle and core, *Science*, 171, 1103, 1971. [493]

ANDERSON, E. M., "The Dynamics of Faulting and Dyke Formation With Applications to Britain," 2d. ed., Oliver & Boyd, 1951. [66, 326-328]

ANDERSON, J. J., Geology of the southwestern high plateaus of Utah: Bear Valley formation, an Oligocene-Miocene volcanic arenite, *Geol. Soc. Am., Bull.*, 82, 1179, 1971. [443]

ANDREWS, J. E., F. P. SHEPARD, and R. J. HURLEY, Great Bahamas Canyon, *Geol. Soc. Am. Bull.*, 81, 1061, 1970. [418]

ANDREWS, J. T., Pattern and cause of variability of postglacial uplift and rate of uplift in Arctic Canada, *J. Geol.*, 76, 404, 1968. [119]

—— and G. FALCONER, Late glacial and post-glacial history and emergence of the Ottawa Islands, Hudson Bay, Northwest Territories: Evidence on the deglaciation of Hudson Bay, *Can. J. Earth Sci.*, 6, 1263, 1969. [119]

ANHAEUSSER, C. R., R. MASON, M. J. VILJOEN, and R. P. VILJOEN, A reappraisal of some aspects of Precambrian shield geology, *Geol. Soc. Am., Bull.*, 80, 2175, 1969. [481-485]

ANTOINE, J. W., and T. E. PYLE, Crustal studies in the Gulf of Mexico, *Tectonophysics*, 10, 477, 1970. [474]

ARCHAMBEAU, C. B., E. A. FLINN, and D. G. LAMBERT, Fine structures of the upper mantle, *J. Geophys. Res.*, 74, 5825, 1969. [307]

ARGAND, E., La tectonique de l'Aise, *Compt. Rend. Cong. Geol. Intern Belgique,* 13, 1922, 5, 171, 1924. [329; 353; 360]

ARMSTRONG, R. L., A model for the evolution of strontium and lead isotopes in a dynamic earth, *Rev. Geophys. Space Phys.*, 6, 175, 1968. [164]

—— and J. BESANCON, A Triassic time scale dilemma: K-Ar dating of Upper Triassic mafic igneous rocks, Eastern U.S.A. and Canada and post-Upper Triassic plutons, western Idaho, U.S.A., *Eclogae Geol. Helv.*, 63, 15, 1970. [417]

ARTEMJEV, M. E., and E. V. ARTYUSHKOV, Structure and isostasy of the Baikal Rift and the mechanism of rifting, *J. Geophys. Res.*, 76, 1197, 1971. [403]

ASAMI, E., On the reverse natural remanent magnetism of basalt of Cape Kawajiri, Yamaguchi Prefecture, *Proc. Japan Acad.*, 30, 102, 1954a. [247]

——, Reverse and normal magnetization of the basalt lavas at Kawajiri-misaki, Japan, *J. Geomagn. Geoelectr.,* 6, 145, 1954b. [247]

ATKINS, F. B., P. E. BAKER, J. D. BELL, and D. G. W. SMITH, Oxford Expedition to Ascension Island, 1964, *Nature,* 204, 722, 1964. [421]

ATWATER, T., Implications of plate tectonics for Cenozoic tectonic evolution of Western North America, *Geol. Soc. Am., Bull*, 81, 3513, 1970. [438; 440; 443; 448]

—— and H. W. MENARD, Magnetic lineations in the Northeast Pacific, *Earth Planet. Sci. Lett.*, 7, 445, 1970. [442]

AUBOUIN, J., "Geosynclines." (Developments in Geotectonics, vol. 1), American Elsevier, 1965. [465; 466]

AULT, W. U., and M. L. JENSEN, Summary of sulfur isotopic standards, in M. L. JENSEN (ed.), "Biogeochemistry of Sulfur Isotopes," *Proc. Nat. Sci. Found. Symp.,* April 1962. [171]

AUMENTO, F., Diorites from the Mid-Atlantic Ridge at 45°N, *Science*, 165, 1112, 1969. [85]

AVERY, O. E., and G. D. BURTON, An aeromagnetic survey of the Norwegian Sea, *J. Geophys. Res.,* 73, 4583, 1968. [409]

AXELROD, D. J., Fossil floras suggest stable not drifting continents, *J. Geophys. Res.*, 68, 3257, 1963; and 69, 1665, 1964. [361]

BABCOCK, H. W., Zeeman effect in stellar spectra, *Astrophys. J.*, 105, 105, 1947. [231]

BACKUS, G. E., A class of self-sustaining dissipative spherical dynamos. *Ann., Phys.*, **4**, 372, 1958. [236]

—— and S. CHANDRASEKHAR, On Cowling's theorem on the impossibility of self-maintained axi-symmetric homogeneous dynamos, *Proc. Nat. Acad. Sci.*, **42**, 105, 1956. [235]

BACON, M., and F. GRAY, Evidence for crust in the deep ocean derived from continental crust, *Nature*, 229, 331, 1971. [406]

BADER, R. G., R. D. GERARD, W. F. BENSON, H. M. BOLLI, W. W. HAY, W. T. ROTHWELL, Jr., M. H. RUEF, W. R. RIEDEL, and F. L. SAYLES, "Initial Reports of the Deep Sea Drilling Project," vol. IV, U.S. Govt. Printing Office, 1970. [370]

BAER, A. J. (ed.), Symposium on basins and geosynclines of the Canadian Shield, *Geol. Surv. Canada Paper*, 70-40, 1970. [483; 487; 488]

BAILEY, D. K. Crustal warping – a possible tectonic control of alkaline magmatism, *J. Geophys. Res.*, 69, 1103 1964. [401]

BAILEY, E. B., Paleozoic mountain systems of Europe and America, *Brit. Assoc. Adv. Sci., Rept. 96th Meet.* 1928, p.57, 1929. [353]

——, "Tectonic Essays, Mainly Alpine," Oxford, 1935. [360]

—— and W. J. MCCALLIEN, Serpentine lavas, the Ankara melange and the Anatolian thrust, *Trans. Edinburgh Roy. Soc.*, 62, pt. II, 403, 1953. [466]

BAILEY, E. H., M. C. BLAKE, Jr., and D. L. JONES, On-land Mesozoic oceanic crust in California coast ranges, *U.S. Geol. Surv., Prof. Paper* 700-C, p. 70, 1970. [85; 439]

BAKANOVA, A. A., I. P. DUDOLADOV, and R. F. TRUNIN, Compression of alkali metals by strong shock waves, *Soviet Phys.-Solid State,* 7, 1307, 1965. [306]

BAKER, B. H., Geology and mineral resources of the Seychelles, *Min. Com. Ind. Geol. Surv. Kenya, Mem.* 3, 1963. [424]

——, The rift systems in Kenya, in A. N. HUNTER (ed.), "East African Rift System," UNESCO Seminar, p. 82, University College, Nairobi, 1965. [398; 401]

—— and J. A. MILLER, Geology and geochronology of the Seychelles Islands and structure of the floor of the Arabian Sea, *Nature*, 199, 346, 1963. [424]

——, L. A. J. WILLIAMS, J. A. MILLER, and F. J. FITCH, Sequence and geochronology of the Kenya rift volcanics, *Tectonophysics*, 11, 191, 1971. [402; 498]

—— and J. WOHLENBERG, Structure and evolution of the Kenya Rift Valley, *Nature*, 229, 538, 1971. [402; 495]

BAKER, I., Petrology of the volcanic rocks of Saint Helena Island, South Atlantic, *Geol. Soc. Am., Bull.*, 80, 1283, 1969. [421]

BAKER, P. E., Historical and geological notes on Bouvetoya, *Br, Antarct. Surv., Bull.* 13, 71, 1967a. [421]

——, Preliminary account of recent geological investigations on Easter Island, *Geol. Mag.*, 104, 116, 1967b. [432]

BALCHAN, A. S., and G. R. COWAN, Shock compression of two iron-silicon alloys to 2.7 megabars, *J. Geophys. Res.*, 71, 3577, 1966. [307]

BALL, M. M., and C. G. A. HARRISON, Crustal plates in the Central Atlantic, *Science*, 167, 1128, 1970. [406]

BALL, R. H., A. B. KAHLE, and E. H. VESTINE, On the determination of surface motions of the Earth's core, *Rand Corp. Mem.* RM-5615-NASA, Dec 1968. [234]

BALLY, A. W., P. L. GORDY, and G. A. STEWART, Structure, seismic data, and orogenic evolution of Southern Canadian Rocky Mountains, *Bull. Can. Pet. Geol.*, 14, 337, 1966. [95; 351]

BAME, S. J., J. R. ASBRIDGE, H. E. FELTHAUSER, E. W. HONES, and I. B. STRONG, Characteristics of the plasma sheet in the Earth's magnetotail, *J. Geophys. Res.*, 72, 113, 1967. [285]

BANGHAR, A. R., and L. R. SYKES, Focal mechanism of earthquakes in the Indian Ocean and adjacent regions, *J. Geophys. Res.*, 74, 632, 1969. [424]

BANKS, R. J., Geomagnetic variations and the electrical conductivity of the upper mantle, *Geophys. J.*, 17, 457, 1969. [263]

BARAGAR, W. R. A., Major element geochemistry of the Yellowknife volcanic rocks, *Can. J. Earth Sci.*, 4, 773, 1968. [96]

BARAZANGI, M., and J. DORMAN, World seismicity maps compiled from ESSA, Coast and Geodetic Survey, Epicenter Data, 1961/1967, *Bull. Seismol. Soc. Am.*, 59, 369, 1969. [347; 375; 392; 427; 444]

——, ——, Seismicity map of the Arctic compiled from ESSA, Coast and Geodetic Survey, Epicenter, Data January 1961 through September 1969, *Bull Seismol. Soc. Am.*, 60, 1741, 1970. [408]

BARKER, P. F., Plate tectonics of the Scotia Sea region, *Nature*, 228, 1293, 1970. [421; 450]

BARR, F. T. (ed.), Geology and Archaeology of Northern Cyrenaica, Libya, Petrol. Explor. Soc. Libya, 10th Ann. Field Conf. 1968, Holland – Breumelhof N. V., Amsterdam, 1968. [462; 466]

BARRETT, D. L., C. E. KEEN, K. S. MANCHESTER, and D. I. ROSS, Baffin Bay–an ocean, *Nature*, 229, 551, 1971. [413]

BARROS, L. A., A Ilha do Principe e a "Linha dos Camaroes," *Mem. Junta Invest Ultram.*, 17, 2d ser., 1960. [420]

BARROW, G., On an intrusion of muscovite-biotite gneiss in the Southeast Highlands of Scotland, *Quart, J. Geol. Soc. London,* 49, 330, 1893. [69; 70]

——, On the geology of lower Dee-side and the Southern Highland Border, *Proc. Geol. Assoc.*, 23, 268, 1912. [69]

BARTLETT, G. A., and L. SMITH, Mesozoic and Cenozoic history of the Grand Banks of Newfoundland, *Am. J. Earth Sci.*, 8, 65, 1971. [415]

BATE, G. L., H. A. POTRATZ, and J. R. HUIZENGA, Thorium in iron meteorites, *Geochim. Cosmochim. Acta,* 14, 118, 1958. [237]

BATEMAN, P. C., and J. P. EATON, Sierra Nevada batholith, *Science*, 158, 1407, 1967. [95]

BATTEY, M. H., The "two-magma theory" and the origin of ignimbrites, *Bull. Volcanol.*, 29, 407, 1966. [496]

BAUER, A., Nouvelle estimation du bilan de masse de l'Indlandsis du Groenland, *Deep Sea Res.*, 14, 13, 1967. [57; 58]

BAUER, M., F. GREY, and D. H. MATTHEWS, Crustal structure studies in the Bay of Biscay, *Earth Planet. Sci. Lett.*, 6, 377, 1969. [415]

BAYLY, B., "Introduction to Petrology," Prentice-Hall, 1968. [67; 71]

BEALS, C. S., and I. HALLIDAY, Terrestrial meteorite craters and their lunar counterparts, in R. FAIRBRIDGE (ed.), "International Dictionary of Geophysics." vol. 2, p. 1520, Pergamon, 1967. [78]

——, M. J. S. INNES, and J. A. ROTTENBERG, Fossil meteorite craters, in B. M. MIDDLEHURST and G. P. KUIPER (eds.), "The Solar System," vol. 4, "The Moon, Meteorites and Comets," The University of Chicago Press, 1963. [18]

BECK, A. E., Techniques of measuring heat-flow on land, in *Am. Geophys. Union Monograph* 8, p. 24, 1965. [182]

——, Energy changes in an expanding Earth, in S. K. RUNCORN (ed.), "The Application of Modern Physics to the Earth and Planetary Interiors," p. 77, Wiley Interscience, 1969. [345]

BECKINSDALE, R. D., and N. H. GALE, A reappraisal of the decay constants and branching ratio of ^{40}K, *Earth Planet, Sci. Lett.*, 6, 289, 1969. [140]

BELLAIR, P., Crozet Islands, in R. ADIE (ed.), "Antarctic Geology," S.C.A.R. Symposium, Elsevier, 1963. [424]

BELOUSSOV, V. V., The origin of folding in the Earth's crust, *J. Geophys. Res.*, 66, 2241, 1961. [340]

——, "Basic Problems in Geotectonics," McGraw-Hill, 1962. [342; 344; 360; 467]

——, The upper mantle project, in H. ODISHAW (ed.) "Research in Geophysics," vol. 2, p. 555, MIT, 1964. [356]

——, Against continental drift, *Sci. J.*, 3, 56, 1967. [353; 366]

——, Debate about the Earth: An open letter to J. Tuzo Wilson, *Geotimes*, 13, 17, 1968a. [353; 366]

——, Against the hypothesis of ocean-floor spreading, *Tectonophysics*, 9, 489, 1970. [353; 367]

—— and I. P. KOSMINSKAYA, Structure and development of transition zones between the continents and oceans, *Can. J. Earth Sci.*, 5, 1011, 1968. [345; 349]

—— and E. M. RUDITCH, Island arcs in the development of the Earth's structure (especially in the region of Japan and the Sea of Okhotsk), *J. Geol.*, 69, 3, 1961. [449]

BENFIELD, A. E., Terrestrial heat flow in Great Britain, *Proc. Roy. Soc. London, Ser. A, 173,* 428, 1939. [198; 372]

——, The temperature in an accreting Earth, *Trans. Am. Geophys. Union*, 31, 53, 1950. [204]

BENIOFF, H., Seismic evidence for the fault origin of oceanic deeps, *Geol. Soc. Am., Bull,* 60, 1837, 1949. [347; 505]

——, Movements on major transcurrent faults, in S. K. RUNCORN (ed.), "Continental Drift," p. 103, Academic, 1962. [388]

BENSON, C. S., and S. EPSTEIN, Stratigraphic studies in the snow and firn of the Greenland Ice Sheet, *S.I.P.R.E. Res. Rept.*, 1960. [173; 174]

BERRY, M. J., and L. KNOPOFF, Structure of the upper mantle under the Western Mediterranean basin, *J. Geophys. Res.,* 72, 3613, 1967. [462]

BETZ, F., Jr., and H. H. HESS, The floor of the North Pacific Ocean, *Geograph Rev.*, 32, 99, 1942. [432]

BEYDOUN, Z. R., Note on the age of the Hadramut Arch, Southern Arabia, *Overseas Geol. Mineral Resources*, 10, 236, 1969. [404]

—— and M. R. BICHAN, The geology of Socotra Island, Gulf of Aden, *Quart. J. Geol. Soc. London,* 125, 413, 1970. [404]

BEZZI, A., and G. B. PICCARDO, Structural features of the Ligurian ophiolites; petrologic evidence for the "oceanic" floor of the Northern Apennines geosyncline; a contribution to the problem of the alpinotype gabbro-peridotite association, *Mem Soc. Geol. Italiana,* 10, 53, 1971. [85; 466]

BHATNAGAR, P. L., *Ind. J. Phys.,* 14, 253, 1940. [9]

BIERMANN, L., Physical processes in comet tails and their relation to solar activity, *Soc. Roy. Sci. de Liège,* 13, 291, 1953. [283]

BIGELEISEN, J., and M. G. MEYER, Calculation of equilibrium constants for isotopic exchange reactions, *J. Chem. Phys.,* 15, 261, 1947. [169]

BIQ, C. C., Role of gravitational gliding in Taiwan tectonogenesis, *Bull. Geol. Soc. Taiwan*, 20, 1, 1969. [454]

BIRCH, F., The variation of seismic velocities within a simplified Earth model in accordance with the theory of finite strain, *Bull. Seismol. Soc. Am.*, 29, 463, 1939. [308]

——, Recent work on the radioactivity of potassium and some related geophysical problems, *J. Geophys. Res.*, 56, 107, 1951. [206]

——, Elasticity and constitution of the Earth's interior, *J. Geophys. Res.*, 57, 227, 1952. [38; 202; 308; 309; 316]

——, Some geophysical applications of high pressure research, in W. PAUL, and D. M. WARSCHAUER (eds.), "Solids Under Pressure," p. 137, McGraw-Hill, 1963. [306; 316; 323]

——, Density and composition of mantle and core, *J. Geophys. Res.*, 69, 4377, 1964. [41; 309]

——, Energetics of core formation, *J. Geophys. Res.*, 70, 6217, 1965. [201; 318]

——, On the possibility of large changes in the Earth's volume, *Phys. Earth Planet. Int.*, 1, 141, 1968. [316; 345]

——, R. F. ROY, and E. R. DECKER, Heat flow and thermal history in New York and New England, in E-AN ZEN, W. S. WHITE, J. B. HADDLEY, and J. B. THOMPSON, Jr. (eds.), "Studies of Appalachian Geology: Northern and Maritimes," Wiley, 1968. [198]

BIRD, J. M., and J. F. DEWEY, Lithosphere plate-continental margin tectonics and the evolution of the Appalachian orogen, *Geol. Soc. Am., Bull.*, 81, 1031, 1970. [473]

BISHOP, W. W., and G. R. CHAPMAN, Early Pliocene sediments and fossils from the Northern Kenya rift valley, *Nature,* 226, 914, 1970. [75]

BJÖRNSSON, S. (ed.), Iceland and mid-ocean ridges, *Visindafelag Islendinga* [Reykjavik], 1967. [381; 412]

BLACK, L. P., N. H. GALE, S. MOORBATH, R. J. PANKHURST, and V. R. MCGREGOR, Isotopic dating of very early Precambrian amphibolite facies gneisses from the Godthaab District, West Greenland, *Earth Planet, Sci. Lett.*, 12, 245, 1971. [153]

BLACKBURN, G., and R. M. TAYLOR, Limestone and red soils of Bermuda, *Geol. Soc. Am., Bull.* 80, 1595, 1969. [418]

BLACKETT, P. M. S., The magnetic field of massive rotating bodies, *Nature*, 159, 658, 1947. [231]

——, A negative experiment relating to magnetism and the Earth's rotation, *Phil. Trans. Roy. Soc. London, Ser. A,* 245, 309, 1952. [231; 383]

——, Lectures on Rock Magnetism, Weizmann Science Press, 1956. [383]

——, Comparison of ancient climates with the ancient latitudes deduced from rock magnetic measurements, *Proc. Roy. Soc. London, Ser. A,* 263, 1, 1961. [360]

——, E. C. BULLARD, and S. K. RUNCORN (eds.), A symposium on continental drift, *Phil. Trans. Roy. Soc. London, Ser. A,* 258, 1965. [412]

BLACKWELL, D. D., Heat-flow determinations in the Northwestern United States, *J. Geophys. Res.*, 74, 992, 1969. [444]

BLACKWELL, J. H., A transient-flow method for determination of thermal constants in insulating materials in bulk, *J. Appl. Phys.*, 25, 137, 1954. [185]

BODVARSSON, G., and G. P. L. WALKER, Crustal drift in Iceland, *Geophys. J.,* 8, 285, 1964. [382]

BOLT, B. A., Gutenberg's early PKP observations, *Nature*, 196, 122, 1962. [36; 40; 323]

——, The velocity of seismic waves near the Earth's core, *Bull. Seismol. Soc. Am.*, 54, 191, 1964. [323]

——, C. LOMNITZ, and T. V. MCEVILLY, Seismological evidence on the tectonics of Central and Northern California and the Mendocino escarpment, *Bull. Seismol. Soc. Am.*, 58, 1725, 1968. [332; 447]

BONATTI, E., C. EMILIANI, G. OSTLUND, and H. RYDELL, Final desiccation of the Afar rift, Ethiopia, *Science,* 172, 468, 1971. [404]

BONHOMMET, N., and J. BABKINE, Sur la présence d'aimantations inversées dans la Chaîne des Puys, *Compt. Rend. Acad. Sci. Paris*, 264, 92, 1967. [248]

BOOKER, J., E. C. BULLARD, and R. L. GRASTY, Palaeomagnetism and age of rocks from Easter Island and Juan Fernandez, *Geophys. J.*, 12, 469, 1967. [432]

BOWIN, C. O., Geophysical study of the Cayman trough, *J. Geophys. Res.*, 73, 5159, 1968. [437]

BRACEY, D. R., Structural implications of magnetic anomalies north of the Bahama-Antilles Islands, *Geophysics*, 33, 950, 1968. [417]

BRAGG, W. H., and W. L. BRAGG (eds.), "The Crystalline State," Bell, 1962. [60]

BRIDEN, J. C., Recurrent continental drift of Gondwanaland, *Nature,* 215, 1334, 1967. [384]

BROOKS, M., A gravity survey of coastal areas of West Finmark, Northern Norway, *Quart. J. Geol. Soc. London*, 125, 171, 1970. [474]
BROWN, A. S., Geology of the Queen Charlotte Islands, *British Columbia Dept. Mines. Pet Res.* 54, 1968. [439]
BROWN, D. A., K. S. W. CAMPBELL, and K. A. W. CROOK, "The Geological Evolution of Australia and New Zealand," Pergamon, 1968. [422; 458; 475]
BRUCKSHAW, J. M., and E. I. ROBERTSON, The magnetic properties of the tholeiite dykes of North England, *Monthly Notices Roy. Astron. Soc., Suppl.* 5, 308, 1949. [243]
BRUHNES, B., Recherches sur le direction d'aimantation des roches volcaniques, *J. Phys.*, 5, 705, 1906. [243]
BRUNDIN, L., On the real nature of transantarctic relationships, *Evolution*, 19, 496, 1965. [361; 363; 459]
——, Transantarctic relationships and their significance, as evidenced by chironomid midges, *Kungl. Swenska Ventenskapsakedemiens Handlingar Fjörde*, Serien 11, 1, 1966. [361]
BRUNE, J. N., Seismic moment, seismicity, and rate of slip along major fault zones, *J. Geophys. Res.*, 73, 777, 1968. [465]
——, Surface waves and crustal structure, in P. J. HART (ed.), "The Earth's Crust and Upper Mantle," *Am. Geophys. Union Monograph* 13, p. 230, 1969. [408]
BRYANT, W. R., J. ANTOINE, M. EWING, and B. JONES, Structure of Mexican continental shelf and slope, Gulf of Mexico, *Am. Assoc. Petrol. Geol. Bull.*, 52, 1204, 1968. [418]
BRYNER, L., Ore deposits of the Philippines – an introduction to their geology, *Econ. Geol.*, 64, 644, 1969. [454; 455]
BUCHER, W. H., "The Deformation of the Earth's Crust," Princeton, 1933. [344; 451; 505]
BULLARD, E. C., The disturbance of the temperature gradient in the Earth's crust by inequalities of height, *Monthly Notices Roy. Astron. Soc., Suppl.* 4, 360, 1938. [183]
——, Heat flow in South Africa, *Proc. Roy. Soc. London, Ser. A.*, 173, 474, 1939. [197; 372]
——, The time necessary for a borehole to attain equilibrium, *Monthly Notices Roy. Astron. Soc., Suppl.* 5, 127, 1947. [183]
——, The magnetic field within the Earth, *Proc. Roy. Soc. London, Ser. A*, 197, 433, 1949a. [236; 238]
——, Electromagnetic induction in a rotating sphere, *Proc. Roy. Soc. London, Ser. A*, 199, 413, 1949b. [236; 238]
——, The flow of heat through the floor of the Atlantic Ocean, *Proc. Roy. Soc. London, Ser. A.*, 222, 408, 1954a. [183; 372]
——, The interior of the Earth, in G. P. KUIPER (ed.), "The Solar System," vol. 2, "The Earth as a Planet," The University of Chicago Press, p. 57, 1954b. [205]
——, Measurement of temperature gradient in the Earth, in S. K. RUNCORN (ed.), "Methods and Techniques in Geophysics," Interscience, vol 1, p. 1, 1960. [182]
——, J. E. EVERETT, and A. G. SMITH, The fit of the continents around the Atlantic, *Phil. Trans. Roy. Soc. London, Ser. A*, 258, 41, 1965. [358; 359; 364; 406; 420]
——, C. FREEDMAN, H. GELLMAN, and J. NIXON, The westward drift of the Earth's magnetic field, *Phil. Trans. Roy. Soc. London, Ser. A*, 243, 67, 1950. [234]
—— and H. GELLMAN, Homogeneous dynamos and terrestrial magnetism, *Phil. Trans. Roy. Soc. London, Ser. A*, 247, 231, 1954. [236]
—— and D.T. GRIGGS, The nature of the Mohorovičić discontinuity, *Geophys. J.* 6, 118, 1961. [297]
——, A. E. MAXWELL, and R. REVELLE, Heat flow through the deep ocean floor, *Advan. Geophys.*, 3, 153, 1956. [183; 208]
BULLEN, K. E., On Rayleigh waves across the Pacific Ocean, *Monthly Notices Roy. Astron. Soc., Geophys. Suppl.* 4, 579, 1939. [372]

——, A hypothesis on compressibility at pressures of the order of a million atmospheres, *Nature*, 157, 405, 1946. [321]

——, Compressibility-pressure hypothesis and the Earth's interior, *Monthly Notices Roy. Astron. Soc., Geophys. Suppl.* 5, 355, 1949. [320]

——, An Earth model based on a compressibility-pressure hypothesis, *Monthly Notices Roy. Astron. Soc., Geophys. Suppl.* 6, 50, 1950. [320]

——, "Seismology," Methuen, 1954. [33]

——, Earth's central density, *Nature*, 196, 973, 1962. [40]

——, "An Introduction to the Theory of Seismology," 3d ed., Cambridge, 1963a. [33; 35; 49; 50; 202; 294]

——, An index of degree of chemical inhomogeneity in the Earth, *Geophys. J.*, 7, 584, 1963b. [40; 321]

——, On compressibility and chemical inhomogeneity in the Earth's core, *Geophys. J.*, 9, 195, 1965a. [322]

——, Models for the density and elasticity of the Earth's lower core, *Geophys. J.*, 9, 233, 1965b. [323]

——, Empirical equations of state for the Earth's lower mantle and core, *Geophys. J.*, 16, 235, 1968. [320]

—— and R. A. W. HADDON, Earth models based on compressibility theory, *Phys. Earth Planet. Int.*, 1, 1, 1967. [321]

—— and ——, Upper bound to change in incompressibility at the Earth's mantle-core boundary, *Geophys. J.*, 17, 179, 1969. [321]

BULMASOV, A. P., Magnetic and gravitational fields of the Baikal region as related to its seismicity, *Byul Soveta po Seismologiyi AN SSSR*, 10, 49 (Def. Res. Bd. Canada Transl. T. 435 R, 1967). [403]

BUNGENSTOCK, H., H. CLOSS, and K. HINZ, Seismische Untersuchungen im nordlichen Teil des Arabischen Meeres (Gulf von Oman), *Erdoel, Kohle, Erdgas, Petrochem*, 19, 237, 1966. [425]

BURBIDGE, G. R., and F. HOYLE, On cosmic rays as an extra-galactic phenomenon, *Proc. Phys. Soc.*, 84, 141, 1964. [289]

BURCKLE, L. H., J. EWING, T. SAITO, and R. LEYDEN, Tertiary sediment from the East Pacific rise, *Science*, 157, 537, 1967. [429]

BUREK, P. J., Tectonic effects of sea-floor spreading on the Arabian Shield, *Geol. Rundschau*, 59, 382, 1970. [405]

BÜRGL, H., The orogenesis in the Andean system of Colombia, *Tectonophysics,* 4, 429, 1967. [433]

BURK, C. A. (ed.), A study of serpentinite, the AMSOC core hole near Mayaguez, Puerto Rico, *Natl. Acad. Sci. – Natl. Res. Council Publ.* 1188, 1964. [85; 294; 437]

——, The geology of the Alaskan Peninsula – island arc and continental margin, *Geol. Soc. Am. Mem.* 99, 1965. [447]

——, The Aleutian arc and Alaska continental margin, in "Continental Margins and Island Arcs," *Geol. Surv. Canada*, Paper 66-15, p. 206, 1966. [447]

BURKE, K., Seismic areas of the Guinea Coast where Atlantic fracture zones reach Africa, *Nature*, 222, 655, 1969. [373]

BURWASH, R. A., Comparative Precambrian geochronology of the North American, European and Siberian shields, *Can. J. Earth Sci.*, 6, 357, 1969. [492]

—— and J. KRUPIČKA, Cratonic reactivation in the Precambrian basement of western Canada, pt II. Metasomatism and isostasy, *Can. J. Earth Sci.*, 7, 1275, 1970. [96]

BUTLER, L. W., Shallow structure of the continental margin, Southern Brazil and Uruguay, *Geol. Soc. Am. Bull.*, 81, 1079, 1970. [420]

BUTLER, S. T., and K. A. SMALL, The excitation of atmospheric oscillations, *Proc. Roy. Soc. London, Ser. A*, 274, 91, 1963. [267]

BUXTORF, A., Prognosen und Befunde beim Hauenstein-basis und Grenchenberg Tunnel und die Bedeutung der letzteren für die Geologie des Juragebriges. *Naturf Gesell.*, Basel, Verh. 27, 1916. [351]

CAHILL, L. H., Investigation of the equatorial electrojet by rocket magnetometer, *J. Geophys. Res.*, 64, 489, 1959. [265]

CAIN, J. C., Models of the earth's magnetic field, in B. M. MCCORMAC (ed.), "Radiation Trapped in the Earth's Magnetic Field," p.7, Gordon & Breach, 1966. [227]

——, W. E. DANIELS, S. J. HENDRICKS, and D. C. JENSEN, An evaluation of the main geomagnetic field, 1940-1962, *J. Geophys. Res.*, 70, 3647, 1965. [226; 227]

——, S. J. HENDRICKS, R. A. LANGEL, and W. V. HUDSON, A proposed model for the International Geomagnetic Reference Field – 1965, *J. Geomagn. Geoelectr.*, 19, 335, 1967. [227]

CALDERON, S., Estudiò petrográfico sobre las rocas volcanicas del Cabo de Gata é Isla de Alboran, *Bol. Comm. Mapa. Geol., Madrid,* 9, 333, 1882. [464]

CAMPBELL, C. D., and S. K. RUNCORN, Magnetization of the Columbia River basalts in Washington and Northern Oregon, *J. Geophys. Res.*, 61, 449, 1956. [243]

CAMPBELL, W. H., Geomagnetic pulsations, in S. MATSUSHITA, and W. H. CAMPBELL (eds.), "Physics of Geomagnetic Phenomena," p. 821, Academic, 1967. [275]

CANITEZ, N., The focal mechanisms in Iran and their relations to tectonics, *Pure Appl. Geophys.*, 75, 76, 1969. [468]

CANN, J. R., Geological processes at mid-ocean ridge crests, *Geophys. J.*, 15, 331, 1968. [85; 99]

——, Spilites from the Carlsberg ridge, Indian Ocean, *J. Petrol.*, 10, 1, 1969. [424]

——, New model for the structure of the ocean crust, *Nature,* 226, 928, 1970. [82; 85; 86; 367]

——, and B. M. FUNNELL, Palmer ridge: A section through the upper part of the ocean crust? *Nature*, 213, 661, 1967. [85; 415]

CAPUTO, M., Gravity in space and the dimensions and mass of the earth, *J. Geophys. Res.*, 68, 4595, 1963. [109]

CAREY, S. W., The Rheid concept in geotectonics, *J. Geol. Soc. Australia*, 1, 67, 1953 [publ. 1954]. [333; 335]

——, Wegener's North America-Africa Assembly, fit or misfit? *Geol. Mag.* 92, 196, 1955. [389]

——, A tectonic approach to continental drift, in "Continental Drift: A Symposium," p. 117, University of Tasmania, 1958. [333; 345; 353; 389; 409; 449]

——, Folding, *J. Alberta Soc. Petrol. Geologists*, 10, 95, 1962. [335]

CAROZZI, A. V., New historical data on the origin of the theory of continental drift, *Geol. Soc. Am. Bull.*, 81, 283, 1970. [352]

CHAMALAUN, F. H., Paleomagnetism of Reunion Island and its bearing on secular variation, *J. Geophys. Res.*, 73, 4674, 1968. [424]

CHANDRASEKHAR, S., The thermal instability of a fluid sphere heated within, *Phil. Mag.*, 7, 1317, 1952. [346]

——, "Hydrodynamic and Hydromagnetic Stability," Oxford, 1961. [319]

CHANEY, R. W., A revision of the fossil Sequoia and Taxadrium in Western North America based on the recent discovery of Metasequoia, *Trans. Am. Phil. Soc.*, New Series 40, 171, 1950. [361]

CH'ANG TA, "The Geology of China," translation, U.S. Dept. Comm., 1963. [475]

CHAO, E. C. T., Pressure and temperature histories of impact metamorphosed rocks – based on petrographic observations, *Neurs. Jahr. Miner. Abh.* 108, 209, 1968. [21]

CHAPMAN, S., and J. BARTELS, "Geomagnetism," Oxford, 1940. [262; 264; 272; 274; 286; 287]

—— and V. C. A. FERRARO, A new theory of magnetic storms, *Terr. Mag.*, 36, 77, 171, 1931; 37, 147, 421, 1932; and 38, 79, 1933. [284; 287; 288]

CHASE, C. G., H. W. MENARD, R. L. LARSON, F. G. SHARMAN III, and S. M. SMITH, History of sea-floor spreading west of Baja, California, *Geol. Soc. Am. Bull.*, 81, 491, 1970. [443]

CHASE, R. L., and E. T. BUNCE, Underthrusting of the eastern margin of the Antilles by the floor of the Western North Atlantic Ocean, and origin of the Barbados ridge, *J. Geophys. Res.*, 74, 1413, 1969. [350; 437]

CHASE, T. E., and H. W. MENARD, "Bathymetric Atlas of the Northwestern Pacific Ocean," U.S. Naval Oceanogr. Office, 1969. [426; 427]

CHAYES, F., The chemical composition of Cenozoic andesite, Proc. Andesite Conference (ed. A. R. MCBIRNEY), *Oregon Dept. Geol. Mineral. Ind. Bull.* 65, 1, 1969. [89]

CHEVALIER, R., L'aimantation des laves de l'Etna et l'orientation du champ terrestre en Sicile du XIIe au XVIIe siecle, *Ann. Phys.*, 4, 5, 1925. [239]

CHILDRESS, S., A class of solutions of the magnetohydrodynamic dynamo problems, in S. K. RUNCORN (ed.), "The Application of Modern Physics to the Earth and Planetary Interiors," p. 629, Wiley, 1969. [236]

CHINNERY, M. A., and M. N. TOKSÖZ, P-wave velocities in the mantle below 700 km, *Bull. Seismol. Soc. Am.*, 57, 199, 1967. [309]

CHRISTOPHER, P. A., Fission track ages of younger intrusions in Southern Maine, *Geol. Soc. Am. Bull.*, 80, 1809, 1969. [415]

CHUBB, L. J., The pattern of some Pacific island chains, *Geol. Mag.*, 94, 221, 1957. [432]

CHURKIN, M., Jr., Paleozoic tectonic history of the Arctic basin north of Alaska, *Science*, 165, 549, 1969. [409]

CLARK, S. P., Effect of radiative transfer on temperatures in the earth, *Geol. Soc. Am. Bull.*, 67, 1123, 1956. [209]

——, Radiative transfer in the earth's mantle, *Trans. Am. Geophys. Union*, 38, 931, 1957. [209]

CLARK, S. P., Jr., and A. E. RINGWOOD, Density distribution and constitution of the mantle, *Rev. Geophys.*, 2, 35, 1964. [43; 302-305; 310]

— and ——, Density, strength, and constitution of the mantle, in T. F. GASKELL (ed.), "The Earth's Mantle," p. 111, Academic, 1967. [309]

CLARKE, D. B., and B. G. J. UPTON, Tertiary basalts of Baffin Island: Field relations and tectonic setting, *Can. J. Earth Sci.*, 8, 248, 1971. [413]

CLAYTON, R. N., I. FRIEDMAN, D. L. GRAF, T. K. MAYEDA, W. F. MEENTS, and N. F. SHIMP, The origin of saline formation waters, 1. Isotopic composition, *J. Geophys. Res.*, 71, 3869, 1966. [172]

CLEARY, J. R., and D. W. SIMPSON, Seismotectonics of the Australian continent, *Nature*, 230, 239, 1971. [393; 403]

CLIFFORD, T. N., and I. G. GASS (eds.), African Magmatism and Tectonics, Oliver & Boyd, 1970. [402-404; 487; 495]

CLOOS, H., Herbung – Spaltung – Vulcanismus, *Geol. Rundschau,* 30, 401, and 637, 1939. [73; 398; 494]

CLOUD, P. E., Jr., Precambrian of North America, the 3rd Penrose Conference, *Geotimes*, 16, 13, 1971. [479; 483; 484]

——, R. G. SCHMIDT, and H. W. BURKE, Geology of Saipan, Mariana Islands, *U.S. Geol. Surv. Prof. Paper* 280A, 1956. [451; 454]

—— and M. A. SEMINKHATOV, Proterozoic stromatolite zonation, *Am. J. Sci.*, 267, 1017, 1969. [477]

CLOUTIER, P. A., and R. C. HAYMES, Vector measurement of the mid-latitude Sq ionospheric current system, *J. Geophys. Res.*, 73, 1771, 1968. [265]

COATS, R. P., The geology and mineralization of the Blinman Dome diapir, *Geol. Surv. South Australia, Rep. Invest.*, 26, 1, 1964. [496]

COATS, R. R., Magma type and crustal structure in the Aleutian arc, in "The Crust of the Pacific Basin," *Am. Geophys. Union Monograph* 6, p. 92, 1962. [447]

COCKS, L. R. M., C. H. C. BRUNTON, A. J. ROWELL, and I. C. RUST, The first lower Palaeozoic fauna proved from South Africa, *Quart, J. Geol. Soc. London*, 125, 583, 1969. [475]
COES, L., Jr., A new dense crystalline silica, *Science*, 118, 131, 1953. [19]
COHEN, M. H., D. L. JAUNCEY, K. I. KELLERMANN, and B. G. CLARK, Radio interferometry at one-thousandth second of arc, *Science*, 162, 88, 1968. [363]
COLEMAN, P. J., Geology of the Solomon and New Hebrides Islands, as part of the Melanesian re-entrant, Southwest Pacific, *Pacific Sci.*, 24, 289, 1970. [456]
COLLE, J. O., W. F. COOKE, Jr., R. L. DENHAM, H. C. FERGUSON, J. H. MCGUIRT, F. REEDY, Jr., and P. WEAVER, Volume of Mesozoic and Cenozoic sediments in Western Gulf Coastal Plain of United States; Pt. 4 of G. E. MURRAY, Sedimentary Volume in Gulf Coastal Plain of United States and Mexico., *Geol. Soc. Am., Bull.*, 63, 1193, 1952. [80]
COLLETTE, B. J., J. I. EWING, R. A. LAGAAY, and M. TRUCHAN, Sediment distribution in the oceans: The Atlantic between 10° and 19° N., *Mar. Geol.*, 7, 279, 1969. [418]
COMPSTON, W., and P. A. ARRIENS, The Precambrian geochronology of Australia, *Can. J. Earth Sci.*, 5, 561, 1968. [487]
CONOLLY, J. R., Western Tasman sea floor, *New Zealand J. Geol. Geophys.*, 12, 310, 1969. [458]
COOK, A. H., The contribution of observations of satellites to the determination of the earth's gravitational potential, *Space Sci. Rev.*, 2, 355, 1963. [111]
—— and T. MURPHY, Geophys. Mem. Dub. Inst. Adv. Stud. 2, pt. IV, 1952. [114]
COOK, K. L., Active rift system in the Basin and Range province, *Tectonophysics*, 8, 469, 1969. [438]
COOKE, H. C., Back to Logan, *Trans. Roy. Soc. Can.*, 42, 29, 1948. [95; 481]
COOMBS, D. S., Some recent work on the lower grades of metamorphism, *Australian J. Sci.*, 24, 203, 1961. [78; 458]
COOPER, J. A., and J. R. RICHARDS, Isotopic and alkali measurements from the Vema seamount of the South Atlantic Ocean, *Nature*, 210, 1245, 1966. [421]
COWIE, J. W., Notes on lower Cambrian stratigraphy in the Boreal region, *Intern. Geol. Congr., 21st, Copenhagen, 1960*, pt. 8, 57, 1960. [474]
COWLING, T. G., The magnetic field of sunspots, *Monthly Notices Roy. Astron. Soc.*, 94, 39, 1934. [235]
COX, A., Lengths of geomagnetic polarity intervals, *J. Geophys. Res.*, 73, 3247, 1969a. [386]
——, Geomagnetic reversals, *Science*, 163, 237, 1969b. [249; 386]
——, R. R. DOELL, and G. B. DALRYMPLE, Reversals of the earth's magnetic field, *Science*, 144, 1537, 1964. [248; 251; 386]
——, D. M. HOPKINS, and G. B. DALRYMPLE, Geomagnetic polarity epochs. Pribilof Islands, Alaska, *Geol. Soc. Am. Bull.*, 77, 883, 1966. [449]
COXELL, H., M. A. POMERANTZ, and S. P. AGARWAL, Survey of cosmic ray intensity in the lower atmosphere, *J. Geophys. Res.*, 71, 143, 1966. [292]
CRAIG, G. Y., "The Geology of Scotland," Oliver & Boyd, 1965. [473; 474]
CRAIG, H., The isotopic geochemistry of water and carbon in geothermal areas, *Proc. Nucl. Geol. on Geotherm. Areas, Spoleto*, Sept. 9-13, 1963. [172; 173]
—— and G. BOATO, Isotopes, *Ann. Rev. Phys. Chem.*, 6, 403, 1955. [178]
CRAWFORD, A. R., and B. DAILY, Probable non-synchroneity of late Precambrian glaciations, *Nature*, 230, 111, 1971. [491]
CREER, K. M., A review of paleomagnetism, *Earth-Sci. Rev.*, 6, 369, 1970. [384]
——, Paleomagnetism of the crust, in R. DEARNLEY (ed.), "Crust of the Earth," Chapman & Hall [1970]. [384]
——, B. J. J. EMBLETON, and D. A. VALENCIO, Comparison between the upper Paleozoic

and Mesozoic paleomagnetic poles for South America, Africa, and Australia, *Earth Planet. Sci. Lett.*, 7, 288, 1969. [383]

——, ——, and ——, Triassic and Permo-Triassic paleomagnetic data for S. America, *Earth Planet. Sci. Lett.*, 8, 173, 1970. [383]

——, E. IRVING, and S. K. RUNCORN, The direction of the geomagnetic pole in remote epochs in Great Britain, *J. Geomagn. Geoelectr.*, 6, 163, 1954. [383]

CULLEN, D. J., Quaternary volcanism at the Antipodes Islands: Its bearing on structural interpretation of the Southwest Pacific, *J. Geophys. Res.*, 74, 4213, 1969. [431]

——, Two-way stretch of sialic crust and plate tectonics in the South-west Pacific, *Nature*, 226, 741, 1970a. [458; 461; 507]

——, On supposed extinct mid-ocean ridges in the Tasman Sea, *Earth Planet. Sci. Lett.*, 9, 446, 1970b. [507]

CURRY, J. R., and D. G. MOORE, Growth of the Bengal deep-sea fan and denudation in the Himalayas, *Geol. Soc. Am. Bull.*, 82, 563, 1971. [425]

DAGLEY, P., R. L. WILSON, J. M. ADE-HALL, G. P. L. WALKER, S. E. HAGGERTY, T. SIGURGEIRSSON, N. D. WATKINS, P. J. SMITH, J. EDWARDS, and R. L. GRASTY, Geomagnetic polarity zones for Icelandic lavas, *Nature*, 216, 25, 1967. [244; 246; 251]

DAINTY, A. M., C. E. KEEN, M. J. KEEN, and J. E. BLANCHARD, Review of geophysical evidence on crust and upper mantle structure on the Eastern Seaboard of Canada, in J. G. STEINHARD and T. J. SMITH (eds.), "The Earth Beneath the Continents," *Am. Geophys. Union Monograph* 10, p. 349, 1966. [415]

DALY, R. A., The geology of American Samoa, *Geophys. Lab. Carnegie Inst. Wash. Publ.* no. 340, Tortugas Lab., 19, 93, 1924. [450]

——, "Our Mobile Earth," Scribner, 1926. [353]

——, "Igneous Rocks and the Depths of the Earth," McGraw-Hill, 1933. [89; 94]

DANA, J. D., On some results of the earth's contraction from cooling, *Am. J. Sci.*, 5, 423, 1873. [81; 472]

DARLINGTON, P. J., "Biogeography at the Southern End of the World," Harvard, 1965. [361]

DARWIN, C., "Structure and Distribution of Coral Reefs," University of California Press, 1962 [original edition, 1842]. [88; 378; 425]

DAVIES, D., When did the Seychelles leave India? *Nature*, 220, 1225, 1968. [424; 425]

—— and T. J. G. FRANCIS, The crustal structure of the Seychelles bank, *Deep-Sea Res.*, 11, 921, 1964. [425]

DAVIES, H. L., Papuan Ultramafic Belt, *Report 23rd Intern. Geol. Congr., Proc. Sec. 1,* 209, 1968. [456]

——, Peridotite-gabbro-basalt complex in Eastern Papua: An overthrust plate of oceanic mantle and crust, *Australia, Bur. Mineral Resources, Bull.* 128, 1970. [457]

DAVIS, G. A., Tectonic correlations, Klamath Mountains and Western Sierra Nevada, California, *Geol. Soc. Am. Bull.*, 80, 1095, 1969. [439]

DAVIS, T. N., K. BURROWS, and J. D. STOLARIK, A latitude survey of the equatorial electrojet with rocket borne magnetometers, *J. Geophys. Res.* 72, 1845, 1967. [265]

DAVIS, W. M., The coral reef problem, *Am. Geol. Soc. Spec. Publ.* 9, 1928. [425]

DE ALMEIDA, F. F. M., Geologia e Petrologia do Arquipelago de Fernando de Noronha, *Brazil Dept. Natl. Prod. Min., Div. Geol. e Min.*, Mon. 13, 1955. [421]

——, Quelques aspects sous-marins au large de la Cote Bresilienne, *Proc. 21st Intern. Geol. Congr.,* pt. X, 23, 1960. [420; 501]

——, Geologia e petrologia de Ilha da Trindade, *Brazil Dept. Natl. Prod. Min., Div. Geol. e Min.*, Mon. 18, 1961. [421]

——, Structure and dynamics of the Brazilian coastal area, in M. MALDONADO-KOERDELL (ed.), "Pan-American Symposium on the Upper Mantle, Mexico, 1968." 2, 29, 1969. [420]

DEARNLEY, R., Orogenic fold-belts and a hypothesis of earth evolution, *Phys. Chem. Earth*, 7, 1, 1966. [345; 492]

DEBELMAS, J., and M. LEMOINE, The Western Alps: Palaegeography and structure, *Earth-Sci. Rev.*, 6, 221, 1970. [360]

DECHOW, E., and M. L. JENSEN, Sulfur isotopes of some Central African sulfide deposits, *Econ. Geol.*, 60, 894, 1965. [171]

DE CSERNA, Z., Tectonic framework of Southern Mexico and its bearing on the problem of continental drift, *Bol. Soc. Geol. Mexicana*, 30, 159, 1967 (1969). [437]

DEEL, S. A., *U.S. Coast Geod. Surv., Paper* 664, 1945. [230]

DEER, W. A., R. A. HOWIE, and J. ZUSSMAN, "An Introduction to the Rock-forming Minerals," Wiley, 1966. [60]

DEFFEYES, K. S., The axial valley: A steady-state feature of the terrain, in H. JOHNSON, and B. L. SMITH (eds.), "The Megatectonics of Continents and Oceans," p. 194, Rutgers, 1970. [71; 439]

DEGENS, E. T., and D. A. ROSS (eds.), "Hot Brines and Recent Heavy Metal Deposits in the Red Sea," Springer-Verlag, 1969. [406]

DEHLINGER, P., Gravity and its relation to topography and geology in the Pacific Ocean in "The Earth's Crust and Upper Mantle," *Am. Geophys. Union Monograph* 13, p. 352, 1969a. [107]

——, Evidence regarding the development of Juan de Fuca and Gorda ridges in the Northeast Pacific, *Trans. N.Y. Acad. Sci.*, 31, 379, 1969b. [439]

—— and S. H. YUNGUL, Experimental determination of the reliability of the La Coste and Romberg surfaceship gravity meter S-9, *J. Geophys. Res.*, 67, 4389, 1962. [107]

DE JONG, K. A., and R. VAN DER VOO, Rotation of Sardinia: Paleomagnetic evidence from Permian rocks, *Nature*, 226, 933, 1970. [464]

DE LEEUW, M. M., New Canadian bathymetric chart of the Western Arctic Ocean north of 72°, *Deep-Sea Res.*, 14, 489, 1967. [408]

DEMENITSKAYA, R. M., and K. L. HUNKINS, Shape and structure of the Arctic Ocean in A. E. MAXWELL (ed.), "The Sea," vol. 4, Wiley-Interscience, 1970. [407; 408]

—— and A. M. KARASIK, The active rift system of the Arctic Ocean, *Tectonophysics*, 8, 345, 1969. [408]

DENGO, G., Geological structure of Central America, in "Studies in Tropical Oceanography 5," *Proc. Intern. Conf. Trop. Oceanogr.*, p. 56, 1967. [437]

——, O. BOHNENBERGER, and S. BONIS, Tectonics and volcanism along the Pacific marginal zone of Central America, *Geol. Rundschau*, 59, 1215, 1970. [437]

DENHAM, D., Distribution of earthquakes in the New Guinea/Solomon Islands region, *J. Geophys., Res.* 74, 4290, 1969. [456]

DENNIS, J. G., "International Tectonics Dictionary," *Am. Assoc. Petrol Geol. Mem.* 7, 1967. [331; 472]

DE ROEVER, W. P., Symposium on the problem of oceanization in the Western Mediterranean, *Verhandel. Ned. Geol. Mijnbouwk. Genoot., Geol. Ser.* 26, 165, 1969. [449; 464]

DE SITTER, L. U., "Structural Geology," 2d ed., McGraw-Hill, 1964. [331; 464]

DEWEY, J. F., and J. M. BIRD, Mountain belts and the new global tectonics, *J. Geophys. Res.*, 75, 2625, 1970a. [395; 396; 398, 403; 469; 472]

—— and ——, Plate tectonics and geosynclines, *Tectonophysics*, 10, 625, 1970b. [395; 398; 403; 469; 472]

—— and B. HORSFIELD, Plate tectonics, orogeny and continental growth, *Nature*, 225, 421, 1970. [439]

DIBNER, V. D., A. J. KRYLOV, M. A. SEDOVA, and V. A. VAKAR, Age and origin of rocks lifted by trawl from the Southwest Greenland shelf, *Medd. Groenland*, bd. 171, 1, 1963. [412]

DICKE, R. H., Average acceleration of the earth's rotation and the viscosity of the deep mantle, *J. Geophys. Res.*, 74, 5895, 1969. [313]

DICKINSON, W. R., Tectonic development of Fiji, *Tectonosphysics*, 4, 543, 1967. [458]
——, Plate tectonic models of geosynclines, *Earth Planet. Sci. Lett.* 10, 165, 1971. [472]
—— and A. GRANTZ (eds.), "Proceedings of Conference on Geologic Problems of San Andreas Fault System," Stanford University Publ. Geol. Sci., no. 11, 1968. [364; 447]
—— and T. HATHERTON, Andesite volcanism and seismicity around the Pacific, *Science*, 157, 801, 1967. [92]
DICKSON, G. O., W. C. PITMAN III, and J. R. HEIRTZLER, Magnetic anomalies in the South Atlantic and ocean floor spreading, *J. Geophys. Res.*, 73, 2087, 1968. [420]
DIETZ, R. S., Possible deep-sea turbidity current channels in the Indian Ocean, *Geol. Soc. Am. Bull.*, 64, 375, 1953. [425]
——, Marine geology of Northwestern Pacific: Description of Japanese bathymetric chart 6901, *Geol. Soc. Am. Bull.*, 65, 1199, 1954. [372; 431; 448]
——, Continent and ocean basin evolution by spreading of the sea floor, *Nature* 190, 854, 1961. [391]
——, Astroblemes; ancient meteorite impact structures on the earth, in B. M. MIDDLEHURST and G. P. KUIPER (eds.), "Moon, Meteorites and Comets," Solar System Series, vol. 4. The University of Chicago Press, 1963a. [18; 21]
——, Cryptoexplosion structures; a discussion, *Am. J. Sci.*, 261, 650, 1963b. [18; 21]
——, A. J. CARSOLA, E. C. BUFFINGTON, and C. J. SHIPEK, Sediments and topography of the Alaskan shelves, in "Papers in Marine Geology – Shepard Commemorative Volume," p. 241, Macmillan, 1964. [449]
—— and J. C. HOLDEN, Reconstruction of Pangaea: Breakup and dispersion of continents, Permian to present, *J. Geophys. Res.*, 75, 4939, 1970a. [358; 417; 418; 420; 497]
—— and ——, The breakup of Pangaea, *Sci. Am.*, 223, 30, 1970b. [358; 417; 420; 497]
—— and ——, Pre-Mesozoic oceanic crust in the Eastern Indian Ocean (Wharton Basin), *Nature*, 229, 309, 1971. [422]
—— and H. J. KNEBEL, Survey of Ross's original deep sea sounding site, *Nature*, 220, 751, 1968. [369]
—— and W. P. SPROLL, East Canary Islands as a microcontinent within the Africa-North America continental drift fit, *Nature*, 226, 1043, 1970a. [418]
—— and ——, Fit between Africa and Antarctica: a continental drift reconstruction, *Science*, 167, 1612, 1970b. [424]
DIMROTH, E., Evolution of the Labrador geosyncline, *Geol. Soc. Am. Bull.*, 81, 2717, 1970. [488; 491]
DIRAC, P. A. M., A new basis for cosmology, *Proc. Roy. Soc. London, Ser. A,* 165, 199, 1938. [345]
DIXEY, F., The geology and geomorphology of Madagascar and a comparison with Eastern Africa, *Quart J. Geol. Soc. London*, 116, 255, 1960. [424]
DOBRIN, M. B., and B. PERKINS, Bikini and nearby atolls, *U.S. Geol. Surv. Prof. Paper* 260J, *Paper* JKL; pt. III, *Geophysics*, 487, 1954. [88]
DOIG, R., An alkaline rock province linking Europe and North America, *Can. J. Earth Sci.*, 7, 22, 1970. [403; 415]
DOLE, M., The oxygen isotope cycle in nature, in "Nuclear Processes in Geologic Settings," Nucl. Sci. Ser. Rep. 19, *Natl. Acad. Sci. – Natl. Res. Counc.* 13, 1955. [172]
DOUGLAS, R. J. W. (ed.), Geology and economic minerals of Canada, *Geol. Surv. Canada, Econ. Geol. Dept.* no. 1, 1970. [446; 482; 490]
DRAKE, C. L., N. J. CAMPBELL, G. SANDER, and J. E. NAFE, A Mid-Labrador sea ridge, *Nature*, 200, 1085, 1963. [413]
——, M. EWING, and G. H. SUTTON, Continental margins and geosynclines: The east coast of North America, north of Cape Hatteras, *Phys. Chem. Earth*, 3, 110, 1959. [417]
—— and R. W. GIRDLER, A geophysical study of the Red Sea, *Geophys. J.*, 8, 473, 1964. [405]

DUBERAL, R. F., and P. RODDA, "Bibliography of the Geology of Fiji," *Dept. Geol. Surv., Govt. of Fiji*, 1968. [458]
DUBERTRET, L., La bordure orientale de la Méditerrané en tant que temoin de l'évolution des accidents de l'Est Africain, *Intern. Geol. Congr., 20th Sess., Mexico Rept. Assoc. Servicios Geologicos Africanos*, p. 377, 1959. [405]
DUFFAUD, F., J.-P. ROTHE, J. DEBRACH, P. ERIMESCO, G. CHOUBERT, and A. FAURE-MURET, Le seisme d'Agadir du 29 Fevrier 1960, Notes et memoires du Service Geologique 154, 1962, *Editions du serv. Geol. du Maroc.* [50]
DUNBAR, C. O., "Historical Geology," 2d ed., Wiley, 1960. [130]
—— and K. M. WAAGE, Historical Geology, 3d ed., Wiley, 1969. [443]
DUNKIN, H. H., Mining for gold in Fiji, *Australian Min.*, 60, 50, 1968. [458]
DU TOIT, A. L., A geological comparison of South America and South Africa, *Carnegie Inst. Wash. Publ.* 381, 1, 1927. [353; 358; 420]
——, "Our Wandering Continents," Oliver & Boyd, 1937. [352; 353; 385; 397; 420; 421]
DYMOND, J. R., N. D. WATKINS, and Y. R. NAYUDU, Age of the Cobb Seamount, *J. Geophys. Res.*, 73, 3977, 1968. [367]
EARDLEY, A. J., "Structural Geology of North America," 2d ed., Harper, 1962. [408]
EATON, J. P., W. H. K. LEE, and L. C. PAKISER, Use of micro-earthquakes in the study of the mechanics of earthquake generation along the San Andreas fault in Central California, *Tectonophysics*, 9, 259, 1970. [392]
—— and K. J. MURATA, How volcanoes grow, *Science*, 132, 925, 1960. [87; 432]
EDWARDS, A. B., Tertiary lavas from the Kerguelen Archipelago, *Brit. Aust. New Zealand Antarct. Res. Exped. Rep.* 2, pt. V, 69, 1938. [424]
EGYED, L., and L. STEGENA, Physical background of a dynamical earth model, *Z. Geophysik.*, 24, 260, 1958. [345]
ELLIOT, D. H., E. H. COLBERT, W. J. BREED, J. A. JENSEN, and J. S. POWELL, Triassic tetrapods from Antarctica: Evidence for continental drift, *Science*, 169, 1197, 1970. [361]
ELSASSER, W. M., On the origin of the Earth's magnetic field, *Phys. Rev.*, 55, 489, 1939. [233]
——, Induction effects in terrestrial magnetism, pt. 1: Theory, *Phys. Rev.*, 69, 106, 1946. [236]
——, Induction effects in terrestrial magnetism, pt. 3: Electric modes, *Phys. Rev.*, 72, 821, 1947. [235, 236]
——, "Early history of the Earth, in Earth Science and Meteoritics," p. 1, North Holland, 1963. [315]
——, Thermal structure of the upper mantle and convection, in "Advances in Earth Science," p. 461, MIT, 1966. [365]
——, The mechanics of continental drift, *Proc. Am. Phil. Soc.*, 112, 344, 1968. [493]
——, Sea-floor spreading as thermal convection, *J. Geophys. Res.*, 76, 1101, 1971. [496]
——, and H. TAKEUCHI, Nonuniform rotation of the Earth and geomagnetic drift, *Trans. Am. Geophys. Union*, 36, 584, 1955. [235]
ELVERS, D. J., G. PETER, and R. MOSES, Analysis of magnetic lineations in the North Pacific, *Trans. Am. Geophys. Union*, 41, 89, 1967. [448]
EMERY, K. O., Marine geology of Johnston Island and its surrounding shallows, Central Pacific Ocean, *Geol. Soc. Am. Bull.*, 67, 1505, 1956. [431]
——, J. I. TRACEY, Jr., and H. S. LADD, Geology of Bikini and nearby atolls, pt. 1, Geology, *U.S. Geol. Surv. Prof. Paper*, 260-A, p. 4, 1954. [450]
——, E. UCHUPI, J. D. PHILLIPS, C. O. BOWIN, E. T. BUNCE, and S. T. KNOTT, Continental rise off Eastern North America, *Am. Assoc. Pet. Geol. Bull.* 54, 44, 1970. [417]

ENGEL, A. E. J., and C. G. ENGEL, Composition of basalts from the mid-Atlantic ridge, *Science*, 144, 1330, 1964. [88;418]
——, C. J. ENGEL, and R. G. HAVENS, Chemical characteristics of oceanic basalts and the upper mantle, *Geol. Soc. Am. Bull.*, 76, 719, 1965. [87]
ENGEL, C. G., and R. L. FISHER, Lherzolite, anorthosite, gabbro and basalt dredged from the mid-Indian Ocean ridge, *Science*, 166, 1136, 1969. [424]
ENGELEN, G. B., A hypothesis on the origin of the Bermuda rise, *Tectonophysics*, 1, 85, 1964. [418;450]
EPSTEIN, S., and T. MAYEDA, Variation of 0^{18} content of waters from natural sources, *Geochim. Cosmochim. Acta*, 4, 213, 1953. [172]
——, R. P. SHARP, and I. GODDARD, Oxygen isotope ratios in Antarctic snow, firn and ice, *J. Geol.*, 71, 698, 1963. [173]
ERICKSON, B. H., and P. J. GRIM, Profiles of magnetic anomalies South of the Aleutian Island arc, *Geol. Soc. Am. Bull.*, 80, 1387, 1969. [448]
——, F. P. NAUGLER, and W. H. LUCAS, Emperor fracture zone: A newly discovered feature in the Central North Pacific, *Nature*, 225, 53, 1970. [431]
ERIKSSON, K. G., Granulométire des sédiments de l'ile d'Alboran, Méditerrané occidentale, *Bull. Geol. Inst. Univ. Upsala*, 40, 269, 1961. [464]
ERNST, W. G., Tectonic contact between the Franciscan melange and the Great Valley Sequence – crustal expression of a late Mesozoic Benioff zone, *J. Geophys. Res.*, 75, 886, 1970. [439]
ESKOLA, P. E., The problem of mantled gneiss domes, *Quart. J. Geol. Soc. London*, 104, 461, 1948. [67; 483]
ESSON, J., Preliminary report on the geology of the Comores Archipelago, *Proc. Geol. Soc. London*, 1649, 1968. [424]
EWING, J., Seismic model of the Atlantic Ocean, in P. J. HART (ed.), "The Earth's Crust and Upper Mantle," *Am. Geophys. Union Monograph* 13, p. 220, 1969. [82]
EWING, J. I., N. T. EDGAR, and J. W. ANTOINE, Structure of the Gulf of Mexico and Caribbean Sea, in A. E. MAXWELL (ed.), "The Sea," vol. 4, pt. 2, 321, Interscience, 1970d. [474]
EWING, J., and M. EWING, Seismic refraction measurements in the Atlantic Ocean basins in the Mediterranean, on the mid-Atlantic ridge, and in the Norwegian Sea, *Geol. Soc. Am. Bull.*, 70, 291, 1959. [370]
—— and ——, Sediment distribution on the mid-ocean ridge with respect to spreading of the sea-floor, *Science*, 156, 1590, 1967. [81;367;370;429]
——, ——, T. AITKEN, and W. J. LUDWIG, North Pacific sediment layers measured by seismic profiling, in L. KNOPOFF, C. L. DRAKE, and P. J. HART (eds.), "The Crust and Upper Mantle of the Pacific Area," *Am. Geophys. Union Monograph* 12, p. 147, 1968. [429]
EWING, J. I., R. E. HOUTZ, and W. J. LUDWIG, Sediment distribution in the Coral sea, *J. Geophys. Res.*, 75, 1963, 1970c. [456]
EWING, J., M. TALWANI, M. EWING, and T. EDGAR, Sediments of the Caribbean, *Trop. Oceanogr.*, 5, 88, 1967. [437]
EWING, J. I., and G. B. TIREY, Seismic profiler, *J. Geophys. Res.*, 66, 2917, 1961. [370]
EWING, J., C. WINDISCH, and M. EWING, Correlation of horizon A with JOIDES core hole results, *J. Geophys. Res.*, 75, 5645, 1970a. [83]
EWING, M., A. P. CRARY, and H. M. RUTHERFORD, Geophysical investigations in the emerged and submerged Atlantic Coastal Plain, pt I: Methods and results, *Geol. Soc. Am. Bull.*, 48, 753, 1937. [370]
——, S. EITTREIM, M. TRUCHAN, and J. I. EWING, Sediment distribution in the Indian Ocean, *Deep-Sea Res.*, 16, 231, 1969b. [424;425]

——, J. I. EWING, R. E. HOUTZ, and R. LEYDEN, Sediment distribution in the Bellingshausen basin, *Symp. Antarct. Oceangr.* 1161, 89, 1966b. [429]
——, and B. C. HEEZEN, Some problems of Antarctic submarine geology, *Am. Geophys. Union Monograph* 1, p. 75, 1956. [369; 374; 389; 427]
—— and ——, Continuity of the mid-oceanic ridge rift valley in the Southwestern Indian Ocean confirmed, *Science*, 131, 1677, 1960. [369]
——, L. V. HAWKINS, and W. J. LUDWIG, Crustal structure of the Coral Sea, *J. Geophys. Res.*, 75, 1953, 1970. [456]
——, R. HOUTZ, and J. EWING, South Pacific sediment distribution, *J. Geophys. Res.*, 74, 2477, 1969c. [429]
——, W. JARDETSKI, and F. PRESS, "Elastic Waves in Layered Media," McGraw-Hill, 1957. [372]
——, X. LE PICHON, and J. EWING, Crustal structure of the mid-ocean ridges, 4, Sediment distribution in the South Atlantic Ocean and the Cenozoic history of the mid-Atlantic ridge, *J. Geophys. Res.*, 71, 1611, 1966a. [418; 420; 450]
——, W. J. LUDWIG, and J. EWING, Oceanic structural history of the Bering Sea, *J. Geophys. Res.*, 70, 4593, 1965. [448]
——, T. SAITO, J. I. EWING, and L. H. BURCKLE, Lower cretaceous sediments from the Northwest Pacific, *Science*, 152, 751, 1966c. [429]
——, J. L. WORZEL, A. O. BEALL et al., Initial reports of the deep sea drilling project, *Scripps Inst. Oceanogr.*, 1, 1969a. [83; 370; 379; 418; 421]
FAHRIG, W. F., and K. E. EADE, The chemical evolution of the Canadian Shield, *Can. J. Earth Sci.*, 5, 1247, 1968. [94; 96]
——, E. IRVING, and G. D. JACKSON, Paleomagnetism of the Franklin diabases, *Can. J. Earth Sci.*, 8, 455, 1971. [481]
FAIL, J. P., L. MONTADERT, J. R. DELTEIL, P. VALERY, PH. PATRIAT, and R. SCHLICH, Prolongation des zones de fractures de l'ocean Atlantique dans le Golfe de Guinee, *Earth Planet. Sci. Lett.*, 7, 413, 1970. [373]
FAIRBAIRN, H. W., et al., A co-operative investigation of precision and accuracy in chemical, spectrochemical and model analysis of silicate rocks, *U.S. Geol. Surv. Bull.* 980, 1951. [63]
FAIRBORN, J. W., Shear wave velocities in the lower mantle, *Bull. Seismol. Soc. Am.*, 59, 1983, 1969. [46]
FAIRBRIDGE, R. W., Early paleozoic South Pole in northwest Africa, *Geol. Soc. Am. Bull.* 80, 113, 1969. [361]
—— and H. D. STEWART, Jr., Alexa bank, a drowned atoll on the Melanesian border plateau, *Deep-Sea Res.*, 7, 100, 1960. [450]
FAIRHEAD, J. D., and R. W. GIRDLER, How far does the rift system extend through Africa?, *Nature*, 221, 1018, 1969. [398]
FALCON, N. L., I. G. GASS, R. W. GIRDLER, and A. S. LAUGHTON, A discussion on the structure and evolution of the Red Sea and the nature of the Red Sea, Gulf of Aden and Ethiopian rift junction, *Phil. Trans. Roy. Soc. London, Ser. A*, 267, 1, 1970. [402-404]
FALLER, J. E., Results of an absolute determination of the acceleration of gravity, *J. Geophys. Res.*, 70, 4035, 1965. [112]
FARRAND, W. R., Postglacial uplift in North America, *Am. J. Sci.*, 260, 181, 1962. [119-121; 346]
FERMI, E., On the origin of cosmic radiation, *Phys. Rev.*, 75, 1169, 1949. [289]
FINCH, H. F., and B. R. LEATON, The earth's main magnetic field-epoch 1955, *Monthly Notices Roy. Astron. Soc., Geophys. Suppl.* 7, 314, 1957. [292]
FISCHER, A. G., and B. C. HEEZEN, Deep sea drilling project leg 6, *Geotimes*, 14, Oct. 13, 1969. [379; 430; 449]

——, ——, R. E. BOYCE, D. BUKRY, R. G. DOUGLAS, R. E. GARRISON, S. A. KLING, V. KRASHENINNIKOV, A. P. LISITZIN, and A. C. PIMM, Geological history of the Western North Pacific, *Science*, 168, 1210, 1970. [430; 431]

FISHER, O., "Physics of the Earth's Crust," 2d ed., Macmillan, 1889. [346; 389]

FISHER, R. A., Dispersion on a sphere, *Proc. Roy. Soc. London Ser. A,* 217, 295, 1953. [383]

FISHER, R. L., Middle America trench: Topography and structure, *Geol. Soc. Am. Bull,* 72, 703, 1961. [433]

——, The median ridge in the South Central Indian Ocean, in T. N. IRVINE (ed.), "The World Rift System," *Geol. Surv. Canada, Paper* 66-14, 135, 1966. [422]

——, G. L. JOHNSON, and B. C. HEEZEN, Mascarene Plateau, Western Indian Ocean, *Geol. Soc. Am. Bull.,* 78, 1247, 1967. [424; 425]

——, J. G. SCLATER, and D. P. McKENZIE, Evolution of the Central Indian ridge, Western Indian Ocean, *Geol. Soc. Am. Bull.,* 82, 553, 1971 [424; 498]

FITCH, F. J., R. L. GRASTY, and J. A. MILLER, Potassium argon ages of rocks from Jan Mayen and an outline of its volcanic history, *Nature*, 207, 1349, 1965. [409]

FITCH, T. J., Earthquake mechanism in the Himalayan, Burmese and Andaman regions and continental tectonics in Central Asia, *J. Geophys. Res.*, 75, 2699, 1970. [470]

FLEISCHER, R. L., and P. B. PRICE, Decay constant for spontaneous fission of U^{238}, *Phys. Rev.*, 133, 63, 1964. [145]

FLORENSOV, N. A., Rifts of the Baikal mountain region, *Tectonophysics*, 8, 443, 1969. [403]

FLORIN, R., The distribution of conifer and taxad genera in time and space, *Acta Horti Bergiani*, 20, 121, 1963. [361]

FLYNN, K. F., and L. E. GLENDENIN, Half-life and beta spectrum of Rb^{87}, *Phys. Rev.*, 116 744, 1959. [143]

FOSBERG, F. R., Description and occurrence of atoll phosphate rock in Micronesia, *Am J. Sci.*, 255, 584, 1957. [450]

FOWLER, W. A., and W. E. STEPHENS, Origin of the elements, Resource Letter OE-1, *Am J. Phys.*, 36, 289, 1968. [179]

FOX, P. J., W. C. PITMAN III, and F. SHEPARD, Crustal plates in the Central Atlantic: Evidence for at least two poles of rotation, *Science*, 165, 487, 1969. [406]

FRANCHETEAU, J., C. G. A. HARRISON, J. G. SCLATER, and M. L. RICHARDS, Magnetization of Pacific seamounts: A preliminary polar curve for the Northeastern Pacific, *J. Geophys. Res.*, 75, 2035, 1970. [428; 432; 448; 450]

FRANCIS, T. J. F., D. DAVIES, and M. N. HILL, Crustal structure between Kenya and the Seychelles, *Phil. Trans. Roy. Soc. London, Ser. A*, 259, 240, 1966. [425]

FRANK, F. C., Curvature of island arcs, *Nature*, 220, 363, 1968. [352; 501]

FRANK, L. A., On the extraterrestrial ring current during geomagnetic storms, *J. Geophys. Res.*, 72, 3753, 1967. [288]

FRASER, J. Z., D. L. HAWKINS, L. HYDOCK, W. L. CROCKER, M. SCHOENBECHLER, D. A. NEWHOUSE, and T. E. CHASE, Surface sediments and topography of the North Pacific, *Geol. Data Center, Scripps Inst. Oceanogr.*, 1972. [426]

FRENCH, B. M., and N. M. SHORT, Shock Metamorphism of Natural Materials, Mono Book Corp., 1968. [21]

FREUND, R., I. ZAK, and Z. GARFUNKEL, Age and rate of the sinistral movement along the Dead Sea rift, *Nature*, 220, 253, 1968. [405]

FREY, F. A., Rare earth and potassium abundances in St. Paul's rocks, *Earth Planet. Sci. Lett.* 7, 351, 1970. [87]

FRIEDMAN, I., Deuterium content of natural waters and other substances, *Geochim. Cosmochim. Acta*, 4, 89, 1953. [172]

FUKAO, Y., On the radioactive heat transfer and the thermal conductivity in the upper mantle, *Bull. Earthquake Res. Inst.*, 47, 549, 1969. [210]
FUNNELL, B. M., and W. B. RIEDEL (eds.), "The Micropalaeontology of Oceans," Cambridge, 1971. [378]
—— and A. G. SMITH, Opening of the Atlantic Ocean, *Nature*, 219, 1328, 1968. [378; 406; 417]
FURON, R., "The Geology of Africa," Oliver & Boyd, 1963. [402; 424; 494]
GAIBAR-PUERTAS, C., *Observ. del Ebro. Mem.* 11, 1953. [234]
GALANOPOULOS, A. G., and E. BACON, "Atlantis: The Truth Behind the Legend," Nelson, 1969. [464]
GALLI-OLIVER, C., Climate: A primary control of sedimentation in the Peru-Chile trench. *Geol. Soc. Am. Bull.*, 80, 1849, 1969. [433]
GANSSER, A., Geology of the Himalayas, Interscience, 1964. [469]
——, The Indian Ocean and the Himalayas – a geological interpretation, *Eclogae Geol. Helv.*, 59, 831, 1966. [395; 425; 466; 468; 469; 473; 477]
——, The large earthquakes of Iran and their geological framework, *Eclogae Geol. Helv.*, 62, 443, 1969. [468]
GANTAR C., C. MORELLI, and M. PISANI, Experimental study of the response of the Graf-Askania Gss 2, No. 13, sea gravity meter, *J. Geophys. Res.*, 67, 4411, 1962. [107]
GAPOSHKIN, E. M., and K. LAMBECK, New geodetic parameters for a standard earth, *J. Geophys. Res.*, 76, 4855, 1971. [123]
GARDINER, J. S., The reefs of the Western Indian Ocean, pt. 1-2, *Trans. Linn. Soc. London*, 19, 393 and 426, 1926-1936. [425]
GARDNER, J. V., Submarine geology of the Western Coral Sea, *Geol. Soc. Am. Bull.*, 81, 2599, 1970. [456]
GARRELS, R. M., and F. T. MACKENZIE, Sedimentary rock types: Relative proportions as a function of geological time, *Science*, 163, 570, 1969. [477]
GASKELL, T. F., and J. C. SWALLOW, Seismic experiments on two Pacific atolls, *Challenger Soc. Occas. Paper* 3, 1, 1954. [450]
GASS, I. G., Geochronology of the Tristan da Cunha group of islands, *Geol. Mag.*, 104, 160, 1967. [421]
——, Is the Troodos Massif of Cyprus a fragment of Mesozoic ocean floor? *Nature*, 220, 39, 1968. [85; 86; 466]
—— and D. MASSON-SMITH, The geology and gravity anomalies of the Troodos Massif, Cyprus, *Phil. Trans. Roy. Soc. London Ser. A*, 225, 417, 1963. [465]
GAST, P. W., Dispersal elements in oceanic volcanic rocks, *Phys. Earth Planct. Int.*, 3, 246, 1970. [87]
GASTIL, G., The distribution of mineral dates in time and space, *Am. J. Sci.*, 258, 1, 1960. [319; 492]
GATES, G. O., and W. M. GIBSON, Interpretation of the configuration of the Aleutian ridge, *Geol. Soc. Am. Bull.*, 67, 127, 1956. [447]
GEES, R. A., K-Ar ages of two basalts from Bermuda, *Eclogae Geol. Helv.*, 63, 93, 1970. [418]
GEIJER, P., Precambrian atmosphere: Evidence from the Precambrian of Sweden, *Geochim Cosmochim, Acta*, 10, 304, 1956. [60]
GELLETICH, H., Über magnetitfuhrende eruptive Gange und Gangesysteme im mittleven Teil des sudlichen Transvaals, *Beitr. Angew. Geophys.*, 6, 337, 1937. [243]
GERARD, R., M. G. LANGSETH, and M. EWING, Thermal gradient measurements in the water and bottom sediments of the Western Atlantic, *J. Geophys. Res.*, 67, 785, 1962. [184; 372]
GERVASIO, F. C., Age and nature of orogenesis of the Philippines, *Tectonophysics*, 4, 379, 1967. [454]

GIBSON, I. L., A comparative account of the flood basalt volcanism of the Columbia Plateau and Eastern Iceland, *Bull. Volcanol.*, 33, pt. II, 419, 1969. [66]

GIBSON, R. D., and P. H. ROBERTS, The Bullard-Gellman dynamo, in S. K. RUNCORN (ed.), "The Application of Modern Physics to the Earth and Planetary Interiors," p. 577, Wiley, 1969. [236]

GIBSON, T. G., and K. M. TOWE, Eocene volcanism and the origin of horizon A, *Science*, 172, 152, 1971. [418]

GIBSON, W. M., and H. NICHOLS, Configuration of the Aleutian Ridge, Rat Islands, Seisopchnoi to West Buildir Island, *Geol. Soc. Am. Bull.*, 64, 1173, 1953. [447]

GIGL, P. D., and F. DACHILLE, Effect of pressure and temperature on the reversal transitions of stishovite, *Meteoritics*, 4, 123, 1968. [19]

GILBERT, G. K., Lake Bonneville, *U.S. Geol. Surv. Monograph* 1, 483, 1890. [472]

GILL, J. E., in M. E. WILSON (ed.), "Structural Geology of Canadian Ore Deposits," Can. Inst. Min. Met., Montreal, p. 29, 1948. [481]

——, Natural divisions of the Canadian Shield, *Trans. Roy. Soc. Canada*, 43, 61, 1949. [481]

GILL, W. D., The Mediterranean Basin, in D. C. ION (ed.), Salt Basins Around Africa, *Inst. Petrol.*, p. 101, 1965. [464]

GILLILAND, W. N., Possible continental continuation of the Mendocino fracture zone, *Science*, 137, 685, 1962. [372; 444]

——, Extension of the theory of zonal rotation to explain global fracturing, *Nature*, 202, 1276, 1964. [444]

GILLULY, J. (ed.), Origin of granite, *Geol. Soc. Am. Mem.*, 28, 1948. [488]

——, Geologic contrasts between continents and ocean basins, in A. POLDERVAART (ed.), "Crust of the Earth," *Geol. Soc. Am. Spec. Paper,* 62, 7, 1955. [90; 156]

——, The tectonic evolution of the Western United States, *Quart. J. Geol. Soc. London*, 119, 133, 1963. [438; 446]

——, Oceanic sediment volumes and continental drift, *Science,* 166, 992, 1969. [438; 439; 446]

——, J. C. CREED, Jr., and W. M. CADY, Sedimentary volumes and their significance, *Geol. Soc. Am. Bull.*, 81, 353, 1970. [446]

GILMAN, R. C., On the composition of circumstellar grains, *Astrophys. J. Letters,* 155, L185, 1969. [314]

GILVARRY, J. J., Equation of the fusion curve, *Phys. Rev.*, 102, 325, 1956. [205]

——, Temperatures in the Earth's interior, *J. Atmospheric Terres. Phys.*, 1, 84, 1957. [205]

GINZBURG, V. L., and S. I. SYROVATSKII, "The Origin of Cosmic Rays," Pergamon, 1964. [289]

GIRDLER, R. W., A review of terrestrial heat flow, in S. K. RUNCORN (ed.), "Mantles of the Earth and Terrestrial Planets," p. 549, Interscience, 1967. [199]

——, J. D. FAIRHEAD, R. C. SEARLE, and W. T. C. SOWERBUTTS, Evolution of rifting in Africa, *Nature*, 224, 1178, 1969. [401]

GJELSVIK, T., Volcano or Jan Mayen alive again, *Nature*, 228, 352, 1970. [409]

GLAESSNER, M. F., Geographic distribution and time range of the Ediocara Precambrian fauna, *Geol. Soc. Am. Bull.*, 82, 509, 1971. [477]

—— and C. TEICHERT, Geosynclines: A fundamental concept in geology, *Am. J. Sci.*, 245, pt. I, 465; pt. II, 571, 1947. [472]

GLANGEAUD, L. J., J. ALINAT, J. POLVECHE, A. GUILLAUME, and O. LEENHARDT, Grandes structures de la mer Ligure, leur évolution et leurs rélations avec les chaines continentales, *Soc. Géol. France*, 7, 921, 1967. [462]

GLENNIE, E. A., Report on the values ot gravity in the Maldive and Laccadive Islands, John Murray Expedition, 1933-1934, *Sci. Reports and Topo.*, 4, 95, 1936. [425]

GLIKSON, A. Y., Geosynclinal evolution and geochemical affinities of early Precambrian systems, *Tectonophysics*, 9, 397, 1970. [483; 487]

GODBY, E. A., P. J. HOOD, and M. E. BOWER, Aeromagnetic profiles across the Reykanes ridge, southwest coast of Iceland, *J. Geophys. Res.*, 73, 7637, 1968. [367]
GOGUEL, J., L'interpretation de l'arc des Alpes occidentales, *Bull. Soc. Geol. France*, 7, 20, 1964. [360]
GOLD, T., Instability of the Earth's axis of rotation, *Nature*, 175, 526, 1955. [252; 253]
——, Irregular changes in the rotation of the Earth, *Observatory*, 76, 96, 1956. [253]
GOLDREICH, P., and A. TOOMRE, Some remarks on polar wandering, *J. Geophys. Res.*, 74, 2555, 1969. [253]
GOLDSCHMIDT, V. M., Norske Videnskaps-Akad. Skrifter, *Math-Naturv. Kl.*, 4, 1937. [16]
GOODWIN, A. M., Evolution of the Canadian Shield, *Proc. Geol. Assoc. Can.*, 19, 1, 1968. [483]
GORAI, M., Some geological problems in the development of Japan and the neighbouring island arcs, in L. KNOPOFF, C. L. DRAKE, and P. J. HART (eds.), "The Crust and Upper Mantle of the Pacific Area," *Am. Geophys. Union Monograph* 12, p. 481, 1968. [451]
GORSHKOV, G. S., Volcanism and the Upper Mantle: Investigations in the Kurile Arc, Plenum, p. 385, 1970. [451]
GOUGH, D. I., and M. PORATH, Long-lived thermal structure under the Southern Rocky Mountains, *Nature*, 226, 837, 1970. [445]
GOULD, S. J., Is uniformitarianism necessary? *Am. J. Sci.*, 263, 223 and 919, 1965. [343]
GRABAU, A. W., "Palaeozoic Formations in the Light of the Pulsation Theory," University Press, National University of Peking, 1936. [128; 344; 361; 473; 474]
GRAESER, S., Isotopic composition of lead in some basic and ultrabasic rocks from the Alps, *Earth Planet. Sci. Lett.*, 6, 491, 1969. [466]
GRAHAM, J. W., The stability and significance of magnetism in sedimentary rocks, *J. Geophys. Res.*, 54, 131, 1949. [244; 383]
——, Evidence of polar shift since Triassic time, *J. Geophys. Res.*, 60, 329, 1955. [383]
GRAHAM, K. W. T., and A. L. HALES, Surface-ship gravity measurements in the Agulhas bank area, south of South Africa, *J. Geophys. Res.*, 70, 4005, 1965. [424]
GRANT, A. C., Recent crustal movements on the Labrador shelf, *Can. J. Earth Sci.*, 7, 571, 1970. [415]
GRASTY, R. L., Orogeny, a cause of world-wide regression of the seas, *Nature*, 216, 779, 1967. [77; 491]
—— and J. T. WILSON, Ages of Florida volcanics and the opening of the Atlantic, *Trans. Am. Geophys. Union*, 48, 212, 1967. [417]
GREEN, D. H., Origin of basaltic magmas, in H. H. HESS and A. POLDERVAART, (eds.), "Basalts: The Poldervaart Treatise on Rocks of Basaltic Composition," vol. 2, p. 835, Interscience, 1968. [85]
—— and A. E. RINGWOOD, Mineral assemblages in a model mantle composition, *J. Geophys. Res.*, 68, 937, 1963. [303]
—— and ——, An experimental investigation of the gabbro to eclogite transformation and its petrological applications, *Geochim. Cosmochim. Acta,* 31, 767, 1967. [298]
GREEN, R. W. E., and A. L. HALES, Seismic refraction measurements in the Southwestern Indian Ocean, *J. Geophys. Res.*, 71, 1637, 1966. [424]
GREEN, T. H., D. H. GREEN, and A. E. RINGWOOD, The origin of high-alumina basalts and their relationship to quartz tholeiites and alkali basalts, *Earth Planet. Sci. Lett.*, 2, 41, 1967. [87]
—— and A. E. RINGWOOD, High pressure experimental studies on the origin of Andesites, in Proceedings of the Andesite Conference, *Intern. Upper Mantle Proj. Sci. Report.*, 16, 13, 1969. [90]
GREGORY, J. W., The African rift valleys, *Geogr. J.*, 56, 13, 1920. [398]

GRIFFITHS, D. H., and P. F. BAKER, Marine geophysics of the Scotia ridge and Scotia sea, *Brit. Antarct. Surv. Bull.*, 12, 93, 1967. [421]

——, R. F. KING, M. A. KHAN, and D. J. BLUNDELL, Seismic refraction line in the Gregory rift, *Nature*, 229, 69, 1971. [401]

——, R. P. RIDDIHOUGH, H. A. CAMERON, and P. KENNETT, Geophysical investigation of the Scotia arc, *Brit. Antarct. Surv., Sci. Rep.* 46, 1964. [421]

GRIGGS, D. T., A theory of mountain building, *Am. J. Sci.*, 237, 611, 1939. [346]

—— and J. W. HANDIN, Rock deformation, *Geol. Soc. Am. Mem.* 79, 1960. [340]

GRIM, P. J., Heat flow measurements in the Tasman Sea, *J. Geophys. Res.*, 74, 3933, 1969. [459]

—— and B. H. ERICKSON, Fracture zones and magnetic anomalies south of the Aleutian trench, *J. Geophys. Res.*, 74, 1488, 1969. [431]

—— and F. P. NAUGLER, Fossil deep-sea channel on the Aleutian abyssal plain, *Science*, 163, 383, 1969. [448]

GRINGAUZ, K. I., V. G. KURT, V. I. MOROG, and I. C. SKLOVSKIY, Sbornik "Iskusstvennye Sputniki Zemli," *Izv. AN. SSSR*, 6, 108, 1961. [282;283]

GROUT, F. F., Origin of Granite, in J. GILLULY (ed.), Origin of Granite, *Geol. Soc. Am. Mem.*, 28, 45, 1948. [95;488]

GROW, J. A., and T. ATWATER, Mid-Tertiary tectonic transition in the Aleutian arc, *Geol. Soc. Am. Bull.*, 81, 3715, 1970. [448]

GUBIN, I. E., Lecture notes on basic problems in seismotectonics, *Intern. Inst. Seismol Earthquake Engn.*, Tokyo, Japan, 1967. [470]

GUIER, W. H., "Recent Progress in Satellite Geodesy," Applied Physics Laboratory, Johns Hopkins, 1965. [199]

GUILBERT, J. M., and J. S. SUMNER, Distribution of porphyry copper deposits in the light of recent tectonic advances, "Southern Arizona Guidebook III," *Ariz. Geol. Soc.*, p. 97, 1963. [446]

GUTARENKO, L. A., K. A. KOSSOVA, A. V. STAKLO, Yu A. TARAKANOV, and V. V. FEDYNSKII, A new method for measuring the force of gravity at sea, *Phys. Solid Earth. Bull. Acad. Sci. USSR*, 12, 800, 1967. [107]

GUTENBERG, B., Changes in sea level, postglacial uplift and mobility of the Earth's interior, *Geol. Soc. Am. Bull.* 52, 721, 1941. [118]

—— and C. F. RICHTER, "Seismicity of the Earth and Associated Phenomena," 2d ed., Princeton, 1954. [373]

HAARMANN, E., "Die Oscillationstheorie," Ferdinand Enke Verlag, 1930. [344]

HADDON, R. A. W., and K. E. BULLEN, An Earth model incorporating free Earth oscillation data, *Phys. Earth Plant. Int.*, 2, 35, 1969. [33;43;47]

HALBOUTY, M. T., "Salt Domes – Gulf Region, United States and Mexico," Gulf, 1967. [418]

HALES, A. L., Gravitational sliding and continental drift, *Earth Planet. Sci. Lett.*, 6, 31, 1969. [493;496]

—— and J. L. ROBERTS, Shear velocities in the lower mantle and the radius of the core, *Bull. Seismol. Soc. Am.*, 60, 1427, 1970. [33]

HALL, J., Paleontology of New York, Natural History Survey, vol. 3, pt. I, *N. Y. Geol. Surv.*, 1859. [81;472]

——, Contribution to the geological history of the American continent, *Am. Assoc. Adv. Sci. Proc.*, 31, 29, 1883. [472]

HAMILTON, E. L., Sunken islands of the Mid-Pacific mountains, *Geol. Soc. Am. Mem.*, 64, 97, 1956. [378;431]

HAMILTON, R. M., and A. W. GALE, Thickness of the mantle seismic zone beneath the North Island of New Zealand, *J. Geophys. Res.*, 74, 1608, 1969. [458]

HAMILTON, W. B., Mesozoic California and the underflow of the Pacific mantle, *Geol. Soc. Am. Bull.*, 80, 2409, 1969. [443]

—— and W. B. MYERS, The nature of batholiths, *U.S. Geol. Surv. Prof. Paper* 554-C, 1967. [67; 95]

—— and ——, Cenozoic tectonic relationships between the Western United States and the Pacific basin, in *Proc. Conf. Geol. Problems, San Andreas Fault System*, Stanford University Publ. Geol. Sci., no. 2, 342, 1968. [443]

HANKS, T. C., and D. L. ANDERSON, The early thermal history of the Earth, *Phys. Earth Planet. Int.*, 2, 19, 1969. [213; 318]

HANSON, W. B., Structure of the ionosphere in F. S. JOHNSON (ed.), "Satellite Environment Handbook," Stanford, 1965. [278]

HARGRAVES, R. B., Paleomagnetic evidence relevant to the origin of the Vredefort ring, *J. Geol.*, 78, 253, 1970. [402]

HARKER, A., The Tertiary igneous rocks of Skye, *Geol. Surv. U.K. Mem.*, Sheets 70, 71, 1904. [413]

HARLAN, R. B., Eötvös corrections for airborne gravimetry, *J. Geophys. Res.*, 73, 4675, 1968. [116]

HARLAND, W. B., Contribution of Spitsbergen to understanding of tectonic evolution of North Atlantic region, in M. KAY (ed.), "North Atlantic – Geology and Continental Drift, a Symposium," *Am. Assoc. Petrol. Geol. Mem.*, 12, 817, 1969. [409]

——, A. G. SMITH, and B. WILCOCK (eds.), Geological Society Phanerozoic Time Scale 1964, *Quart. J. Geol. Soc. London*, 120S (suppl. vol.), 260, 1964. [126; 131]

HARRINGTON, HILARY J., Geology and morphology of Antarctica, in "Biogeography and Ecology in Antarctica," *Monograph Biol.*, 15, 1, 1965. [422]

——, B. L. WOOD, I. C. McKELLAR, and G. L. LENSEN, Topography and geology of the Cape Hallett district, Victoria Land, Antarctica, *New Zealand Geol. Surv. Bull.*, 80, 1967. [422]

HARRINGTON, HORACIO J., Paleogeographic development of South America, *Am. Assoc. Petrol. Geol. Bull.*, 46, 1773, 1962. [434; 435]

——, Deep focus earthquakes in South America and their possible relation to continental drift in "Polar Wandering and Continental Drift," *Soc. Econ. Paleon. Mineral. Spec. Publ.* 10, 55, 1963. [435]

HARRIS, P. G., Basalt type and Africa rift valley tectonism, *Tectonophysics*, 8, 427, 1969. [402; 496]

HARRISON, J. C., An interpretation of gravity anomalies in the Eastern Mediterranean, *Phil. Trans. Roy. Soc. London Ser. A*, 248, 283, 1955. [465]

HART, P. J. (ed.), "The Earth's Crust and Upper Mantle," *Am. Geophys. Union Monograph* 13, 1969. [427]

HART, S. R., Isotope geochemistry of crust-mantle processes, in P. J. HART (ed.), "The Earth's Crust and Upper Mantle," *Am. Geophys. Union Monograph* 13, p. 58, 1969. [85]

——, K, Rb, Cs, Sr and Ba contents and Sn isotope ratios of ocean floor basalts, *Phil. Trans. Roy. Soc. London Ser. A*, 268, 573, 1971. [87]

——, C. BROOKS, T. E. KROGH, G. L. DAVIS, and D. NAVA, Ancient and modern volcanic rocks: a trace element model, *Earth Planet. Sci. Lett.*, 10, 17, 1970. [483; 497]

—— and G. L. DAVIS, Zircon U-Pb and whole rock Rb-Sr ages and early crustal development near Rainy Lake, Ontario, *Geol. Soc. Am. Bull.*, 80, 595, 1969. [150]

—— and J. S. STEINHART, Terrestrial heat-flow: Measurement in lake bottoms, *Science*, 149, 1499, 1965. [184]

HASKELL, N. A., The viscosity of the asthenosphere, *Am. J. Sci.*, 33, 22, 1937. [346]

HATHERTON, T., E. W. DAWSON, and F. C. KINSKY, Balleny Islands reconnaissance expedition 1964, *New Zealand J. Geol. Geophys.*, 8, 164, 1965. [422; 432]

HATTEN, C. W., and A. A. MEYERHOFF, The Caribbean area: A case of destruction and regeneration of a continent: Discussion and reply by V. Skvor, *Geol. Soc. Am. Bull.*, 81, 1855, 1970. [437]

HAUGHTON, S. H., "Geological History of Southern Africa," *Geol. Soc. South Africa*, Cape Town, 1969. [475]

HAWKINS, G. S., "Meteors, Comets and Meteorites," McGraw-Hill, 1964. [13]

HAYES, D. E., A geophysical investigation of the Peru-Chile trench, *Mar. Geol.*, 4, 309, 1966. [433; 444]

—— and W. J. LUDWIG, The Manila trench and West Luzon trough/gravity and magnetic measurements, *Deep-Sea Res.*, 14, 545, 1967. [454]

—— and W. C. PITMAN III, Marine geophysics and sea-floor spreading in the Pacific-Antarctic area: A review, *Antarct. J. U.S.*, 5, 70, 1970. [427]

HAYS, J. D., Deep sea drilling project leg 9, *Geotimes*, 15, 10, 1970. [81; 429]

—— (ed.), Geological Investigations of the North Pacific, *Geol. Soc. Am. Mem.* 126, 1971. [447]

——, T. SAITO, N. D. OPDYKE, and L. H. BURCKLE, Pliocene-Pleistocene sediments of the equatorial Pacific: Their paleomagnetic, biostratigraphic and climatic record, *Geol. Soc. Am. Bull.*, 80, 1481, 1969. [429]

HEATH, G. R., Carbonate sedimentation in the abyssal equatorial Pacific during the past 50 million years, *Geol. Soc. Am. Bull.*, 80, 689, 1969. [428]

HEDBERG, H. D., Continental margins from the viewpoint of the petroleum geologist, *Amer. Assoc. Pet. Geol. Bull.*, 54, 3, 1970. [79; 417]

HEEZEN, B. C., The rift in the ocean floor, *Sci. Am.*, 203, 98, 1960. [345; 382]

——, The deep-sea floor, in S. K. RUNCORN (ed.), "Continental Drift," chap. 9, p. 256, fig. 16, Academic, 1962. [369]

——, The Atlantic continental margin, *Univ. Missouri at Rolla J.*, 1, 5, 1968. [417]

——, E. T. BUNCE, J. B. HERSEY, and M. THARP, Chain and Romanche fracture zones, *Deep-Sea Res.*, 11, 11, 1964. [372; 421]

—— and C. L. DRAKE, Grand Banks slump, *Bull. Am. Assoc. Pet. Geol.*, 48, 221, 1964. [83]

——, D. B. ERICSON, and W. M. EWING. Turbidity currents and submarine slumps and the 1929 Grand Banks (Newfoundland): Further evidence, *Deep-Sea Res.*, 1, 193, 1954. [84]

——, and M. EWING, Orleansville earthquake and turbidity currents, *Bull. Am. Assoc. Pet. Geol.*, 39, 2504, 1955. [462]

—— and ——, The mid-oceanic ridge and its extension through the Arctic basin in G. O. RAASCH (ed.), "Geology of the Arctic," University of Toronto Press, p. 622, 1961. [408]

——, ——, and E. T. MILLER, Trans-Atlantic profile of total magnetic intensity and topography, Dakar to Barbados, *Deep-Sea Res.*, 1, 25, 1953. [386]

——, B. GLASS, and H. W. MENARD, The Manihiki plateau, *Deep-Sea Res.*, 13, 445, 1966b. [450]

——, C. GRAY, A. G. SEGRE, and E. F. K. ZARUDSKI, Evidence of foundered continental crust beneath the Central Tyrrhenian Sea, *Nature*, 229, 327, 1971. [464]

——, C. D. HOLLISTER, and W. F. RUDDIMAN, Shaping of the continental rise by deep geostrophic contour currents, *Science*, 152, 502, 1966a. [82; 83]

—— and G. L. JOHNSON, The south Sandwich trench, *Deep-Sea Res.*, 12, 185, 1965. [421]

——, ——, and C. D. HOLLISTER, The Northwest Atlantic mid-ocean canyon, *Can. J. Earth Sci.*, 7, 1441, 1970a. [413]

—— and I. P. KOSMINSKAYA (eds.), The structure of the crust and mantle beneath inland and marginal seas, *Tectonophysics*, 10, 473, 1970b. [462]

—— and A. S. LAUGHTON, Abyssal plains, in M. N. HILL (ed.), "The Sea"; Ideas and observations on progress in the study of the seas, vol. 3, p. 312, Interscience, 1963. [83]

—— and J. E. NAFE, Vema trench: Western Indian Ocean, *Deep-Sea Res.*, 11, 79, 1964. [423; 425]

—— and M. THARP, Physiographic diagram of the Indian Ocean, *Geol. Soc. Am. Spec. Publ.*, 1965. [421; 422]

—— and ——, Physiography of the Indian Ocean, *Phil. Trans. Roy Soc. London, Ser. A*, 259, 137, 1966. [421]

—— and ——, Indian Ocean Floor, map supplement, *Natl. Georgr. Mag.*, 132, no. 4, 1967. [421]

—— and ——, Pacific Ocean Floor, map supplement, *Natl. Geogr. Mag.*, 136, no. 4, 1969. [426]

—— and ——, Physiographic diagram of the Western Pacific Ocean, *Geol. Soc. Am.*, 1971. [426]

——, ——, and M. EWING, The floors of the ocean: I, the North Atlantic, *Geol. Soc. Am. Spec. Paper*, 65, 1959. [367; 373; 413]

HEIM, A., "Untersuchungen über den Mechanismus der Gebirgsbildung in Anschluss an die geologische Monographie der Todi-Windgallen Gruppe," Schwabe, 1878. [360]

HEIRTZLER, J. R., G. O. DICKSON, E. M. HERRON, W. C. PITMANN III, and X. LE PICHON, Marine magnetic anomalies, geomagnetic field reversals and motions of the ocean floor and continents, *J. Geophys. Res.*, 73, 2119, 1968. [81; 388; 389; 392; 413]

—— and M. L. HADLEY, Magnetic anomaly over Vema seamount, *Nature*, 212, 912, 1966. [421]

—— and D. E. HAYES, Magnetic boundaries in the North Atlantic Ocean, *Science*, 157, 185, 1967. [417]

——, X. LE PICHON, and J. G. BARON, Magnetic anomalies over the Reykjanes ridge, *Deep-Sea Res.*, 13, 427, 1966. [388]

HEKINIAN, R., Rocks from the mid-oceanic ridge in the Indian Ocean, *Deep-Sea Res.*, 15, 195, 1968. [424]

HELMERT, F. R., "Die mathematischen und physikalischen Theorien der höheren Geodäsie," vol. II, Teubner, 1884. [108, 116; 514]

HELSLEY, C. H., and M. B. STEINER, Evidence for long intervals of normal polarity during Cretaceous period, *Earth Planet. Sci. Lett.*, 5, 325, 1969. [389]

HENDRICKS, S. J., and J. C. CAIN, Magnetic field data for trapped particle evaluations, *J. Geophys. Res.*, 71, 346, 1966. [227]

HEPPLE, P. (ed.), The Exploration for Petroleum in Europe and North Africa, *Inst. Pet. London.*, 1969. [462; 464; 466]

HEPPNER, J. P., The world magnetic survey, *Space Sci. Rev.*, 11, 315, 1963. [227]

HERCZEG, T., Planetary cosmogonics, in A. BEER (ed.), "Vistas in Astronomy," vol. 10, p. 175, Pergamon, 1968. [13]

HERMAN, Y., Cretaceous, Paleocene and Pleistocene sediments from the Indian Ocean, *Science*, 140, 1316, 1963. [424]

HERMES, J. J., The Papuan geosyncline and the concept of geosynclines, *Geol. Mijnbouw.*, 47, 81, 1968. [456]

HERRIN, E., Regional variations of P-wave velocity in the upper mantle beneath North America, in P. J. HART (ed.), "The Earth's Crust and Upper Mantle," *Am. Geophys. Union Monograph* 13, p. 242, 1969. [445]

—— and J. TAGGART, Regional variations in P_n velocity and their effect on the location of epicenters, *Bull. Seismol. Soc. Am.*, 52, 1037, 1966. [445]

HERRON, E. M., and D. E. HAYES, A geophysical study of the Chile Ridge, *Earth Planet. Sci. Lett.*, 6, 77, 1969. [435]

—— and J. R. HEIRTZLER, Sea-floor spreading near the Galapagos, *Science*, 158, 775, 1967. [437]

HERSEY, J. B., Sedimentary basins of the Mediterranean Sea, in W. F. WHITTARD and R. BRADSHAW (eds.), "Submarine Geology and Geophysics" (Colston Papers, vol. 17), 75, 1965. [462]

—— (ed.), "Deep-Sea Photography," Johns Hopkins, 1967. [370]

—— and M. EWING, Seismic reflections from beneath the ocean floor, *Trans. Am. Geophys. Union*, 30, 5, 1949. [370]

HERZENBERG, A., Geomagnetic dynamos, *Phil. Trans. Roy. Soc. London, Ser. A,* 250, 543, 1958. [236]

HESS, H. H., Drowned ancient islands of the Pacific basin, *Am. J. Sci.*, 244, 772, 1946. [372; 377; 430; 450]

——, Major structural features of the Western North Pacific, an interpretation of H. O. 5485, bathymetric chart, Korea to New Guinea, *Geol. Soc. Am. Bull.* 59, 417, 1948. [372]

——, Serpentine orogeny and epeirogeny, in A. POLDERVAART (ed.), "Crust of the Earth," *Geol. Soc. Am. Spec. Paper* 62, 391, 1955. [456]

——, The AMSOC hole to the Earth's mantle, *Trans. Am. Geophys. Union*, 40, 340, 1959. [295]

——, History of ocean basins, in A. E. J. ENGEL, H. L. JAMES, and B. F. LEONARD (eds.), "Petrological Studies – a Volume in Honour of A. F. Buddington," *Geol. Soc. Am.*, 1962. [85; 295; 331; 366; 378; 391; 427; 436; 449]

——, Seismic anistropy of the uppermost mantle under oceans, *Nature*, 203, 629, 1964. [430]

——, Mid-oceanic ridges and tectonics of the sea-floor, in W. F. WHITTARD, and R. BRADSHAW (eds.), "Submarine Geology and Geophysics" (Colston Papers, vol. 17), 317, 1965. [88; 449]

—— (ed.), Caribbean Geological Investigations, *Geol. Soc. Am., Mem.* 98, 1966. [437]

—— and A. POLDERVAART (eds.), "Basalts: The Poldervaart Treatise on Rocks of Basaltic Composition," 2 vols., Interscience, 1967 and 1968. [73]

HIDE, R., Free hydromagnetic oscillations of the Earth's core and the theory of the geomagnetic secular variation, *Phil. Trans. Roy. Soc. London, Ser. A*, 259, 615, 1966. [238; 248]

HIGGS, D. V., Quantitative areal geology of the United States, *Am. J. Sci.*, 247, 575, 1949. [95; 156; 477]

HILL, M. L., and T. W. DIBBLEE, Jr., San Andreas, Garlock and Big Pine faults, California – A study of the character, history and tectonic significance of their displacements, *Geol. Soc. Am. Bull.*, 64, 443, 1953. [331; 447]

HILL, M. N. (ed.), "The Sea: Ideas and Observations on Progress in the Study of the Seas," vol. 3, Interscience, 1963. [370]

HO, C. S., Structural evolution of Taiwan, *Tectonophysics*, 4, 367, 1967. [454]

HOFMANN, R. B., Recent changes in California fault movement, in W. R. DICKINSON, and A. GRANTZ (eds.), "Proceedings of Conference on Geologic Problems of the San Andreas Fault System," Stanford University Publ. Geol. Sci. no. 11, 89, 1968. [364]

HOLMBERG, E. R. R., A suggested explanation of the present value of the velocity of rotation of the earth, *Monthly Notices Roy. Astron. Soc., Suppl.*, 6, 325, 1952. [103]

HOLMES, A., Radioactivity in the Earth and the Earth's thermal history, *Geol. Mag.*, 2, 60 and 102, 1915. [209]

——, Radioactivity and earth movements, *Trans. Geol. Soc. Glasgow*, 1928-1929, 18, 559, 1931. [346; 353; 389]

——, The sequence of Precambrian orogenic belts in South and Central Africa, *18th Intern. Geol. Cong. Gt. Brit.* pt. XIV, 254, 1948. [481]
——, The oldest dated minerals of the Rhodesian shield, *Nature*, 173, 612, 1954. [137]
——, A revised geological time-scale, *Trans. Edinburgh Geol. Soc.*, 17, pt. III, 183, 1960. [126; 131; 156]
——, "Principles of Physical Geology," 2d ed., Nelson, 1965. [61; 342; 353; 425; 492]
HOLSER, W. T., and I. R. KAPLAN, Isotope geochemistry of sedimentary sulphates, *Chem. Geol.*, 1, 93, 1966. [171]
HOLTEDAHL, O., On the morphology of the West Greenland shelf with general remarks on the "marginal channel" problem, *Mar. Geol.*, 8, 155, 1970. [412]
HOOD, P., Flemish cap, Galacia bank and continental drift, *Earth Planet. Sci. Lett.*, 1, 205, 1966. [415]
HOPKINS, D. M., "The Bering Land Bridge," Stanford, 1967. [408]
——, D. W. SCHOLL, W. O. ADDICOTT, R. L. PIERCE, P. B. SMITH, J. A. WOLE, D. GERSHANOVICH, B. KOTENEV, K. E. LOHMAN, J. H. LIPPS, and J. OBRADOVICH, Cretaceous, Tertiary and Early Pleistocene rocks from the continental margin in the Bering Sea, *Geol. Soc. Am. Bull.*, 80, 1471, 1969. [448]
HORAI, K., M. CHESSMAN, and G. SIMMONS, Heat flow measurements on the Reykjanes ridge, *Nature*, 225, 264, 1970. [193; 194]
—— and G. SIMMONS, Spherical harmonic analysis of terrestrial heat flow, *Earth Planet. Sci. Lett.*, 6, 386, 1969. [186-188; 199]
—— and S. UYEDA, Terrestrial heat flow in Japan, *J. Geophys. Res.*, 69, 2121, 1964. [193; 195]
HORN, D. R., M. N. DERLACH, and B. M. HORN, Distribution of volcanic ash layers and turbidites in the North Pacific, *Geol. Soc. Am. Bull.*, 80, 1715, 1969. [448]
HORSFIELD, W. T., and P. I. MATOU, Transform faulting along the De Geer line, *Nature*, 226, 256, 1970. [408]
HOSPERS, J., Remanent magnetism of rocks and the history of the geomagnetic field, *Nature*, 168, 1111, 1951. [243; 383]
——, Reversals of the main geomagnetic field, *Koninkl. Ned. Akad. Wetenschap. Proc., Ser. B*, pt. I: 56, 467, 1953; pt. II: 56, 477, 1953; pt. III: 57; 112, 1954. [243]
—— and S. I. VAN ANDEL, Paleomagnetic data from Europe and Northern America and their bearing on the origin of the North Atlantic Ocean, *Tectonophysics*, 6, 475, 1968. [413]
—— and ——, Paleomagnetic and tectonics: A review, *Earth Sci. Rev.*, 5. 5. 1969. [386; 451; 464]
HOTZ, E. F. (ed.), Guide to the geology and culture of Greece, *Petrol. Explor. Soc. Libya, 7th Ann. Field Conf.* 1965, Drukkerij, Holland N.V., Amsterdam, 1965. [462; 465]
HOUTZ, R. E., J. I. EWING, and X. LE PICHON, Velocities of deep-sea sediments and sonobuoy data, *J. Geophys. Res.*, 73, 2615, 1968. [83]
HOWELL, F. C., Remains of hominidae from Pliocene/Pleistocene formations in the lower basin, Ethiopia, *Nature*, 223, 1234, 1969. [75]
HOYLE, F., "Frontiers of Astronomy," Heinemann, 1955. [11]
——, On the origin of the solar nebula, *Quart. J. Roy. Astr. Soc.*, 1, 28, 1960. [11]
HUBBERT, M. K., Theory of scale models as applied to the study of geologic structures, *Geol. Soc. Am. Bull.*, 48, 1459, 1937. [339]
——, Mechanical basis for certain familiar geologic structures, *Geol. Soc. Am. Bull.*, 62, 355, 1951. [339]
HUNKINS, K., T. HERRON, H. KUTSCHALE, and G. PETER, Geophysical studies of the Chukchi cap, Arctic Ocean, *J. Geophys. Res.*, 67, 235, 1962. [408]
HURLEY, P. M., The confirmation of continental drift, *Sci. Am.*, 218, 52, 1968. [358; 359]
——, Distribution of age provinces in Laurasia, *Earth Planet. Sci. Lett.*, 8, 189, 1970. [358]
—— and J. R. RAND, Pre-drift continental nuclei, *Science*, 164, 1229, 1969. [420]

HURWITZ, L., D. G. KNAPP, J. H. NELSON, and D. E. WATSON, Mathematical model of the geomagnetic field, *J. Geophys. Res.*, 71, 2373, 1966. [226]
HUTCHINSON, R. D., Stratigraphy and trilobite faunas of the Cambrian sedimentary rocks of Cape Breton Island, Nova Scotia, *Geol. Surv. Canada Mem.* 263, 52, 1952. [473;474]
ICHIYE, T., Marine geological research and exploration, in "Undersea Technology Handbook/Directory 1968, Section A, Technical and Scientific Review," Compass, Arlington, Va., 1968. [370]
ILLIES, J. H., An intercontinental belt of the world rift system, *Tectonophysics*, 8, 5, 1969. [403]
—— and ST. MUELLER, Graben problems, "Proceedings of an International Rift Symposium held in Karlsruhe, October 10-12, 1968," *Intern. Upper Mantle Proj. Sci. Rep.*, 27, 1970. [403]
INGHRAM, M. G., Manhattan Project, Tech. Series, div. II, vol. 14, chap. 4, p. 35, McGraw-Hill, 1947.
IRVING, E., "Paleomagnetism and Its Application to Geological and Geophysical Problems," Wiley, 1964. [251;383]
——, Paleomagnetic evidence for shear along the Tethys, in C. G. ADAMS and D. V. AGER (eds.), "Aspects of Tethyan Biogeography," *Systematics Assoc. Publ.*, 7, 59, 1967. [464]
——, The Mid-Atlantic ridge at 45°N. XIV. Oxidation and magnetic properties of basalt; review and discussion, *Can. J. Earth Sci.*, 7, 1528, 1970. [415]
——, J. K. PARK, S. E. HAGGERTY, F. AUMENTO, and B. LONCAREVIC, Magnetism and opaque mineralogy of basalts from the Mid-Atlantic ridge at 45°N, *Nature*, 228, 974, 1970. [367;391]
—— and W. A. ROBERTSON, Test for polar wandering and some possible implications, *J. Geophys. Res.*, 74, 1026, 1969. [422]
——, P. J. STEPHENSON, and A. MAJOR, Magnetism in Heard Island rocks, *J. Geophys. Res.*, 70, 3421, 1965. [424]
ISACKS, B., J. OLIVER, and L. R. SYKES, Seismology and the new global tectonics, *J. Geophys. Res.*, 73, 5855, 1968. [123;305;392-394;433;458]
——, L. R. SYKES, and J. OLIVER, Focal mechanisms of deep and shallow earthquakes in the Tonga-Kermadec region and the tectonics of island arcs, *Geol. Soc. Am. Bull.* 80, 1443, 1969. [395;433;458-460]
JACOBS, J. A., The Earth's inner core, *Nature*, 172, 297, 1953a. [202;203;205]
——, Temperature-pressure hypothesis and the Earth's interior, *Can. J. Phys.*, 31, 370, 1953b. [237]
——, Temperature distribution within the Earth's core, *Nature*, 173, 258, 1954. [202;205]
——, "Geomagnetic Micropulsations," Springer-Verlag, 1970. [275]
—— and D. W. ALLAN, Temperatures and heat flow within the Earth, *Trans. Roy. Soc. Can.*, 48, 33, 1954. [209]
—— and ——, The thermal history of the Earth, *Nature*, 177, 155, 1956. [209]
JACOBY, W. R., Instability in the upper mantle and global plate movements, *J. Geophys. Res.*, 75, 5671, 1970. [493;496]
JAEGER, J. C., Conduction of heat in an infinite region bounded internally by a circular cylinder of a perfect conductor, *Australian J. Phys.*, 9, 167, 1956. [185]
——, The effect of the drilling fluid on temperatures measured in bore holes, *J. Geophys. Res.*, 66, 563, 1961. [183]
JEANS, J. H., "Problems of Cosmogony and Stellar Dynamics," Cambridge, 1919. [8]
JEFFREY, J. G., "Chemical Methods of Rock Analysis," Pergamon, 1970. [63]
JEFFREYS, H., On the early history of the solar system, *Monthly Notices Roy. Astron. Soc.*, 78, 424, 1918. [8]
——, The times of the core waves, *Monthly Notices Roy. Astron. Soc. Geophys. Suppl.* 4, 594, 1939. [307]

——, "The Earth," 4th ed., Cambridge, 1962. [33]
——, How soft is the Earth? *Quart. J. Roy. Astr. Soc.*, 5, 10, 1964. [364; 365]
——, Continental drift, *Nature*, 222, 706, 1969. [353; 364; 365]
——, "The Earth, Its Origin, History and Physical Constitution," 5th ed., Cambridge, 1970. [333; 344; 346; 353; 358; 364; 365]
JENSEN, D. C., and J. C. CAIN, An interim geomagnetic field, *J. Geophys. Res.*, 67, 3568, 1962. [226]
JENSEN, M. L., Biogenic sulfur and sulfide deposits, in M. L. JENSEN (ed.), Biogeochemistry of Sulfur Isotopes, Proc. Nat. Sci. Found. Symp., April 1962. [171]
——, Sulfur isotopes and biogenic origin of uraniferous deposits of the Grants and Laguna districts, New Mexico, *New Mexico Bur. Mines Mining Res., Mem.* 15, 182, 1963. [171]
JOHNSON, E. A., T. MURPHY, and O. W. TORRESON, Pre-history of the Earth's magnetic field, *Terr. Magn. Atmos. Elec.*, 53, 349, 1948. [239; 240; 383]
JOHNSON, F. S., The ion distribution above the F_2 maximum, *J. Geophys. Res.*, 65, 577, 1960. [279]
——, Structure of the upper atmosphere, in F. S. JOHNSON (ed.), "Satellite Environment Handbook," Stanford, 1965. [276]
JOHNSON, G. L., Peter 1 Island, *Norsk Polarinstitutt. Arbok. 1965*, p. 85, 1966. [432]
——, Morphology of the Eurasian Arctic basin, *Polar Record*, 14, 619, 1969. [408; 410]
—— and B. C. HEEZEN, Morphology and evolution of the Norwegian-Greenland Sea, *Deep-Sea Res.*, 14, 755, 1967. [409]
—— and M. K. JUGEL, Recent developments in hydrography, "Under sea Technology Handbook/Directory 1968. Section A, Technical and Scientific Review," Compass, Arlington, Va., 1968. [370]
JOHNSON, L. R., Array measurements of P velocities in the upper mantle, *J. Geophys. Res.*, 72, 6309, 1967. [37]
——, Array measurements of P velocities in the lower mantle, *Bull. Seismol. Soc. Am.*, 59, 973, 1969. [33; 37]
JOHNSON, R. H., and A. MALAHOFF, Relation of Macdonald Volcano to migration of volcanism along the Austral chain, *J. Geophys. Res.*, 76, 3282, 1971. [432]
JOLLIFFE, A. W., The Northwestern part of the Canadian Shield, *Intern. Geol. Congr. Rep. 18th Sess., Gr. Brit. 1948*, pt. XII, 1952. [481]
JOLY, J., "Radioactivity and Geology," Constable, 1909. [344]
JONES, E. J. W., and B. M. FUNNELL, Association of a seismic reflector and upper Cretaceous sediment in the Bay of Biscay, *Deep-Sea Res.*, 15, 701, 1968. [415]
JORDAN, P., "The Expanding Earth," Interscience, 1969. [345]
JOYNER, W. B., Basalt-eclogite transition as a cause for subsidence and uplift, *J. Geophys. Res.*, 72, 4977, 1967. [298]
KAHLE, A. B., R. H. BALL, and E. H. VESTINE, Comparison of estimates of fluid motions at the surface of the Earth's core for various epochs, *J. Geophys. Res.*, 72, 4917, 1967a. [234]
——, E. H. VESTINE, and R. H. BALL, Estimated surface motions of the Earth's core, *J. Geophys. Res.*, 72, 1095, 1967b. [234]
KAJIWARA, Y., H. R. KROUSE, and A. SASAKI, Experimental study of sulfur isotope fractionation between coexistent sulfide minerals, *Earth Planet. Sci. Lett.*, 7, 271, 1969. [171]
KALININ, N. A., Problems of oil and gas geology in India, *Intern. Geol. Congr., Rept. 22nd Sess., India*, pt. I, p. 244, 1964. [403]
KANAMORI, H., and F. PRESS, How thick is the lithosphere? *Nature*, 226, 330, 1970. [493]

KANASEWICH, E. R., Seismicity and other properties of geological provinces, *Nature*, 208, 1275, 1965. [438]

KAPLAN, I. R., and J. R. HULSTON, The isotope abundance and content of sulfur in meteorites, *Geochim. Cosmochim. Acta*, 30, 479, 1966. [171]

KARIG, D. E., Structural history of the Mariana Island arc system, *Geol. Soc. Am. Bull.*, 82, 323, 1971. [425; 454; 504; 506]

KATILI, J. A., Large transcurrent faults in Southeast Asia with special reference to Indonesia, *Geol. Rundschau*, 59, 581, 1970. [425]

KATSUMATA, M., and L. R. SYKES, Seismicity and tectonics of the Western Pacific: Izu-Mariana-Caroline and Ryukyu-Taiwan regions, *J. Geophys. Res.*, 74, 5923, 1969. [454; 505]

KATZ, M. B., The nature and origin of the granulites of Mont Tremblant Park, Quebec, *Geol. Soc. Am. Bull.*, 80, 2019, 1969. [488]

KAULA, W. M., Tests and combinations of satellite determinations of the gravity field with gravimetry, *J. Geophys. Res.*, 71, 5303, 1966. [111]

——, Geophysical implications of satellite determinations of the Earth's gravitational field, *Space Sci. Rev.*, 7, 769, 1967. [199]

——, A tectonic classification of the main features of the Earth's gravitational field, *J. Geophys. Res.*, 74, 4807, 1969a. [123]

——, The gravitational field of the moon, *Science*, 166, 1581, 1969b. [124]

——, Global gravity and tectonics, *Proc. Francis Birch Symposium, Harvard Univ.*, 1970. [123]

KAUSEL, E., and C. LOMNITZ, Tectonics of Chile, Pan American Symposium on the Upper Mantle, Mexico, 1968, vol. 2, 47, 1969. [435; 505]

KAY, M., Geosynclinical nomenclature and the craton, *Bull. Am. Assoc. Petrol. Geologists,* 31, 1289, 1947. [472]

——, North American geosynclines, *Geol. Soc. Am. Mem.* 48, 1951. [78; 95; 491]

——, On geosynclinal nomenclature, *Geol. Mag.*, 104, 311, 1967. [472]

—— (ed.), "North Atlantic – Geology and Continental Drift, a Symposium," *Am. Assoc. Petrol. Geologists Mem.* 12, 1969. [413; 474]

KAY, R., N. J. HUBBARD, and P. W. GAST, Chemical characteristics and origin of oceanic ridge volcanic rocks, *J. Geophys. Res.*, 75, 1585, 1970. [87; 418]

KEAR, D., and B. L. WOOD, The geology and hydrology of Western Samoa, *New Zealand Geol. Surv. Bull.*, 63, 1959. [450]

KEEN, C., and C. TRAMONTINI, A seismic refraction survey on the Mid-Atlantic ridge, *Geophys. J.*, 20, 473, 1970. [377; 430]

KEEN, M. J., "An Introduction to Marine Geology," p. 218, Pergamon, 1968. [370; 415]

KEITH, M. L., Ocean floor convergence: A contrary view of global tectonics, *Trans. Am. Geophys. Union*, 51, 430, 1970. [378]

KELLOGG, P. J., Calculations of cosmic-ray trajectories near the equator, *J. Geophys. Res.*, 65, 2701, 1960. [292]

KENNEDY, G. C., The origin of continents, mountain ranges and ocean basins, *Am. Sci.*, 47, 491, 1959. [296]

KENNEDY, W. Q., The Great Glen fault, *Quart. J. Geol. Soc. London*, 102, 41, 1946. [329]

—— et al., Salt basins around Africa, *Proc. Joint Mtg. Inst. Petrol. and Geol. Soc.*, London, 1965, Institute Petroleum, 1965. [420]

KENT, P. E., G. E. SATTERTHWAITE, and A. M. SPENGLER, "Time and Place in Orogeny," *Geol. Soc. London Spec. Publ.* 3, 1969. [479]

KERR, J. W., A submerged continental remnant beneath the Labrador Sea, *Earth Planet., Sci, Lett.*, 2, 283, 1967. [413]

——, Today's topography and tectonics in North-eastern Canada, *Can. J. Earth Sci.*, 7, 570, 1970. [413]

KHAN, M. A., and J. MANSFIELD, Gravity measurement in the Gregory rift, *Nature*, 229, 72, 1971. [401]
KHLOPIN, V. G., and E. K. GERLING, *Dokl. Akad. Nauk SSSR*, 58, 1415, 1947. [144]
KHRAMOV, A. N., V. P. RODINOV, and R. A. KOMISSAROVA, New data on the Paleozoic history of the geomagnetic field in the U.S.S.R.; translated by E. R. Hope, *Canada Def. Res. Bd.*, T46 OR, 1966. [365;474]
KIEFT, C., Mission de peridotites: chromite eluvial de la region des Pirogues, Carte geologique B. R. G. M., 1962. [456]
KIMURA, T., Structural division of Japan and the Honshu arc, *Japan J. Geol. Geography Trans.*, 38, 117, 1967. [451]
KING, B. C., and D. S. SUTHERLAND, Alkaline rocks of Eastern and Southern Africa, *Sci. Prog.*, 48, 298, 1960. [401]
KING, L. C., Canons of landscape evolution, *Geol. Soc. Am. Bull.*, 64, 721, 1953. [353]
——, "The Morphology of the Earth," 2d. ed., Hafner, 1967. [353;385;402]
KING, L. H., and B. MacLEAN, Continuous seismic-refraction study of orpheus gravity anomaly, *Am. Assoc. Pet. Geol.*, 54, 2007, 1970a. [415]
—— and ——, Origin of the outer part of the Laurentian channel, *Can. J. Earth Sci.*, 7, 1470, 1970b. [415]
KING, P. B., "The Evolution of North America," Princeton, 1959. [95;474]
——, New tectonic map of North America, *Geol. Soc. Am. Bull.*, 80, 2039, 1969a. [395]
——, The tectonics of North America – a discussion to accompany the tectonic map of North America, Scale 1:5,000,000, *U.S. Geol. Surv. Prof. Paper* 628, 1969b. [395]
KING-HELE, D. G., The effect of the Earth's oblateness on the orbit of a near satellite, *Proc. Roy. Soc. London, Ser. A* 247, 49, 1958. [109]
——, G. E. COOK, and D. W. SCOTT, New evaluation of odd zonal harmonics in the geopotential, *Nature*, 219, 1143, 1968. [110]
KISTLER, R. W., J. F. EVERNDEN, and H. R. SHAW, Sierra Nevada plutonic cycle: pt. 1, Origin of composite granitic batholiths, *Geol. Soc. Am. Bull.*, 82, 853, 1971. [64;95]
KLEIN, G. de V., Deposition of Triassic sedimentary rocks in separate basins, Eastern North America, *Geol. Soc. Am. Bull.*, 80, 1825, 1969. [417]
KLEMME, H. D., Regional geology of Circum-Mediterranean region, *Bull. Am. Assoc. Petrol. Geol.*, 47, 477, 1958. [462] Cf. Kay, M. (Geologists).
KNACKE, R. F., J. E. GAUSTAD, F. C. GILLETT, and W. A. STEIN, A possible identification of interstellar silicate absorption in the infrared spectrum of 119 Tauri, *Astrophys. J. Letter*, 155, L189, 1969. [314]
KNOPOFF, L., The convection current hypothesis, *Rev. Geophys.*, 2, 89, 1964. [210]
——, International Upper Mantle Project Report 6, Physics Dept., UCLA, 1968. [427]
——, The Upper Mantle of the Earth, *Science*, 1963, 1277, 1969. [365]
——, C. L. DRAKE, and P. J. HART, "The Crust and Upper Mantle of the Pacific Area," *Am. Geophys. Union Monograph* 12,, 1968. [449;456]
——, B. C. HEEZEN, and G. J. F. MacDONALD, The world rift system, a symposium, Zurich, 1967, *Tectonophysics*, 8, 265, 1969. [427]
—— and J. N. SHAPIRO, Comments on the interrelations between Grüneisen's parameter and shock and isothermal equations of state, *J. Geophys. Res.*, 74, 1439, 1969. [306]
KOBER, L., "Der Bau der Erde," 2d ed., Borntraeger, 1928. [360]
KOENIGSBERGER, J. G., Natural residual magnetism of eruptive rocks, pts. I and II, *Terr. Mag.*, 43, 119, 1938. [241]
KOGAN, S. D., Travel times of longitudinal and transverse waves, calculated from data on nuclear explosions made in the region of the Marshall Islands, *Bull. Acad. Sci. USSR, Geophys. Ser.*, no. 3, 246, 1960. [40]
KORMER, S. B., and A. I. FUNTIKOV, A study of impact compression of ferrosilicon and a possible composition of the Earth's core, *Bull. Acad. Sci. USSR, Phys. Solid Earth, Ser. 5*, 285, 1965. [307]

KOVACH, A., A re-determination of the half-life of rubidium-87, *Acta Phys. Acad. Sci. Hung.*, XVII, pt. 3, 1964. [143]

KOVACH, R. L., and D. L. ANDERSON, Attenuation of shear waves in the upper and lower mantle, *Bull. Seismol. Soc. Am.*, 54, 1855, 1964. [311]

KRASNY, L. I., Structure and geologic history of the Northwestern Pacific mobile belt, *Tectonophysics*, 4, 339, 1967. [449]

KRAUSE, D. C., Guinea fracture zone in the equatorial Atlantic, *Science*, 146, 57, 1964. [373]

——, Submarine geology north of New Guinea, *Geol. Soc. Am. Bull.*, 76, 27, 1965. [456]

——, Tectonics, marine geology and bathymetry of Celebes Sea – Sulu Sea region, *Geol. Soc. Am. Bull.*, 77, 813, 1966. [456]

——, Bathymetry and geologic structure of the Northwestern Tasman Sea – Coral Sea – South Solomon Sea area of the Southwestern Pacific Ocean, *New Zealand Dept. Sci. Ind. Res. Bull.*, 183, *New Zealand Oceanogr. Inst. Mem.* 41, 1967. [456]

—— and N. D. WATKINS, North Atlantic crustal genesis in the vicinity of the Azores, *Geophys. J.*, 19, 261, 1970. [415]

——, W. C. WHITE, D. J. W. PIPER, and B. C. HEEZEN, Turbidity currents and cable breaks in the Western New Britain trench, *Geol. Soc. Am. Bull.*, 81, 2153, 1970. [83]

KRAUSKOPF, K. B., Lava movement at Paricutin Volcano, Mexico, *Geol. Soc. Am. Bull.*, 59, 1267, 1948. [66]

KRESTINIKOV, V. N., and I. L. NERSESOV, Relations of the deep structure of the Pamirs and Tien Shan to their tectonics, *Tectonophysics*, 1, 183, 1964. [470]

KROPOTKIN, P. N., Eurasia as a composite continent, *Tectonophysics*, 12, 261, 1971. [474; 480]

KRYNINE, P. D., Sediments and the search for oil, *Producers Monthly*, 9, 12, 1945. [67; 68]

——, The megascopic study and field classification of the sedimentary rocks, *J. Geol.*, 56, 130, 1948. [67; 71]

KU, T.-L., and W. S. BROECKER, Rates of sedimentation in the Arctic Ocean, *Progr. Oceanogr.*, 4, 95, 1967. [84]

KUHN, T. S., "The Structure of Scientific Revolutions," 2d ed., The Univ. of Chicago Press, 1970. [355]

KUIPER, G. P., The law of planetary and satellite distances, *Astrophys. J.*, 109, 308, 1949. [10]

——, On the origin of the solar system, *Proc. Nat. Acad. Sci. Washington*, 37, 1, 1951a. [10]

——, On the origin of the solar system, in H. A. HYNEK (ed.), "Astrophysics," McGraw-Hill, 1951b. [10]

——, The formation of the planets, *J. Roy. Astron. Soc. Can.* L. 57, 105, 158, 1956. [7]

KULLENBERG, B., The piston core sampler, *Svenska Hydrograf Biol. Komm. Skrifter (Ser 3, Hydrograf.)*, 1, 1, 1947. [370]

KUMAGAI, N., and H. ITO, Creep of granite observed in a laboratory for 10 years, in S. ONOGI (ed.), *Proc. Fifth Intern. Congr. Rheol.*, 2, 579, 1970. [346]

KUMARAPELI, P. S., and V. A. SAULL, The St. Lawrence valley system: A North American equivalent of the East African rift valley system, *Can. J. Earth Sci.*, 3, 639, 1966. [403]

KUMMEL, B., "History of the Earth: An Introduction to Historical Geology." 2d ed., Freeman, 1970. [361; 451; 468]

KUNO, H., Literal variation of basalt magma across continental margins and island arcs in W. H. POOLE (ed.), "Continental Margins and Island Arcs," *Geol. Surv. Canada Paper*, 66-15, 317, 1966a. [92; 451]

——, Literal variation of basalt magma type across continental margins and island arcs, *Bull. Volcanol.*, 29, 195, 1966b. [451]

——, R. L. FISHER, and N. MASU, Rock fragments and pebbles dredged near Jimma seamount, Northwest Pacific, *Deep-Sea Res.*, 3, 126, 1956. [431]
KURTH, G., "Evolution and Hominisation," Fischer, Stuttgart and Abel, 1968. [75]
LACHENBRUCH, A. H., Preliminary geothermal model of the Sierra Nevada, *J. Geophys. Res.*, 73, 6977, 1968. [198]
——, Crustal temperature and heat production: Implications of the linear heat-flow relation, *J. Geophys. Res.*, 75, 3291, 1970. [198]
—— and B. V. MARSHALL, Heat flow from the Arctic Ocean basin, preliminary results, *Trans. Am. Geophys. Union*, 45, 123, 1964. [196]
—— and ——, Heat-flow through the Arctic Ocean floor: The Canada Basin – Alpha Rise boundary, *J. Geophys. Res.*, 71, 1223, 1966. [196]
—— and ——, Heat flow in the Arctic, *Arctic*, 22, 300, 1969. [197]
LACOSTE, L. J. B., Measurement of gravity at sea and in the air, *Rev. Geophys.*, 5, 477, 1967. [107]
——, N. CLARKSON, and G. HAMILTON, LaCoste and Romberg stabilized platform shipboard gravity meter, *Geophysics*, 32, 99, 1967. [107]
LADD, H. S., and S. O. SCHLANGER, Drilling operations on Eniwetok atoll, *U.S. Geol. Surv. Prof. Paper* 260-Y, 863, 1960. [450]
——, J. I. TRACEY, Jr., and M. G. GROSS, Deep drilling on Midway atoll, *U.S. Geol. Surv. Prof. Paper* 680A, 1970. [432]
LAFEHR, T. R., and L. L. NETTLETON, Quantitative evaluation of a stabilized platform shipboard gravity meter, *Geophysics*, 32, 110, 1967. [107]
LAHIRI, B. N., and A. T. PRICE, Electromagnetic induction in non-uniform conductors, and the determination of the conductivity of the Earth from terrestrial magnetic variations, *Phil. Trans. Roy. Soc. London, Ser. A*, 237, 509, 1938. [263]
LAMAKIN, V. V., The origin of the Baikal depression, *Can. Def. Res. Bd. Transl.* T504-R, 1969; *Priroda*, 4, 48, 1968. [403]
LAMB, H. H., Volcanic dust in the atmosphere; with a chronology and assessment of its meteorological significance, *Phil. Trans. Roy. Soc. London, Ser. A*, 266, 425, 1970. [57; 66; 90; 421]
LAMBERT, I. B., and P. J. WYLLIE, Low velocity zone of the earth's mantle incipient melting caused by water, *Science* 169, 764, 1970. [303]
LANDISMAN, M., Y. SATO, and J. NAFE, Free vibrations of the Earth and the properties of its deep interior regions, I, Density, *Geophys. J.*, 9, 439, 1965. [40; 43; 48]
LANGSETH, M. G., Jr., Techniques of measuring heat-flow through the ocean floor, *Am. Geophys. Union Monograph 8*, p. 58, 1965. [184]
——, P. J. GRIM, and M. EWING, Heat flow measurements in the East Pacific Ocean, *J. Geophys. Res.*, 70, 367, 1965. [189; 191]
——, X. LE PICHON, and M. EWING, Crustal structure of the mid-ocean ridges 5. Heat flow through the Atlantic Ocean floor and convection currents, *J. Geophys. Res.*, 71, 5321, 1966. [190; 192; 193]
—— and P. T. TAYLOR, Recent heat-flow measurements in the Indian Ocean, *J. Geophys. Res.*, 72, 6249, 1967. [191; 192]
—— and R. P. VON HERZEN, Heat flow through the floors of the world oceans, in A. E. MAXWELL (ed.), "The Sea," pt. 1, vol. 4, Interscience, 1970. [187]
LARIMER, J. W., Chemical fractionations in meteorites 1: Condensations of the elements, *Geochim. Cosmochim. Acta*, 31, 1215, 1967. [314]
LARMOR, J., How could a rotating body such as the sun become a magnet? *Rep. Brit. Assoc.*, 159, 1919. [232; 235]
LAROCHELLE, A., Paleomagnetism of the Monteregian Hills: New results, *J. Geophys. Res.*, 73, 3239, 1968. [448]
LARSON, R. L., and W. C. PITMAN III, World-wide correlation of Mesozoic magnetic

anomalies, and its implications, *Geol. Soc. Am. Bull.*, 83, 3645, 1972. [417; 420;428;431]

——, H. W. MENARD, and S. M. SMITH, Gulf of California: A result of ocean-floor spreading and transform faulting, *Science*, 161, 781, 1968. [443]

LATIMER, W. M., Astrochemical problems in the formation of the Earth, *Science*, 112, 101, 1950. [314]

LATTER, J. H., Active volcanoes and fumarole fields of the world on punched cards (with their eruptions since January 1963), *Bull. Volcanol.*, 32, 299, 1968. [66]

——, Natural disasters, *Advanc. Sci.*, June 1969. [66; 347]

LAUGHTON, A. S., The Gulf of Aden, in relation to the Red Sea and the Afar depression of Ethiopia, in "The World Rift System," *Geol. Surv. Canada Paper* 66-14, 78, 1966. [373]

—— and W. A. BERGGREN, Deep sea drilling project, leg 12, *Geotimes*, 15, 10, 1970. [412]

—— and C. TRAMONTINI, Recent studies of the crustal structure in the Gulf of Aden, *Tectonophysics*, 8, 359, 1969. [404; 405]

LEAKEY, L. S. B., "Fossil Vertebrates of Africa," Academic, 1969. [75]

LEATON, B. R., S. R. C. MALIN, and M. J. EVANS, An analytical representation of the estimated geomagnetic field and its secular change for the epoch 1965.0, *J. Geomagn. Geoelectr.*, 17, 187, 1965. [226]

LEBEDEV, M. M., I. A. TARARIN, and E. A. LAGOVSKAYA, Metamorphic zones of Kamchatka as an example of the metamorphic assemblages of the inner part of the Pacific belt, *Tectonophysics*, 4, 445, 1967. [449]

LEE, W. H. K., Heat flow data analysis, *Rev. Geophys.*, 1, 449, 1963. [187]

——, The thermal history of the Earth; doctoral dissertation, UCLA, 1967. [209; 210; 212-214]

——, Effects of selective fusion on the thermal history of the Earth's mantle, *Earth Planet. Sci. Lett.*, 4, 270, 1968. [210]

——, On the global variations of terrestrial heat flow, *Phys. Earth Planet. Int.*, 2, 332, 1970. [494]

—— and S. P. CLARK, Jr., Heat flow and volcanic temperatures, Sect. 22, in "Handbook of Physical Constants," *Geol. Soc. Am. Mem.*, 97, 481, 1966. [187]

—— and G. J. F. MACDONALD, The global variation of terrestrial heat flow, *J. Geophys. Res.*, 68, 6481, 1963. [199]

—— and S. UYEDA, Review of heat flow data, *Am. Geophys. Union Monograph 8*, p. 87, 1965. [187; 189; 191]

——, S. UYEDA, and P. T. TAYLOR, Geothermal studies of continental margins and island arcs, *Geol. Surv. Canada Paper*, 66-15, 398, 1966. [349]

LEGGO, P. J., W. COMPSTON, and A. F. TRENDALL, Radiometric ages of some Precambrian rocks from the northwest division of Western Australia, *J. Geol. Soc. Australia*, 12, 53, 1965. [489]

LEHMANN, I., P', *Bur. Cent. Séism. Int., Ser. A*, 14, 3, 1936. [32]

LE MAǏTRE, R. W., Petrology of volcanic rocks, Gough Island, South Atlantic, *Geol. Soc. Am. Bull.*, 73, 1309, 1962. [421]

LE PICHON, X., Sea-floor spreading and continental drift, *J. Geophys. Res.*, 73, 3661, 1968; and corrections in 75, 2793, 1970. [305; 371; 392; 394; 427; 468; 470; 498; 500]

—— and J. R. HEIRTZLER, Magnetic anomalies in the Indian Ocean and sea-floor spreading, *J. Geophys. Res.*, 73, 2101, 1968. [406; 422; 423]

—— and M. TALWANI, Regional gravity anomalies in the Indian Ocean, *Deep-Sea Res.*, 16, 263, 1969. [107; 422]

LEVIN, B. Y., S. V. KOZLOVSKAIA, and A. G. STARKOVA, *Meteoritika*, 14, 38, 1956. [16]

LEVINSON, A. A. (ed.), "Proceedings of the Apollo II Lunar Science Conference, Houston, Texas, January 5-8, 1970." 3 vols., Pergamon, 1970 (supplement to *Geochim. Cosmochim. Acta*). [56]
LEWIS, J. S., and H. R. KROUSE, Isotopic composition of sulfur and sulfate produced by oxidation of FeS, *Earth Planet. Sci. Lett.*, 5, 425, 1969. [171]
LILLEY, F. E. M., On kinematic dynamos, *Proc. Roy. Soc. London, Ser. A,* 316, 153, 1970. [236]
LILLIE, A. R., and R. N. BROTHERS, The Geology of New Caledonia, *New Zealand J. Geol. Geophys.*, 13, 145, 1970. [456]
LINDZEN, R. S., Thermally driven diurnal tide in the atmosphere, *Quart. J. Roy. Meteorol. Soc.*, 93, 18, 1967. [267]
——, The application of classical atmospheric tidal theory, *Proc. Roy. Soc. London. Ser. A*, 303, 299, 1968. [267]
LIPMAN, P. W., Alkalic and tholeiitic basaltic volcanism related to the Rio Grande depression, Southern Colorado and Northern New Mexico, *Geol. Soc. Am. Bull.*, 80, 1343, 1969. [496]
LISTER, C. R. B., Measurement of in situ sediment conductivity by means of a Bullard-type probe, *Geophys. J.*, 19, 521, 1970a. [185]
——, Heat flow west of the Juan de Fuca ridge, *J. Geophys. Res.*, 75, 2648, 1970b. [439]
LIUBIMOVA, YE. A., G. A. TOMARA, and A. L. ALEKSANDROV, Heat flux through the floor of the Arctic basin in the region of the Lomonosov ridge, *Dokl. Akad. Nauk SSSR*, 184, 403, 1969. [196]
LIVINGSTON, D. E., R. L. MAUGER, and P. E. DAMON, Geochronology of the emplacement, enrichment and preservation of Arizona porphyry copper deposits, *Econ. Geol.*, 63, 31, 1968. [446]
LOGAN, W., Geology of Canada, *Geol. Surv. Canada*, 1863. [95; 487]
LOMNITZ, C., Sea floor spreading as a factor of tectonic evolution in Southern Chile, *Nature*, 222, 366, 1969. [433; 435]
LOWDER, G. G., and I. S. E. CARMICHAEL, The volcanoes and caldera of Talasea, New Britain, geology and petrology, *Geol. Soc. Am. Bull.*, 81, 17, 1970. [456]
LOWES, F. J., and I. WILKINSON, Geomagnetic dynamo: A laboratory model, *Nature*, 198, 1158, 1963. [237]
LUBIMOVA, E. A., Thermal history of the Earth with consideration of the variable thermal conductivity of its mantle, *Geophys. J*, 1, 115, 1958. [209]
——, Theory of thermal state of the Earth's mantle, in T. F. GASKELL (ed.), "The Earth's Mantle," p. 272, Academic, 1967. [209; 345]
—— and B. G. POLYCK, Heat flow map of Eurasia, in "The Earth's Crust and Upper Mantle," *Am. Geophys. Union Monograph* 13, p. 82, 1969. [198]
LUDWIG, W. J., J. I. EWING, M. EWING, S. MURAUCHI, N. DEN, S. ASANO, H. HOTTA, M. HAYAKAWA, T. ASANUMA, K. ICHIKAWA, and I. NOGUCHI, Sediments and structure of the Japan trench, *J. Geophys. Res.*, 71, 2121, 1966. [394; 433; 454]
——, D. E. HAYES, and J. I. EWING, The Manila trench and West Luzon trough, I. Bathymetry and sediment distribution, *Deep-Sea Res.*, 14, 533, 1967. [454]
LUSKIN, B., B. C. HEEZEN, M. EWING, and M. LANDISMAN, Precision measurement of depth, *Deep-Sea Res.*, 1, 131, 1954. [370]
LYTTLETON, R. A., The origin of the solar system, *Monthly Notices Roy. Astron. Soc.*, 96, 559, 1936. [9]
——, On the origin of binary stars, *Monthly Notices Roy. Astron. Soc.*, 98, 646, 1938. [9]
——, On the origin of the solar system, *Monthly Notices Roy. Astron. Soc.*, 101, 216, 1941a. [9]
——, Note on the origin of planets and satellites, *Monthly Notices Roy. Astron. Soc.*, 101, 349, 1941b. [9]

LYUSTIKH, E. N., The calculation of the rheological properties of the asthenosphere from the "floating" of Fennoscandia, *Izv. Akad, Nauk SSSR, Ser. Geofiz*, 360, 1956. [119]

——, Criticism of hypotheses of convection and continental drift, *Geophys. J.*, 14, 347, 1967. [353]

MA, T.-Y. H., The cause of late Paleozoic glaciation in Australia and South America, *Intern. Geol. Congr. Rep. 21st Sess., Mexico 1956*, pt. 12, 111, 1960. [361]

MAACK, R., Kontinentaldrift und Geologie des Sudatlantischen Ozeans, Walter de Cruyter, 1969. [353; 358]

MACDONALD, G. A., Petrography of the Wallis Islands, *Geol. Soc. Am. Bull.*, 56, 861, 1945. [450]

——, Composition and origin of Hawaiian lavas, *Geol. Soc. Am. Mem.* 116, 477, 1968. [432]

—— and A. T. ABBOTT, "Volcanoes in the Sea – The Geology of Hawaii," University of Hawaii Press, 1970. [432]

MACDONALD, G. J. F., Calculations on the thermal history of the Earth, *J. Geophys. Res.*, 64, 1967, 1959. [200; 209; 210]

——, The deep structure of continents, *Rev. Geophys.*, 1, 587, 1963. [210; 311]

——, Dependence of the surface heat flow on the radioactivity of the Earth, *J. Geophys. Res.*, 69, 2933, 1964. [210]

—— and L. KNOPOFF, On the chemical composition of the outer core, *Geophys. J.*, 1, 284, 1958. [307]

MACELWANE, J. B., Evidence on the interior of the Earth derived from seismic sources, in B. GUTENBERG (ed.), "The Internal Composition of the Earth," Dover, 1951. [33]

MACGREGOR, A. M., Some milestones in the Precambrian of Southern Rhodesia, *Geol. Soc. South Africa Proc., Pres. Addr.*, 54, 27, 1951. [481]

MACHADO, F., W. H. PARSONS, A. F. RICHARDS, and J. W. MULFORD, Capelinhos eruption of Fayal Volcano, Azores, 1957-58, *J. Geophys. Res.*, 67, 3519, 1962. [415]

MACLEOD, W. N., L. E. de la HUNTY, W. R. JONES, and R. HOLLINGER, A preliminary report on the Hamersley Iron Province, Northwest division, *Ann. Rept. Geol. Surv. W. Australia, 1962*, p. 44, 1963. [488]

MADIRAZZA, I., and S. FREGERSLEV, Lower Eocene tuffs at Mønsted, North Jutland, *Bull. Geol. Soc. Denmark*, 19, 283, 1969. [413]

MAGNITSKY, V. A., and I. V. KALASKNIKOVA, Problem of phase transitions in the upper mantle and its connection with the earth's crustal structure, *J. Geophys. Res.*, 75, 877, 1970. [493]

MALAHOFF, A., Gravity and magnetic studies of the New Hebrides Island arc, *New Hebrides Condominium Geol. Surv.*, 1970. [456]

—— and G. P. WOOLLARD, Magnetic and tectonic trends over the Hawaiian ridge, in L. KNOPOFF, C. L. DRAKE, and P. J. HART (eds.), "The Crust and Upper Mantle of the Pacific Area," *Am. Geophys. Union Monograph,* 12, p. 241, 1968. [431]

MALDONADO-KOERDELL, M., Geological and geophysical studies in the Gulf of Fonseca-Nicaragua depression area, Central America, in "The World Rift System," *Geol. Surv. Canada,* 66-14, 220, 1966. [437]

MALKUS, W. V. R., Precession of the Earth as the cause of geomagnetism, *Science*, 160, 259, 1968. [238]

MALLICK, D. I. J., Annual Report of the Geological Survey for the year 1969, *New Hebrides Anglo-Fr. Condominium*, 1970. [456]

MANSINHA, L., and D. E. SMYLIE, Effect of earthquakes in the Chandler wobble and the secular polar shift, *J. Geophys. Res.*, 72, 4731, 1967. [105]

—— and ——, Earthquakes and the Earth's wobble, *Science*, 161, 1127, 1968. [105]

MANSON, V., Geochemistry of basaltic rocks: major elements in H. H. HESS and A. POLDERVAART (eds.), "Basalts," vol. 1, p. 215, Interscience, 1967. [73; 74]

MARKL, R. G., G. M. BRYAN, and J. I. EWING, Structure of the Blake-Bahama outer ridge, *J. Geophys. Res.*, 75, 4539, 1970. [417]
MARKOVSKY, A. P., Structure géologique de l'U.R.S.S. (translated by P. de Saint-Aubin and J. Roget), Centre Nat. Res. Sci., Paris, 1961. [408; 475]
MARKOWITZ, W., and B. GUINOT (eds.), "Continental Drift, Secular Motion of the Pole and Rotation of the Earth," Springer-Verlag, 1968. [363]
MARSHALL, P., Geology of Mangaia, *Bernice P. Bishop Mus. Bull.*, 72, 1, 1930. [90; 450]
——, Geology of Rarotonga and Atui, *Bernice P. Bishop Mus. Bull.*, 72, 1, 1930. [450]
MARTIN, H., The hypothesis of continental drift in the light of recent advances of geological knowledge in Brazil and Southwest Africa, *Geol. Soc. South Africa Annexare*, 64, 47, 1961. [358]
MARTIN, L., (ed.), Guidebook to the geology and history of Tunisia, *Petrol. Explor. Soc. Libya, 9th Ann. Field. Conf.*, 1967. [462; 464]
MARTSINYAK, A. J., *Izmer. Tekh. Poverochn. Delo*, 5, 11, 1956. [112]
MARTYN, D. F., The theory of magnetic storms and aurorae, *Nature*, 167, 92, 1951. [287]
MASON, B., "Meteorites," Wiley, 1962. [14; 16]
——, "The Principles of Geochemistry," 3d ed., Wiley, 1966. [58; 59; 98]
—— and W. G. MELSON, "The Lunar Rocks," Wiley-Interscience, 1970. [56]
MASON, R. G., A magnetic survey off the west coast of the United States, *Geophys. J.*, 1, 320, 1958. [427]
MATSUDA, T., and S. UYEDA, On the Pacific-type orogeny and its model – extension of the paired belts concept and possible origin of marginal seas. *Tectonophysics*, 11, 5, 1971. [471; 505]
MATSUMOTO, T. (ed.), Age and nature of the Circum-Pacific orogenesis, *Tectonophysics*, 4, 317, 1967. [427]
MATTAUER, M., Sur la rotation de l'Espagne, *Earth Planet. Sci. Lett.*, 7, 87, 1969. [415]
MATTHEWS, D. H., Aspects of the geology of the Scotia arc, *Geol. Mag.*, 95, 425, 1959. [421]
——, The northern end of the Carlsberg ridge, in T. N. IRVINE (ed.), "The World Rift System," *Geol. Surv. Canada Paper*, 66-14, 124, 1966. [424]
—— and C. A. WILLIAMS, Linear magnetic anomalies in the Bay of Biscay, *Earth Planet. Sci. Lett.*, 4, 315, 1968. [415]
——, ——, and A. S. LAUGHTON, Mid-ocean ridge in the mouth of the Gulf of Aden, *Nature*, 215, 1052, 1967. [404]
MATTSON, P. H., R. S. HURLEY, J. B. HERSEY, E. T. BUNCE, M. TALWANI, and W. HAMILTON, Session on Caribbean Island arcs, in "Continental Margins and Island Arcs," *Geol. Surv. Canada Paper*, 66-15, 124, 1966. [437]
MAXWELL, A. E., R. P. VON HERZEN, K. JINGHWA HSU, J. E. ANDREWS, T. SAITO, S. F. PERCIVAL, Jr., E. D. MILOW, and R. E. BOYCE, Deep sea drilling in the South Atlantic, *Science*, 168, 1047, 1970a. [81; 367; 389; 420]
—— et al., Initial reports of the Deep Sea Drilling Project, vol. 3, U.S. Govt. Printing Office, 1970b. [81; 367; 370; 378; 389; 420]
——, E. C. BULLARD, E. GOLDBERG, and J. L. WORZEL (eds.), "The Sea: Ideas and Observations on Progress in the Study of the Seas," vol. 4, Interscience, 1970c. [370; 427; 429]
MAXWELL, J. A., "Rock and Mineral Analysis" (Chemical Analysis Series), p. 27, Interscience, 1968. [63]
MAXWELL, J. C., Continental drift and a dynamic Earth, *Am. Sci.*, 56, 35, 1968. [494]
MAYHEW, M. A., C. L. DRAKE, and J. E. NAFE, Marine geophysical measurements on the continental margins of the Labrador Sea, *Can. J. Earth Sci.*, 7, 199, 1970. [415]
MAYNARD, G. L., Crustal layer of seismic velocity of 6.9 to 7.6 kilometers per second under the deep oceans, *Science*, 168, 120, 1970. [81; 82; 85; 370; 428]
MAYNE, K. I., Terrestrial helium, *Geochim. Cosmochim. Acta*, 9, 174, 1956. [59]

MCBIRNEY, A. R., Conductivity variations and terrestrial heat flow distribution, *J. Geophys. Res.*, 68, 6323, 1963. [209]
—— (ed.), Proceedings of the Andesite Conference, *Oregon, Dept. Geol. Mineral Ind. Bull.*, 65, 1969. [88]
—— and K.-I. AOKI, Petrology of the Island of Tahiti, *Geol. Soc. Am. Mem.*, 116, 523, 1968. [432]
—— and I. G. GASS, Relations of oceanic volcanic rocks to mid-ocean rises and heat flow, *Earth Planet. Sci. Lett.*, 2, 265, 1967. [87; 88; 418; 432; 499; 500]
—— and H. WILLIAMS, Volcanic history of Nicaragua, Univ. California Publ. Geol. Sci. no. 55, 1, 1965. [437]
—— and ——, Geology and petrology of the Galápagos Islands, *Geol. Soc. Am. Mem.*, 118, 1969. [432; 437]
MCCABE, H. R., and B. B. BANNATYRE, The Lake St. Martin crypto-explosion crater and geology of the surrounding area, *Manitoba Geol. Surv. Geol. Paper*, 3-70, 1970. [78]
MCCONNELL, R. B., The East African rift system, *Nature*, 215, 578, 1967. [400; 402]
——, Fundamental fault zones in the Guiana West African shields in relation to presumed axes of Atlantic spreading, *Geol. Soc. Am. Bull.*, 80, 1775, 1969. [420]
MCCONNELL, R. K., Jr., Comments on letter by H. Takeuchi, "Time scales of isostatic compensations," *J. Geophys. Res.*, 68, 4397, 1963. [311]
——, Isostatic adjustment in a layered earth, *J. Geophys. Res.*, 70, 5171, 1965. [346]
——, Viscosity of the mantle from relaxation time spectra of isostatic adjustment, *J. Geophys. Res.*, 73, 7089, 1968. [313]
MCCREA, J. M., On the isotopic chemistry of carbonates and a paleotemperature scale, *J. Chem. Phys.*, 18, 849, 1950. [175]
MCCREA, W. H., The origin of the solar system, *Proc. Roy. Soc. London, Ser. A*, 256, 245, 1960. [11; 12]
MCDOUGALL, I., Potassium-argon measurements on dolerites from Antarctica and South Africa, *J. Geophys. Res.*, 68, 1535, 1963. [402]
—— and F. H. CHAMALAUN, Geomagnetic polarity scale of time, *Nature*, 212, 1415, 1966. [386]
—— and ——, Isotopic dating and geomagnetic polarity studies on volcanic rocks from Mauritius, Indian Ocean, *Geol. Soc. Am. Bull.*, 80, 1419, 1969. [424]
——, P. R. DUNN, W. COMPSTON, A. W. WEBB, J. R. RICHARDS, and V. M. BOFINGER, Isotopic age determinations on Precambrian rocks of the Carpentaria region Northern Territory, Australia, *J. Geol. Soc. Australia*, 12, 67, 1965. [489]
—— and N. R. RUEGG, Potassium-argon dates on the Serra Geral formation of South America, *Geochim. Cosmochim. Acta*, 30, 191, 1966. [420]
MCELHINNY, M. W., Northwood drift of India – examination of recent paleomagnetic results, *Nature*, 217, 342, 1968. [425]
——, Formation of the Indian Ocean, *Nature*, 228, 977, 1970. [384; 385; 422; 470]
——, J. C. BRIDEN, D. L. JONES, and A. BROCK, Geological and geophysical implications of paleomagnetic results from Africa, *Rev. Geophys.*, 6, 201, 1968. [383-385; 488]
—— and P. WELLMAN, Polar wandering and sea-floor spreading in the Southern Indian Ocean, *Earth Planet. Sci. Lett.*, 6, 198, 1969. [424]
MCFARLANE, D. T., and W. I. RIDLEY, An interpretation of gravity data for Lanzarote, Canary Islands, *Earth Planet. Sci. Lett.*, 6, 431, 1969. [418]
MCGINNIS, L. D., Gravity fields and plate tectonics in the Hindu Kush, *J. Geophys. Res.*, 76, 1894, 1971. [470]
MCKEE, E. D., J. CHRONIC, and E. B. LEOPOLD, Sedimentary belts in lagoon in Kapingamaringi atoll, *Bull. Am. Assoc. Pet. Geol.* 43, 501, 1959. [450]
—— and E. H. MCKEE, Pliocene uplift – the Grand Canyon region – time of drainage adjustment, *Geol. Soc. Am. Bull.*, 83, 1923, 1972. [448]

MCKENZIE, D. P., The viscosity of the lower mantle, *J. Geophys. Res.*, 71, 3995, 1966. [311]
——, Plate tectonics of the Mediterranean region, *Nature*, 226, 239, 1970. [393; 395; 464; 465; 468]
——, D. DAVIES, and P. MOLNAR, Plate tectonics of the Red Sea and East Africa. *Nature*, 226, 243, 1970. [404]
—— and W. J. MORGAN, Evolution of triple junctions, *Nature*, 224, 125, 1969. [333; 337; 435; 438; 443; 447; 458]
—— and R. L. PARKER, The North Pacific: An example of tectonics on a sphere, *Nature*, 216, 527, 1967. [392]
—— and J. G. SCLATER, Heat flow inside the island arcs of the Northwestern Pacific, *J. Geophys. Res.*, 73, 3173, 1968. [458]
—— and ——, Evolution of the Indian Ocean, *Geophys, J.*, 24, 437, 1971. [424]
MCKINSTRY, H. E., Precambrian problems in Western Australia, *Am. J. Sci.*, 243A, 448, 1945. [481]
MCMAHON, B. E., and D. W. STRANGWAY, Kiaman magnetic interval in the Western United States, *Science*, 155, 1012, 1967. [389]
MCMANUS, D. A., et al., Initial reports of the deep sea drilling project, vol. V, U.S. Govt. Printing Office, 1970. [370]
MCMULLEN, C. C., K. FRITZE, and R. H. TOMLINSON, The half-life of rubidium-87, *Can. J. Phys.*, 44, 3033, 1966. [143]
MCQUEEN, R. G., J. N. FRITZ, and S. P. MARSH, On the equation of state of stishovite, *J. Geophys. Res.*, 68, 2319, 1963. [306]
——, ——, and ——, On the composition of the Earth's interior, *J. Geophys. Res.*, 69, 2947, 1964. [43; 306]
—— and S. P. MARSH, Equation of state for nineteen metallic elements from shock wave measurements to two megabars, *J. Appl. Phys.*, 31, 1253, 1960. [306]
—— and ——, Shock wave compression of iron nickel alloys and the Earth's core, *J. Geophys. Res.*, 71, 1751, 1966. [306]
——, ——, and J. N. FRITZ, Hugoniot equation of state of twelve rocks, *J. Geophys. Res.*, 72, 4999, 1967. [306]
MEHNERT, K. R., "Migmatites and the Origin of Granitic Rocks," Elsevier, 1968. [64]
MEINEL, A. B., Doppler-shifted auroral hydrogen emission, *Astrophys. J.*, 113, 50, 1951. [286]
MELSON, W. G., S. R. HART, and G. THOMPSON, St. Paul's rocks, equatorial Atlantic: Petrogenesis, radiometric ages and implications on sea-floor spreading [unpublished manuscript], Woods Hole Oceanogr. Inst., Ref. 71-20, 1971. [421]
——, E. JAROSEWICH, V. T. BOWEN, and G. THOMPSON, St. Peter and St. Paul rocks: A high temperature, mantle-derived intrusion, *Science*, 155, 1532, 1967a. [421]
——, ——, and R. CIFELLI, Alkali olivine basalt dredged near St. Paul's rocks, mid-Atlantic ridge, *Nature*, 215, 381, 1967b. [421]
—— and G. THOMPSON, Layered basic complex in oceanic crust. Romanche fracture, equatorial Atlantic ocean, *Geophys. J.*, 19, 817, 1970. [85; 497]
——, ——, and T. H. VAN ANDEL, Volcanism and metamorphism in the mid-Atlantic ridge, 22°N latitude, *J. Geophys. Res.*, 73, 5925, 1968. [418]
MELVILLE, R., Continental drift, Mesozoic continents, and the migration of the angiosperms, *Nature*, 211, 116, 1966. [362; 363]
MENARD, H. W., The East Pacific rise, *Science*, 132, 1737, 1960. [373; 427; 428; 438]
——, "Marine Geology of the Pacific," McGraw-Hill, 1964. [88; 372; 378; 379; 388; 426; 427; 480; 449]
——, Extension of Northeastern Pacific fracture zones, *Science*, 155, 72, 1967. [372]

——, Elevation and subsidence of oceanic crust, *Earth Planet. Sci. Lett.*, 6, 275, 1969a. [367; 376; 418, 449; 450]
——, The deep-ocean floor, *Sci. Am.*, 221, 126, 1969b. [373; 394; 442]
——, Growth of drifting volcanoes, *J. Geophys. Res.*, 74, 4827, 1969c. [449]
—— and T. M. ATWATER, Changes in direction of sea-floor spreading, *Nature*, 219, 463, 1968. [373; 375; 428]
—— and ——, Origin of fracture zone topography, *Nature*, 222, 1037, 1969. [373; 394]
—— and R. S. DIETZ, Submarine geology of the Gulf of Alaska, *Geol. Soc. Am. Bull.*, 62, 1263, 1951. [372; 432; 448]
—— and ——, Mendocino submarine escarpment, *J. Geol.*, 60, 266, 1952. [331]
—— and J. MAMMERICKX, Abyssal hills magnetic anomalies and the East Pacific rise, *Earth Planet. Sci. Lett.*, 2, 465, 1967. [427]
—— and S. M. SMITH, Hypsometry of ocean basin provinces. *J. Geophys. Res.*, 17, 4305, 1966. [57; 58]
——, ——, and R. M. PRATT, The Rhone deep-sea fan, in W. E. WHITTARD and R. BRADSHAW (eds.), "Submarine Geology and Geophysics" (Colston Papers vol. 17), 271, 1965. [462]
MENDIGUREN, J. A., Focal mechanism of a shock in the middle of the Nazca Plate, *J. Geophys. Res.*, 76, 3861, 1971. [496]
MERCANTON, P. L., Inversion de l'inclination magnetique terrestre aux ages geologiques, *Terres. Magn. Atmos. Elec.*, 31, 187, 1926. [383]
MEYERHOFF, A. A., Continental drift: Implications of paleomagnetic studies, meteorology, physical oceanography, and climatology, *J. Geol.*, 78, 1, 1970a. [353; 361]
—— Continental drift, II, High-latitude evaporite deposits and geologic history of Arctic and North Atlantic Oceans, *J. Geol.*, 78, 406, 1970b. [353; 361]
MILLER, E. T., and M. EWING, Geomagnetic measurement in the Gulf of Mexico and in the vicinity of Caryn Peak, *Geophysics*, 21, 406, 1956. [370; 418]
MILLER, S. L., and H. C. UREY, Organic compound synthesis on the primitive Earth, *Science*, 130, 245, 1959. [59]
MILNE, W. G., W. E. T. SMITH, and G. C. ROGERS, Canadian seismicity and microearthquake research in Canada, *Can. J. Earth Sci.*, 7, 591, 1970. [413]
MINATO, M., M. GORAI, and M. HUNAHASHI, Geologic development of the Japanese Islands, *Assoc. of Geol. Collab.*, p. 442, Tokyo 1965. [451]
MISENER, A. D., and A. E. BECK, The measurement of heat flow over land, in S. K. RUNCORN (ed.), "Methods and Techniques in Geophysics," vol. 1, p. 10, Interscience, 1960. [182; 185]
MITRONOVAS, W., B. ISACKS, and L. SEEBER, Earthquake locations and seismic wave propagation in the upper 250 km of the Tonga Island arc, *Bull. Seismol. Soc. Am.*, 59, 1115, 1969. [458]
MIYASHIRO, A., Evolution of metamorphic belts, *J. Petrol.* 2, 277, 1961. [454]
—— and F. SHIDO, Progressive metamorphism in zeolite assemblages, *Lithos.*, 3, 251, 1970. [78]
——, ——, and M. EWING, Diversity and origin of abyssal tholeiite from the Mid-Atlantic ridge near 24° and 30° north latitude, *Contr. Min. Petrol.*, 23, 38, 1969. [418]
——, ——, and ——, Crystallization and differentiation in abyssal tholeiites and gabbros from mid-ocean ridges, *Earth Planet. Sci. Lett.*, 7, 361, 1970. [85]
MOHR, P. A., "The Geology of Ethiopia," University College of Addis Ababa Press, 1962. [73]
——, Major Volcano – Tectonic lineament in the Ethiopian rift system, *Nature*. 213, 664, 1967. [404]
——, Plate tectonics of the Red Sea and East Africa, *Nature*, 228, 547, 1970. [404]

——, Ethiopian rift and plateaus: Some volcanic petrochemical differences, *J. Geophys. Res..*, 76, 1967, 1971. [402;496]

MOLNAR, P., and J. OLIVER, Lateral variations in the upper mantle and discontinuities in the lithosphere, *J. Geophys. Res.*, 74, 2648, 1969. [395;445]

—— and L. R. SYKES, Tectonics of the Caribbean and Middle America regions from focal mechanisms and seismicity, *Geol. Soc. Am. Bull.*, 80, 1639, 1969. [393;435; 437]

MONGER, J. W. H., and C. A. ROSS, Distribution of Fusulinaceans in the Western Canadian Cordillera, *Can. J. Earth Sci.*, 8, 259, 1971. [460;474;479]

MONROE, W. H., The age of the Puerto Rico trench, *Geol. Soc. Am. Bull.*, 79, 487, 1968. [437]

MOORBATH, S., and G. P. L. WALKER, Strontium isotope investigations of igneous rocks from Iceland, *Nature*. 207, 837, 1965. [381]

—— and H. WELKE, Isotopic evidence for the continental affinity of the Rockall bank, North Atlantic, *Earth Planet. Sci. Lett.*, 15, 211, 1968. [412]

——, R. K. O'NIONS, R. J. PANKHURST, N. H. SALE, and V. R. MCGREGOR, Further rubidium-strontium age determinations on the very early Precambrian rocks of the Godthab District, West Greenland, *Nature*, 240, 78, 1972. [483]

MOORE, D. G., and E. C. BUFFINGTON, Transform faulting and growth of the Gulf of California since the late Pliocene, *Science*, 161, 1238, 1968. [443]

MOORE, R. C., "Introduction to Historical Geology," 2d ed., McGraw-Hill, 1958. [361; 476]

——, Stability of the Earth's crust, *Geol. Soc. Am. Bull.*, 81, 1285, 1970. [361]

MOORES, E. M., Petrology and structure of the Vourinos ophiolitic complex of Northern Greece, *Geol. Soc. Am. Spec. Paper*, 118, 1969. [464]

——, Patterns of continental fragmentation and reassembly: Some implications, *Geol. Soc. Am. Abstracts with Programs,* 2, 629, 1970a. [77;491]

——, Ultramafics and orogeny, with models of the U.S. Cordillera and the Tethys, *Nature*, 228, 837, 1970b. [439]

MOOSER, F., The Mexican volcanic belt: Structure and development, "Pan-American Symposium on the Upper Mantle, Mexico, 1968," vol. 2, group 2, p. 15, 1969. [437]

MORGAN, W. J., Rises, trenches, great faults and crustal blocks, *J. Geophys. Res.*, 73, 1959, 1968. [372; 392; 393;406;494]

——, Convection plumes in the lower mantle, *Nature*, 230, 42, 1971. [87; 88;494;496]

——, P. R. VOGT, and D. F. FALLS, Magnetic anomalies and sea-floor spreading on the Chile rise, *Nature*, 222, 137, 1969. [435]

MORITZ, H., The geodetic reference system, 1967, *Allgem. Vermessungs-Nachrich*, 2-7, 1968. [109]

MORLEY, L. W., and A. LAROCHELLE, Paleomagnetism as a means of dating geological events, *Roy. Soc. Canada Spec. Publ.* 8, 39, 1964. [387]

MORLIER, P., Les fluge des roches, *Ann. Paris Inst. Tech. Bat. Trer.*, 19, 89, 1966. [338]

MORRIS, G. B., R. W. RAITT, and G. G. SHOR, Jr., Velocity anisotropy and delay-time maps of the mantle near Hawaii, *J. Geophys. Res.*, 74, 4300, 1969. [428; 430]

MORRISON, R. P., "Structural Geology of South America," Longmans, 1973. [403, 435; 475]

MOURAD, A. G., New techniques for geodetic measurements at sea, *Am. Geophys. Union Trans.*, 1, 864, 1970. [370]

MUEHLBERGER, W. R., R. E. DENISON, and E. G. LIDIAK, Basement rocks in continental interior of United States, *Am. Assoc. Petrol. Geol. Bull.*, 51, 2351, 1967. [96]

MULLER, P. M., and W. J. SJOGREN, Mascons: Lunar mass concentrations, *Science*, 161, 680, 1968. [124]

MUNK, W. H., and G. J. F. MACDONALD, "The Rotation of the Earth," Cambridge, 1960. [102]

—— and R. REVELLE, On the geophysical interpretation of irregularities in the rotation of the earth, *Monthly Notices Roy. Astron. Soc., Geophys. Suppl.*, 6, 331, 1952. [234]

MURAUCHI, S., N. DEN, D. ASANO, H. HOTTA, T. YOSHII, T. ASANUMA, K. HAGIWARA, K. ICHIKAWA, T. SATO, W. J. LUDWIG, J. I. EWING, N. T. EDGAR, and R. E. HOUTZ, Crustal structure of the Philippine Sea, *J. Geophys. Res.*, 73, 3143, 1968. [454]

MURPHY, L. M., World-wide seismic network, in ESSA Symposium on Earthquake Prediction, February, 1966, p. 53, U.S. Govt. Printing Office, 1966. [392]

MURRAY, C. G., Magma genesis and heat flow: Differences between mid-oceanic ridges and African rift valleys, *Earth Planet. Sci. Lett.*, 9, 34, 1970. [402]

MURRAY, G. E., "Geology of the Atlantic and Gulf Coastal Province of North America," Harper, 1961. [79]

MURTHY, V. R., and C. C. PATTERSON, Primary isochron of zero age for meteorites and the earth, *J. Geophys. Res.*, 67, 1161, 1962. [152]

NAFE, J. E., and C. L. DRAKE, Floor of the North Atlantic – summary of geophysical data, in M. KAY (ed.), "North Atlantic – Geology and Continental Drift, a Symposium," *Am. Assoc. Pet. Geol. Mem.*, 12, 59, 1969. [389; 417; 418]

NAIRN, A. E. M. (ed.), "Descriptive Palaeoclimatology," Interscience, 1961. [360]

NALIVKIN, D. V., "The Geology of the U.S.S.R." English translation, Pergamon, 1960. [403; 447; 475]

NAUGLER, F. P., Aleutian deep-sea channel on the Aleutian abyssal plain, *Nature,* 228, 1081, 1970. [448]

NÉEL, L., L'inversion de l'aimantation permanente des roches, *Ann, Geophys.,* 7, 90, 1951. [244]

——, Some theoretical aspects of rock magnetism, *Phil. Mag. Supp. Adv. Phys.*, 4, 191, 1955. [244]

NELSON, J. H., L. HURWITZ, and D. G. KNAPP, "Magnetism of the Earth," Publ. 40-1, U.S. Dept. Comm. Coast Geod. Surv., 1962. [219-222; 258-261]

NEPROCHNOV, YU. P., I. P. KOSMINISKAYA, and YA. P. MALOVITSKY, Structure of the crust and upper mantle of the Black and Caspian Seas, *Tectonophysics*, 10, 517, 1970. [466]

NESS, N. F., C. S. SCEARCE, and J. B. SEEK, Initial results of the IMP 1 magnetic field experiment, *J. Geophys. Res.*, 69, 3531, 1964. [285]

NETTLETON, L. L., L. J. B. LACOSTE, and M. GLICKEN, Quantitative evaluation of precision of airborne gravity meter, *J. Geophys. Res.*, 67, 4395, 1962. [107]

NIER, A. O., Variations in the relative abundances of the isotopes of common lead from various sources, *J. Am. Chem. Soc.*, 60, 1571, 1938. [160]

——, A re-determination of the relative abundances of the isotopes of carbon, nitrogen, oxygen, argon and potassium, *Phys. Rev.*, 77, 789, 1950. [142]

NINKOVICH, D., and B. C. HEEZEN, Santorini tephra, in "Submarine Geology and Geophysics," (Colston Paper, vol. 17), 413, 1965. [464]

——, N. OPDYKE, B. C. HEEZEN, and J. H. FOSTER, Paleomagnetic stratigraphy rates of deposition and tephrachronology in North Pacific deep-sea sediments, *Earth Planet. Sci. Lett.*, 1, 476, 1966. [388; 429]

NOBLE, J. A., Metal provinces of the Western United States, *Geol. Soc. Am. Bull.*, 81, 1607, 1970. [445; 446]

NOE-NYGAARD, A., On extrusion forms in plateau basalts, *Sci. Iceland*, anniv. vol. 10, 1968. [412]

NÖLCKE, F., Der Entwicklungsgang unseres Planetensystems, Bonn, 1930. [9]

NORTHROP, J. W., III, and A. A. MEYERHOFF, Validity of polar continental movement hypothesis based on paleomagnetic studies, *Bull. Am. Assoc. Pet. Geol.*, 47, 575, 1963. [365]

NORTHROP, J., R. A. FROSCH, and R. FRASSETTO, Bermuda–New England seamount arc, *Geol. Soc. Am. Bull.*, 73, 587, 1962. [418]

NORTHRUP, D. A., and R. N. CLAYTON, Oxygen isotope fractionations in systems containing dolomites, *J. Geol.*, 71, 174, 1966. [177]

NOUGIER, J., Contribution à l'étude géologique et géomorphologique des îles Kerguelen, Thèse présentée à la Faculté des Sciences de Paris, 2 vols., no. 5, d'enregistrement au C.N.R.S., A.O. 3755, 1969. [424]

NUGENT, L. E., Jr., Coral reefs in the Gilbert, Marshall and Caroline Islands, *Geol. Soc. Am. Bull.*, 57, 735, 1946. [450]

OBERMÜLLER, A. G., Contribution à l'étude géologique et minérale de île Clipperton (Polynesie Français), *Recherche Geol. Min. Polynesie Franc*, 45, 1959. [432]

O'BRIEN, B. J., Lifetimes of outer zone electrons and their precipitation into the atmosphere, *J. Geophys. Res.*, 67, 3687, 1962. [483]

——, J. A. VAN ALLEN, C. D. LAUGHLIN, and L. A. FRANK, Absolute electron intensities in the heart of the Earth's outer radiation zone, *J. Geophys. Res.*, 67, 397, 1962. [283]

O'DRISCOLL, E. S., Inference patterns from inclined shear fold systems, *Bull. Can. Pet. Geol.*, 12, 279, 1962. [335]

OFFICER, C. B., J. I. EWING, J. F. HENNION, D. G. HARKRIDER, and D. E. MILLAR, Geophysical investigations in the Eastern Caribbean: summary of 1955 and 1956 cruises, in L. H. AHRENS et al. (eds.), "Physics and Chemistry of the Earth," vol. 3, 17, 1959. [439]

OLIVER, J., M. EWING, and F. PRESS, Crustal structure of the Arctic regions from the L_g phase, *Geol. Soc. Am. Bull.*, 66, 1063, 1955. [408]

——, T. JOHNSON, and J. DORMAN, Postglacial faulting and seismicity in New York and Quebec, *Can. J. Earth Sci.*, 7, 879, 1970. [363; 393]

OMODEO, P., Distribution of the terricolous oligochaetes on the two shores of the Atlantic, in A. LÖVE, and D. LÖVE (eds.), "North Atlantic Biota and Their History," p. 127, Pergamon, 1963. [363]

OMOTE, S., Y. OSAWA, I. SKINNER, and Y. YOSHIMI, Philippines: Luzon earthquake of 2 August 1968, UNESCO Serial No. 977, Paris 1969. [454]

O'NEIL, J. R., and R. N. CLAYTON, Oxygen isotope geothermometry, in H. CRAIG, et al. (eds.), "Isotopic and Cosmic Chemistry" (Urey Volume), p. 157, North Holland Publ., 1964. [177]

—— and S. EPSTEIN, Oxygen isotope fractionation in the system dolomite-calcite-carbon dioxide, *Science*, 152, 198, 1966. [177]

—— and H. P. TAYLOR, Jr., Oxygen isotope fractionation between mascovite and water, *Trans. Am. Geophys. Union*, 47, 212, 1966. [177]

—— and ——, The oxygen isotope and cation exchange chemistry of feldspars, *Am. Mineralogist*, 52, 1414, 1967. [177]

OPDYKE, N. D., Palaeoclimatology and continental drift, in S. K. RUNCORN (ed.), "Continental Drift," p. 41, Academic, 1962. [361]

—— and B. P. GLASS, The paleomagnetism of sediment cores from the Indian Ocean, *Deep-Sea Res.*, 16, 249, 1969. [422]

——, ——, J. D. HAYS, and J. FOSTER. A paleomagnetic study of Antarctic deep-sea sediments, *Science*, 154, 349, 1966. [83; 250; 251; 388; 429]

—— and K. W. HENRY, A test of the dipole hypothesis, *Earth Planet. Sci. Lett.*, 6, 139, 1969. [365]

OROWAN, E., Density of the moon and nucleation of planets, *Nature*, 222, 867, 1969a. [314]

——, The origin of the oceanic ridges, *Sci. Am.*, 221, 102, 1969b. [493]

OSTIC, R. G., R. D. RUSSELL, and R. L. STANTON, Additional measurements of the isotopic composition of lead from stratiform deposits, *Can. J. Earth Sci.*, 4, 245, 1967. [162]

OTTMANN, F., L'atol das Rocas, dans l'Atlantique sud tropical, *Rev. Geogr. Phys. et Geol. Dynamique*, 2d ser., 5, 101, 1962. [421]

OVERSBY, V. M., and P. W. GAST, Isotopic composition of lead from oceanic islands, *J. Geophys. Res.*, 75, 2097, 1970. [497]

——, J. LANCELOT, and P. W. GAST, Isotopic composition of lead in volcanic rocks from Tenerife, Canary Islands, *J. Geophys. Res.*, 76, 3402, 1971. [499]

OXBURGH, E. R., and D. L. TURCOTTE, Mid-ocean ridges and geothermal distribution during mantle convection, *J. Geophys. Res.*, 73, 2643, 1968. [85; 494]

—— and ——, Origin of paired metamorphic belts and crustal dilation in island arc regions, *J. Geophys. Res.*, 76, 1315, 1971. [505]

PACKHAM, G. H., and D. A. FALVEY, An hypothesis for the formation of marginal seas in the Western Pacific, *Tectonophysics*, 11, 79, 1971. [504]

PAGE, B. M., Sur-Nacimiento fault zone of California: Continental margin tectonics, *Geol. Soc. Am. Bull.*, 81, 667, 1970. [364; 447]

PAGE, R., Late Cenozoic movement on the Fairweather fault in Southeastern Alaska, *Geol. Soc. Am. Bull.*, 80, 1873, 1969. [448]

PAKISER, L. C., J. P. EATON, J. H. HEALY, and C. B. RALEIGH, Earthquake prediction and control, *Science*, 166, 1467, 1969. [447]

—— and J. S. STEINHART, Explosion seismology in the western hemisphere, in H. ODISHAW (ed.), "Research in Geophysics," vol. 2, "Solid Earth and Interface Phenomena," p. 123, M.I.T., 1964. [445]

PALMER, H., East Pacific rise and westward drift of North America, *Nature*, 220, 341, 1968. [438]

PAPANASTASSIOU, D. A., and G. J. WASSERBURG, Initial strontium isotopic abundances and the resolution of small time differences in the formation of planetary objects, *Earth Planet. Sci. Lett.*, 5, 361, 1969. [166]

—— and ——, Rb-Sr ages from the Ocean of Storms, *Earth Planet. Sci. Lett.*, 8, 269, 1970. [153]

—— and ——, Lunar chronology and evolution from Rb-Sr studies of Apollo 11 and 12 samples, *Earth Planet. Sci. Lett.*, 11, 37, 1971. [153]

PARKER, E. N., "Interplanetary Dynamical Processes," vol. 8 of Monographs and Texts in Physics and Astronomy, Interscience, 1963. [283]

PARSONS, R. C., and D. G. F. HEDLEY, The analysis of the various properties of rocks for classification, *Int. J. Rock Mech. Min. Sci.*, 3, 325, 1966. [338]

PATERSON, W. S. B., and L. K. LAW, Additional heat flow determinations in the area of Mould Bay, Arctic Canada, *Can. J. Earth Sci.*, 3, 237, 1966. [197]

PATTERSON, B., K. K. BEHRENSWEGER, and W. O. LILL, Geology and fauna of a new Pliocene locality in Northwestern Kenya, *Nature*, 226, 918, 1970. [75]

PAVONI, N., Tectonic interpretation of the magnetic anomalies southwest of Vancouver Island, *Rev. Pure, Appl. Geophys.*, 63, 172, 1966. [439]

PEARCE, G. W., and G. N. FREDA, Magnetization of the Perry formation of New Brunswick, and the rotation of Newfoundland: Discussion, *Can. J. Earth Sci.*, 6, 353, 1969. [386]

PEASE, R. W., Normal faulting and lateral shear in Northeastern California, *Geol. Soc. Am. Bull.*, 80, 715, 1969. [439]

PEEBLES, J., and R. H. DICKE, The temperature of meteorites and Dirac's cosmology and Mach's principle, *J. Geophys. Res.*, 67, 4063, 1962. [345]

PEKERIS, C. L., Thermal convections in the interior of the Earth, *Monthly Notices Roy. Astron. Soc., Geophys. Suppl.* 3, 343, 1935. [346]

——, Atmospheric oscillations, *Proc. Roy. Soc. London, Ser. A,* 158, 650, 1937. [266; 267]
——, The internal constitution of the Earth, *Geophys. J.*, 11, 85, 1966. [41]
PELLETIER, B. R., Development of submarine physiography in the Canadian Arctic and its relation to crustal movements, *Roy. Soc. Canada Spec. Publ.*, 9, 77, 1966. [413]
PETER, G., Preliminary results of a systematic geophysical survey south of the Alaska Peninsula, in "Continental Margins and Island Arcs," *Geol. Surv. Canada Paper*, 66-15, 223, 1966. [447]
—— and O. E. DEWALD, Geophysical reconnaissance in the Gulf of Tadjura, *Geol. Soc. Am. Bull.*, 80, 2313, 1969. [404]
—— and R. LATTIMORE, Magnetic structure of the Juan de Fuca – Gorda ridge area, *J. Geophys. Res.*, 74, 586, 1969. [438]
PETERMAN, Z. E., and C. E. HEDGE, Related strontium isotopic and chemical variations in oceanic basalts, *Geol. Soc. Am. Bull.*, 82, 493, 1971. [497]
PETERSON, M. N. A., et al., Initial reports of the Deep Sea Drilling Project, vol. 2, U.S. Govt. Printing Office, 1970. [370; 379; 418]
PETTIJOHN, F. J., Archean sedimentation, *Geol. Soc. Am. Bull.*, 54, 1955, 1943. [481]
——, "Sedimentary Rocks," 2d ed., Harper, 1957. [67; 68; 71; 76; 91]
——, The Canadian Shield – a status report, 1970, in A. J. BAER (ed.), "Symposium on Basins and Geosynclines of the Canadian Shield," *Geol. Surv. Canada Paper*, 70-40, 239, 1970. [479; 488]
PEYVE, A. V., Oceanic crust of the geologic past, *Geotectonika 1969*, English translation 4, 210, 1969. [345; 467]
PHILLIPS, J. D., and D. FORSYTH, Plate tectonics, paleomagnetism and the opening of the Atlantic, *Geol. Soc. Am. Bull.*, 83, 1579, 1972. [383; 406]
—— and B. G. LUYENDYK, Central North Atlantic plate motions over the last 40 million years, *Science.* 70, 727, 1970. [81; 392; 406]
PICARD, L., On Afro-Arabian graben tectonics, *Geol. Rundschau,* 59, 337, 1970. [405]
PICHAMUTHU, C. S., The Precambrian of India, in K. RANKAMA (ed.), "The Precambrian," vol. 3, Interscience, 1968. [483]
PIRES SOARES, J. M., Observations, géologique sur les îles du Cap Vert, *Soc. Geol. France*, B55 tl18 f6-7, 383, 1949. [378]
PITMAN, W. C., III, and D. E. HAYES, Sea-floor spreading in the Gulf of Alaska, *J. Geophys. Res.*, 73, 6571, 1968. [428; 448]
—— and J. R. HEIRTZLER, Magnetic anomalies over the Pacific-Antarctic ridge, *Science*, 154, 1164, 1966. [388; 427]
——, E. M. HERRON, and J. R. HEIRTZLER, Magnetic anomalies in the Pacific and sea-floor spreading, *J. Geophys. Res.*, 73, 2069, 1968. [423]
PLAFKER, G. P., and J. C. SAVAGE, Mechanism of the Chilean earthquakes of May 21 and 22, 1960, *Geol. Soc. Am. Bull.*, 81, 1001, 1970. [435]
PLUMSTEAD, E. P., Palaeobotany of Antarctica, in R. J. ADIE (ed.), "Antarctic Geology," North-Holland 637, 1964. [363]
POLDERVAART, A., Chemistry of the Earth's crust, in A. POLDERVAART (ed.), "The Crust of the Earth," *Geol. Soc. Am. Spec. Paper*, 62, 119, 1955. [98]
POLYCK, B. G., and YA. B. SMIRNOV, Relationship between terrestrial heat flow and the tectonics of continents, *Geotectonics*, 4, 205, 1968. [187]
PORATH, H., and A. DZIEWONSKI, Crustal resistivity anomalies from geomagnetic deep sounding studies, *Rev. Geophys.*, 9, 891, 1971. [446]
—— and D. I. GOUGH, Mantle conductive structures in the western United States from magnetometer array studies, *Geophys. J.*, 22, 161, 1971. [446]
PRESS, F., Displacements, strains and tilts at teleseismic distances, *J. Geophys. Res.*, 70, 2395, 1965. [105]
——, Density distribution in Earth, *Science*, 160, 1218, 1968a. [43; 46; 47; 304]

——, Earth models obtained by Monte Carlo inversion, *J. Geophys. Res.*, 73, 5223, 1968b. [43; 46; 47; 304]
——, The sub-oceanic mantle, *Science*, 165, 174, 1969. [304; 310]
——, Earth models consistent with geophysical data, *Phys. Earth Planet. Int.*, 3, 3, 1970a. [44-47; 304]
——, Regionalized earth models, *J. Geophys. Res.*, 75, 6575, 1970b. [304]
—— and M. EWING, Waves with Pn and Sn velocity at great distances, *Proc. Nat. Acad. Sci.*, 41, 24, 1955. [372]
PRESTON-THOMAS, H., L. G. TURNBULL, E. GREEN, T. M. DAUPHINEE, and S. N. KALRA, An absolute measurement of the acceleration due to gravity at Ottawa, *Can. J. Phys.*, 38, 824, 1960. [112]
PRICE, A. T., Magnetic variations and telluric currents, in T. F. GASKELL (ed.), "The Earth's Mantle," p. 125, Academic, 1967. [263]
——, Daily variations of the geomagnetic field, *Space Sci. Rev.*, 9, 151, 1969. [263]
——, The electrical conductivity of the Earth, *Quart. J. Roy. Astron. Soc.*, 11, 23, 1970. [263]
PRICE, P. B., and R. M. WALKER, Fossil tracks of charged particles in mica and the age of minerals, *J. Geophys. Res.*, 68, 4847, 1963. [144; 145]
PUMINOV, A. P., A chart of recent tectonics of the Arctic, *Dokl. Akad. Nauk. SSR*, 175, 901, 1967. (Canada Def. Res. Bd., Transl. T 487 R) [408]
QUENBY, J. J., and W. R. WEBBER, Cosmic ray cut-off rigidities and the Earth's magnetic field, *Phil. Mag.*, 4, 90, 1959. [292]
—— and G. J. WENK, Cosmic ray threshold rigidities and the Earth's magnetic field, *Phil. Mag.*, 7, 1457, 1962. [292]
QUENNELL, A. M., The structural and geomorphic evolution of the Dead Sea rift, *Quart. J. Geol. Soc. London.* 114, 1, 1958. [405]
QURESHY, M. N., Thickening of a basalt layer as a possible cause for the uplift of the Himalayas – a suggestion based on gravity data, *Tectonophysics,* 7, 137, 1969, [470]
RABINOWITZ, P. D., and W. B. F. RYAN, Gravity anomalies and crustal shortening in the Eastern Mediterranean, *Tectonophysics*, 10, 585, 1970. [466]
RADFORTH, N. W., The ancient flora and continental drift, *Roy. Soc. Canada Spec. Publ.* 9, 53, 1966. [363; 469]
RAFF, A. D., Sea-floor spreading – another rift, *J. Geophys. Res.*, 73, 3699, 1968. [437]
—— and R. G. MASON, Magnetic survey off the west coast of North America, 40° N latitude to 52° N latitude, *Geol. Soc. Am. Bull.*, 72, 1267, 1961. [386; 387; 438]
RAITT, R. W., G. G. SHOR, Jr., T. J. G. FRANCIS, and G. B. MORRIS, Anisotropy of the Pacific upper mantle, *J. Geophys. Res.*, 74, 3095, 1969. [430]
RALEIGH, C. B., and W. H. K. LEE, Sea-floor spreading and island arc tectonics, in A. R. MCBIRNEY (ed.), "Proceedings of the Andesite Conference," *Oregon Dep. Geol. Mineral Res., Bull.*, 65, 99, 1969. [90]
—— and M. S. PATERSON, Experimental deformation of serpentinite and its tectonic implications, *J. Geophys. Res.*, 70, 3965, 1965. [90; 294]
RAMBERG, H., "Gravity, Deformation of the Earth's Crust as Studied by Centrifuged Models," Academic, 1967. [340]
——, Mantle diapirism and its tectonic and magmagenetic consequences, *Phys. Earth Planet. Interiors*, 5, 45-60, 1972. [340]
RAMSEY, A. T. S., The Pre-Pleistocene stratigraphy and paleontology of the Palmer ridge area (Northeast Atlantic), *Mar. Geol.*, 9, 261, 1970. [415]
RAMSEY, W. H., On the nature of the Earth's core, *Monthly Notices Roy. Astron. Soc., Geophys. Suppl.*, 5, 409, 1949. [316]
RANALLI, G., and A. E. SCHEIDEGGER, Rheology of the tectonosphere as inferred from seismic aftershock sequences, *Ann. Geophys.*, 22, 293, 1969. [341]

RAPP, R. H., Gravitational potential of the Earth determined from a combination of satellite, observed and model anomalies, *J. Geophys. Res.*, 73, 6555, 1968. [111]
RASMUSSEN, J., and A. NOE-NYGAARD, Geology of the Faeroe Islands, *Geol. Surv. Denmark, I. Series*, no. 25, 1970. [412]
RAYLEIGH, J. W. S., On convection currents in a horizontal layer of fluid when the higher temperature is on the underside, *Phil. Mag.*, 32, 529, 1916. [346]
REA, D. K., Changes in structure and trend of fracture zones north of the Hawaiian ridge and relation to sea-floor spreading, *J. Geophys. Res.*, 75, 1421, 1970. [431]
READ, H. H., Granites and granites, in J. GILLULY (ed.), "Origin of Granite," *Geol. Soc. Am. Mem.* 28, 1, 1947. [64]
——, "The Granite Controversy," Murby, 1957. [64; 95; 488]
REDFIELD, A. C., and I. FRIEDMAN, Factors affecting the distribution of deuterium in the oceans, in "Symposium of Marine Geochemistry," Rhode Island Univ. Narragansett Marine Lab. Occas. Publ. 3, 149, 1965. [172]
REED, G. W., and A. TURKEVICH, The uranium content of two iron meteorites, *Natl. Acad. Sci. – Natl. Res. Council,* Publ. 400, 97, 1956. [237]
REID, J. B., Jr., and F. A. FREY, Rare Earth distributions in lherzolite and garnet pyroxenite and the constitution of the upper mantle, *J. Geophys. Res.*, 76, 1184, 1971. [497]
REILLY, W. I., Gravity map of New Zealand 1:4,000,000 Bouguer anomalies, *New Zealand Dept. Sci. Ind. Res.*, 1955a. [458]
——, Gravity map of New Zealand 1:4,000,000 isostatic anomalies, *New Zealand Dept. Sci. Ind. Res.*, 1965b. [458]
REINER, M., Twelve Lectures on Theoretical Rheology, North-Holland, 1949. [338]
REINHARDT, B. M., On the genesis and emplacement of ophiolites in the Oman Mountains' geosyncline, *Schweiz. Mineral. Petrog. Mitt.*, 49, 1, 1969. [85; 469]
REVELLE, R., and A. E. MAXWELL, Heat flow through the floor of the Eastern North Pacific Ocean, *Nature*, 170, 199, 1952. [372]
REYMENT, R. A., A brief review of the stratigraphic thickness of West Africa (Angola to Senegal), *Proc. 2d West African Micropalaeontolog. Colloquium*, p. 162, Holland, 1966. [420]
REYNOLDS, J. H., Xenology, *J. Geophys. Res.*, 68, 2939, 1963. [180]
REYNOLDS, P. H., and R. D. RUSSELL, Isotopic composition of lead from Balmat, New York, *Can. J. Earth Sci.*, 5, 1239, 1968. [163]
REYNOLDS, R. T., P. E. FRICKER, and A. L. SUMMERS, Effect of melting upon thermal models of the Earth, *J. Geophys. Res.*, 71, 573, 1966. [210]
REZANOV, I. A., and S. S. CHAMO, Reasons for absence of a "granitic" layer in basins of the South Caspian and Black Sea type, *Can. J. Earth Sci.*, 6, 671, 1969. [466; 467]
RICE, M. H., R. G. MCQUEEN, and J. M. WALSH, Compression of solids by strong shock waves, *Solid State Phys.*, 6, 1, 1958. [306]
RICHARDS, J. R., J. A. COOPER, A. W. WEBB, and P. J. COLEMAN, Potassium-argon measurements of the age of basal schists in the British Solomon Islands, *Nature*, 211, 1241, 1966. [456]
RICHARDSON, A., and N. D. WATKINS, Palaeomagnetism of Atlantic Islands: Fernando Noronha, *Nature*, 215, 1470, 1967. [471]
RICHTER, C. F., An instrumental earthquake magnitude scale, *Bull. Seismol. Soc. Am.*, 25, 1, 1935. [51]
——, "Elementary Seismology," Freeman, 1958. [52]
RIDD, M. F., Faults in South-East Asia, and the Andaman Rhombochasm, *Nature*, 229, 51, 1971. [425]
RIKITAKE, T., and Y. HAGIWARA, Non-steady state of a Herzenberg dynamo, *J. Geomagn. Geoelectr.*, 18, 393, 1966. [237]
——, S. MIYAMURA, I. TSUBOKAWA, S. MURAUCHI, S UYEDA, H. KUNO, and

M. GORAI, Geophysical and geological data in and around the Japan arc, *Can. J. Earth Sci.*, 5, 1101, 1968. [451]
RINGWOOD, A. E., On the chemical evolution and densities of the planets, *Geochim. Cosmochim. Acta*, 15, 257, 1959. [307]
——, Some aspects of the thermal evolution of the Earth, *Geochim. Cosmochim. Acta*, 20, 241, 1960. [201; 315]
——, Mineralogical constitution of the deep mantle, *J. Geophys. Res.*, 67, 4005, 1962a. [302; 303]
——, A model for the upper mantle, *J. Geophys. Res.*, 67, 857, 1962b. [303]
——, Chemical evolution of the terrestrial planets, *Geochim. Cosmochim. Acta*, 30, 41, 1966a. [313]
——, The chemical composition and origin of the Earth, in P. M. HURLEY (ed.), "Advances in Earth Science," 287, MIT, 1966b. [210; 211; 318]
——, Phase transformations in the mantle, *Earth Planet. Sci. Lett.*, 5, 401, 1969. [308; 309]
—— and D. H. GREEN, An experimental investigation of the gabbro-eclogite transformation and some geophysical implications, *Tectonophysics*, 3, 383, 1966. [298-301; 305; 487]
—— and A. MAJOR, Synthesis of $Mg_2SiO_4 - Fe_2SiO_4$ spinel solid solutions, *Earth Planet. Sci. Lett.*, 1, 241, 1966. [309]
—— and ——, Apparatus for phase transformation studies at high pressures and temperatures, *Phys. Earth Planet. Int.*, 1, 164, 1968. [309]
RITSEMA, A. R., The fault-plane solutions of earthquakes of the Hindu Kush Centre, *Tectonophysics*, 3, 147, 1966. [470]
——, On the origin of the Western Mediterranean Sea basins, *Tectonophysics*, 10, 609, 1970. [467]
ROBERTS, D. G., Structural evolution of the rift zones in the Middle East, *Nature*, 223, 55, 1969. [404]
ROBERTS, G. O., Spatially periodic dynamos, *Phil. Trans. Roy. Soc. London, Ser. A*, 266, 535, 1970. [235; 236]
ROCHE, A., Sur les caràcteres magnétiques du système éruptif de Gerovie, *Compt. Rend. Acad. Sci.*, 230, 113, 1950a. [243]
——, Anomalies magnétiques accompagnant les massifs de massifs de pépérites de la Limagne d'Auvergne, *Compt. Rend. Acad. Sci.*, 230, 1603, 1950b. [243]
——, Sur les inversions de l'aimantation remanente des roches volcaniques dans les monts d'Auvergne, *Compt. Rend. Acad. Sci.*, 230, 1132, 1951. [243]
——, Sur l'origne des inversions de l'aimantation constantées dans les roches d'Auvergne. *Compt. Rend. Acad. Sci.*, 236, 107, 1953. [243]
ROCHESTER, M. G., Geomagnetic westward drift and irregularities in the Earth's rotation, *Phil. Trans. Roy. Soc., London, Ser. A*, 252, 531, 1960. [235]
ROD, E., Strike-slip faults of Northern Venezuela, *Am. Assoc. Pet. Geol. Bull.*, 40, 457, 1956. [437]
RODGERS, J., "The Tectonics of the Appalachians," Interscience, 1970. [474]
RODOLFO, K. S., Bathymetry and marine geology of the Andaman basin, and tectonic implications for Southeast Asia, *Geol. Soc. Am. Bull.*, 80, 1203, 1969. [425]
ROMER, A. S., Fossils and Gondwanaland, in G. PIEL (ed.), "Gondwanaland Revisited: New Evidence for Continental Drift," *Proc. Am. Phil. Soc.*, 112, 335, 1968. [361]
RONA, P. A., Possible salt domes in the deep Atlantic off Northwest Africa, *Nature,* 224, 141, 1969. [417]
——, Comparison of continental margins of Eastern North America at Cape Hatteras and Northwestern Africa at Cap Blanc, *Am. Assoc. Pet. Geol. Bull.*, 54, 109, 1970. [417]
——, Depth distribution in ocean basins and plate tectonics, *Nature,* 231, 179, 1971. [496]

RONOV, A. B., and A. A. YAROSHEVSKY, Chemical composition of the Earth's crust, in P. J. HART (ed.), "The Earth's Crust and Upper Mantle," *Am. Geophys. Union Monograph* 13, 37, 1969. [83; 97-99]

ROSCOE, S. M., Huronian rocks and uraniferous conglomerates in the Canadian Shield, *Canada. Geol. Surv. Paper* 68-40, 1969. [488]

ROSE, D. C., K. B. FENTON, J. KATZMAN, and J. A. SIMPSON, Latitude effect of the cosmic ray nuclear and meson components at sea level from the Arctic to the Antarctic, *Can. J. Phys.*, 34, 968, 1956. [291]

ROSE, R. D., H. M. PARKER, R. A. LOWRY, A. R. KUHLTHAU, and G. W. BEAMS, Determination of the gravitational constant G. *Phys. Rev. Lett.*, 23, 655, 1969. [105]

ROSS, D. A., Sediments of the Northern Middle America trench, *Geol. Soc. Am. Bull.*, 82, 303, 1971. [437]

ROSS, R. J., Jr., and J. K. INGHAM, Distribution of the Toquimatable head (Middle Ordovician Whiterock) faunal realm in the Northern Hemisphere, *Geol. Soc. Am. Bull.*, 81, 393, 1970. [473]

ROTHE, P., and H.-U. SCHMINCKE, Contrasting origins of the Eastern and Western Islands of the Canarian Archipelago, *Nature*, 218, 1152, 1968. [418]

ROY, R. F., and D. D. BLACKWELL, Heat generation of plutonic rocks and continental heat flow provinces, *Earth Planet. Sci. Lett.*, 5, 4, 1968b. [198]

——, E. R. DECKER, D. D. BLACKWELL, and F. BIRCH, Heat flow in the United States, *J. Geophys. Res.*, 73, 5207, 1968a. [198]

RUBEY, W. W., Geologic history of sea water, an attempt to state the problems, *Geol. Soc. Am. Bull.*, 62, 1111, 1951. [59]

—— and M. K. HUBBERT, Role of fluid pressures in mechanics of overthrust faulting, *Geol. Soc. Am. Bull.*, 70, 115, 1959. [328]

RÜEGG, W., An intra-Pacific ridge, its continuation on to the Permian mainland and its bearing on the hypothetical Pacific landmass, *Proc. 21st Intern. Geol. Congr.*, pt. 10, 29, 1960. [437]

RUNCORN, S. K., The Earth's core, *Trans. Am. Geophys. Union*, 35, 49, 1954. [233; 234]

——, Paleomagnetic comparisons between Europe and North America, *Proc. Geol. Assoc. Can.*, 8, 301, 1956. [383]

——, Towards a theory of continental drift, *Nature*, 193, 311, 1962a. [318; 383]

——, Paleomagnetic evidence for continental drift and its geophysical cause, in S. K. RUNCORN (ed.), "Continental Drift," p. 1, Academic, 1962b. [318; 320; 383]

——, "Continental Drift," Academic, 1962c. [346]

——, Changes in the convection pattern in the Earth's mantle and continental drift: Evidence for a cold origin of the Earth, *Phil. Trans. Roy. Soc. London, Ser. A*, 258, 228, 1965. [346; 492]

——, A. C. BENSON, A. F. MOORE, and D. H. GRIFFITHS, Measurements of the variation with depth of the main geomagnetic field, *Phil. Trans. Roy. Soc. London, Ser. A*, 244, 113, 1951. [231]

RUSSELL, H. N., "The Solar System and Its Origin," Macmillan, 1935. [8]

RUSSELL, R. D., The systematics of double spiking, *J. Geophys. Res.*, 76, no. 20, 4949, 1971. [152]

——, Evolutionary model for lead isotopes in conformable ores and in ocean volcanics, *Rev. Geophys. Space Phys.*, 10, 529–549 (1972). [164]

—— and L. H. AHRENS, Additional regularities among discordant lead-uranium ages, *Geochim. Cosmochim. Acta*, 11, 213, 1957. [150]

——, W. F. SLAWSON, T. J. ULRYCH, and P. H. REYNOLDS, Further applications of concordia plots to rock lead isotope abundances, *Earth Planet. Sci. Lett.*, 3, 284, 1968. [164]

RUTLAND, R. W. R., A tectonic study of part of the Philippine fault zone, *Quart. J. Geol. Soc. London*, 123, 193, 1968. [454]

RUTTEN, M. G., "The Geology of Western Europe," Elsevier, 1969. [360; 462; 464; 474]

RYAN, W. F. B., and B. C. HEEZEN, Ionian Sea submarine canyons and the 1908 Messina turbidity current, *Geol. Soc. Am. Bull.*, 76, 915, 1965. [83; 464]

SABINE, P. A., Rockall, an unusual occurrence of Tertiary granite, *Proc. Geol. Soc. London*, 1621, 51, 1965. [412]

SACHET, M.-H., Geography and land ecology of Clipperton Island, *Atoll. Res. Bull.* 86, NAS-NRC, 1962. [432]

SACKS, S., Diffracted wave studies of the Earth's core. 1. Amplitude, core size, and rigidity, *J. Geophys. Res.*, 71, 1173, 1966. [32]

——, Diffracted P-wave studies of the Earth's core. 2. Lower mantle velocity, core size, lower mantle structure, *J. Geophys. Res.*, 72, 2589, 1967. [32]

SAINT-AMAND, P., Los terremotos de Mayo-Chile 1960, U.S. Naval Ordnance Test Station, Michelson Laboratories, Tech. Art. 14, NOTS TP 2701, 1961. [435]

SAITO, T., The occurrence of Prequaternary planktonic foraminifera in the oceans, S. C. O. R. Working Group 19, Symposium on Micropaleontology of Marine Bottom Sediments, 1967. [423]

——, L. BURCKLE, and M. EWING, Lithology and Paleontology of the reflective layer horizon A, *Science*, 154, 1173, 1966b. [418]

——, M. EWING, and L. H. BURCKLE, Tertiary sediment from the Mid-Atlantic ridge, *Science*, 151, 1075, 1966a. [378]

SANTO, T. A., Dispersion of surface waves along various faults to Uppsala, Sweden, Pt. II, Arctic and Atlantic Oceans, *Ann. Geofis.*, 15, 277, 1962. [408]

SASAJIMA, S., J. NISHIDA, and M. SHIMADA, Paleomagnetic evidence of a drift of the Japanese main island during the Paleogene period, *Earth Planet. Sci. Lett.*, 5, 135, 1968. [451]

SAUNDERS, J. R., (ed.), "Transactions of the Fourth Caribbean Geological Conference," Caribbean Printers, p. 457, 1968. [437]

SCHAFER, C., and J. BROOKE, Cores from the crest of the Mid-Atlantic ridge, *Geotimes*, 15, 14, 1970. [370]

SCHEIDEGGER, A. E., Rheology of the Earth, the basic problem of geodynamics, *Can. J. Phys.*, 35, 383, 1957. [340; 341]

——, "Principles of Geodynamics," 2d ed., Springer-Verlag, 1963. [342; 344; 345; 347; 501]

——, On the rheology of rock creep, *Rock Mech.*, 2, 138, 1970. [338]

SCHENCK, H. G., and S. W. MULLER, Stratigraphic terminology, *Geol. Soc. Am. Bull.*, 52, 1419, 1941. [129]

SCHLANGER, S. O., and J. W. BROOKHART, Geology and water resources of Falalop Island, Ulithi Atoll, Western Caroline Islands, *Am. J. Sci.*, 253, 553, 1955. [450]

SCHMIDT, O., A meteoric theory of the origin of the Earth and planets, *Dokl. Akad. Nauk SSSR*, 45, 6, 229, 1944. [10]

——, "A Theory of the Origin of the Earth; Four Lectures," Lawrence and Wishart, 1959. [10]

SCHNEIDER, E. D., The deep sea – a habitat for petroleum? *Under Sea Technology,* 10, 32, 1969. [75; 398]

——, P. J. FOX, C. D. HOLLISTER, H. D. NEEDHAM, and B. C. HEEZEN, Further evidence of contour currents in the Western North Atlantic, *Earth Planet. Sci. Lett.*, 2, 351, 1967. [81; 83]

SCHOLL, D. W., E. C. BUFFINGTON, and D. M. HOPKINS, Geologic history of the continental margin of North America in the Bering Sea, *Mar. Geol.*, 6, 297, 1968. [449]

——, M. N. CHRISTENSEN, R. VON HUENE, and M. S. MARLOW, Peru-Chile trench sediments and sea-floor spreading, *Geol. Soc. Am, Bull.*, 81, 1339, 1970. [433]

—— and R. VON HUENE, Comments on paper by R. L. CHASE and E. T. BUNCE, Underthrusting of the eastern margin of the Antilles by the floor of the Western North Atlantic Ocean and origin of the Barbados ridge, *J. Geophys. Res.*, 75, 488, 1970. [395]

——, ——, and J. B. RIDLON, Spreading of the ocean-floor: Undeformed sediments in the Peru-Chile trench, *Science*, 159, 869, 1968. [433; 435]

SCHOPF, J. M., Gondwanaland paleobotany, *Antarct. J.*, 5, 62, 1970. [363]

SCHREIBER, B. C., New evidence concerning the age of the Hawaiian ridge, *Geol. Soc. Am. Bull.*, 80, 2601, 1969. [432]

SCHUCHERT, C., Sites and nature of the North American geosynclines, *Geol. Soc. Am. Bull.*, 34, 151, 1923. [472]

SCHUILING, R. D., Continental drift and oceanic heat-flow, *Nature*, 210, 1027, 1966. [366]

SCHWARZBACH, M., "Climates of the Past," Van Nostrand, 1963. [361]

SCLATER, J. G., C. E. CORRY, and V. VACQUIER, In situ measurement of the thermal conductivity of ocean floor sediments, *J. Geophys. Res.*, 74, 1070, 1969. [185]

—— and J. FRANCHETEAU, The implications of terrestrial heat flow observations on current tectonic and geochemical models of the crust and upper mantle of the earth, *Geophys. J.*, 20, 509, 1970. [187; 366]

—— and C. G. A. HARRISON, Elevation of mid-ocean ridges and the evolution of the Southwest Indian ridge, *Nature*, 230, 175, 1971. [425; 496]

—— and H. W. MENARD, Topography and heat flow of the Fiji Plateau, *Nature*, 216, 991, 1967. [196; 458]

SEARLE, R. C., Evidence from gravity anomalies for thinning of the lithosphere beneath the rift valley in Kenya, *Geophys. J.*, 21, 13, 1970. [401]

SEDERHOLM, J. J., "Selected Works: Granites and Migmatites," Wiley, 1967. [64; 95]

SEIGEL, H. O., A theory of fracture of material and its application to geology, *Trans. Am. Geophys. Union,* 31, 611, 1950. [329]

SEN, H. K., and M. L. WHITE, Thermal and gravitational excitation of atmospheric oscillations, *J. Geophys. Res.* 60, 483, 1955. [267]

SERRALHEIRO, A., Formacoes sedimentares do Arquipelago de Cabo Verde, Agrupamento Cientifico de Geologia da Universidade de Lisboa da Junta de Investigacoes do Ultramar Lisboa, 1968. [378]

——, Geologia da Ilha Maio (Cabo Verde), *Junta de Investig. do Ultramar*, 1970. [418]

SERSON, P. H., S. E. MACK, and K. WHITHAM, A three-component airborne magnetometer, *Publ. Dominion. Obs., Canada*, 19, 1957. [370]

SEYFERT, C. K., Undeformed sediments in oceanic trenches with sea floor spreading, *Nature*, 222, 70, 1969. [395; 433]

SHAND, S. J., "Eruptive Rocks," Wiley, 1943. [63]

SHAPIRO, J. N., and L. KNOPOFF, Reduction of shock wave equations of state to isothermal equations of state, *J. Geophys. Res.*, 74, 1435, 1969. [306]

SHARP, R. P., "Glaciers" (Condon Lectures), University of Oregon Press, 1960. [173; 175]

SHAW, D. M., Radioactive elements in the Canadian Precambrian Shield and the interior of the Earth, in L. H. AHRENS (ed.), "Origin and Distribution of the Elements," p. 855, Pergamon, 1968. [208]

——, G. A. REILLY, J. R. MUYSSON, G. E. PATTENDEN, and F. E. CAMPBELL, An estimate of the chemical composition of the Canadian Precambrian Shield, *Can. J. Earth Sci.*, 4, 829, 1967. [96; 98]

SHEA, M. A., A comparison of theoretical and experimental cosmic-ray equators, *J. Geophys. Res.*, 74, 2407, 1969. [293]

SHELTON, J. S., "Geology Illustrated," Freeman, 1966. [16]

SHERIDAN, R. E., R. E. HOUTZ, C. L. DRAKE, and M. EWING, Structure of continental margin off Sierra Leone, West Africa, *J. Geophys. Res.*, 74, 2512, 1969. [418; 419]

SHILLIBEER, H. A., and R. D. RUSSELL, The argon-40 content of the atmosphere and the age of the Earth, *Geochim. Cosmochim. Acta*, 8, 16, 1955. [59]

SHOR, G. G., Structure of the Bering Sea and the Aleutian ridge, *Mar. Geol.*, 1, 213, 1964. [447]

—— and R. L. FISHER, Middle America trench: Seismic-refraction studies, *Geol. Soc. Am. Bull.*, 72, 721, 1961. [433; 437]

—— and R. W. RAITT, Explosion seismic reflection studies of the crust and upper mantle in the Pacific and Indian oceans, in P. J. HART (ed.), "The Earth's Crust and Upper Mantle," *Am. Geophys. Union Monograph* 13, 225, 1969. [82; 428]

SIEBERT, M., Atmospheric tides, *Adv. Geophys.*, 7, 105, 1961. [267]

SIEDNER, G., and J. A. MILLER, K-Ar age determinations on basaltic rocks from Southwest Africa and their bearing on continental drift, *Earth Planet. Sci. Lett.*, 4, 451, 1968. [420]

SILVER, E. A., Transitional tectonics and late Cenozoic structure of the continental margin off northernmost California, *Geol. Soc. Am. Bull.*, 82, 1, 1971. [439]

SILVER, L. T., Radioactive Dating Symposium held by the Internatl. Atomic Energy Agency in co-operation with the Joint Commission on Applied Radioactivity, 1963. [150]

SIMKIN, T., and K. A. HOWARD, Caldera collapse in the Galapagos Islands, 1968, *Science*, 169, 429, 1970. [437]

SIMMONS, G., and K. HORAI, Heat flow data 2, *J. Geophys. Res.*, 73, 6608, 1968. [187]

—— and A. NUR, Granites: Relation of properties in situ to laboratory measurements, *Science*, 162, 789, 1968. [185]

—— and R. F. ROY, Heat flow in North America, in P. J. HART (ed.), "The Earth's Crust and Upper Mantle," *Am. Geophys. Union Monograph* 13, 78, 1969. [444]

SIMON, F. E., The melting of iron at high pressures, *Nature*, 172, 746, 1953. [205]

SIMPSON, E. S. W., The geology and mineral resources of Mauritius, *Colonial Geol. Mineral Resources*, 1, 217, 1951. [424]

SINGER, S. F., E. MAPLE, and W. A. BOWEN, Evidence for ionosphere currents from rocket experiments near the geomagnetic equator, *J. Geophys. Res.*, 56, 265, 1951. [265]

SITTLER, C., The sedimentary trough of the Rhinegraben, *Tectonophysics*, 8, 543, 1969. [403]

SKVOR, V., The Caribbean Area: A case of destruction and regeneration of continent, *Geol. Soc. Am. Bull.*, 80, 961, 1969. [437]

SLICHTER, L. B., Cooling of the Earth, *Geol. Soc. Am. Bull.*, 52, 561, 1941. [209]

SLOSS, L. L., Sequences in cratonic interior of North America, *Geol. Soc. Am. Bull.*, 74, 93, 1963. [77; 492]

——, Orogeny and epeirogeny: The view from the craton, *Trans. N. Y. Acad. Sci.*, 28, 579, 1966. [77]

SMITH, A. G., Potassium-argon decay constants and age tables, in "The Phanerozoic Time Scale," A Symposium Dedicated to Professor Arthur Holmes, *J. Geol. Soc. London*, 120 S, 129, 1964. [140]

—— and A. HALLAM, The fit of the southern continents, *Nature*, 225, 139, 1970. [385; 422]

SMITH, J. D., and J. H. FOSTER, Geomagnetic reversal in Brunhe normal polarity epoch, *Science*, 163, 565, 1969. [248]

SMITH, P. J., The intensity of the ancient geomagnetic field: A review and analysis, *Geophys. J.*, 12, 321, 1967. [241-243; 247]

SOUTHER, J. G., Volcanism and its relationship to recent crustal movements in the Canadian Cordillera, *Can. J. Earth Sci.*, 7, 553, 1970. [439]

SOWERBUTTS, W. T. C., Crustal structure of the East African plateau and rift valleys from gravity measurements, *Nature*, 223, 143, 1969. [398]
SPALL, H., Precambrian apparent polar wandering: Evidence from North America, *Earth Planet. Sci. Lett.*, 10, 273, 1971. [488]
SPITZER, L., The dissipation of planetary filaments, *Astrophys. J.*, 90, 675, 1939. [9]
SPRIGG, R. C., "Sedimentation in the Adelaide Geosyncline and the Formation of the Continental Terrace," Sir Douglas Mawson Anniversary Volume, p. 153, University of Adelaide, 1952. [422]
SPROLL, W. P., and R. S. DIETZ, Morphological continental drift fit of Australia and Antarctica, *Nature*, 222, 345, 1969. [422]
STANLEY, D. J., C. E. GEHIN, and C. BARTOLINI, Flysch-type sedimentation in the Alboran Sea, Western Mediterranean, *Nature*, 228, 979, 1970. [462]
STANTON, R. L., and R. D. RUSSELL, Anomalous leads and the emplacement of lead sulfide ores, *Econ. Geol.*, 54, 588, 1959. [162]
STARK, J. T., and R. L. HAY, Zenoliths in pyroclastic breccia of Truk Islands, Western Pacific, *Geol. Soc. Am., Bull.*, 66, 1621, 1955. [450]
STAUB, R., "Der Bewegungsmechanismus der Erde," Gebruder Borntraeger, 1928. [353; 360]
STAUDER, W., Mechanism of the Rat Island earthquake sequence of February 4, 1965, with relation to island arcs and sea floor spreading, *J. Geophys. Res.*, 73, 3847, 1968. [448]
STEARNS, H. T., Geology of the Wallis Islands, *Geol. Soc. Am. Bull.*, 56, 840, 1945. [450]
——, "Geology of the State of Hawaii," Pacific Books, 1966. [432]
STEHLI, F. G., A paleoclimatic test of the hypothesis of an axial dipolar magnetic field, in R. A. PHINNEY (ed.), "The History of the Earth's Crust," p. 195, Princeton, 1968. [366]
——, R. G. DOUGLAS, and N. D. NEWELL, Generation and maintenance of gradients in taxonomic diversity, *Science*, 164, 947, 1969. [361]
STEIGER, R. H., and G. J. WASSERBURG, Comparative U-Th-Pb systematics in 2.7 x 10^9 yr plutons of different geologic histories, *Geochim. Cosmochim. Acta*, 33, 1213, 1969. [150]
STEIN, W. A., and F. C. GILLETT, Spectral distribution of infrared radiation from the trapezium region of the Orion nebula, *Astrophys. J. Lett.*, 155, L197, 1969. [314]
STEINBERG, B. S., and L. A. RIVOSH, Geophysical study of the Kamchatka volcanoes, *J. Geophys. Res.*, 70, 3341, 1965. [447]
STEINHART, J. S., S. R. HART, and T. J. SMITH, "Heat Flow," *Carnegie Inst. Annual Report, 1967-68,* 360, 1969. [184]
STEINMANN, G., Geologische Beobachtungen in den Alpen II. Die Scharttsche überfaltungs-theorie und die Geologische Bedeutung des Tiefseeabsatz und der Ophiolithischen Massengesteine, *Ber. Naturforsch. Ges. Freiburg Breisgau*, 1, Bd. 16, 44, 1905. [466]
STEPHENSON, P. J., Some geological observations on Heard Island, *Antarct. Geol. SCAR Proc.*, p. 14, 1963. [422]
STEVENS, R. K., Cambro-Ordovician flysch sedimentation and tectonics in West Newfound-land and their possible bearing on a proto-Atlantic Ocean, *Geol. Assoc. Canada Spec. Paper* 7, 165, 1970. [85]
STEWART, J. H., Basin and range structures: A system of horsts and graben produced by deep-seated extension, *Geol. Soc. Am. Bull.*, 82, 1019, 1971. [443]
——, J. P. ALBERS, and F. G. POOLE, Reply to paper by F. A. WRIGHT and B. W. TROXEL, *Geol. Soc. Am. Bull.*, 81, 2175, 1970. [438]
STILLE, H., "Einfuhrung in den Bau Amerikas," p. 83, Borntraeger, 1940. [472]
STISHOV, S. M., and S. V. POPOVA, New dense polymorphic modification of silica, *Geokhimiya*, 10, 837, 1961. [19; 308]

STÖCKLIN, J., Structural history and tectonics of Iran; a review, *Am. Assoc. Pet. Geol. Bull.*, 52, 1229, 1968. [468]

STOCKWELL, C. H., Age determinations and geological studies, *Geol. Surv. Canada Paper* 64-17, pt. 2, p. 1, 1964. [490]

STONE, D. B., Geophysics in the Bering Sea and surrounding areas: A review, *Tectonophysics*, 6, 433, 1968. [448]

STONELEY, R., The structural development of the Gulf of Alaska sedimentary province in Southern Alaska, *Quart. J. Geol. Soc. London*, 123, 25, 1967. [448]

STOREY, L. R. O., An investigation of whistling atmospherics, *Phil. Trans. Roy. Soc. London, Ser. A*, 246, 113, 1953. [280]

STÖRMER, C., "The Polar Aurora," Oxford, 1955. [282; 286]

STØRMER, L., Some aspects of the Caledonian geosyncline and foreland west of the Baltic Shield, *Quart. J. Geol. Soc. London*, 123, 183, 1967. [474]

STRANGE, W. E., and M. A. KHAN, On the relation between satellite gravity results, heat flow and convection currents, *Trans. Am. Geophys. Union*, 46, 544, 1965. [199]

STRANGWAY, D. W., "History of the Earth's Magnetic Field," McGraw-Hill, 1970. [383]

STRIDE, A. H., J. R. CARREY, D. G. MOORE, and R. H. BEDERSON, Marine Geology of the Atlantic continental margin of Europe, *Phil. Trans. Roy. Soc. London, Ser. A*, 264, 31, 1969. [413]

STRONG, D. F., and M. F. J. FLOWER, The significance of sandstone inclusions in lavas of the Comores Archipelago, *Earth Planet. Sci. Lett.*, 7, 47, 1969. [424]

—— and C. JACQUOT, The Karthala Caldera, Grande Comore, *Bull. Volcanol.*, 34, 663, 1970. [424]

STRONG, H. M., The experimental fusion curve of iron to 96,000 atmospheres, *J. Geophys. Res.*, 64, 653, 1959. [205]

SUESS, E., "Das Antilitz der Erde," 3 vols., Prague, 1885-1909; "The Face of the Earth," translated by H. B. C. Sollas, 5 vols., Oxford, 1904-1924. [347; 397; 398; 469]

SUGGATE, R. P., The alpine fault, *Trans. Roy. Soc. New Zealand*, 2, 105, 1963. [458]

SUGIURA, M., and S. CHAPMAN, The average morphology of geomagnetic storms with sudden commencement, *Abhandl. Akad. Wiss. Göttingen, Math-Physik. Kl., Sonderh.* 4, 1960. [268]

—— and J. P. HEPPNER, The Earth's magnetic field, in W. N. HESS (ed.), "Introduction to Space Science," Gordon and Breach, 1965. [270]

SUMMERHAYES, C. P., Marine geology of the New Zealand Subantarctic sea floor, *New Zealand Dep. Sci. Ind. Res. Bull.*, 190, 1969. (New Zealand Oceanogr. Inst. Mem. 50, 1969.) [431; 458]

SUTTON, J., Long-term cycles in the evolution of the continents, *Nature*, 198, 731, 1963. [492]

SWANSON, D. A., D. B. JACKSON, W. A. DUFFIELD, and D. W. PETERSON, Maura Ulu Eruption, Kilauea Volcano, *Geotimes*, 16, 12, 1971. [432]

SYKES, L. R., Deep-focus earthquakes in the New Hebrides region, *J. Geophys. Res.*, 69, 5353, 1964. [456]

——, The seismicity of the Arctic, *Bull. Seismol. Soc. Am.*, 55, 519, 1965. [408]

——, Mechanism of earthquakes and nature of faulting on the mid-oceanic ridges, *J. Geophys. Res.*, 72, 2131, 1967. [332; 374; 421; 427]

——, Seismological evidence for transform faults, sea floor spreading and continental drift in R. A. PHINNEY (ed.), "The History of the Earth's Crust," p. 120, Princeton, 1968. [332; 404]

——, Seismicity of the Indian Ocean and a possible nascent island arc between Ceylon and Australia, *J. Geophys. Res.*, 75, 5041, 1970. [396; 421; 425]

——, B. L. ISACKS, and J. OLIVER, Spatial distribution of deep and shallow earthquakes of small magnitudes in the Fiji-Tonga region, *Bull. Seismol. Soc. Am.*, 59, 1093, 1969. [458]

TAKAI, F., T. MATSUMOTO, and R. TORIYAMA, "Geology of Japan," University of Tokyo Press, 1963. [451]
TAKEUCHI, H., On the Earth tide in the compressible Earth of varying density and elasticity, *Trans. Am. Geophys. Union*, 31, 651, 1950. [35]
——, Time scales of isostatic compensations, *J. Geophys. Res.*, 68, 2357, 1963. [310]
—— and Y. HASEGAWA, Viscosity distribution within the Earth, *Geophys. J.*, 9, 503, 1965. [311]
—— and H. KANAMORI, Equations of state of matter from shock wave experiments, *J. Geophys. Res.*, 71, 3985, 1966. [306; 321]
——, S. UYEDA, and H. KANAMORI, "Debate about the Earth," 2d ed., Freeman, Cooper, 1970. [342]
TALWANI, M., A review of marine geophysics, *Mar. Geol.*, 2, 29, 1964. [370]
—— and X. LE PICHON, Gravity field over the Atlantic Ocean, in P. J. HART (ed.), "The Earth's Crust and Upper Mantle," *Am. Geophys. Union Monograph* 13, p. 341, 1969. [107]
——, ——, and M. EWING, Crustal structure of the mid-ocean ridges, *J. Geophys. Res.*, 70, 341, 1965. [376; 377]
——, J. L. WORZEL, and M. EWING, Gravity anomalies and crustal section across the Tonga trench, *J. Geophys. Res.*, 66, 1265, 1961. [458]
TARLING, D. H., The paleomagnetism of the Samoan and Tongan Islands, *Geophys. J.*, 10, 497, 1966. [450]
——, Results of a palaeomagnetic reconnaissance of the New Hebrides and New Caledonia, *Tectonophysics*, 4, 55, 1967a. [456]
——, The paleomagnetism of some rock samples from Viti Levu, Fiji, *New Zealand J. Geol. Geophys.*, 10, 1235, 1967b. [458]
——, Gondwanaland, paleomagnetism and continental drift, *Nature*, 229, 17, 1971. [422]
TARPLEY, J. D., The ionospheric wind dynamo – I Lunar tide, *Planetary Space Sci.*, 18, 1075, 1970a. [267]
——, The ionospheric wind dynamo – II Solar tides, *Planetary Space Sci.*, 18, 1091, 1970b. [267]
TATSUMOTO, M., Genetic relations of oceanic basalts as indicated by lead isotopes, *Science*, 153, 1094, 1966. [163]
——, Lead isotopes in volcanic rocks and possible ocean-floor thrusting beneath island arcs, *Earth Planet. Sci. Lett.*, 6, 369, 1969. [454]
TAUBENECK, W. H., An evaluation of tectonic rotation in the Pacific Northwest, *J. Geophys. Res.*, 71, 2113, 1966. [438]
TAYLOR, F. B., Bearing of the Tertiary mountain belt on the origin of the Earth's plan, *Geol. Soc. Am. Bull.*, 21, 179, 1910. [352]
TAYLOR, G. I., The oscillations of the atmosphere, *Proc. Roy. Soc. London, Ser. A*, 156, 318, 1936. [266]
TAYLOR, H. P., JR., and S. EPSTEIN, $0^{18}/0^{16}$ ratios in rocks and coexisting minerals of the Skaergaard intrusion, East Greenland, *J. Petrol.*, 4, 51, 1963. [177]
TAYLOR, P. T., I. ZIETZ, and L. S. DENNIS, Geologic implications of aeromagnetic data for the eastern continental margin of the United States, *Geophysics*, 33, 755, 1968. [417]
TAYLOR, S. R., Trace element chemistry of andesites and associated calc-alkaline rocks, in "Proceedings of the Andesite Conference," *Intern. Upper Mantle Proj., Sci. Rept.* 16, 43, 1969. [95]
—— and A. J. R. WHITE, Geochemistry of andesites and the growth of continents, *Nature*, 208, 271, 1965. [95]
TAZIEFF, H., Potash-bearing evaporites – Danakil area, *Econ. Geol.*, 64, 228, 1969. [404]

——, The Afar triangle, *Sci. Am.*, 222, 32, 1970. [404]

TER HAAR, D., On the origin of the solar system, *Ann. Rev. Astron. Astrophys.*, 5, 267, 1967. [13]

TERMIER, P., Les nappes des Alpes orientales et la synthese des Alpes, *Bull. Soc. Geol. France*, 3, 711, 1903. [360]

TEVES, J. S., Philippine structural history and relation with neighbouring areas, *Philippine Geologist*, 9, 18, 1955. [454;455]

THELLIER, E., and O. THELLIER, Sur l'intensité du champ magnétique terrestre dans le passé historique et géologique. *Ann. Geophys.*, 15, 285, 1959 [This paper contains full references to the authors' earlier work.] [241]

THOMPSON, G. A., The rift system of the Western United States, in T. N. IRVINE (ed.), "The World Rift System," *Geol. Surv. Canada Paper* 66-14, 280, 1966. [435]

THOMPSON, J. B., Jr., P. ROBINSON, T. N. CLIFFORD, and N. J. TRASK, Jr., Nappes and gneiss domes in west-central New England, in E.-AN ZEN, and J. B. THOMPSON (eds.), "Studies of Appalachian Geology," Interscience, 1968. [483]

THOMPSON, J. E., and N. H. FISHER, Mineral deposits of New Guinea and Papua and their tectonic setting, *Australia Dept. Nat. Dev., Bur. Mineral Resources, Geol. Geophys.*, Records 1965/10, 1965. [456]

THOMSON, B. P., Precambrian rock groups in the Adelaide geosyncline: a new subdivision, *Geol. Surv. South Australia Q. Geol. Notes* 9, 1, 1964. [489]

THORARINSSON, S. (ed.), On the geology and geophysics of Iceland (Guide to Excursion No. A2), *Intern. Geol. Congr., 21st,* Norway, 1960. [381]

——, The median zone of Iceland, in T. N. IRVINE (ed.), "The World Rift System," *Geol. Surv. Canada Paper* 66-14, 187, 1966. [381]

——, Some problems of volcanism in Iceland, *Geol. Rundschau*, Bd. 57, 1, 1967. [381]

THULIN, B. A. Résultat d'une nouvelle détermination absolue de l'accélération due à la pesanteau, au Parillon de Breteaul, *Compt. Rend. Acad. Sci.*, 246, 3322, 1958. [112]

THYER, R. F., Geophysical exploration – Australia, *Geophysics*, 28, 273, 1963. [403]

TILTON, G. R., Volume diffusion as a mechanism for discordant lead ages, *J. Geophys. Res.*, 65, 2933, 1960. [150]

——, C. PATTERSON, H. BROWN, M. INGHRAM, R. HAYDEN, D. HESS, and E. LARSEN, Jr., Isotopic composition and distribution of lead, uranium and thorium in a Precambrian granite, *Geol. Soc. Am. Bull.*, 66, 1131, 1955. [139]

TISSOT, B., and A. NOESMOEN, Les bassins de Noumea et de Bourail (Nouvelle-Caledonie), *Rev. Inst. Franc. Petrole*, 13, 739, 1958. [456]

TOBIN, D. G., and L. R. SYKES, Seismicity and tectonics of the Northeast Pacific Ocean, *J. Geophys. Res.*, 73, 3821, 1968. [332; 447]

——, P. L. WARD, and C. L. DRAKE, Microearthquakes in the rift valley of Kenya, *Geol. Soc. Am. Bull.*, 80, 2043, 1969. [398]

TOKSÖZ, M. N., J. ARKANI-HAMED, and C. A. KNIGHT, Geophysical data and long-wave heterogeneities of the Earth's mantle, *J. Geophys. Res.*, 74, 3751, 1969. [200]

——, M. A. CHINNERY, and D. L. ANDERSON, Inhomogeneities in the Earth's mantle, *Geophys. J.*, 13, 31, 1967. [309]

TOKUDA, S., On the echelon structure of the Japanese Archipelagoes, *Japan. J. Geol. Geography,* Eng. trans., 5, 41, 1926-1927. [451;453;505]

TORRANCE, K. E., and D. L. TURCOTTE, Structure of convection cells in the mantle, *J. Geophys. Res.*, 76, 1154, 1971, [494]

TORRESON, O. W., T. MURPHY, and J. W. GRAHAM, Magnetic polarization of sedimentary rocks and the Earth's magnetic history, *J. Geophys. Res.*, 54, 111, 1949. [239; 241]

TOZER, D. C., The electrical properties of the Earth's interior, in L. H. AHRENS (ed.), "Physics and Chemistry of the Earth," vol. 3, p. 414, Pergamon, 1959. [262]

——, Thermal history of the Earth I. The formation of the core, *Geophys. J.*, 9, 95, 1965. [316]
——, Towards a theory of thermal convection in the mantle, in T. F. GASKELL (ed.), "The Earth's Mantle," p. 325, Academic, 1967. [365]
TOZER, E. T., and R. THORSTEINSSON, Western Queen Elizabeth Islands, Arctic Archipelago, *Geol. Surv. Canada Mem.* 332, 1964. [409]
TRAMONTINI, C., and D. DAVIES, A seismic refraction survey in the Red Sea, *Geophys. J.*, 17, 225, 1969. [405]
TRENDALL, A. F., Towards rationalism in Precambrian stratigraphy, *J. Geol. Soc. Australia*, 13, 517, 1966. [157; 479; 489]
TROITSKAYA, V. A., Rapid variations of the electromagnetic field of the Earth in H. ODISHAW (ed.), "Research in Geophysics," vol. 1, "Sun: Upper Atmosphere and Space," p. 485, MIT, 1964. [275]
——, Micropulsations and the state of the magnetosphere, in J. W. KING, and W. S. NEWMAN (eds.), "Solar-Terrestrial Physics," p. 213, Academic, 1967. [275]
TRUEMAN, N. A., The phosphate, volcanic and carbonate rocks of Christmas Islands, Indian Ocean, *J. Geol. Soc. Australia*, 12, 261, 1965. [425]
TRÜMY, R., Paleotectonic evolution of the Central and Western Alps, *Geol. Soc. Am. Bull.*, 71, 843, 1960. [360; 466]
TRUNIN, R. F., V. I. GONSHAKOVA, G. V. SIMAKOV, and N. E. GALDIN, A study of rocks under the high pressures and temperatures created by shock compression, *Bull. Phys. Solid Earth*, Eng. trans., 9, 579, 1965. [306; 317]
TRUSWELL, J. F., Marion Island, South Indian Ocean, *Nature*, 205, 65, 1965. [424]
TUDGE, A. P., and H. G. THODE, Thermodynamic properties of isotopic compounds of sulphur, *Can. J. Phys.*, 28b, 567, 1950. [169; 515]
TURCOTTE, D. L., and E. R. OXBURGH, Thermal structure of island arcs, *Geol. Soc. Am. Bull.*, 81, 1665, 1970. [349]
TUREKIAN, K. K., and S. P. CLARK, Jr., Inhomogeneous accumulation of the Earth from the primitive solar nebula, *Earth Planet. Sci. Lett.*, 6, 346, 1969. [314]
TURNER, F. J., and J. VERHOOGEN, "Igneous and Metamorphic Petrology," McGraw-Hill, 1960. [69; 71; 89]
TUTTLE, O. F., and N. L. BOWEN, Origin of granite in the light of experimental studies in the system $NaAlSi_3O_8-KAlSi_3O_8-SiO_2-H_2O$, *Geol. Soc. Am. Mem.* 74, 1958. [95; 488]
TYRRELL, G. W., "The Principles of Petrology," Methuen, 1926. [71]
TYURMINA, L. O., and T. N. CHEREVKO, Analytical representation of the geomagnetic field according to Cosmos 49 Data 1, *Geomagnetism i Aeronomiya*, 7, 207, 1967. [226]
UDINTSEV, G. B. (ed.), Map of the Indian Ocean, *Soviet Inst. Oceanogr.*, 1962. [421]
——, Results of upper mantle project studies in the Indian Ocean by the research vessel Vityaz, in T. N. IRVINE (ed.), "The World Rift System," *Geol. Surv. Canada Paper* 66-14, 148, 1966. [422]
—— et al. (ed.), Bathymetric chart of the Pacific Ocean, *Inst. of Oceanol., Moscow*, 1964. [426]
UFFEN, R. J., A method of estimating the melting point gradient in the Earth's mantle, *Trans. Am. Geophys. Union*, 33, 893, 1952. [205]
ULRYCH, T. J., Oceanic basalt leads: A new interpretation and an independent age for the earth, *Science*, 158, 252, 1967. [163]
—— and P. H. REYNOLDS, Whole-rock and mineral leads from the Llano uplift Texas, *J. Geophys. Res.*, 71, 3089, 1966. [148]
UMBGROVE, J. H. F., "The Pulse of the Earth," 2d ed., Martinus Nijhoff, 1947. [344; 350; 492]
——, "Structural History of the East Indies," Cambridge, 1949. [350; 425]

UNSÖLD, A. O. J., Stellar abundances and the origin of the elements, *Science*, 163, 1015, 1969a. [55]

——, "The New Cosmos," English translation, Springer-Verlag, 1969b. [55]

UPTON, B. G. J., and W. J. WADSWORTH, Intravolcanic intrusions of Reunion, *J. Geol. Spec. Issue* no. 2, 141, 1970. [424]

——, ——, and T. C. NEWMAN, The petrology of Rodriguez Island, Indian Ocean, *Geol. Soc. Am. Bull.*, 78, 1495, 1967. [424]

UREY, H. C., The thermodynamic properties of isotopic substances, *J. Chem. Soc.*, p. 562, 1947. [515]

——, The abundances of the elements, *Phys. Rev.*, 88, 248, 1952a. [16]

——, "The Planets, Their Origin and Development," Yale, 1952b. [12; 102; 314; 318]

——, The origin of the Earth, *Sci. Am.*, 187, 53, 1952c. [59]

——, Evidence regarding the origin of the Earth, *Geochim. Cosmochim. Acta*, 26, 1, 1962. [201]

——, The origin and evolution of the solar system, in D. P. LEGALLEY (ed.), "Space Science," Wiley, 1963. [12]

——, F. G. BRICKWEDDE, and G. M. MURPHY, Hydrogen isotope of mass Z and its concentration, *Phys. Rev.*, 39, 164, 1932; 40, 1, 1932. [167]

——, H. A. LOWENSTAM, S. EPSTEIN, and C. R. MCKINNEY, Measurement of paleotemperatures and temperatures of the upper Cretaceous of England, Denmark and the Southeastern United States, *Geol. Soc. Am. Bull.*, 62, 399, 1951. [174; 176; 177]

UYEDA, S., and A. M. JESSOP (eds.), Geothermal problems, *Tectonophysics*, 10, 1, 1970. [504]

—— and T. RIKITAKE, Electrical conductivity anomaly and terrestrial heat flow, *J. Geomagn. Geoelectn.*, 22, 75, 1970. [445]

—— and V. VACQUIER, Geothermal and geomagnetic data in and around the island arc of Japan, in L. KNOPOFF, C. L. DRAKE, and P. J. HART (eds.), "The Crust and Upper Mantle of the Pacific Area," *Am. Geophys. Union Monograph*, 12, 349, 1968. [430; 452; 454]

VACQUIER, V., Measurement of horizontal displacements along faults in the ocean floor, *Nature*, 183, 452, 1959. [331; 372; 386]

——, Magnetic evidence for horizontal displacements in the floor of the Pacific Ocean, in S. K. RUNCORN (ed.), "Continental Drift," p. 135, Academic, 1962. [372; 373; 387; 427; 450]

——, J. G. SCLATER, and C. E. CORRY, Studies of the thermal state of the Earth. The 21st Paper: Heat flow, Eastern Pacific, *Bull. Earthquake Res. Inst.*, 45, 375, 1967. [189]

——, S. UYEDA, M. YASUI, J. SCLATER, C. CORRY, and T. WATANABE, Studies of the thermal state of the Earth. The 19th Paper, Heat-flow measurements in the Northwestern Pacific, *Bull. Earthquake Res. Inst.*, 44, 1519, 1966. [190; 196; 197]

VAIL, J. R., An outline of the geochronology of the late Precambrian formation of Eastern Central Africa, *Proc. Roy. Soc. London, Ser. A*, 284, 354, 1965. [398]

——, The southern extension of the East African rift system and related igneous activity, *Geol. Rundschau*, 57, 601, 1967. [398; 400]

VALENTINE, J. W., Plate tectonics and the history of marine life, *Geol. Soc. Am. Abstracts with Programs*, 2, 710, 1970. [491]

VAN ALLEN, J. A., and L. A. FRANK, The mission of Mariner 2, Preliminary observations; the Iowa radiation experiment, *Science*, 138, 1097, 1962. [282]

VAN ANDEL, T. H., and J. J. VEEVERS, Submarine morphology of the Sahul shelf, Northwestern Australia, *Geol. Soc. Am. Bull.*, 76, 695, 1965. [422]

VAN BEMMELEN, R. W., The undation theory of development of the Earth's crust, *16th Sess. Intern. Geol. Congr.*, 2, 965, 1933 (1935). [344]

——, "The Geology of Indonesia," vol. 1A, Govt. Printing Office, The Hague, 1949. [425]

——, On the origin and evolution of the Earth's crust and magmas, *Geol. Rundschau,* 57, 657, 1968. [344]
——, Notes on the history and future use of the term "geonomy," *Earth-Sci. Rev. Atlas,* 5, 85, 1969. [356]
VAN DER LINDEN, W. J. M., Extinct mid-ocean ridges in the Tasman Sea and in the Western Pacific, *Earth Planet. Sci. Lett.,* 6, 483, 1969. [458]
VAN DE VOO, R., Paleomagnetic evidence for the rotation of the Iberian Peninsula, *Tectonophysics,* 7, 5, 1969. [415]
VAN PADANG, M. N. (ed.), Catalogue of the active volcanoes of the world including Solfatara fields, *Intern. Assoc. Volc.,* pts. 1-21, 1951-1967. [66]
VAN STEENIS, C. G. G. J., The land-bridge theory in botany, *Blumea,* 11, 235, 1962. [363]
VAN WATERSCHOOT VAN DER GRACHT, W. A. J. M., et al., Theory of Continental Drift, *Am. Assoc. Pet. Geol. Publ.,* 1928. [353; 366]
VARNE, R., R. D. GEE, and P. G. J. QUILTY, Macquarie Island and the cause of oceanic linear magnetic anomalies, *Science,* 166, 230, 1969. [432]
VEEVERS, J. J., J. G. JONES, and J. A. TALENT, Indo-Australian stratigraphy and the configuration and dispersal of Gondwanaland, *Nature,* 229, 383, 1971. [422]
VENING MEINESZ, F. A., Theory and Practice of Pendulum Observations at Sea, *Publ. Netherlands Geod. Comm.,* pts. 1 and 2, Waltman, 1929 and 1941. [347; 370; 394; 425]
——, Gravity expeditions at sea, 1923-1938. Vol. 4, *Netherlands Geod. Comm.,* Delft, 1948a. [107]
——, Major tectonic phenomena and the hypothesis of convection currents in the Earth, *Quart. J. Geol. Soc. London,* 103, 191, 1948b. [346]
——, Convection currents in the Earth and the origin of the continents I, *Koninkl. Ned. Akad. Wetenschap.,* 55, 527, 1952. [319]
——, "The Earth's Crust and Mantle," Elsevier, 1964. [346]
VERHOOGEN, J., F. J. TURNER, L. E. WEISS, C. WAHRHAFTING, and W. S. FYFE, "The Earth, an Introduction to Physical Geology," Holt, 1970. [342]
VERMA, R. K., and H. NARAIN, Terrestrial heat flow in India, in "The Crust and Upper Mantle of the Pacific Area," *Am. Geophys. Union Monograph,* 12, 1968. [198]
VERWOERD, W. J., and O. LANGENEGGER, Marion and Prince Edward Islands: Geological studies, *Nature,* 213, 231, 1967. [424]
VESTINE, E. H., The geographical incidence of aurora and magnetic disturbance, Northern Hemisphere, *Terr. Mag.,* 49, 77, 1944. [282]
——, I. LANGE, L. LAPORTE, and W. E. SCOTT, The geomagnetic field, its description and analysis, *Carnegie Inst. Wash. Publ.,* 580, 1947b. [226; 234]
——, I. LAPORTE, C. COOPER, I. LANGE, and W. C. HENDRIX, Description of the Earth's main magnetic field and its secular change, 1905-1945, *Carnegie Inst. Wash. Publ.,* 578, 1947a. [224; 225]
VILJOEN, M. J., and R. P. VILJOEN, Archaean vulcanicity and continental evolution in the Barberton region, Transvaal, in T. N. CLIFFORD, and I. G. GASS (eds.), "African Magmatism and Tectonics," Oliver & Boyd, p. 27, 1970. [483]
VINE, F. J., Spreading of the ocean floor; new evidence, *Science,* 154, 1405, 1966. [367; 392; 413; 427]
——, Evidence from submarine geology, *Proc. Am. Phil. Soc.,* 112, 325, 1968. [422; 424; 429]
——, The Geophysical Year, *Nature,* 227, 1013, 1970. [330]
—— and H. H. HESS, Sea-floor spreading, in A. E. MAXWELL (ed.), "The Sea," vol. 4, pt 2, p. 587, Interscience, 1970. [345; 367; 389; 427; 431; 448]
—— and D. H. MATTHEWS, Magnetic anomalies over oceanic ridges, *Nature,* 199, 947, 1963. [386]

—— and J. T. WILSON, Magnetic anomalies over a young oceanic ridge off Vancouver Island, *Science*, 150, 485, 1965. [388; 438]

VINOGRADOV, A. P., and A. I. TUGARINOV, The geologic age of Precambrian rocks of the Ukrainian and Baltic shields, *Ann. N. Y. Acad. Sci.*, 91, 500, 1961. [492]

VISSER, W. A., and J. J. HERMES, Geological results of the exploration for oil in Netherlands, New Guinea, *Verhandel. Ned. Geol. Mijnbouwk Genoot., Geol. Ser.*, 20, 1, 1962. [456]

VOGT, P. R., Magnetized basement outcrops on the Southeast Greenland continental shelf, *Nature*, 226, 743, 1970. [75; 398; 413]

——, C. N. ANDERSON, D. R. BRACEY, and E. D. SCHNEIDER, North Atlantic magnetic smooth zones, *J. Geophys. Res.*, 75, 3955, 1970. [418]

——, O. E. AVERY, E. D. SCHNEIDER, C. N. ANDERSON, and D. R. BRACEY, Discontinuities in sea-floor spreading, *Tectonophysics*, 8, 285, 1969a. [85; 367; 389; 394; 406; 413; 414; 417]

—— and R. H. HIGGS, An aeromagnetic survey of the Eastern Mediterranean and its interpretation, *Earth Planet. Sci. Lett.*, 5, 439, 1969. [465]

——, ——, and G. L. JOHNSON, Hypotheses on the origin of the Mediterranean basin: Magnetic data, *J. Geophys. Res.*, 76, 3207, 1971. [462]

—— and N. A. OSTENSO, Magnetic and gravity profiles across the Alpha Cordillera and their relation to Arctic sea-floor spreading, *J. Geophys. Res.*, 75, 4925, 1970b. [408; 475]

——, ——, and G. L. JOHNSON, Magnetic and bathymetric data bearing on sea-floor spreading north of Iceland, *J. Geophys. Res.*, 75, 903, 1970a. [373; 409; 410; 411; 417]

——, E. D. SCHNEIDER, and G. L. JOHNSON, The crust and upper mantle beneath the sea, in P. J. HART (ed.), "The Earth's Crust and Upper Mantle," *Am. Geophys. Union Monograph* 13, 556, 1969b. [82; 85; 390; 393; 429]

VOLET, C., Sur la mesure absolue de la gravité, *Compt. Rend. Acad. Sci.*, 222, 373, 1946. [111]

VON HERZEN, R. P., and M. G. LANGSETH, Present status of oceanic heat flow measurements, in L. H. AHRENS (ed.), "Physics and Chemistry of the Earth," vol. VI, 365, Pergamon, 1966. [187]

—— and W. H. K. LEE, Heat flow in oceanic regions, in P. J. HART (ed.), "The Earth's Crust and Upper Mantle," *Am. Geophys. Union Monograph* 13, 88, 1969. [422]

—— and A. E. MAXWELL, The measurement of thermal conductivity of deep-sea sediments by a needle-probe method, *J. Geophys. Res.*, 64, 1557, 1959. [185]

—— and S. UYEDA, Heat flow through the eastern Pacific Ocean floor, *J. Geophys. Res.*, 68, 4219, 1963. [187; 190; 191]

—— and V. VACQUIER, Terrestrial heat flow in Lake Malawi, Africa, *J. Geophys. Res.*, 72, 4221, 1967. [185]

VON HUENE, R., Geologic structure between the Murray fracture zone and the Transverse ranges, *Mar. Geol.*, 7, 475, 1969. [438]

—— and G. G. SHOR, Jr., The structure and tectonic history of the Eastern Aleutian trench, *Geol. Soc. Am. Bull.*, 80, 1889, 1969. [448]

VON WEIZSÄCKER, C. V., Über die Entstehung der Planetensystems, *Z. Astrophys.*, 22, 319, 1944. [9]

WACKERLE, J., Shock-wave compression of quartz, *J. Appl. Phys.*, 33, 922, 1962. [306]

WAGER, L. R., and W. A. DEER, A dyke swarm and crustal flexure in East Greenland, *Geol. Mag.*, 75, 39, 1938. [412]

—— and ——, Geological investigations in East Greenland, Part III, The petrology of the Skaergaard intrusion, Kangerdlugssuaq, East Greenland, *Medd. Groenland,* Bd. 105, no. 4, pt. 3, p. 35, 1939 [reissue 1962]. [412; 413]

WALCOTT, R. I., Flexure of the lithosphere at Hawaii, *Tectonophysics*, 9, 397, 1970. [432; 498]

WALCZAK, J. E., and T. CARTER, A bathymetric and geomagnetic survey of the New England seamount chain, *Int. Hydrogr. Rev.*, 41, 59, 1964. [418]

WALKER, F., and L. O. NICOLAYSEN, The petrology of Mauritius, *Colonial Geol. Mineral Resources,* 4, 3, 1954. [424]

WALKER, G. P. L., Zeolite zones and dike distribution in relation to the structure of the basalts of Eastern Iceland, *J. Geol.*, 68, 515, 1960. [381]

——, Evidence of crustal drift from Icelandic geology, *Phil. Trans. Roy. Soc. London, Ser. A.* 258, 199, 1965. [381]

WALKER, K. R., The Palisades sill, New Jersey: A reinvestigation, *Geol. Soc. Am. Spec. Paper* 111, 1969. [417]

WALSH, J. B., and E. R. DECKER, Effect of pressure and saturating fluid on the thermal conductivity of compact rock, *J. Geophys. Res.*, 71, 3053, 1966. [185]

WALSH, J. M., and R. H. CHRISTIAN, Equation of state of metals from shock wave measurements, *Phys. Rev.*, 97, 1544, 1955. [306]

——, M. H. RICE, R. G. MCQUEEN, and F. L. YARGER, Shock wave compressions of twenty-seven metals. Equations of state of metals, *Phys. Rev.*, 108, 196, 1957. [306]

WANG, C.-Y., Some geophysical implications from gravity and heat flow data, *J. Geophys. Res.*, 70, 5629, 1965. [199]

——, Phase transitions in rocks under shock compression, *Earth Planet. Sci. Lett.*, 2, 107, 1967. [306]

——, Constitution of the lower mantle as evidenced from shock wave data for some rocks, *J. Geophys. Res.*, 73, 6459, 1968. [43; 306]

——, Equation of state of periclase and some of its geophysical implications, *J. Geophys. Res.*, 74, 1451, 1969. [43; 306]

——, Density and constitution of the mantle, *J. Geophys. Res.*, 75, 3264, 1970. [43]

WANLESS, R. K., R. D. STEVENS, and W. D. LOVERIDGE, Excess radiogenic argon in biotites, *Earth Planet. Sci. Lett.*, 7, 167, 1969. [151]

WARD, P. L., and T. MATSUMOTO, A summary of volcanic and seismic activity in Katmai National Monument, Alaska, *Bull. Volcanol.* 31, 107, 1967. [447]

——, G. PALMASON, and C. DRAKE, Microearthquake survey and Mid-Atlantic ridge in Iceland, *J. Geophys. Res.*, 74, 665, 1969. [382]

WASSERBURG, G. J., Geochronology, and isotopic data bearing on development of the continental crust, in P. M. HURLEY (ed.), "Advances in Earth Science," p. 431, MIT, 1966. [165]

——, G. J. F. MACDONALD, F. HOYLE, and W. A. FOWLER, Relative contributions of uranium, thorium and potassium to heat production in the Earth, *Science*, 143, 465, 1964. [211]

WATERHOUSE, J. B., A new Permian fauna from New Caledonia and its relationship to Gondwana and the Tethys, UNESCO Gondwana Stratigraphy, *IUGS Sympos. Buenos Aires*, p. 249, 1967. [366; 456]

—— and S. PIYASIN, Mid-Permian brachiopods from Khao Phrik, Thailand, *Sonder-Abdruck Palaeontographica*, Bd. 135, Abt. A, 83, 1970 [458]

—— and P. VELLA, A Permian fauna from Northwest Nelson, New Zealand, *Trans. Roy. Soc. New Zealand, Geol.*, 3, 57, 1965. [458; 459]

WATKINS, N. D., Behavior of the geomagnetic field during the Miocene period in Southeastern Oregon, *Nature*, 197, 126, 1963. [246]

——, Comments on a paper by William Taubeneck, An evaluation of tectonic rotation in the Pacific Northwest, *J. Geophys. Res.*, 72, 1411, 1967. [438]

WATSON, J. A., and G. L. JOHNSON, Mediterranean diapiric structures, *Am. Assoc. Pet. Geol. Bull.* 52, 2247, 1968. [462]

—— and ——, The marine geophysical survey in the Mediterranean, *Int. Hydrogr. Rev.*, 46, 81, 1969. [462;463]

WATT, W. S., The coast-parallel dike swarm of Southwest Greenland in relation to the opening of the Labrador Sea, *Can. J. Earth Sci.*, 6, 1320, 1969. [413]

WAYLAND, E. J., Rift valleys and Lake Victoria, *Rept. Intern. Geol. Congr. 1929*, 2, 323, 1930. [398]

WEERTMAN, J., Mechanism for continental drift, *J. Geophys. Res.*, 67, 1133, 1962. [365]

WEGENER, A., "The Origin of Continents and Oceans," 4th ed. (1929 trans.), Dover, 1966. [329; 352; 353; 355; 365; 368]

WEGMANN, C. E., Geological tests of the hypothesis of continental drift in the Arctic regions, *Medd. Groenland*, 114, 1, 1948. [408]

WELLMAN, H. W., The Alpine fault in detail: River terrace displacement at Mauruia River. *New Zealand J. Sci. Technol.*, 33, 409, 1952. [458]

——, Data for the study of recent and late Pleistocene faulting in the south island of New Zealand, *New Zealand J. Sci. Technol., B*, 34, 270, 1953. [331; 363]

——, New Zealand quaternary tectonics, *Geol. Rundschau*, 43, 248, 1955. [331]

——, Structural outline of New Zealand, *New Zealand Dept. Sci. and Ind. Res. Bull.*, 71, 1956. [458]

——, Active wrench faults of Iran, Afghanistan and Pakistan, *Geol. Rundschau*, 55, 716, 1966. [468]

——, Wrench (transcurrent) fault systems, in P. J. HART (ed.), "The Earth's Crust and Upper Mantle," *Am. Geophys. Union Monograph* 13, 544, 1969. [331]

WELLMAN, P., M. W. MCELHINNY, and I. MCDOUGALL, On the polar-wander path for Australia during the Cenozoic, *Geophys. J.*, 18, 371, 1969. [422]

WELLS, J. W., Coral growth and geochronometry, *Nature*, 197, 948, 1963. [103]

WESSON, P. S., The position against continental drift, *Quart. J. Roy. Astron. Soc.*, 11, 312, 1970. [353]

WESTOLL, T. S., and D. R. STODDART (eds.), A discussion on the results of the Royal Society Expedition to Aldabra 1967-68, *Phil. Trans. Roy. Soc. London, Ser. B.* 260, 1971. [425]

WETHERILL, G. W., Discordant lead-uranium ages, I. *Trans. Am. Geophys. Union*, 37, 320, 1956. [149]

——, Steady-state calculations bearing on geological implications of a phase transition Mohorovičić discontinuity, *J. Geophys. Res.*, 66, 2983, 1961. [298]

——, The beginning of continental evolution, in A. R. RITSEMA (ed.), "The Upper Mantle," *Tectonophysics*, 13, 31, 1972. [153]

WHEELER, J. O. (ed.), Structure of the Southern Canadian Cordillera, *Geol. Assoc. Canada Spec. Paper* 6, 166, 1970. [95]

WHITE, D. A., D. H. ROEDER, T. H. NELSON, and J. C. CROWELL, Subduction, *Geol. Soc. Am. Bull.*, 81, 3431, 1970. [325; 394]

WHITE, W. C., and O. N. WARIN, A survey of phosphate deposits in the Southwest Pacific and Australian waters, *Australia, Bur. Mineral Resources, Geol. Geophys. Bull.* 69, 1964. [450]

WHITESIDE, H. C. M., Volcanic rocks of the Witwatersrand Triad, in T. N. CLIFFORD, and I. G. GASS (eds.), "African Magmatism and Tectonics," Oliver & Boyd, 1970. [488]

WHITTINGTON, H. B., and C. P. HUGHES, Ordovician geography and faunal provinces deduced from trilobite distribution, *Phil. Trans. Roy. Soc. London, Ser. B*, 263, 235, 1972. [474]

WILLIAMS, I. P., and A. W. CREMIN, A survey of theories relating to the origin of the solar system, *Quart. J. Roy. Astron. Soc.*, 9, 40, 1968. [13]

WILLIAMS, J. J. (ed.), South-Central Libya and Northern Chad, *Pet. Explor. Soc. Libya 8th Ann. Field Conf., 1966*, Drukkerij, Holland N. V., Amsterdam, 1966. [462; 466]

WILLIAMSON, E. D., and L. H. ADAMS, Density distribution in the Earth, *J. Wash. Acad. Sci.*, 13, 413, 1923. [39]

WILLIS, B., East African plateaus and rift valleys, *Carnegie Inst. Wash.*, Publ. 470, 1936. [398]

WILSON, H. D. B., Volcanology, *Geol. Surv. Canada, Paper* 67-41, 155, 1967. [483]

——, Superior Province, Geological Provinces of Canada, Exploration and Outlook, *Can. Inst. Mining, Bull.* 63, 195, 1970. [483]

WILSON, J. TINSLEY, The Love waves of the South Atlantic earthquakes of August 28, 1933, *Bull. Seismol. Soc. Am.*, 38, 273, 1940. [372]

WILSON, J. TUZO, On the growth of continents, R. M. Johnston Memorial Lecture, *Papers and Proc. Roy. Soc. Tasmania (1950).* p. 85, 1951. [99]

——, The development and structure of the crust, in G. P. KUIPER (ed.), "The Earth as a Planet," p. 138, The University of Chicago Press, 1954. [344; 351; 403; 451; 462]

—— et al., "Handbuch der Physik," 47, 1956. [133-135]

——, Some consequences of expansion of the Earth, *Nature*, 185, 880, 1960. [111]

——, Evidence from islands on the spreading of ocean floors, *Nature*, 197, 536, 1963a. [378; 425; 494]

——, Continental drift, *Sci. Am.*, p. 2, April, 1963b. [409; 494]

——, Hypothesis of the Earth's behavior, *Nature*, 198, 925, 1963c. [409; 422; 425; 475; 494]

——, Pattern of uplifted islands in the main ocean basins, *Science*, 139, 592, 1963d. [450]

——, A possible origin of the Hawaiian Islands, *Can. J. Phys.*, 41, 863, 1963e. [88; 432; 498; 504]

——, Submarine fracture zones, a seismic ridge and the ICSU rise, a proposed western margin of the East Pacific rise, *Nature*, 207, 907, 1965a. [88; 372; 378; 379; 420; 431; 432; 494; 497]

——, A new class of faults and their bearing on continental drift, *Nature*, 207, 343, 1965b. [333; 421; 424]

——, Convection currents and continental drift, *Phil. Trans. Roy. Soc. London, Ser. A*, 258, 145, 1965c. [408]

——, Did the Atlantic close and then reopen? *Nature*, 211, 676, 1966. [396; 468; 473; 475-477]

——, Theories of building of continents, in T. F. GASKELL (ed.), "The Earth's Mantle," p. 445, Academic, 1967. [90; 487]

——, Static or mobile Earth: The current scientific revolution, *Proc. Amer. Phil. Soc.*, 112, 309, 1968. [398; 403; 409; 474; 478]

——, Tectonics, in "International Upper Mantle Project, Report No. 6," p. 32, U.C.L.A., 1969. [331]

——, Mantle plumes and plate motion, *Tectonophysics* (in press). [403; 415; 494; 495; 501; 505]

——, New insights into old shields, *Tectonophysics*, 13, 73, 1972. [483]

——, R. D. RUSSELL, and R. M. FARQUHAR, Radioactivity and age of minerals, in S. FLÜGGE (ed.), "Encyclopedia of Physics," 47, 288, 1956. [132; 134; 135; 479; 481]

WILSON, R. L., Palaeomagnetism, *Die Naturwiss.*, 11, 286, 1965. [252]

—— and N. D. WATKINS, Correlation of petrology and natural magnetic polarity in Columbia plateau basalts, *Geophys. J.*, 12, 405, 1967. [247]

WINDHAUSEN, A., Elementos de estructura en el subauclo de Patagonia, *Bol. Acad. Nac. Cienc.*, Corboda, Tomo 25, 125, 1921. [435]

WINDISCH, C. C., R. J. LEYDEN, J. L. WORZEL, T. SAITO, and J. EWING, Investigations of Horizon Beta, *Science*, 162, 1473, 1968. [418]

WINKLER, H. G. F., Petrogenesis of Metamorphic Rocks, Springer-Verlag, 1965. [71]

WISE, D. U., An outrageous hypothesis for the tectonic pattern of North American Cordillera, *Geol. Soc. Am. Bull.*, 74, 357, 1963. [439]

WISEMAN, J. D. H., The petrography and significance of a rock dredged from a depth of 744 fathoms, near the Providence reef, Indian Ocean, *Trans. Linn. Soc. London*, 19, 437, 1926/1936, 1937. [425]

WOHLENBERG, J., Remarks on the seismicity of East Africa between 4°N–12°S and 23°E–40°E, *Tectonophysics*, 8, 567, 1969. [398]

WONG, K., and E. F. K. ZARUDZKI, Thickness of unconsolidated sediments in the Eastern Mediterranean Sea, *Geol. Soc. Am. Bull.*, 80, 2611, 1969. [466]

WOOD, J. A., On the origin of chondrules and chondrites, *Icarus*, 2, 152, 1963a. [22]

——, Physics and chemistry of meteorites, in B. M. MIDDLEHURST, and G. P. KUIPER (eds.), "The Solar System," vol. IV, The University of Chicago Press, 1963b. [22]

WOODSIDE, J., and C. BOWIN, Gravity anomalies and inferred crustal structure in the Eastern Mediterranean Sea, *Geol. Soc. Am. Bull.*, 81, 1107, 1970. [466]

WOOLF, N. J., and E. P. NEY, Circumstellar infrared emission from cool stars, *Astrophys. J. Lett.*, 155, L181, 1969. [314]

WOOLLARD, G. P., The interrelationship of the crust, the upper mantle, and isostatic gravity anomalies in the United States, in L. KNOPOFF, C. L. DRAKE, and P. J. HART (eds.), "The Crust and Upper Mantle of the Pacific Area," *Am. Geophys. Union Monograph* 12, 312, 1968. [445]

——, Standardization of gravity measurements, in P. J. HART (ed.), "The Earth's Crust and Upper Mantle," *Am. Geophys. Union Monograph* 13, 283, 1969. [113; 114; 121]

WORSLEY, T. R., Terminal Cretaceous events, *Nature*, 230, 318, 1971. [490]

WORZEL, J. L., Continuous gravity measurements on a surface-ship with the Graf sea gravimeter, *J. Geophys. Res.*, 64, 1299, 1959. [107]

——, "Pendulum Gravity Measurements at Sea 1936-1959," Interscience, 1965. [107]

——, Survey of continental margins, in D. T. DONOVAN (ed.), "Geology of the Shelf Seas," Oliver & Boyd, p. 117, 1968. [417]

WRIGHT, F. A., and B. W. TROXEL, Summary of regional evidence for right-lateral displacement in the Western Great Basin discussion, *Geol. Soc. Am. Bull.*, 81, 2167, 1970. [438]

WRIGHT, J. B., South Atlantic continental drift and the Benue trough, *Tectonophysics*, 6, 301, 1968. [403; 420]

——, Re-interpretation of a mixed petrographic province – the Sintra intrusive complex (Portugal) and related rocks, *Geol. Rundschau*, 58, 538, 1969. [415]

—— and P. RIX, Evidence for trough faulting in Eastern Central Kenya, *Overseas Geol. Mineral Resources* 10, 30, 1967. [398]

WÜST, G., The major deep-sea expeditions and research vessels, 1873-1960, *Progr. Oceanogr.*, 2, 1, 1964. [369]

WYLLIE, P. J., The nature of the Mohorovičić discontinuity, a compromise, *J. Geophys. Res.*, 68, 4611, 1963. [298]

——, "Ultramafic and Related Rocks," Wiley, 1967. [73; 92]

WYNNE-EDWARDS, H. R., Age relations in high-grade metamorphic terrains, *Geol. Assoc. Can. Spec. Paper* 5, 1969. [487]

YASUI, M., and Y. HASHIMOTO, Geomagnetic studies of the Japan Sea – Anomaly pattern in the Japan Sea, *Oceangr. Mag.*, 19, 221, 1967. [454]

——, T. KISHII, and K. SUDO, Terrestrial heat-flow in the Okhotsk Sea 1, *Oceanogr. Mag.*, 19, 87, 1967. [195]

——, ——, T. WATANABE, and S. UYEDA, Studies of the thermal state of the Earth. 18th Paper: Terrestrial heat-flow in the Japan Seas (2), *Bull. Earthquake Res. Inst.*, 44, 1501, 1966. [194]

——, ——, ——, and ——, Heat-flow in the Sea of Japan, in L. KNOPOFF, C. DRAKE, and P. J. HART (eds.), "The Crust and Upper Mantle of the Pacific Area," *Am. Geophys. Union Monograph* 12, 3, 1968b. [194;451;454]

——, K. NAGASAKA, Y. HASHIMOTO, and K. ANMA, Geomagnetic and bathymetric study of the Okhotsk Sea, *Oceanogr. Mag.*, 20, 65, 1968c. [451]

——, ——, and T. KISHII, Terrestrial heat-flow in the Okhotsk Sea 2, *Oceanogr. Mag.*, 20, 73, 1968a. [195;451]

——, ——, ——, and A. J. HALUNEN, Terrestrial heat-flow in the Okhotsk Sea, *Oceanogr. Mag.*, 20, 73, 1968d. [451]

YATES, R. G., The Trans-Idaho discontinuity, *23d Intern. Geol. Congr., Proc. Sect. 1*, p. 117, 1968. [439]

YEN, T. P., Crustal features of the Taiwan region, *Proc. Geol. Soc. China*, 11, 130, 1968. [454]

ZAMBRANO, J. J., and C. M. URIEN, Geological outline of the basins in Southern Argentina and their continuation off the Atlantic shore, *J. Geophys. Res.*, 75, 1363, 1970. [83; 420]

ZARTMAN, R. E., P. M. HURLEY, H. W. KRUEGER, and B. J. GILETTI, A Permian disturbance of K-Ar radiometric ages in New England: Its occurrence and cause, *Geol. Soc. Am. Bull.*, 81, 3359, 1970. [417]

ZBYSZEWSKI, G., and O. DA VEIGA FERREIRA, Carta Geològica de Portugal, na Escala 1/50,000: Notícia Explicativa da Folha de Ilha Santa Maria (Açores), *Serviços Géol. de Portugal*, 1961. [415]

—— and ——, La Faune Miocène de l'Île de Santa Maria (Açores), *Comm. dos Serviocos Géol de Portugal*, 46, 247, 1962. [415]

ZIELINSKI, R. A., and F. A. FREY, Gough Island: Evaluation of a fractional crystallization model, *Contr. Mineral and Petrol.*, 29, 242, 1970. [496]

ZWART, H. J. (ed.), Symposium on metamorphic facies and facies series, *Publ. Dept. Geol.*, Aarhus University, Copenhagen, 53, 1967. [487]

ZIJDERVELD, J. D. A., G. J. A. HAGEN, M. NERDIN, and R. VAN DER VOO, Shear in the Tethys and the Permian paleomagnetism in the Southern Alps, including new results, *Tectonophysics*, 10, 639, 1970. [464]

Name Index

Names appearing in the text *exclusive* of *first* authors listed in the bibliography.

Geographic Index

Subject Index